AF769796

Grundlehren der mathematischen Wissenschaften 340

A Series of Comprehensive Studies in Mathematics

Series editors

M. Berger P. de la Harpe F. Hirzebruch
N.J. Hitchin L. Hörmander M. Kashiwara
A. Kupiainen G. Lebeau F.-H. Lin B.C. Ngô
M. Ratner D. Serre Ya.G. Sinai
N.J.A. Sloane A.M. Vershik M. Waldschmidt

Editor-in-Chief

A. Chenciner J. Coates S.R.S. Varadhan

Half of an analog to Costa's surface which is stationary in a configuration consisting of four nearly semicircular arcs and a plane. Courtesy of K. Polthier

For further volumes:
http://www.springer.com/series/138

Ulrich Dierkes • Stefan Hildebrandt •
Anthony J. Tromba

Regularity of Minimal Surfaces

With assistance and contributions by A. Küster

Revised and enlarged 2nd edition

With 62 Figures and 4 Color Plates

Springer

Ulrich Dierkes
Fakultät für Mathematik
Universität Duisburg-Essen
Campus Duisburg
Forsthausweg 2
47057 Duisburg
Germany
ulrich.dierkes@uni-due.de

Stefan Hildebrandt
Mathematical Institute
University of Bonn
Endenicher Allee 60
53115 Bonn
Germany

Anthony J. Tromba
Department of Mathematics
University of California at Santa Cruz
Baskin 621B
CA 95064 Santa Cruz
USA
tromba@ucsc.edu

With contributions by
Albrecht Küster
Walter-Flex-Str. 19
47809 Krefeld
Germany

This volume is the second part of a treatise on *Minimal Surfaces* in the series *Grundlehren der mathematischen Wissenschaften*. Part One is Vol. 339 ISBN 978-3-642-11697-1, Part Three is Vol. 341 ISBN 978-3-642-11705-3. A 1st edition of the treatise appeared as Vols. 295 and 296 of the same series.

ISSN 0072-7830
ISBN 978-3-642-11699-5 e-ISBN 978-3-642-11700-8
DOI 10.1007/978-3-642-11700-8
Springer Heidelberg Dordrecht London New York

Library of Congress Control Number: 2010930923

Mathematics Subject Classification (2010): 49Q05, 53A05, 53A07, 53B20, 35J20, 35J47, 35J50, 35J75, 49Q20, 30C20

© Springer-Verlag Berlin Heidelberg 1992, 2010
This work is subject to copyright. All rights are reserved, whether the whole or part of the material is concerned, specifically the rights of translation, reprinting, reuse of illustrations, recitation, broadcasting, reproduction on microfilm or in any other way, and storage in data banks. Duplication of this publication or parts thereof is permitted only under the provisions of the German Copyright Law of September 9, 1965, in its current version, and permission for use must always be obtained from Springer. Violations are liable to prosecution under the German Copyright Law.
The use of general descriptive names, registered names, trademarks, etc. in this publication does not imply, even in the absence of a specific statement, that such names are exempt from the relevant protective laws and regulations and therefore free for general use.

Cover design: VTEX, Vilnius

Printed on acid-free paper

Springer is part of Springer Science+Business Media (www.springer.com)

Preface

This book is the second volume of a treatise on minimal surfaces consisting of altogether three volumes which can be read and studied independently of each other. The central theme is *boundary value problems for minimal surfaces* such as Plateau's problem. The present treatise forms a greatly extended version of the monograph *Minimal Surfaces I, II* by U. Dierkes, S. Hildebrandt, A. Küster, and O. Wohlrab, published in 1992, which is often cited in the literature as [DHKW]. New coauthors are Friedrich Sauvigny for the first volume and Anthony J. Tromba for the second and third volume.

The four main topics of this second volume are *free boundary value problems*, *regularity of minimal surfaces* and their *geometric properties*, and finally a new method is introduced to show that minimizers of area are immersed. Since minimal surfaces in $\mathbb{R}^3$ are understood as harmonic, conformally parametrized mappings $X : \Omega \to \mathbb{R}^3$ of an open domain Ω in $\mathbb{R}^2$, they are real analytic in Ω, and so the *problem of smoothness* for X is the question how smooth X is at the boundary $\partial\Omega$ if X is subject to certain boundary conditions. However, even if X is "analytically regular", it might not be "geometrically regular" since it could have branch points. We investigate how X behaves in the neighbourhood of branch points, and secondly whether such points actually exist. In addition we describe geometric properties of minimal surfaces in $\mathbb{R}^3$ or, more generally, of H-surfaces in an n-dimensional Riemannian manifold. This book can be read independently from the preceding volume of this treatise although we use some terminology and results from the previous material.

We thank E. Kuwert, F. Müller, D. Schwab, H. von der Mosel, D. Wienholtz, and S. Winklmann for pointing out errors and misprints in [DHKW] which are corrected here. Particularly we are indebted to Frank Müller for some penetrating contributions to Chapter 3, and to Albrecht Küster who supplied a considerable part of Chapter 1 (which was taken from Vol. 1 of [DHKW]). Special thanks also to Ruben Jakob who carefully read and corrected Chapters 4 and 6, and to Klaus Steffen and Friedrich Tomi for their valuable comments to the Scholia of Chapter 6. Furthermore we thank Klaus

Bach, Frei Otto, and Eric Pitts for providing us with photographs of various soap film experiments. Thanks also to M. Bourgart, D. Hoffman, J.T. Hoffman, and K. Polthier for permitting us to reproduce some of their computer generated figures.

The continued support of our work by the Sonderforschungsbereich 611 at Bonn University as well as by the Hausdorff Institute for Mathematics in Bonn and its director Matthias Kreck was invaluable. We also thank the Centro di Ricerca Matematica Ennio De Giorgi in Pisa and its director Mariano Giaquinta for generous support of our work.

We are especially grateful to Anke Thiedemann and Birgit Dunkel who professionally and with untiring patience typed many versions of the new text.

Last but not least we should like to thank our publisher and in particular our very patient editors, Catriona Byrne, Marina Reizakis, and Angela Schulze-Thomin, for their encouragement and support.

Duisburg — *Ulrich Dierkes*
Bonn — *Stefan Hildebrandt*
Santa Cruz — *Anthony Tromba*

Contents

Part I. Boundary Behaviour of Minimal Surfaces

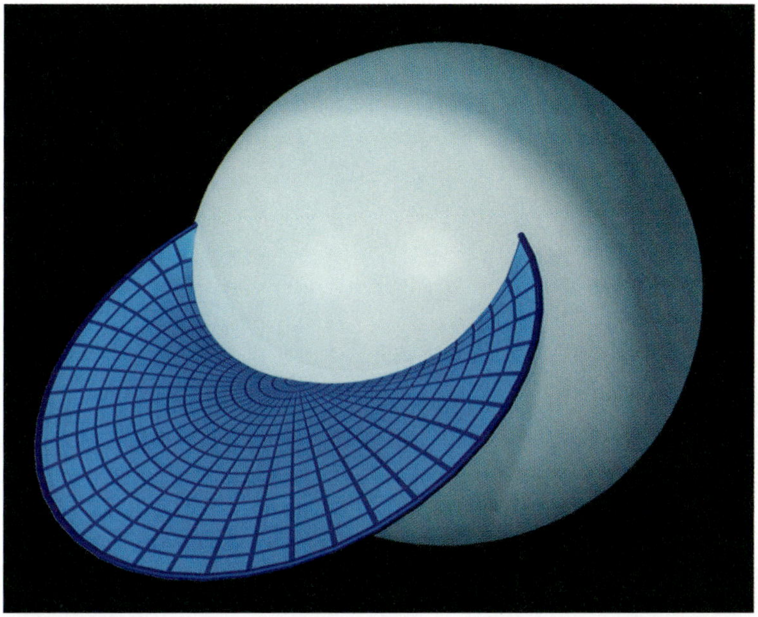

Plate Ia. A minimal surface which intersects a sphere perpendicularly. Courtesy of M. Bourgart

Plate Ib. A boundary configuration $\langle \Gamma_1, \Gamma_2, S \rangle$ spanning a minimal surface with a free boundary. Courtesy of M. Bourgart

Plate II. Four soap film experiments providing solutions of partially free boundary value problems. Courtesy of ILF Stuttgart

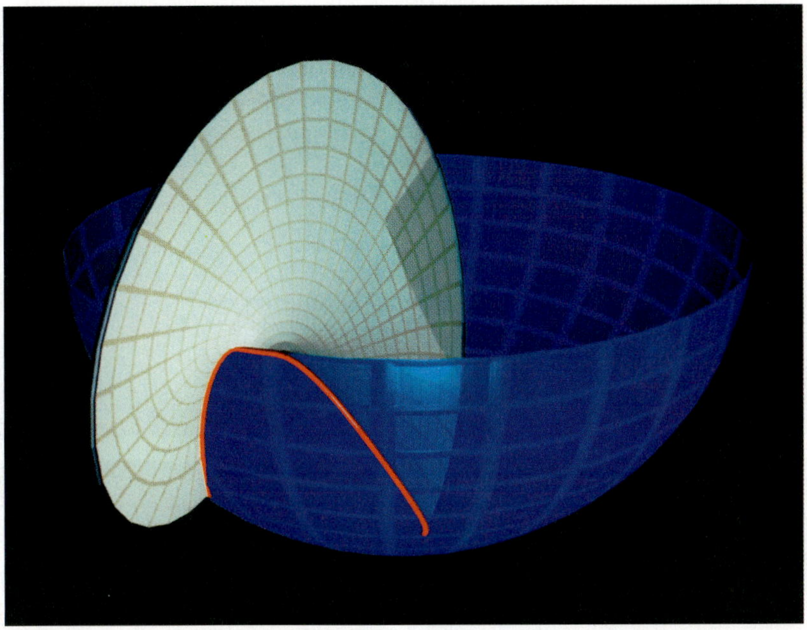

Plate III. A soap film (bright colour) spanned by a circular arc and a (blue) halfsphere. The film intersects the halfsphere perpendicularly along its (red) trace, except where the trace touches the equator, the boundary of the half sphere. Courtesy of M. Bourgart

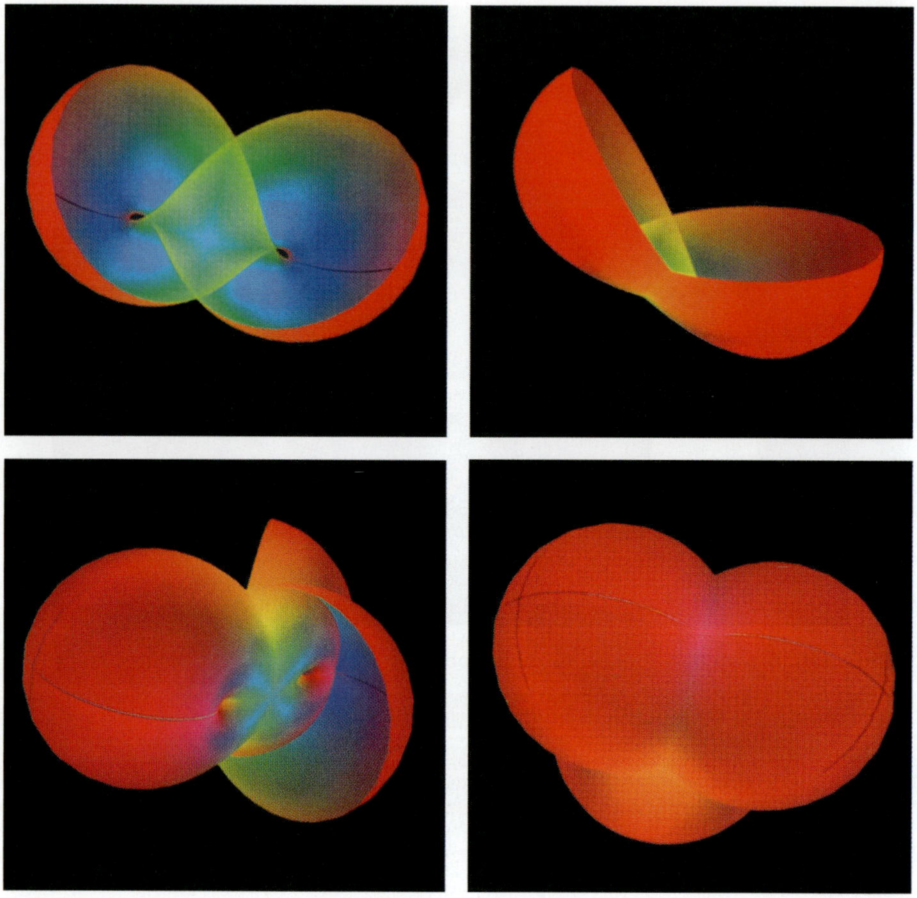

Plate IV. Construction of Wente's compact H-surface from the three building blocks. Courtesy of D. Hoffman and J.T. Hoffman

Introduction

We begin this volume with a survey on minimal surfaces with entirely free boundaries. In the Sections 1.1–1.3 Courant's existence result is described. It follows the derivation of the *transversality condition* at the free boundary in Section 1.4 and of the *condition of balancedness* that are to be satisfied by stationary surfaces. Further results and examples in Sections 1.6–1.10 conclude this chapter.

In Chapter 2 we investigate the boundary behaviour of minimal surfaces subject to Plateau boundary conditions or to free boundary conditions. Roughly speaking we show that a minimal surface is as smooth at the boundary as the data of the boundary conditions to which it is subject. There is a basic difference between grappling with the regularity problem for area minimizing surfaces or for merely stationary solutions of boundary value problems. For disk-type minimizers it is always possible to derive a priori estimates, while examples show that it is generally impossible to establish a priori estimates for stationary solutions of free boundary problems. Thus it becomes necessary to apply indirect methods if one wants to prove boundary regularity of minimal surfaces subject to free boundary conditions.

For a more complete understanding of the boundary behaviour of minimal surfaces one has not only to investigate their class of smoothness at the boundary, but it is also necessary to find out whether singular points occur at the boundary and, if so, how a minimal surface behaves in the neighbourhood of such points. This question is tackled in Chapter 3. If a minimal surface $X(w), w = u + iv$, is given in conformal parameters u, v, then its singular (= nonregular) points are exactly its branch points w_0, which are characterized by the relation $X_w(w_0) = 0$. In Chapter 3 we derive asymptotic expansions of minimal surfaces at boundary branch points which can be seen as a generalization of Taylor's formula to the nonanalytic case. Moreover, we also derive expansions of minimal surfaces with nonsmooth boundaries (e.g. polygons) at boundary points which are mapped onto vertices of the nonsmooth boundary frame.

Asymptotic expansions and boundary smoothness are very useful if one wants to treat subtle geometric and analytic problems. Furthermore they are indispensable for the derivation of *index theorems* and for the investigation of the *Euler characteristic* of minimal surfaces. Topics of this kind will be discussed in Volume 3.

The long Chapter 4 could have been labeled as *geometric properties of minimal surfaces*. First we derive *inclusion theorems* for minimal surfaces in dependence of their boundary data. Such results, obtained in Sections 4.1 and 4.2, are more or less sophisticated versions of the maximum principle. They lead to interesting nonexistence results for connected minimal surfaces and H-surfaces whose boundaries consist of several disjoint components, as it is seen in Sections 4.3–4.5. Here we even discuss the situation for higher dimensional surfaces and for solutions of variational inequalities, obtained from *obstacle* problems. Inclusion principles for such solutions are the fundament for results ensuring the existence of minimal surfaces and H-surfaces solving Plateau's problem in Euclidean space or in a Riemannian manifold respectively, see Sections 4.7 and 4.8. Of particular interest are the *Jacobi field estimates* obtained in Section 4.8.

Isoperimetric inequalities for minimal surfaces solving either Plateau's problem or a free boundary value problem are derived in Sections 4.5 and 4.6. The simplest kind of such an inequality was already stated in Section 4.14 of Vol. 1; for the sake of completeness we repeat here the derivation. Furthermore, in Section 6.4 of Vol. 1 an isoperimetric inequality for harmonic mappings $X : \Omega \to \mathbb{R}^3$, due to Morse & Tompkins, was derived, which plays an essential role in Courant's theory of unstable minimal surfaces.

In Chapter 5 we investigate an extension of the isoperimetric problem, the so-called *thread problem*, and prove the existence and regularity of minimal surfaces with *movable boundary parts of fixed lengths*, which in soap film experiments are formed by very thin threads.

The last chapter contains a new approach to the celebrated result that a minimizer of area in a given contour has no interior branch points. The novelty consists particularly in the fact that, in certain cases, relative minimizers of Dirichlet's integral are shown to be free of nonexceptional branch points, and this is achieved by a purely analytical reasoning.

The Scholia serve as sources of additional information. In particular we try to give credit to the authorship of the results presented in the main text, and we sketch some of the main lines of the historical development. References to the literature and brief surveys of relevant topics not treated in our text complete the picture.

Our *notation* is essentially the same as in the treatises of Morrey [8] and of Gilbarg and Trudinger [1]. Sobolev spaces are denoted by H_p^k instead of $W^{k,p}$; the definition of the classes C^0, C^k, C^∞ and $C^{k,\alpha}$ is the same as in Gilbarg and Trudinger [1]; C^ω denotes the class of real analytic functions; $C_c^\infty(\Omega)$ stands for the set of C^∞-functions with compact support in Ω. For greater precision we write $C^k(\Omega, \mathbb{R}^3)$ for the class of C^k-mappings $X : \Omega \to \mathbb{R}^3$, whereas the corresponding class of scalar functions is denoted by $C^k(\Omega)$, and similarly for the other classes of differentiability. Another standard symbol is $B_r(w_0)$ for the disk $\{w = u + iv \in \mathbb{C} : |w - w_0| < r\}$ in the complex plane. On some occasions it is convenient to switch several times from this meaning of B to another one. Moreover, some definitions based on one meaning of B have to be transformed mutatis mutandis to the other one. This may sometimes require slight changes but we have refrained from pedantic adjustments which the reader can easily supply himself.

Part I

Boundary Behaviour of Minimal Surfaces

Chapter 1

Minimal Surfaces with Free Boundaries

This chapter is centered on the proof of existence theorems for minimal surfaces with completely free boundaries. We approach the problem by applying the direct methods of the calculus of variations, thus establishing the existence of minimizers with a boundary on a given supporting surface S. However, this method does not yield the existence of stationary minimal surfaces which are not area minimizing. As certain kinds of supporting surfaces are not able to hold nontrivial minimizers, our method is restricted by serious topological limitations. For example, it does not furnish existence of nontrivial stationary minimal surfaces within a closed convex surface. It seems that the techniques of geometrical measure theory are best suited to handle this problem. Unfortunately they are beyond the scope of our lecture notes, but we shall at least present a survey of the pertinent results in Section 1.8 as well as an existence result for the particular case of S being a tetrahedron. There the reader will also find references to the literature.

In the following we shall describe Courant's method for proving the existence of a nontrivial and minimizing minimal surface whose boundary lies on a given closed supporting surface. This problem is more difficult than the Plateau problem or the semifree problem treated in Chapter 4 of Vol. 1 because an arbitrary minimal sequence will shrink to a single point. In order to exclude this phenomenon, we have to impose suitable topological conditions on the boundary values of admissible surfaces. For instance, one could assume that the boundary values are continuous curves on S which are contained in a prescribed homotopy class. This approach would, however, lead to a rather difficult problem. One would first have to prove that a suitable minimizing sequence tends to a limit with continuous boundary values, and then one would have to show that these boundary values lie in a prescribed homotopy class. Therefore we abandon this idea.

Instead we show in Section 1.1 how a kind of homotopy class can be set up for surfaces X which are of class $H_2^1(B, \mathbb{R}^3)$ and have their boundary values on S. We shall also prove by way of example that the problem of prescribed

U. Dierkes, S. Hildebrandt, A.J. Tromba, *Regularity of Minimal Surfaces*,
Grundlehren der mathematischen Wissenschaften 340,
DOI 10.1007/978-3-642-11700-8_1, © Springer-Verlag Berlin Heidelberg 2010

homotopy class need not have a solution. In Section 1.2 we set up the classes of admissible functions for which we can solve the minimum problem and in which we are able to find nondegenerate solutions.

The free boundary problem will be solved in Section 1.3; the supporting set S may look as bizarre as the one in Fig. 1 or as simple as the catenoid. The gist of our reasoning consists in an indirect argument showing that the limit of a suitable minimizing sequence satisfies the prescribed topological condition, and therefore it will be a nondegenerate solution of the minimum problem.

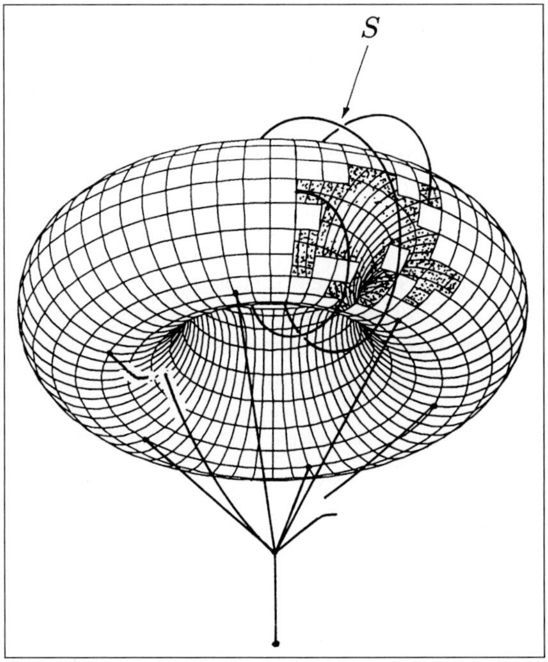

Fig. 1. A bizarre supporting set

The remaining part of the chapter will deal with additional properties of minimal surfaces with free boundaries.

In Section 1.4 we give a precise definition of a *stationary minimal surface* X whose free boundary lies on a given support surface S. Here we do not require X to be a minimizer. It will be investigated how the condition of being stationary is linked with the condition that X intersects S perpendicularly at its free trace Σ, provided Σ does not touch the boundary of S. This discussion is used in Section 1.5 to set up necessary conditions for the existence of stationary minimal surfaces with boundary on S. This will lead us to a class of non-existence results which explain, for example, why soap films in a funnel always run to its narrow end.

In Section 1.6 we prove the existence of three embedded stationary surfaces with their boundaries on a tetrahedron, following the discussion of B. Smyth. This is a case where the minimizing approach cannot be used.

Section 1.7 is concerned with stationary surfaces whose boundaries lie on a sphere. We shall prove Nitsche's result that flat disks are the only solutions to this problem that are of the type of the disk.

After a report on the existence of stationary minimal surfaces with boundaries on a convex surface (Section 1.8), in Section 1.9 we shall present some results concerning uniqueness and nonuniqueness of minimal surfaces with a free boundary on a given support surface. In particular, we construct a family of minimizing minimal surfaces with boundaries on a regular, real analytic surface of the topological type of a torus which are nonisometric to each other. Moreover, we discuss some finiteness results of Alt & Tomi for minimizers with boundaries on a real analytic supporting surface.

1.1 Surfaces of Class H_2^1 and Homotopy Classes of Their Boundary Curves. Nonsolvability of the Free Boundary Problem with Fixed Homotopy Type of the Boundary Traces

Let us fix some closed set S in $\mathbb{R}^3$. Then we want to define the class $\mathcal{C}(S)$ of surfaces $X \in H_2^1(B, \mathbb{R}^3)$ with boundary values $X|_{\partial B}$ on S. The parameter domain B will be chosen as the unit disk:

$$B = \{w = u + iv\colon |w| < 1\}.$$

In the following we shall usually pick an ACM-representative[1] for a given Sobolev mapping X. If we work with polar coordinates r, θ about the origin, i.e., $w = re^{i\theta}$, this means that we choose a representative $X(r, \theta)$ such that $X(r, \cdot)$ is absolutely continuous for almost all $r \in (0, 1)$, and that $X(\cdot, \theta)$ is absolutely continuous for almost all $\theta \in (0, 2\pi)$. Thus X is in particular a continuous function on almost all circles $C_r = \{w \in \mathbb{C}\colon |w| = r\}$.

Any function $X \in H_2^1(B, \mathbb{R}^3)$ possesses a *trace* (or *boundary values*) ξ on ∂B which is of class $L_2(C, \mathbb{R}^3), C := \partial B$, and we have both

(1) $$\lim_{r \to 1-0} X(r, \varphi) = \xi(\varphi) \quad \text{for almost all } \varphi \in [0, 2\pi]$$

and

(2) $$\lim_{r \to 1-0} \int_0^{2\pi} |X(r, \varphi) - \xi(\varphi)|^2 \, d\varphi = 0.$$

[1] ACM stands for absolutely continuous in the sense of Morrey; cf. Morrey [8], Lemma 3.1.1.

However, the trace $\Sigma = \{\xi(\varphi)\colon \varphi \in [0, 2\pi]\}$ of an arbitrary Sobolev function $X \in H_2^1(B, \mathbb{R}^3)$ will in general not be a continuous curve, whereas the curves

$$\Sigma_r := \{X(r, \varphi)\colon 0 \le \varphi \le 2\pi\}$$

are absolutely continuous for a.a. $r \in (0, 1)$. As we cannot formulate topological conditions for a possibly noncontinuous curve Σ, we shall use the continuous curves Σ_r as a substitute. In view of (1) and (2) we can expect that conditions on curves Σ_r close to Σ express conditions on Σ in an appropriate sense.

We begin, however, by defining the class $\mathcal{C}(S)$ of surfaces with boundary values on a supporting set S. *We assume once and for all that supporting sets S are closed, proper, and nonempty subsets of $\mathbb{R}^3$. However, if a boundary configuration contains other parts besides S, we allow S to be empty.*

Definition 1. *Let S be a supporting set in $\mathbb{R}^3$. Then we denote by $\mathcal{C}(S)$ the class of functions $X \in H_2^1(B, \mathbb{R}^3)$ whose L_2-trace $\xi := X|_C$ sends almost every $w \in C = \partial B$ into S.*

For any closed set S in $\mathbb{R}^3$, $S \ne \emptyset$, and for any number $\mu > 0$, we define the *tubular μ-neighbourhood T_μ* $= T_\mu(S)$ of S by

$$T_\mu(S) := \{x \in \mathbb{R}^3 : \operatorname{dist}(x, S) < \mu\}. \tag{3}$$

Then we can formulate our first result on surfaces of class $\mathcal{C}(S)$ which will shed some light on their boundary behaviour.

Theorem 1. *Let S be a supporting set in $\mathbb{R}^3$, and suppose that X belongs to $\mathcal{C}(S)$. Then, for every $\mu > 0$ and every $\varepsilon > 0$, there is a subset $\mathcal{I} \subset (1-\varepsilon, 1)$ of positive measure such that, for all $r \in \mathcal{I}$, the curve $\Sigma_r = \{X(r, \varphi)\colon 0 \le \varphi \le 2\pi\}$ is a closed continuous curve which is contained in the tubular neighbourhood $T_\mu(S)$ of S.*

Note that other curves $\Sigma_r, r \in (1-\varepsilon, 1) \setminus \mathcal{I}$, may stay arbitrarily far from $T_\mu(S)$ as can be shown by simple examples; cf. Fig. 1.

We shall prove Theorem 1 in several steps.

Lemma 1. *For any closed set S in $\mathbb{R}^3$, the function $d_s := \operatorname{dist}(\cdot, S)$ is Lipschitz continuous on $\mathbb{R}^3$ with a Lipschitz constant less than or equal to one.*

Proof. For arbitrary points $P_1, P_2 \in \mathbb{R}^3$ there exist points $Q_1, Q_2 \in S$ such that

$$d_s(P_1) = |P_1 - Q_1| = \inf_{Q \in S} |P_1 - Q|,$$
$$d_s(P_2) = |P_2 - Q_2| = \inf_{Q \in S} |P_2 - Q|.$$

Therefore we obtain

$$d_S(P_2) \le |P_2 - Q_1|$$

and

$$d_S(P_2) - d_S(P_1) \le |P_2 - Q_1| - |P_1 - Q_1| \le |P_2 - P_1|,$$

and analogously

$$d_S(P_1) - d_S(P_2) \le |P_1 - P_2|.$$

Therefore we have

$$|d_S(P_1) - d_S(P_2)| \le |P_1 - P_2| \quad \text{for all } P_1, P_2 \in \mathbb{R}^3. \qquad \square$$

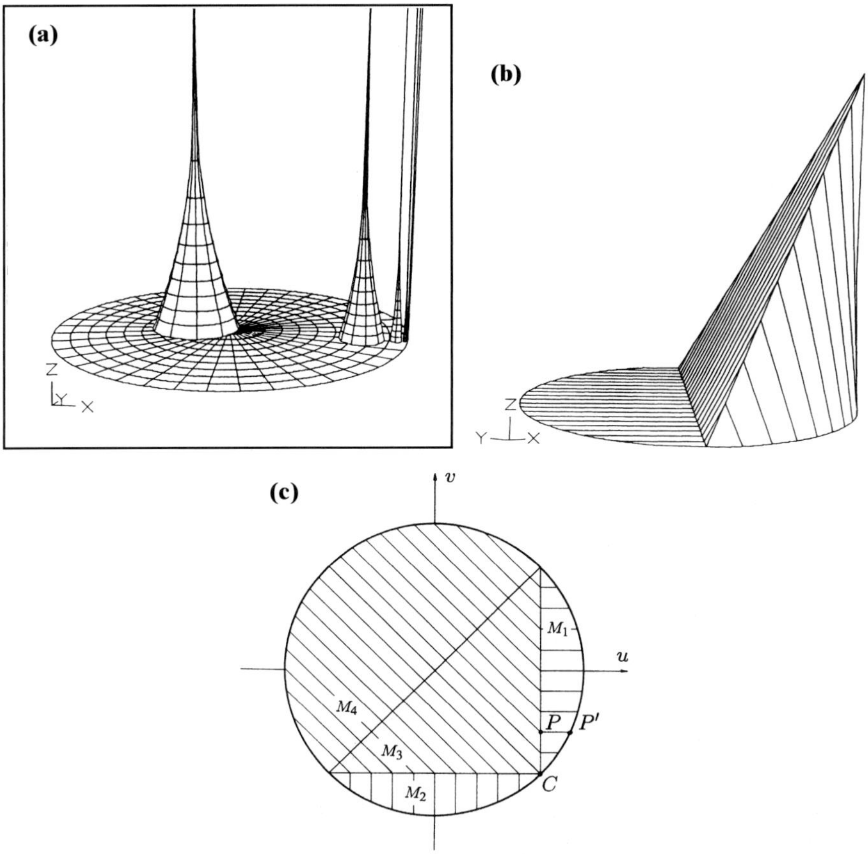

Fig. 1. **(a)** The graph of a bizarre function $f \in \mathring{H}_2^1(B)$ which has infinitely many peaks congruent to a part of the graph of $\log|\log|w||$. These peaks converge to the point $w = 1$ on ∂B. Given $\varepsilon > 0$ and $\delta > 0$, there is a set of values $r \in (1-\delta, 1)$ of positive measure such that the absolute values of f on C_r remain less that ε; see Lemma 4. This is a borderline case of the boundary behaviour of functions of class $\mathring{H}_p^1(B)$. For $p > 2$ they are continuous up to ∂B, and therefore their values on *all* circles sufficiently close to ∂B remain close to zero. For $p < 2$ there may be *no* such circle, as is shown by the function depicted in **(b)** and **(c)** which belongs to $\mathring{H}_p^1(B)$ for all $p \in (1, 2)$ and has a discontinuity at $C \in \partial B$

Lemma 2. *A function $X \in H_2^1(B, \mathbb{R}^3)$ belongs to $\mathcal{C}(S)$ if and only if the scalar function $d_S \circ X$ is an element of the space $\mathring{H}_2^1(B)$ of functions $f \in H_2^1(B)$ with generalized boundary values zero.*

Proof. Note that $X \in H_2^1(B, \mathbb{R}^3)$ implies $d_S \circ X \in H_2^1(B)$. Then the assertion follows from well-known properties of functions of class $\mathring{H}_2^1(B)$ (see Gilbarg and Trudinger [1]). □

Lemma 3. *Let X belong to $H_2^1(B, \mathbb{R}^N), N \geq 1$. Then, for any two numbers $\mu > 0$ and $\delta > 0$, there is an $\varepsilon > 0$ with the following property:*

If $\mathcal{I}' = [\theta_1, \theta_2]$ is an angular interval with $\theta_2 - \theta_1 = \delta$, then there exists a subset $\sigma \subset \mathcal{I}'$ of positive measure such that

$$|X(1,\theta) - X(r,\theta)| \leq \mu$$

holds for all $\theta \in \sigma$ and for all $r \in (1-\varepsilon, 1)$. In fact, we can choose ε as

$$\varepsilon = \min\left\{\frac{1}{2}, \frac{1}{4}\frac{\mu^2\delta}{D(X)}\right\}. \tag{4}$$

Proof. From

$$r\int_{\theta_1}^{\theta_2}\int_r^1 |X_\rho(\rho,\theta)|^2\, d\rho\, d\theta \leq 2D(X)$$

we conclude that there is a subset $\sigma \subset [\theta_1, \theta_2]$ of positive measure such that

$$\int_r^1 |X_\rho(\rho,\theta)|^2\, d\rho \leq \frac{2}{r\delta}D(X)$$

holds for all $\theta \in \sigma$ and for $\delta = \theta_2 - \theta_1$. Moreover, we have

$$\begin{aligned} |X(1,\theta) - X(r,\theta)| &\leq \int_r^1 |X_\rho(\rho,\theta)|\, d\rho \\ &\leq \sqrt{1-r}\left(\int_r^1 |X_\rho(\rho,\theta)|^2\, d\rho\right)^{1/2} \end{aligned}$$

for $\theta \in \sigma$ and $0 < r < 1$, whence

$$|X(1,\theta) - X(r,\theta)| \leq \{2r^{-1}(1-r)\delta^{-1}D(X)\}^{1/2} \quad \text{for } \theta \in \sigma.$$

Choosing ε as in (4), the assertion follows at once. □

Lemma 4. *Let f belong to $\mathring{H}_2^1(B)$. Then, for any $\mu > 0$ and any $\varepsilon > 0$, the set $\mathcal{I} := \{r\colon 1-\varepsilon < r < 1, |f|_{0,C_r} < \mu\}$ has positive measure.*

Proof. Suppose that the assertion were false. Then we would have $D(f) > 0$, and there were numbers $\varepsilon > 0$ and $\mu > 0$ such that

$$|f|_{0,C_r} \geq \mu \tag{5}$$

for almost all $r \in (1-\varepsilon, 1)$. Without loss of generality we can assume that

$$0 < \mu < \sqrt{D(f)}$$

holds true.

Because of (6) we infer that, for almost all $r \in (1-\varepsilon, 1)$, there is an angle $\theta(r)$ such that

$$|f(re^{i\theta(r)})| \geq \mu.$$

Furthermore we choose some $\delta \in (0, 1)$ such that

$$\varepsilon' := \min\left\{\frac{1}{2}, \frac{\mu^2\delta}{16D(f)}\right\} \tag{6}$$

satisfies $0 < \varepsilon' < \varepsilon$. By Lemma 3, every angular interval $\mathcal{I}'$ of width δ contains an angle θ' such that $f(\cdot, \theta')$ is absolutely continuous and that

$$|f(re^{i\theta'})| < \tfrac{1}{2}\mu \quad \text{for all } r \in (1-\varepsilon', 1).$$

Conclusion: For almost all $r \in (1-\varepsilon', 1)$, there exist angles $\theta(r)$ and $\theta'(r)$ with $|\theta(r) - \theta'(r)| < \delta$ and

$$|f(re^{i\theta(r)})| \geq \mu, \quad |f(re^{i\theta'(r)})| \leq \frac{\mu}{2}.$$

Thus

$$\frac{\mu}{2} \leq \left| \int_{\theta(r)}^{\theta'(r)} |f_\theta(re^{i\theta})| \, d\theta \right|$$

and consequently

$$\frac{\mu^2}{4\delta} \leq \int_0^{2\pi} f_\theta^2(re^{i\theta}) \, d\theta.$$

Thus

$$\begin{aligned}\int_{\{1-\varepsilon'<|w|<1\}} |\nabla f|^2 \, du \, dv &\geq \int_{1-\varepsilon'}^1 \int_0^{2\pi} \frac{1}{r^2} f_\theta^2(re^{i\theta}) r \, d\theta \, dr \\ &\geq \int_{1-\varepsilon'}^1 \left(\int_0^{2\pi} f_\theta^2(re^{i\theta}) \, d\theta \right) dr \geq \frac{\varepsilon'\mu^2}{4\delta}.\end{aligned}$$

Because of (6), we have

$$\int_{\{1-\varepsilon'<|w|<1\}} |\nabla f|^2 \, du \, dv \geq \frac{\mu^4}{64D(f)}$$

for $0 < \delta \ll 1$, and $\varepsilon' \to 0$ as $\delta \to +0$. This is impossible for an H_2^1-function. □

Proof of Theorem 1. The assertion of Theorem 1 is now an immediate consequence of the Lemmata 1–4. □

Remark 1. The assertion of Lemma 4 holds for trivial reasons if $f \in \mathring{H}_p^1(B)$ and $p > 2$, because Sobolev's embedding theorem yields that $f \in C^0(\overline{B})$ and $f = 0$ on ∂B. The assertion turns out to be false if $p < 2$, as one can find examples of functions $f \in \mathring{H}_p^1(B), p < 2$, such that near ∂B the function $|f(w)|$ is bounded away from zero by an arbitrary constant (cf. Fig. 1).

Now we want to give a reasonable definition for a homotopy class of a boundary mapping $\xi(\theta) = X(1,\theta)$ of a surface X of class $\mathcal{C}(S)$ which is not necessarily continuous on B. To this end we consider the curves $\Sigma_r = \{X(r,\theta) \colon 0 \le \theta \le 2\pi\}$ for r close to one which are absolutely continuous and lie in a tubular neighbourhood T_μ of S. By Theorem 1, there exist sufficiently many of them: In fact, for any number $\varepsilon \in (0,1)$ there is a set $\mathcal{I} \subset (1-\varepsilon, 1)$ of positive measure such that, for every $r \in \mathcal{I}$, the mapping $X(r,\cdot)$ is absolutely continuous and $\Sigma_r \subset T_\mu$.

Now we can state the following result:

Theorem 2. *Let T_μ be the μ-neighbourhood of some closed set S in $\mathbb{R}^3$, and suppose that $X \in \mathcal{C}(S)$. Then for $\delta := \frac{1}{4}\pi\mu^2 > 0$, the following holds true:*

If $r_1, r_2 \in (0,1)$ are two radii such that

(i) *the Dirichlet integral of X over the annulus*

$$\Omega(r_1, r_2) := \{w \in \mathbb{C} \colon r_1 < |w| < r_2\}$$

is at most δ;

(ii) *the curves $X|_{C_1}$ and $X|_{C_2}$ with $C_k := C_{r_k} = \{w \colon |w| = r_k\}$ are absolutely continuous, and their traces $\Sigma_k := X(C_k)$ are contained in $T_{\mu/2}$;*

(iii) *there is an angle θ such that the curve $X(r,\theta), r_1 \le r \le r_2$, connecting Σ_1 and Σ_2 is absolutely continuous and that its trace lies in $T_{\mu/2}$; then the curves $X|_{C_1}$ and $X|_{C_2}$ are homotopic in T_μ.*

Recall that two closed continuous curves $\gamma_1 \colon C \to T_{\mu/2}$ and $\gamma_2 \colon C \to T_{\mu/2}$ are *homotopic in* T_μ if there is a continuous map $H \colon C \times [0,1] \to T_\mu$ such that $H(\cdot, 0) = \gamma_1$ and $H(\cdot, 1) = \gamma_2$. The mapping H is called a *homotopy.*

Furthermore, a closed curve $\gamma \colon C \to T_{\mu/2}$ is *contractible in* T_μ if it is homotopic in T_μ to a constant map or, equivalently, if it extends to a continuous map $\overline{B} \to T_\mu$.

Remark 2. Close to $C = \partial B$, the angle θ appearing in condition (iii) can be found by virtue of Lemma 3.

The *proof of Theorem 2* can be reduced to proving the following

Lemma 5. *Let T_μ be the μ-neighbourhood of some closed set S in $\mathbb{R}^3$, and set $\delta := \frac{1}{4}\pi\mu^2$. Suppose, moreover, that X is a mapping of class $H_2^1(B, \mathbb{R}^3) \cap C^0(\partial B, \mathbb{R}^3)$ whose boundary curve $X|_{\partial B}$ is contained in $T_{\mu/2}$ and which satisfies $D(X) < \delta(\mu)$. Then the curve $X|_{\partial B}$ is contractible in T_μ.*

In fact, let r_1 and r_2 be two radii as in Theorem 2, and let $\theta \in [0, 2\pi)$ be an angle as in (iii) of the theorem. Then we consider a conformal map τ of B onto the slit annulus

$$\{w = re^{i\varphi}: r_1 < r < r_2, \varphi \in [0, 2\pi), \varphi \neq \theta\}$$

and apply Lemma 5 to the surface $Z := X \circ \tau$, thus obtaining that $Z|_{\partial B}$ is contractible in T_μ. A straightforward reasoning now implies that the curves $X|_{C_1}$ and $X|_{C_2}$ are homotopic in T_μ.

Proof of Lemma 5. We begin by choosing a mapping $Y: \overline{B} \to \mathbb{R}^3$ which is harmonic in B, continuous on $\overline{B}$, of class $H_2^1(B, \mathbb{R}^3)$ and satisfies $Y - X \in \mathring{H}_2^1(B, \mathbb{R}^3)$ and $Y = X$ on ∂B. We know that $D(Y) \leq D(X)$. Since $X(\partial B) = Y(\partial B)$ is contained in $T_{\mu/2}$, there exists a strip $U = \{w: 1 - \varepsilon < |w| \leq 1\}$ about the boundary $C = \partial B$ such that $Y(U) \subset T_{\mu/2}$. Then we can find a regular, real analytic curve $Y|_{C_r}, r \in (1 - \varepsilon, 1)$, which is homotopic to $X|_C = Y|_C$ in $T_{\mu/2}$. Thereafter we can find a sequence $\{\Gamma_k\}$ of smooth closed Jordan curves Γ_k given by smooth topological mappings $\Phi_k: C \to \Gamma_k$ such that $|\Phi_k - \Phi|_{2,C} \to 0$ as $k \to \infty$ holds for the mapping $\Phi: C \to \mathbb{R}^3$ defined by $\Phi(e^{i\theta}) := Y(re^{i\theta})$.

Now let $Z(w) := Y(rw)$ and $Z_k(w)$ be the harmonic extensions to B of the boundary values Φ and Φ_k respectively, and let X_k be a solution of the variational problem $\mathcal{P}(\Gamma_k)$. Then the maximum principle implies $|Z_k - Z|_{0,\overline{B}} \to 0$ as $k \to \infty$ and, applying the estimate of Lemma 7 in Section 2.1 together with the Arzelà–Ascoli theorem, we also obtain $|Z_k - Z|_{1,\overline{B}} \to 0$ as $k \to \infty$. This implies

$$\lim_{k \to \infty} D(Z_k) = D(Z).$$

Consequently we have

$$A(X_k) = D(X_k) \leq D(Z_k) \to D(Z) = D_{B_r}(Y) \leq D(Y) \leq D(X).$$

By assumption, we have also

$$D(X) < \delta(\mu) = \tfrac{1}{4}\pi\mu^2,$$

whence

$$A(X_k) = D(X_k) < \pi(\mu/2)^2 \tag{7}$$

is satisfied for k sufficiently large.

If for one of these k the minimal surface X_k were not contained in T_μ, then there would exist some $w \in B$ such that $X_k(w) \notin T_\mu$. We choose a conformal selfmapping of B satisfying $\tau(0) = w$ and note that all the boundary values of $X_k \circ \tau$ lie outside the ball of radius $\mu/2$ centered at $X_k(w) = X_k(\tau(0))$. Then we infer from Vol. 1, Section 3.2, Proposition 2 that

$$A(X_k) \geq \pi(\mu/2)^2$$

which contradicts (7). Thus we have shown that

(8) $$X_k(\overline{B}) \subset T_\mu \quad \text{for all } k \gg 1.$$

Moreover, every minimal surface X_k furnishes a topological mapping of C onto Γ_k (see Vol. 1, Section 4.5, Theorem 3). Thus $X_k|_C$ furnishes a parameter representation of Γ_k equivalent to Φ_k, and we infer from (9) that Φ_k is contractible in T_μ for $k \gg 1$. Since $X|_C$ is homotopic in T_μ to all of the Γ_k with $k \gg 1$, we infer that $X|_C$ is contractible in T_μ. □

Recall now that $\mathcal{C}(S)$ has been defined as the class of all surfaces $X \in H_2^1(B, \mathbb{R}^3)$ having their boundary values $X|_C$ on a closed subset S in $\mathbb{R}^3$ (see Definition 1).

We now denote by $\tilde{\Pi}_1(S)$ the *set of all homotopy classes of closed paths* in S. (For details, we refer for instance to Schubert [1], or to Greenberg [1].)

Assumption (A). *Suppose that there is a number $\mu > 0$ such that the inclusion map $S \to T_\mu$ of the closed set S into its μ-neighbourhood T_μ induces a bijection from $\tilde{\Pi}_1(S)$ to $\tilde{\Pi}_1(T_\mu)$.*

For example, this assumption is fulfilled for sufficiently small $\mu > 0$ if S is a smooth compact submanifold of $\mathbb{R}^3$.

Let $\mu > 0$ be a number as in Assumption (A), and recall that the curves $X|_{C_r}$ are absolutely continuous for almost all $r \in (0, 1)$.

If $X \in \mathcal{C}(S)$, then there is a number $\varepsilon > 0$ such that any two curves $X|_{C_r}$ and $X|_{C_{r'}}$ contained in $T_{\mu/2}$ and with $r, r' \in (1 - \varepsilon, 1)$ define the same homotopy class in $\tilde{\Pi}_1(T_\mu)$; this homotopy class will be viewed as *homotopy class of the boundary values* $X|_C$. It is denoted by $[X|_C]$ and will be called *the boundary class of a surface* $X \in \mathcal{C}(S)$. Because we have a bijection

$$\tilde{\Pi}_1(T_\mu) \leftrightarrow \tilde{\Pi}_1(S),$$

we can view the class $[X|_C]$ as an element of $\tilde{\Pi}_1(S)$. If the mapping $X: C \to \mathbb{R}^3$ is continuous, then $[X|_C]$ coincides with the usual homotopy class of $X|_C$.

Note that the definition of the homotopy class $[X|_{\partial B}]$ does not depend on the particular *ACM*-representative of X that we have chosen since any two of them coincide on almost all circles C_r.

Moreover, the definition $[X|_C]$ is even independent of μ in the following sense: Suppose that the inclusion maps $S \to T_\mu$ and $S \to T_{\mu'}$ induce two bijections $\tilde{\Pi}_1(S) \leftrightarrow \tilde{\Pi}_1(T_\mu)$ and $\tilde{\Pi}_1(S) \leftrightarrow \tilde{\Pi}_1(T_{\mu'})$. Then both constructions with respect to μ and μ' lead to the same class $[X|_C]$ in $\tilde{\Pi}_1(S)$.

Indeed, according to the definition we first have to choose an $\varepsilon > 0$ such that any two of the curves $X|_{C_r}$, $r \in (1 - \varepsilon, 1)$, lying completely in $T_{\mu/2}$ (or in $T_{\mu'/2}$) are homotopic in T_μ (or in $T_{\mu'}$). This ε may be the same for μ and μ' because decreasing ε does not change the class $[X|_{\partial B}]$. If, say, $\mu' \le \mu$, then we find in $(1 - \varepsilon, 1)$ a subset $\mathfrak{J}'$ of positive measure or radii r such that the curves $X|_{c_r}$, $r \in \mathfrak{J}'$, are completely contained in $T_{\mu'/2}$ and that any two of them are homotopic in $T_{\mu'}$. Therefore all these curves $X|_{C_r}$ define a homotopy class α' in $\tilde{\Pi}_1(T_{\mu'})$ which corresponds to the boundary class $[X|_{\partial B}]' \in \tilde{\Pi}_1(S)$ which is constructed by means of $T_{\mu'}$.

On the other hand, all curves $X|_{C_r}, r \in \mathcal{J}'$, are contained in $T_{\mu/2} \supset T_{\mu'/2}$, and any two of them are homotopic in $T_\mu \supset T_{\mu'}$. Therefore all these curves $X|_{C_r}, r \in \mathcal{J}'$, define a homotopy class $\alpha \in \tilde{\Pi}_1(T_\mu)$ which by the definition of ε corresponds to the boundary class $[X|_{\partial B}] \in \tilde{\Pi}_1(S)$ defined by means of T_μ. Since the inclusion $T_{\mu'} \to T_\mu$ induces a bijection $\tilde{\Pi}_1(T_{\mu'}) \to \tilde{\Pi}_1(T_\mu)$ which maps α' to α, the boundary classes $[X|_{\partial B}]$ and $[X|_{\partial B}]'$ and identical. □

Collecting our results and inspecting Chapter 4 of Vol. 1, we obtain the following

Theorem 3 (Natural boundary classes). *Let S be a subset of $\mathbb{R}^3$ such that for some $\mu > 0$ the inclusion $S \to T_\mu$ induces a bijection $\tilde{\Pi}_1(S) \to \tilde{\Pi}_1(T_\mu)$ between the corresponding sets $\tilde{\Pi}_1$ of homotopy classes of closed paths in S and T_μ respectively.*

(i) *Then for every surface $X \in \mathcal{C}(S)$ a boundary homotopy class $[X|_{\partial B}] \in \tilde{\Pi}_1(S)$ is defined in a natural way.*

(ii) *If σ is a closed curve in S which is not contractible in S and if $[\sigma] \in \tilde{\Pi}_1(S)$ denotes its homotopy class, then every minimizer of the Dirichlet integral $D(X) = \frac{1}{2}\int_B |\nabla X|^2\, du\, dv$ in the class*

$$\mathcal{C}(\sigma, S) := \{X \in \mathcal{C}(S) \colon [X|_{\partial B}] = [\sigma]\} \tag{9}$$

is a minimal surface.

Let us denote the minimum problem

$$D(X) \to \min \quad \text{in } \mathcal{C}(\sigma, S) \tag{10}$$

by $\mathcal{P}(\sigma, S)$.

In general one encounters serious difficulties if one tries to solve the problem $\mathcal{P}(\sigma, S)$. For instance, the classes $\mathcal{C}(\sigma, S)$ are not necessarily closed with respect to weak convergence in H_2^1; yet this fact was crucial for the existence proof carried out in Section 4.6 of Vol. 1.

All basic difficulties of this problem can already be seen in the comparatively simple case that we shall consider next. The reader who is not interested in the details of the following discussion may very well skip it since it is not anymore needed in the later sections.

Let us choose a *torus* T in $\mathbb{R}^3$ as the prescribed supporting surface, and consider the corresponding variational problem

$$\mathcal{P}(\sigma, T): \qquad D(X) \to \min \text{ in } \mathcal{C}(\sigma, T).$$

To be precise, let T be the torus in $\mathbb{R}^3$ which is obtained by revolving the circle

$$\{(x,y,z) : y = 0, (x-R)^2 + z^2 = r^2\}, \quad 0 < r < R,$$

about the z-axis (see Fig. 2). Denote by $\sigma_1, \sigma_2 \colon [0, 2\pi] \to T$ the two circles

$$\sigma_1(t) = (R - r\cos t, 0, -r\sin t)$$

and

$$\sigma_2(t) = ((R-r)\cos t, (R-r)\sin t, 0).$$

Finally let $P = \sigma_1(0) = \sigma_2(0) = (R-r, 0, 0)$ be the base point of T.

Note that in this case the assumption made in the construction of the boundary classes $[X|_{\partial B}]$ of a surface $X \in \mathcal{C}(T)$, namely that the inclusion map $T \to T_\mu$ induces a bijection $\tilde{\Pi}_1(T) \leftrightarrow \tilde{\Pi}_1(T_\mu)$, is satisfied for all sufficiently small μ since for these μ the above inclusion $T \to T_\mu$ is a homotopy equivalence.

In general the set $\tilde{\Pi}_1(M)$ of all equivalence classes of (freely) homotopic closed curves in a topological space M is different from its fundamental group $\Pi_1(M, *)$; but if Π_1 is Abelian and if M is connected, then the canonical map $\Pi_1(M, *) \to \tilde{\Pi}_1(M), [\sigma] \to [\sigma]$, is indeed a bijection (cf. Schubert [1]).

The fundamental group of the torus T is isomorphic to $\mathbb{Z} \oplus \mathbb{Z}$, and it is freely generated by $[\sigma_1]$ and $[\sigma_2]$. Therefore, in the case of the torus, the class $\mathcal{C}(T)$ of all H_2^1-surfaces with boundary values on T is the disjoint union of the classes $\mathcal{C}^{k,l}, k, l \in \mathbb{Z}$, of surfaces $X \in \mathcal{C}(T)$ whose boundary class $[X|_{\partial B}]$ can be represented by the closed path $\sigma_1^k \cdot \sigma_2^l$. (First k-times along σ_1, then l times along σ_2, negative powers denote reversal of orientation.)

Now we can state our *nonexistence result.*

Theorem 4. *Let T be the torus defined before.*

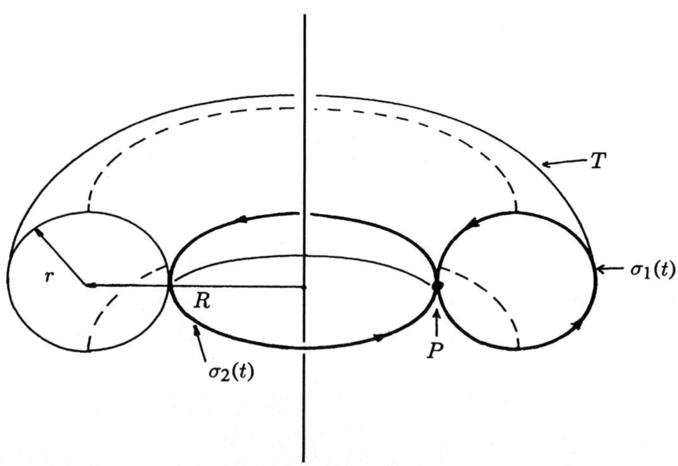

Fig. 2. The points and curves on a torus T used in the study of minimizing sequences for the Dirichlet integral of surfaces with free boundaries on T whose boundary curves have a prescribed homotopy class

(i) *For all* $k, l \in \mathbb{Z}$ *the numbers* $d_{k,l} := \inf\{D(X) \colon X \in \mathcal{C}^{k,l}\}$ *are given by*

$$d_{k,l} = \pi\{|k|r^2 + |l|(R-r)^2\}.$$

(ii) *The variational problem*

$$D(X) \to \min \quad \text{in } \mathcal{C}^{\,k,l}$$

has a solution if and only if $k = 0$ *or* $l = 0$.

For the proof of Theorem 4 we shall need the following

Lemma 6 (A formula for the oriented area). *Assume that the boundary values of a mapping* $X = (X^1, X^2) \in H_2^1(B, \mathbb{R}^2)$ *are contained in* $\mathbb{R}^2 \setminus B_\rho(w_0)$. *Then the boundary class* $[X|_{\partial B}] \in \tilde{\Pi}_1(\mathbb{R}^2 \setminus B_\rho(w_0))$ *is well defined, and it is characterized by the winding number* $U([X|_{\partial B}], w_0)$. *If* $\Omega := \{w \in B \colon X(w) \in B_\rho(w_0)\}$, *then we have for the oriented area*

$$A^0_\Omega(X) := \int_\Omega X_u \wedge X_v \, du\, dv$$

of the mapping X *the formula*

$$A^0_\Omega(X) := \int_\Omega \{X_v^1 X_v^2 - X_v^1 X_u^2\}\, du\, dv = \pi\rho^2 U([X|_{\partial B}], w_0).$$

Proof of Lemma 6. Approximating H_2^1-mappings $Z \in H_2^1(B, \mathbb{R}^2)$ by smooth mappings, we obtain the following two formulas that are well known for smooth maps:

(i) For almost all $R \in (0,1)$, the oriented surface area of Z is given by

$$A^0_{B_R}(Z) = \frac{1}{2}\int_0^{2\pi} \{Z^1 Z^2_\theta - Z^2 Z^1_\theta\}\, d\theta.$$

(ii) If Z is absolutely continuous on ∂B_R, then

$$U(Z|_{\partial B_R}, 0) = \frac{1}{2\pi}\int_0^{2\pi} \frac{Z^1 Z^2_\theta - Z^2 Z^1_\theta}{|Z|^2}\, d\theta$$

unless $Z = 0$ somewhere on ∂B_R. Of course, $0 < R < 1$ and $Z = Z(Re^{i\theta})$, etc.

Let us now prove the lemma. We may assume without loss of generality that $w_0 = 0$. Moreover, for $0 < \varepsilon < \rho$, let $\pi_\varepsilon \colon \mathbb{R}^2 \to \overline{B}_{\rho-\varepsilon}(0)$ denote the radial projection

$$Z \mapsto \begin{cases} Z & \text{if } |Z| < \rho - \varepsilon, \\ \frac{Z}{|Z|}(\rho - \varepsilon) & \text{otherwise,} \end{cases}$$

and set $Y^\varepsilon := \pi_\varepsilon \circ X$, which is again of class $H_2^1(B, \mathbb{R}^2)$ since π_ε is Lipschitz continuous. The boundary values of Y^ε are contained in $\mathbb{R}^2 \setminus B_{\rho-\varepsilon}(0)$, and we have

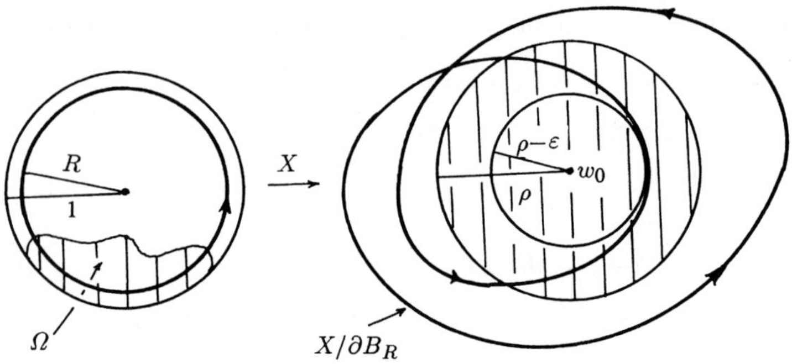

Fig. 3. The area functional $A(X)$ of a map $X\colon B \to \mathbb{R}^2$, whose boundary curve winds around a disk in $\mathbb{R}^2$, can be calculated from the radius of the disk and the winding number of the boundary curve, cf. Lemma 6

$$U([X|_{\partial B}], 0) = U([Y^\varepsilon|_{\partial B}], 0).$$

Now we choose R so close to 1 that $X|_{\partial B_R} \subset \mathbb{R}^2 \setminus B_{\rho-\varepsilon}(0)$ represents the boundary class $[X|_{\partial B}]$ and that the integration-by-parts-formula (i) holds true. Then we conclude that

$$\begin{aligned} U([X|_{\partial B}], 0) &= U(X|_{\partial B_R}, 0) = U(Y^\varepsilon|_{\partial B_R}, 0) \\ &= \frac{1}{2\pi} \int_0^{2\pi} \frac{Y^\varepsilon \wedge Y^\varepsilon_\theta}{|Y^\varepsilon|^2}\bigg|_{r=R} d\theta = \frac{1}{\pi(\rho-\varepsilon)^2} \frac{1}{2} \int_0^{2\pi} Y^\varepsilon \wedge Y^\varepsilon_\theta \, d\theta \\ &= \frac{1}{\pi(\rho-\varepsilon)^2} A^0_{B_R}(Y^\varepsilon). \end{aligned}$$

Now, on the one hand, $Y^\varepsilon_r \wedge Y^\varepsilon_\theta = 0$ almost everywhere on $\Omega_\varepsilon = \{|X| \ge \rho - \varepsilon\}$ since both Y^ε_r and Y^ε_θ are tangential to $\partial B_{\rho-\varepsilon}(0)$. On the other hand, we have $Y^\varepsilon_r \wedge Y^\varepsilon_\theta = X_r \wedge X_\theta$ almost everywhere on $\Omega'_\varepsilon = \{|X| < \rho - \varepsilon\}$. Therefore

$$A^0_{B_R}(Y^\varepsilon) = A^0_{\Omega'_\varepsilon \cap B_R}(X).$$

Thus at last, if ε decreases to zero, the radii R chosen above tend to one whence

$$A^0_{B_R}(Y^\varepsilon) \to A^0_\Omega(X),$$

and the lemma is proved. □

Now we turn to the

Proof of Theorem 4. Let $\pi_1 : \mathbb{R}^3 \to \mathbb{R}^2$ be the orthogonal projection onto the x, y-plane given by

$$(x, y, z) \to (x, y),$$

and denote by $\pi_2 : \mathbb{R}^3 \to \mathbb{R}^2$ the projection mapping

$$(x, y, z) \to (\rho, z) \quad \text{with } \rho = \sqrt{x^2 + y^2}$$

which maps (x, y, z) onto the point $(\rho, z) \in \mathbb{R}^2$ defined by the two cylinder coordinates ρ and z. Note that π_2 is Lipschitz continuous and, for $\rho \neq 0$, even real analytic.

Then, for any $X \in H_2^1(\Omega, \mathbb{R}^3)$, we have the following inequalities:[2]

(I) $D(\pi_1 \circ X) \le D(X)$, and the equality sign holds if and only if $\nabla z(w) = 0$ a.e. on Ω, where $z(w)$ is the third component of $X(w)$.

(II) $D(\pi_2 \circ X) \le D(X)$, and the equality sign holds if and only if $\nabla \varphi(w) = 0$ a.e. in Ω, where $\varphi(w) := \arctan \frac{y(w)}{x(w)}$ is the angle belonging to the cylinder coordinates ρ, φ, z. For the assertion of (II) to hold we have to assume that $X(\Omega) \Subset \mathbb{R}^3 \setminus \mathcal{H}$ where $\mathcal{H}$ is some halfplane in $\mathbb{R}^3$ having the z-axis as its boundary.

Now, given $X = (x, y, z) \in \mathcal{C}^{k,l}$, let us consider the sets

$$\Omega_1 := \{w \in B \colon x^2(w) + y^2(w) < (R - r)^2\}$$

and

$$\Omega_2 := \{w \in B \colon |\pi_2(X(w)) - (R, 0)| < r\}$$

which are the pre-images of the cylinder $\{0 \le \rho < R - r\}$ and of the open solid torus T, respectively. The sets Ω_1 and Ω_2 are disjoint.

From (I), (II) and Lemma 6 we infer

$$\begin{aligned} D(X) &\ge D_{\Omega_1}(X) + D_{\Omega_2}(X) \ge D_{\Omega_1}(\pi_1 \circ X) + D_{\Omega_2}(\pi_2 \circ X) \\ &\ge |A^0_{\Omega_1}(\pi_1 \circ X)| + |A^0_{\Omega_2}(\pi_2 \circ X)| \ge \pi |l| (R - r)^2 + \pi |k| r^2, \end{aligned}$$

that is,

$$\text{(III)} \qquad \pi(|l|(R - r)^2 + |k| r^2) \le D(X).$$

In order to complete the proof of the first part of the theorem, we construct a minimizing sequence as follows. For $0 < \rho \ll 1$, we introduce the set

$$\Omega_\rho := B_1(0) \cup B_\rho(1) \cup B_1(2).$$

As Ω_ρ is conformally equivalent to the unit disk B (see Fig. 4), we can choose Ω_ρ as parameter domain. For $k, l \ge 0$, we define

$$X_{\rho(w)} := \begin{cases} ((R - r)\mathrm{Re}\, w^l, (R - r)\mathrm{Im}\, w^l, 0) & \text{if } w \in \overline{B}_1(0) \setminus B_\rho(1), \\ (R - r\, \mathrm{Re}(2 - w)^k, 0, -r\, \mathrm{Im}(2 - w)^k) & \text{if } w \in \overline{B}_2(0) \setminus B_\rho(1). \end{cases}$$

If $k < 0$, we replace in the definition of X_ρ the variable $w \in \overline{B}_1(0) \setminus B_\rho(1)$ by $\overline{w}$, and for $l < 0$ we substitute $w \in \overline{B}_2(0) \setminus B_\rho(1)$ by $\overline{w}$.

---

[2] The proof of the second fact is not totally trivial. It can be derived by choosing an ACM-representation of $\pi_2 \circ X$ in conjunction with Fubini's theorem.

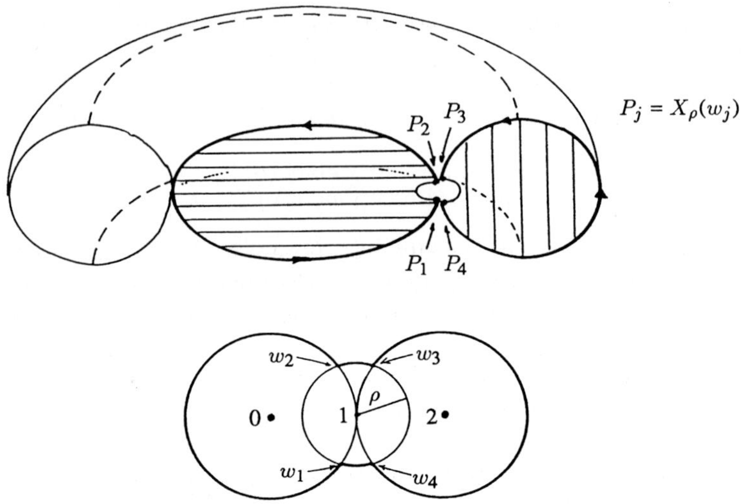

Fig. 4. Construction of a minimizing sequence for the Dirichlet integral for surfaces with free boundaries on T and a boundary class homotopic to σ_1 followed by σ_2

Now let $w_1, \ldots, w_4$ be the four vertices in Ω_ρ. Then we connect every two of the points $P_j := X(w_j)$ by geodesic lines on the torus T such that the curve $X_\rho|_{\partial\Omega_\rho}$ is homotopic to $\sigma_1^k \cdot \sigma_2^l$. These geodesics are parametrized in proportion to the arc length by means of the boundary pieces of $\partial\Omega_\rho$ between w_1 and w_4, w_2 and w_3.

Having thus defined $X|_{\partial B_\rho(1)}$, one completes the construction by filling in a harmonic surface in $B_\rho(1)$ with the boundary values X on $\partial B_\rho(1)$. Since $D_B(w^n) = \pi n$ and $D_{B_\rho(1)}(X_\rho)$ tends to zero with ρ, we have found a minimizing sequence.

In order to show part (ii) of the theorem, we consider a minimizer X in $\mathcal{C}^{k,l}$. Then X is harmonic in B and equality holds in (III). Our initial remarks (I) and (II) imply that $X(B)$ lies in a plane which either contains the z-axis (in which case $l = 0$) or is orthogonal to the z-axis (implying $k = 0$). Finally, minimizers in $\mathcal{C}^{k,0}$ and $\mathcal{C}^{0,l}$ can be constructed again using powers of w. □

1.2 Classes of Admissible Functions. Linking Condition

If we enlarge the class of admissible functions in a suitable way, the minimum problem becomes solvable. The difficulty consists in finding a proper class $\tilde{\mathcal{C}}$ of surfaces between $\mathcal{C}(\sigma, S)$ and $\mathcal{C}(S)$ such that the Dirichlet integral has a nondegenerate minimizer in $\tilde{\mathcal{C}}$. In this section we want to set up several of such classes $\tilde{\mathcal{C}}$ which serve this purpose.

To this end we shall assume throughout that S is a closed, proper, nonempty subset of $\mathbb{R}^3$ satisfying *Assumption* (A) of Section 1.1: There is a $\mu > 0$ such that the inclusion $S \to T_\mu$ induces a bijection $\tilde{\Pi}_1(S) \leftrightarrow \tilde{\Pi}_1(T_\mu)$. Then we can define

$$\mathcal{C}^+(S) := \bigcup_{[\sigma]\neq[\text{const}]} \mathcal{C}(\sigma, S), \tag{1}$$

where the union is to be taken over all closed curves σ in S which are not homotopic in S to a constant map. In other words, $\mathcal{C}^+(S)$ consists of all those surfaces $X \in \mathcal{C}(S)$ whose boundary class $[X|_{\partial B}]$ is not represented by a constant map.

Clearly, the position of the competing surfaces $X \in \mathcal{C}^+(S)$ is not particularly restricted. Therefore the minimizer in $\mathcal{C}^+(S)$ will always fill the smallest hole in S.

In order to specify the position of the boundary values of the competing surfaces more precisely, we choose some polygon Π (that is, a piecewise linear image of ∂B) which does not meet the tubular neighbourhood T_μ of S.

Then we introduce the variational class $\mathcal{C}(\Pi, S)$ of all surfaces $X \in \mathcal{C}(S)$ whose boundary class $[X|_{\partial B}]$ is linked with the polygon Π, that is, whose *linking number* $\mathcal{L}([X|_{\partial B}], \Pi)$ is nonzero:

$$\mathcal{C}(\Pi, S) := \{X \in \mathcal{C}(S) \colon \mathcal{L}([X|_{\partial B}], \Pi) \neq 0\}.$$

The classes $\mathcal{C}^+(S)$ and $\mathcal{C}(\Pi, S)$ will be the two sets on which we want to minimize the Dirichlet integral in order to obtain nondegenerate minimal surfaces with a free boundary on S. The minimizing procedures will be carried out in the next section.

For the convenience of the reader we shall in the following sketch the main features of the linking number. For proofs and further details we refer to the treatise of Alexandroff and Hopf [1].

Definition and Properties of the Linking Number

(I) First we define the intersection number of two oriented simplices $e^p = (a_0, \dots, a_p)$ and $f^q = (b_0, \dots, b_q)$ for two particular cases.

(α) If the corresponding geometric simplices furnished by the convex hulls of $\{a_0, \dots, a_p\}$ and $\{b_0, \dots, b_q\}$ are disjoint, then we define the intersection number $\emptyset(e^p, f^q)$ to be zero.

(β) If $p + q = 3$, and if the intersection of the corresponding geometrical simplices is neither empty nor does it contain any vertex of e^p, f^q, we define the intersection number $\emptyset(e^p, f^q)$ to be one if the ordered base $(a_1 - a_0, \dots, a_p - a_0, b_1 - a_0, \dots, b_q - a_0)$ has the same orientation as the standard simplex $e_1 = (1,0,0), e_2 = (0,1,0), e_3 = (0,0,1)$ of $\mathbb{R}^3$, and we set $\emptyset(e^p, f^q) = -1$ if the orientations are different.

(II) Secondly we define the linking number of two disjoint closed polygons Π_1 and Π_2.

Assume that Π_1 and Π_2 have r (resp. s) corners $P_{r+1} = P_1, \dots, P_r$ and $Q_{s+1} = Q_1, \dots, Q_s$, and choose a point $P \in \mathbb{R}^3$ such that any pair of simplices $e_j := (P, P_j, P_{j+1})$ and $f_k = (Q_k, Q_{k+1}), j = 1, \dots, r; k = 1, \dots, s$, satisfies one of the above conditions $(\alpha), (\beta)$ in (I). Then we define

$$\mathfrak{L}(\Pi_1, \Pi_2) := \sum_{j=1}^{r} \sum_{k=1}^{s} \emptyset(e_j, f_k)$$

as the *linking number of the two polygons* Π_1 *and* Π_2.

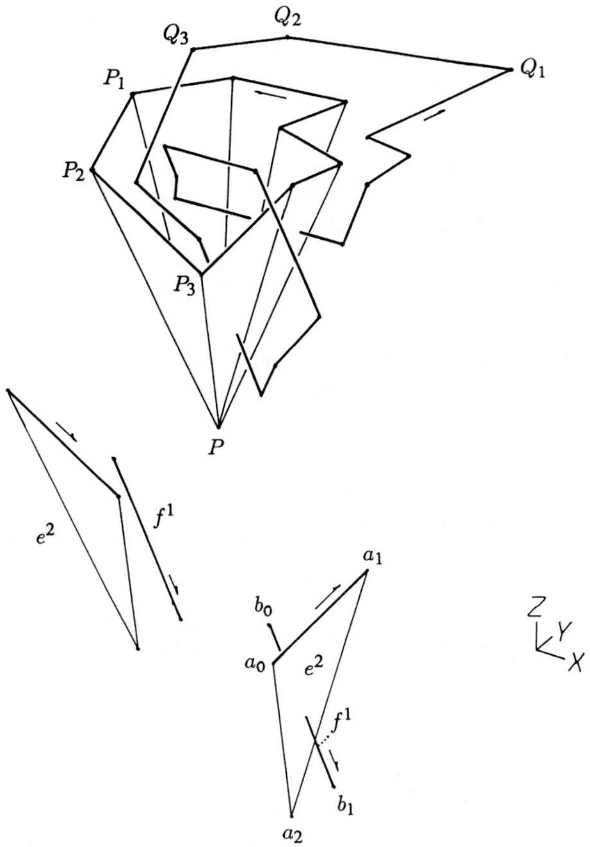

Fig. 1. The definition of the linking number of two closed polygons $P_1, P_2, \dots$ and $Q_1, Q_2, \dots$ is reduced to the intersection numbers of the faces (2-dimensional simplices) of a cone erected over the first polygon with the line segments (1-dimensional simplices) of the second. The intersection number is 0 if the simplices are disjoint, and +1 or −1 otherwise depending on their orientations. The resulting linking number for the polygons shown here is −2

(III) Finally, if c_1 and c_2 are two closed curves $\partial B \to \mathbb{R}^3$ with disjoint traces $c_i(\partial B)$, say, $\operatorname{dist}(c_1(\partial B), c_2(\partial B)) = \delta > 0$, then we choose two closed polygons Π_1 and Π_2 such that

$$|c_1 - \Pi_1|_{0,\partial B}, \quad |c_2 - \Pi_2|_{0,\partial B} < \frac{\delta}{2},$$

and define the *linking number of* c_1 *and* c_2 as

$$\mathfrak{L}(c_1, c_2) := \mathfrak{L}(\Pi_1, \Pi_2).$$

(IV) Some of its *properties* are:

(i) The definition of the linking number of two disjoint closed curves is independent of all choices made above (see Alexandroff and Hopf [1], p. 423).

(ii) *Deformation invariance.* If $h_1(t,\theta)$ and $h_2(t,\theta)\colon [0,1] \times \partial B \to \mathbb{R}^3$ are two homotopies of closed curves such that for every $t \in [0,1]$ the supports of the deformed curves are disjoint, then

$$\mathfrak{L}(h_1(0,\cdot), h_2(0,\cdot)) = \mathfrak{L}(h_1(1,\cdot), h_2(1,\cdot))$$

(see Alexandroff and Hopf [1], p. 424).

(iii) *Additivity of linking numbers.* If c_1, c_2 and c are three closed curves such that c_1 and c_2 have the same end points and that

$$c_i(\partial B) \cap c(\partial B) = \emptyset \quad \text{for } i = 1, 2,$$

then we have for the composite curve $c_1 \cdot c_2$

$$\mathfrak{L}(c_1 \cdot c_2, c) = \mathfrak{L}(c_1, c) + \mathfrak{L}(c_2, c).$$

This follows immediately from the construction (see Alexandroff and Hopf [1], p. 418).

(V) In view of the homotopy invariance of the linking numbers, the linking number of a boundary class $[X|_{\partial B}]$ with a polygon Π at a distance greater than μ from S is well defined:

$$\mathfrak{L}([X|_{\partial B}], \Pi) := \mathfrak{L}(X|_{C_R}, \Pi),$$

where $X|_{C_R}$, $R \in (0,1)$, is any curve in $T_{\mu/2}$ which represents the boundary class $[X|_{\partial B}]$.

1.3 Existence of Minimizers for the Free Boundary Problem

Let us now treat some free boundary problems for minimal surfaces with a prescribed supporting surface. We shall minimize the Dirichlet integral

$$D(X) = \frac{1}{2} \int_B |\nabla X|^2 \, du \, dv$$

both in $\mathcal{C}^+(S)$ and $\mathcal{C}(\Pi, S)$, the classes introduced in the previous sections. We shall describe some geometric conditions on S such that the two variational problems

$$\mathcal{P}(\Pi, S): \qquad D(X) \to \min \quad \text{in } \mathcal{C}(\Pi, S)$$

and

$$\mathcal{P}^+(S): \qquad D(X) \to \min \quad \text{in } \mathcal{C}^+(S)$$

are solvable.

By definition of the classes $\mathcal{C}^+(S)$ and $\mathcal{C}(\Pi, S)$, every solution of $\mathcal{P}(\Pi, S)$ and $\mathcal{P}^+(S)$ is nondegenerate.

Theorem 1. *Let S be a supporting set in $\mathbb{R}^3$ satisfying Assumption* (A) *of Section* 1.1, *i.e. there is some $\mu > 0$ such that the inclusion map $S \to T_\mu$ of S into its μ-neighbourhood T_μ induces a bijection from $\tilde{\Pi}_1(S)$ to $\tilde{\Pi}_1(T_\mu)$. Then we have:*

(i) *If there is a closed polygon Π in $\mathbb{R}^3$ which does not meet T_μ and for which $\mathcal{C}(\Pi, S)$ is nonempty, then there exists a solution of $\mathcal{P}(\Pi, S)$.*

(ii) *If S is compact and $\mathcal{C}^+(S)$ is nonempty, then there is a solution of $\mathcal{P}^+(S)$.*

(iii) *Any solution X of $\mathcal{P}(\Pi, S)$ or of $\mathcal{P}^+(S)$ is a minimal surface. That is, X is a nonconstant mapping of class $C^2(B, \mathbb{R}^3)$ and satisfies the equations*

$$\Delta X = 0, \tag{1}$$

$$|X_u|^2 = |X_v|^2, \quad \langle X_u, X_v \rangle = 0 \tag{2}$$

in B.

It is of great importance to investigate the boundary behaviour of solutions of $\mathcal{P}(\Pi, S)$ and $\mathcal{P}^+(S)$. If X is a solution of one of these problems that is smooth up to its boundary (say, $X \in C^1(\overline{B}, \mathbb{R}^3)$), and if S is a smooth surface with an empty boundary ∂S, then we shall prove in the next section that X meets S perpendicularly along its free trace $\Sigma = X(\partial B)$ on the supporting surface S. However, if ∂S is nonempty, then it may very well happen that Σ touches ∂S (this phenomenon is studied in Chapter 2, and in the Chapters 1, 2 of Vol. 3); then one cannot anymore expect that X meets S perpendicularly everywhere along Σ. In fact, a right angle between X and S is generally formed only at those parts of Σ which do not coincide with ∂S.

Moreover, we have to answer the question as to whether a solution of $\mathcal{P}(\Pi, S)$ or of $\mathcal{P}^+(S)$ is smooth on the closure $\overline{B}$ of its parameter domain B, so that we can apply the succeeding results of Section 1.4. A detailed discussion of this and related problems is given in Chapter 2. There and in

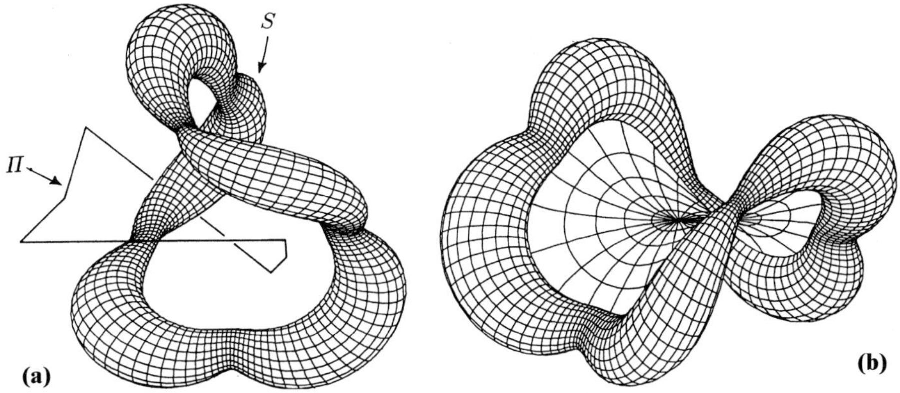

Fig. 1. **(a)** A closed smooth surface S linked with a polygon Π for which the class of surfaces $\mathcal{C}(\Pi, S)$ is non-empty. **(b)** A solution of the corresponding free boundary value problem $\mathcal{P}(\Pi, S)$

Chapter 3, we also investigate how a solution X and its trace curve Σ behave in the neighbourhood of a boundary branch point.

Let us now turn to the proof of Theorem 1. We need the notion of the *greatest distance* $g(A, B)$ of a closed set A of $\mathbb{R}^3$ to another closed set B of $\mathbb{R}^3$ which is defined by

$$(3) \qquad g(A, B) := \sup\{\text{dist}(x, B)\colon x \in A\}.$$

Clearly, we have $0 \le g(A, B) \le \infty$.

Lemma 1. *Let S_k and S be closed sets in $\mathbb{R}^3$ such that $\lim_{k\to\infty} g(S_k, S) = 0$, and suppose that $\{X_k\}$ is a sequence of surfaces $X_k \in \mathcal{C}(S_k)$ which tends weakly in $H_2^1(B, \mathbb{R}^3)$ to some surface X. Then X is of class $\mathcal{C}(S)$.*

Proof. By passing to a suitable subsequence of $\{X_k\}$ and renumbering, we can assume that the $L_2(\partial B, \mathbb{R}^3)$-boundary values converge pointwise almost everywhere on ∂B to $X|_{\partial B}$ (cf. Morrey [8], Theorem 3.4.5). Then we obtain

$$\text{dist}(X(1, \theta), S) \le |X(1, \theta) - X_k(1, \theta)| + g(S_k, S) \to 0$$

as $k \to \infty$, for almost all $\theta \in [0, 2\pi]$. □

Proof of Theorem 1. (i) Suppose that X is a surface of class $\mathcal{C}(S')$, where S' is a closed set with $g(S', S) < \mu/4$. Because of Assumption (A), we can define a boundary class $[X|_{\partial B}]$ which can be viewed as element of $\tilde{\Pi}_1(S)$.

Definition 1. *A sequence of surfaces $X_k \in H_2^1(B, \mathbb{R}^3)$ is said to be a* generalized admissible sequence *for the problem $\mathcal{P}(\Pi, S)$ if there is a sequence of closed sets $S_k \subset \mathbb{R}^3$ such that $\lim_{k\to\infty} g(S_k, S) = 0$ and $X_k \in \mathcal{C}(\Pi, S_k), k \in \mathbb{N}$, holds true.*

We set

$$(4) \qquad e := \inf\{D(X) \colon X \in \mathcal{C}(\Pi, S)\}$$

and

$$(5) \qquad e^* := \inf\{\liminf_{k\to\infty} D(X_k) \colon \{X_k\} \text{ is a generalized admissible sequence for } \mathcal{P}(\Pi,S)\}.$$

Evidently we have

$$(6) \qquad e^* \le e.$$

Now we pick a sequence $\{\mathcal{S}^l\}$ of generalized admissible sequences $\mathcal{S}^l = \{Z_k^l\}_{k\in\mathbb{N}}$ for $\mathcal{P}(\Pi, S)$ such that

$$\lim_{l\to\infty} \liminf_{k\to\infty} D(Z_k^l) = e^*.$$

From the sequences $\mathcal{S}^l$ we can extract a sequence $\mathcal{S} = \{Z_k\}$ of surfaces Z_k which is a generalized admissible sequence for $\mathcal{P}(\Pi, S)$ and satisfies

$$\lim_{k\to\infty} D(Z_k) = e^*.$$

Definition 2. *Such a sequence* $\mathcal{S}$ *of surfaces* $Z_k \in \mathcal{C}(\Pi, S_k)$ *is said to be a* generalized minimizing sequence *for the minimum problem* $\mathcal{P}(\Pi, S)$.

Next we choose radii $\rho_k \in (0,1)$ with $\rho_k \to 1$ having the following properties on the circles $C_k := C_{\rho_k}$:

(α) The curve $Z_k|_{C_k}$ is absolutely continuous, $Z_k(C_k)$ lies in $T_{\mu/2}$ and is linked with the polygon Π, i.e. $\mathcal{L}(Z_k|_{C_k}, \Pi) \neq 0$.

(β) The sequence of surfaces $Y_k(w) := Z_k(\rho_k w), w \in \overline{B}$, with boundary values on $S_k := Z_k(C_k)$ is a generalized minimizing sequence for $\mathcal{P}(\Pi, S)$. Thus we have in particular

$$\lim_{k\to\infty} D(Y_k) = e^*.$$

In addition, all $Y_k|_{\partial B}$ are continuous curves whose greatest distance from S converges to zero as k tends to infinity.

Now we pass from the sequence $\{Y_k\}$ to the sequence of harmonic mappings $X_k \colon B \to \mathbb{R}^3$ which are continuous on $\overline{B}$ and have the boundary values $Y_k|_{\partial B}$ on ∂B. We know that $X_k - Y_k \in \mathring{H}_2^1(B, \mathbb{R}^3)$ and

$$D(X_k) \le D(Y_k).$$

Therefore, also $\{X_k\}$ is a generalized minimizing sequence, and we have in particular

$$(7) \qquad \lim_{k\to\infty} D(X_k) = e^*,$$

whence there is a constant M such that

(8) $$D(X_k) \le M \quad \text{for all } k \in \mathbb{N}.$$

By virtue of the mean value theorem for harmonic functions, there is a constant c such that

(9) $$|\nabla X_k(w)| \le c\sqrt{M}\rho^{-1} \quad \text{for all } k \in \mathbb{N} \text{ and for } |w| \le 1-\rho,$$

where $\rho \in (0,1)$.

Without loss of generality we can also assume that $X_k(0)$ lies on the closed polygon Π for all $k \in \mathbb{N}$, since we can replace X_k by $X_k \circ \tau_k$, where τ_k is a conformal selfmapping of B that maps $w = 0$ onto some point $w_k^* \in B$ with $X_k(w_k^*) \in \Pi$, and such a point can always be found since the polygon Π and the curve $X_k|_{\partial B}$ are linked.

In conjunction with (9) we infer that the harmonic mappings $X_k, k \in \mathbb{N}$, are uniformly bounded on every subset $\Omega \Subset B$. Applying a standard compactness result for harmonic mappings, there is a subsequence of $\{X_k\}$ that converges uniformly on every set $\Omega \Subset B$. By renumbering this subsequence we can achieve that the sequence X_k tends to a harmonic mapping $X\colon B \to \mathbb{R}^3$ on every compact subset of B. In conjunction with (8), we obtain that the $H_2^1(B)$-norms of the surfaces X_k are uniformly bounded, and thus we may also assume that the X_k tend weakly in $H_2^1(B,\mathbb{R}^3)$ and strongly in $L_2(\partial B,\mathbb{R}^3)$ to X which then is of class $H_2^1(B,\mathbb{R}^3)$.

From Lemma 1 we infer that $X \in \mathcal{C}(S)$, and the relations $X_k(0) \in \Pi$ imply in the limit that $X(0) \in \Pi$. Thus the harmonic mapping X is certainly not a constant, and therefore $X(w) \not\equiv \text{const}$ on any open subset Ω of B. Hence

(10) $$D_\Omega(X) > 0 \quad \text{for every nonempty open set } \Omega \Subset B.$$

The lower semicontinuity of the Dirichlet integral with respect to weak convergence in $H_2^1(B,\mathbb{R}^3)$ yields

$$D(X) \le \liminf_{k\to\infty} D(X_k),$$

and together with (6) and (7) we arrive at

$$D(X) \le e^* \le e.$$

As X is of class $\mathcal{C}(S)$, we shall expect X to be a solution of $\mathcal{P}(\Pi, S)$. However, it remains to be shown that X lies in $\mathcal{P}(\Pi, S)$. To this end we have to prove that the linking number of the polygon Π with the boundary class $X|_{\partial B}$ does not vanish. This will be proved by contradiction.

Hence we suppose that $\mathcal{L}([X|_{\partial B}], \Pi) = 0$. Then a sequence of radii $r_k \in (1/2, 1)$ with $r_k \to 1$ can be found such that

$$\xi_k(\theta) := X(r_k, \theta), \quad 0 \le \theta \le 2\pi,$$

represents the boundary class $[X|_{\partial B}]$ of X, and that both the conditions

$$\mathcal{L}(\xi_k, \Pi) = 0, \quad k \in \mathbb{N},$$

and

$$\lim_{k\to\infty} g(\xi_k, S) = 0$$

hold true.

Recall that $\{X_k\}$ converges to X uniformly on every $\Omega \Subset B$. Then, by passing to another subsequence of X_k and renumbering it, we may assume that

$$\max_{0\le\theta\le 2\pi} |X_k(r_k, \theta) - \xi_k(\theta)| \to 0 \quad \text{as } k \to \infty.$$

Set

$$\xi_k^*(\theta) := X_k(r_k, \theta), \quad 0 \le \theta \le 2\pi.$$

Then we infer that

$$\mathcal{L}(\xi_k^*, \Pi) = 0 \quad \text{for } k \in \mathbb{N} \tag{11}$$

and

$$\lim_{k\to\infty} g(\xi_k^*, S) = 0. \tag{12}$$

Moreover, it follows as in the proof of Lemma 3 in Section 1.1 that there is an angle $\theta_k \in [0, 2\pi]$ such that

$$|X_k(r, \theta_k) - X_k(1, \theta_k)| \le \left(\frac{2M}{\pi}\right)^{1/2} \sqrt{1-r} \tag{13}$$

is satisfied for $1/2 \le r \le 1$ and for all $k \in \mathbb{N}$.

Finally we choose conformal mappings τ_k from B onto the slit annuli

$$\{w = re^{i\theta} \in B \colon r_k < r < 1, \theta \ne \theta_k\}.$$

We use these mappings to define a new sequence of surfaces $\hat{X}_k := X_k \circ \tau_k$. On account of the additivity of linking numbers, of (11), (12), and of $\mathcal{L}([X_k|_{\partial B}], \Pi) \ne 0$, it follows that

$$\mathcal{L}([\hat{X}_k|_{\partial B}], \Pi) \ne 0 \quad \text{for } k \in \mathbb{N}.$$

Moreover, we infer from (12) and (13) that the surfaces $\hat{X}_k$ are of class $C^0(\overline{B}, \mathbb{R}^3) \cap H_2^1(B, \mathbb{R}^3)$ and have boundary values on closed sets $\Sigma_k := \hat{X}_k(\partial B)$ with $g(\Sigma_k, S) \to 0$ as $k \to \infty$. Consequently, $\{\hat{X}_k\}$ is a generalized admissible sequence for the problem $\mathcal{P}(\Pi, S)$, and we obtain from (5) that

$$e^* \le \liminf_{k\to\infty} D(\hat{X}_k). \tag{14}$$

On the other hand, the conformal invariance of the Dirichlet integral yields

$$D(\hat{X}_k) = D(X_k) - D_{B_{r_k}}(X_k) \le D(X_k) - D_{B_{1/2}}(X_k),$$

where $B_r := \{w \in \mathbb{C}\colon |w| < r\}$. By a classical result on harmonic mappings, we infer from

$$\lim_{k\to 0} |X - X_k|_{0,\Omega} = 0 \quad \text{for any } \Omega \Subset B$$

that also

$$\lim_{k\to 0} |\nabla X - \nabla X_k|_{0,\Omega} = 0 \quad \text{for any } \Omega \Subset B$$

holds true, whence

$$\lim_{k\to\infty} D_{B_{1/2}}(X_k) = D_{B_{1/2}}(X).$$

In conjunction with (7), we conclude that

$$\liminf_{k\to\infty} D(\hat{X}_k) \le e^* - D_{B_{1/2}}(X),$$

and now (14) yields

$$e^* \le e^* - D_{B_{1/2}}(X).$$

This is a contradiction to (10). Consequently we obtain $\mathcal{L}([X|_{\partial B}], \Pi) \neq 0$, whence $X \in \mathcal{C}(\Pi, S)$ and therefore

$$e \le D(X).$$

In view of (6) and (7) it follows that

$$D(X) = e = e^* \tag{15}$$

which shows that X is a solution of the minimum problem $\mathcal{P}(\Pi, S)$. This completes the proof of part (i) of the theorem.

(ii) The proof of part (ii) essentially follows the same lines of reasoning if we replace the conditions "$\mathcal{L}(\ldots, \Pi) \neq 0$" by "the boundary class of ... is not contractible", and if the relations "$X_k(0) \in \Pi$" are substituted by the assumption "S is compact". Then we have

$$|X_k|_{0,\partial B} \le M' \quad \text{for all } k \in \mathbb{N},$$

and the maximum principle for harmonic functions yields

$$|X_k|_{0,\overline{B}} \le M' \quad \text{for all } k \in \mathbb{N}.$$

Now we may carry on as before.

(iii) The third assertion of the theorem follows in the same way as for solutions of the Plateau problem; cf. Chapter 4 of Vol. 1. This completes the proof of Theorem 1. □

Theorem 2. *Every minimizer of the Dirichlet integral $D(X)$ in the class $\mathcal{C}(\Pi, S)$ (or in $\mathcal{C}^+(S)$) is a surface of least area in $\mathcal{C}(\Pi, S)$ (or in $\mathcal{C}^+(S)$).*

The *proof* of this result can be carried out in the same way as that of Theorem 4 in Section 4.5 of Vol. 1. An alternative method to establish

$$\inf_{\mathcal{C}(\Pi,S)} A = \inf_{\mathcal{C}(\Pi,S)} D$$

and

$$\inf_{\mathcal{C}^+(S)} A = \inf_{\mathcal{C}^+(S)} D$$

without using Morrey's lemma on ϵ-conformal mappings consists in applying the technique of Section 4.10 in Vol. 1, namely, to minimize $A^\epsilon := (1 - \epsilon)A + \epsilon D$. □

1.4 Stationary Minimal Surfaces with Free or Partially Free Boundaries and the Transversality Condition

In the preceding chapters we have considered minimal surfaces which minimize Dirichlet's integral in suitable classes of admissible surfaces. However, the definition of minimal surfaces does not require them to be minimizers, and thus we are led to study also minimal surfaces that are only stationary within a given free boundary configuration. This, roughly speaking, means that the first order change of Dirichlet's integral is zero if we change the stationary surface in such a way that the boundary values remain on the prescribed supporting surface S. It will turn out that stationary minimal surfaces essentially are minimal surfaces which intersect S perpendicularly at their trace curves on S, provided that S is smooth and the boundary of S is empty. However, we also want to consider the case when ∂S is nonempty and consists of smooth regular curves.

By definition we want two distinguish two types of stationary minimal surfaces. The *first type* is defined by the differential equations

$$\Delta X = 0 \tag{1}$$

$$|X_u|^2 = |X_v|^2, \quad \langle X_u, X_v \rangle = 0 \tag{2}$$

which have to hold in B, and by a *natural boundary condition* which is to be satisfied on the free part of ∂B.

The *second type* will be described as critical points of the Dirichlet integral with respect to *inner* and *outer variations.*

Then we shall prove that both types of stationary minimal surfaces are the same provided that both S and ∂S are sufficiently smooth.

Let us begin by defining minimal surfaces in a partially free boundary configuration $\langle \Gamma, S \rangle$. In the following we use the notation of Section 4.6 of Vol. 1; in particular we define the class $\mathcal{C}(\Gamma, S)$ of admissible surfaces for the partially free problem as in that Section.

At present we assume that S and ∂S are of class C^1 and that Γ is a rectifiable arc.

For any point P on S, we denote by $T_p(S)$ the tangent plane of S at P. If $P \in \partial S$, then $T_p(S)$ is divided by the tangent $T_p(\partial S)$ of ∂S at P into two halfplanes. If $N_{\partial S}(P) \in T_P(S)$ denotes the outward unit normal of ∂S at $P \in \partial S$, then we call all tangent vectors $V \in T_P(S)$ with $\langle V, N_{\partial S}(P) \rangle \leq 0$ *interior tangent vectors of S at $P \in \partial S$.*

Definition 1. *A* stationary minimal surface *in $\mathcal{C}(\Gamma, S)$ is an element of $\mathcal{C}(\Gamma, S)$ satisfying*

(i) *$X \in C^1(B \cup I, \mathbb{R}^3) \cap C^2(B, \mathbb{R}^3)$, where $I = \partial B \cap \{\operatorname{Im} w < 0\}$ denotes the free boundary (of the parameter domain $B := B_1(0)$) of X.*

(ii) *In B we have the equations* (1) *and* (2).

(iii) Along $I_1 := \{w \in I \colon X(w) \in \operatorname{int} S\}$, the exterior normal derivative $\frac{\partial X}{\partial \nu}$ is perpendicular to S. (Using polar coordinates r, θ about the origin $w = 0$, we have $\frac{\partial X}{\partial \nu} = \frac{\partial X}{\partial r}$.)

(iv) *For any w belonging to $I_2 := \{w \in I \colon X(w) \in \partial S\}$ and every interior tangent vector $V \in T_{X(w)}S$, we have $\langle \frac{\partial X}{\partial \nu}(w), V \rangle \geq 0$.*

Definition 2. *An element $X \in \mathcal{C}(\Gamma, S)$ is called a* critical *(or* stationary*) point of Dirichlet's integral in the class $\mathcal{C}(\Gamma, S)$ if*

$$(3) \qquad \lim_{\varepsilon \to +0} \frac{1}{\varepsilon} \{D(X_\varepsilon) - D(X)\} \geq 0$$

holds for all admissible variations *$X_\varepsilon, |\varepsilon| < \varepsilon_0$, of X. A family $\{X_\varepsilon\}_{|\varepsilon| < \varepsilon_0}$ of surfaces $X_\varepsilon \in \mathcal{C}(\Gamma, S)$ is said to be an* admissible variation *of X, if it is of one of the following two types.*

Type I (inner variations). *$X_\varepsilon = X \circ \sigma_\varepsilon$ where $\{\sigma_\varepsilon\}_{|\varepsilon| < \varepsilon_0}$, $\varepsilon_0 > 0$, is a differentiable family of diffeomorphisms $\sigma_\varepsilon \colon \overline{B}^*_\varepsilon \to \overline{B}$ which are defined as inverse mappings of the diffeomorphisms $\tau_\varepsilon \colon \overline{B} \to \overline{B}^*_\varepsilon$ defined by*

$$\tau_\varepsilon(w) = w - \varepsilon \lambda(w), \quad \lambda \in C^1(\mathbb{R}^2, \mathbb{R}^2),$$

cf. Section 4.5 *of Vol.* 1.

Type II (outer variations). *$X_\varepsilon = X + \varepsilon \phi(\cdot, \varepsilon)$ for all $\varepsilon \in (0, \varepsilon_0)$ and some $\varepsilon_0 > 0$, where the following holds:*

(α) *the Dirichlet integrals of the mappings $\phi(\cdot, \varepsilon)$ are uniformly bounded, i.e.,*

$$D(\phi(\cdot, \varepsilon)) \leq \text{const} \quad \text{for all } \varepsilon \in (0, \varepsilon_0);$$

(β) *the functions* $\phi(\cdot, \varepsilon)$ *converge pointwise a.e. in* B *to some function* $\phi_0 \in H_2^1 \cap L_\infty(B, \mathbb{R}^3)$ *as* $\varepsilon \to +0$.

By Proposition 2 of Vol. 1, Section 4.5, we obtain from the inner variations that a critical point X of Dirichlet's integral in the class $\mathcal{C}(\Gamma, S)$ satisfies

$$\partial D(X, \lambda) = 0 \quad \text{for all } \lambda = (\mu, \nu) \in C^1(\overline{B}, \mathbb{R}^2). \tag{4}$$

Here $\partial D(X, \lambda)$ denotes the first inner variation of the Dirichlet integral, given by

$$2\partial D(X, \lambda) = \int_B \{a(\mu_u - \nu_v) + b(\mu_v + \nu_u)\}\, du\, dv, \tag{5}$$

where a and b denote the functions

$$a = |X_u|^2 - |X_v|^2, \quad b = 2\langle X_u, X_v \rangle. \tag{6}$$

Moreover, applying outer variations, it follows that

$$\lim_{\varepsilon \to +0} \frac{1}{\varepsilon}\{D(X_\varepsilon) - D(X)\} = \int_B \langle \nabla X, \nabla \phi_0 \rangle\, du\, dv. \tag{7}$$

In fact, if $\{\varepsilon_n\}$ is a sequence of positive numbers tending to zero, then $\phi_n(w) := \phi(w, \varepsilon_n) \to \phi_0(w)$ a.e. on B.

By Egorov's theorem, for any $\delta > 0$ there is a compact subset B_δ of B with meas $(B \setminus B_\delta) < \delta$ such that $\lim_{n\to\infty} |\phi_0 - \phi_n|_{0,B_\delta} = 0$. By virtue of ($\alpha$) and of Poincaré's inequality (see Morrey [8], Theorem 3.6.4) we then infer that the $H_2^1(B)$-norms of the ϕ_n are uniformly bounded. From this we deduce that the sequence ϕ_n converges weakly in $H_2^1(B, \mathbb{R}^3)$ to ϕ_0, and this implies relation (7).

The following result states that the two kinds of stationary minimal surfaces are identical if S and ∂S are sufficiently smooth.

Theorem 1. *Assume that* S *and* ∂S *are of class* C^1 *and that* Γ *is rectifiable. Then every stationary minimal surface in* $\mathcal{C}(\Gamma, S)$ *is a stationary point of Dirichlet's integral in* $\mathcal{C}(\Gamma, S)$. *If* S *and* ∂S *are of class* $C^{3,\beta}, \beta \in (0, 1)$, *then also the converse holds true, that is, every stationary point of Dirichlet's integral in* $\mathcal{C}(\Gamma, S)$ *furnishes a stationary minimal surface in* $\mathcal{C}(\Gamma, S)$.

For the proof we need the following auxiliary result:

Lemma 1. *Let* $X_\varepsilon = X + \varepsilon\phi(\cdot, \varepsilon), 0 \le \varepsilon \le \varepsilon_0$, *be an outer variation (i.e. an admissible variation of type II) of a surface* $X \in \mathcal{C}(\Gamma, S)$, *and let* $\phi_0 = \phi(\cdot, 0)$. *Then we have:*

(i) *For almost all* $w \in I$, *the vector* $\phi_0(w)$ *is a tangent vector of* S *at* $X(w)$. *If* $X(w)$ *lies on* ∂S, *then* $\phi_0(w)$ *is an interior tangent vector.*

(ii) *If, in addition to our general assumption, the arc* Γ *is of class* C^1 *or if* X *is a stationary minimal surface, then* $\phi_0(w)$ *is tangent to* Γ *at* $X(w)$ *for almost all* $w \in C = \partial B \setminus I$.

Proof. Choose a sequence $\{\varepsilon_n\}$ with $\varepsilon_n \to +0$ such that $\phi(w, \varepsilon_n) \to \phi_0(w)$ for a.e. $w \in \partial B$. Then assertion (i) follows from

$$\phi(w, \varepsilon_n) = \frac{1}{\varepsilon_n}\{X_{\varepsilon_n}(w) - X(w)\}$$

and from $X(w), X_{\varepsilon_n}(w) \in S$ for a.a. $w \in \partial B$.

(ii) is verified in the same way. (If Γ is only rectifiable we note that it has a tangent everywhere except at countably many points. Moreover, if X is a minimal surface then it follows from Theorem 1 in Vol. 1, Section 4.7 that, for almost all $w \in C = \partial B \setminus I$, the curve Γ has a tangent at the point $X(w)$.) □

Now we turn to the

Proof of Theorem 1. (i) Let X be a stationary minimal surface in $\mathcal{C}(\Gamma, S)$. In order to show that X is a stationary point of the Dirichlet integral, we have to verify (3) for all admissible variations $\{X_\varepsilon\}_{|\varepsilon|<\varepsilon_0}$ of X. Since the case of inner variations (type I) has already been settled in Section 4.5 of Vol. 1, it suffices to consider variations of type II. In view of (7) we have to show

$$\int_B \langle \nabla X, \nabla \phi_0 \rangle \, du \, dv \geq 0. \tag{8}$$

The Courant–Lebesgue lemma (see Section 4.4 of Vol. 1) shows that, given any $\delta \in (0, 1)$, there are two radii $r_1(\delta)$ and $r_2(\delta)$ with $\delta \leq r_1, r_2 \leq \sqrt{\delta}$ such that

$$\int_{\gamma_k} \left|\frac{\partial X}{\partial \nu}\right| ds = \int_{\gamma_k} \left|\frac{\partial X}{\partial t}\right| ds \leq \frac{M}{\{\log \frac{1}{\delta}\}^{1/2}} \tag{9}$$

holds true for $\gamma_1 := \overline{B} \cap \partial B_{r_1}(1)$ and $\gamma_2 := \overline{B} \cap \partial B_{r_2}(-1)$, where $M = \text{const}\,\sqrt{D(X)}$. (Here ν and t denote unit normal and unit tangent to γ_1 and γ_2, respectively.)

On account of Theorem 2 in Section 4.7 of Vol. 1 we have

$$\int_{\Omega_\delta} \langle \nabla X, \nabla \phi_0 \rangle \, du \, dv = \int_{\partial\Omega_\delta} \left\langle \frac{\partial X}{\partial \nu}, \phi_0 \right\rangle ds, \tag{10}$$

where $\Omega_\delta := B \setminus \{\overline{B}_{r_1}(1) \cup \overline{B}_{r_2}(-1)\}$. Letting δ tend to zero, we infer from (9) and (10) that

$$\int_B \langle \nabla X, \nabla \phi_0 \rangle \, du \, dv = \int_{\partial B} \left\langle \frac{\partial X}{\partial r}, \phi_0 \right\rangle ds. \tag{11}$$

By (iii) and (iv) of Definition 1 it follows that

$$\left\langle \frac{\partial X}{\partial r}, \phi_0 \right\rangle \geq 0 \quad \text{on } I = \partial B \cap \{\operatorname{Im} w < 0\}$$

holds true, whereas in view of Lemma 1 we obtain

$$\left\langle \frac{\partial X}{\partial r}, \phi_0 \right\rangle = 0 \quad \text{a.e. on } C = \partial B \cap \{\operatorname{Im} w \geq 0\}.$$

Consequently we have

$$\int_{\partial B} \left\langle \frac{\partial X}{\partial r}, \phi_0 \right\rangle ds \geq 0,$$

whence the identity (11) implies (8).

(ii) Let us now consider a stationary point X of Dirichlet's integral. By the results of Chapter 4 such a mapping X is a minimal surface, that is, equations (1) and (2) are satisfied in B (cf. equations (5)–(7)). The regularity results of 2.4 imply that $X \in C^1(B \cup I, \mathbb{R}^3)$. Thus it remains to prove conditions (iii) and (iv) of Definition 1. This will be carried out by applying the fundamental lemma of the calculus of variations to the equation

$$(12) \qquad \int_{\partial B} \left\langle \frac{\partial X}{\partial r}, \phi_0 \right\rangle ds \geq 0$$

which follows from (3), (7) and (11). As we shall see it will be enough to consider outer variations

$$X_\varepsilon = X + \varepsilon\phi(\cdot, \varepsilon), \quad 0 \leq \varepsilon < \varepsilon_0,$$

with

$$\text{support } \phi(\cdot, \varepsilon) \subset B \cup I.$$

Then also supp $\phi_0 \subset B \cup I$, and (12) reduces to

$$(13) \qquad \int_I \langle X_r, \phi_0 \rangle \, ds \geq 0.$$

Consider now an arbitrary function $V \in C_c^1(I, \mathbb{R}^3)$ with $V(w) \in T_{X(w)}S$ for all $w \in I$ and

$$(14) \qquad \langle V(w), N_{\partial S}(X(w)) \rangle < 0 \quad \text{for all } w \in I_2.$$

Here and in the sequel, the subsets I_1 and I_2 of I be defined in the same way as in Definition 1.

Then we solve the initial value problem

$$\frac{D}{d\varepsilon}\frac{d}{d\varepsilon} Z(w, \varepsilon) = 0,$$

$$Z(w, 0) = X(w), \quad \frac{dZ}{d\varepsilon}(w, 0) = V(w)$$

for fixed $w \in I$ and $0 \leq \varepsilon < \varepsilon_0$ with $0 < \varepsilon_0 \ll 1$, where $\frac{D}{d\varepsilon}$ denotes the covariant derivative on S. In other words, we define $Z(w, \varepsilon)$ as the geodesic flow on S starting at $X(w)$ in direction of $V(w), w \in I$. This flow exists for $0 \le \varepsilon < \varepsilon_0$ and $w \in I$, where ε_0 denotes a sufficiently small positive number, and we have

$$Z(w,\varepsilon) = X(w) + \varepsilon\Psi(w,\varepsilon) = X(w) + \varepsilon V(w) + o(\varepsilon)$$

for any $w \in I$, as well as

$$Z(w,\varepsilon) = X(w) \quad \text{for } w \in I \setminus \operatorname{supp} V \text{ and } 0 \le \varepsilon < \varepsilon_0.$$

Therefore we obtain

$$\Psi(\cdot,\varepsilon) \in C_c^1(I,\mathbb{R}^3), \quad \Psi(w,0) = V(w) \quad \text{for } w \in I$$

and

$$Z(w,\varepsilon) \in S \quad \text{for } (w,\varepsilon) \in I \times [0,\varepsilon_0).$$

Then we extend $\Psi(\cdot,\varepsilon)$ to functions $\phi(\cdot,\varepsilon)$ for class $C_c^1(B \cup I,\mathbb{R}^3)$ which depend smoothly on $(w,\varepsilon) \in B \cup I \times [0,\varepsilon_0)$, and set

$$X_\varepsilon(w) := X(w) + \varepsilon\phi(w,\varepsilon) \quad \text{for } w \in \overline{B} \text{ and } 0 \le \varepsilon < \varepsilon_0.$$

By construction we have $X_\varepsilon \in \mathcal{C}(\Gamma,S)$, and the function $\phi_0 := \phi(\cdot,0)$ satisfies $\phi_0(w) = V(w)$ for all $w \in I$. Consequently, the relation (13) holds true. This implies

$$\int_I \langle X_r, V\rangle\, ds \ge 0 \tag{15}$$

for every $V \in C_c^1(I,\mathbb{R}^3)$ with $V(w) \in T_{X(w)}S$ for all $w \in I$ which, in addition, satisfies $\langle V(w), N_{\partial S}(X(w))\rangle \le 0$ for $w \in I_2$. (In contrast to (14), we may admit the equality sign in the last inequality as can be proved by a straightforward approximation argument.)

Let us write

$$X_r = X_r' + X_r'', \quad X_r' \in T_X S, \quad X_r'' \perp T_X S;$$

then (15) is equivalent to

$$\int_I \langle X_r', V\rangle\, ds \ge 0. \tag{16}$$

Suppose now that $w_0 \in I_1$. Then there exists some $\rho > 0$ such that $I_\rho(w_0) := I \cap B_\rho(w_0)$ is contained in I_1, and we infer that

$$\int_{I_\rho(w_0)} \langle X_r', V\rangle\, ds \ge 0$$

is satisfied for every $V \in C_c^1(I_\rho(w_0),\mathbb{R}^3)$ with $V(w) \in T_{X(w)}S, w \in I_\rho(w_0)$, and since the same inequality holds if we replace V by $-V$, we even have

$$\int_{I_\rho(w_0)} \langle X_r', V\rangle \, ds = 0. \tag{17}$$

The fundamental lemma of the calculus of variations yields $X_r' = 0$ on $I_\rho(w_0)$, whence $X_r(w_0) = X_r''(w_0)$. Consequently the normal derivative $X_r(w_0)$ is perpendicular to $T_{X(w_0)}S$ for every $w_0 \in I_1$, and we have verified property (iii) of Definition 1.

Similarly we infer from (15) that condition (iv) of Definition 1 is fulfilled. We leave it as an exercise for the reader to carry out the details. □

Remarks and Generalizations

(i) Analogous to the Definitions 1 and 2 one can define stationary minimal surfaces in $\mathcal{C}(S)$ (or $\mathcal{C}^+(S)$ or $\mathcal{C}(\Pi, S)$) as well as *stationary points of the Dirichlet integral in* $\mathcal{C}(S)$ (or in $\mathcal{C}^+(S)$, or $\mathcal{C}(\Pi, S)$, respectively). We only have to replace I by $\partial B, C$ by the empty set, and $\mathcal{C}(\Gamma, S)$ by $\mathcal{C}(S)$ (or by $\mathcal{C}^+(S)$, or $\mathcal{C}(\Pi, S)$); all statements about Γ are now to be omitted. Then, analogous to Theorem 1, we obtain

Theorem 2. *Assume that S and ∂S are of class C^1. Then every stationary minimal surface in $\mathcal{C}(S)$ (or in $\mathcal{C}^+(S)$ or $\mathcal{C}(\Pi, S)$) is a stationary point of Dirichlet's integral in $\mathcal{C}(S)$ (or in $\mathcal{C}^+(S)$ or $\mathcal{C}(\Pi, S)$). If S and ∂S are of class $C^{3+\beta}, \beta \in (0,1)$, also the converse holds true.*

A stationary minimal surface in $\mathcal{C}(S)$ will also be called *stationary minimal surface with respect to S, or: with a free boundary on S.*

If $\partial S = \emptyset$ and $S \in C^{2,\beta}$, $0 < \beta < 1$, then a stationary point of Dirichlet's integral is even of class $C^{2,\beta}$ up to its free boundary, according to results by Dziuk and Jost. In this case, the second statements of the Theorems 1 and 2 also hold under the assumption $S \in C^{2,\beta}$.

(ii) If we want to define *stationary minimal surfaces $X\colon \Omega \to \mathbb{R}^3$ with a free boundary on S and critical points of the Dirichlet integral with a free boundary on S* which are defined on multiply connected parameter domains Ω or even on Riemann surfaces, the matter is slightly more complicated. We are not anymore allowed to fix Ω, but only the *conformal type of Ω* can be prescribed. Then the boundary behaviour of the minimal surface does not only depend on S but also on the boundary $\partial\Omega$ of the parameter domain. However, we never have a real problem. For instance, a theorem of Koebe [1] states that every k-fold connected domain in $\mathbb{C}$ is conformally equivalent to a bounded domain in $\mathbb{C}$ whose boundary consists of k disjoint circles. Therefore we can essentially proceed as before.

(iii) In Section 1.6 we shall also consider stationary minimal surfaces $X\colon \Omega \to \mathbb{R}^3$ having their boundaries on a *simplex* or, more generally, on a *polyhedron*. In this case we shall call a *minimal surface X stationary* if for some finite subset M of ∂B the surface X is of class $H_2^1(B, \mathbb{R}^3) \cap C^1(\overline{B} \setminus M, \mathbb{R}^3)$ and if X meets the interiors of the faces of the polyhedron orthogonally along $\partial B \setminus M$.

(iv) In the special case that Γ is a rectifiable Jordan curve and that $\mathcal{C}(\Gamma)$ is defined as in Chapter 4 of Vol. 1, we obtain the following result: The solutions of Plateau's problem within the class $\mathcal{C}(\Gamma)$ which are not necessarily minimizers of the area are precisely the stationary points of the Dirichlet integral in $\mathcal{C}(\Gamma)$, i.e., those elements X of $\mathcal{C}(\Gamma)$ which satisfy

$$\lim_{\varepsilon \to 0} \frac{1}{\varepsilon} \{D(X_\varepsilon) - D(X)\} = 0$$

for all admissible variations $\{X_\varepsilon\}_{|\varepsilon| < \varepsilon_0}$ of X which are of class $\mathcal{C}(\Gamma)$. These admissible variations are defined as in Definition 2 if we replace S and I by the empty set, and C by ∂B.

Let us close this section with a simple

Example. If S is the boundary of an open, convex and bounded subset $\mathcal{K}$ of $\mathbb{R}^3$ of class C^1, and if E is a plane which intersects S orthogonally (e.g., a plane of symmetry of $\mathcal{K}$), then a conformal map X from B onto $E \cap \mathcal{K}$ defines a stationary minimal surface having $S \cap E$ as its trace. Therefore *the plane disks bounded by the great circles of a sphere S are stationary minimal surfaces in S. As we shall see in Section 1.7, they are the only stationary disk-type surfaces in the sphere.*

Moreover, the ellipses in an *ellipsoid* $\mathcal{E}$, having two of the axes of $\mathcal{E}$ as their principal axes, are *three stationary minimal surfaces in* $\mathcal{E}$. It is unknown whether they are the only stationary surfaces in $\mathcal{E}$ which are of the type of the disk.

1.5 Necessary Conditions for Stationary Minimal Surfaces

Let us agree that throughout this section we consider minimal surfaces $X : \Omega \to \mathbb{R}^3$, the parameter domain Ω of which will be bounded by finitely many disjoint circles $C_1, C_2, \ldots, C_k$. Moreover, the surfaces X will be stationary minimal surfaces with a free boundary on a polyhedron S or on a closed, orientable, regular C^1-surface S.

Now, if V is an arbitrary constant vector in $\mathbb{R}^3$, then an integration by parts shows that

$$\int_{\partial\Omega} \left\langle \frac{\partial X}{\partial \nu}, V \right\rangle ds = \int_{\Omega} \langle \Delta X, V \rangle \, ds = 0.$$

(If S is a polyhedron this can be justified as in Section 1.4.) Since X is stationary, we know that (almost) everywhere on $\partial\Omega$

$$\begin{aligned}\frac{\partial X}{\partial \nu}(w) &= \pm N_S(X(w)) \cdot \left|\frac{\partial X}{\partial \nu}(w)\right| \\ &= \pm N_S(X(w)) \cdot \left|\frac{\partial X}{\partial t}(w)\right|,\end{aligned}$$

where N_S is the surface normal of S, and $\frac{\partial}{\partial \nu}(\frac{\partial}{\partial t})$ denotes the normal (tangential) derivative along $\partial\Omega$. Furthermore, s denotes the parameter of the arc length on $\partial\Omega$.

Therefore we have obtained the following

Proposition 1. *If $X\colon \Omega \to \mathbb{R}^3$ is a stationary minimal surface with respect to a surface S, X and S satisfying the general assumption, then*

$$\int_{\partial\Omega} N_S(X(w)) \cdot \left|\frac{\partial X}{\partial t}(w)\right| \, ds = 0 \tag{1}$$

unless $\langle \frac{\partial X}{\partial \nu}, N_S(X)\rangle$ changes its sign on $\partial\Omega$.

Remarks. (i) If we denote by $\Sigma := X|_{\partial\Omega}$ the trace of a stationary minimal surface X, then formula (1) could also be written as

$$\int_\Sigma \mu(P) N_S(P) \, d\mathcal{H}^1 = 0,$$

where $\mathcal{H}^1$ is the one-dimensional Hausdorff-measure in $\mathbb{R}^3$ and $\mu(P)$ the number of points $w \in \partial\Omega$ such that $X(w) = P$.

(ii) The geometric interpretation of formula (1) is that the integral of the normal N_S over the trace Σ of the stationary minimal surface X vanishes.

(iii) Here are some conditions implying that

$$\left\langle \frac{\partial X}{\partial \nu}, N_S(X) \right\rangle \quad \text{does not change its sign on } \partial\Omega. \tag{2}$$

A first condition is

(I) *S is smooth and X has no branch points on $\partial\Omega$.*

Recall that $w \in \overline{\Omega}$ is a branch point of X if $|\nabla X(w)|^2 = 0$. Since by conformality we have the identity $|\frac{\partial X}{\partial \nu}|^2 = \frac{1}{2}|\nabla X|^2$ on $\partial\Omega$, property (2) follows from (I).

(II) More generally, *for surfaces S of class C^4 (or $C^{3,\beta}$, $\beta \in (0,1)$), the absence of branch points of odd order on $\partial\Omega$ is also sufficient for* (2).
This follows from the expansion formula in Section 2.10 which describes the asymptotic behavior of the minimal surface near a boundary branch point.

(III) *The surface $X(\overline{\Omega})$ stays on one side of S.*
Clearly, (2) follows from this property of X which, in turn, is true if

(IV) *S is the boundary of a convex body $\mathcal{K} \in \mathbb{R}^3$*, as we can infer from the maximum principle, or if

(V) *S is the graph of a C^2-function defined on the 2-sphere S^2 such that the mean curvature of S with respect to the inward unit normal N_S is nowhere negative.*
This follows from a maximum principle to be stated in the Chapter 4.

(iv) A system $\Gamma = \{\Gamma_1, \dots, \Gamma_k\}$ of rectifiable curves Γ_j on a surface S is sometimes called a *system of balanced curves on S* if

$$\int_\Gamma N_S\, d\boldsymbol{s} = 0,$$

that is, if

$$\sum_{j=1}^{k} \int_{\Gamma_j} N_S\, d\boldsymbol{s} = 0$$

holds true (here $\boldsymbol{s}$ denotes the parameter of the arc length of Γ).

According to Proposition 1, formula (1), we have

$$\int_\Sigma N_S\, d\boldsymbol{s} = 0$$

for the free trace Σ of a minimal surface X with a free boundary on S and satisfying (2), since $d\boldsymbol{s} = |X_t|\, ds$ if s denotes the arc length on $\partial\Omega$. Consequently, we can read Proposition 1 as follows:

The free trace Σ of a stationary minimal surface in $\mathcal{C}(S)$ is a system of balanced curves.

We shall now draw several conclusions from Proposition 1.

Corollary 1. *Let $X\colon \Omega \to \mathbb{R}^3$ be a stationary minimal surface with respect to a support surface S which is the boundary of an open set $\mathcal{K}$ in $\mathbb{R}^3$, and assume that $X(\overline{\Omega})$ is contained in $\overline{\mathcal{K}}$. Then the free trace $X(\partial\Omega)$ of X on S cannot be contained in a subset of S which is mapped by the Gauss map $N_S\colon S \to S^2$ of S into an open hemisphere of S^2.*

Corollary 2. *In particular there are no stationary minimal surfaces which have their boundaries on a paraboloid or on one sheet of a hyperboloid of two sheets. Likewise there is no stationary minimal surface with respect to a simplex whose trace intersects only three of the four faces.*

Corollary 3. *If X and S are as in Corollary 1 and if $X(\partial\Omega)$ is contained in a subset $\mathcal{U}$ of S whose Gauss image $N_S(\mathcal{U})$ is contained in a closed hemisphere H of S^2, then $N_S(X(\partial\Omega))$ is the great circle ∂H.*

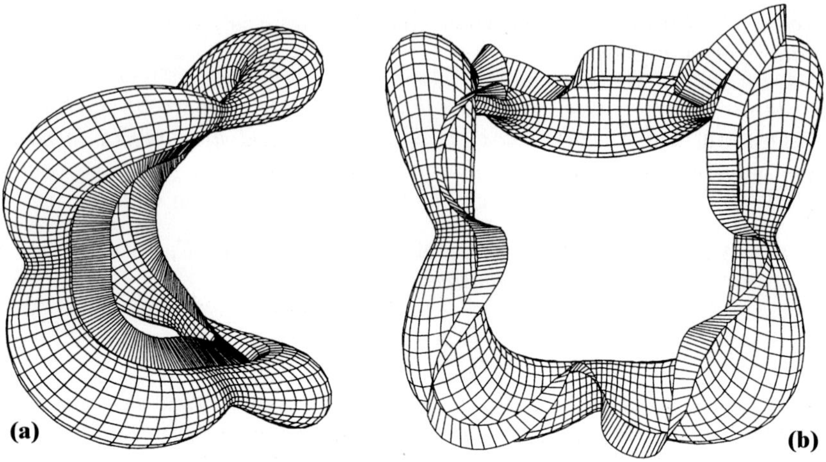

(a) (b)

Fig. 1. **(a)** The integral of the unit normal bundle along a balanced curve on a smooth surface vanishes. **(b)** The integral of the unit normal bundle along this unbalanced curve on the same surface has a non-vanishing component pointing to the reader

Corollary 4. *If X and S are as in Corollary 1, and if Ω is simply connected, S is of class C^1, and if the image of the trace $X(\partial\Omega)$ under the Gauss map $N_S\colon S \to S^2$ is a great circle, then Σ is a plane curve, and X is a plane minimal surface.*

Proof. We can now assume that Ω is the unit disk B. Let X^* be the adjoint minimal surface of X. Then we have

$$(3) \qquad X_u = X_v^* \quad \text{and} \quad X_v = -X_u^* \quad \text{in } B,$$

or, in polar coordinates,

$$X_r^* = -\frac{1}{r}X_\theta \quad \text{and} \quad X_\theta^* = rX_r,$$

hence for $0 \le \theta \le 2\pi$

$$(4) \qquad \begin{aligned} X^*(1,\theta) &= X^*(1,0) + \int_0^\theta X_\theta^*(1,\varphi)\,d\varphi \\ &= X^*(1,0) + \int_0^\theta X_r(1,\varphi)\,d\varphi. \end{aligned}$$

By assumption, the normals of S along Σ are contained in a plane. Consequently the vectors $X_r(w) = \pm|X_r(w)|N_S(X(w)), w \in \partial B$, lie in a plane, and (4) implies that $X^*(1,\theta)$ is contained in a parallel plane. The maximum principle now yields that $X^*(\overline{\Omega})$ lies in this plane, and the assertion follows from (3). □

Corollary 5. *Stationary minimal surfaces of the type of the disk (i.e., $\Omega = B$) with their free boundary on a cylinder are plane disks orthogonal to the cylinder axis.*

This is an immediate consequence of the preceding corollary. Note that the infinite strip is excluded, as our definition of "stationary" implies "finite area".

1.6 Existence of Stationary Minimal Surfaces in a Simplex

The examples of stationary minimal surfaces with a free boundary on a supporting surface S have been rather trivial since all of them were planar surfaces. The first nontrivial example of a minimal surface with a free boundary on a tetrahedron was found by H.A. Schwarz in 1872 (cf. Math. Abhandlungen [2], vol. 1, pp. 149–150); we have copied Schwarz's drawing in Fig. 1. Schwarz obtained this surface as an adjoint of the minimal surface bounded by four consecutive edges of a regular tetrahedron. In the following we describe a result of B. Smyth [1] which may be viewed as a generalization of the Schwarz surface to arbitrary simplices in $\mathbb{R}^3$.

Theorem 1. *Let S be the boundary of a simplex in $\mathbb{R}^3$. Then there are exactly three stationary minimal surfaces of disk-type having connected intersections with each of the four faces of S. They neither have branch points in B nor on the arcs of ∂B which are mapped into the faces of S.*

Remark. Exactly three means, of course, exactly three except for reparametrizations.

Proof of Theorem 1. First of all, in order to prove existence, choose a fixed order $H_1, \dots, H_4$ of the faces H_i of S and let $N_1, \dots, N_4$ be their outward unit normals. Next choose four real numbers l_i such that $\sum_{i=1}^4 l_i N_i = 0$ (note that all l_i are different from zero). Now let Γ be the quadrilateral which is determined by the four vectors $l_1 N_1, \dots, l_4 N_4$ just in this order, i.e., $\Gamma(t) := 4 l_1 N_1 \cdot t$ for $0 \le t \le \frac{1}{4}$, $\Gamma(t) = l_1 N_1 + 4(t - \frac{1}{4}) \cdot l_2 N_2$ for $\frac{1}{4} \le t \le \frac{1}{2}$, etc. Since Γ can be projected onto a convex curve in a plane, there is exactly one solution Y of the Plateau problem $\mathcal{P}(\Gamma)$ (see Section 4.9 of the Vol. 1). By the reflection principle (cf. Vol. 1, Section 4.8), Y is of class $H_2^1(B, \mathbb{R}^3) \cap C^\omega(\overline{B} \setminus M, \mathbb{R}^3)$, where M contains the four points of ∂B corresponding to the corners of Γ. Then the minimal surface $\hat{X} := -Y^*$ is stationary with respect to a simplex $\hat{S}$ similar to S. After a suitable choice of $a > 0$ and $A_0 \in \mathbb{R}^3$, the surface $X := A_0 + a\hat{X}$ is a stationary minimal surface with respect to the given simplex S, which crosses the faces $H_1, \dots, H_4$ in this order.

Now note that, since a stationary minimal surface has to cross all four faces of the simplex (Corollary 2 of Section 1.5), one can select any of them

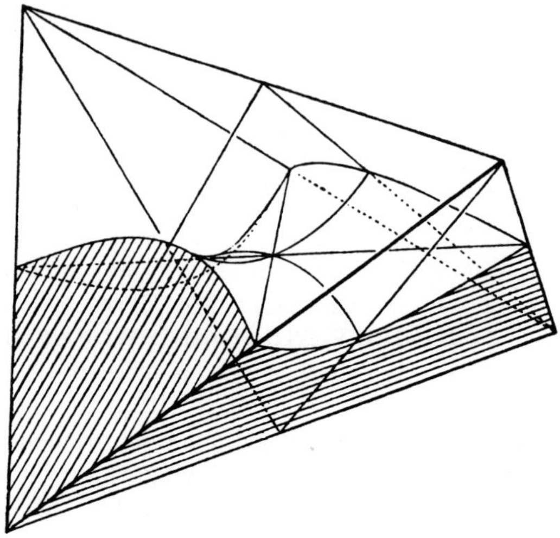

Fig. 1. Schwarz's stationary minimal surface in a tetrahedron

as the first to be crossed. But then the three possible choices of the face to be crossed next but one lead to three geometrically different stationary minimal surfaces. This proves the existence of at least three stationary minimal surfaces.

Before we show the uniqueness part of the theorem let us show that *the length of the trace $X|_{\partial B}$ of any stationary minimal surface in the simplex S having connected intersections with the faces is finite.*

Denote by $C_1, \dots, C_4$ the four open subarcs of ∂B which are mapped by X into the interiors of the faces H_1, H_2, H_3, H_4 of the simplex. Next, note that the adjoint minimal surface X^* of X also belongs to $H_2^1(B, \mathbb{R}^3) \cap C^\omega(\overline{B} \setminus M, \mathbb{R}^3)$, where $M = \partial B \setminus \bigcup_1^4 C_i$. By virtue of the maximum principle and the boundary condition, we obtain

$$\frac{\partial X^*}{\partial \theta} = \frac{\partial X}{\partial r} = \left| \frac{\partial X^*}{\partial r} \right| \cdot N_i \quad \text{on } C_i,$$

whence we see that X^* maps the four arcs C_i monotonically onto four mutually nonparallel straight lines $\hat{L}_i$ parallel to N_i. The Courant–Lebesgue lemma now implies that $\hat{L}_i$ intersects $\hat{L}_{i+1(\text{mod } 4)}$, and that X^* is continuous on $\overline{B}$ and bounded by the quadrilateral Γ given by the line segments L_i on $\hat{L}_i$ between the intersections of $\hat{L}_i$ with $\hat{L}_{i-1(\text{mod } 4)}$ and $\hat{L}_{i+1(\text{mod } 4)}$.

In particular we have for $i = 1, \dots, 4$ that

$$l_i = \int_{C_i} |X_\theta| \, ds = \int_{C_i} |X_\theta^*| \, ds < \infty$$

(one can now also show that X is continuous in $\overline{B}$). Since the boundary curve $X|_{\partial B}$ is balanced (see Section 1.5), we have

$$\sum_{i=1}^{4} l_i N_i = 0.$$

This equation shows that the lengths of the intercepts of X with the faces are determined up to a constant, since there is only one linear relation between four vectors in $\mathbb{R}^3$, no three of which are dependent.

Consequently, if Y is another stationary minimal surface in the simplex S which intersects the faces in the same order as X, then the bounding quadrilaterals of X^* and Y^* are homothetic, hence also $X^*(B)$ and $Y^*(B)$, as follows from the uniqueness theorem in Section 4.9 of Vol. 1. Therefore $X(B)$ and $Y(B)$ are homothetic too, and they even coincide since they are bounded by the same simplex.

Hence we have shown that a particular choice of the order in which the faces are crossed determines the minimal surface uniquely. Hence only three essentially different stationary minimal surfaces remain. This proves the assertion. □

A stationary minimal surface X in the simplex S has no interior branch points since X^* has none (Theorem 1 in Vol. 1, Section 4.9). The simplex is convex; therefore X stays on one side of each of the faces H_i. Hence we may first continue X by reflection across H_i as a minimal surface and then exclude branch points on C_i by means of the expansion formulas stated in Section 3.2 of Vol. 1.

Remark. By means of the *theorem of Krust* presented at the end of Section 3.3 of Vol. 1, it follows that the three stationary solutions of Smyth are graphs, since their adjoints are graphs; thus, in particular, they are embedded minimal surfaces.

1.7 Stationary Minimal Surfaces of Disk-Type in a Sphere

In this section we shall prove that plane disks are the only stationary minimal surfaces of disk type that have their boundaries on a sphere.

Theorem. *Let* $X \in C^1(\overline{B}, \mathbb{R}^3) \cap C^2(B, \mathbb{R}^3)$ *satisfy*

(1) $$\Delta X = 0 \quad \textit{in } B,$$

(2) $$|X_u|^2 = |X_v|^2, \quad \langle X_u, X_v \rangle = 0 \quad \textit{in } B.$$

Moreover, assume that $X(\partial B)$ *is contained in a sphere* S *and that the normal derivative* $\frac{\partial X}{\partial \nu}$ *is orthogonal to* S *along* ∂B. *Then* $X(\overline{B})$ *is a plane disk.*

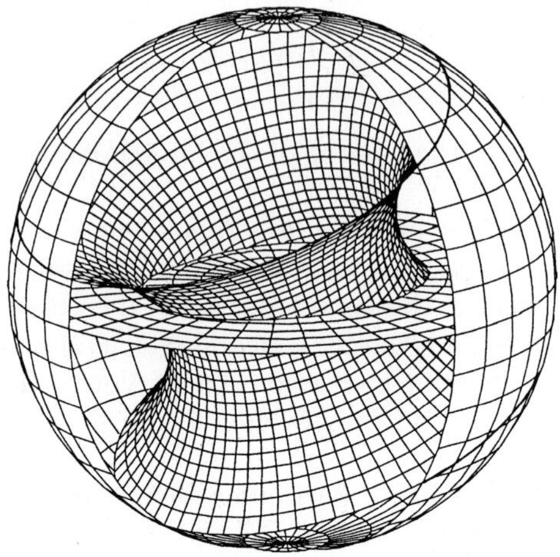

Fig. 1. The catenoid yields a doubly connected non-planar minimal surface intersecting a sphere perpendicularly. On the other hand, all simply connected stationary minimal surfaces in a sphere are planar surfaces

Remark. Note that the theorem is false if we admit minimal surfaces of a different topological type. For example, any sphere S bounds a catenoid intersecting S orthogonally along its trace.

Proof of the Theorem. We shall prove in Chapter 2 that

(3) $$X \textit{ is real analytic in } \overline{B}.$$

Then the arguments used in Chapter 3 of Vol. 1 show the following:

(4) *X has only finitely many isolated branch points in $\overline{B}$.* The surface normal $N(w)$ of $X(w)$and hence also the coefficients of the second fundamental form of X can be extended continuously to all of $\overline{B}$.

Let $M \subset \overline{B}$ denote the set of branch points, i.e. of points w with $|X_u(w)| = |X_v(w)| = 0$. Then, by H. Hopf's observation (cf. Vol. 1, Section 1.3), the function $f(w) = \frac{1}{2}(\mathcal{L} - \mathcal{N}) - i\mathcal{M}$ is holomorphic in $B \setminus M$ and continuous on $\bar{B}$. Consequently all interior singularities of f are removable and f is holomorphic in all of B. Let us now assume without loss of generality that $S = S^2$, and consider the boundary condition, which is equivalent to

$$\begin{aligned} X_\rho(e^{i\theta}) &= \lambda(e^{i\theta}) N_S(X(e^{i\theta})) \\ &= \lambda(e^{i\theta}) X(e^{i\theta}), \end{aligned}$$

where $N_S(X)$ denotes the outward unit normal of S at X and $\lambda(\theta) := \sqrt{\mathcal{E}(e^{i\theta})}$. Next we differentiate this equation in $\partial B \setminus M$ with respect to θ. We obtain ($' = \frac{\partial}{\partial\theta}$): $X_{\rho\theta} = \lambda' X + \lambda X_\theta = \frac{\lambda'}{\lambda} X_\rho + \lambda X_\theta$, and a comparison with formula (36) of Section 1.3 in Vol. 1 shows that the boundary values of the imaginary part β of $g := w^2 f$ vanish. Hence $\beta \equiv 0$ in $\overline{B}$, and therefore $\alpha := \text{Re}\, g$ is identically constant in B, whence $\alpha \equiv \alpha(0) = 0$. Thus $\mathcal{L} = \mathcal{N}$ and $\mathcal{M} = 0$ in B. Now Weingarten's equations (cf. Vol. 1, Section 1.2, (38) ff. and (42)) imply that

$$\nabla N = -H \nabla X = 0 \quad \text{in } B \setminus M.$$

Therefore $N \equiv$ const in B and $X(B)$ is contained in a plane orthogonal to N. $\square$

Remark. Suppose that $X \in C^1(\overline{B}, \mathbb{R}^3) \cap C^2(B, \mathbb{R}^3)$ is a disk-type surface of constant mean curvature H. Then our previous reasoning shows that $f(w) = \frac{1}{2}(\mathcal{L} - \mathcal{N}) - i\mathcal{M}$ is again holomorphic, and the same arguments as before yield that $\mathcal{L} = \mathcal{N}$ and $\mathcal{M} = 0$, as one has the same asymptotic expansion about branch points as for minimal surfaces (see Heinz [15]). Then it is fairly easy to prove that X *is a parametrization of a spherical cap.* This result as well as Theorem 1 are due to Nitsche [35]. Furthermore, one can construct an example where this spherical cap actually covers a whole sphere of radius $1/|H|$.

1.8 Report on the Existence of Stationary Minimal Surfaces in Convex Bodies

Let $S \subset \mathbb{R}^3$ be an embedded submanifold of $\mathbb{R}^3$ without boundary and of genus $g \geq 1$, that is, S has at least one hole to be spanned. Then there exists a closed polygon Π such that the class $\mathcal{C}(\Pi, S)$ is nonempty, and we can prove the existence of a stationary minimal surface X which has its boundary on S and such that $X|_{\partial B}$ is not contractible in $\mathbb{R}^3 \setminus \Pi$, see Theorem 1 in Section 1.3. Such a surface X is constructed as a solution to the variational problem $\mathcal{P}(\Pi, S)$: $D_B(X) \to \min$ in $\mathcal{C}(\Pi, S)$. In Section 1.6, on the other hand, we have considered the case of a simplex S, and we have proved the existence of three (distinct) stationary minimal surfaces in $\mathcal{C}(S)$. Clearly these surfaces cannot be solutions of the minimum problem $\mathcal{P}(S)$: $D_B(\cdot) \to$ *minimum in* $\mathcal{C}(S)$ as the constant surfaces have a smaller Dirichlet integral, and the classes $\mathcal{C}^+(S)$ and $\mathcal{C}(\Pi, S)$ are void. Thus we cannot use a minimizing procedure to obtain nondegenerate minimal surfaces in S.

Consider now the ellipsoid E given by $\frac{x^2}{a^2} + \frac{y^2}{b^2} + \frac{z^2}{c^2} = 1$, with $a > b > c$. As we have already noted, there exist at least three geometrically distinct, stationary minimal surfaces inside E which are of the type of the disk, namely, the parts of the coordinate planes $\{x = 0\}$, $\{y = 0\}$, and $\{z = 0\}$ lying in the interior of E.

Thus, if S is the boundary of a convex body $\mathcal{K} \subset \mathbb{R}^3$, it is tempting to conjecture that there exist at least three geometrically different stationary minimal surfaces with boundary on S. As mentioned before, we cannot obtain these surfaces by a minimum procedure. Hence more refined *minimax procedures* (or *saddle-point methods*) have to be used if we want to find such surfaces which are not minimizers. As a first result in this direction Struwe [3] proved the following

Theorem 1. *For any embedded surface S of class C^4 which is diffeomorphic to the unit sphere S^2 in $\mathbb{R}^3$, there exists a stationary minimal surface $X \in \mathcal{C}(S)$ of the type of the disk which has its free boundary on S.*

Struwe's proof applies a minimax principle from Palais [1] to a modified class of variational problems $\mathcal{P}_\alpha, \alpha > 1$, which satisfy the Palais–Smale condition and hence admit a saddle-type solution X_α. A nonconstant stationary minimal surface is obtained by passing to the limit $\alpha \to 1$ via a suitable subsequence of the surfaces X_α. This approach can be viewed as an adaptation of a method due to K. Uhlenbeck (see, for instance, Sacks and Uhlenbeck [1]).

Struwe's theorem does not answer the question as to whether one can find an *embedded* stationary minimal surface with its free boundary on the surface S of some convex body $\mathcal{K}$, or if there is at least an *immersed* stationary minimal surface in $\mathcal{C}(S)$. In case that S is the boundary of a strictly convex subset $\mathcal{K} \subset \mathbb{R}^3$ of class C^4, Grüter and Jost [1] have found the following stronger result.

Theorem 2. *There exists an embedded, stationary disk-type minimal surface having its free boundary on S (and values in $\mathcal{K}$).*

The proof of this theorem uses methods from geometric measure theory which have not been treated in these notes. Let us only mention some main ingredients of the arguments used by Grüter and Jost. First the minimax methods from Pitts [1] are employed to obtain a so-called *almost minimizing varifold* in the sense of Pitts [1] and Simon and Smith [1], which meets S transversally along its trace. The regularity of this varifold at its free boundary relies on an extension of Allard's regularity results to free boundary value problems due to Grüter and Jost [2]. Finally, Simon and Smith proved the existence of a minimally embedded two-sphere in any manifold diffeomorphic to the three-sphere. The methods of these authors are used in an essential way to show that the above varifold is both embedded and simply connected, that is, the minimizing varifold is of the type of the disk or of a collection of disks.

Theorem 2 also extends to Riemannian manifolds if one adapts methods by Pitts [1] and by Meeks, Simon, and Yau [2].

The following theorem due to Jost [15] (cf. also [9] for an earlier, more restricted result) shows that a closed convex surface S bounds in fact three different stationary minimal surfaces.

Theorem 3. *Let S be the boundary of a strictly convex body $\mathcal{K} \subset \mathbb{R}^3$ of class C^5. Then there exist three geometrically different, stationary, embedded*

minimal surfaces in $\mathcal{K}$ which are of disk type and have their free boundaries on S.

In fact, Jost [15] proved that the assertion still holds true if S is merely *H-convex*.

A generalization of Theorem 2 to convex polyhedral surfaces S was established by Jost [15]. His result to be stated next contains a part of the Theorem of B. Smyth as a special case.

Theorem 4. *Let S be a compact convex polyhedron in $\mathbb{R}^3$. Then there exists an embedded minimal surface X of the type of the disk meeting S perpendicularly along its boundary such that no segment of any edge of S is contained in the boundary of X.*

1.9 Nonuniqueness of Solutions to a Free Boundary Problem. Families of Solutions

Examples of minimal surfaces with free or partially free boundaries on a prescribed supporting surface S were already investigated during the last century. The first geometric problem leading to minimal surfaces with free boundaries was posed by the French mathematician Gergonne [1] in 1816, but a correct solution was only found by H.A. Schwarz in 1872 (see [2], pp. 126–148, and Tafel 4 at the end of vol. I).

Gergonne's problem consists in finding a minimal surface spanning a frame $\langle \Gamma_1, \Gamma_2, S_1, S_2 \rangle$ that consists of two parallel faces S_1 and S_2 of some cube and of two straight arcs Γ_1 and Γ_2 lying on opposite faces of the cube.[3] As depicted in Fig. 1, we assume that the two diagonals Γ_1 and Γ_2 are perpendicular to each other. In contrast to his predecessors, Schwarz arrived at correct stationary surfaces spanning the configuration $\langle \Gamma_1, \Gamma_2, S_1, S_2 \rangle$ since he had discovered the proper free boundary condition: each stationary surface has to meet the two supporting surfaces at a right angle. In addition to an area minimizing solution which is depicted in Fig. 1, Schwarz discovered infinitely many other *non-congruent* stationary minimal surfaces in the frame $\langle \Gamma_1, \Gamma_2, S_1, S_2, S_3, S_4 \rangle$ consisting of the four vertical faces S_i and the two horizontal arcs Γ_1, Γ_2. In other words, *a partially free boundary problem may have infinitely many distinct (i.e. noncongruent) solutions.*

Let us set up the definition of free or partially free boundary problems in some more generality than in Section 1.4. We consider boundary configurations $\langle \Gamma, S \rangle$ in $\mathbb{R}^3$ consisting of a system Γ of Jordan curves $\Gamma_1, \dots, \Gamma_m$ and of a system S of surfaces $S_1, \dots, S_n$. Each of the curves Γ_i is either a closed curve or else a Jordan arc with end points on S. We shall call S the *free part*

[3] In fact, the original form of this problem is somewhat different; it was stated as a *partition problem* for the cube.

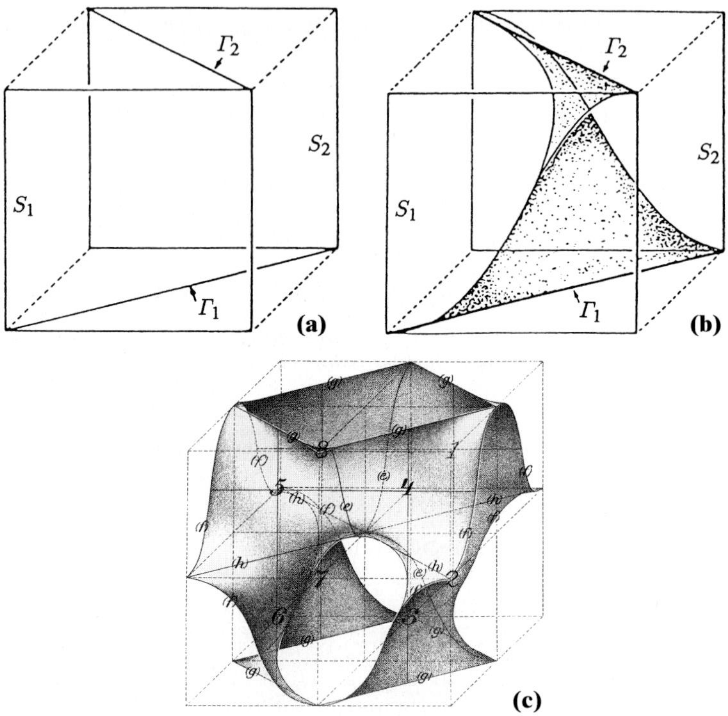

Fig. 1. (a) The Schwarzian chain $\langle \Gamma_1, \Gamma_2, S_1, S_2 \rangle$ forming the boundary frame for Gergonne's problem. **(b)** Gergonne's surface, the area minimizing solution of Gergonne's problem discovered by Schwarz. **(c)** Gergonne's surface generates the fifth periodic minimal surface known to Schwarz (Lithograph by H.A. Schwarz)

of the configuration $\langle \Gamma, S \rangle$. The fixed part Γ of the boundary frame may be empty.

A minimal surface $\mathcal{M}$ is said to be *stationary within the configuration* $\langle \Gamma, S \rangle$ if the boundary of $\mathcal{M}$ lies on $\Gamma \cup S$ and, moreover, if $\mathcal{M}$ meets S orthogonally at the part $\Sigma = \partial\mathcal{M} \cap S$ of its boundary. As usual, we shall call Σ the *free trace* of $\mathcal{M}$ on S.

Remark. If this definition is to make sense we have to assume that each of the support surfaces S_j is a regular surface of class C^1. Furthermore we shall suppose the each Γ_k is a piecewise smooth regular arc. Similarly, $\mathcal{M}$ is supposed to be smooth except for finitely many points. Note that in this section we assume that $\mathcal{M}$ meets S *everywhere* at a right angle (except for at most finitely many points). In other words, we essentially exclude the case that $\partial\mathcal{M}$ attaches in segments (i.e. intervals) to ∂S since in this case the two surfaces $\mathcal{M}$ and S need not include an angle of ninety degrees.

The *free-boundary problem* of a configuration $\langle \Gamma, S \rangle$ is the problem to determine a stationary minimal surface within $\langle \Gamma, S \rangle$. As before, such a problem

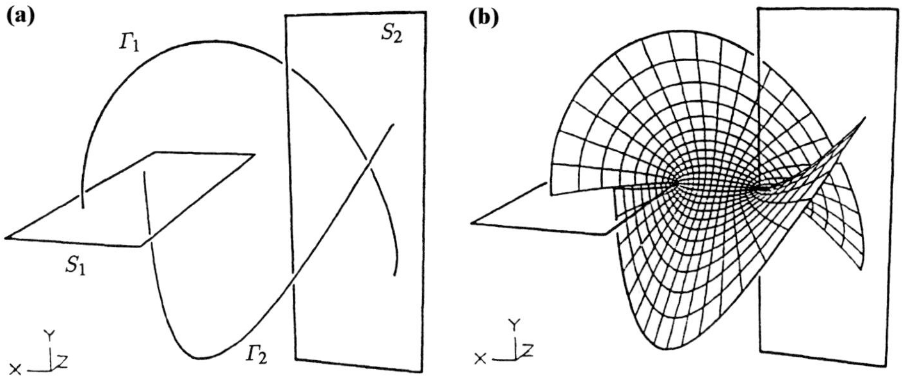

Fig. 2. **(a)** A partially free boundary problem for a frame $\langle \Gamma_1, \Gamma_1, S_1, S_2 \rangle$. **(b)** A part of Henneberg's surface forms a disk-type solution of the problem. Note that S_1 and S_2 are surfaces with boundary. As in the present case, this can lead to singularities of the free boundary of a solution (see Chapters 1 and 2 of Vol. 3)

is said to be *partially free* if Γ is nonvoid; otherwise we call it *completely free* or simply *free*.

As usual we describe minimal surfaces $\mathcal{M}$ by mappings X from a planar parameter domain Ω or from a Riemann surface $\mathcal{R}$ into $\mathbb{R}^3$; $\partial\Omega$ and $\partial\mathcal{R}$ are assumed to be piecewise smooth, and X will be smooth in $\overline{\Omega}$ or $\overline{\mathcal{R}}$ except for at most finitely many points on $\partial\Omega$ or $\partial\mathcal{R}$.

It is trivial to find supporting surfaces S which bound continua of stationary minimal surfaces. The sphere, the cylinder, or the torus furnish simple examples. In these cases, however, all minimal surfaces belonging to the same continuum are congruent to each other.

Therefore it is of interest to see that there are free or even partially free boundary problems which possess denumerably many noncongruent solutions, or even continua of noncongruent solutions.

As we have mentioned above, Gergonne's problem furnishes an example of such a free boundary problem. In fact, using the helicoids, Schwarz was able to exhibit an even simpler and completely elementary example of such a boundary configuration. Consider a boundary frame $\langle \Gamma_1, \Gamma_2, S \rangle$ consisting of a cylinder surface S and two straight arcs Γ_1 and Γ_2 which are perpendicular to each other as well as to the cylinder axis and pass through the axis at different heights. This configuration bounds denumerably many left and right winding helicoids which meet the cylinder S at a right angle (Fig. 3). Only two of these helicoids are area minimizing, the others are only stationary.

A slight modification of the previous example yields a boundary frame $\langle \Gamma, S \rangle$ consisting of a cylinder S as surface of support and of a polygon Γ made of a piece A of a cylinder axis and of two straight segments A_1 and A_2 which connect A with S; we assume that A_1 and A_2 are perpendicular to each other. There are again infinitely many stationary surfaces for $\langle \Gamma, S \rangle$, all of which are helicoidal surfaces (cf. Fig. 4).

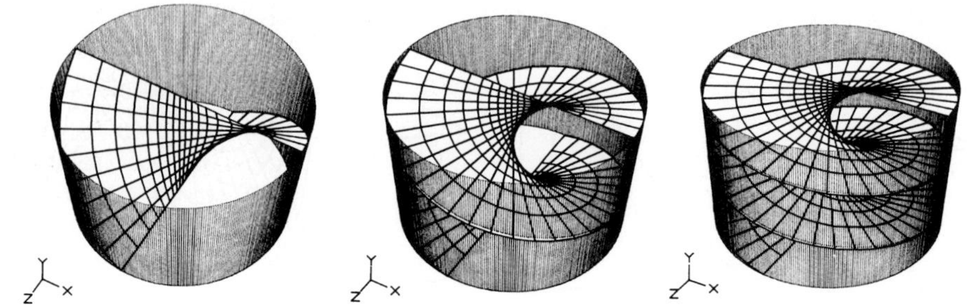

Fig. 3. Three of infinitely many noncongruent minimal surfaces that are stationary within a configuration $\langle \Gamma_1, \Gamma_2, S \rangle$

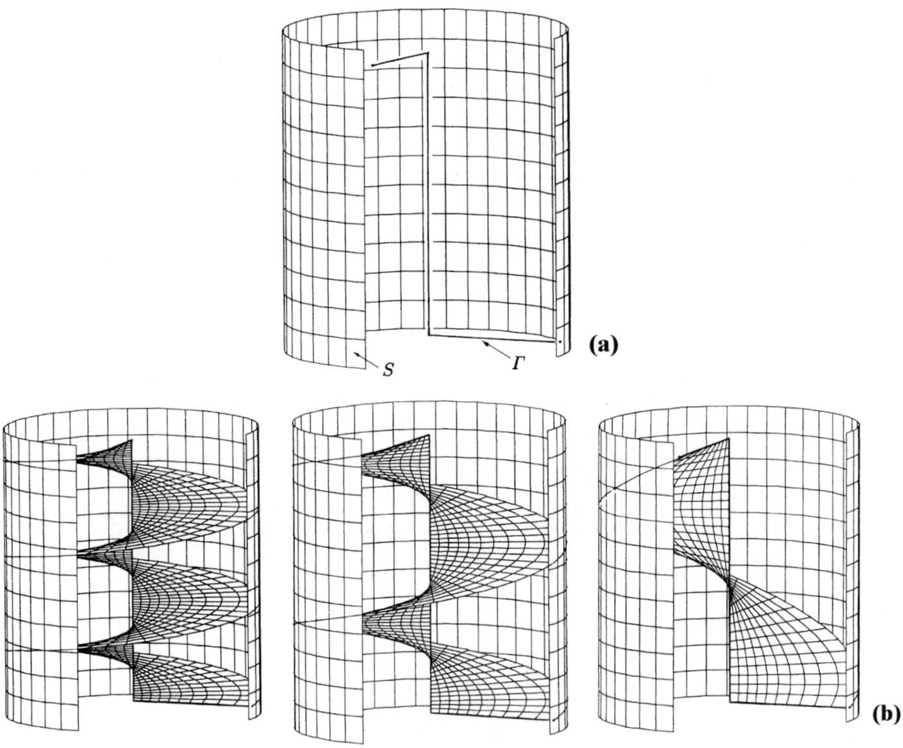

Fig. 4. A boundary configuration $\langle \Gamma, S \rangle$ **(a)** bounding infinitely many stationary minimal surfaces of the type of the disk; these are pieces of helicoids **(b)**

Next we consider a configuration $\langle \Gamma, S \rangle$ consisting of a circle Γ and of a supporting surface S which bounds a continuum of noncongruent and even area minimizing minimal surfaces. It turns out that such an example can be derived from the classical calculus of variation. In the following we freely use some of these results, see Bolza [1] (Beispiel I); Bliss [1], pp. 85–127; Carathèodory [3], pp. 340–341, 360–367; Giaquinta and Hildebrandt [1].

Let $x(t), y(t), t_1 \leq t \leq t_2$, be the parameter representation of a curve contained in the upper half plane $\{y > 0\}$. The surface area of its surface of revolution about the x-axis is given by the integral $2\pi \int_{t_1}^{t_2} y\sqrt{dx^2 + dy^2}$. Thus the minimal surfaces of revolution are described by the extremals of the functional $\int y\sqrt{dx^2 + dy^2}$, $y > 0$, which are the parallels to the positive y-axis,

$$x = x_0, \quad y > 0,$$

and the catenaries

(1) $$y = a\cosh\left(\frac{x - x_0}{a}\right), \quad -\infty < x < \infty,$$

which form a 2-parameter family of nonparametric curves, $a > 0, -\infty < x_0 < \infty$. The point (x_0, a) is the vertex of the catenary (1).

Let us consider all catenaries passing through some fixed point $P = (0, b), b > 0$, on the y-axis. They must satisfy $b = a\cosh(\frac{x_0}{a})$ or $b = a\cosh\lambda$, if we introduce the new parameter $\lambda = -\frac{x_0}{a}$. Then there is a 1-1 correspondence between all real values of the parameter λ and all catenaries passing through $(0, b)$ which is given by

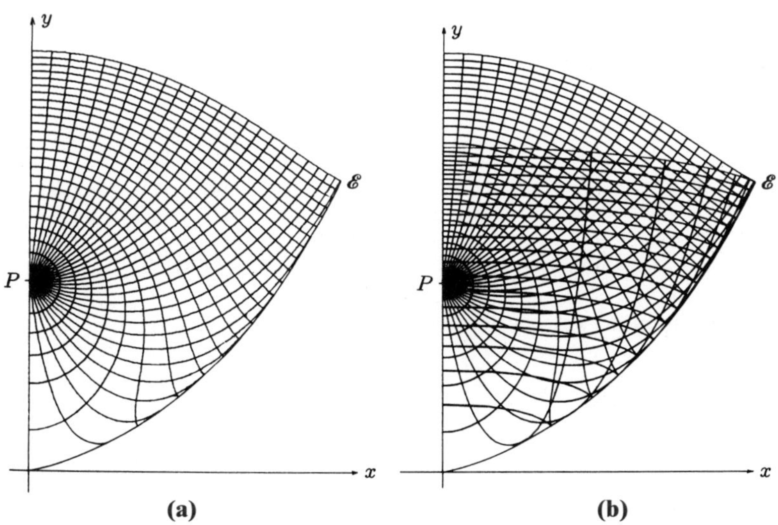

Fig. 5. (a) Catenaries emanating from P to the right, and their wave fronts. **(b)** A complete figure: The stable catenaries emanating from P and terminating at their envelope $\mathcal{E}$, together with their wave fronts

(2)
$$y = g(x,\lambda) := a(\lambda)\cosh\left(\lambda + \frac{x}{a(\lambda)}\right), \quad x \in \mathbb{R},$$
$$a(\lambda) := \frac{b}{\cosh\lambda}, \quad \lambda \in \mathbb{R}.$$

We can also write

$$g(x,\lambda) = b\cosh\frac{x}{a(\lambda)} + a(\lambda)\sinh\lambda\sinh\frac{x}{a(\lambda)},$$

and $\sinh\lambda = \pm\sqrt{b^2 - a^2(\lambda)}/a(\lambda)$.

We now consider the branches $y = g(x,\lambda), x \geq 0$, lying in the first quadrant of the x,y-plane. There exists exactly one conjugate point $Q(\lambda) = (\xi(\lambda), \eta(\lambda))$ with respect to P on each catenary (2). The points $Q(\lambda), \lambda \in \mathbb{R}$, form a real-analytic curve $\mathcal{E}$ that resembles a branch of a parabola extending from the origin to infinity. The curve $\mathcal{E}$ is given by the condition

$$\frac{\partial}{\partial\lambda}g(x,\lambda) = 0$$

and describes the envelope of the catenary arcs

$$C_\lambda = \{(x, g(x,\lambda)) : 0 \leq x \leq \xi(\lambda)\}, \quad \lambda \in \mathbb{R}.$$

The domain $\Omega = \{(x,y) : 0 < x < \xi(\lambda), y > \eta(\lambda) \text{ for some } \lambda\}$ is simply covered by the open arcs $\overset{\circ}{C}_\lambda = C_\lambda \setminus \{P, Q(\lambda)\}$.

Consider the *wavefronts* $W_c, c > 0$, emanating from P. The curves W_c are the real analytic level lines $\{\mathcal{S}(x,y) = c\}$ of the wave function $\mathcal{S}(x,y)$ that satisfies the Hamilton–Jacobi equation

$$\mathcal{S}_x^2 + \mathcal{S}_y^2 = y^2$$

and is given by

$$\mathcal{S}(x, g(x,\lambda)) = \mathcal{I}(x,\lambda), \quad 0 \leq x \leq \xi(\lambda),$$

where the right-hand side is defined by

$$\mathcal{I}(x,\lambda) = \int_0^x g(u,\lambda)\sqrt{1 + g'(u,\lambda)^2}\,du,$$

and $g'(u,\lambda) = \frac{\partial}{\partial u}g(u,\lambda)$.

The two families of curves $C_\lambda, \lambda \in \mathbb{R}$, and $W_c, c > 0$, form the *complete figure* (in sense of Carathéodory) associated with the variational problem

$$\int y\sqrt{dx^2 + dy^2} \to \text{Extr}, \quad y(0) = b,$$

in $x \geq 0, y > 0$, see Fig. 5.

By Adolf Kneser's transversality theorem, the curves W_c intersect the catenaries C_λ orthogonally. Two curves W_{c_1} and W_{c_2}, $c_1 < c_2$, cut a piece $C_\lambda(c_1, c_2)$ out of each curve C_λ such that

$$\int_{C_\lambda(c_1,c_2)} y\sqrt{dx^2 + dy^2} = c_2 - c_1,$$

and $c_2 - c_1$ is the infimum of the integral $\int y\sqrt{dx^2 + dy^2}$ along all paths joining W_{c_1} and W_{c_2} within Ω. In particular, if $C_{\lambda,c} = \{(x, g(x,\lambda)) \colon 0 \leq x \leq x_0(\lambda, c)\}$ denotes the subarc of the catenary that connects P with W_c, then $\mathcal{I}(x_0(\lambda, c), \lambda)$ is the infimum of the integral $\int y\sqrt{dx^2 + dy^2}$ taken along all curves joining P and W_c within Ω. If we now rotate the whole configuration shown in Fig. 5 about the x-axis, the wavefront W_c generates a surface of revolution S_c, and each catenary $C_{\lambda,c}$ produces a minimal catenoid $K_{\lambda,c}$ with the area $2\pi c$. The catenoid $K_{\lambda,c}$ is bounded by two parallel coaxial circles Γ and $\Sigma_{\lambda,c}$ centered on the x-axis. Γ is generated by the rotation of P, and $\Sigma_{\lambda,c}$ by the rotation of the intersection point of C_λ with W_c. Each catenoid $K_{\lambda,c}$ intersects S_c orthogonally and, therefore, is a stationary minimal surface within the configuration $\langle \Gamma, S_c \rangle$. All catenoids $K_{\lambda,c}$, c fixed, have the same area and minimize area among all surfaces of revolution bounded by $\langle \Gamma, S \rangle$ which lie in the open set $\mathcal{H}$ generated by rotating $\Omega \cup \Omega^* \cup \{x = 0, y > 0\}$ about the x-axis. Here Ω^* is the mirror image of Ω at the y-axis in the x, y-plane (cf. Fig. 6).

In fact, it turns out that the catenoids $K_{\lambda,c}$ even minimize area among all orientable surfaces $\mathcal{F}$ bounded by $\langle \Gamma, S_c \rangle$ that are contained in $\mathcal{H}$. A well-known projection argument shows that it suffices to prove Area$(K_{\lambda,c}) \leq$ Area$(\mathcal{F})$ for all oriented surfaces $\mathcal{F}$ with boundary on $\Gamma \cup S_c$ that are contained in $\mathcal{H}^+ = \mathcal{H} \cap \{x \geq 0\}$.

Let now $\mathcal{F}$ be such a surface with $\gamma = \partial\mathcal{F} \cap S_c$. Then there is a region T in the surface S_c with integer multiplicities, the boundary of which equals $\gamma - \Sigma_{\lambda,c}$. Therefore $K_{\lambda,c} - \mathcal{F} + T$ is a cycle, and it follows that there is a three-dimensional region $\mathcal{R}$ with integer multiplicities such that the boundary of $\mathcal{R}$ is $K_{\lambda,c} - \mathcal{F} + T$. Gauss's theorem yields

$$\int_{\mathcal{R}} \operatorname{div} X \,\mathrm{dvol} = \int_{\partial\mathcal{R}} \langle X, N_{\partial\mathcal{R}} \rangle \, dA, \tag{3}$$

where $N_{\partial R}$ is the oriented unit normal to $\partial\mathcal{R}$. Let $X = X(x, y, z)$ be a field of unit vectors normal to the foliation formed by the catenoids $K_{\lambda,c}$. Then we infer from Vol. 1, Section 2.7, in particular from formula (3) that

$$\operatorname{div} X = -2H,$$

H being the mean curvature of the leaves of the foliation. Since $H \equiv 0$, the vector field X is divergence free. Since $\langle X, N_T \rangle = 0$ and X can be chosen in such a way that $\langle X, N_{K_{\lambda,c}} \rangle = 1$, we obtain

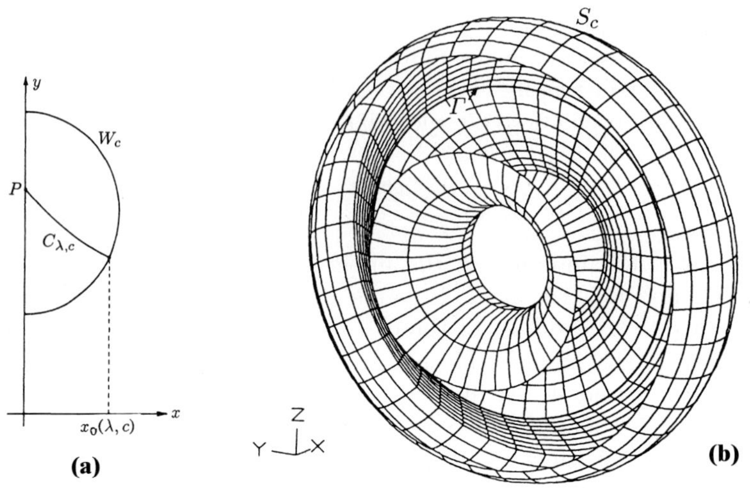

Fig. 6. **(a)** Rotation of the wave front W_c about the x-axis yields half the surface S_c; the whole surface S_c is then obtained by reflection at the plane $x = 0$. The curve Γ is a circle obtained by rotating P about the x-axis. **(b)** This drawing depicts the configuration $\langle \Gamma, S_c \rangle$ and two of the minimal leaves within $\langle \Gamma, S_c \rangle$. A part of S_c is removed to permit a glimpse into the interior

$$\text{Area}(K_{\lambda,c}) = \int_{K_{\lambda,c}} \langle X, N_{K_{\lambda,c}} \rangle \, dA = \int_{\mathcal{F}} \langle X, N_{\mathcal{F}} \rangle \, dA.$$

Because of $\langle X, N_{\mathcal{F}} \rangle \le 1$, the term on the right-hand side is estimated from above by Area($\mathcal{F}$). Thus we have proved:

Theorem 1. *There exists a configuration $\langle \Gamma, S_c \rangle$ consisting of a circle Γ and a real analytic surface of revolution S_c that bounds a family $\{K_{\lambda,c}\}$ of stationary and even area-minimizing minimal surfaces of annulus-type that are really distinct in the sense that, for any two different values λ_1, λ_2, the surfaces $K_{\lambda_1,c}$ and $K_{\lambda_2,c}$ are not congruent.*

A simple modification of the previous example leads to *boundary configurations S as shown in Fig. 7 that bound continua $\mathcal{C}$ of noncongruent stationary surfaces of annulus type which have a completely free boundary on S. The surfaces of $\mathcal{C}$ are even area minimizing within the class $\mathcal{C}^*$ of annulus type surfaces whose free boundaries are homologous to those of the surfaces of $\mathcal{C}$.*

For this purpose, we take two wavefront curves W_{c_1} and $W_{c_2}, c_1, c_2 > 0$, contained in $x > 0, y > 0$. If c_1 and c_2 are chosen sufficiently small, both curves terminate at the positive y-axis and meet this axis orthogonally. Reflecting both arcs at the y-axis, we obtain two closed real analytic curves Γ_{c_1} and Γ_{c_2}, and their rotation about the x-axis leads to two closed torus-type surfaces S_1 and S_2 that are orthogonally met by a family of catenoids, generated by the catenary arcs $C_\lambda(c_1, c_2)$. These catenoids are stationary annulus-type

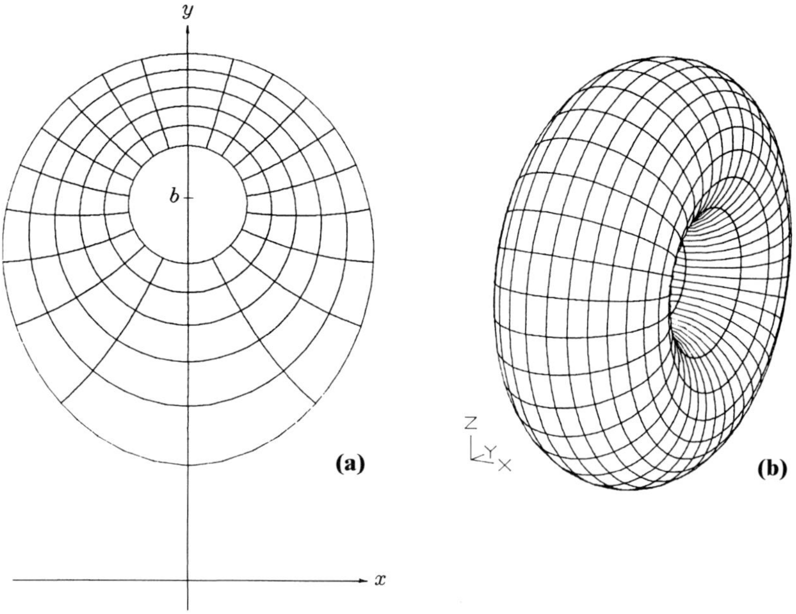

Fig. 7. A modification of the example depicted in Fig. 6. Any of the closed curves W_c in **(a)** generates a real analytic and rotationally symmetric surface S_c as depicted in **(b)**. Any configuration $\langle S_1, S_2 \rangle$ with $S_i := S_{c_i}$ bounds a one parameter family $\mathcal{C}$ of annulus-type minimal leaves which are parts of catenoids. **(c)**, **(d)** Parts of the configuration $\langle S_1, S_2 \rangle$. **(e)**, **(f)** Three surfaces of the family $\mathcal{C}$ outside and within $\langle S_1, S_2 \rangle$

minimal surfaces within the configuration $\langle S_1, S_2 \rangle$, and a reasoning similar to the previous one shows that they even minimize area within $\mathcal{C}^*$ (cf. Fig. 7).

A somewhat different example, which is not rotationally symmetric, leads to a free-boundary problem for minimal surfaces of the type of the disk, with their boundary lying on a given real analytic torus-like surface. Let $K_\lambda, \lambda \in \mathbb{R}$, be the catenoids obtained by rotating the arc C_λ about the x-axis, and let K_λ^* be the surface obtained from K_λ by reflection at the y, z-plane. Moreover, let $K_{-\infty}$ be the disk interior to the circle Γ in the y, z-plane, and let K_∞ be the plane domain exterior to Γ. We may think of $K_{\pm\infty}$ as degenerate catenoids obtained for $\lambda \to \pm\infty$. Then the surfaces $K_\lambda, K_\lambda^*, -\infty \le \lambda \le \infty$, describe a minimal foliation, singular at Γ, of the rotationally symmetric domain $\mathcal{H}$.

We now introduce cylindrical coordinates (x, r, θ), where $y = r\cos\theta, z = r\sin\theta$. For each $r \in (0, b)$, there exists exactly one value $c(r) > 0$ such that the closed, real analytic curve $\Gamma_{c(r)}$ in the plane $\theta = 0$, obtained from the wavefront $W_{c(r)}$ as described before, passes through $(0, r, 0)$.

Denote by $L_{r,\theta}$ the closed curve that is obtained by rotating $\Gamma_{c(r)}$ about the angle θ around the x-axis. The curves $L_{r,\theta}, 0 < r < b, 0 \le \theta < 2\pi$, meet the plane $x = 0$ orthogonally at the points $(0, r, \theta)$ and sweep out an open subdomain $\mathcal{H}_0$ of $\mathcal{H}$.

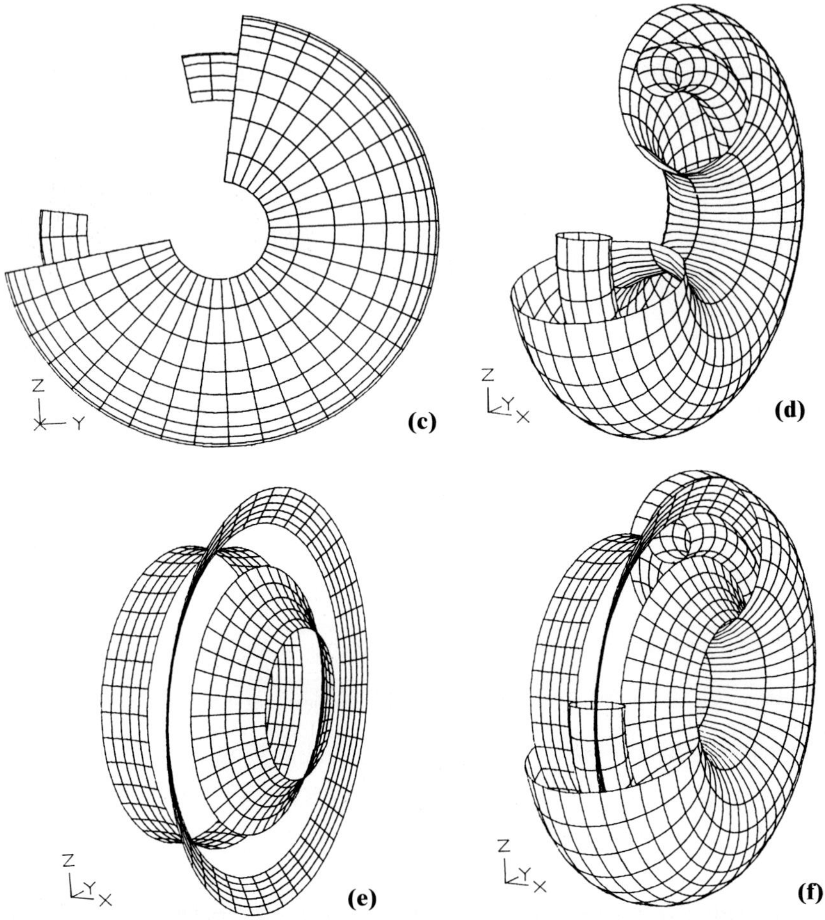

Fig. 7. (Continued. Captions see preceding page)

Let γ_0 be a real analytic Jordan curve in the plane $x = 0$, say, a circle, which is contained in the open disk $K_{-\infty}$ (the interior of Γ) and does not wind about the origin. As the point $(0, r, \theta)$ traverses the curve γ_0, the curves $L_{r,\theta}$ sweep out a toruslike surface S which bounds a tube G. This tube is foliated by a family $\mathcal{M}_\lambda, \mathcal{M}^*_\lambda, -\infty \le \lambda \le \infty$, of minimal surfaces that are cut by S out of the catenoids K_λ, K^*_λ. The surfaces $\mathcal{M}_\lambda, \mathcal{M}^*_\lambda$ are of the type of the disk and meet S perpendicularly; hence they are stationary within S (cf. Figs. 8 and 9). Moreover, the unit normal vectors to $\mathcal{M}_\lambda, \mathcal{M}^*_\lambda$ form a divergence free vector field on the set $\mathcal{H} \setminus \Gamma$ containing G which is tangent to S. Then, by an argument parallel to the previous reasoning, all surfaces $\mathcal{M}_\lambda, \mathcal{M}^*_\lambda$ have equal area, and each oriented surface $\mathcal{F}$ contained in $\mathcal{H} \setminus \Gamma$ and with a boundary γ homologous in S to γ_0 has area larger than the leaves $\mathcal{M}_\lambda, \mathcal{M}^*_\lambda$ unless it coincides with one of these surfaces. Thus we have shown:

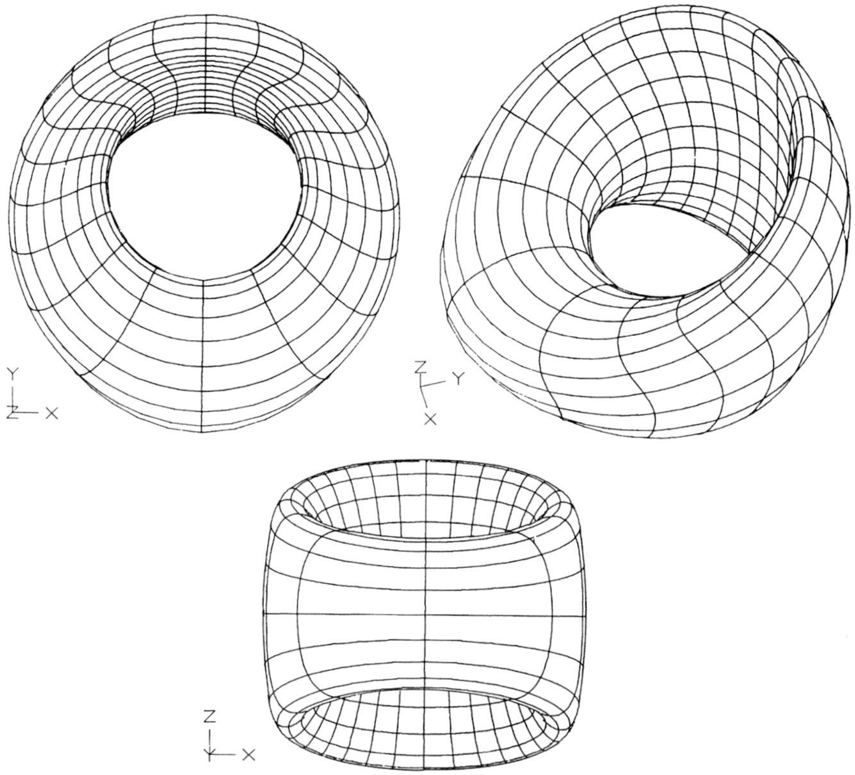

Fig. 8. Three views of a real analytic topological torus bounding a 1-parameter family of minimal disks whose traces on S are depicted in the figures by curvy lines

Theorem 2. *There exists a real analytic, embedded surface S of the type of the torus, and a homology class $[\gamma_0]$ in $H_1(S;\mathbb{Z})$, so that S bounds a family of stationary minimal surfaces of the type of the disk which have smallest area among all oriented surfaces in $\mathcal{H} \setminus \Gamma$ having their boundaries lying on S and homologous in S to γ_0.*

In view of the two examples described in Theorems 1 and 2, the following two *theorems* will be rather surprising.

Theorem 3 (F. Tomi). *If a compact analytic H-convex body M in $\mathbb{R}^3$ has the property that there is closed Jordan curve in M which cannot be contracted in M and, secondly, that the free boundary problem for ∂M admits infinitely many minimizing solutions of disk-type contained in M, then M must be homeomorphic to a solid torus, and the set of all such solutions is an analytic S^1-family of minimal embeddings of the disk.*

For the proof of Theorem 3, we refer the reader to Tomi's paper [10]. There one also finds the following interesting observation:

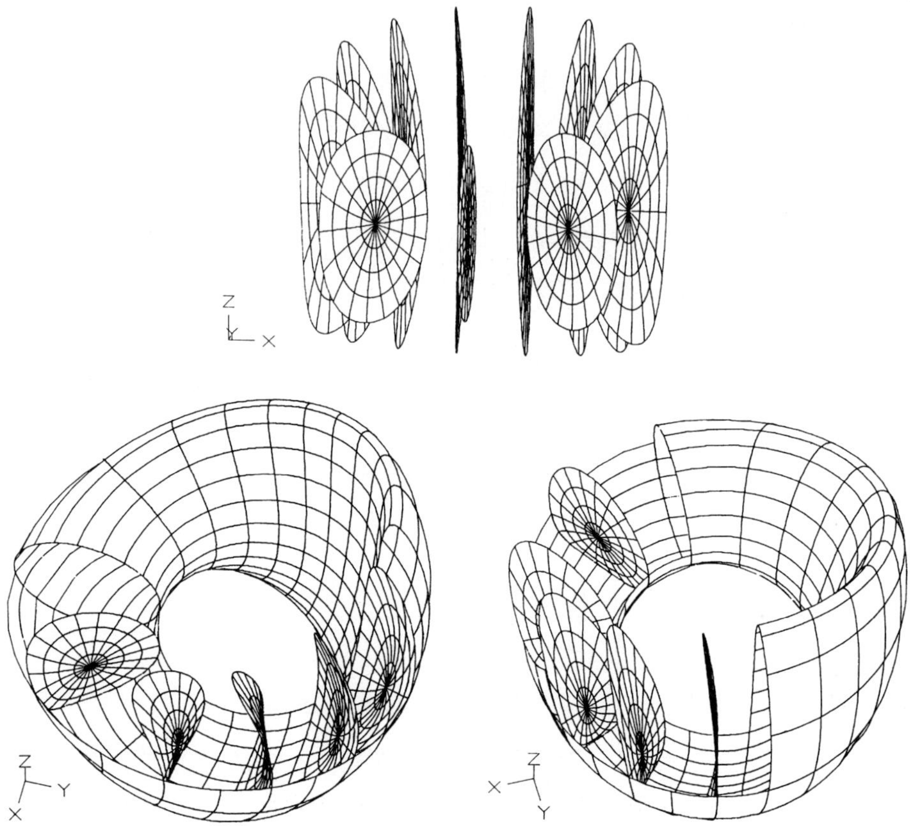

Fig. 9. Samples of minimal leaves $\mathcal{M}_\lambda, \mathcal{M}^*_\lambda$, two of which are flat, and how they fit into the surface S shown in Fig. 8. A part of S has been removed to permit a glimpse into the interior of S

If a torus M is foliated by a smooth S^1-family of plane disk-type minimal surfaces being orthogonal to ∂M, then all surfaces in the family are congruent.

In contrast, we obtain from Theorem 2 the following result:

There exist real analytic (topological) tori admitting families of non-flat disk-type minimal surfaces which intersect the tori at a right angle, and secondly, the surfaces in such a family need not be congruent (nor isometric).

Related to Theorem 3 there is a *finiteness theorem* due to Alt and Tomi [1] that will be stated as Theorem 4. We shall outline a proof of this result. As their techniques are closely related to the methods used for proving Theorem 3, the reader will obtain a good idea of how such results are proved.

Let S be a compact, embedded, real analytic surface in $\mathbb{R}^3$, and let Π be a homotopically nontrivial closed polygon in the unbounded component of $\mathbb{R}^3 \setminus S$.

As in Section 1.2 we define the class $\mathcal{C}(\Pi, S)$ by

$$\mathcal{C}(\Pi, S) := \{X \in \mathcal{C}(S) \colon \mathcal{L}([X|_{\partial B}], \Pi) \neq 0\};$$

that is, $\mathcal{C}(\Pi, S)$ is defined as the set of all $X \in \mathcal{C}(S)$ whose boundary values are not contractible in $\mathbb{R}^3 \setminus \Pi$. Then we obtain the following finiteness result contrasting with Theorem 2:

Theorem 4. *There are only finitely many geometrically different minimal surfaces which are minimizers of Dirichlet's integral in $\mathcal{C}(\Pi, S)$.*

As a by-product of the proof of Theorem 4 we obtain the following result which is of independent interest.

Theorem 5. *Let $X \in \mathcal{C}(\Pi, S)$ be a strong relative minimum of $\mathcal{P}(\Pi, S)$, i.e. we have $D(X) \le D(Y)$ for all surfaces $Y \in \mathcal{C}(\Pi, S)$ with $Y(\bar{B}) \subset \mathcal{U}, \mathcal{U}$ being an open neighbourhood of $X(\bar{B})$. Then X is immersed up to the boundary, that is, $|X_u(w)| = |X_v(w)| \neq 0$ for all $w \in \bar{B}$.*

Anticipating the regularity results of Chapter 2 we may assume that each minimizer X – and even each stationary point – can be continued analytically across the boundary ∂B. Moreover, if $w_0 \in \bar{B}$ is a branch point (i.e., $X_u(w_0) = X_v(w_0) = 0$), then we obtain as in Section 3.2 of Vol. 1 and Section 2.10 of this volume in suitable (new) coordinates x^1, x^2, x^3 the representation

$$\begin{aligned} x^1(w) + ix^2(w) &= A(w - w_0)^m + O(|w - w_0|^{m+1}), \\ x^3(w) &= O(|w - w_0|^{m+1}) \end{aligned} \tag{4}$$

with $A \in \mathbb{C}^3 \setminus \{0\}$. Next we infer from Lemma 5 of Section 5.3 the existence of a C^1-diffeomorphism $F \colon \mathcal{U} \to V$ defined on a neighbourhood $\mathcal{U}$ of w_0 such that for some function $\varphi \in C^2(V)$ we have

$$\begin{aligned} x^1(w) + ix^2(w) &= [F(w)]^m \\ x^3(w) &= \varphi(F(w)). \end{aligned} \tag{5}$$

Moreover it follows from the proof of Lemma 5 of Section 5.3 that $F(w_0) = 0, F \in C^\omega(\mathcal{U} \setminus \{w_0\})$ and $\varphi \in C^\omega(V \setminus \{0\})$, V being a suitable neighbourhood of $0 \in \mathbb{R}^2 \mathrel{\widehat{=}} \mathbb{C}$. Of course we may assume that V is a disk $B_r(0)$ of a sufficiently small radius $r > 0$. The representation (5) permits us to introduce the new variable $\tilde{w} = F(w) \in V$, and we have

$$\nabla^k \varphi(\tilde{w}) = O(|\tilde{w}|^{m+1-k}) \quad \text{as } \tilde{w} \to 0$$

for $k = 0, 1, 2$.

We have to distinguish true and false branch points of a given minimal surface $X(w)$; the surface has different geometric properties in a neighbourhood of different kinds of branch points.

We call a branch point w_0 of X a *false branch point* if in some neighbourhood of w_0 the surface $X(w)$ can be reparametrized as an immersed surface. This is true if and only if φ is a function of $(\tilde{w})^m$. Otherwise w_0 is called a *true branch point*. It is shown in Chapter 5 how to exclude true branch points on the boundary by using only the minimum property of X. Since the argument is similar for true interior branch points, we refrain from repeating the procedure and refer to Section 5.3 as well as to the original papers by Osserman [12], Alt [1] and Gulliver [2]. Another possibility could be to apply Tromba's technique, which is presented in Chapter 6. We are going to outline the discussion for false branch points. Note that by analytic continuation we may assume X to be defined on some open neighbourhood $B_R, R > 1$, of the closed unit disk $\bar{B}$. Denoting the new function again by X, we may in addition assume that all branch points of X lie in $\bar{B}$ and that (5) continues to hold. Moreover, we can define a continuous unit normal $N(w)$ for $X(w)$ on all of $\bar{B}$.

Definition 1. *Two points $z, w \in B_R$ are called* equivalent, *$z \sim w$, if there are fundamental systems of open neighbourhoods $\mathfrak{U}_n(z), V_n(w), n \in \mathbb{N}$ such that $X(\mathfrak{U}_n) = X(V_n)$ for all n. We also define the* equivalent boundary *$\tilde{\partial} B$ by $\tilde{\partial} B = \{z \in \overline{B} : z \sim w \text{ for some } w \in \partial B\}$.*

Proposition 1. *Suppose $z_k \to z, w_k \to w$ and $z_k \sim w_k$. Then $z \sim w$. In particular, the equivalent boundary $\tilde{\partial} B$ is closed.*

Proposition 1 is a consequence of

Lemma 1. *Let z and w be two points in $B_R, R > 1$, such that $X(z) = X(w)$ and $N(z) = \pm N(w)$. Furthermore denote by $\mathfrak{U}$ and V coordinate neighbourhoods of z and w such that a representation (5) holds, and suppose that there is an open subset $\mathfrak{U}'$ of $\mathfrak{U}$ with the property that $X(\mathfrak{U}') \subset X(V)$. Then it follows that $z \sim w$.*

Proof. From (5) we infer the existence of small positive number r and s such that,

$$X(\mathfrak{U}) = \{(x^1, x^2, x^3) \colon x^1 + ix^2 = (\tilde{w})^m, x^3 = \varphi(\tilde{w}), |\tilde{w}| < r\},$$
$$X(V) = \{(x^1, x^2, x^3) \colon x^1 + ix^2 = (\tilde{\omega})^n, x^3 = \psi(\tilde{\omega}), |\tilde{\omega}| < s\}.$$

Since $X(\mathfrak{U}) \subset X(V)$, it follows that for some open set of numbers z contained in $\{w \colon 0 < |w| < \min(\sqrt[n]{r}, \sqrt[m]{s})\}$ the relation $\varphi(z^n) = \psi(\eta_n z^m)$ holds true where η_n denotes some n-th root of unity. By the analyticity of φ and ψ in $0 < |z| < r$ and $0 < |z| < s$ respectively we conclude that $\varphi(z^n) = \psi(\eta_n z^m)$ holds for all z with $|z| < \min(\sqrt[n]{r}, \sqrt[m]{s})$. Next define

$$\mathfrak{U}_\varepsilon := \{z \in \mathfrak{U} \colon |\tilde{w}(z)| < \sqrt[m]{\varepsilon}\}, \quad V_\varepsilon := \{z \in V \colon |\tilde{\omega}(z)| < \sqrt[n]{\varepsilon}\}.$$

Then it follows that $X(\mathfrak{U}_\varepsilon) = X(V_\varepsilon)$ for suitably small $\varepsilon > 0$. □

For the formulation of the next result the following definition will be clarifying.

Definition 2. *An analytic arc in* $\mathbb{R}^n, n \geq 2$, *emanating from a point* p *of* $\mathbb{R}^n$, *is the image of a closed interval* $[0, \delta]$ *under a nonconstant, real analytic map* α *which is defined on an open interval containing* $[0, \delta]$ *and satisfies* $\alpha(0) = p$.

Lemma 2. *Let* $\varphi \colon \mathfrak{U} \to \mathbb{C}$ *be an analytic function defined on some neighbourhood* $\mathfrak{U}$ *of the origin* 0 *in* $\mathbb{C}$, *and suppose that*

$$\varphi(z) = az^m + O(|z|^{m+1}) \quad \text{as } z \to 0, \tag{6}$$

for some $m \in \mathbb{N}, a \in \mathbb{C}, a \neq 0$. *Furthermore, let* $\alpha \colon [0, \tau] \to \mathbb{C}$ *be a regular analytic arc emanating from* $0 \in \mathbb{C}$. *Then there exists some* $\tau_0 \in (0, \tau]$ *such that* $\varphi^{-1}(\alpha[0, \tau_0])$ *consists of* m *analytic arcs emanating from* 0.

Proof. Choose some neighbourhood $B_\delta = B_\delta(0) \subset \mathbb{C}$ with $\varphi \neq 0$ for all $z \in B_\delta \setminus \{0\}$, and introduce polar coordinates $(r, \xi) \in [0, \delta] \times S^1$. Without loss of generality we assume that $a = 1$. Then (6) implies that

$$\varphi(r, \xi) = r^m \xi^m \{1 + \varphi_1(r, \xi)\} \tag{7}$$

with some analytic function φ_1 satisfying $\varphi_1(0, \xi) = 0$. Let $\hat{\Phi} \colon [0, \delta] \times S^1 \to \mathbb{R}^+ \times S^1$ be a mapping so that the following diagram commutes:

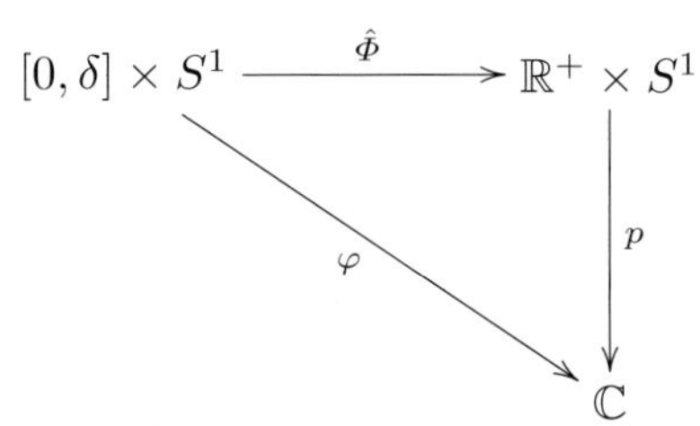

where $p(r, \xi) := r \cdot \xi$. Hence $\hat{\Phi}(r, \xi) = (|\varphi(r, \xi)|, \frac{\varphi(r,\xi)}{|\varphi(r,\xi)|})$. Similarly, let $\hat{\alpha} : [0, \tau] \to \mathbb{R}^+ \times S^1$ be chosen in such a way that $p \circ \hat{\alpha}(t) = \alpha(t)$, i.e., $\hat{\alpha}(t) = (\rho(t), \gamma(t))$ with real analytic functions $\rho(t) \geq 0$ and $\gamma(t) \in S^1$. In fact, replacing $\alpha(t)$ by the mapping $\tilde{\alpha}(t) := \alpha(t^{2m})$ which parametrizes the same arc, we may even assume that $\hat{\alpha}(t) = (\rho_1^m(t), \gamma(t))$ with analytic functions $\rho_1(t)$ and $\gamma(t)$. Note that $\gamma(0) = |\dot{\alpha}(0)|^{-1}\dot{\alpha}(0)$ is the direction of α at zero. We infer from (7) that

$$\hat{\Phi}(0, \xi) = (0, \xi^m) \tag{8}$$

and that $\hat{\Phi}$ can be continued analytically onto $[-\delta, \delta] \times S^1$ for some suitable $\delta > 0$. We can now define an analytic map $\tilde{\Phi}$ by

$$\tilde{\Phi}(r, \xi) := (\sqrt[m]{|\varphi(r, \xi)|}, |\varphi(r, \xi)|^{-1}\varphi(r, \xi)).$$

We are interested in the pre-image of $\hat{\alpha}$ under $\hat{\Phi}$, or equivalently, in the pre-image of $(\rho_1(t), \gamma(t))$ under $\tilde{\Phi}$. By virtue of (8) we infer that $\tilde{\Phi}^{-1}(0, \gamma(0))$ consists of the m points $(0, \gamma_1), \ldots, (0, \gamma_m)$ where $\gamma_1, \ldots, \gamma_m$ denote the m-th roots of $\gamma(0)$.

From the properness of $\tilde{\Phi}$ we infer that, for any given ε_0-neighbourhood $U_{\varepsilon_0}(0, \gamma_j)$ of $(0, \gamma_j)$ in $[0, \delta] \times S^1$, there exists a number $\varepsilon > 0$ such that

$$\tilde{\Phi}^{-1}(U_\varepsilon(0, \gamma(0))) \subset \bigcup_{j=1}^{m} U_{\varepsilon_0}(0, \gamma_j).$$

We choose ε_0 in such a way that $\tilde{\Phi}$ is an analytic diffeomorphism on each rectangle $\{(r, \xi) \colon |r| < \varepsilon_0, |\xi - \gamma_j| < \varepsilon_0\}$. Finally we select $\tau_0 > 0$ so small that $\rho_1(t) < \varepsilon$ and $|\gamma(t) - \gamma(0)| < \varepsilon$ holds for all $t \in [0, \tau_0]$. Then $\tilde{\Phi}^{-1}(\rho_1(t), \gamma(t))|_{[0,\tau_0]}$ consists of m analytic arcs emanating from $(0, \gamma_1), \ldots,$ $(0, \gamma_m)$. Therefore the set $\varphi^{-1}(\alpha[0, \tau_0])$ consists of m disjoint arcs starting at 0 with the directions $\gamma_1, \ldots, \gamma_m$. □

Lemma 3. *The equivalent boundary $\tilde{\partial} B$ is the union of finitely many analytic arcs.*

Proof. By Proposition 1, the set $\tilde{\partial} B$ is compact, and hence we may argue locally. First we claim that, for arbitrary $z_0 \in \bar{B}$, the pre-image of $P_0 := X(z_0)$ consists of only finitely many points. In fact, assuming the contrary, we would obtain a sequence $\{Z_j\}_{j \in \mathbb{N}} \in X^{-1}(P_0)$ with $z_j \to w$ whence, by continuity of X, we would have $w \in X^{-1}(P_0)$. However this would contradict (5) since any neighbourhood of w would contain points z_j with $X(z_j) = X(w) = X(z_0)$. Thus there are only finitely many points $z_1, \ldots, z_n \in \partial B$ which are equivalent to a given $z_0 \in \tilde{\partial} B$. For given (small) neighbourhoods $\mathfrak{U}_j = \mathfrak{U}_j(z_j)$ we can find a neighbourhood $\mathfrak{U}$ of z_0 with

$$\{w \in \partial B \colon w \sim z \in \mathfrak{U}\} \subset \bigcup_{j=1}^{m} \mathfrak{U}_j(z_j) \cap \partial B. \tag{9}$$

Otherwise there would exist a sequence of points $\xi_k \in B_R, R > 1$, with $\xi_k \to z_0$, and another sequence of points $w_k \in \partial B$ with $w_k \sim \xi_k$ but $w_k \notin \bigcup_{j=1}^{n} \mathfrak{U}_j(z_j)$. Passing to a sequence, we could assume that $w_k \to w \in \partial B \setminus \bigcup_{j=1}^{n} \mathfrak{U}_j(z_j)$. Because of Proposition 1 we would have $w \sim z_0$ or $w = z_j$ for some $j \in \{1, \ldots, n\}$, an obvious contradiction. Since $z_0, z_1, \ldots, z_n$ are equivalent, we may assume that $X(z_0) = X(z_1) = \cdots = X(z_n) = 0$ and that the common tangent plane is the (x^1, x^2)-plane. Denote by φ the mapping

$P \circ X$, where $P\colon \mathbb{R}^3 \to \mathbb{R}^2$ is the orthogonal projection onto the x^1, x^2-plane. Then (9) implies

$$\tilde{\partial} B \cap \mathfrak{U} \subset \bar{B} \cap \mathfrak{U} \cap \varphi^{-1}\left(\bigcup_{j=1}^{n} \varphi(\mathfrak{U}_j \cap \partial B)\right).$$

The set $\mathfrak{U}_j \cap \partial B$ consists of two circular arcs γ_j^+, γ_j^- emanating from z_j in opposite directions. Also, $\bigcup_{j=1}^n \varphi(\mathfrak{U}_j \cap \partial B)$ is a finite union of analytic arcs starting from the origin. Now we apply Lemma 2, choosing possibly smaller neighbourhoods $\mathfrak{U}_j$ and $\mathfrak{U}$, and conclude that $\varphi^{-1}(\bigcup_{j=1}^n \varphi(\mathfrak{U}_j \cap \partial B)) \cap \mathfrak{U}$ is a collection of analytic arcs $\alpha_1, \ldots, \alpha_N$, all starting at z_0. The lemma is proved if we can show that every arc α_k containing one point $z \in \tilde{\partial} B$ different from z_0, already belongs to $\tilde{\partial} B$. To this end let $\varphi(\alpha_k) \subset \varphi(\gamma_j^+)$ for some j and suppose that $z \in \alpha_k \setminus \{z_0\}$ is equivalent to $w \in \gamma_j^+ \setminus \{z_j\}$. We infer from (6) that we can write $X(\sigma \cap \mathfrak{U}_j)$ as a graph over the plane domain $\varphi(\sigma \cap \mathfrak{U}_j)$, where σ denotes the open sector

$$\{z_j + re^{i\theta}\colon r > 0, |\theta - \theta_j| < \varepsilon\}, \quad \theta_j = \arg z_j \pm \pi/2.$$

Since $\varphi(\alpha_k) \subset \varphi(\gamma_j^+)$ and X is continuous, we can find another open sector

$$\sigma_0 = \{z_0 + re^{i\theta}\colon r > 0, |\theta - \theta_0| < \delta\},$$

$e^{i\theta_0}$ being the direction of α_k at z_0, such that we have $\varphi(\sigma_0 \cap \mathfrak{U}) \subset \varphi(\sigma \cap \mathfrak{U}_j)$ and $\alpha_k \setminus \{z_0\} \subset \sigma_0$ for sufficiently small $\mathfrak{U}$.

Also $X|_{\sigma_0 \cap \mathfrak{U}s}$ is a graph over $\varphi(\sigma_0 \cap \mathfrak{U})$. Since $z \in \alpha_k \setminus \{z_0\}$ and $w \in \gamma_j^+ \setminus \{z_j\}$ are equivalent, we infer from the analyticity of minimal graphs that $X(\sigma_0 \cap \mathfrak{U}) \subset X(\sigma \cap \mathfrak{U}_j)$. In particular, we have $\alpha_k \subset \tilde{\partial} B$. □

Lemma 4. *Denote by $\tilde{\partial}_1 B$ the connected component of $\tilde{\partial} B$ which contains ∂B. Then $B \setminus \tilde{\partial}_1 B$ is connected.*

Proof. Lemma 3 implies that $B \setminus \tilde{\partial}_1 B$ consists of finitely many connected components $B_1, \ldots, B_n$, having piecewise analytic boundaries, whence

$$X|_{\partial B} = \sum_{k=1}^{n} X|_{\partial B_k}$$

and

$$X(\partial B_k) \subset X(\tilde{\partial} B) \subset X(\partial B) \subset S.$$

Choose some j so that $X|_{\partial B_j}$ is linked with Π, and then select some conformal map $\tau\colon B \to B_j$ of B onto B_j. If n were greater than 1, we would have

$$D(X \circ \tau) = D(X|_{B_j}) < D(X),$$

which contradicts the minimality of X. □

Consider now a (relative) minimizer X to the variational problem $\mathcal{P}(\Pi, S)$. We claim that, for a suitable reparametrization $\tilde{X} = X \circ \tau$ of X, we obtain another minimizer $\tilde{X}$ with $\tilde{\partial}_1 B = \partial B$. In fact, Lemmata 3 and 4 imply that $\tilde{\partial}_1 B$ consists of ∂B together with a finite number of trees growing out of certain points on ∂B. Let $\tau\colon B \to B \setminus \tilde{\partial}_1 B$ be a conformal map. Then the loop $X \circ \tau|_{\partial B}$ is homotopic to $X|_{\partial B}$ on S, whence $\tilde{X} = X \circ \tau \in \mathcal{C}(\Pi, S)$. We also have $D(\widetilde{X}) = D(X)$ and $\tilde{\partial}_1 B = \partial B$.

Note that the conformal reparametrization $\tau\colon B \to B \setminus \tilde{\partial}_1 B$ produces boundary branch points for the surface $\tilde{X}\colon B \to \mathbb{R}^3$ at those points $w \in \partial B$ which correspond to an endpoint $z \in \tilde{\partial}_1 B \cap B$ since, at these points, the boundary mapping runs back and forth in its own trace. Thus we have proved the following

Proposition 2. *Suppose that each strong relative minimizer $X \in \mathcal{C}(\Pi, S)$ which in addition satisfies $\tilde{\partial}_1 B = \partial B$, is immersed up to the boundary. Then the relation $\tilde{\partial}_1 B = \partial B$ holds for any strong relative minimizer $X \in \mathcal{C}(\Pi, S)$ of the variational problem $\mathcal{P}(\Pi, S)$.*

Let us now consider a minimizer X which satisfies $\tilde{\partial}_1 B = \partial B$.

Lemma 5. *Suppose that for a strong relative minimizer $X \in \mathcal{C}(\Pi, S)$ the relation $\tilde{\partial}_1 B = \partial B$ holds true. Then it follows that $\tilde{\partial} B = \partial B$.*

Proof. We argue by contradiction. Assume that the set

$$\partial_0 B := \{z \in \partial B\colon z \sim z_0 \in B\}$$

were not empty. From the definition of $\sim$ we then infer that $\partial_0 B$ is open in ∂B. The set $\partial_0 B$ in ∂B is also closed because of Proposition 1 and the assumption $\tilde{\partial}_1 B = \partial B$. In fact, let $z_n \in \partial_0 B$ be a sequence with $z_n \to z \in \partial B$ and $z_n \sim z_{0_n} \in B$. Without loss of generality, let $z_{0_n} \to z_0$. Because of Proposition 1 we obtain that $z_0 \in \tilde{\partial} B$, and since $\tilde{\partial}_1 B = \partial B$ it follows that $z_0 \in B$. Clearly, we have $z_0 \sim z$, whence $z \in \partial_0 B$. We conclude that $\partial B = \partial_0 B$ which means that X maps some neighbourhood of ∂B into $X(B)$. Thus $X(B)$ would be a compact minimal surface in $\mathbb{R}^3$, which is impossible because of the maximum principle. □

Proposition 3. *Let $X \in \mathcal{C}(\Pi, S)$ be a strong relative minimizer of $\mathcal{P}(\Pi, S)$ such that $\tilde{\partial}_1 B = \partial B$ holds true. Then X is immersed up to the boundary.*

Sketch of the proof. As we have already mentioned before, we only show the absence of false branch points. We argue by contradiction and assume first that $z_0 \in B$ is a false (interior) branch point of order m. Let $\gamma_1(t), t \in [0, 1]$, be an analytic Jordan arc which avoids branch points and points equivalent to branch points and has the following properties:

$$\gamma_1(0) = z_0, \quad \gamma_1(1) \in \partial B, \quad \gamma_1([0, 1)) \subset B.$$

We claim that there exist Jordan arcs $\gamma_k(t), t \in [0,1], k = 2, \dots, m$, with

$$\begin{aligned} &\gamma_k([0,1)) \subset B, \quad \gamma_k(0) = z_0, \quad \gamma_k(1) \in \partial B, \\ &\gamma_k(t) \sim \gamma_l(t) \qquad\qquad\qquad\qquad \text{for } 1 \le k \le l \le m, \\ &\gamma_k((0,1)) \cap \gamma_l((0,1)) = \emptyset \qquad\quad \text{for } k \neq l. \end{aligned}$$

In fact, the local existence of γ_k follows from the representation (5), while global existence is secured by analytic continuation. The inclusion $\gamma_k(1) \in \partial B$ follows from Lemma 1. An additional argument is required to show that γ_1 can be chosen in such a way that all γ_k are free of intersections; for details, see Alt [1] and Alt and Tomi [1]. Now let $z_1 = \gamma_1(1), \dots, z_m = \gamma_m(1)$ denote consecutive points on ∂B in positive orientation. For convenience, we put $z_{m+1} = z_1$. If σ_k denotes the arc of ∂B bounded by z_k and z_{k+1}, we see that $X(\sigma_k)$ is a closed loop on S and that $X|_{\partial B} = \sum_{k=1}^m X|_{\sigma_k}$. We choose k such that $X|_{\sigma_k}$ is not contractible in $\mathbb{R}^3 \setminus \Pi$ and denote by B_k the subdomain of B bounded by γ_k, σ_k and γ_{k+1}. There exists a conformal map $\tau\colon \bar{B} \setminus [0,1] \to B_k \cup \overset{\circ}{\sigma}_k$ with the property that

$$\lim_{\substack{z \to t \\ \operatorname{im} z > 0}} \tau(z) = \gamma_k(t) \quad \text{for all } t \in [0,1],$$

and that

$$\lim_{\substack{z \to t \\ \operatorname{im} z < 0}} \tau(z) = \gamma_{k+1}(t) \quad \text{for all } t \in [0,1].$$

Since $\gamma_k(t) \sim \gamma_{k+1}(t)$, we infer that $X \circ \tau$ is continuous in $\bar{B}$ and that $X \circ \tau$ is contained in $\mathcal{C}(\Pi, S)$. If m were larger than 1, we had $D(X \circ \tau) < D(X)$, a contradiction to the minimum property of X.

Next we consider a false branch point z_0 on the boundary ∂B which is of order $m \ge 2$. Here it is convenient to map the closed disk $\bar{B}$ conformally onto the half plane $(\operatorname{im} z \ge 0\} \cup \{\infty\}$ and z_0 onto 0. Denote the open half plane by B, and let X be the corresponding minimal surface. Then we may also assume that $X(0) = 0$, and that the tangent plane at $X(0)$ is the x^1, x^2-plane, applying a suitable motion in $\mathbb{R}^3$. Suppose also that the direction of the curve $X(\mathbb{R}^+) \subset S$ at 0 is given by $(1,0,0)$. We want to show the existence of a curve $\alpha\colon [0,1] \to \bar{B}$ with $\alpha(0) = z_0 = 0, \alpha((0,1)) \subset B$, and $X(\alpha[0,1]) \subset S$. From the representation formula (5) we infer the existence of numbers $r, R > 0$ and $\theta \in (\frac{\pi}{m}, \frac{2\pi}{m})$ such that the image of the sector $S_{r,\theta} := \{\rho e^{i\varphi}\colon 0 < \rho < r, 0 < \varphi < \theta\}$ under the mapping $\phi = P \circ X$ covers the half disk

$$H_R = \left\{\rho e^{i\varphi}\colon \frac{\pi}{2} < \varphi < \frac{3\pi}{2}, 0 < \rho < R\right\},$$

and $X(S_{r,\theta})$ is a graph over $\phi(S_{r,\theta})$. Then $X(S_{r,\theta})$ intersects S along an analytic arc $\hat{\alpha}\colon [0,1] \to \mathbb{R}^3$ with $\hat{\alpha}(t) \to 0$ as $t \to 0$ and $\frac{\hat{\alpha}'(t)}{|\hat{\alpha}'(t)|} \to (-1,0,0)$ as $t \to 0$. Thus the arc $\alpha := X^{-1}_{|S_{r,\theta}}(\hat{\alpha})$ has all the desired properties.

Next let $B' = B \setminus \alpha([0,1])$, and consider some conformal map $\tau\colon B \to B'$ of B onto B'. If we put $X' := X \circ \tau$, then $D(X') = D(X)$ and $X' \in \mathcal{C}(\Pi, S)$, whence X' is a solution to $\mathcal{P}(\Pi, S)$. It follows that X' intersects S orthogonally along ∂B, which means that X and S intersect perpendicularly along α. The curve α can be continued analytically until it hits ∂B. Moreover, by analyticity, the surface X remains orthogonal to S along α. We also note that α cannot have any double points in B, since X is a local embedding in B and hence intersects S in an embedded arc. Thus we have shown that $B \setminus \alpha$ consists of two simply connected domains B_1 and B_2 such that

$$X|_{\partial B} = X|_{\partial B_1} + X|_{\partial B_2}$$

holds true.

Suppose that $X|_{\partial B_1}$ is not contractible in $\mathbb{R}^3 \setminus \Pi$, and let $\tau\colon B \to B_1$ be a conformal equivalence. Then we obtain $X \circ \tau \in \mathcal{C}(\Pi, S)$ and $D(X \circ \tau) < D(X)$, which contradicts the minimality of X, and Proposition 3 is proved. □

Proof of Theorem 5. By virtue of Proposition 3 we only have to show that $\tilde{\partial}_1 B = \partial B$ holds for any strong relative minimizer $X \in \mathcal{C}(\Pi, S)$. But this immediately follows from Proposition 3 in conjunction with Proposition 2. □

Sketch of the proof of Theorem 4. By Theorem 5 we can assume that each minimizer $X \in \mathcal{C}(\Pi, S)$ is an immersion of $\bar{B}$ into $\mathbb{R}^3$.

If we apply a suitable conformal selfmapping of the disk B, we can also achieve the normalization $X(0) \in \Pi$. This condition ensures compactness of minimizers in C^k. In fact, we have

Proposition 4. *Let $\mathcal{C}^* \subset \mathcal{C}(\Pi, S)$ be the set of all minimizing minimal surfaces with $X(0) \in \Pi$. Then $\mathcal{C}^*$ is uniformly bounded in $C^{k,\alpha}(\bar{B})$, for any $k \geq 2, \alpha \in (0,1)$.*

Proof. In order to apply the results of Chapter 2, in particular Theorem 1 of Section 2.5, we wish to verify the following condition which is to hold uniformly in $\mathcal{C}^*$:

For each $\delta > 0$ there exists some $\varepsilon > 0$ such that

$$D_{B \setminus B_{1-\varepsilon}(0)}(X) < \delta \quad \text{for all } X \in \mathcal{C}^*. \tag{10}$$

Suppose on the contrary that there exist $\delta > 0$ and sequences $X_n \in \mathcal{C}^*, \varepsilon_n \to 0$ with $D_{B \setminus B_{1-\varepsilon_n}(0)}(X_n) \geq \delta$. Then all X_n are harmonic and bounded and hence a subsequence, again denoted by X_n, converges to some harmonic X uniformly in $C^k(\Omega)$, for all $\Omega \Subset B, k \in \mathbb{N}$. Because of

$$D(X_n) = d := \inf_{Y \in \mathcal{C}(\Pi, S)} D(Y),$$

we infer that

$$D(X) \leq d - \delta \quad \text{and} \quad X(0) \in \Pi. \tag{11}$$

Recalling the argument in the proof of Theorem 1 of Section 1.3, we conclude that $X \in \mathcal{C}(\Pi, S)$, whence $D(X) \geq d$, contradicting (11). Hence the relation (10) holds true.

On the other hand, (10) enables us to employ the regularity results of Chapter 2. First we see from the proof of Theorem 1 of Section 2.5 that the elements X in $\mathcal{C}^*$ satisfy a uniform global Hölder condition. Once having established a uniform Hölder condition, one can easily derive the higher order estimates by applying Theorem 1′ in Section 2.8. □

Suppose now that there are infinitely many geometrically different minimizing surfaces in $\mathcal{C}(\Pi, S)$. By Proposition 4, we can select a sequence $\{X_n\}$ that converges in $C^k(\bar{B})$ to some $X^* \in \mathcal{C}(\Pi, S)$ which must again be minimizing. By virtue of the immersed character of X^*, it can be shown as in Tomi [10] that there even exists a one-parameter family $F(t)$, $|t| < \varepsilon$, of area minimizing surfaces in $\mathcal{C}(\Pi, S)$ with $F(0) = X^*$, and $F'(0)$ is a nonvanishing normal field along X^*. Furthermore, each solution of $\mathcal{P}(\Pi, S)$ sufficiently close to X^* belongs to the family F (after a suitable reparametrization).

Now let Σ^* denote the connected component of X^* in the set of minimizing surfaces. Then the set

$$U^* := U \cap \left\{ \bigcup_{X \in \Sigma^*} X(B) \right\}$$

must be open and nonempty in the unbounded component U of $\mathbb{R}^3 \setminus S$. On the other hand, the set U^* must be bounded and closed in U according to Proposition 4. Thus we infer $U = U^*$ which clearly is impossible. □

1.10 Scholia

1. The first existence theorem for minimal surfaces with free boundaries was given by Courant [6] and [9] in the years 1938–40. At that time these results were considerable mathematical achievements comparable to the solution of Plateau's problem by Douglas and Radó. We also mention a paper by Courant and Davids [1] as well as a generalization of these results to *generalized Schwarzian chains* $\langle \Gamma_1, \ldots, \Gamma_k, S_1, \ldots, S_m \rangle$ given by Ritter [1]. A comprehensive treatment can be found in Courant [15] and in Nitsche [28].

2. Our exposition in the Sections 1.1–1.3 follows Küster [1]. The reader who is familiar with Courant's treatise [15] will have noticed that we have replaced Courant's condition

(1) $$\lim_{w \to w_0} \operatorname{dist}(X(w), S) = 0 \quad \text{for all } w_0 \in \partial B$$

by the simpler condition $X \in \mathcal{C}(S)$. It is somewhat easier to define the linking condition

$$\mathfrak{L}(X|_{\partial B}, \Pi) \neq 0$$

for surfaces satisfying (1). However, one then has to verify a compactness theorem that will ensure the condition (1) to hold in the limit, whereas our Lemma in Section 1.3 is close to trivial.

Moreover, our approach has the additional advantage that it can easily be carried over to *obstacle* problems with only modest smoothness assumptions on the obstructions, and it can also be used to handle more general functionals than the Dirichlet integral.

The proof that the curves $X|_{C_r}$ and $X|_{C_{r'}}$ are homotopic if r and r' are sufficiently close together (Section 1.1) has been adapted from an analogous theorem due to Schoen and Yau [2].

3. Let us mention some related existence results. Davids [1] proved the existence of multiply connected minimal surfaces with free boundaries. Hildebrandt [6] and Küster [1] treated surfaces of prescribed mean curvature, Lipkin [1] studied 2-dimensional parametric integrals, and F.P. Harth [1] proved existence of minimal surfaces with free boundaries in Riemannian manifolds. Meeks and Yau [1] dealt with Riemannian manifolds as ambient spaces.

P. Tolksdorf [2] stated that any non-trivial homotopy class in $\tilde{\Pi}_1(S)$ can be decomposed into finitely many nontrivial homotopy classes for which the problem of prescribed homotopy class has a solution, assuming that S is a smooth compact surface in $\mathbb{R}^3$. However, R. Ye [6] has pointed but that Tolksdorf's reasoning is faulty.

R. Ye [5,6] proved the existence of a minimal surface with prescribed boundary homotopy class α provided that α satisfies some Douglas-type condition. His method generalizes to Riemannian manifolds as well.

We also refer to remarks by E. Kuwert [5], p. 6, concerning the papers of Tolksdorf and Ye.

4. The existence proof of three different stationary minimal surfaces in a simplex presented in Section 1.6 is due to Smyth [1]. Smyth also stated that each of the three stationary surfaces possesses a non-parametric representation with respect to suitably chosen coordinates.

5. The uniqueness result for stationary minimal surfaces of disk-type in a sphere proved in Section 1.7 is due to Nitsche [35].

6. The examples in Section 1.9 of foliations given by 1-parameter families of minimizing minimal surfaces with their boundaries on a real analytic supporting surface S of the topological type of the torus are due to Gulliver and Hildebrandt [1], and we have followed their exposition quite closely.

7. Concerning detailed proofs of the *finiteness results* of Tomi [10] and Alt and Tomi [1] described in Section 1.9 we refer the reader to the original papers.

8. The most exciting recent development in the theory of minimal surfaces with free boundaries are the beautiful existence results for stationary minimal surfaces in convex bodies some of which we have listed in Section 1.8. We emphasize the importance of the contributions by Sacks and Uhlenbeck [1,2],

Struwe [3], Grüter and Jost [1], Pitts [1], Simon and Smith [1], and Jost [9, 13,15].

9. We also mention a paper by Karcher, Pinkall, and Sterling [1] on new examples of compact embedded minimal surfaces in the 3-sphere which generalizes the important earlier work by Lawson [4]. The Karcher–Pinkall–Sterling approach is closely related to the ideas of Smyth presented in Section 1.6, as their main construction consists in solving free boundary problems in S^3 (instead of $\mathbb{R}^3$).

10. Finally we shall briefly describe the work of E. Kuwert [5–7] on minimizers of Dirichlet's integral among disk-type surfaces $X \in H_2^1(B, \mathbb{R}^n)$ *whose boundary curves* $X|_{\partial B}$ *represent a given homotopy class* α of free loops on a closed configuration S in $\mathbb{R}^n$, $n \geq 3$.

Kuwert's work is an important and far reaching generalization of the theory presented before in this chapter. It deals with the problem of minimizing Dirichlet's integral among all disk-type surfaces $X : B \to \mathbb{R}^n, n \geq 2$, whose boundary values lie on a given configuration S and satisfy certain homotopy constraints. Here one observes *degeneration*, just as in the Douglas problem, and this causes *concentrations of the parametrization*, which in Kuwert's setting can occur only at the boundary of the disk B and leads to a separation of disks. It is proved that *any minimizing sequence has a subsequence which decomposes in the limit into a finite or countably infinite collection of disk-type surfaces, each of which is a minimizer with respect to its own homotopy class.* Here S can be any compact set in $\mathbb{R}^n$, or an unbounded closed set satisfying a suitable condition that prevents the escape of components to infinity.

Kuwert takes the view of Jesse Douglas and considers minimal surfaces as critical points of Dirichlet's integral within the class of *harmonic surfaces* $X \in H_2^1(B, R^n)$ satisfying the prescribed boundary conditions. Since such surfaces are uniquely determined by their "boundary values" $x = X|_{\partial B}$ (i.e. by their "Sobolev trace" on ∂B), the minimum problem is reduced to the minimization of Douglas's functional $A_0(x)$ among all admissible boundary curves x, since for any harmonic extension X of x one has $A_0(x) = D(X)$. However, we have seen before that, for free boundary value problems, it is not feasible to work with continuous boundary values $x(\theta) = X(e^{i\theta})$, since there is no a priori certainty that the minimization procedure leads to a continuous minimizer. To overcome this difficulty, Kuwert applies Courant's artifice of using sequences $\boldsymbol{x} = \{x_k\}$ with $x_k(\theta) := X(r_k e^{i\theta})$, $r_k \to 1-0$, which approximate x in $H_2^1(\partial B, \mathbb{R}^n)$ and satisfy $x_k \in H_2^1(\partial B, \mathbb{R}^n) \cap C^0(\partial B, \mathbb{R}^n)$; for the x_k it is possible to impose homotopy conditions. Finally, one altogether forgets the origin of $\boldsymbol{x}$ and operates with suitable sequences $\boldsymbol{x} = \{x_k\}$ of continuous curves x_k. Keeping this idea in mind, we turn to the technicalities needed to formulate Kuwert's results.

Let S be a nonempty closed set in $\mathbb{R}^n$, and denote by $U_\delta(S)$ the δ-neighbourhood of S in $\mathbb{R}^n$, $\delta > 0$:

$$U_\delta(S) := \{p \in \mathbb{R}^n : d(p,S) < \delta\}, d(p,S) := \mathrm{dist}(p,S).$$

Throughout we assume that $U_\delta(S)$ *is connected for* any $\delta > 0$.

Let $\pi_1(S)$ be the set of homotopy classes $[x]$ of free loops $x \in C^0(\mathbb{R}/2\pi, S)$. In order to define the sequence space $\Pi_1(S)$, we have to introduce the equivalence relations "$x \overset{\delta}{\sim} y$" between two curves $x, y \in C^0(\mathbb{R}/2\pi, \mathbb{R}^n)$, which means: *Both curves lie in* $U_\delta(S)$ *and are freely homotopic to each other in* $U_\delta(S)$. Then we set

$$\Pi_1(S) := \{\boldsymbol{x} = \{x_k\} : x_k \in C^0(\mathbb{R}/2\pi, \mathbb{R}^n), \text{ and for any } \delta > 0$$
$$\text{there is a } k_0 \in \mathbb{N} \text{ with } x_k \overset{\delta}{\sim} x_l \text{ for all } k, l > k_0\}.$$

For any $\boldsymbol{x} \in \Pi_1(S)$ we denote the smallest possible $k_0 \in \mathbb{N}$ by $k(\boldsymbol{x}, \delta)$.

On $\Pi_1(S)$ we introduce the equivalence relation "$\boldsymbol{x} \sim \boldsymbol{y}$" by:

For any $\delta > 0$ *there is a* $k_0 \in \mathbb{N}$ *such that* $x_{\sim} y_l$ *for all* $k, l \geq k_0$.

The quotient

$$\hat{\pi}_1(S) := \Pi_1(S)/\sim$$

will be the substitute for $\pi_1(S)$, if we operate with sequences $\boldsymbol{x} = \{x_k\}$ of loops x_k close to S instead of loops x on S. It turns out that $\hat{\pi}_1(S)$ is the inverse limit of the set $\pi_1(U_\delta(S))$ of homotopy classes of free loops in $U_\delta(S)$, i.e.

$$\hat{\pi}_1(S) = \lim_{\delta \to 0} \pi_1(U_\delta(S)).$$

We obtain the maps

$$i : \pi_1(S) \to \hat{\pi}_1(S) \quad \text{with } [x] \mapsto [\{x_k \equiv x\}]$$

and

$$i_\delta : \hat{\pi}_1(S) \to \pi_1(U_\delta(S)) \quad \text{with } [\boldsymbol{x}] \mapsto [x_k], k = k(\boldsymbol{x}, \delta).$$

For $\alpha, \beta \in \hat{\pi}_1(S)$ we define $\delta(\alpha, \beta) \in [0, \infty]$ by

$$\delta(\alpha, \beta) := \inf\{\delta > 0 : i_\delta(\alpha) = i_\delta(\beta)\};$$

this is a complete generalized metric on $\hat{\pi}_1(S)$, except that $\delta(\alpha, \beta) = \infty$ if S is unbounded, and $\hat{\pi}_1(S)$ is arcwise totally disconnected. Moreover i is injective if S is a retract of $U_{\delta_0}(S)$ for some $\delta_0 > 0$, and if S is a uniform deformation retract of $U_{\delta_0}(S)$, then i, i_{δ_0} are bijective and $\delta(\alpha, \beta) \geq \delta_0$ for $\alpha, \beta \in \hat{\pi}_1(S)$ with $\alpha \neq \beta$.

Finally we define $|\alpha|$ for $\alpha \in \hat{\pi}_1(S)$ by

$$|\alpha| := \inf\{\delta > 0 : \ i_\delta(\alpha) \text{ contains a constant map}\},$$

and we call $\alpha \in \hat{\pi}_1(S)$ *trivial* if $|\alpha| = 0$, otherwise *nontrivial.*

Now we consider the space H of Fourier series

$$x \sim a_0 + \sum_{m=1}^{\infty} (a_m \cos m\theta + b_m \sin m\theta), \quad a_m, b_m \in \mathbb{R}^n,$$

satisfying

$$E(x) := \frac{\pi}{2} \sum_{m=1}^{\infty} m(|a_m|^2 + |b_m|^2) < \infty.$$

The harmonic extension of X with $X(e^{i\theta}) = x(\theta)$ satisfies

$$D(X) = E(x) = A_0(x).$$

The space H with the norm $\|x\|_H$, defined

$$\|x\|_H^2 := |a_0|^2 + E(x),$$

can be identified with the Hilbert space $H^{1/2,2}(\mathbb{R}/2\pi, \mathbb{R}^n)$. Set

$$\begin{aligned}\mathcal{H}(S) := \{\boldsymbol{x} = \{x_k\} : x_k \in H \cap C^0(\partial B, \mathbb{R}^n), d(x_k, S) \to 0,\\ \{x_k\} \text{ is a Cauchy sequence in } H\},\end{aligned}$$

where $d(x_k, S) := \sup\{\alpha(x_k(\theta), S) : 0 \le \theta \le 2\pi\}$.

For $\boldsymbol{x}, \boldsymbol{y} \in \mathcal{H}(S)$ we write

$$\boldsymbol{x} \sim \boldsymbol{y} \quad \text{if and only if} \quad \|x_k - y_k\|_H \to 0.$$

We define the quotient space

$$H(S) := \mathcal{H}(S)/\sim$$

and note that $H(S)$ can be identified isometrically with $(W(S), \|\cdot\|_H)$, where

$$W(S) := \{x \in H : x(\theta) \in S \text{ for a.e. } \theta \in [0, 2\pi]\},$$

i.e.

$$W(S) = H(S),$$

and for $\boldsymbol{x} = \{x_k\} \in \mathcal{H}(S)$ we have $\|x_k - x\|_H \to 0$ for some $x \in W(S)$; then the equivalence class of $\boldsymbol{x}$ is identified with x, and $\|x\|_H = \lim_{k\to\infty} \|x_k\|_H =: \|\boldsymbol{x}\|_{\mathcal{H}(S)}$.

One obtains the following topological substitute for a Sobolev embedding of H into $C^0(\partial B, \mathbb{R}^n)$ which is essentially due to Courant (cf. Section 1.1); a proof can be found in B. White [7] and Kuwert [5,7].

Theorem A. *The set $\mathcal{H}(S)$ is a subset of $\Pi_1(S)$, and the inclusion $\mathcal{H}(S) \subset \Pi_1(S)$ induces a well-defined assignment from any $x \in W(S)$ to a homotopy class in $\hat{\pi}_1(S)$ which will be denoted by $[x]$. The mapping $x \mapsto [x]$ from $W(S) = H(S)$ into $\hat{\pi}_1(S)$ is continuous.*

The following can be seen: *For any $x \in W(S)$, the harmonic extension X satisfies*

$$\lim_{|w|\to 1} d(X(w), S) = 0,$$

and for any sequence $\{r_k\}$ with $r_k \to 1-0$ the sequence $x_k(\theta) := X(r_k e^{i\theta})$, $k \in \mathbb{N}$, can be used for the definition of $[x] \in \hat{\pi}_1(S)$.

Now we formulate the *minimization problem for a given class $\alpha \in \hat{\pi}_1(S)$.* We set

$$E_*(\alpha) := \inf\Big\{\liminf_{k\to\infty} E(x_k) : \boldsymbol{x} = \{x_k\} \text{ with } [\boldsymbol{x}] \in \alpha\Big\}.$$

The function $E_* : \hat{\pi}_1(S) \to [0, \infty]$ is lower semicontinuous and satisfies $E_*(\alpha) \geq \pi|\alpha|^2$ as well as: $E_*(\alpha) = 0 \Leftrightarrow \alpha$ is trivial. Furthermore:

For any $\alpha \in \hat{\pi}_1(S)$ there is always a sequence $\boldsymbol{x} = \{x_k\}$ with $[\boldsymbol{x}] \in \alpha$ and $x_k \in C^\infty(\mathbb{R}/2\pi, \mathbb{R}^n)$ such that

$$\lim_{k\to\infty} E(x_k) = E_*(\alpha).$$

Definition 1. (i) *A minimizing sequence for $\alpha \in \hat{\pi}_1(S)$ is a sequence $\boldsymbol{x} = \{x_k\} \in \Pi_1(S)$ with $[x] \in \alpha$ satisfying $E(x_k) \to E_*(\alpha)$.*

(ii) *Any $x \in W(S)$ with $[x] = \alpha$ and $E(x) = E_*(\alpha)$ is called a minimizer of E.*

We set

$$\mathcal{F}(S) := \{\boldsymbol{x} = \{x_k\} : x_k \in C^0(\mathbb{R}/2\pi, \mathbb{R}^n), d(x_k, S) \to 0\};$$

in particular we have $\Pi_1(S) \subset \mathcal{F}(S)$.

A sequence $\boldsymbol{x} = \{x_k\}$ is said to be *trivial*, if for any $\delta > 0$ there is a $k_1(\delta) \in \mathbb{N}$ such that x_k is contractible in $U_\delta(S)$ for all $k \geq k_1(\delta)$; otherwise $\boldsymbol{x}$ is called *nontrivial*. Then we introduce $\epsilon_*(S)$ and $\epsilon_0(S) \in \mathbb{R}$ with $0 \leq \epsilon_*(S) < \epsilon_0(S)$ by

$$\begin{aligned}
\epsilon_*(S) &:= \inf\Big\{\epsilon > 0 : \text{ There is a nontrivial sequence } \boldsymbol{x} = \{x_k\} \in \mathcal{F}(S) \\
&\qquad\qquad \text{with } \limsup_{k\to\infty} E(x_k) \leq \epsilon\Big\}, \\
\epsilon_0(S) &:= \inf\{E_*(\alpha) : \alpha \in \hat{\pi}_1(S) \text{ is nontrivial}\}.
\end{aligned}$$

Note that $\alpha \in \hat{\pi}_1(S)$ is nontrivial if $\alpha = [\boldsymbol{x}]$ with $\boldsymbol{x} \in \Pi_1(S)$ and $\boldsymbol{x}$ is nontrivial. Observe also that $\epsilon_*(S)$ is only defined if there is a nontrivial sequence in $\mathcal{F}(S)$, and $\epsilon_0(S)$ is only defined if there is a nontrivial $\alpha \in \hat{\pi}_1(S)$.

For a sequence $\boldsymbol{M} = \{M_k\}$ of sets $M_k \subset \mathbb{R}^n$ we define the closed set of accumulation points of $\boldsymbol{M}$ by

$$\begin{aligned}
\mathcal{A}(\boldsymbol{M}) := \{p \in \mathbb{R}^n :\ &\text{There is a subsequence } \{k_l\} \text{ and a sequence} \\
&\text{of point } p_l \in M_{k_l} \text{ with } p_l \to p\}.
\end{aligned}$$

Furthermore, define the subset $\mathcal{F}'$ of $\mathcal{F}$ by

$$\mathcal{F}' := \{\boldsymbol{x} = \{x_k\} \in \mathcal{F} : x_k \in H, E(x_k) \to e \text{ for some } e \in [0, \infty)\}$$

and set

$$\mathcal{A}(\boldsymbol{x}) := \mathcal{A}(\{\operatorname{im} x_k\}), \quad \mathcal{A}(\boldsymbol{X}) := \mathcal{A}(\{\operatorname{im} X_k\})$$

for $\boldsymbol{x} = \{x_k\} \in \mathcal{F}'$ and $\boldsymbol{X} := \{X_k\}$, $X_k =$ harmonic extension of x_k,

$$\operatorname{im} x_k = \text{ image of } x_k, \quad \operatorname{im} X_k = \text{image of } X_k.$$

One has $\mathcal{A}(\boldsymbol{x}) \subset S \cap \mathcal{A}(X)$ and

$$\mathcal{A}(\boldsymbol{X}) \subset \operatorname{clos}(U_{\delta(e)}(\mathcal{A}(\boldsymbol{x}) \cap S)) \quad \text{for } \boldsymbol{x} = \{x_k\} \in \mathcal{F}'$$

where $\delta(e) := \sqrt{e/\pi}$, and $e = \lim_{k\to\infty} E(x_k)$. Furthermore, $\mathcal{A}(\boldsymbol{X}) = \emptyset$ if and only if $\mathcal{A}(\boldsymbol{x}) = \emptyset$.

One of Kuwert's main tools is a decomposition result for sequences $\boldsymbol{x} = \{x_k\} \in \mathcal{F}'$ which is formulated as Lemma 3 in Section 2 of his paper [5], but is too involved to be stated here.

Now the following compactness question is raised: *Given a sequence $\boldsymbol{z} = \{z_k\} \in \mathcal{F}'$, is there always a subsequence $\boldsymbol{x} = \{x_l\}, x_l = z_{k_l}$ with $\boldsymbol{x} \in \Pi_1(S)$?*

It turns out that the answer is negative in general. To clarify the situation, some topological notions are needed. Recall first that $U_\delta(S)$ is assumed to be connected for any $\delta > 0$, which is the case if S is connected. Therefore there is a unique trivial element o in $\hat{\pi}_1(S)$ which is represented by any sequence $\{x_k\}$ of constant loops $x_k(\theta) \equiv p_k$ with $p_k \to S$, and $|\alpha| = \delta(\alpha, 0)$.

Now we consider m-tupel $\underline{\alpha} = (\alpha^1, \alpha^2, \dots, \alpha^m)$ of homotopy classes $\alpha^j \in \hat{\pi}_1(S)$ which are either finite, $m \in \mathbb{N}$, or infinite, $m = \infty$. We also require that, for any $S > 0$, the set $I(\underline{\alpha}, \delta) = \{j : |\alpha^j| > \delta\}$ is finite with the number $m(\underline{\alpha}, \delta)$ of elements (which may be zero). Next we introduce the *sequence space* $\Pi_1(\underline{\alpha})$ as follows:

$$\begin{aligned} \Pi_1(\underline{\alpha}) := \{\boldsymbol{x} = \{x_k\} \in \mathcal{F} : \ & \text{For any } \delta > 0 \text{ there is a smallest possible} \\ & k(\boldsymbol{x}, \delta) \in \mathbb{N} \text{ such that the homotopy class } [x_k] \in \pi_1(U_\delta(S)) \\ & \text{belongs to the composition set of the classes } i_\delta(\alpha^j), j \in I(\underline{\alpha}, \delta), \\ & \text{for all } k \geq k(\boldsymbol{x}, \delta)\}. \end{aligned}$$

We say that $\alpha \in \hat{\pi}_1(S)$ belongs to the *composition set* $\mathcal{C}(\underline{\alpha})$ of $\underline{\alpha} = (\alpha^1, \dots, \alpha^m)$ if and only if $i_\delta(\alpha)$ belongs to the composition set of the finite $\boldsymbol{m}(\alpha, \delta)$-tupel of the $i_\delta(\alpha^j)$ with $j \in I(\underline{\alpha}, \delta)$, for all $\delta > 0$. Equivalently we say: $\underline{\alpha}$ is a decomposition of $\alpha, \underline{\alpha} \in \mathcal{D}(\alpha)$.

This means: Given $\boldsymbol{x} = \{x_k\} \in \Pi_1(\underline{\alpha})$, $\delta > 0$, and representatives $\boldsymbol{x}^j = \{x_k^j\}$ of α^j, then the boundary data $x_k, x^j_{k(\boldsymbol{x},\delta)}$ with $j \in I(\underline{\alpha}, \delta)$ can be extended to an $(m(\underline{\alpha}, \delta) + 1)$-fold connected domain by a map into $U_\delta(S)$ for $k \geq k(\boldsymbol{x}, \delta)$. If α is finite then $m(\alpha, \delta) \equiv \text{const}$ for $0 < \delta \ll 1$.

For unbounded S it can happen that disks escape to infinity. This will be excluded by imposing an energy condition. For this purpose we define

$$\epsilon_\infty(S) := \inf\Big\{\epsilon > 0 : \text{There is a nontrivial sequence}$$
$$\boldsymbol{x} = \{x_k\} \in \mathcal{F}(S) \text{ with } \mathcal{A}(X) \subset S$$
$$\text{and } \limsup_{k\to\infty} E(x_k) \le \epsilon\Big\}.$$

Clearly, $\epsilon_*(S) \le \epsilon_\infty(S)$, and $\epsilon_\infty(S) = \epsilon_0(S)$ if S is compact.

We have the following answer to the "compactness question" raised above:

Theorem B. *Let $z = \{z_k\} \in \mathcal{F}'$ be a sequence with $E(x_k) \to e < \epsilon_\infty$. Then there exist an m-tupel $\underline{\alpha} = (\alpha^1, \ldots, \alpha^m)$ of $\alpha^j \in \hat{\pi}_1(S)$, $|\alpha^j| > 0$, with $|\alpha^j| \to 0$ as $j \to \infty$ if $m = \infty$, and a subsequence $\boldsymbol{x} = \{x_l\}, x_l = z_{k_l}$, such that $\boldsymbol{x} \in \Pi_1(\underline{\alpha})$, and it addition*

$$\sum_j E_*(\alpha^j) \le e \quad \textit{and} \quad m \le e/\epsilon_0(S).$$

Moreover, if $e < \min\{2\epsilon_0(S), \epsilon_\infty(S)\}$ then $\boldsymbol{x} \in \Pi_1(S)$, i.e. $\boldsymbol{x}$ defines a homotopy class. Finally, $\epsilon_*(S) = \epsilon_0(S)$ provided that $\epsilon_*(S) < \epsilon_0(S)$.

Theorem C. *Let $\alpha \in \hat{\pi}_1(S)$ be a nontrivial homotopy class with $E_*(\alpha) < \infty$, and $\boldsymbol{x} = \{x_k\}$ be a minimizing sequence for $\alpha \in \hat{\pi}_1(S)$ which converges weakly in H to $x \in W(S)$. Then x is a minimizer with respect to its own homotopy class.*

The hypothesis on α can be verified if α can be represented by a sequence of equibounded length. While $\{x_k\}$ will not converge strongly in general, it is often possible to extract a nonconstant weak limit.

The next theorem is the main result of Kuwert [5–7]. It states that *any minimizing sequence contains a subsequence which decomposes in the limit both in homotopy and in energy into a union of minimizing disks.*

Theorem D. *Let $\alpha \in \hat{\pi}_1(S)$ be a given nontrivial homotopy class with $E_*(\alpha) < \epsilon_\infty(S)$, and let $\boldsymbol{z} = \{z_k\}$ with $[z] \in \alpha$ be a given minimizing sequence. Then there are a subsequence $\{z_{k_l}\}$, a number $m \in \mathbb{N} \cup \{\infty\}$, a sequence $\{h_l^1\}$ of conformal automorphisms of B, topological disks $D_l^j, l \in \mathbb{N}, 1 \le j \le m$, and Riemann mapping functions $g_l^j : B^j \to D_l^j$ such that the loops $x_l := z_{k_l} \circ h_l^1$ satisfy:*

(i) *$D_l^j \Subset B, \overline{D}_l^j \cap \overline{D}_j^k = \emptyset$ for $j \ne k$; ∂D_l^j is regular and real analytic; $D_j^1 = \{w \in B : |w| < r_l^1\}$ with $r_l^1 \to 1 - 0$; $g_l^1(w) = r_l^1 w$.*

(ii) *For any j, the sequence $\{X_l \circ g_l^j\}_{l\in\mathbb{N}}$ converges strongly in $H_2^1(B^j, \mathbb{R}^n)$ to a nontrivial minimizer $X^j : B^j \to \mathbb{R}^n$ with the boundary values $x^j \in W(S)$,*

(B_j identified with B), and $\alpha^j := \{x^j\}$ is an element of $\hat{\pi}_1(S) \setminus \{0\}$. Each mapping X^j is a (possibly branched) minimal surface.

(iii) $E_*(\alpha) = \lim_{k\to\infty} E(z_k)$ *can be written as*

$$E_*(\alpha) = \sum_{j=1}^{m} E_*(\alpha^j) = \sum_{j=1}^{\infty} E(x^j).$$

(iv) *The m-tupel $\underline{\alpha} := (\alpha^1, \alpha^2, \dots, \alpha^m)$ is a decomposition of the given class α.*

(v) *For $M_l := B \setminus \bigcup_{j=1}^m D_l^j$ we have $d(X_l|_{M_l}, S) \to 0$, and the Dirichlet integrals $D_{M_l}(X_l)$ of X_l over M_l tend to zero as $l \to \infty$.*

(vi) *If $\epsilon_0(S) > 0$ then $m \le E_*(\alpha)/\epsilon_0(S) < \infty$.*

We mention that this result can be used to generalize H.W. Alt's solution of the so-called *thread problem*, treated in Chapter 5 of this volume; cf. Kuwert [5], pp. 51–52. Kuwert's approach allows to consider threads whose endpoints are fixed at support surfaces (instead of arcs).

Now we want to collect several applications of Theorem D in case that $\alpha \in \hat{\pi}_1(S)$ satisfies a *sufficient Douglas condition*. This means: *For any proper decomposition $\underline{\alpha} = (\alpha^1, \alpha^2, \dots, \alpha^m) \in \mathcal{D}(\alpha)$ we have the strict inequality*

$$E_*(\alpha) < \sum_{j=1}^{m} E_*(\alpha^j).$$

Note however that, so far, this condition can only be verified for specific homotopy classes.

Theorem E. *Let $\alpha \in \hat{\pi}_1(S)$ satisfy the sufficient Douglas condition, and assume that $E_*(\alpha) < \epsilon_\infty(S)$. Then we have:*

(i) *For any minimizing sequence $\boldsymbol{z} = \{z_k\}$ representing α there are a subsequence $\{z_{k_l}\}$ as well as conformal automorphisms h_l of B such that $\boldsymbol{x} = \{x_l\}$ with $x_l := z_{k_l} \circ h_l$ converges strongly in H to some minimizer $x \in W(S)$ for α, and $E_*(\alpha)$ is attained.*

(ii) *If a minimizing sequence converges weakly to a nonconstant map $x \in W(S)$, then it converges strongly to x.*

(iii) *The nonempty set $\mathcal{M}(\alpha)$ of minimizers for α is compact in $W(S)$ modulo the conformal automorphism group* $\mathrm{Aut}(B)$ *of B.*

Kuwert also shows ([5], pp. 57–60) how the results presented in the following Chapter 2 can be used to show regularity of minimal surfaces X defined by minimizers x for α and to prove compactness for $\mathcal{M}(\alpha)$ with respect to $C^{0,\beta}$ or $C^{k,\beta}$.

Chapter 2

The Boundary Behaviour of Minimal Surfaces

In this chapter we deal with the boundary behaviour of minimal surfaces, with particular emphasis on the behaviour of stationary surfaces at their free boundaries. This and the following chapter will be the most technical and least geometric parts of our lectures. They can be viewed as a section of the regularity theory for nonlinear elliptic systems of partial differential equations. Yet these results are crucial for a rigorous treatment of many geometrical questions, and thus they will again illustrate what role the study of partial differential equations plays in differential geometry.

The first part of this chapter, comprising Sections 2.1–2.3, deals with the boundary behaviour of minimal surfaces at a fixed boundary. Consider for example a minimal surface $X\colon B \to \mathbb{R}^3$ which is continuous on $\bar{B}$ and maps ∂B onto some closed Jordan curve Γ. Then we shall prove that X is as smooth on $\bar{B}$ as Γ, more precisely, that X is of class $C^\infty(\bar{B}, \mathbb{R}^3)$ (or $X \in C^\omega(\bar{B}, \mathbb{R}^3)$, or $X \in C^{m,\alpha}(\bar{B}, \mathbb{R}^3)$) if Γ is of class C^∞ (or $\Gamma \in C^\omega$, or $\Gamma \in C^{m,\alpha}$, respectively). These results are worked out in Section 2.3. In Section 2.1 we shall supply some results from potential theory that will be needed, and in Section 2.2 we shall derive various regularity results and estimates for vector-valued solutions X of differential inequalities of the kind

$$|\Delta X| \le a|\nabla X|^2$$

which will be crucial for our considerations in Section 2.3.

The central part of this chapter consists of Sections 2.4–2.9 where we prove analogous regularity results for minimal surfaces with free boundaries on a support surface S. If the boundary ∂S of S is empty, the reasoning is considerably simpler than for $\partial S \neq \emptyset$; in fact this second case has to be viewed as a Signorini problem (or else, as a thin obstacle problem). For a survey of the results on the boundary behaviour of minimal surfaces with free boundaries we refer the reader to Section 2.4.

Finally, in Section 2.10, we shall derive an asymptotic expansion for any minimal surface at a boundary branch point which is analogous to the expan-

U. Dierkes, S. Hildebrandt, A.J. Tromba, *Regularity of Minimal Surfaces*,
Grundlehren der mathematischen Wissenschaften 340,
DOI 10.1007/978-3-642-11700-8_2, © Springer-Verlag Berlin Heidelberg 2010

sion at an interior branch point that was obtained in Section 3.2 of Vol. 1. The results of Section 2.10 are based on the discussion in Chapter 3.

2.1 Potential-Theoretic Preparations

In this section we want to supply some results from potential theory which will be needed in Section 2.3 for investigating the boundary behaviour of minimal surfaces which are bounded by smooth Jordan arcs. The reader who is well acquainted with Schauder estimates may skip this part at a first reading. Although a large part of the material can be found in the treatise of Gilbarg and Trudinger [1], a brief presentation may be welcome because it will enable the reader to study the essential results of Section 2.2 on solutions of differential inequalities without consulting additional sources.

In what follows we shall use the following notation: We write $w = u + iv$, $\zeta = \xi + i\eta$, $d\zeta = d\xi + i\,d\eta$, and $d^2\zeta = d\xi\,d\eta$ denotes the two-dimensional area element. Moreover, we set

$$\begin{aligned} \frac{\partial}{\partial w} &= \frac{1}{2}\left(\frac{\partial}{\partial u} - i\frac{\partial}{\partial v}\right), \quad \frac{\partial}{\partial \overline{w}} = \frac{1}{2}\left(\frac{\partial}{\partial u} + i\frac{\partial}{\partial v}\right), \\ \Delta &= \frac{\partial^2}{\partial u^2} + \frac{\partial^2}{\partial v^2} = 4\frac{\partial}{\partial w}\frac{\partial}{\partial \overline{w}}, \\ B_R &= B_R(0) = \{w \in \mathbb{C}\colon |w| < R\}, \quad B := B_1(0). \end{aligned}$$

Green's function $G_R(w,\zeta)$ for the disk B_R is given by

$$G_R(w,\zeta) = \frac{1}{2\pi}\log\left|\frac{R^2 - \overline{w}\zeta}{R(\zeta - w)}\right|, \tag{1}$$

and the Poisson kernel $\mathcal{P}_R(w,\varphi) = \mathcal{P}_R^*(w,\zeta), w = re^{i\theta}, \zeta = \mathrm{Re}^{i\varphi} \in \partial B_R$, is defined[1] by

$$\begin{aligned} \mathcal{P}_R(w,\varphi) &= \frac{1}{2\pi}\frac{R^2 - r^2}{R^2 - 2rR\,\cos(\theta - \varphi) + r^2} = \frac{1}{2\pi}\mathrm{Re}\frac{R + re^{i(\theta-\varphi)}}{R - re^{i(\theta-\varphi)}} \\ &= \frac{1}{2\pi}\mathrm{Re}\frac{\zeta + w}{\zeta - w} = \frac{1}{2\pi}\frac{R^2 - |w|^2}{|\zeta - w|^2} = -R\frac{\partial}{\partial \nu_\zeta}G_R(w,\zeta), \end{aligned} \tag{2}$$

where ν_ζ denotes the exterior normal to ∂B_R at ζ. One computes that

$$\frac{\partial}{\partial w}G_R(w,\zeta) = \frac{1}{4\pi}\left(\frac{1}{\zeta - w} - \frac{\overline{\zeta}}{R^2 - w\overline{\zeta}}\right), \tag{3}$$

[1] Note that often the expression $\frac{1}{R}\mathcal{P}_R(w,\varphi) = -\frac{\partial}{\partial \nu_\zeta}G_R(w,\zeta)$ is called *Poisson kernel*; cf. for instance Gilbarg and Trudinger [1], formula (2.29).

whence it follows that

$$\frac{\partial^s}{\partial w^s} G_R(w,\zeta) = \frac{(s-1)!}{4\pi}\left[\frac{1}{(\zeta-w)^s} - \frac{\overline{\zeta}^s}{(R^2 - w\overline{\zeta})^s}\right]. \tag{4}$$

A straight-forward estimation shows that

$$R|\zeta - w| \le |R^2 - w\overline{\zeta}| \quad \text{for all } w, \zeta \in \overline{B}_R,$$

which implies

$$\left|\frac{\partial^s}{\partial w^s} G_R(w,\zeta)\right| \le \frac{(s-1)!}{2\pi}\,\frac{1}{|\zeta - w|^s} \quad \text{for all } \zeta, w \in \overline{B}_R \text{ with } w \ne \zeta. \tag{5}$$

The following results is a direct consequence of Green's formula and can be found in any textbook on partial differential equations.[2]

Proposition 1. *Any function $x \in C^0(\overline{B}_R) \cap C^2(B_R)$ with $q := \Delta x \in L_\infty(B_R)$ and $\boldsymbol{x}(\varphi) := x(Re^{i\varphi})$ can be written in the form*

$$x(w) = h(w) - \int_{B_R} G_R(w,\zeta) q(\zeta)\, d^2\zeta, \tag{6}$$

where

$$h(w) := \int_0^{2\pi} \mathcal{P}_R(w,\varphi)\boldsymbol{x}(\varphi)\, d\varphi \tag{7}$$

denotes the harmonic function in B_R which is continuous on $\overline{B}_R$ and satisfies $h = x$ on ∂B_R.

Proposition 2. *Suppose that $\boldsymbol{x}(\varphi)$ is of class $C^2(\mathbb{R})$ and periodic with the period 2π, and let $q(w)$ be of class $L_\infty(B)$. Assume also that*

$$\sup_B |q| \le \alpha, \quad \sup_{\mathbb{R}} |\boldsymbol{x}''| \le \beta$$

holds for some numbers α, β. Then the function $x(w), w \in B$, defined by (6) and (7) for $R = 1$, can be extended to $\overline{B}$ as a function which is of class $C^{1,\mu}(\overline{B})$ for any $\mu \in (0,1)$ and satisfies $x(e^{i\varphi}) = \boldsymbol{x}(\varphi)$. For suitable numbers $c_1(\alpha,\beta)$ and $c_2(\alpha,\beta,\mu)$ depending only on the indicated parameters and not on q and $\boldsymbol{x}$, we have

$$|\nabla x|_{0,\overline{B}} \le c_1(\alpha,\beta), \quad [\nabla x]_{\mu,\overline{B}} \le c_2(\alpha,\beta,\mu). \tag{8}$$

If $q \in C^{0,\sigma}(B)$ holds for some $\sigma \in (0,1)$, then we have $x \in C^{2,\sigma}(B)$, and the equation $\Delta x = q$ is satisfied on B. Moreover, for any R, R' with $0 < R' < R \le 1$ the function $y(w)$ defined by

[2] Cf. for instance Gilbarg and Trudinger [1], p. 18; John [1], p. 96.

$$y(w) := \int_{B_R} G_R(w,\zeta) q(\zeta)\, d^2\zeta \tag{9}$$

is of class $C^{2,\sigma}(B_R)$ *and satisfies*

$$|y|_{2+\sigma, B_{R'}} \le c(R, R', \sigma) |q|_{0+\sigma, B_R}. \tag{10}$$

Here and in the following we use the notation

$$|x|_{0,B} = \sup_B |x|, \quad [x]_{\mu,B} = \sup \left\{ \frac{|x(w) - x(w')|}{|w - w'|^\mu} : w, w' \in B, w \ne w' \right\},$$

$$|x|_{s,B} = \sum_{k=0}^{s} |\nabla^k x|_{0,B}, \quad |x|_{s+\mu,B} + |x|_{s,B} + [\nabla^s x]_{\mu,B}.$$

Moreover, we shall use the notation $\nabla_w = (\frac{\partial}{\partial u}, \frac{\partial}{\partial v})$ in order to distinguish the real gradient $\nabla_w f = (f_u, f_v)$ of a function $f(u,v)$ from its complex derivative $f_w = \frac{1}{2}(f_u - i f_v)$.

The reader will find more complete results on Schauder estimates in Gilbarg and Trudinger [1], Chapters 2–4 and 6; Morrey [8], Chapters 2 and 6; Stein [1]; Agmon, Douglis, and Nirenberg [1,2]. We shall use some of these refined results later on. For the present the reader might welcome to see how one can obtain Schauder estimates in the simple situation at hand. Proposition 2 and related results will be proved by a sequence of auxiliary results.

Lemma 1. *Let* $H(w,\zeta)$ *be a* C^2*-kernel on the set* $\{w, \zeta \in \overline{B}_R : w \ne \zeta\}$ *such that*

$$|H(w,\zeta)| \le b \left| \log \frac{1}{r} \right|, \quad |\nabla_w H(w,\zeta)| \le \frac{b}{r}, \quad |\nabla_w^2 H(w,\zeta)| \le \frac{b}{r^2} \tag{11}$$

holds for $r = |w - \zeta|$ *and some constant* $b > 0$. *In addition we assume that* $q \in L_\infty(B_R)$. *Then the function* $y(w)$ *defined by*

$$y(w) = \int_{B_R} H(w,\zeta) q(\zeta)\, d^2\zeta, \quad w \in B_R, \tag{12}$$

can be extended to a function of class $C^{1,\mu}(\overline{B}_R)$ *satisfying*

$$|y|_{1,\overline{B}_R} \le c_1(b, R) |q|_{0,\overline{B}_R}, \tag{13}$$

$$[\nabla y]_{\mu,\overline{B}_R} \le c_2(b, R, \mu) |q|_{0,\overline{B}_R}, \tag{14}$$

with constants c_1, c_2 *depending on the parameters* b, R *and* b, R, μ, *respectively. Moreover, we have*

$$\nabla_w y(w) = \int_{B_R} \nabla_w H(w,\zeta) q(\zeta)\, d^2\zeta \quad \text{for } w \in B_R. \tag{15}$$

Proof. As $\frac{1}{r} \in L_1(B_R)$ and $q \in L_\infty(B_R)$, the integrals (12) and (15) are well defined. We choose a cut-off function $\eta_h \in C^\infty(\mathbb{R})$ with $0 \le \eta_h \le 1, \eta_h(r) = 0$ for $r \le h, \eta_h(r) = 1$ for $r \ge 2h$, and $\eta_h'(r) \le \frac{2}{h}$. Then we set

$$H_h(w,\zeta) := \eta_h(r)H(w,\zeta) \quad \text{with } r = |w-\zeta|.$$

Then we have $|H_h| \le |H|$ and $H = H_h$ for $r \ge 2h$. The function

$$y_h(w) := \int_{B_R} H_h(w,\zeta)q(\zeta)\,d^2\zeta, \quad w \in B_R, \tag{16}$$

is of class C^2 and, setting $a := |q|_{0,B_R}$, we obtain

$$\begin{aligned}|y(w) - y_h(w)| &\le a \int_{B_R \cap B_{2h}(w)} \{|H| + |H_h|\}\,d^2\zeta \le 2a \int_{B_R \cap B_{2h}(w)} |H|\,d^2\zeta \\ &\le \text{const} \cdot h \to 0 \quad \text{as } h \to 0.\end{aligned}$$

Thus we infer that $y \in C^0(B_R)$.

Now we define

$$z(w) := \int_{B_R} \nabla_w H(w,\zeta)q(\zeta)\,d^2\zeta, \quad w \in B_R.$$

We want to show that $y \in C^1(B_R)$ and $\nabla_w y = z$. In fact, we have

$$\nabla y_h(w) = I_1^h(w) + I_2^h(w)$$

with

$$\begin{aligned}I_1^h(w) &:= \int_{B_R} \eta_h(r)\nabla_w H(w,\zeta)q(\zeta)\,d^2\zeta, \\ I_2^h(w) &:= \int_{B_R} \nabla_w \eta_h(r)H(w,\zeta)q(\zeta)\,d^2\zeta.\end{aligned}$$

As before we show

$$|z(w) - I_1^h(w)| \le \text{ const } \cdot h \to 0 \quad \text{as } h \to 0,$$

and a straight-forward estimate yields

$$|I_2^h(w)| \le \text{const } h^{-1}h^{2-\alpha} \quad \text{for any } \alpha > 0$$

whence we infer that ∇y_h tends uniformly to z on every $\Omega \Subset B_R$. Together with the uniform convergence of y_h to y on $\Omega \Subset B_R$ as $h \to 0$ we infer that $y \in C^1(B_R)$ and $\nabla y(w) = z(w)$ for any $w \in B_R$. Consequently

$$\begin{aligned}|y(w)| + |\nabla_w y(w)| &\le \int_{B_R} \{|H(w,\zeta)| + |\nabla_w H(w,\zeta)|\}|q(\zeta)|\,d^2\zeta \\ &\le c(b,R)a \quad \text{for all } w \in B_R;\end{aligned}$$

thus (13) is also verified.

Now let $w_1, w_2 \in B_R$, and set $\rho := |w_1 - w_2|$. Then we infer from (15) that

$$\begin{aligned} \left|\frac{\partial y}{\partial u}(w_1) - \frac{\partial y}{\partial u}(w_2)\right| &= \left|\int_{B_R} \left\{\frac{\partial H}{\partial u}(w_1,\zeta) - \frac{\partial H}{\partial u}(w_2,\zeta)\right\} q(\zeta)\, d^2\zeta\right| \\ &\le a \int_{B_R\cap B_{2\rho}(w_1)} \left|\frac{\partial H}{\partial u}(w_1,\zeta)\right| d^2\zeta + a\int_{B_R\cap B_{2\rho}(w_1)} \left|\frac{\partial H}{\partial u}(w_2,\zeta)\right| d^2\zeta \\ &\quad + a\int_{B_R\setminus B_{2\rho}(w_1)} \left|\frac{\partial H}{\partial u}(w_1,\zeta) - \frac{\partial H}{\partial u}(w_2,\zeta)\right| d^2\zeta. \end{aligned}$$

Note that

$$\left|\frac{\partial H}{\partial u}(w_1,\zeta)\right| \le b|w_1-\zeta|^{-1}, \qquad \left|\frac{\partial H}{\partial u}(w_2,\zeta)\right| \le b|w_2-\zeta|^{-1},$$

and the mean value theorem implies

$$\left|\frac{\partial H}{\partial u}(w_1,\zeta) - \frac{\partial H}{\partial u}(w_2,\zeta)\right| \le \frac{2b\rho}{|w^*-\zeta|^2}$$

for some $w^* = (1-t)w_1 + tw_2, 0 < t < 1$. If $|\zeta - w_1| \ge 2\rho$, we infer that

$$|\zeta - w^*| \ge |\zeta - w_1| - |w_1 - w^*| \ge \frac{1}{2}|\zeta - w_1|,$$

and therefore

$$\left|\frac{\partial H}{\partial u}(w_1,\zeta) - \frac{\partial H}{\partial u}(w_2,\zeta)\right| \le \frac{8b\rho}{|\zeta - w_1|^2} \quad \text{for } |\zeta - w_1| \ge 2\rho.$$

Thus we arrive at

$$\begin{aligned} &\left|\frac{\partial y}{\partial u}(w_1) - \frac{\partial y}{\partial u}(w_2)\right| \\ &\quad \le ab\bigg[\int_{B_{2\rho}(w_1)} |w_1-\zeta|^{-1}\, d^2\zeta \\ &\qquad + \int_{B_{3\rho}(w_2)} |w_2-\zeta|^{-1}\, d^2\zeta + 8\rho\int_{B_R\setminus B_{2\rho}(w_1)} |w_1-\zeta|^{-2}\, d^2\zeta\bigg] \\ &\quad \le ab\left[4\pi\rho + 6\pi\rho + 16\pi\rho\log\frac{R}{\rho}\right] \le ac(b,R,\mu)\rho^\mu \end{aligned}$$

for any $\mu \in (0,1)$ and $\rho = |w_1 - w_2|$, and (14) is proved. The estimates (13) and (14) imply that y can be extended to $\overline{B}_R$ as a function of class $C^{1,\mu}(\overline{B}_R)$ for any $\mu \in (0,1)$. □

Lemma 2. *Let $H(w,\zeta)$ be a kernel of the form*

$$H(w,\zeta) = K(w-\zeta) = K(u-\xi, v-\eta)$$

for some function $K(\zeta)$ which is of class C^2 on $\{\zeta \neq 0\}$, and suppose that $H(w,\zeta)$ satisfies the growth condition (11). Furthermore we assume that $q(w)$ is of class $C^{0,\mu}(\overline{B}_R), 0 < \mu < 1$. Then the function $y(w)$ defined by (12) is of class $C^2(B_R)$, and we have

$$(17) \qquad |\nabla^2 y|_{0,B_{R'}} \leq c_1 |q|_{\mu,B_R}, \quad |y|_{2,B_{R'}} \leq c_2 |q|_{\mu,B_R}$$

for $0 < R' < R$. Here c_1 and c_2 denote constants depending solely on b, μ, R and R'.

Moreover, if $K(\zeta)$ is of class C^3 for $\zeta \neq 0$ and if also

$$(11^*) \qquad |\nabla_w^3 H(w,\zeta)| \leq b|w-\zeta|^{-3},$$

then $y(w)$ is of class $C^{2,\mu}(B_R)$ and satisfies

$$(18) \qquad |\nabla^2 y|_{\mu,B_{R'}} \leq c_3 |q|_{\mu,B_R}, \quad |y|_{2+\mu,B_{R'}} \leq c_4 |q|_{\mu,B_R}$$

for $0 < R' < R$. Here the numbers c_3 and c_4 only depend on b, μ, R', and R.

Proof. We set again $H_h = \eta_h H$ where η_h is chosen as in the proof of Lemma 1; but in addition we arrange that $|\eta_h''(r)| \leq \gamma h^{-2}$ for some constant $\gamma > 0$. Then

$$z_h(w) := \int_{B_R} \nabla_w H_h(w,\zeta) q(\zeta)\, d^2\zeta$$

is of class $C^1(\overline{B}_R, \mathbb{R}^2)$, and for $D = \frac{\partial}{\partial u}$ or $\frac{\partial}{\partial v}$ we can write

$$\begin{aligned} Dz_h(w) &= \int_{B_R} D\nabla_w H_h(w,\zeta) q(\zeta)\, d^2\zeta \\ &= \int_{B_R} D\nabla_w H_h(w,\zeta)[q(\zeta) - q(w)]\, d^2\zeta + q(w) \int_{B_R} D\nabla_w H_h(w,\zeta)\, d^2\zeta. \end{aligned}$$

By integration by parts, we obtain

$$\begin{aligned} \int_{B_R} D_u \nabla_w H_h(w,\zeta)\, d^2\zeta &= -\int_{B_R} D_\xi \nabla_w H_h(w,\zeta)\, d^2\zeta \\ &= -\int_{\partial B_R} \nabla_w H_h(w,\zeta) \cos\alpha\, ds(\zeta), \end{aligned}$$

where ds is the line element on ∂B_R and $\cos\alpha = \frac{\xi}{|\zeta|}$. If $2h < |w-\zeta|$ we obtain

$$Dz_h(w) = \phi_h(w) - q(w) \int_{\partial B_R} \nabla_w H(w,\zeta) \cos\alpha\, ds(\zeta)$$

with $\cos\alpha = \frac{\xi}{|\zeta|}$ or $= \frac{\eta}{|\zeta|}$ and

$$\phi_h(w) := \int_{B_R} D\nabla_w H_h(w,\zeta)[q(\zeta) - q(w)]\, d^2\zeta.$$

Similarly we set

$$\phi(w) := \int_{B_R} D\nabla_w H(w,\zeta)[q(\zeta) - q(w)]\, d^2\zeta. \tag{19}$$

For $r = |w - \zeta|$ and $a := |q|_{\mu, B_R}$ we have

$$|D\nabla_w H(w,\zeta)| \le br^{-2} \quad \text{and} \quad |q(\zeta) - q(w)| \le ar^{\mu},$$

whence

$$|\phi(w)|, |\phi_h(w)| \le 2\pi ab\mu^{-1}R^{\mu}. \tag{20}$$

By a similar reasoning we obtain ($\nabla = \nabla_w$ and $0 < h \ll 1$):

$$\begin{aligned}|\phi_h(w) - \phi(w)| &\le \int_{B_R \cap B_{2h}(w)} |\eta_h(w) - 1||D\nabla H(w,\zeta)||q(\zeta) - q(w)|\, d^2\zeta \\ &\quad + \int_{B_R} \{|\nabla^2 \eta_h||H| + 2|\nabla \eta_h||\nabla H|\}|q(\zeta) - q(w)|\, d^2\zeta \\ &\le \text{const} \cdot h^{\mu}\left(1 + \log\frac{1}{h}\right) \to 0 \quad \text{as } h \to 0.\end{aligned}$$

Thus $Dz_h(w)$ tends uniformly to

$$\phi(w) - q(w) \int_{\partial B_R} \nabla_w H(w,\zeta) \cos\alpha(\zeta)\, ds(\zeta)$$

as $h \to 0$, for $w \in B_{R'}$ and $0 < R' < R$. On the other hand, if y_h is defined by (16), we know that

$$z_h = \nabla y_h, \quad Dz_h = D\nabla y_h, \quad z_h \in C^1,$$

and, as shown in the proof of Lemma 1, we also have

$$\lim_{h \to 0} |y - y_h|_{1, B_{R'}} = 0 \quad \text{for } 0 < R' < R.$$

Consequently we have $y \in C^2(B_R)$ and

$$D\nabla y(w) = \phi(w) - q(w) \int_{\partial B_R} \nabla_w H(w,\zeta) \cos\alpha(\zeta)\, ds(\zeta) \quad \text{for } |w| < R. \tag{21}$$

Now inequalities (17) follow from (20) and (21).

Finally, taking assumption (11*) into account, we derive from the representation formulas (19) and (21) that y is of class $C^{2,\mu}(B_R)$ and in conjunction with (17) that the estimates (18) are satisfied. Since we may proceed in the same way as in the last part of the proof of Lemma 1, we shall skip this part of the proof. □

Proposition 3. *Suppose that* $x \in C^0(\overline{B}_R) \cap C^2(B_R)$ *and that* $\nabla x \in L_2(B_R)$ *and* $\Delta x \in L_\infty(B_R)$. *Then we obtain the following representation formulas which are satisfied for* $w \in B_R$:

$$x(w) = \frac{1}{2\pi} \int_{\partial B_R} \left(\mathrm{Re} \frac{\zeta + w}{\zeta - w} \right) x(\zeta) \frac{d\zeta}{i\zeta} - \int_{B_R} G_R(w, \zeta) \Delta x(\zeta) \, d^2\zeta, \tag{22}$$

$$\frac{\partial}{\partial w} x(w) = \frac{1}{2\pi i} \int_{\partial B_R} \frac{x(\zeta)}{(\zeta - w)^2} \, d\zeta - \int_{B_R} \frac{\partial}{\partial w} G_R(w, \zeta) \Delta x(\zeta) \, d^2\zeta, \tag{23}$$

$$x_u(0) = \frac{1}{\pi R^2} \int_{B_R} x_u(u,v) \, du \, dv - \frac{1}{2\pi} \int_{B_R} u \left[\frac{1}{r^2} - \frac{1}{R^2} \right] \Delta x(u,v) \, du \, dv, \tag{24}$$

$$x_v(0) = \frac{1}{\pi R^2} \int_{B_R} x_v(u,v) \, du \, dv - \frac{1}{2\pi} \int_{B_R} v \left[\frac{1}{r^2} - \frac{1}{R^2} \right] \Delta x(u,v) \, du \, dv, \tag{25}$$

$$r = |w| = \sqrt{u^2 + v^2}.$$

Proof. Formula (22) is merely a reformulation of (6) and (7). Differentiating (22), it follows in conjunction with Lemma 2 (in particular, with (15)) that (23) holds if we take

$$\frac{2\zeta}{(\zeta - w)^2} = \frac{\partial}{\partial w} \frac{\zeta + w}{\zeta - w} = 2 \frac{\partial}{\partial w} \mathrm{Re} \frac{\zeta + w}{\zeta - w} = 2 \frac{\partial}{\partial w} \frac{R^2 - |w|^2}{|\zeta - w|^2}$$

for $\zeta \in \partial B_R$ into account.

By applying (23) to $w = 0$ and noting that

$$\frac{\partial}{\partial w} G_R(0, \zeta) = \frac{1}{4\pi} \left(\frac{1}{\zeta} - \frac{\overline{\zeta}}{R^2} \right)$$

we infer that

$$x_w(0) = \frac{1}{2\pi i} \int_{\partial B_R} \frac{x(\zeta)}{\zeta^2} \, d\zeta - \int_{B_R} \frac{\overline{\zeta}}{4\pi} \left(\frac{1}{|\zeta|^2} - \frac{1}{R^2} \right) \Delta x(\zeta) \, d^2\zeta.$$

Because of $\zeta^{-2} \, d\zeta = -R^{-2} \, d\overline{\zeta}$, it follows that

$$\begin{aligned} \frac{1}{2\pi i} \int_{\partial B_R} \zeta^{-2} x(\zeta) \, d\zeta &= -\frac{1}{2\pi i R^2} \int_{\partial B_R} x(\zeta) \, d\overline{\zeta} \\ &= -\frac{1}{2\pi i R^2} \int_{\partial B_R} x(\zeta)(d\xi - i \, d\eta) \\ &= -\frac{1}{2\pi i R^2} \int_{B_R} (-i x_\xi - x_\eta) \, d\xi \, d\eta \\ &= \frac{1}{\pi R^2} \int_{B_R} x_\zeta \, d^2\zeta. \end{aligned}$$

Replacing ζ by w, we arrive at (24) and (25) by separating the real and imaginary parts. Actually we first prove (24) and (25) for $B_{R'}, R' < R$, instead for B_R, and then we let $R' \to R$. $\square$

Now we prove Schwarz's result concerning the boundary continuity of Poisson's integral.

Lemma 3. *Let $\boldsymbol{x}(\varphi)$ be a continuous, 2π-periodic function on $\mathbb{R}$, and let $h(w) := \int_0^{2\pi} \mathcal{P}_R(w,\varphi)\boldsymbol{x}(\varphi)\,d\varphi$ be the corresponding Poisson integral, which is a harmonic function of $w \in B_R$. Then we obtain $h(w) \to \boldsymbol{x}(\varphi)$ as $w \to Re^{i\varphi}$. Thus $h(w)$ can be extended to a continuous function on $\overline{B}_R$ such that $h(Re^{i\varphi}) = \boldsymbol{x}(\varphi)$ for all $\varphi \in \mathbb{R}$.*

Proof. It suffices to treat the case $R = 1$. Then we have to prove $\lim_{r\to 1-0} h(re^{i\theta}) = \boldsymbol{x}(\theta)$ uniformly in $\theta \in \mathbb{R}$. We can write

$$h(re^{i\theta}) = \frac{1}{2\pi}\int_{-\pi}^{\pi} \frac{1-r^2}{|e^{i\varphi}-r|^2}\boldsymbol{x}(\theta+\varphi)\,d\varphi.$$

Because of the identity

$$\frac{1}{2\pi}\int_{-\pi}^{\pi} \frac{1-r^2}{|e^{i\varphi}-r|^2}\,d\varphi = 1$$

it follows that

$$h(re^{i\theta}) - \boldsymbol{x}(\theta) = \frac{1}{2\pi}\int_{-\pi}^{\pi} \frac{1-r^2}{|e^{i\varphi}-r|^2}[\boldsymbol{x}(\theta+\varphi)-\boldsymbol{x}(\theta)]\,d\varphi$$

whence

$$\begin{aligned} |h(re^{i\theta}) - \boldsymbol{x}(\theta)| &\le \frac{1}{2\pi}\int_{-\pi}^{\pi} \frac{1-r^2}{|e^{i\varphi}-r|^2}|\boldsymbol{x}(\theta+\varphi)-\boldsymbol{x}(\theta)|\,d\varphi \\ &\le \frac{1}{2\pi}\int_{-\pi}^{-\delta}\cdots + \frac{1}{2\pi}\int_{-\delta}^{\delta}\cdots + \frac{1}{2\pi}\int_{\delta}^{\pi}\cdots = I_1 + I_2 + I_3 \end{aligned}$$

for any $\delta \in (0, \frac{\pi}{2})$. Fix some $\varepsilon > 0$ and choose $\delta > 0$ so small that $|\boldsymbol{x}(\varphi) - \boldsymbol{x}(\theta)| < \varepsilon$ for all φ and θ with $|\varphi - \theta| < \delta$. Then we obtain

$$I_2 \le \varepsilon \cdot \frac{1}{2\pi}\int_{-\delta}^{\delta} \frac{1-r^2}{|e^{i\varphi}-r|^2}\,d\varphi \le \varepsilon.$$

Moreover, by setting $M := \max_{\mathbb{R}} |\boldsymbol{x}|$ we obtain

$$I_1, I_3 \le \frac{1}{2\pi}(\pi-\delta)(1-r)(1+r)\frac{2M}{\sin^2\delta} \le \frac{2M}{\sin^2\delta}(1-r)$$

since $|e^{i\varphi} - r|^2 \ge \sin^2\delta$ for $\delta \le |\varphi| \le \pi$. Thus we arrive at

$$|h(re^{i\theta}) - \boldsymbol{x}(\theta)| \le \varepsilon + \frac{4M}{\sin^2 \delta(\varepsilon)}(1-r) \quad \text{for } r \in (0,1)$$

and therefore

$$\lim_{r \to 1-0} h(re^{i\theta}) = \boldsymbol{x}(\theta) \quad \text{uniformly in } \theta. \qquad \square$$

As a by-product of this proof we have found:

Lemma 4. *Let* $h \in C^0(\overline{B}) \cap C^2(B)$ *be harmonic in* B *and suppose that* $|h(e^{i\varphi}) - h(e^{i\theta})| < \varepsilon$ *holds for all* φ *with* $|\varphi - \theta| \le \delta, \delta \in \left(0, \frac{\pi}{2}\right)$*. Then it follows that*

$$|h(re^{i\theta}) - h(e^{i\theta})| \le \varepsilon + \frac{4|h|_{0,\partial B}}{\sin^2 \delta}(1-r) \tag{26}$$

holds for all $r \in (0,1)$.

Lemma 5. *Let* $h \in C^0(\overline{B}_R) \cap C^2(B_R)$ *be harmonic in* B_R*, and suppose that the boundary values* $\boldsymbol{x}(\varphi)$ *of* h *defined by* $\boldsymbol{x}(\varphi) := h(Re^{i\theta})$ *are of class* $C^2(\mathbb{R})$ *and satisfy* $|\boldsymbol{x}''(\varphi)| \le k$ *for all* $\varphi \in \mathbb{R}$*. Then we obtain*

$$|\nabla h(w)| \le cR^{-1}k \quad \textit{for all } w \in B_R, \tag{27}$$

where c *is an absolute constant independent of* h *and* R.

Proof. By virtue of an obvious scaling argument we can restrict our attention to the case $R = 1$. Then we have to prove

$$|\nabla h(w)| \le \text{const } k \quad \text{for } w \in B.$$

Let h^* be the conjugate harmonic function to h. Then $f(w) := h(w) + ih^*(w)$ is a holomorphic function of $w = u + iv$, and we have the convergent power series expansion

$$f(w) = \sum_{l=0}^{\infty} c_l w^l \quad \text{for } |w| < 1.$$

Set $c_0 = \frac{1}{2}(a_0 - ib_0), c_l = a_l - ib_l$ if $l \ge 1, a_l, b_l \in \mathbb{R}$. Then we have for $w = re^{i\varphi}$ that

$$h(w) = \frac{a_0}{2} + \sum_{l=1}^{\infty} r^l (a_l \cos l\varphi + b_l \sin l\varphi)$$

whence

$$\begin{aligned}
\boldsymbol{x}(\varphi) &= \frac{a_0}{2} + \sum_{l=1}^{\infty} (a_l \cos l\varphi + b_l \sin l\varphi), \\
a_l &= \frac{1}{\pi} \int_0^{2\pi} \boldsymbol{x}(\varphi) \cos l\varphi \, d\varphi = -\frac{1}{\pi l^2} \int_0^{2\pi} \boldsymbol{x}''(\varphi) \cos l\varphi \, d\varphi, \\
b_l &= \frac{1}{\pi} \int_0^{2\pi} \boldsymbol{x}(\varphi) \sin l\varphi \, d\varphi = -\frac{1}{\pi l^2} \int_0^{2\pi} \boldsymbol{x}''(\varphi) \sin l\varphi \, d\varphi.
\end{aligned}$$

Because of $f' = h_u + ih_u^* = h_u - ih_v$ we infer that

$$\begin{aligned}|\nabla h(w)| &= |f'(w)| = \left|\sum_{l=1}^{\infty} l(a_l - ib_l)w^{l-1}\right| \\ &\le \sum_{l=1}^{\infty} l\sqrt{a_l^2 + b_l^2} \le \left\{\sum_{l=1}^{\infty} \frac{1}{\pi l^2}\right\}^{1/2} \left[\int_0^{2\pi} |\boldsymbol{x}''(\varphi)|^2 \, d\varphi\right]^{1/2} \\ &\le \frac{\pi}{\sqrt{3}} k \quad \text{if } |w| < 1,\end{aligned}$$

taking Schwarz's inequality into consideration as well as Parseval's relation for the Fourier series expansion of $\boldsymbol{x}''$. □

The next result is known in the literature as *Theorem of Korn and Privalov.*

Lemma 6. *Let $f(w) = h(w) + ih^*(w)$ be holomorphic in $B_R, h = \operatorname{Re} f$, $h^* = \operatorname{Im} f$, and suppose that $h^* \in C^0(\overline{B}_R) \cap C^{0,\mu}(\partial B_R)$ holds for some $\mu \in (0,1)$. Then f is of class $C^{0,\mu}(\overline{B}_R)$ and we have*

$$[f]_{\mu,\overline{B}_R} \le c(\mu)[h^*]_{\mu,\partial B_R}. \tag{28}$$

Proof. We can assume that $R = 1$ applying a scaling argument. Set $H := [h^*]_{\mu,\partial B}$. Then we have

$$|h^*(e^{i\theta}) - h^*(e^{i\varphi})| \le H|e^{i\theta} - e^{i\varphi}|^\mu \tag{29}$$

for all $\theta, \varphi \in \mathbb{R}$. Fix some $\varphi \in [0, 2\pi)$ and consider the function

$$\psi(w) = \operatorname{Re}(1 - we^{-i\varphi})^\mu$$

which can be viewed as a univalent harmonic function of $w \in B$. Introducing the angle α between the rays $\{te^{i\varphi} : t \ge 0\}$ and $\{t(e^{i\varphi} - w) : t \ge 0\}$, we obtain

$$\psi(w) = |w - e^{i\varphi}|^\mu \cos(\mu\alpha(w)),$$

where $|\alpha(w)| \le \frac{\pi}{2}$. Thus we infer from (29) that

$$-\frac{H\psi(w)}{\cos\frac{\mu\pi}{2}} \le h^*(w) - h^*(e^{i\varphi}) \le \frac{H\psi(w)}{\cos\frac{\mu\pi}{2}} \tag{30}$$

holds for all $w \in \partial B$. Applying the maximum principle, we obtain that (31) holds for all $w \in \overline{B}$ and in particular for all $w \in B_{1-r}(re^{i\varphi})$ if $0 < r < 1$. Hence we infer that

$$|h^*(w) - h^*(e^{i\varphi})| \le \frac{2^\mu H}{\cos\frac{\mu\pi}{2}}(1-r)^\mu \quad \text{for } |w - re^{i\varphi}| < 1 - r \tag{31}$$

is satisfied.

Now we set $w_0 := re^{i\varphi}, 0 < r < 1$, and $h_0^* := h^*(w_0)$. Applying Gauss's mean value theorem to the harmonic function h_u^* and to the ball $B_\rho(w_0)$ with some radius $\rho \in (0, 1-r)$, we obtain

$$h_u^*(w_0) = \frac{1}{\pi\rho^2}\int_{B_\rho(w_0)} h_u^*(w)\,d^2w = \frac{1}{\pi\rho^2}\int_{B_\rho(w_0)} (h^* - h_0^*)_u\,d^2w,$$

and an integration by parts yields

$$h_u^*(w_0) = \frac{1}{\pi\rho^2}\int_{\partial B_\rho(w_0)} \frac{u-u_0}{\rho}(h^* - h_0^*)\,ds.$$

Analogously,

$$h_v^*(w_0) = \frac{1}{\pi\rho^2}\int_{\partial B_\rho(w_0)} \frac{v-v_0}{\rho}(h^* - h_0^*)\,ds.$$

Thus we obtain

$$(32) \qquad |\nabla h^*(w_0)| \le 2\rho^{-1}|h^* - h_0^*|_{0,\partial B_\rho(w_0)}.$$

If we let $\rho \to 1-r$ and combine the resulting inequality with (31), it follows that

$$|\nabla h^*(w_0)| \le \frac{4H}{\cos\frac{\mu\pi}{2}}(1-r)^{-1+\mu}.$$

Since we can choose φ arbitrarily, we obtain that

$$(33) \qquad |f'(w)| \le c(\mu)H(1-|w|)^{-1+\mu} \quad \text{for all } w \in B,$$

if we set $c(\mu) := 4(\cos\frac{\mu\pi}{2})^{-1}$. Then it follows from (33) for $0 \le r < 1$ that

$$(7.33') \qquad |f'(w)| \le c(\mu)H(r-|w|)^{-1+\mu} \quad \text{for all } w \in B_r.$$

For any r and r' with $0 \le r < r' < 1$ we now conclude

$$|f(r'e^{i\theta}) - f(re^{i\theta})| = \left|\int_{re^{i\theta}}^{r'e^{i\theta}} f'(w)\,dw\right| \le c(\mu)H\int_r^{r'} (r'-\rho)^{-1+\mu}\,d\rho$$

whence

$$(34) \qquad |f(r'e^{i\theta}) - f(re^{i\theta})| \le c(\mu)\mu^{-1}H(r'-r)^\mu \quad \text{for } 0 \le r < r' < 1.$$

We infer that $\lim_{r\to 1-0} f(re^{i\theta})$ exists for any $\theta \in \mathbb{R}$. Setting $\xi(\varphi) := \lim_{r\to 1-0} f(re^{i\varphi})$ we extend $f(w)$ from B to $\overline{B}$ by defining $f(e^{i\varphi}) := \xi(\varphi)$. We now want to show that $f \in C^{0,\mu}(\overline{B})$. In fact, setting $c^*(\mu) := \mu^{-1}c(\mu)$ we obtain from (34) that

$$(34') \qquad |f(r'e^{i\theta}) - f(re^{i\theta})| \le c^*(\mu)H(r'-r)^\mu \quad \text{for } 0 \le r \le r' \le 1, \theta \in \mathbb{R},$$

and in particular

$$|\xi(\theta) - f(re^{i\theta})| \le c^*(\mu) H (1-r)^\mu \quad \text{for } 0 < r < 1 \text{ and } \theta \in \mathbb{R}.$$

Then it follows for $\theta_1 < \theta_2$ that

$$\begin{aligned} |\xi(\theta_1) - \xi(\theta_2)| &\le |\xi(\theta_1) - f(re^{i\theta_1})| + |\xi(\theta_2) - f(re^{i\theta_1})| \\ &\le |\xi(\theta_1) - f(re^{i\theta_1})| + |\xi(\theta_2) - f(re^{i\theta_2})| + |f(re^{i\theta_1}) - f(re^{i\theta_2})| \\ &\le 2c^*(\mu) H (1-r)^\mu + \int_{\theta_1}^{\theta_2} |f'(re^{i\theta})| r \, d\theta. \end{aligned}$$

Moreover, we derive from (33) that

$$\int_{\theta_1}^{\theta_2} |f'(re^{i\theta})| r \, d\theta \le H c(\mu) r (1-r)^{-1+\mu} (\theta_2 - \theta_1).$$

Suppose that $0 < \theta_2 - \theta_1 < 1$, and choose $r = 1 - (\theta_2 - \theta_1)$. Then it follows that

$$|\xi(\theta_1) - \xi(\theta_2)| \le \{2c^*(\mu) + c(\mu)\} H |\theta_2 - \theta_1|^\mu$$

if $|\theta_1 - \theta_2| \le 1$. Renaming $8c^*(\mu) + 4c(\mu)$ by $c(\mu)$, we arrive at

$$(35) \qquad |\xi(\theta_1) - \xi(\theta_2)| \le c(\mu) H |\theta_1 - \theta_2|^\mu \quad \text{for all } \theta_1, \theta_2 \in \mathbb{R}.$$

Applying the maximum principle to the modulus of the holomorphic mapping $f(e^{i\alpha}w) - f(w), w \in B$, we see that

$$\max_{w \in \overline{B}} |f(e^{i\alpha}w) - f(w)| \le \max_{w \in \partial B} |f(e^{i\alpha}w) - f(w)|$$

holds for all $\alpha \in \mathbb{R}$, and in view of (35) we obtain

$$|f(e^{i\alpha}w) - f(w)| \le c(\mu) H |\alpha|^\mu \quad \text{for } w \in \overline{B} \text{ and } \alpha \in \mathbb{R}.$$

This estimate is equivalent to

$$(35') \qquad |f(re^{i\theta_2}) - f(re^{i\theta_1})| \le c(\mu) H |\theta_1 - \theta_2|^\mu \quad \text{for } 0 \le r \le 1, \ \theta_1, \theta_2 \in \mathbb{R}.$$

Combining the estimates (34′) and (35′) we arrive at

$$\begin{aligned} |f(r_1 e^{i\theta_1}) - f(r_2 e^{i\theta_2})| &\le |f(r_1 e^{i\theta_1}) - f(r_2 e^{i\theta_1})| + |f(r_2 e^{i\theta_1}) - f(r_2 e^{i\theta_2})| \\ &\le c^*(\mu) H |r_1 - r_2|^\mu + c(\mu) H |\theta_1 - \theta_2|^\mu \end{aligned}$$

for arbitrary $r_1, r_2 \in [0,1]$ and $\theta_1, \theta_2 \in \mathbb{R}$.

If $w_1 = r_1 e^{i\theta_1}, w_2 = r_2 e^{i\theta_2}, \frac{1}{2} \le r_1, r_2 \le 1, |\theta_2 - \theta_1| \le \pi$, then there is a constant K such that

$$|r_1 - r_2|^\mu + |\theta_1 - \theta_2|^\mu \le K |w_1 - w_2|^\mu.$$

Consequently we have

$$|f(w_1) - f(w_2)| \le c(\mu)H|w_1 - w_2|^\mu$$

for all $w_1, w_2 \in \overline{B} \setminus B_{1/2}$ and for some constant $c(\mu)$, and because of (33) the same estimate holds for any $w_1, w_2 \in \overline{B}_{1/2}$. Then we easily infer that

$$|f(w_1) - f(w_2)| \le c(\mu)H|w_1 - w_2|^\mu \quad \text{for all } w_1, w_2 \in \overline{B}$$

holds true. □

Lemma 7. *Suppose that $h \in C^0(\overline{B}_R) \cap C^2(B_R)$ is harmonic in B_R and that its boundary values $\boldsymbol{x}(\varphi) := h(Re^{i\varphi})$ satisfy $\boldsymbol{x} \in C^2(\mathbb{R})$ and $|\boldsymbol{x}''(\varphi)| \le k$ for all $\varphi \in \mathbb{R}$. Then we obtain $h \in C^{1,\mu}(\overline{B}_R)$ for every $\mu \in (0,1)$ and*

$$[\nabla h]_{\mu,\overline{B}_R} \le c(\mu) R^{-1-\mu} k, \tag{36}$$

where the number $c(\mu)$ only depends on μ.

Proof. It is sufficient to prove the result for $R = 1$. Let us introduce the tangential difference quotient

$$(T_\theta h)(re^{i\varphi}) := \frac{1}{\theta}[h(re^{i(\varphi+\theta)}) - h(re^{i\varphi})]$$

and note that $(T_\theta h)(w)$ is a harmonic function of $w \in B$ which is continuous on $\overline{B}$ and has the boundary values

$$(\tau_\theta \boldsymbol{x})(\varphi) := \frac{1}{\theta}[\boldsymbol{x}(\varphi + \theta) - \boldsymbol{x}(\varphi)].$$

By assumption the boundary values $(\tau_\theta \boldsymbol{x})(\varphi)$ tend uniformly to $\boldsymbol{x}'(\varphi)$ as $\theta \to 0$. Then, on account of Harnack's first convergence theorem, we easily infer that the functions $(T_\theta h)(w)$ tend uniformly on $\overline{B}$ to the harmonic function $h_\varphi(w)$ with the boundary values $\boldsymbol{x}'(\varphi) = \frac{\partial}{\partial\varphi} h(e^{i\varphi})$ on ∂B which, by assumption, are Hölder continuous for any exponent $\mu < 1$, and

$$[\boldsymbol{x}']_{\mu,\mathbb{R}} \le 2\pi k. \tag{37}$$

Consider a holomorphic function $f(w)$ on B with $f = h + ih^*$, that is, $h = \operatorname{Re} f, h^* = \operatorname{Im} f$. Then $g(w) := iwf'(w) = \frac{\partial f}{\partial \varphi}(w), w = re^{i\varphi}$, is another holomorphic function on B with $\frac{\partial h}{\partial\varphi} = \operatorname{Re} g$ and $\boldsymbol{x}'(\varphi) = \frac{\partial h}{\partial \varphi}(e^{i\varphi}), \boldsymbol{x}' \in C^{0,\mu}(\mathbb{R})$ for any $\mu \in (0,1)$. Hence we can apply Lemma 6 to the holomorphic function $ig(w) = -wf'(w), w \in B$, and we obtain that $ig(w)$ is of class $C^{0,\mu}(\overline{B})$. This implies

$$f' \in C^{0,\mu}(\overline{B} \setminus B_{1/2}),$$

and inequalities (28) and (37) yield

(38) $$[f']_{\mu,B\setminus B_{1/2}} \leq \text{const} \cdot k.$$

Moreover, (27) implies

$$|f'|_{0,B} \leq \text{const} \cdot k,$$

and Cauchy's estimate for holomorphic functions then gives

$$|f''|_{0,B_{1/2}} \leq \text{const} \cdot k,$$

whence

$$[f']_{\mu,B_{1/2}} \leq \text{const} \cdot k.$$

Combining this estimate with (38), we arrive at the desired inequality

$$[f']_{\mu,B} \leq \text{const} \cdot k.$$

If we now recall that $f' = h_u - ih_v$, we find that the lemma is proved. □

Proof of Proposition 2. We now see that Proposition 2 is a direct consequence of Lemmata 1–7 in conjunction with Proposition 1 and with formulas (1)–(5). □

Remark. We have formulated the estimates and the regularity results of Proposition 1 in a global way. Analogous local results can be derived by similar methods, but certain changes will be necessary to obtain local estimates at the boundary. A very simple approach to local $C^{1,\mu}$-estimates is based on a reflection method: it will be described in the next section.

2.2 Solutions of Differential Inequalities

In this section we want to derive a priori estimates for solutions $X(u,v) = X(w) = (x^1(w), x^2(w), \dots, x^N(w))$ of differential inequalities

(1) $$|\Delta X| \leq a|\nabla X|^2,$$

which can equivalently be written as

(1′) $$|X_{w\overline{w}}| \leq a|X_w|^2.$$

Here a denotes a fixed nonnegative constant.

Lemma 1. *Let $X \in C^2(\Omega, \mathbb{R}^N)$ be a solution of (1) in the open set Ω of $\mathbb{R}^2$ which satisfies $|X|_{0,\Omega} \leq M$. Then we obtain*

(2) $$\Delta|X|^2 \geq 2(1 - aM)|\nabla X|^2$$

in Ω. In particular, if $aM < 1$, then $|X|^2$ is subharmonic in Ω.

Proof. Because of

$$|\langle X, \Delta X\rangle| \le |X||\Delta X| \le aM|\nabla X|^2,$$

the inequality (2) is an immediate consequence of the identity

$$\Delta|X|^2 = 2|\nabla X|^2 + 2\langle X, \Delta X\rangle \tag{3}$$

which holds for every mapping X of class C^2. □

Lemma 2. *Suppose that $X \in C^0(\overline{B}_R(w_0), \mathbb{R}^N) \cap C^2(B_R(w_0), \mathbb{R}^N)$ satisfies (1) in $B_R(w_0)$. Assume also that $|X(w)| \le M$ for $w \in \overline{B}_R(w_0)$ and $aM < 1$ are satisfied. Then for any $\rho \in (0, R)$ we have*

$$\int_{B_\rho(w_0)} |\nabla X|^2\, du\, dv \le \frac{1}{\log \frac{R}{\rho}} \frac{2\pi M}{1 - aM} \max_{w \in \partial B_R(w_0)} |X(w) - X(w_0)| \tag{4}$$

and

$$\int_{B_\rho(w_0)} |\nabla X|^2\, du\, dv \le \frac{1}{\log \frac{R}{\rho}} \frac{4\pi M^2}{1 - aM}. \tag{5}$$

Proof. Choose some $\rho \in (0, R)$ and apply Proposition 1 of Section 2.1 to the function $x(w) := |X(w)|^2$ and to the domain $B_R(w_0)$ instead of $B_R = B_R(0)$, assuming in addition that X is of class C^2 on $\overline{B}_R(w_0)$. Then formula (6) of Section 2.2 yields

$$\frac{1}{2\pi} \int_0^{2\pi} [x(w_0 + Re^{i\varphi}) - x(w_0)]\, d\varphi = \frac{1}{2\pi} \int_{B_R(w_0)} \log \frac{R}{|w - w_0|} \Delta x\, d^2 w.$$

Because of

$$|x(w) - x(w_0)| \le 2M|X(w) - X(w_0)|$$

we infer that

$$\frac{1}{2\pi} \int_{B_R(w_0)} \log \frac{R}{|w - w_0|} \Delta x\, d^2 w \le 2M \max_{w \in \partial B_R(w_0)} |X(w) - X(w_0)|.$$

On the other hand, Lemma 1 gives

$$2(1 - aM)|\nabla X|^2 \le \Delta x,$$

whence

$$(1 - aM)\pi^{-1} \int_{B_R(w_0)} \log \frac{R}{|w - w_0|} |\nabla X|^2\, d^2 w \le 2M \max_{w \in B_R(w_0)} |X(w) - X(w_0)|.$$

Moreover,

$$\log \frac{R}{\rho} \int_{B_\rho(w_0)} |\nabla X|^2\, d^2 w \le \int_{B_R(w_0)} \log \frac{R}{|w - w_0|} |\nabla X|^2\, d^2 w$$

if $0 < \rho < R$, and (4) is proved. The additional hypothesis can be removed if we first apply the reasoning to ρ and R' with $0 < \rho < R' < R$, and then let $R' \to R - 0$. Inequality (5) is a direct consequence of (4). □

Proposition 1. *There is a continuous function $\kappa(t), 0 \le t < 1$, with the following property: For any solution $X \in C^2(B_R(w_0), \mathbb{R}^N)$ of the differential inequality* (1) *in $B_R(w_0)$ satisfying*

$$|X(w)| \le M \quad \textit{for } w \in B_R(w_0) \tag{6}$$

and for some constant M with $aM < 1$, the estimates

$$|\nabla X(w_0)| \le \kappa(aM)\frac{M}{R} \tag{7}$$

and

$$|\nabla X(w_0)| \le \frac{\kappa(aM)}{R} \sup_{w \in B_R(w_0)} |X(w) - X(w_0)| \tag{8}$$

hold true.

Proof. Fix any $R' \in (0, R)$, and consider the nonnegative function

$$f(w) := (R' - |w - w_0|)|\nabla X(w)|$$

on $\overline{B}_{R'}(w_0)$ which vanishes on $\partial B_{R'}(w_0)$. Then there is some point $w_1 \in B_{R'}(w_0)$ where $f(w)$ assumes its maximum K, i.e.,

$$f(w_1) = K := \max\{f(w) \colon w \in \overline{B}_{R'}(w_0)\}.$$

Set $r = |w - w_1|$ for $w \in \overline{B}_{R'}(w_0)$ and $\rho := R' - |w_1 - w_0|$. Clearly, we have $0 < \rho < R'$.

By formulas (24) and (25) of Section 2.1, we obtain for any $\theta \in (0,1)$ that

$$\begin{aligned} X_u(w_1) &= \frac{1}{\pi\rho^2\theta^2}\int_{B_{\rho\theta}(w_1)} X_u \, du\, dv \\ &\quad - \frac{1}{2\pi}\int_{B_{\rho\theta}(w_1)} (u - u_1)\left(\frac{1}{r^2} - \frac{1}{\rho^2\theta^2}\right) \Delta X \, du\, dv, \end{aligned}$$

and an analogous formula holds for $X_v(w_1)$. By means of Schwarz's inequality we infer that

$$|\nabla X(w_1)| \le \frac{1}{\sqrt{\pi}\rho\theta}\left\{\int_{B_{\rho\theta}(w_1)} |\nabla X|^2 \, du\, dv\right\}^{1/2} + \frac{1}{2\pi}\int_{B_{\rho\theta}(w_1)} \frac{|\Delta X|}{r}\, du\, dv.$$

Applying Lemma 2, (5) to $\rho\theta$ and ρ instead of ρ and R, we also obtain

$$\int_{B_{\rho\theta}(w_1)} |\nabla X|^2 \, du\, dv \le \frac{c(a, M)}{\log\frac{1}{\theta}}.$$

Taking $|\Delta X| \le a|\nabla X|^2$ into account, we arrive at

$$|\nabla X(w_1)| \le \frac{\sqrt{c}}{\sqrt{\pi}\rho\theta\big(\log\frac{1}{\theta}\big)^{1/2}} + \frac{a}{2\pi}\int_{B_{\rho\theta}(w_1)} \frac{1}{r}|\nabla X|^2\,du\,dv$$

and

$$\frac{a}{2\pi}\int_{B_{\rho\theta}(w_1)} \frac{1}{r}|\nabla X|^2\,du\,dv \le a\rho\theta \sup_{B_{\rho\theta}(w_1)} |\nabla X|^2.$$

On account of

$$K = f(w_1) = \rho|\nabla X(w_1)|$$

we obtain

$$|\nabla X(w_1)| = K/\rho.$$

Moreover, if $r = |w - w_1| < \rho\theta$, it follows that

$$R' - |w - w_0| \ge R' - |w_0 - w_1| - |w - w_1| = \rho - r > (1-\theta)\rho.$$

Thus we infer from

$$|\nabla X(w)|(R' - |w - w_0|) \le K \quad \text{for all } w \in B_{R'}(w_0)$$

that

$$|\nabla X(w)| \le \frac{K}{(1-\theta)\rho} \quad \text{for all } w \in B_{\rho\theta}(w_1)$$

holds true, and we conclude that

$$\frac{a}{2\pi}\int_{B_{\rho\theta}(w_1)} \frac{1}{r}|\nabla X|^2\,du\,dv \le \frac{a\theta K^2}{(1-\theta)^2\rho}$$

whence

$$\frac{K}{\rho} \le \frac{\sqrt{c}}{\sqrt{\pi}\rho\theta\big(\log\frac{1}{\theta}\big)^{1/2}} + \frac{a\theta K^2}{(1-\theta)^2\rho},$$

and finally

$$K \le \frac{\sqrt{c/\pi}}{\theta\big(\log\frac{1}{\theta}\big)^{1/2}} + \frac{a\theta K^2}{(1-\theta)^2}.$$

Set

$$\alpha(\theta) := \frac{a\theta}{(1-\theta)^2}, \quad \beta(\theta) := \frac{\sqrt{c/\pi}}{\theta\big(\log\frac{1}{\theta}\big)^{1/2}}.$$

Then we have

$$\alpha K^2 - K + \beta \ge 0,$$

or equivalently

$$\left(K - \frac{1}{2\alpha}\right)^2 \ge \frac{1 - 4\alpha\beta}{4\alpha^2}.$$

Note that

$$\alpha(\theta)\beta(\theta) = \frac{a\sqrt{c/\pi}}{(1-\theta)^2\left(\log\frac{1}{\theta}\right)^{1/2}}, \qquad c = c(a, M).$$

Hence there exists a number $\theta_0(a, M) \in (0, 1)$ such that

$$4\alpha(\theta)\beta(\theta) \le \frac{3}{4} \quad \text{if } 0 < \theta \le \theta_0,$$

that is,

$$\sqrt{1 - 4\alpha(\theta)\beta(\theta)} \ge \frac{1}{2} \quad \text{if } 0 < \theta \le \theta_0.$$

Set

$$m^-(\theta) := \frac{1 - \sqrt{1 - 4\alpha(\theta)\beta(\theta)}}{2\alpha(\theta)}, \qquad m^+(\theta) := \frac{1 + \sqrt{1 - 4\alpha(\theta)\beta(\theta)}}{2\alpha(\theta)}.$$

Then we infer for any $\theta \in (0, \theta_0]$ that either

$$\text{(i) } K \le m^-(\theta), \quad \text{or} \quad \text{(ii) } K \ge m^+(\theta)$$

holds true.

Moreover, the functions $m^-(\theta)$ and $m^+(\theta)$ are continuous on $(0, \theta_0]$ and satisfy

$$m^-(\theta) < m^+(\theta) \quad \text{for } 0 < \theta \le \theta_0$$

and

$$\lim_{\theta \to +0} m^+(\theta) = \infty.$$

The last relation yields that case (ii) cannot occur for θ close to zero; hence we have $K \le m^-(\theta)$ for θ near zero, and a continuity argument then implies

$$K \le m^-(\theta) \quad \text{for all } \theta \in (0, \theta_0],$$

in particular, $K \le m^-(\theta_0)$. Finally, for $\theta \in (0, \theta_0]$, we also obtain that

$$\begin{aligned} m^-(\theta) &= \frac{1 - \sqrt{1 - 4\alpha(\theta)\beta(\theta)}}{2\alpha(\theta)} = \frac{1}{2\alpha(\theta)} \frac{1 - (1 - 4\alpha(\theta)\beta(\theta))}{1 + \sqrt{1 - 4\alpha(\theta)\beta(\theta)}} \\ &= \frac{2\beta(\theta)}{1 + \sqrt{1 - 4\alpha(\theta)\beta(\theta)}} \le \frac{2\beta(\theta)}{1 + 1/2} = \frac{4}{3}\beta(\theta) \end{aligned}$$

whence

$$m^-(\theta_0) < \tfrac{4}{3}\beta(\theta_0).$$

Consequently,

$$K \le \frac{4}{3}\beta(\theta_0) = \frac{4\sqrt{c/\pi}}{3\theta_0\left(\log\frac{1}{\theta_0}\right)^{1/2}} := c^*(a, M).$$

Because of

$$R'|\nabla X(w_0)| = f(w_0) \le f(w_1) = K \le c^*(a, M)$$

we arrive at

$$|\nabla X(w_0)| \le c^*(a, M)/R' \quad \text{for any } R' \in (0, R)$$

whence

$$(7^*) \qquad |\nabla X(w_0)| \le c^*(a, M)/R.$$

This estimate is close to (7). We now introduce the function $\kappa(t) := c^*(t, 1)$. A close inspection of the previous computations shows that $\kappa(t)$ can assumed to be an increasing and continuous function on the interval $[0, 1)$.

In order to prove (7) we assume that $M > 0$ because that inequality trivially holds true if $M = 0$. Then $Z(w) := M^{-1}X(w)$ satisfies both $|Z(w)| \le 1$ and

$$|\Delta Z| \le aM|\nabla Z|^2.$$

Applying the estimate (7^*) to Z, we arrive at

$$|\nabla Z(w_0)| \le \kappa(aM)/R.$$

Multiplying this inequality by M, we obtain (7).

Estimate (8) is now an easy consequence of (7). To see this we introduce the quantity

$$m := \sup\{|X(w) - X(w_0)| \colon w \in B_R(w_0)\}.$$

If $m = 0$ or $m = \infty$, the estimate (8) is true for trivial reasons. If $M \le m < \infty$, (8) follows directly from (7). If $0 < m < M$, we introduce $Z := m^{-1}[X - X(w_0)]$ and obtain as before

$$|\nabla Z(w_0)| \le \kappa(am)/R \le \kappa(aM)/R,$$

and this implies (8). □

Corollary 1. *Suppose that $X \in C^2(B_R(w_0), \mathbb{R}^N)$ is a solution of (1) in $B_R(w_0)$ satisfying*

$$|X(w)| \le M \quad \textit{for } w \in B_R(w_0)$$

and for some constant M with $aM < 1$. Then we have

$$(9) \qquad |\nabla X(w)| \le \kappa(aM)\frac{M}{\rho} \quad \textit{for all } w \in B_{R-\rho}(w_0),\ 0 < \rho < R.$$

Proposition 2. *Let $X \in C^0(\overline{B}_R(w_0), \mathbb{R}^N) \cap C^2(B_R(w_0), \mathbb{R}^N)$ be a solution of the differential inequality* (1) *in $B_R(w_0)$, and suppose that $|X(w)| \le M$ holds for all $w \in \overline{B}_R(w_0)$ and for some number M with $2aM < 1$. Moreover, set $x(w) := |X(w)|^2$, and let $H \in C^0(\overline{B}_R(w_0), \mathbb{R}^N) \cap C^2(B_R(w_0), \mathbb{R}^N)$ and $h \in C^0(\overline{B}_R(w_0)) \cap C^2(B_R(w_0))$ be the solutions of the boundary value problems*

$$\Delta H = 0 \quad \text{in } B_R(w_0), \qquad H = X \quad \text{on } \partial B_R(w_0), \tag{10}$$

$$\Delta h = 0 \quad \text{in } B_R(w_0), \qquad h = x \quad \text{on } \partial B_R(w_0). \tag{11}$$

Then for any $w \in B_R(w_0)$ and $w^* \in \partial B_R(w_0)$ we have the inequality

$$\begin{aligned} |X(w) - X(w^*)| \le {} & \frac{a}{2(1-2aM)}|h(w) - h(w^*)| \\ & + \frac{1-aM}{1-2aM}|H(w) - H(w^*)|. \end{aligned} \tag{12}$$

Proof. Inequality (2) implies

$$|\nabla X|^2 \le \frac{1}{2(1-aM)}\Delta x,$$

which in conjunction with

$$|\Delta X| \le a|\nabla X|^2$$

yields

$$|\Delta X| \le \frac{a}{2(1-aM)}\Delta x.$$

Pick some constant vector $E \in \mathbb{R}^N$ with $|E| = 1$ and consider the auxiliary function $z \in C^0(\overline{B}_R(w_0)) \cap C^2(B_R(w_0))$ which is defined by

$$z(w) := \frac{a}{2(1-aM)}[x(w) - h(w)] + \langle H(w) - X(w), E\rangle$$

and vanishes on $\partial B_R(w_0)$. Because of

$$\begin{aligned} \Delta z &= \frac{a}{2(1-aM)}\Delta x - \langle \Delta X, E\rangle \\ &\ge \frac{a}{2(1-aM)}\Delta x - |\Delta X| \ge 0, \end{aligned}$$

we see that z is subharmonic on $B_R(w_0)$. Then the maximum principle yields

$$\langle H - X, E\rangle \le \frac{a}{2(1-aM)}[h - x] \quad \text{on } \overline{B}_R(w_0)$$

for any unit vector E of $\mathbb{R}^N$, and we conclude that

$$|H - X| \leq \frac{a}{2(1 - aM)}[h - x] \tag{13}$$

holds on $\overline{B}_R(w_0)$. Moreover, the inequality $|X(w)| \leq M$ in conjunction with the maximum principle gives

$$|H(w)| \leq M \quad \text{for all } w \in \overline{B}_R(w_0).$$

Then we obtain

$$\begin{aligned} h - x &= (|H|^2 - |X|^2) + (h - |H|^2) \\ &\leq 2M(|H| - |X|) + (h - |H|^2), \end{aligned}$$

whence

$$|H - X| \leq \frac{aM}{1 - aM}|H - X| + \frac{a}{2(1 - aM)}(h - |H|^2).$$

Since

$$0 < 1 - \frac{aM}{1 - aM} = \frac{1 - 2aM}{1 - aM} < 1,$$

it follows that

$$|H - X| \leq \frac{a}{2(1 - 2aM)}(h - |H|^2) \quad \text{on } \overline{B}_R(w_0). \tag{14}$$

For $w^* \in \partial B_R(w_0)$ we have

$$|X(w^*)|^2 = x(w^*) = |H(w^*)|^2 = h(w^*),$$

and therefore

$$\begin{aligned} |X(w) - X(w^*)| &\leq |X(w) - H(w)| + |H(w) - H(w^*)| \\ &\leq \frac{a}{2(1 - 2aM)}(h(w) - |H(w)|^2) + |H(w) - H(w^*)|. \end{aligned}$$

Because of

$$\begin{aligned} h(w) - |H(w)|^2 &= h(w) - h(w^*) + |H(w^*)|^2 - |H(w)|^2 \\ &\leq |h(w) - h(w^*)| + 2M|H(w) - H(w^*)|, \end{aligned}$$

we may now conclude that (12) holds for any $w \in \overline{B}_R(w_0)$ and for any $w^* \in \partial B_R(w_0)$. □

Remark 1. Note that the differential inequality (1) remains invariant with respect to conformal transformations of the parameter domain. Thus we can carry over Proposition 2 from $B_R(w_0)$ to any bounded domain Ω in $\mathbb{C}$ which is of the conformal type of the disk and has a closed Jordan curve as its boundary.

Proposition 3. *For any $a \ge 0, R > 0, M \ge 0$, and $k \ge 0$ with $2aM < 1$, there is a number $c = c(a, R, M, k) \ge 0$ having the following property:*

Let $X \in C^0(\overline{B}_R(w_0), \mathbb{R}^N) \cap C^2(B_R(w_0), \mathbb{R}^N)$ be a solution of (1) in $B_R(w_0)$ satisfying $|X(w)| \le M$ for all $w \in \overline{B}_R(w_0)$. Suppose also that the boundary values $\mathcal{X}(\varphi) := X(w_0 + Re^{i\varphi})$ are of class $C^2(\mathbb{R})$ and satisfy $|\mathcal{X}''(\varphi)| \le k$ for all $\varphi \in \mathbb{R}$. Then we have

$$(15) \qquad |\nabla X(w)| \le c(a, R, M, k) \quad \text{for all } w \in B_R(w_0).$$

Proof. It suffices to treat the case $w_0 = 0$ and $R = 1$, that is, we consider the parameter domain $B = B_1(0)$. Let $w = re^{i\theta}, 0 < r < 1$, be an arbitrary point of B. By formula (8) of Proposition 1 we have

$$(16) \qquad |\nabla X(w)| \le \frac{c(a, M)}{1-r} \sup\{|X(w') - X(w)| : w' \in B_{1-r}(w)\}.$$

Moreover, for $w, w' \in B$ and $w^* \in \partial B$ it follows from Proposition 2 that

$$(17) \qquad \begin{aligned} |X(w) - X(w')| &\le |X(w) - X(w^*)| + |X(w') - X(w^*)| \\ &\le \frac{a}{2(1-2aM)}\{|h(w) - h(w^*)| + |h(w') - h(w^*)|\} \\ &\quad + \frac{1-aM}{1-2aM}[|H(w) - H(w^*)| + |H(w') - H(w^*)|] \end{aligned}$$

holds true where H and h are harmonic in B and have the boundary values X and $x := |X|^2$ respectively on ∂B. By Lemma 5 of Section 2.1 we obtain that

$$|\nabla H(w)| \le ck \quad \text{for all } w \in B$$

whence

$$|H(w_1) - H(w_2)| \le ck|w_1 - w_2| \quad \text{for all } w_1, w_2 \in \overline{B}.$$

Therefore we have

$$(18) \qquad |H(w) - H(w^*)| + |H(w') - H(w^*)| \le 3ck(1-r)$$

for $w = re^{i\theta}, w^* = e^{i\theta}, w' \in B_{1-r}(w)$.

Furthermore, the boundary values $\eta(\varphi) := |\mathcal{X}(\varphi)|^2$ of $x(e^{i\varphi})$ satisfy $\eta'' = 2|\mathcal{X}'|^2 + 2\langle \mathcal{X}, \mathcal{X}'' \rangle$, hence

$$|\eta''| \le 2|\mathcal{X}'|^2 + 2|\langle \mathcal{X}, \mathcal{X}'' \rangle| \le 2|\mathcal{X}'|^2 + 2Mk.$$

Let $E \in \mathbb{R}^N$ be a constant unit vector. Then we have

$$\int_0^{2\pi} \langle E, \mathcal{X}'(\varphi) \rangle \, d\varphi = 0.$$

Consequently, there is some $\varphi_0 \in [0, 2\pi]$ such that

$$\langle E, \mathfrak{X}'(\varphi_0)\rangle = 0$$

and therefore

$$\langle E, \mathfrak{X}'(\varphi)\rangle = \int_{\varphi_0}^{\varphi} \langle E, \mathfrak{X}''(\varphi)\rangle \, d\varphi.$$

Hence we obtain

$$|\langle E, \mathfrak{X}'(\varphi)\rangle| \le 2\pi k \quad \text{for all } \varphi \in [0, 2\pi].$$

Since E can be chosen as an arbitrary vector of $\mathbb{R}^N$, we conclude that

$$|\mathfrak{X}'(\varphi)| \le 2\pi k \quad \text{for all } \varphi \in \mathbb{R},$$

and therefore

$$|\eta''| \le 8\pi^2 k^2 + 2Mk.$$

Then we infer from Lemma 5 of Section 2.1 that

$$|\nabla h(w)| \le c^*(1+k^2) \quad \text{for all } w \in B$$

whence

$$|h(w_1) - h(w_2)| \le c^*(1+k^2)|w_1 - w_2| \quad \text{for all } w_1, w_2 \in \overline{B}$$

and consequently

$$(19) \qquad |h(w) - h(w^*)| + |h(w') - h(w^*)| \le 3c^*(1+k^2)(1-r)$$

for $w = re^{i\theta}, w^* = e^{i\theta}, w' \in B_{1-r}(w)$.

Combining (17), (18), and (19), we arrive at

$$|X(w) - X(w')| \le c(a, M, k)(1-r) \quad \text{for } w = re^{i\theta}, 0 < r < 1, \text{ and } w' \in B_{1-r}(w),$$

and this implies

$$|\nabla X(w)| \le c(a, M, k) \quad \text{for all } w \in B,$$

taking (16) into account. □

Theorem 1. *Suppose that the assumptions of Proposition 3 are satisfied. Then X is of class $C^{1,\mu}(\overline{B}_R(w_0), \mathbb{R}^N)$ for all $\mu \in (0,1)$ and we have*

$$(20) \qquad [\nabla X]_{\mu, \overline{B}_R(w_0)} \le c(a, R, M, k, \mu).$$

Proof. This result is an immediate consequence of Proposition 2 of Section 2.1 in conjunction with Proposition 3 that we have just proved. □

Now we come to the proof of the most important result with regard to the next section.

Theorem 2. *For $w_0 \in \partial B$, we introduce the set $S_\rho(w_0) := B \cap B_\rho(w_0)$. Assume that, for some $\rho \in (0,1)$, $X \in C^0(\overline{S}_\rho(w_0), \mathbb{R}^N) \cap C^2(S_\rho(w_0), \mathbb{R}^N)$ is a solution of the differential inequality* (1) *in $S_\rho(w_0)$ that vanishes on $\partial S_\rho(w_0) \cap \partial B$. Then we obtain $X \in C^{1,\mu}(\overline{S}_{\rho'}(w_0), \mathbb{R}^N)$ for every $\mu \in (0,1)$ and every $\rho' \in (0, \rho)$.*

Proof. It suffices to show that for any $w^* = e^{i\theta} \in \partial S_\rho(w_0) \cap \partial B$ there is a $\delta > 0$ such that $X \in C^{1,\mu}(\overline{S}_\delta(w^*), \mathbb{R}^N)$, where $S_\delta(w^*)$ denotes the circular two-gon $B \cap B_\delta(w^*)$. We may also assume that $a > 0$.

Thus, having fixed an arbitrary $w^* = e^{i\theta} \in \partial B \cap \partial S_\rho(w_0)$, we first choose an $\varepsilon > 0$ such that $S_{3\varepsilon}(w^*) \subset S_\rho(w_0)$ holds and that

$$\sup\{|X(w)| : w \in S_{3\varepsilon}(w^*)\} \le \frac{1}{4a}.$$

Then the mapping

$$Z(w) := 4aX, \quad w \in S_{3\varepsilon}(w^*),$$

satisfies the inequalities

$$|Z| \le 1 \quad \text{and} \quad |\Delta Z| \le \tfrac{1}{4}|\nabla Z|^2$$

in $S_{3\varepsilon}(w^*)$.

We now consider the functions $H(w)$ and $h(w)$ which are harmonic in $S_{3\varepsilon}(w^*)$ and which have the boundary values X and $|X|^2$ respectively on $\partial S_{3\varepsilon}$. As h and H vanish on the circular arc

$$C := \partial B \cap \partial S_{3\varepsilon}(w^*)$$

we can extend h and H to harmonic functions in $B_{3\varepsilon}(w^*)$ by reflection at C, applying Schwarz's reflection principle. Hence there is a number $c(\varepsilon)$ such that

$$|\nabla H(w)| + |\nabla h(w)| \le c(\varepsilon) \quad \text{for all } w \in B_{2\varepsilon}(w^*)$$

whence

$$(21) \qquad \begin{aligned} &|H(w_1) - H(w_2)| + |h(w_1) - h(w_2)| \le c(\varepsilon)|w_1 - w_2| \\ &\quad \text{for all } w_1, w_2 \in B_{2\varepsilon}(w^*). \end{aligned}$$

Fix some $w = re^{i\varphi} \in S_\varepsilon(w^*)$. Then we have $|w| = r > 1 - \varepsilon$ and, for $|w - w'| < 1 - r$, we have $|w' - w^*| \le |w' - w| + |w - w^*| < \varepsilon + \varepsilon = 2\varepsilon$, and consequently

$$B_{1-r}(w) \subset S_{2\varepsilon}(w^*) \subset S_{3\varepsilon}(w^*) \subset S_\rho(w_0).$$

By Proposition 2 and the subsequent Remark 1 we obtain for $\Omega := S_{3\varepsilon}(w^*)$, $w' \in \Omega$ and $e^{i\varphi} \in \partial B \cap \partial S_{3\varepsilon}(w^*)$ that

$$|Z(w')| = |Z(w') - Z(e^{i\varphi})| \le \tfrac{1}{4}|h(w') - h(e^{i\varphi})| + \tfrac{3}{2}|H(w') - H(e^{i\varphi})|.$$

In connection with (21) we infer for any $w' \in B_{1-r}(w)$ that

$$|Z(w')| \le \tfrac{7}{4}c(\varepsilon)|w' - e^{i\varphi}| \le \tfrac{7}{2}c(\varepsilon)(1-r) < 4c(\varepsilon)(1-r).$$

In other words, we have

$$\sup\{|Z(w')|\colon w' \in B_{1-r}(w)\} \le 4c(\varepsilon)(1-r)$$

for any $w \in S_\varepsilon(w^*)$ with $|w| = r$.

Moreover, we infer from Proposition 1, (8) that

$$|\nabla Z(w)| \le \frac{2\kappa(1/4)}{1-r}\sup\{|Z(w')|\colon w' \in B_{1-r}(w)\}$$

for any $w \in S_\varepsilon(w^*)$ with $|w| = r$.

This implies

$$|\nabla Z(w)| \le c^*(\varepsilon) \quad \text{for all } w \in S_\varepsilon(w^*).$$

Since $X = \frac{1}{4a}Z$, we conclude that

$$|\Delta X|_{0,S_\varepsilon(w^*)} \le \text{const}, \quad \text{and} \quad |\nabla X|_{0,S_\varepsilon(w^*)} \le \text{const}.$$

Now we choose a cut-off function $\eta \in C_c^\infty(\mathbb{R}^2)$ with $\eta(w) = 1$ for $w \in B_\delta(w^*), \delta := \frac{\varepsilon}{2}$, and with $\eta(w) = 0$ for $|w - w^*| \ge \frac{3}{4}\varepsilon$. Then the mapping $Y := \eta X$ on $S_\varepsilon(w^*)$ satisfies

$$\Delta Y = \eta\Delta X + 2\nabla\eta\cdot\nabla X + \Delta\eta X$$

and therefore

(22) $\quad |\Delta Y(w)| \le \text{const} \quad \text{for all } w \in S_\varepsilon(w^*),$

(23) $\quad Y(w) = 0 \quad \text{on } \partial S_\varepsilon(w^*),$

(24) $\quad Y(w) = 0 \quad \text{for all } w \in S_\varepsilon(w^*) \text{ with } \frac{3}{4}\varepsilon < |w - w^*| < \varepsilon.$

Consider a conformal mapping τ of the unit disk B onto the two-gon $S_\varepsilon(w^*)$. We can extend τ to a homeomorphism of $\overline{B}$ onto $\overline{S}_\varepsilon(w^*)$, and it can be assumed that $\zeta = \pm 1$ are mapped onto the two vertices of the two-gon. By the reflection principle the mapping $\tau(\zeta)$ is holomorphic on $\overline{B} \setminus \{-1, 1\}$. Then it follows from (22) and (24) that the mapping $Y^*(\zeta) := Y(\tau(\zeta))$ is of class $C^0(\overline{B}, \mathbb{R}^N) \cap C^2(B, \mathbb{R}^N)$ and satisfies

$$|\Delta Y^*(\zeta)| \le \text{const} \quad \text{for all } \zeta \in B.$$

Moreover, we infer from (23) that

$$Y^*(\zeta) = 0 \quad \text{for all } \zeta \in \partial B.$$

Thus we can apply Proposition 2 of Section 2.1 and obtain that $Y^* \in C^{1,\mu}(\overline{B}, \mathbb{R}^N)$ for any $\mu \in (0,1)$. It follows that $Y \in C^{1,\mu}(\overline{S}_\varepsilon(w^*), \mathbb{R}^N)$, and therefore $X \in C^{1,\mu}(\overline{S}_\delta(w^*), \mathbb{R}^N)$ as $Y(w) = X(w)$ for all $w \in \overline{S}_\delta(w^*)$, $\delta = \frac{\varepsilon}{2}$. □

2.3 The Boundary Regularity of Minimal Surfaces Bounded by Jordan Arcs

In this section we want to investigate the boundary behaviour of minimal surfaces at smooth Jordan arcs. The results can be applied to minimal surfaces bounded by one or several Jordan curves, to solutions of the partially free boundary problem, and to solutions of the thread problem (see Chapters 1 and 5 as well as Vol. 3, Chapters 1, 2).

As all results are of local nature, it suffices to formulate them on simply connected boundary domains, say, for minimal surfaces $X\colon B \to \mathbb{R}^3$ defined on the unit disk $B = \{w \in \mathbb{C}\colon |w| < 1\}$. The same results can be carried over without any problem to minimal surfaces $X\colon B \to \mathbb{R}^N, N \geq 2$. At the end of this section we shall sketch analogous results for minimal surfaces $X\colon B \to \mathcal{M}$ in an n-dimensional Riemannian manifold $\mathcal{M}$.

The main theorem of this section is the following result.

Theorem 1. *Consider a minimal surface $X\colon B \to \mathbb{R}^3$ of class $C^0(B \cup \gamma, \mathbb{R}^3) \cap C^2(B, \mathbb{R}^3)$ which maps an open subarc γ of ∂B into an open Jordan arc Γ of $\mathbb{R}^3$ which is a regular curve of class $C^{m,\mu}$ for some integer $m \geq 1$ and some $\mu \in (0,1)$. Then X is of class $C^{m,\mu}(B \cup \gamma, \mathbb{R}^3)$. Moreover, if Γ is a regular real analytic Jordan arc, then X can be extended as a minimal surface across γ.*

In fact, we shall only prove a slightly weaker result. We want to show that the statement of the theorem holds under the assumption $\Gamma \in C^{m,\mu}$ with $m \geq 2$ and $0 < \mu < 1$. It remains to verify that the assumption $\Gamma \in C^{1,\mu}$ implies $X \in C^{1,\mu}(B \cup \gamma, \mathbb{R}^3)$. This can be carried out by employing a reflection method combined with refined potential-theoretic estimates. A version of this reasoning was invented by W. Jäger [3]. Other methods to prove this initial step can be found in Nitsche [16,20] and [28] (see Kapitel V, 2.1), Kinderlehrer [1], and Warschawski [5].

It will turn out that the method to be described also covers the boundary behaviour of surfaces of prescribed mean curvature at a smooth arc. Thus we shall deal with this more general result.

Theorem 2. *Let $X \in C^0(B \cup \gamma, \mathbb{R}^3) \cap C^2(B, \mathbb{R}^3)$ be a solution of the equations*

$$\Delta X = 2H(X) X_u \wedge X_v, \tag{1}$$

$$|X_u|^2 = |X_v|^2, \quad \langle X_u, X_v \rangle = 0 \tag{2}$$

in B which maps an open subarc $\gamma = \{e^{i\theta}\colon \theta_1 < \theta < \theta_2\}$ of ∂B into some open regular Jordan arc Γ of $\mathbb{R}^3$, i.e. $X(w) \in \Gamma$ for all $w \in \gamma$. Then the following holds:

(i) *If $\mathcal{H}(w) := H(X(w))$ is of class $L_\infty(B)$, and if $\Gamma \in C^2$, then we obtain that $X \in C^{1,\mu}(B \cup \gamma, \mathbb{R}^3)$ for any $\mu \in (0,1)$.*

(ii) *If H is of class $C^{0,\mu}$ on $\mathbb{R}^3$, and if $\Gamma \in C^{2,\mu}, 0 < \mu < 1$, then $X(w)$ is of class $C^{2,\mu}(B \cup \gamma, \mathbb{R}^3)$.*

Proof. (i) It suffices to show that for any $w_0 \in \gamma$ there is some $\delta > 0$ such that $X \in C^{1,\mu}(\overline{S}_\delta(w_0), \mathbb{R}^3), 0 < \mu < 1$, provided that $\mathcal{H}(w) := H(X(w))$ is of class $L_\infty(B)$ and that $\Gamma \in C^2$. Here $S_\delta(w_0)$ denotes as usual the two-gon $B \cap B_\delta(w_0)$.

Thus we fix some $w_0 \in \gamma$. Without loss of generality we may assume that $X(w_0) = 0$. For sufficiently small $\rho > 0$ we can represent $\Gamma \cap \mathcal{K}_\rho(0)$ in the form

$$(3) \qquad x^1 = g^1(t), \quad x^2 = g^2(t), \quad x^3 = t, \quad |t| < 2t_0,$$

where the functions $g^1(t)$ and $g^2(t)$ are of class C^2, and by a suitable motion in $\mathbb{R}^3$ we can arrange that

$$(4) \qquad g^k(0) = 0, \quad \dot{g}^k(0) = 0, \quad k = 1, 2,$$

choosing the parameter t appropriately.

We may also assume that $w_0 = 1$ and that $\gamma = \{w \in \partial B\colon |w - 1| < R_0\}$ for some $R_0 \in (0, 1)$. Choosing $t_0 > 0$ and $R \in (0, R_0]$ sufficiently small we can achieve that

$$(5) \qquad |\dot{g}^1(t)|^2 + |\dot{g}^2(t)|^2 \le \tfrac{1}{8} \quad \text{for } |t| < t_0$$

and

$$(6) \qquad |x^3(w)| < t_0 \quad \text{for } w \in \overline{S}_R(1).$$

Consider the auxiliary function $Y(w) = (y^1(w), y^2(w))$ which is defined by

$$(7) \qquad y^k(w) := x^k(w) - g^k(x^3(w)), \quad k = 1, 2, \; w \in \overline{S}_R(1),$$

where $X(w) = (x^1(w), x^2(w), x^3(w))$. Clearly, we have $Y \in C^0(\overline{S}_R(1), \mathbb{R}^2) \cap C^2(S_R(1), \mathbb{R}^2)$ and $Y(w) = 0$ for $w \in \partial B \cap \partial S_R(1)$. Moreover, we infer from (1) and from the relations

$$(8) \qquad \Delta y^k = \Delta x^k - \dot{g}^k(x^3)\Delta x^3 - \ddot{g}^k(x^3)|\nabla x^3|^2, \quad k = 1, 2,$$

that

$$(8^*) \qquad |\Delta Y| \le \alpha |\nabla X|^2$$

holds for some constant $\alpha > 0$.

In addition, we have

$$(9) \qquad x^k_w = y^k_w + \dot{g}^k(x^3)x^3_w, \quad k = 1, 2,$$

and therefore

$$(10) \qquad \begin{aligned} |x^1_w|^2 + |x^2_w|^2 &\le 2|y^1_w|^2 + 2|y^2_w|^2 + 2|x^3_w|^2 \sum_{k=1}^{2} |\dot{g}^k(x^3)|^2 \\ &\le 2|y^1_w|^2 + 2|y^2_w|^2 + \tfrac{1}{4}|x^3_w|^2 \end{aligned}$$

since (5) and (6) imply $\sum_{k=1}^2 |\dot{g}^k(x^3)|^2 \le \frac{1}{8}$.

Now we write the conformality relations (2) as

$$0 = \langle X_w, X_w \rangle = (x_w^1)^2 + (x_w^2)^2 + (x_w^3)^2 \tag{11}$$

whence

$$|x_w^3|^2 \le |x_w^1|^2 + |x_w^2|^2$$

and therefore

$$\tfrac{1}{2}|X_w|^2 \le |x_w^1|^2 + |x_w^2|^2. \tag{12}$$

From (10) and (11) we infer

$$\tfrac{1}{4}|X_w|^2 \le 2|Y_w|^2$$

whence

$$|\nabla X|^2 \le 8|\nabla Y|^2. \tag{13}$$

From (8*) and (12) we derive the differential inequality

$$|\Delta Y| \le 8\alpha|\nabla Y|^2 \quad \text{on } S_R(1), \tag{14}$$

and we know already that

$$Y = 0 \quad \text{on } \partial B \cap \partial S_R(1). \tag{15}$$

Thus we can apply Theorem 2 of Section 2.2 to $Y \colon \overline{S}_R(1) \to \mathbb{R}^2$, and we obtain $Y \in C^{1,\mu}(\overline{S}_\varepsilon(1), \mathbb{R}^2)$ for any $\varepsilon \in (0, R)$ and any $\mu \in (0, 1)$.

Combining (9) and (11) it follows that

$$0 = \sum_{k=1}^2 (y_w^k)^2 + 2\sum_{k=1}^2 \dot{g}^k(x^3) y_w^k x_w^3 + \left\{1 + \sum_{k=1}^2 |\dot{g}^k(x^3)|^2\right\}(x_w^3)^2. \tag{16}$$

If we introduce

$$p^k(t) := \frac{\dot{g}^k(t)}{q(t)}, \quad q(t) := 1 + \sum_{k=1}^2 |\dot{g}^k(t)|^2, \tag{17}$$

this relation can be rewritten as

$$\left[x_w^3 + \sum_{k=1}^2 p^k(x^3) y_w^k\right]^2 = \left\{\sum_{k=1}^2 p^k(x^3) y_w^k\right\}^2 - \frac{1}{q(x^3)} \sum_{k=1}^2 (y_w^k)^2. \tag{18}$$

As the right-hand side of (18) is continuous in $\overline{S}_\varepsilon(1)$, it follows that [...] and therefore also x_w^3 are continuous. Thus we arrive at $X \in C^1(\overline{S}_\varepsilon(1), \mathbb{R}^3)$ for any $\varepsilon \in (0, R)$.

Multiplying (18) by $-w^2$, we obtain

$$(19)\quad \left[iwx_w^3 + \sum_{k=1}^{2} p^k(x^3)iwy_w^k\right]^2 = \frac{1}{q(x^3)}\sum_{k=1}^{2}(wy_w^k)^2 - \left\{\sum_{k=1}^{2} p^k(x^3)wy_w^k\right\}^2 .$$

Introducing polar coordinates r, φ with $w = re^{i\varphi}$, we find that

$$\begin{array}{cc} wx_w^l = \frac{1}{2}(rx_r^l - ix_\varphi^l), & wy_w^k = \frac{1}{2}(ry_r^k - iy_\varphi^k) \\ (l = 1, 2, 3) & (k = 1, 2). \end{array}$$

For $w \in \gamma' := \partial B \cap \partial S_\varepsilon(1)$, we infer from (15) that the right-hand side of (19) is equal to

$$\frac{1}{4}|q(x^3)|^{-2}\left\{\sum_{k=1}^{2} q(x^3)|y_r^k|^2 - \left(\sum_{k=1}^{2}\dot{g}^k(x^3)y_r^k\right)^2\right\}$$

and this expression is real and nonnegative on account of (5) and (6). The left-hand side of (19) is of the form

$$(a + ib)^2 = (a^2 - b^2) + 2iab$$

with

$$a := \frac{r}{2}\left(x_\varphi^3 + \sum_{k=1}^{2} p^k(x^3)y_\varphi^k\right), \quad b := \frac{1}{2}\left(x_r^3 + \sum_{k=1}^{2} p^k(x^3)y_r^k\right).$$

From the relations

$$a^2 - b^2 \ge 0 \quad \text{and} \quad ab = 0$$

we infer that $b = 0$, that is,

$$(20)\qquad x_r^3 + \sum_{k=1}^{2} p^k(x^3)y_r^k = 0 \quad \text{on } \gamma'.$$

Thus we have found:

$$(21)\qquad |\Delta x^3| + |\nabla x^3| \le \text{const} \quad \text{on } S_\varepsilon(1), \qquad \frac{\partial}{\partial r}x^3 \in C^{0,\mu}(\gamma').$$

Now we choose a cut-off function $\eta \in C_c^\infty$ which is rotationally symmetric with respect to the pole $w = 1$ and satisfies $\eta(w) = 1$ for $w \in B_\delta(1), \delta := \frac{\varepsilon}{2}, \eta(w) = 0$ for $|w - 1| \ge \frac{3}{4}\varepsilon$. Set

$$(22)\qquad y(w) := \eta(w)x^3(w), \quad w \in \overline{S}_\varepsilon(1).$$

We have $y_r = \eta x_r^3 + \eta_r x^3$, and therefore

$$y_r \in C^{0,\mu}(\gamma'). \tag{23}$$

From the identity

$$\Delta y = \eta \Delta x^3 + x^3 \Delta \eta + 2\nabla\eta \cdot \nabla x^3$$

and from (21) we infer that

$$|\Delta y| \le \text{const} \quad \text{on } S_\varepsilon(1). \tag{24}$$

Finally we have

$$y(w) = 0 \quad \text{for all } w \in S_\varepsilon(1) \text{ with } \tfrac{3}{4}\varepsilon < |w-1| < \varepsilon. \tag{25}$$

Consider a conformal mapping τ of the unit disk B onto the two-gon $S_\varepsilon(1)$. We can extend τ to a homeomorphism of $\overline{B}$ onto $\overline{S}_\varepsilon(1)$, and it can be assumed that $\zeta = \pm 1$ are mapped onto the two vertices of the two-gon. By the reflection principle the mapping $\tau(\zeta)$ is holomorphic on $\overline{B} \setminus \{-1, 1\}$. Then it follows from (23) and (25) that the function $y^*(\zeta) := y(\tau(\zeta)), \zeta \in \overline{B}$, is of class $C^1(\overline{B}) \cap C^2(B)$ and satisfies

$$|\Delta y^*| \le \text{const} \quad \text{on } B,$$

$$\frac{\partial y^*}{\partial r} \in C^{0,\mu}(\partial B);$$

here the radial derivative y^*_r is the normal derivative of y^* on ∂B.

According to Section 2.1, Proposition 2, the solution $p(\zeta)$ of the boundary value problem

$$\Delta p = \Delta y^* \quad \text{in } B, \quad p = 0 \quad \text{on } \partial B$$

is of class $C^{1,\mu}(\overline{B})$ for any $\mu \in (0,1)$. Hence $h := y^* - p$ is of class $C^1(\overline{B}) \cap C^2(B)$, harmonic in B, and h_r is of class $C^{0,\mu}$ on ∂B. Therefore the conjugate harmonic function h^* with respect to h is of class $C^1(\overline{B})$ too, and the equation $h_r = h^*_\varphi$ on ∂B implies that $h^*|_{\partial B}$ is of class $C^{1,\mu}$. Applying the Korn–Privalov theorem (see Section 2.1, Lemmata 6 and 7) we infer that $h \in C^{1,\mu}(\overline{B})$, and therefore also $y^* \in C^{1,\mu}(\overline{B})$. Returning to $y = y^* \circ \tau^{-1}$ it follows that $y \in C^{1,\mu}(\overline{S}_\varepsilon(1))$. Since $y(w) = x^3(w)$ holds true for $w \in \overline{S}_\delta(1)$, $\delta = \frac{\varepsilon}{2}$, we finally arrive at $x^3 \in C^{1,\mu}(\overline{S}_\delta(1))$, and therefore $X \in C^{1,\mu}(\overline{S}_\delta(1), \mathbb{R}^3)$ for any $\mu \in (0,1)$. This concludes the proof of the first part of the theorem.

(ii) The initial step (i) is the crucial part of our investigation whereas the further proof is essentially potential-theoretic routine. However, as our estimates in Section 2.1 are not quite complete, we only want to indicate how one can proceed. The reader should use the Schauder estimates (described for example in Gilbarg and Trudinger [1]) to derive higher regularity by bootstrapping.

Thus let us assume that $H \in C^{0,\mu}(\mathbb{R}^3)$ and that $\Gamma \in C^{2,\mu}$ for some $\mu \in (0,1)$. Then $\mathcal{H}(w) := H(X(w))$ is of class $C^{0,\mu}(B \cup \gamma)$ and, using the notation of (i), the functions $g^1(t)$ and $g^2(t)$, $|t| < 2t_0$, are of class $C^{2,\mu}$. On account of (8) and (2) the mapping Y satisfies

$$\begin{aligned} \Delta Y &= Q \quad \text{in } S_R(1), \\ Y &= 0 \quad \text{on } \partial B \cap \partial S_R(1) \end{aligned} \tag{26}$$

with $Q \in C^{0,\mu}(\overline{S}_R(1), \mathbb{R}^2)$.

Then a potential-theoretic reasoning yields $Y \in C^{2,\mu}(\overline{S}_\varepsilon(1), \mathbb{R}^2)$ for $0 < \varepsilon < R$. Now we use the equations (cf. (1) and (20))

$$\begin{aligned} \Delta x^3 &= 2H(X)(x_u^1 x_v^2 - x_v^1 x_u^2) \quad \text{in } S_R(1), \\ x_r^3 &= -\sum_{k=1}^{2} p^k(x^3) y_r^k \quad \text{on } \partial B \cap \partial S_R(1) \end{aligned} \tag{27}$$

to prove by a potential-theoretic argument that $x^3 \in C^{2,\mu}(\overline{S}_\varepsilon(1))$ for $0 < \varepsilon < R$. We only have to note that $p^1(x^3)y_r^1 + p^2(x^3)y_r^2$ is of class $C^{1,\mu}(\gamma')$ for $\gamma' = \partial B \cap \partial S_R(1)$ because of the result for Y that we obtained before. By virtue of (7) we then infer $X \in C^{2,\mu}(\overline{S}_\varepsilon(1), \mathbb{R}^3)$, and therefore $X \in C^{2,\mu}(B \cup \gamma, \mathbb{R}^3)$. □

Remark 1. Similarly one proves $X \in C^{m,\mu}(B \cup \gamma, \mathbb{R}^3)$ as claimed in Theorem 1 if $\Gamma \in C^{m,\mu}$ and $H \in C^{m-2,\mu}(\mathbb{R}^3)$. The proof is carried out by a bootstrap reasoning, considering the boundary value problems alternatingly. Since a similar idea is developed in detail in the following sections, we want to omit the proof of higher boundary regularity of X except for proving analyticity in the case that Γ is a real analytic, regular arc. This will be done next for a minimal surface. We shall present *H. Lewy's regularity theorem.*

In the following we shall suppose B to be the semidisk $\{w = u + iv \colon |w| < 1, v > 0\}$, and I will denote the straight segment $\{u \in \mathbb{R} \colon |u| < 1\}$ on the boundary of B.

Theorem 3. *Let $X \in C^0(B \cup I, \mathbb{R}^3) \cap C^2(B, \mathbb{R}^3)$ be a minimal surface which maps I into a real-analytic and regular Jordan arc Γ in $\mathbb{R}^3$. Then X can be extended analytically across I as a minimal surface.*

Proof. Let $X^*(w)$ be an adjoint minimal surface to $X(w) = (x^1(w), x^2(w), x^3(w))$ which is assumed to satisfy $\Delta X = 0$ and (2) in B, and let

$$f(w) = X(w) + iX^*(w) = (f^1(w), f^2(w), f^3(w)) \tag{28}$$

be the holomorphic curve in $\mathbb{C}^3$ with $X = \operatorname{Re} f$ and $X^* = \operatorname{Im} f$ satisfying

$$\langle f'(w), f'(w) \rangle = 0. \tag{29}$$

By Theorem 2 we know already that $X \in C^2(B \cup I, \mathbb{R}^3)$ holds true. We have to show that for any $u_0 \in I$ there is some $\delta > 0$ such that $f(w)$ can be extended as a holomorphic curve from $S_\delta(u_0) := B \cap B_\delta(u_0)$ to $B_\delta(u_0)$. Without loss of generality we can assume that $u_0 = 0$. Set $B_\delta := B_\delta(0)$ and $S_\delta = S_\delta(0) = B \cap B_\delta$. As in the proof of Theorem 2 we can arrange for the following:

$$x^1(0) = x^2(0) = x^3(0) = 0, \quad \text{i.e.} \quad X(0) = 0.$$

For sufficiently small $\rho > 0$, we can represent $\Gamma \cap \mathcal{K}_\rho(0)$ in the form

$$x^1 = g^1(t), \quad x^2 = g^2(t), \quad x^3 = t, \quad t \in I_{2R_0} \tag{30}$$

where $I_\delta := \{t \in \mathbb{R} \colon |t| < \delta\}$, and $g^1(t)$ and $g^2(t)$ are real analytic functions on I_{2R_0}, $R_0 > 0$. Hence, choosing R_0 sufficiently small, we can assume that $g^1(\zeta)$ and $g^2(\zeta)$ are holomorphic functions of $\zeta \in B_{2R_0}$; hence

$$g(\zeta) := (g^1(\zeta), g^2(\zeta), \zeta), \quad |\zeta| < 2R_0$$

is a holomorphic curve on B_{2R_0}. In addition, we may (in accordance with $X(0) = 0$) assume that

$$g^1(0) = g^2(0) = \dot{g}^1(0) = \dot{g}^2(0) = 0$$

and

$$\left|\frac{dg^1}{d\zeta}(\zeta)\right|^2 + \left|\frac{dg^2}{d\zeta}(\zeta)\right|^2 \le \frac{1}{2} \quad \text{for } |\zeta| < 2R_0$$

are satisfied, and that

$$|f^3(w)| < R_0 \quad \text{holds for } w \in \overline{S}_R,$$

where R is a sufficiently small positive number.

Consider now the holomorphic function

$$F(w, \zeta) := \frac{\langle g'(\zeta), f'(w)\rangle}{\langle g'(\zeta), g'(\zeta)\rangle} \tag{31}$$

of $(w, \zeta) \in S_R \times B_{R_0}$, and note that F is of class C^1 on $\overline{S}_R \times \overline{B}_{R_0}$ (of course, differentiability in the second statement is real differentiability).

We claim that the differential equation

$$x^3_u(u) = F(u, x^3(u)) \quad \text{for } u \in I_R \tag{32}$$

holds true. In fact, the boundary condition $X(I) \subset \Gamma$ together with the above normalization implies that

$$X(u) = g(x^3(u)) \quad \text{for } u \in I_R, \tag{33}$$

hence

$$X_u(u) = g'(x^3(u))x^3_u(u) \quad \text{for } u \in I_R. \tag{34}$$

Therefore $x_u^3(u) = 0$ for some $u \in I_R$ yields $X_u(u) = 0$, and (2) gives $X_v(u) = 0$ or $X_u^*(u) = 0$ and consequently $f'(u) = 0$; therefore we also have $F(u, x^3(u)) = 0$, and (32) is trivially satisfied. Thus we may now assume that $x_u^3(u) \neq 0$. Because of (34) and (2), we obtain on I_R:

$$\begin{aligned} x_u^3 \langle g'(x^3), f' \rangle &= \langle g'(x^3) x_u^3, f' \rangle = \langle X_u, f' \rangle \\ &= \langle X_u, X_u + iX_u^* \rangle = |X_u|^2 - i\langle X_u, X_v \rangle = |X_u|^2 \\ &= \langle g'(x^3), g'(x^3) \rangle (x_u^3)^2 \end{aligned}$$

whence

$$x_u^3(u) = \frac{\langle g'(x^3), f' \rangle}{\langle g'(x^3), g'(x^3) \rangle}(u) = F(u, x^3(u)),$$

and (32) is verified. Thus $\zeta(u) = x^3(u)$ is a solution of the integral equation

$$\zeta(u) = \int_0^u F(\underline{u}, \zeta(\underline{u}))\, d\underline{u}. \tag{35}$$

It can easily be shown that there is some constant $M > 0$ such that

$$|F(w, \zeta) - F(w, \zeta')| \le M |\zeta - \zeta'|$$

holds for all $w \in \overline{S}_R$ and $\zeta, \zeta' \in \overline{B}_{R_0}$. Then it follows from a standard fixed point argument that there is a number $\delta \in (0, R)$ such that the integral equation

$$z(w) = \int_0^w F(\underline{w}, z(\underline{w}))\, d\underline{w}, \quad w \in \overline{S}_\delta, \tag{36}$$

has exactly one solution $z(w), w \in \overline{S}_\delta$, in the Banach space $\mathcal{A}(\overline{S}_\delta)$ of functions $z\colon \overline{S}_\delta \to \mathbb{C}$ which are holomorphic in S_δ and continuous on $\overline{S}_\delta$. (As usual, the proof of this fact can easily be carried out by Picard's iteration method.[3]) Similarly one sees that the real integral equation (35) has (for $u \in \overline{I}_\delta$) exactly one solution $\zeta(u), u \in \overline{I}_\delta$, whence we obtain $\zeta(u) = x^3(u)$ and $\zeta(u) = z(u)$ for $|u| \le \delta$, that is,

$$z(u) = x^3(u) \quad \text{for } u \in \overline{I}_\delta. \tag{37}$$

Consequently $z(w)$ is real-valued on I_δ, and by Schwarz's reflection principle we can extend $z(w)$ to a holomorphic function B_δ.

Now we consider the mapping $\phi\colon \overline{S}_\delta \to \mathbb{C}^3$, defined by

$$\phi(w) := f(w) - g(z(w)), \tag{38}$$

which is continuous on $\overline{S}_\delta$, holomorphic in S_δ, and purely imaginary on I_δ, since we have

[3] The integral in (36) is a complex line integral independent of the path from 0 to w within $\overline{S}_\delta$.

$$(39) \qquad \phi(u) = f(u) - g(z(u)) = X(u) + iX^*(u) - X(u) = iX^*(u)$$

on account of (33). Applying the reflection principle once again, we can extend $\phi(w)$ to a holomorphic function on B_δ, and therefore also

$$(40) \qquad f(w) = \phi(w) + g(z(w))$$

is extended to a holomorphic mapping on B_δ. □

We conclude this section by sketching the proof of a generalization of Theorem 1, employing the method of the proof of Theorem 2.

Theorem 4. *Let $\mathcal{M}$ be a Riemannian manifold of class C^2, and let Γ be an open regular Jordan arc in $\mathcal{M}$ which is of class C^2. Moreover let $X \in C^2(B, \mathcal{M}), B = \{w \in \mathbb{C}\colon |w| < 1\}$ be a minimal surface in $\mathcal{M}$. Finally we assume that γ is an open subarc of ∂B such that $X \in C^0(B \cup \gamma, \mathcal{M})$ and that $X(\gamma) \subset \Gamma$. Then we have:*

(i) *$X \in C^{1,\mu}(B \cup \gamma, \mathcal{M})$ for any $\mu \in (0, 1)$.*

(ii) *If $\mathcal{M}$ and Γ are of class $C^{m,\mu}, m \geq 2, 0 < \mu < 1$, then $X \in C^{m,\mu}(B \cup \gamma, \mathcal{M})$.*

(iii) *If $\mathcal{M}$ and Γ are real analytic, then X is real analytic in $B \cup \gamma$ and can be extended as a minimal surface across γ.*

Proof. We shall sketch a proof of (i). The results of (ii) can be derived from (i) by employing a bootstrap reasoning together with potential-theoretic estimates, as described in the proof of Theorem 2 and in Remark 1. The proof of (iii) now follows from a general theorem by Morrey [8] (cf. Theorem 6.8.2, pp. 278–279). We refer the reader to Hildebrandt [3], p. 80, for an indication how Morrey's result can be used to prove (iii). Another proof (in the spirit of H. Lewy) can be obtained by the method of F. Müller [1–3].

Let us now turn to step (i). We fix some point $w_0 \in \gamma$. Then there is some $R > 0$ such that X maps $\overline{S}_R(w_0) := \overline{B \cap B_R(w_0)}$ into some coordinate patch on the manifold $\mathcal{M}$ since X is continuous on $B \cup \gamma$. Introducing local coordinates $(x^1, x^2, \ldots, x^n)$ on this patch, we can represent X in the form

$$X(w) = (x^1(w), x^2(w), \ldots, x^n(w)) \quad \text{for } w \in S_R(w_0)$$

with $X \in C^0(\overline{S}_R(w_0) \cup \gamma, \mathbb{R}^n) \cap C^1(S_R(w_0), \mathbb{R}^n)$.

Suppose that the line element ds of $\mathcal{M}$ on the patch is given by

$$(41) \qquad ds^2 = g_{kl}(x)\, dx^k\, dx^l,$$

where repeated Latin indices are to be summed from 1 to n, and let

$$(42) \qquad \Gamma^l_{jk} = \tfrac{1}{2} g^{rl}(g_{jr,k} + g_{rk,j} - g_{jk,r})$$

be the Christoffel symbols corresponding to g_{kl}, where $(g^{rl}) = (g_{jk})^{-1}$. Then we have the equations

$$\Delta x^l + \Gamma^l_{jk}(X)\{x^j_u x^k_u + x^j_v x^k_v\} = 0, \quad 1 \le l \le n, \tag{43}$$

and

$$g_{kl}(X) x^k_u x^l_u = g_{kl}(X) x^k_v x^l_v, \quad g_{kl}(X) x^k_u x^l_v = 0. \tag{44}$$

(Equations (43) replace the equations $\Delta x^l = 0$ holding in the Euclidean case, and equations (44) are the Riemannian substitute of the conformality relations (2).)

Without loss of generality we may assume that $w_0 = 1$, and we set $S_R := S_R(1), 0 < R < 1$, and $\gamma' = \partial B \cap \partial S_R$. We can also assume that the coordinate patch containing $X(\overline{S}_R)$ is described by $\{x \in \mathbb{R}^n : |x| < 1\}$ and that $X(1) = 0$. Furthermore, we can assume that Γ in $\{|x| < 1\}$ is described by $x^1 = x^2 = \cdots = x^{n-1} = 0$, and that $g_{kl} \in C^1, g_{kl}(0) = \delta_{kl}$. Thus we have

$$|X(w)| < 1 \quad \text{for } w \in \overline{S}_R$$

and

$$x^\alpha(w) = 0 \quad \text{for } \alpha = 1, \dots, n-1 \text{ and } w \in \gamma'.$$

We write (44) as

$$g_{kl}(X) x^k_w x^l_w = 0, \quad w \in \overline{S}_R,$$

which can be transformed into

$$\left(x^n_w + \frac{g_{\alpha n}(X)}{g_{nn}(X)} x^\alpha_w\right)^2 = \left(\frac{g_{\alpha n}(X)}{g_{nn}(X)} x^\alpha_w\right)^2 - \frac{g_{\alpha\beta}(X)}{g_{nn}(X)} x^\alpha_w x^\beta_w \tag{45}$$

(summation with respect to repeated Greek indices is supposed to run from 1 to $n-1$).

The definiteness of the matrix (g_{kl}) implies

$$m_1 \le g_{nn}(x) \quad \text{and} \quad |g_{kl}(x)| \le m_2 \quad \text{for } |x| < 1$$

where m_1 and m_2 denote two positive constants. Then we obtain from (45) that there is some constant $m_3 > 0$ such that

$$|\nabla x^n|^2 \le m_3 \sum_{\alpha=1}^{n-1} |\nabla x^\alpha|^2 \quad \text{on } \overline{S}_R.$$

As in the proof of Theorem 1 we infer from (43) and (45) that the mapping

$$Y(w) := (x^1(w), x^2(w), \dots, x^{n-1}(w))$$

is of class $C^0(\overline{S}_R) \cap C^2(S_R)$ and satisfies the relations

$$\begin{aligned} |\Delta Y| &\le m_4 |\nabla Y|^2 \quad \text{on } S_R, \\ Y &= 0 \quad \text{on } \gamma'. \end{aligned} \tag{46}$$

Now we proceed as in the proof of Theorem 2. In fact, from (46) we infer that $Y \in C^{1,\nu}(\overline{S}_\varepsilon)$ for any $\nu \in (0,1)$ and $\varepsilon \in (0,R)$, whence (45) implies that x^n_w is of class $C^0(\overline{S}_\varepsilon)$. Therefore we obtain $X \in C^1(\overline{S}_\varepsilon)$.

Moreover, from (45) and (46_2) it follows that

$$\left(iwx^n_w + \frac{g_{\alpha n}}{g_{nn}} iwx^\alpha_w\right)^2 = \left\{\frac{i}{2}\left(x^n_r + \frac{g_{\alpha n}}{g_{nn}} x^\alpha_r\right) + \frac{1}{2}\left(x^n_\varphi + \frac{g_{\alpha n}}{g_{nn}} x^\alpha_\varphi\right)\right\}^2$$
$$= \frac{g_{\alpha\beta}}{g_{nn}}(wx^\alpha_w)(wx^\beta_w) - \left(\frac{g_{\alpha n}}{g_{nn}} wx^\alpha_w\right)^2 \geq 0$$

on $\gamma'' := \gamma' \cap \partial S_\varepsilon$ (cf. the computations leading to (20)). Hence we have

$$x^n_r(e^{i\varphi}) = -\frac{g_{\alpha n}(0,\dots,0,x^n(e^{i\varphi}))}{g_{nn}(0,\dots,0,x^n(e^{i\varphi}))} x^\alpha_r(e^{i\varphi}) \quad \text{on } \gamma''. \tag{47}$$

Setting

$$p = -\Gamma^n_{kl}(X)\{x^k_u x^l_u + x^k_v x^l_v\}, \tag{48}$$

it follows that

$$\Delta x^n = p \quad \text{in } S_\varepsilon, \quad x^n_r = f \quad \text{on } \gamma'', \tag{49}$$

where x^n is of class C^1 on $\overline{S}_\varepsilon$, of class C^2 on $S_\varepsilon, p \in L_\infty(S_\varepsilon), f \in C^{0,\nu}(\gamma'')$. Then a potential-theoretic reasoning yields $x^n \in C^{1,\nu}(\overline{S}_{\varepsilon'})$ for $0 < \varepsilon' < \varepsilon$ and therefore $X \in C^{1,\nu}(\overline{S}_{\varepsilon'})$.

Alternating between (49) and

$$\Delta Y = Q \quad \text{in } S_\varepsilon, \quad Y = 0 \quad \text{on } \gamma'', \tag{50}$$

where

$$Q^\alpha := -\Gamma^\alpha_{kl}(X)(x^k_u x^l_u + x^k_v x^l_v),$$

we obtain higher regularity of X at the boundary part γ. This completes the sketch of the proof. □

2.4 The Boundary Behaviour of Minimal Surfaces at Their Free Boundary: A Survey of the Results and an Outline of Their Proofs

The boundary behaviour of minimal surfaces with free boundaries is somewhat more difficult to treat than that of solutions of Plateau's problem. In fact, Courant [9,15] has exhibited a number of examples indicating that the trace of a minimal surface with a free boundary on a continuous support surface S need not be continuous. One of his examples even shows that the trace curve can be unbounded although S is smooth (but not compact). Unfortunately Courant's examples are not rigorous as their construction is based on a heuristic principle, the *bridge theorem*, which has not yet been established for solutions of free boundary problems, and therefore we shall describe Courant's idea only in the Scholia. However, one of Courant's constructions is not based on the bridge theorem and has been made perfectly rigorous by Cheung [1].

We consider here a modification of Cheung's example. The supporting surface S (see Fig. 1) in our example will be defined as follows. Let us define sets $B_1, B_2, C, E_\pm, G$ and curves $\gamma_\pm, \beta_\pm$ by

$$\begin{aligned}
B_1 &:= \{(x,y,z): x = 0, -1 \le y \le 1, -3 \le z \le 0\},\\
B_2 &:= \{(x,y,z): x = 0, -1 \le y \le 1, -5 \le z \le -3\},\\
C &:= \{(x,y,z): z = 0, x \ge 0, -e^{-x} \le y \le e^{-x}\},\\
E_\pm &:= \{(x,y,z): x \ge 0, y = \pm 1, -5 \le z \le -3\},\\
G &:= \{(x,y,z): x \ge 0, -1 \le y \le 1, z = -5\},\\
\gamma_\pm &:= \{(x,y,0): x \ge 0, y = \pm e^{-x}\},\\
\beta_\pm &:= \{(x,y,-3): x \ge 0, y = \pm 1\}.
\end{aligned}$$

Now we connect each point $(x, \pm e^{-x}, 0)$ on $\gamma_\pm$ by straight segments with the corresponding point $(x, \pm 1, -3)$ on $\beta_\pm$, thus obtaining two ruled surfaces $F_\pm$.

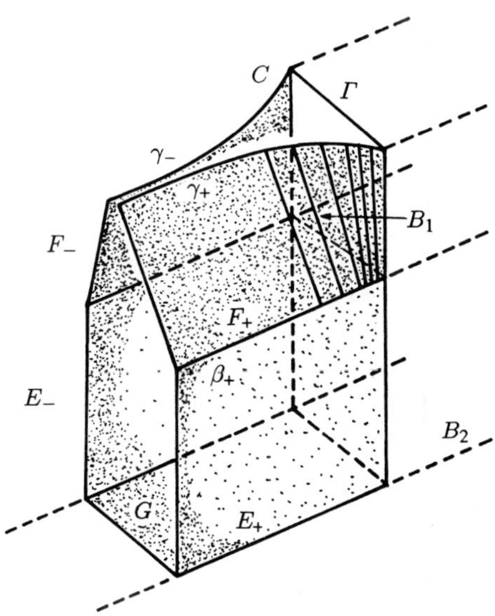

Fig. 1. A noncompact, Lipschitz continuous, nonclosed supporting surface S which satisfies no chord-arc condition. The configuration $\langle \Gamma, S\rangle$ bounds an unbounded minimal surface of the type of the disk

Let

$$S_1 \colon = E_+ \cup E_- \cup F_+ \cup F_- \cup G$$

and denote by S_1^* the reflection of S_1 at the plane $\{x = 0\}$. Then we define

$$S := S_1 \cup S_1^*$$

and

$$\Gamma := \{(x, y, z) \colon x = 0, z = 0, -1 \le y \le 1\}.$$

Claim. *Every solution $Y \in \mathcal{C}(\Gamma, S)$ of the corresponding free boundary problem $\mathcal{P}(\Gamma, S)$ has an unbounded trace on S; in particular Y is discontinuous along the interval I.*

In fact, suppose that $Y \in \mathcal{C}(\Gamma, S)$ is a solution of $\mathcal{P}(\Gamma, S)$ which is continuous on $B \cup \overline{I}$. Then the trace $Y(\overline{I})$ is compact and has to pass G continuously as $Y(\overline{I}) \subset S$. By a projection argument we infer that the area of the part of $Y(B)$ below the plane $\{z = -3\}$ is greater than or equal to the area of B_2 which is 4, and thus it is larger than the area of C which is 2. Thus each solution Y of $\mathcal{P}(\Gamma, S)$ must have a discontinuous trace $Y|_{\overline{I}}$. In fact, $Y|_{\overline{I}}$ cannot be contained in the subregion $S_R := \mathcal{K}_R(0) \cap S$ for any $R > 0$. (Otherwise, by Theorem 2 of Section 2.5, we would obtain $Y \in C^{0,\mu}(\overline{B}, \mathbb{R}^3)$ for some $\mu > 0$.) Hence it follows that the trace $Y|_{\overline{I}}$ is unbounded, and a projection argument shows that Y has to be a parametrization of C or of its reflection C^* at $\{x = 0\}$. In other words, if there is a solution Y of $\mathcal{P}(\Gamma, S)$, it will be given either by C or by C^*. As the existence theory of Vol. 1, Chapter 4 yields the existence of a solution of $\mathcal{P}(\Gamma, S)$, we infer that *C and C^* are the two solutions of $\mathcal{P}(\Gamma, S)$ and that there is no other solution of this minimum problem.*

By reflecting S at the plane $\{z = 0\}$ we can extend it to a Lipschitz continuous noncompact surface $\tilde{S}$ without boundary. Furthermore, by rounding off the edges of S and of $\tilde{S}$, we can even construct examples of smooth supporting surfaces, with or without boundary, having the desired property that there is no solution $Y \in C^0(B \cup \overline{I}, \mathbb{R}^3)$ of the minimum problem $\mathcal{P}(\Gamma, S)$.

Thus we have an example of a boundary configuration $\langle \Gamma, S \rangle$ consisting of a smooth arc Γ and a smooth support surface S for which the minima of area in $\mathcal{C}(\Gamma, S)$ have a discontinuous (and even unbounded) trace on S. However, the reader will note that the surface S in the Courant example does not satisfy a uniform extrinsic Lipschitz condition in $\mathbb{R}^3$, i.e., the quotient of the distance of two points on S divided by their air distance is unbounded. We say that S does not satisfy a *chord-arc condition* (the precise definition of this condition will be given in Section 2.5).

Surprisingly, the chord-arc condition suffices to enforce that all minima of a free or partially free boundary problem have a continuous free trace. In fact, we shall prove:

(i) *Suppose that* X *minimizes Dirichlet's integral in the class* $\mathcal{C}(\Gamma, S)$ *and that* $D(X) > 0$. *Assume also that the support surface* S *satisfies a chord-arc condition. Then* X *is of class* $C^{0,\mu}(B \cup I, \mathbb{R}^3)$ *for some* $\mu \in (0,1)$.

This result is the main statement of Theorem 1 in Section 2.5. The proof is based on an adaptation of Morrey's idea to compare any minimizer locally with a suitable harmonic mapping. To make this idea effective one constructs such a mapping by exploiting the chord-arc condition in order to set up its boundary values on S.

Several variants of the assertion (i) are given in Theorems 2–4 of Section 2.5. In particular, Theorem 4 of Section 2.5 provides a regularity theorem analogous to (i), holding for minimizers of a completely free boundary problem.

In Section 2.6 we shall prove regularity of stationary points of Dirichlet's integral at their free boundaries. At present it is not known whether the free trace Σ of any such surface X is a continuous curve provided that the support surface S satisfies merely a chord-arc condition. However, assuming that S is of class C^2 we obtain the desired result. More precisely, we have:

(ii) *Let* S *be an admissible support surface of class* C^2, *and suppose that* X *is a stationary point of Dirichlet's integral in the class* $\mathcal{C}(\Gamma, S)$. *Then there is some* $\alpha \in (0,1)$ *such that* $X \in C^{0,\alpha}(B \cup I, \mathbb{R}^3)$.

This result is the content of Theorem 2 in Section 2.6; a similar statement can be obtained for solutions of completely free boundary problems (cf. Section 2.6, Remark 2).

The proof of (ii) is quite different from that of (i). Whereas in (i) we shall proceed by deriving a priori estimates for X, the approach in (ii) is indirect. Using the finiteness of Dirichlet integral of X we shall first derive suitable *monotonicity results* for functionals that are closely related to Dirichlet's integral. Combining these results we shall infer that X has to be continuous on $B \cup I$ if $D(X) < \infty$.

Once the boundary values $X|_I$ are shown to be continuous, we can apply suitable techniques from the theory of nonlinear elliptic equations to obtain $X \in C^{0,\alpha}(B \cup I, \mathbb{R}^3), \alpha \in (0,1)$. For instance, Widman's hole-filling method (cf. Lemma 5 in Section 2.6) yields a direct way to this result; the details are carried out in the proof of Theorem 2 in Section 2.6.

Note that in all these cases the support surface S may have a nonempty boundary. If ∂S is void, we can say much more on the free trace $\Sigma = \{X(w) \colon w \in I\}$ of X on S. Roughly speaking Σ will turn out to be as good as the support surface S. We shall, in fact, obtain:

(iii) *Let* S *be an admissible support surface with* $\partial S = \emptyset$ *which is of class* $C^{m,\beta}, m \geq 3, \beta \in (0,1)$. *Then any stationary point* X *of Dirichlet's integral in* $\mathcal{C}(\Gamma, S)$ *is of class* $C^{m,\beta}(B \cup I, \mathbb{R}^3)$. *If* S *is real analytic, then* X *is real analytic on* $B \cup I$ *and can be continued analytically across* I.

This result is formulated in Theorems 1 and 2 of Section 2.8. Starting from (ii), we shall first verify that X is contained in $C^1(B \cup I, \mathbb{R}^3)$. This can either be achieved by transforming the boundary problem for X locally into an interior regularity question for some weak solution Z of an elliptic system

$$\Delta Z = F(w)|\nabla Z|^2,$$

which is derived by a reflection argument, and then applying Tomi's regularity theorem, or by playing the full regularity machinery for nonlinear elliptic boundary value problems. The first possibility is sketched in Remark 1 of Section 2.8, whereas the second approach is discussed in Section 2.7 in great detail and in a wider context (see in particular Theorem 4 of Section 2.7).

Having proved that X is of class $C^1(B \cup I, \mathbb{R}^3)$, we use classical results from potential theory to derive $X \in C^{m,\beta}(B \cup I, \mathbb{R}^3)$ by employing a suitable bootstrap argument. The reader can find this reasoning in the proof of Proposition 1 in Section 2.8.

In Theorem 2 of Section 2.8 we show that X can be continued analytically across its free boundary if the support surface S is real analytic. To this end, we set up a Volterra integral equation

$$Z(w) = \int_0^w F(\omega, Z(\omega))\, d\omega$$

which has exactly one solution Z in the space $\mathcal{A}(\overline{S}_\delta)$ of mappings $Z\colon \overline{S}_\delta \to \mathbb{C}^3$ which are continuous on $\overline{S}_\delta$ and holomorphic in $S_\delta := \{w\colon |w| < \delta < 1, \operatorname{Im} w > 0\}$, and F is constructed in such a way that

$$Z(u) = X(u) \quad \text{for } u \in \mathbb{R} \text{ with } |u| < \delta$$

(assuming a suitable normalization of X).

Let X^* be an adjoint surface of X and $f = X + iX^*$. Then both f and $g := f - Z$ are of class $\mathcal{A}(\overline{S}_\delta)$, and we have

$$\operatorname{Im} Z = 0 \quad \text{and} \quad \operatorname{Re} g = 0 \quad \text{on } I_\delta.$$

By Schwarz's reflection principle, we can continue both Z and g across I_δ as holomorphic functions, whence also $f = g + Z$ and $X = \operatorname{Re} f$ are continued analytically across I_δ.

This approach to analyticity at the boundary, due to H. Lewy, is by far the easiest, but it cannot be carried over to H-surfaces or to minimal surfaces in a Riemannian manifold as it uses the holomorphic function $f = X + iX^*$. This tool is, however, not available in those other cases. Here one can apply a general regularity theorem due to Morrey [5] (cf. also Morrey and Nirenberg [1] and Morrey [8]), or the work of Frank Müller which extends Lewy's method to more general situations.

Let us now turn to the case when the support surface S has a nonempty boundary. Then we shall establish the following result (cf. Section 2.7, Theorem 1):

(iv) *Let S be an admissible support surface of class C^4* (by definition, this implies $\partial S \in C^4$; cf. Section 2.6, Definitions 1 and 2). *Moreover, let X be a stationary point of Dirichlet's integral in $\mathcal{C}(\Gamma, S)$. Then X is of class $C^{1,1/2}(B \cup I, \mathbb{R}^3)$.*

According to Remark 1 in Section 1.8 of Vol. 3, this is the best possible result which can, in general, be expected. This follows from the asymptotic expansions (1) and (2) in Section 2.10 around points u_1 and u_2 on I where the free trace $X|_I$ of X on S lifts off the boundary ∂S of the support surface S. We could interpret (iv) as a regularity result for a *Signorini problem* (or else for a *thin obstacle problem*). The proof of (iv) will be carried out in three steps. First, by applying Nirenberg's difference quotient technique, we shall derive L_2-estimates for the second derivatives $\nabla^2 X$ up to the free boundary. For this purpose we need the Hölder continuity of X on $B \cup I$, established in (ii), as well as an important calculus inequality due to Morrey (cf. Section 2.7, Lemma 2) which implies that the Morrey seminorm is reproducible.

As a second step it will be shown that X is of class $C^1(B \cup I, \mathbb{R}^3)$. This follows from L_p-estimates for solutions of the Poisson equation. In order to apply these estimates we introduce suitable local coordinates $\{\mathcal{U}, g\}$ on S such that the boundary conditions for $Y(w) = (y^1(w), y^2(w), y^3(w)) = g(X(w))$ become uncoupled. For y^2 we derive a Neumann condition and for y^3 a Dirichlet condition. Then we apply the L_p-estimates to y^2 and y^3, thus obtaining Hölder continuity of ∇y^2 and ∇y^3 up to the free boundary. Finally, the continuity of y^1 up to the free boundary will be derived from the conformality relations.

As a third step we devise an iteration scheme, which allows us to attain $X \in C^{1,1/2}(B \cup I, \mathbb{R}^3)$, by exploiting once again the conformality relations.

The use of L_p-estimates can be circumvented by a method which is developed in Section 2.9. Here one derives directly that $\nabla^2 X$ satisfies a Dirichlet growth condition (i.e., has a finite Morrey seminorm) up to the free boundary by alternatively applying one of two possible Poincaré inequalities. This method is nothing but a skilful improvement of the estimates derived in step 1.

Note that the regularity results (i)–(iv) are not directly meaningful for differential geometry, as the free boundary I may contain branch points. This can, at least partially, be remedied in the following way. First, by applying a technique due to Hartman and Wintner, we show that, for every branch point $w_0 \in I$, we have an asymptotic expansion

$$X_w(w) = A(w - w_0)^\nu + o(|w - w_0|^\nu) \quad \text{as } w \to w_0$$

with some $\nu \in \mathbb{N}$ and $A \in \mathbb{C}^3, A \neq 0, \langle A, A \rangle = 0$.

This implies that there exists a limit tangent plane of X as $w \to w_0$ with the normal $N_0 = \lim_{w \to w_0} N(w)$, where

$$N(w) = |X_u|^{-2}(X_u \wedge X_v),$$

and that the oriented tangent

$$t(u) := |X_u(u)|^{-1} X_v(u) \quad \textit{as } u \to w_0 = u_0 \in I$$

of the free trace $X|_I$ *is either continuous or jumps by 180 degrees; the first case occurs if the order* ν *of the branch point* $w_0 = u_0 \in I$ *is even, the second case, if* ν *is odd.*

Hence, if ν is even, the representation $\boldsymbol{x}(s)$ of the trace $\Sigma = X|_I$ with respect to its arc length

$$s = \int_{u_0}^{u} |X_u(u)|\, du$$

is of class C^1, and therefore the trace Σ can be viewed as a regular C^1-curve in the neighbourhood of $x_0 := X(u_0)$. If ν is odd, then Σ has a *cusp at* x_0, and only the unoriented tangent is continuous at x_0.

We sketch the derivation of this result in Section 2.10; the details of the Hartman–Wintner technique are given in Chapter 3. In the first two chapters of Vol. 3 we shall study cases where boundary branch points can entirely be excluded.

Most of our results will be stated and proved merely for stationary points of Dirichlet's integral in $\mathcal{C}(\Gamma, S)$, that is, for solutions of a partially free boundary problem. Similar results hold mutatis mutandis for minimal surfaces with completely free boundaries, or for minimal surfaces of higher topological type spanned in a general boundary configuration $\langle \Gamma_1, \dots, \Gamma_l, S_1, \dots, S_m \rangle$, and their proofs can be carried out in essentially the same way. In fact, the considerations in Sections 2.7–2.10 are strictly local and require only changes in notation, and the reasoning of Sections 2.5 and 2.6 can be adjusted without major difficulties. We leave it as an exercise to the reader to carry out the details.

2.5 Hölder Continuity for Minima

Courant's examples indicate that one cannot expect a solution of a free or a semifree boundary problem to be regular at its free boundary, even if the support surface S is of class C^∞. On the other hand, we shall see that a minimal surface is continuous up to its free boundary if it is minimizing and if S satisfies a kind of uniform (local) Lipschitz condition. Such a condition on S will be called a *chord-arc* condition.

Definition. *A set* S *in* $\mathbb{R}^3$ *is said to fulfil a chord-arc condition with constants* M *and* δ, $M \geq 1$ *and* $\delta > 0$, *if it is closed and if any two points* P *and* Q *of* S *whose distance* $|P - Q|$ *is less than or equal to* δ *can be connected in* S *by a rectifiable arc* Γ^* *whose length* $L(\Gamma^*)$ *satisfies*

$$L(\Gamma^*) \leq M|P - Q|.$$

For example, every compact regular C^1-surface S without boundary satisfies a chord-arc condition, and the same holds true if the boundary ∂S is nonempty but smooth.

Let us first deal with the semifree problem. We now denote by B the parameter domain

$$B = \{w = u + iv \colon |w| < 1, v > 0\}$$

the boundary of which consists of the circular arc

$$C = \{w = u + iv \colon |w| = 1, v \geq 0\}$$

and of the interval

$$I = \{u \in \mathbb{R} \colon |u| < 1\}$$

on the real axis.

Consider a boundary configuration $\langle \Gamma, S \rangle$ consisting of a closed set S in $\mathbb{R}^3$ satisfying a chord-arc condition and of a Jordan curve Γ in $\mathbb{R}^3$ whose endpoints P_1 and P_2 lie on S, $P_1 \neq P_2$. As in Section 4.6 of Vol. 1 we define the class of admissible surfaces for the semifree problem as the set $\mathcal{C}(\Gamma, S)$ of mappings $X \in H_2^1(B, \mathbb{R}^3)$ satisfying

(i) $X(w) \in S$ $\mathcal{H}^1$-a.e. on I;

(ii) $X \colon C \to \Gamma$ is a continuous, weakly monotonic mapping of C onto Γ such that $X(1) = P_1, X(-1) = P_2$.

Let us also introduce the sets

$$\begin{aligned} Z_d &:= \{w \in B \colon |w| < 1 - d\} = \{w \in B \colon \operatorname{dist}(w, C) > d\}, \quad (0 < d < 1), \\ S_r(w_0) &:= B \cap B_r(w_0). \end{aligned}$$

Then we can prove:

Theorem 1. *Suppose that $X \in \mathcal{C}(\Gamma, S)$ minimizes the Dirichlet integral $D(X)$ within the class $\mathcal{C}(\Gamma, S)$, and let $e = e(\Gamma, S) := \inf\{D(Y) \colon Y \in \mathcal{C}(\Gamma, S)\}$ be positive. Moreover, assume that S satisfies a chord-arc condition with constants M and δ. Then, for any $d \in (0,1)$, and $w_0 \in \overline{Z}_d$, and for any $r > 0$, we have*

$$(1) \qquad \int_{S_r(w_0)} |\nabla X|^2 \, du \, dv \leq \left(\frac{2r}{d}\right)^{2\mu} \int_B |\nabla X|^2 \, du \, dv,$$

where

$$(2) \qquad \mu := \min\{(1 + M^2)^{-1}, \delta^2/(2e\pi)\}.$$

It follows that X is of class $C^{0,\mu}(\overline{Z}_d, \mathbb{R}^3)$ and that

$$(3) \qquad [X]_{\mu, \overline{Z}_d} \leq c(\mu)(1 - d)^{-\mu} \sqrt{D(X)} = c(\mu)(1 - d)^{-\mu} \sqrt{e(\Gamma, S)}$$

holds true for some constant $c(\mu) > 0$.

Proof. Let X be a minimizer of the Dirichlet integral in $\mathcal{C}(\Gamma, S)$. Then X is harmonic in B, satisfies the conformality relations

$$(4) \qquad |X_u| = |X_v|, \quad \langle X_u, X_v \rangle = 0,$$

and

$$D(X) = e.$$

For any point $w_0 \in \overline{B}$ we define

$$(5) \qquad \Phi(r, w_0) := \int_{S_r(w_0)} |\nabla X|^2 \, du \, dv.$$

We begin by proving that for any $d \in (0,1)$ and for any $w_0 \in I$ with $|w_0| \le 1-d$ the inequality

$$(6) \qquad \Phi(r, w_0) \le (r/d)^{2\mu} \Phi(d, w_0)$$

holds true for all $r \in (0, d]$. To this end we fix $w_0 \in I$ with $|w_0| \le 1 - d$ and set $B_r := B_r(w_0), S_r := S_r(w_0)$, and $\Phi(r) := \Phi(r, w_0)$. Introducing polar coordinates ρ, θ around w_0 by $w = w_0 + \rho e^{i\theta}$ and writing somewhat sloppily

$$X(w) = X(w_0 + \rho e^{i\theta}) = X(\rho, \theta),$$

we obtain

$$(7) \qquad \Phi(r) = \int_0^r \int_0^\pi \{|X_\rho(\rho, \theta)|^2 + \rho^{-2} |X_\theta(\rho, \theta)|^2\} \rho \, d\theta \, d\rho.$$

From (4) we infer

$$(8) \qquad |X_\rho|^2 = \rho^{-2} |X_\theta|^2, \quad \langle X_\rho, X_\theta \rangle = 0,$$

hence

$$(9) \qquad \Phi(r) = 2 \int_0^r \rho^{-1} \int_0^\pi |X_\theta(\rho, \theta)|^2 \, d\theta \, d\rho.$$

There is a set $\mathcal{N} \subset [0, d]$ of 1-dimensional measure zero such that

$$(10) \qquad \int_0^\pi |X_\theta(r, \theta)|^2 \, d\theta < \infty \quad \text{for } r \in (0, d) \setminus \mathcal{N}$$

and that the absolutely continuous function $\Phi(r)$ is differentiable at the values $r \in (0, d) \setminus \mathcal{N}$ and satisfies

$$(11) \qquad \Phi'(r) = 2r^{-1} \int_0^\pi |X_\theta(r, \theta)|^2 \, d\theta.$$

We can therefore assume that, for $r \in (0, d) \setminus \mathcal{N}$, the function $X(r, \theta)$ is an absolutely continuous function of $\theta \in [0, \pi]$; in particular, the limits

$$Q_1(r) := \lim_{\theta \to \pi - 0} X(r, \theta), \quad Q_2(r) := \lim_{\theta \to +0} X(r, \theta)$$

exist for $r \in (0, d) \setminus \mathcal{N}$.

Consider now any $r \in (0, d) \setminus \mathcal{N}$ for which

$$\int_0^\pi |X_\theta(r, \theta)|^2 \, d\theta \le \pi^{-1} \delta^2 \tag{12}$$

holds true. Then we infer from

$$|Q_1(r) - Q_2(r)| \le \int_0^\pi |X_\theta(r, \theta)| \, d\theta \le \sqrt{\pi} \left\{ \int_0^\pi |X_\theta(r, \theta)|^2 \, d\theta \right\}^{1/2} \tag{13}$$

the inequality

$$|Q_1(r) - Q_2(r)| \le \delta.$$

Since S satisfies a chord-arc condition with constants M and δ, there exists a rectifiable arc

$$\Gamma^* = \{\xi(s) \colon 0 \le s \le l^*\}$$

of length $l^* = L(\Gamma^*)$ on S which connects the points $Q_1(r)$ and $Q_2(r)$, and whose length $L(\Gamma^*)$ satisfies

$$l^* = L(\Gamma^*) \le M |Q_1(r) - Q_2(r)|. \tag{14}$$

We assume s to be chosen as parameter of the arc length on Γ^*. Then it follows that $|\xi'(s)| = 1$ a.e. on $[0, l^*]$. Introducing the reparametrization $\zeta(\theta), \pi \le \theta \le 2\pi$, of Γ^* which is defined by

$$\zeta(\theta) := \xi(\pi^{-1}(\theta - \pi) l^*),$$

we obtain

$$|\zeta_\theta(\theta)| = \text{const} = l^*/\pi \quad \text{a.e. on } [\pi, 2\pi]$$

and

$$l^* = \int_\pi^{2\pi} |\zeta_\theta| \, d\theta;$$

therefore also

$$\pi \int_\pi^{2\pi} |\zeta_\theta|^2 \, d\theta = \left(\int_\pi^{2\pi} |\zeta_\theta| \, d\theta \right)^2 = l^{*2}. \tag{15}$$

From (13)–(15) we conclude that

$$\int_\pi^{2\pi} |\zeta_\theta|^2 \, d\theta \le M^2 \int_0^\pi |X_\theta(r, \theta)|^2 \, d\theta. \tag{16}$$

Consider now the harmonic vector function $H(w)$ in B_r whose boundary values $\eta(\theta) = H(w_0 + re^{i\theta})$ are defined by

$$\eta(\theta) := \begin{cases} X(r,\theta) & \text{for } 0 \le \theta \le \pi \\ \zeta(\theta) & \text{for } \pi \le \theta \le 2\pi \end{cases} .$$

Because of (16), we have

$$\int_0^{2\pi} |\eta_\theta|^2 \, d\theta \le (1+M^2) \int_0^{\pi} |X_\theta(r,\theta)|^2 \, d\theta. \tag{17}$$

Expanding H and ζ in Fourier series we obtain

$$H(w) = \frac{1}{2}A_0 + \sum_{n=1}^{\infty} \left(\frac{\rho}{r}\right)^n (A_n \cos n\theta + B_n \sin n\theta)$$

and

$$\eta(\theta) = \frac{1}{2}A_0 + \sum_{n=1}^{\infty} (A_n \cos n\theta + B_n \sin n\theta).$$

From these expressions, we derive

$$\int_{B_r} |\nabla H|^2 \, du \, dv = \pi \sum_{n=1}^{\infty} n(|A_n|^2 + |B_n|^2),$$

$$\int_0^{2\pi} |\eta_\theta|^2 \, d\theta = \pi \sum_{n=1}^{\infty} n^2(|A_n|^2 + |B_n|^2),$$

and therefore

$$\int_{B_r} |\nabla H|^2 \, du \, dv \le \int_0^{2\pi} |\eta_\theta|^2 \, d\theta. \tag{18}$$

Relations (11), (17) and (18) imply that

$$\int_{B_r} |\nabla H|^2 \, du \, dv \le \frac{1}{2}(1+M^2) r \Phi'(r). \tag{19}$$

Next we consider the mapping $Y(w)$ on $B \cup B_r$ which is defined as

$$Y(w) := \begin{cases} H(w) & \text{for } w \in B_r \\ X(w) & \text{for } w \in B \setminus B_r \end{cases} .$$

Clearly Y is continuous and of class H_2^1 on $B \cup B_r$. Let τ be the homeomorphism of $\overline{B}$ onto $\overline{B \cup B_r}$ which maps B conformally onto $B \cup B_r$, keeping the points $1, -1, i$ fixed. Then the mapping $Z := Y \circ \tau$ is contained in $\mathcal{C}(\Gamma, S)$, and the minimum property of X implies

$$\int_B |\nabla X|^2 \, du \, dv \le \int_B |\nabla Z|^2 \, du \, dv.$$

On account of the conformal invariance of the Dirichlet integral we have

$$\int_B |\nabla X|^2 \, du \, dv \le \int_{B \cup B_r} |\nabla Y|^2 \, du \, dv,$$

and the definition of Y now implies

$$\int_{S_r} |\nabla X|^2 \, du \, dv \le \int_{B_r} |\nabla H|^2 \, du \, dv. \tag{20}$$

By virtue of (5), (19), and (20) we obtain the relation

$$\Phi(r) \le \tfrac{1}{2}(1 + M^2) r \Phi'(r) \tag{21}$$

for every $r \in (0, d) \setminus \mathcal{N}$ satisfying equation (12).

On the other hand, if the equation

$$\int_0^\pi |X_\theta(r, \theta)|^2 \, d\theta > \pi^{-1} \delta^2$$

holds for some $r \in (0, d) \setminus \mathcal{N}$, then we trivially have

$$\Phi(r) \le 2D(X) = 2e < 2e\pi\delta^{-2} \int_0^\pi |X_\theta(r, \theta)|^2 \, d\theta,$$

and the identity (11) yields

$$\Phi(r) \le \pi e \delta^{-2} r \Phi'(r). \tag{22}$$

Defining the number $\mu \in (0, 1)$ as in (2), it follows that

$$2\mu\Phi(r) \le r\Phi'(r) \quad \text{for all } r \text{ in } (0, d) \setminus \mathcal{N}, \tag{23}$$

and an integration yields

$$\Phi(r) \le (r/d)^{2\mu} \Phi(d) \quad \text{for all } r \in [0, d].$$

Thus we have established (6) for any $d \in (0, 1), w_0 \in I$ with $|w_0| < 1 - d$, and $r \in [0, d]$.

Consider any w_0 with $|w_0| \le 1 - R$ and $\operatorname{Im} w_0 \ge R$ for some $R \in (0, 1)$. Then we have $B_r(w_0) \subset B$ for any $r \in (0, R)$, and analogously to (18) we obtain

$$\int_{B_r(w_0)} |\nabla X|^2 \, du \, dv \le \int_0^{2\pi} |X_\theta(r, \theta)|^2 \, d\theta$$

for almost all $r \in (0, R)$. By (5) and (11) we therefore infer

$$\Phi(r, w_0) \le \frac{1}{2} r \frac{d}{dr} \Phi(r, w_0)$$

for a.a. $r \in (0, R)$, and an integration yields

$$\Phi(r, w_0) \le (r/R)^2 \Phi(R, w_0) \quad \text{for all } r \in [0, R]. \tag{24}$$

Finally we fix some $d \in (0, 1)$ and choose an arbitrary point $w_0 \in \overline{Z}_d = \overline{B} \cap \{|w| \le 1 - d\}$. Set $u_0 = \operatorname{Re} w_0$ and $v_0 = \operatorname{Im} w_0$. We distinguish three cases:

(i) $v_0 \ge d/2$.
Choosing $R = d/2$ we infer from (24) that

$$\Phi(r, w_0) \le (2r/d)^2 \int_B |\nabla X|^2 \, du \, dv \quad \text{for } 0 \le r \le d/2 \tag{25}$$

holds true.

(ii) $0 \le v_0 \le d/2$ and $v_0 \le r \le d/2$.
Then we have $B_r(w_0) \subset B_{2r}(u_0)$, and it follows that

$$\Phi(r, w_0) \le \Phi(2r, u_0).$$

Applying (6) we have also

$$\Phi(2r, u_0) \le (2r/d)^{2\mu} \Phi(d, u_0),$$

and therefore

$$\Phi(r, w_0) \le (2r/d)^{2\mu} \int_B |\nabla X|^2 \, du \, dv. \tag{26}$$

In particular we have

$$\Phi(v_0, w_0) \le (2v_0/d)^{2\mu} \int_B |\nabla X|^2 \, du \, dv \quad \text{for any } v_0 \in [0, d/2]. \tag{27}$$

(iii) $0 \le v_0 \le d/2$ and $0 \le r \le v_0$.
Applying (24) to the case $R = v_0$ we obtain

$$\Phi(r, w_0) \le (r/v_0)^2 \Phi(v_0, w_0).$$

Combining this inequality with (27) it follows that

$$\begin{aligned} \Phi(r, w_0) &\le (r/v_0)^2 (2v_0/d)^{2\mu} \int_B |\nabla X|^2 \, du \, dv \\ &\le (2r/d)^{2\mu} \int_B |\nabla X|^2 \, du \, dv. \end{aligned} \tag{28}$$

On account of (25), (26) and (28) inequality (1) holds true for any $r \in [0, d/2]$, and for $r > d/2$ estimate (1) is satisfied for trivial reasons. The bound (3) and $X \in C^{0,\mu}(\overline{Z}_d, \mathbb{R}^3)$ now follow from Morrey's Dirichlet growth theorem (see Morrey [8], p. 79). □

Remark. Note that the assumptions of Theorem 1 do not require S to be a regular surface. In fact, S is allowed to degenerate to a rectifiable arc. Thus several variants of Theorem 1 can be proved. For instance we get:

Theorem 2. *Suppose that $S \cup \Gamma$ satisfies a chord-arc condition with constants M and δ, and let $X \in \mathcal{C}(\Gamma, S)$ be a minimizer of the Dirichlet integral in the class $\mathcal{C}(\Gamma, S)$, that is, a solution of the minimum problem $\mathcal{P}(\Gamma, S)$ considered in Section 4.6 of Vol. 1, which satisfies $D(X) > 0$. Then X is of class $C^{0,\mu}(\overline{B}, \mathbb{R}^3)$ for some $\mu \in (0, 1)$.*

Fixing a third point $P_3 \in \Gamma$ and requiring $X(i) = P_3$ we can even derive an a priori estimate for $[X]_{\mu,\overline{B}}$ analogous to (3).

In particular, the chord-arc condition for $S \cup \Gamma$ implies the Hölder continuity of any minimizer X in the corners $w = \pm 1$ which are mapped by X on the points P_1 and P_2 where the arc Γ is attached to S.

If we consider minimal surfaces bounded by a preassigned closed Jordan curve Γ of finite length, we can even drop the minimizing property of X since we then can avoid the detour via the comparison surface $Z = Y \circ \tau$ obtained from X and H. Instead we derive an inequality of the type (21) directly by applying the isoperimetric inequality to the part $X|_{S_r(w_0)}$ of the minimal surfaces. Leaving a detailed discussion to the reader we just formulate the final result:

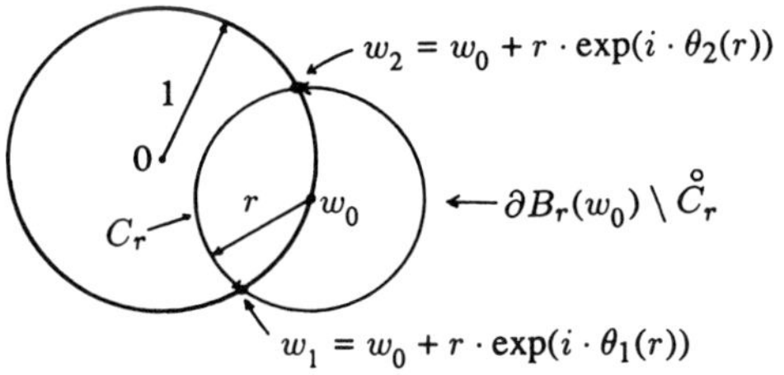

Fig. 1.

Theorem 3. *Let Γ be a closed rectifiable Jordan arc in $\mathbb{R}^3$ of the length $L(\Gamma)$ satisfying a chord-arc condition with constants M and δ. Denote by $\mathfrak{F}(\Gamma)$ a family of minimal surfaces $Y \in \mathcal{C}(\Gamma)$ bounded by Γ which maps three fixed points on $C = \partial B$ onto three fixed points on Γ. Then there exists a number $R > 0$ such that for all $X \in \mathfrak{F}(\Gamma)$ we have*

(29) $$\int_{S_r(w_0)} |\nabla X|^2 \, du \, dv \le (r/R)^{2\mu} D(X) \quad \textit{for all } r > 0$$

with the exponent $\mu = (1 + M)^{-2}$, and

(30) $$[X]_{0+\mu,\overline{B}} \le cL(\Gamma),$$

where the constant c only depends on M, δ and on the chosen three-point condition of the family $\mathfrak{F}(\Gamma)$.[4]

Similar results hold for solutions of minimum problems with a completely free boundary, i.e., for the minimizers of the Dirichlet integral within one of the classes $\mathcal{C}(\sigma, S)$, $\mathcal{C}^+(S)$, and $\mathcal{C}(\Pi, S)$ introduced in Sections 1.1 and 1.2. As in Chapter 1 we now choose the parameter domain B as the unit disk in $\mathbb{C}$,

$$B = \{w \in \mathbb{C}\colon |w| < 1\},$$

and

$$C = \partial B = \{w \in \mathbb{C}\colon |w| = 1\}.$$

Moreover, we set

$$S_r(w_0) := B \cap B_r(w_0), \quad C_r(w_0) := \overline{B} \cap \partial B_r(w_0).$$

Theorem 4. *Let S be a closed, nonempty, proper subset of $\mathbb{R}^3$ satisfying a chord-arc condition with constants M, δ. Moreover assume that for some $\mu > 0$ the inclusion $S \to T_\mu$ of S in T_μ induces a bijection of the corresponding homotopy classes: $\tilde{\pi}_1(S) \leftrightarrow \tilde{\pi}_1(T_\mu)$.*[5] *Finally, suppose that $\mathcal{C}$ denotes one of the classes $\mathcal{C}(\sigma, S), \mathcal{C}^+(S), \mathcal{C}(\Pi, S)$. Then for every minimizer X of the Dirichlet integral in the class $\mathcal{C}$ there is a constant c such that*

(31) $$\int_{S_r(w_0)} |\nabla X|^2 \, du \, dv \le cr^{2\nu}$$

holds for any $w_0 \in \overline{B}$ and any $r > 0$, where

(32) $$\nu = (1 + M^2)^{-1}.$$

In particular, we have $X \in C^{0,\nu}(\overline{B}, \mathbb{R}^3)$ and

(33) $$\lim_{w \to w_0} \operatorname{dist}(X(w), S) = 0 \quad \text{for all } w_0 \in \partial B.$$

Sketch of the proof. Set $\delta_0 := \frac{1}{4}\pi\mu^2$; this constant is nothing but the number δ which appears in Theorem 2 of Section 1.1. Then there is a number $R_0 \in (0, 1)$ such that

(34) $$\int_{\Omega_0} |\nabla X|^2 \, du \, dv < \delta_0$$

holds true for the annular domain $\Omega_0 := \{w \in \mathbb{C}\colon 1 - R_0 < |w| < 1\}$.

For any point $w_0 \in \overline{B}$ we define

[4] See Hildebrandt [3], pp. 55–59, for a sketch of the proof.

[5] This is Assumption (A) of Section 1.1.

$$\Phi(r) = \Phi(r, w_0) = \int_{S_r(w_0)} |\nabla X|^2 \, du \, dv. \tag{35}$$

Introducing polar coordinates ρ, θ around w_0 by $w = w_0 + \rho e^{i\theta}$ and writing $X(w) = X(w_0 + \rho e^{i\theta}) = X(\rho, \theta)$, we obtain analogously to (9) that

$$\Phi(r) = 2 \int_0^r \int_{\theta_1(\rho)}^{\theta_2(\rho)} \rho^{-1} |X_\theta(\rho, \theta)|^2 \, d\theta \, d\rho \tag{35'}$$

holds for two angles θ_1, θ_2 with $0 \le \theta_2(\rho) - \theta_1(\rho) \le 2\pi$. Consequently the absolutely continuous function $\Phi(r)$ satisfies

$$\int_{\theta_1(r)}^{\theta_2(r)} |X_\theta(r, \theta)|^2 \, d\theta = \tfrac{1}{2} r \Phi'(r) \tag{36}$$

for all $r \in (0, \infty) \setminus \mathcal{N}$ where $\mathcal{N}$ is a one-dimensional null set.

Let $w_0 \in C$ and consider some positive number β which will be specified later. Moreover, let $r \in (0, R_0) \setminus \mathcal{N}$.

Case 1.

$$\int_{\theta_1(r)}^{\theta_2(r)} |X_\theta(r, \theta)|^2 \, d\theta \ge \pi^{-1} \beta^2.$$

Then we obtain the trivial inequality

$$\Phi(r) \le 2\pi\beta^{-2} D(X) \int_{\theta_1(r)}^{\theta_2(r)} |X_\theta(r, \theta)|^2 \, d\theta = \pi\beta^{-2} D(X) r \Phi'(r). \tag{37}$$

Case 2.

$$\int_{\theta_1(r)}^{\theta_2(r)} |X_\theta(r, \theta)|^2 \, d\theta < \pi^{-1} \beta^2.$$

Then for any two points $P := X(r, \theta)$ and $P' := X(r, \theta')$ on $X(C_r(w_0))$ we have

$$|P' - P| \le \int_\theta^{\theta'} |X_\theta(r, \theta)| \, d\theta \le |\theta' - \theta|^{1/2} \left\{ \int_\theta^{\theta'} |X_\theta(r, \theta)|^2 \, d\theta \right\}^{1/2}$$

whence

$$|P' - P| \le \int_\theta^{\theta'} |X_\theta(r, \theta)| \, d\theta \le \beta. \tag{38}$$

In particular, this estimate holds true for the two endpoints $Q_1(r)$ and $Q_2(r)$ of the arc $X \colon C_r(w_0) \to \mathbb{R}^3$ which lie on S. Choosing β less than or equal to δ (M and δ being the constants of the chord-arc condition of S), we have

$$|Q_1(r) - Q_2(r)| \leq \delta.$$

Thus there is a rectifiable arc $\Gamma^* = \{\xi(s) \colon 0 \leq s \leq l^*\}$ of length $l^* = L(\Gamma^*)$ which connects the points $Q_1(r)$ and $Q_2(r)$, and whose length satisfies

$$l^* = L(\Gamma^*) < M|Q_1(r) - Q_2(r)|. \tag{39}$$

Consider now the harmonic vector function $H(w)$ in $B_r = B_r(w_0)$, the boundary values $\eta(\theta) = H(w_0 + re^{i\theta})$ of which are defined by

$$\eta(\theta) := \begin{cases} X(r,\theta) & \\ & \text{for} \\ \zeta(\theta) & \end{cases} \quad \begin{matrix} \theta_1(r) \leq \theta \leq \theta_2(r) \\ \\ \theta \in [0, 2\pi] \setminus [\theta_1(r), \theta_2(\theta)] \end{matrix}$$

where $\zeta(\theta)$ is a suitable reparametrization of Γ^* proportional to the arc length. Then analogously to (19) we obtain

$$\int_{B_r} |\nabla H|^2 \, du \, dv \leq \frac{1}{2}(1 + M^2) r \Phi'(r). \tag{40}$$

We now define the mapping $Y(w)$ on $B \cup B_r$ by

$$Y(w) := \begin{cases} H(w) & \\ & \text{for} \\ X(w) & \end{cases} \quad \begin{matrix} w \in B_r \\ \\ w \in B \setminus B_r \end{matrix}$$

and set

$$Z := Y \circ \tau,$$

where τ is a homeomorphism of $\overline{B}$ onto $\overline{B \cup B_r}$ which maps B conformally onto $B \cup B_r$.

Claim. *The mapping Z is an admissible comparison surface, i.e. $Z \in \mathcal{C}$, if we choose β as*

$$\beta := \min\{\delta, \mu, [(1 + M^2)^{-1} \pi \delta_0]^{1/2}\}. \tag{41}$$

Then analogously to (21) we arrive at

$$\Phi(r) \leq \tfrac{1}{2}(1 + M^2) r \Phi'(r). \tag{42}$$

Combining the discussion of the cases 1 and 2 we infer from (37) and (40) that

$$\Phi(r) \leq \tfrac{1}{2} c r \Phi'(r) \quad \text{a.e. on } (0, R_0), \tag{43}$$

where

$$c := \max\{1 + M^2, 2\pi\beta^{-2} D(X)\}. \tag{44}$$

Now we can proceed as in the proof of Theorem 1, and we obtain the Dirichlet growth condition (31) with $\nu = 1/c$. As this implies $X \in C^{0,\nu}(\overline{B}, \mathbb{R}^3)$, we can repeat the previous discussion in such a way that case 1 becomes void. For this purpose we only have to choose $R_0 > 0$ so small that

$$|Q_1(r) - Q_2(r)| < \delta \quad \text{for } r \in (0, R_0) \setminus \mathcal{N}.$$

Then we obtain condition (31) with the desired exponent $\nu = (1 + M^2)^{-1}$.

It remains to verify the claim.

First of all we choose a radius $\rho \in (0,1)$ so close to 1 that the closed curve $Z\colon \partial B_\rho(0) \to \mathbb{R}^3$ is completely contained in $T_{\mu/2}$ and represents the boundary class $[Z_{|\partial B}]$. We shall show that this curve is homotopic in T_μ to some curve $X\colon \partial B_{\rho'}(0) \to \mathbb{R}^3$ which represents the boundary class of X.

Let $Q_1(r) = X(w_1), Q_2(r) = X(w_2), w_1, w_2 \in \partial B$. Since τ is a conformal mapping of B onto $B \cup B_r(w_0)$, the tangent of the curve $\tilde{C}_r(w_0) := \tau^{-1}(C_r(w_0))$ tends to a limit as w tends along $C_r(w_0)$ to one of the endpoints w_1 and w_2 of $C_r(w_0)$, and this limit is different from the tangent of ∂B. This can either be seen by an explicit computation of τ or from a general theorem of the theory of conformal mappings (cf. Carathéodory [4], p. 91). Therefore the above number ρ can be selected in such a way that $\partial B_\rho(0)$ intersects $\tilde{C}_r(w_0)$ in exactly two points z_3 and z_4, and that the curve $\tau(\partial B_\rho(0))$ is completely contained in $\Omega_\varepsilon := B_r(w_0) \cup B \setminus B_{1-\varepsilon}(0)$ where ε is chosen to satisfy $0 < \varepsilon < r \le R_0$. Because of (34), it follows that

$$2D_{\Omega_\varepsilon}(X) < \delta_0. \tag{45}$$

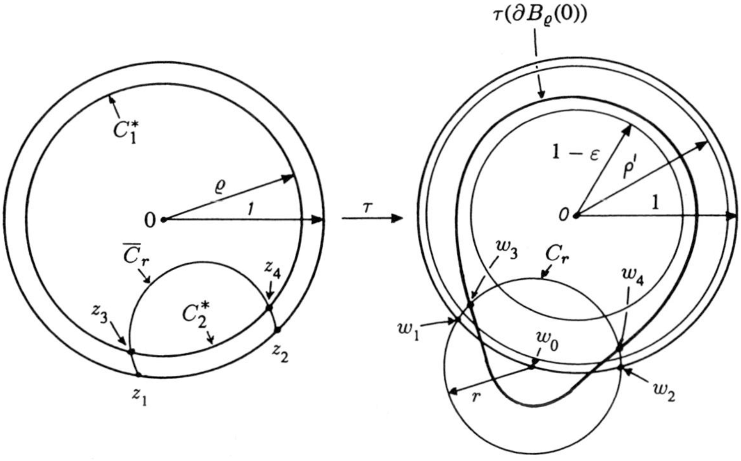

Fig. 2. This sketch illustrates the proof that the comparison surface used in the regularity proof is admissible

We can find a number $\rho' \in (1-\varepsilon, 1)$ such that the trace of the curve

$$X\colon \partial B_{\rho'}(0) \to \mathbb{R}^3$$

is completely contained in $T_{\mu/2}$ and represents the boundary class $[X|_{\partial B}]$ of X. We can also achieve that $\partial B_{\rho'}(0) \setminus B_r(w_0)$ lies between ∂B and $\tau(C_1^*)$, where C_1^* denotes that part of $\partial B_\rho(0)$ which is mapped by τ into $B \setminus B_r(w_0)$. Set $C_2^* := \partial B_\rho(0) \setminus C_1^*$. Moreover, note that the curve $X\colon C_r(w_0) \to \mathbb{R}^3$ remains completely in $T_{\mu/2}$ since its endpoints lie on S and its length is less than or equal to β (cf. (38)), and $\beta \le \mu$ on account of (41).

Finally we infer from (36) and (40) that

$$\int_{B_r(w_0)} |\nabla Y|^2 \, du \, dv \le (1+M^2) \int_{\theta_1(r)}^{\theta_2(r)} |X_\theta(r,\theta)|^2 \, d\theta,$$

and the right-hand side of this inequality is bounded from above by

$$(1+M^2)\pi^{-1}\beta^2.$$

By virtue of (41) we arrive at

$$\int_{B_r(w_0)} |\nabla Y|^2 \, du \, dv \le \delta_0. \tag{46}$$

We infer from (34), (45) and (46) as well as from Theorem 2 of Section 1.1 that all curves to be considered in the following are contained in $T_{\mu/2}$, and that we obtain the following homotopies ($\simeq$). Here C_r' will denote the subarc of $C_r(w_0)$ which connects the intersection points w_3 and w_4 of $C_r(w_0)$ with $\tau(\partial B_\rho(0))$:

$$\begin{aligned} X|_{\partial B_{\rho'}(0)} &\simeq X|_{\partial B_{\rho'}(0)\setminus B_r(w_0)} \cdot X|_{C_r(w_0)\cap \overline{B}_{\rho'}(0)} \\ &\simeq X|_{\tau(C_1^*)} \cdot X|_{C_r'} \\ &\simeq Y \circ \tau|_{C_1^*} \cdot Y|_{C_r'} \\ &\simeq Y \circ \tau|_{C_1^*} \cdot Y \circ \tau|_{C_2^*} = Z|_{\partial B_\rho(0)}. \end{aligned}$$

This completes the proof of the *claim* and thus of the theorem. □

Remark. An Inspection of the proof of Theorem 4 shows that the constant c in (31) will depend on the number R_0 which in turn depends on X. Hence (31) does not yield an a priori estimate of the Morrey seminorm or of the Hölder seminorm of X.

2.6 Hölder Continuity for Stationary Surfaces

In the previous section we have proved that minimizers of the Dirichlet integral in various classes of admissible surfaces corresponding to free boundary problems are Hölder continuous up to their free boundary. The proof has made essential use of the minimum property of the solution of the free boundary problem. In case of partially free problems we have even derived a priori estimates for the Hölder seminorm up to the free boundary. Now we want to establish Hölder continuity of stationary minimal surfaces up to the free boundary. However, we shall have to use a completely different approach in this case as we are not able to derive a priori estimates for the Hölder seminorm or even for the modulus of continuity. In fact, such estimates do not exist, as an inspection of the Schwarz examples discussed in Section 1.9 will show. Consider, for instance, the boundary configuration $\langle \Gamma, S \rangle$ depicted in Fig. 1 which consists of a cylinder surface S and of a polygon Γ with its endpoints on S. For this particular configuration the corresponding semi-free boundary problem possesses infinitely many stationary solutions, all of which are simply connected parts of helicoids, and it is fairly obvious that there is neither an upper bound for their areas (Dirichlet integrals), nor for the length of their free traces, nor for their moduli of continuity.

For this reason we shall not approach the regularity problem by deriving estimates. Instead we want to use an indirect reasoning, first proving continuity up to the boundary by a contradiction argument. We shall constrain our attention to stationary surfaces in the class $\mathcal{C}(\Gamma, S)$ defined for semi-free problems. Similar results can be obtained for stationary solutions of completely free problems without any essential alterations.

We begin by defining Assumption (B) and the notion of admissible support surfaces.

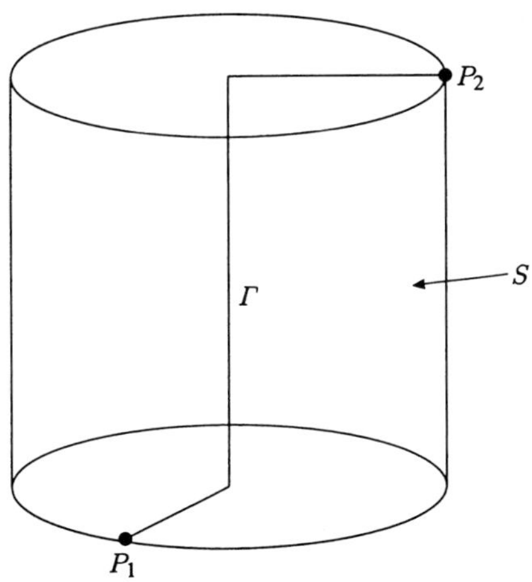

Fig. 1. The stationary solutions of the boundary value problem for the configuration $\langle \Gamma, S \rangle$ cannot be estimated a priori

Definition 1. *An* admissible support surface S *of class* $C^m, m \geq 2$, *(or of class* $C^{m,\beta}$ *with* $0 < \beta \leq 1$*) is a two-dimensional manifold of class* C^m *(or of*

class $C^{m,\beta}$*) embedded in* $\mathbb{R}^3$*, with or without boundary, which has the following two properties:*

(i) *The boundary* ∂S *of the manifold* S *is a regular one-dimensional submanifold of class* C^m *(or* $C^{m,\beta}$*) which can be empty.*

(ii) *Assumption* (B) *is fulfilled.*

Assumption (B), a uniformity condition at infinity, is defined next. We write $x = (x^1, x^2, x^3), y = (y^1, y^2, y^3), \dots$ for points $x, y, \dots$ in $\mathbb{R}^3$.

Definition 2. *A support surface* S *is said to fulfil* **Assumption (B)** *if the following holds true: For each* $x_0 \in S$ *there exist a neighbourhood* $\mathfrak{U}$ *of* x_0 *in* $\mathbb{R}^3$ *and a* C^2*-diffeomorphism* h *of* $\mathbb{R}^3$ *onto itself such that* h *and its inverse* $g = h^{-1}$ *satisfy:*

(i) *The inverse* g *maps* $\mathfrak{U}$ *onto some open ball* $B_R(0) = \{y \in \mathbb{R}^3 : |y| < R\}$ *such that* $g(x_0) = 0; 0 < R < 1$.

(ii) *If* ∂S *is empty, then*

$$g(S \cap \mathfrak{U}) = \{y \in B_R(0) \colon y^3 = 0\}.$$

If ∂S *is nonvoid, then there exists some number* $\sigma = \sigma(x_0) \in [-1, 0]$ *such that*

$$\begin{aligned} g(S \cap \mathfrak{U}) &= \{y \in B_R(0) \colon y^3 = 0, y^1 \geq \sigma\}, \\ g(\partial S \cap \mathfrak{U}) &= \{y \in B_R(0) \colon y^3 = 0, y^1 = \sigma\} \end{aligned}$$

holds true. If $x_0 \in \partial S$*, then* $\sigma = 0$*, and* $\sigma \leq -R$ *if* $\partial S \cap \mathfrak{U}$ *is empty.*

(iii) *There are numbers* m_1 *and* m_2 *with* $0 < m_1 \leq m_2$ *such that the components*

$$g_{ik}(y) = h^l_{y^i}(y) h^l_{y^k}(y)$$

of the fundamental tensor of $\mathbb{R}^3$ *with respect to the curvilinear coordinates* y *satisfies*

$$m_1|\xi|^2 \leq g_{ik}(y)\xi^i\xi^k \leq m_2|\xi|^2 \quad \textit{for all } y, \xi \in \mathbb{R}^3.$$

(iv) *There exists a number* $K > 0$ *such that*

$$\left| \frac{\partial g_{ik}}{\partial y^l}(y) \right| \leq K$$

is satisfied on $\mathbb{R}^3$ *for* $i, k, l = 1, 2, 3$.

We call the pair $\{\mathfrak{U}, g\}$ *an* **admissible boundary coordinate system centered at** x_0.

Let us recall the standard notation used for semifree problems and for the definition of $\mathcal{C}(\Gamma, S)$: The parameter domain B is the semidisk

$$B = \{w = u + iv \colon |w| < 1, v > 0\},$$

the boundary of which consists of the circular arc

$$C = \{w = u + iv\colon |w| = 1, v \geq 0\}$$

and of the segment

$$I = \{w \in \mathbb{R}\colon |w| < 1\}.$$

Moreover, we set

$$Z_d = \{w = u + iv\colon |w| < 1 - d, v > 0\}, \quad d \in (0,1),$$

$$S_r(w_0) = B \cap B_r(w_0), \quad I_r(w_0) = I \cap B_r(w_0),$$

$$C_r(w_0) = \overline{B} \cap \partial B_r(w_0).$$

Next we introduce some terminology with respect to a fixed *admissible boundary coordinate system* $\{\mathfrak{U}, g\}$. Given a minimal surface $X\colon B \to \mathbb{R}^3$, we use the diffeomorphism $g\colon \mathbb{R}^3 \to \mathbb{R}^3$ to define a new mapping $Y \in C^3(B, \mathbb{R}^3)$ by

$$Y(u,v) := g(X(u,v)), \tag{1}$$

whence also

$$X(u,v) = h(Y(u,v)). \tag{1'}$$

In other words, we have

$$Y = g \circ X \quad \text{and} \quad X = h \circ Y.$$

Quite often we use the following *normalization:*

$$\begin{cases} \textit{Let } w_0 \in I, \textit{ and set } x_0 := X(w_0). \textit{ Suppose that } \{\mathfrak{U}, g\} \textit{ is an} \\ \textit{admissible boundary coordinate system for } S \textit{ centered at } x_0. \\ \textit{Then } Y(w_0) = 0. \end{cases} \tag{2}$$

In case of this normalization, the following holds true:

$$\begin{cases} \textit{Let } \Omega \textit{ be a subset of } B \textit{ such that } X(\Omega) \subset \mathfrak{U}. \textit{ Then we have} \\ |Y(w)| < R. \textit{ If } w_0 \in I,\ d = 1 - |w_0|,\ r < d,\ X \in C^0(\overline{S}_r(w_0), \mathbb{R}^3), \\ \textit{and } X\colon I_r(w_0) \to S, \textit{ then we have } y^3(w) = 0 \textit{ for } w \in I_r(w_0). \\ \textit{If } \partial S \textit{ is nonempty and } x_0 \in \partial S, \textit{ then we have } y^1(w) \geq \sigma \\ \textit{for all } w \in I_r(w_0). \end{cases} \tag{2'}$$

For any $Z = (z^1, z^2, z^3) \in H_2^1(\Omega, \mathbb{R}^3), \Omega \subset \mathbb{C}$, we define the *transformed Dirichlet integral* (or: *energy functional*) $E_\Omega(Z)$ by

$$E_\Omega(Z) := \frac{1}{2} \int_\Omega g_{ik}(Z)[z_u^i z_u^k + z_v^i z_v^k]\, du\, dv \tag{3}$$

and we set

$$(3') \qquad E(Z) := E_B(Z).$$

We note that

$$(4) \qquad E_\Omega(Z) = D_\Omega(h \circ Z) \quad \text{for all } Z \in H_2^1(\Omega, \mathbb{R}^3),$$

whence, by (1′),

$$(5) \qquad E_\Omega(Y) = D_\Omega(X), \quad E(Y) = D(X).$$

For every $\phi = (\varphi^1, \varphi^2, \varphi^3) \in H_2^1 \cap L_\infty(B, \mathbb{R}^3)$ and for $X_\varepsilon := h(Y + \varepsilon\phi)$ we have

$$\lim_{\varepsilon \to +0} \frac{1}{\varepsilon}\{E(Y + \varepsilon\phi) - E(Y)\} = \lim_{\varepsilon \to +0} \frac{1}{\varepsilon}\{D(X_\varepsilon) - D(X)\}.$$

The left-hand side is equal to the first variation $\delta E(Y, \phi)$ of E at Y in direction of ϕ, and a straightforward computation yields

$$(6) \qquad \begin{aligned} \delta E(Y, \phi) &= \int_B g_{ik}(Y)\{y_u^i \varphi_u^k + y_v^i \varphi_v^k\}\, du\, dv \\ &\quad + \int_B \frac{1}{2} g_{ik,l}(Y)\{y_u^i y_u^k + y_v^i y_v^k\}\varphi^l\, du\, dv \end{aligned}$$

while the right-hand side tends to

$$(7) \qquad \delta D(X, \Psi_0) = \int_B \langle \nabla X, \nabla \Psi_0 \rangle\, du\, dv, \quad \Psi_0 := h_y(Y)\phi$$

because of

$$X = h(Y), \quad X_\varepsilon = h(Y + \varepsilon\phi) = h(Y) + \varepsilon\Psi(\cdot, \varepsilon) = X + \varepsilon\Psi(\cdot, \varepsilon)$$

with

$$\Psi_0 := \Psi(\cdot, 0) = \lim_{\varepsilon \to 0} \frac{1}{\varepsilon}\{X_\varepsilon - X\} = h_y(Y)\phi.$$

Thus we have

$$(8) \qquad \delta E(Y, \phi) = \delta D(X, \Psi_0).$$

Now we can reformulate the conditions which define stationary points X of the Dirichlet integral in terms of the transformed surfaces $Y = g(X)$. Recall Definition 2 in Section 1.4:

If X is a stationary point of Dirichlet's integral in $\mathcal{C}(\Gamma, S)$ and if $X_\varepsilon = X + \varepsilon\Psi(\cdot, \varepsilon)$ is an outer variation (type II*) of X with $X_\varepsilon \in \mathcal{C}(\Gamma, S)$ for $0 \le \varepsilon < \varepsilon_0$, we have*

$$\lim_{\varepsilon \to +0} \frac{1}{\varepsilon}\{D(X_\varepsilon) - D(X)\} = \int_B \langle \nabla X, \nabla \Psi_0 \rangle\, du\, dv \ge 0$$

for $\Psi_0 := \Psi(\cdot, 0)$, *and this is equivalent to*

$$\delta E(Y, \phi) \geq 0. \tag{9}$$

This holds true in particular for every $\phi \in C_c^\infty(B, \mathbb{R}^3)$ and thus we have both

$$\delta E(Y, \phi) \geq 0 \quad \text{and} \quad \delta E(Y, -\phi) \geq 0$$

whence

$$\delta E(Y, \phi) = 0 \quad \text{for all } \phi \in C_c^\infty(B, \mathbb{R}^3). \tag{10}$$

An integration by parts yields

$$\begin{aligned} &-\int_B g_{il}(Y)\{y_u^i \varphi_u^l + y_v^i \varphi_v^l\}\, du\, dv \\ &= \int_B [g_{il}(Y)\nabla y^i \varphi^l + g_{il,k}(Y)(y_u^i y_u^k + y_v^i y_v^k)\varphi^l]\, du\, dv \end{aligned}$$

for any $\phi \in C_c^\infty(B, \mathbb{R}^3)$, and we infer from (6) and (10) that

$$\int_B [g_{il}(Y)\Delta y^i + \{g_{il,k}(Y) - \tfrac{1}{2} g_{ik,l}(Y)\}(y_u^i y_u^k + y_v^i y_v^k)]\varphi^l\, du\, dv = 0 \quad \text{for all } \phi \in C_c^\infty(B, \mathbb{R}^3). \tag{11}$$

Then the fundamental lemma of the calculus of variations yields

$$g_{il}(Y)\Delta y^i + \{g_{il,k}(Y) - \tfrac{1}{2} g_{ik,l}(Y)\}(y_u^i y_u^k + y_v^i y_v^k) = 0. \tag{12}$$

Introducing the Christoffel symbols of the first kind,

$$\Gamma_{ilk} = \tfrac{1}{2}\{g_{lk,i} - g_{ik,l} + g_{il,k}\}$$

we can rewrite (12) in the form

$$g_{ik}(Y)\Delta y^i + \Gamma_{ilk}(Y)(y_u^i y_u^k + y_v^i y_v^k) = 0 \tag{13}$$

using the symmetry relation $\Gamma_{ilk} = \Gamma_{kli}$, and this implies

$$\Delta y^l + \Gamma_{jk}^l(Y)(y_u^j y_u^k + y_v^j y_v^k) = 0, \quad l = 1, 2, 3, \tag{14}$$

if, as usual, $\Gamma_{jk}^l = g^{lm}\Gamma_{jmk}$ and $(g^{lm}) = (g_{jk})^{-1}$. As one can reverse the previous computations, we have found:

The equation $\Delta X = 0$ *is equivalent to the system* (14).

Moreover, we infer by a straight-forward computation from (1′) and from $g_{ik} = h_{y^i}^l h_{y^k}^l$:

The conformality relations

$$|X_u|^2 = |X_v|^2, \quad \langle X_u, X_v \rangle = 0$$

are equivalent to

$$g_{jk}(Y)y_u^j y_u^k = g_{jk}(Y)y_v^j y_v^k, \quad g_{jk}(Y)y_u^j y_v^k = 0. \tag{15}$$

The advantage of the new coordinate representation $Y(w)$ over the old representation $X(w)$ is that we have transformed the nonlinear boundary condition $X(I) \subset S$ into linear conditions as described in (2′). We pay, however, by having to replace the linear Euler equation $\Delta X = 0$ by the nonlinear system (14). The *variational inequality* (9) will be the key to all regularity results. Together with the conformality relations (15) it expresses the fact that $X = h \circ Y$ is a stationary point of the Dirichlet integral in the class $\mathcal{C}(\Gamma, S)$. (Here Γ can even be empty if X is a stationary point for a completely free boundary configuration; however, to have a clear-cut situation, we restrict our attention to partially free problems.)

The two main steps of this section are:

(i) First we prove continuity in $B \cup I$, that is, up to the free boundary I, using an indirect reasoning. The corresponding result will be formulated as Theorem 1.

(ii) In the second step we establish Hölder continuity on $B \cup I$ employing the hole-filling technique. The corresponding result is stated as Theorem 2.

Let us begin with the *first step* by formulating

Theorem 1. *Let S be an admissible support surface of class C^2, and suppose that $X(w)$ is a stationary point of Dirichlet's integral in the class $\mathcal{C}(\Gamma, S)$. Then $X(w)$ is continuous on $B \cup I$.*

The proof of this result will be based on four lemmata which we are now going to discuss.

Lemma 1. *Let $X: B \to \mathbb{R}^3$ be a minimal surface. For any point $w^* \in B$ we introduce $x^* := X(w^*)$ and the set*

$$K_\rho(x^*) := \{w \in B\colon |X(w) - x^*| < \rho\}.$$

Then, for each open subset Ω of B with $w^* \in \Omega$, we obtain

$$\limsup_{\rho \to +0} \frac{1}{\pi \rho^2} \int_{\Omega \cap K_\rho(x^*)} |\nabla X|^2 \, du \, dv \ge 2.$$

Proof. Fix some $w^* \in B$ and some Ω in B with $w^* \in \Omega$. We can assume that $x^* = X(w^*) = 0$. Then we introduce the set

$$\mathcal{U}_\rho := \{w\colon w = w^* + te^{i\theta}, t \ge 0, \theta \in \mathbb{R}, |X(w^* + re^{i\theta})| < \rho \text{ for all } r \in [0, t]\}.$$

Clearly $\mathcal{U}_\rho$ is an open set with $w^* \in \mathcal{U}_\rho$, and we have

$$\mathcal{U}_\rho \Subset \Omega \quad \text{for } 0 < \rho \ll 1$$

and therefore

$$\mathcal{U}_\rho \Subset \Omega \cap K_\rho(x^*) \quad \text{for } 0 < \rho \ll 1.$$

Hence it suffices to prove

$$\limsup_{\rho \to +0} \frac{1}{\pi \rho^2} \int_{\mathcal{U}_\rho} |\nabla X|^2 \, du \, dv \geq 2.$$

This relation is, however, an immediate consequence of Proposition 2 in Section 3.2 of Vol. 1. □

Lemma 2. *For each $X \in C^1(B, \mathbb{R}^3)$, every $w_0 \in I$, and each $\rho \in (0, 1 - |w_0|)$, there is a number r with $\rho/2 \leq r \leq \rho$ such that*

$$\operatorname{osc}_{C_r(w_0)} X \leq (\pi / \log 2)^{1/2} \left\{ \int_{S_\rho(w_0)} |\nabla X|^2 \, du \, dv \right\}^{1/2}.$$

Proof. Let us introduce polar coordinates r, θ about w_0 setting $w = w_0 + re^{i\theta}$ and $X(r, \theta) = X(w)$. Then, for $0 \leq \theta_1 \leq \theta_2 \leq \pi$, we obtain

$$|X(r, \theta_2) - X(r, \theta_1)| \leq \int_{\theta_1}^{\theta_2} |X_\theta(r, \theta)| \, d\theta \leq \sqrt{\pi p(r)}$$

where we have set

$$p(r) := \int_0^\pi |X_\theta(r, \theta)|^2 \, d\theta.$$

If $\rho/2 \leq r \leq \rho$, it follows that

$$\int_{\rho/2}^{\rho} p(r) \frac{dr}{r} \leq \int_{S_\rho(w_0)} |\nabla X|^2 \, du \, dv.$$

Consequently, there is a number $r \in [\rho/2, \rho]$ such that

$$\left(\int_{\rho/2}^{\rho} \frac{dt}{r} \right) p(r) \leq \int_{S_\rho(w_0)} |\nabla X|^2 \, du \, dv$$

or

$$p(r) \leq \frac{1}{\log 2} \int_{S_\rho(w_0)} |\nabla X|^2 \, du \, dv,$$

and the assertion is proved. □

Lemma 3. *Let $w_0 \in I, r \in (0, 1 - |w_0|)$, and $X \in C^1(B, \mathbb{R}^3)$. Assume also that there are positive numbers α_1 and α_2 such that*

$$\operatorname{osc}_{C_r(w_0)} X \leq \alpha_1$$

and

$$\sup_{w^* \in S_r(w_0)} \inf_{w \in C_r(w_0)} |X(w) - X(w^*)| \le \alpha_2.$$

Then we obtain

$$\text{osc}_{S_r(w_0)} X \le 2\alpha_1 + 2\alpha_2.$$

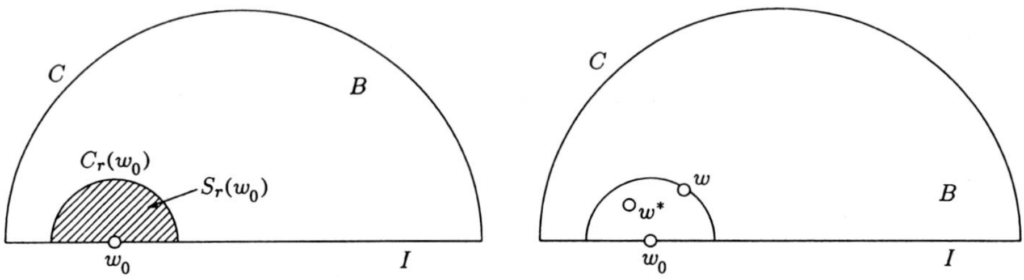

Fig. 2. A domain used in Lemma 3

Proof. Let $w \in S_r(w_0)$ and $w', w'' \in C_r(w_0)$. Then we infer from

$$|X(w) - X(w')| \le |X(w) - X(w'')| + |X(w'') - X(w')|$$

that

$$|X(w) - X(w')| \le \inf_{w'' \in C_r(w_0)} |X(w) - X(w'')| + \text{osc}_{C_r(w_0)} X.$$

Thus we have

$$|X(w) - X(w')| \le \alpha_1 + \alpha_2 \quad \text{for all } w \in S_r(w_0) \text{ and } w' \in C_r(w_0).$$

This yields for arbitrary $w_1, w_2 \in S_r(w_0)$ and $w' \in C_r(w_0)$ the inequalities

$$|X(w_1) - X(w_2)| \le |X(w_1) - X(w')| + |X(w_2) - X(w')| \le 2\alpha_1 + 2\alpha_2,$$

and the assertion is proved. □

Lemma 4. *Let X be a stationary point of Dirichlet's integral in the class $\mathcal{C}(\Gamma, S)$. Suppose also that the support surface S is of class C^2, and let R, K, m_1, m_2 be the constants appearing in Assumption (B) that is to be satisfied by S. Then, for $R_1 := R\sqrt{m_2}$ and for some number $c > 0$ depending only on R, K, m_1, m_2, we have: If for some $r \in (0, 1 - |w_0|)$ and for some number $R_2 \in (0, R_1)$ the inequality*

$$\left[\int_{S_r(w_0)} |\nabla X|^2 \, du \, dv \right]^{1/2} < R_2/c$$

holds true, then it follows that

$$\sup_{w^* \in S_r(w_0)} \inf_{w \in C_r(w_0)} |X(w) - X(w^*)| \le R_2.$$

Before we come to the proof of Lemma 4 which is the main result in step 1 of our discussion, let us turn to the *Proof of Theorem 1.* Since $\int_B |\nabla X|^2 \, du \, dv < \infty$, we have

$$\lim_{r \to +0} \int_{S_r(w_0)} |\nabla X|^2 \, du \, dv = 0$$

for every $w_0 \in I$. Then Lemmata 2, 3, and 4 immediately imply that

$$\lim_{r \to 0} \operatorname{osc}_{S_r(w_0)} X = 0$$

for $w_0 \in I$. In conjunction with $X \in C^0(B, \mathbb{R}^3)$ we then infer that X is continuous on $B \cup I$. □

Proof of Lemma 4. Let $w_0 \in I$ and $0 < r < 1 - |w_0|$. Then we have to prove the following statement:

There is a number $c = c(R, K, m_1, m_2)$ with the property that for any R_2 with $0 < R_2 < R_1$ and for any $w^* \in S_r(w_0)$ with

$$\inf_{w \in C_r(w_0)} |X(w) - X(w^*)| > R_2 \tag{16}$$

the inequality

$$R_2 \le c \left[\int_{S_r(w_0)} |\nabla X|^2 \, du \, dv \right]^{1/2} \tag{17}$$

holds true.

Thus let us consider some $w^* \in S_r(w_0), w_0 \in I, 0 < r < 1 - |w_0|$, and set

$$x^* := X(w^*), \quad \delta(x^*) := \operatorname{dist}(x^*, S).$$

We shall distinguish between two cases, $\delta(x^*) > 0$ and $\delta(x^*) = 0$.

Case (i): $\delta(x^*) > 0$.

Then we proceed as follows: Choose some function $\lambda \in C^1(\mathbb{R})$ with $\lambda' \ge 0$ and with $\lambda(t) = 0$ for $t \le 0$, and introduce the real valued function

$$\varphi(\rho) := \frac{1}{2} \int_{S_r(w_0)} \lambda(\rho - |X - x^*|) |\nabla X|^2 \, du \, dv,$$

for $0 < \rho < \min\{\delta(x^*), d^2 R_2\}$, where R_2 is some number with $0 < R_2 < R_1 := R\sqrt{m_2}$, and where we have set

$$d := \frac{1}{2} \sqrt{\frac{m_1}{m_2}}, \quad 0 < d \le 1/2.$$

Define a test function $\eta(w)$ as

$$\eta(w) := \begin{cases} \lambda(\rho - |X(w) - x^*|)[X(w) - x^*] & \text{for } w \in \overline{S_r}(w_0), \\ 0 & \text{for } w \in B \setminus \overline{S_r}(w_0). \end{cases}$$

We use η to define a family $\{X_\varepsilon\}_{0 \le \varepsilon < \varepsilon_0}$ of outer variations

$$X_\varepsilon(w) := X(w) - \varepsilon \eta(w).$$

On account of

$$|X(w) - x^*| \ge R_2 > \rho \quad \text{for } w \in C_r(w_0)$$

we find that X_ε is of class $H_2^1(B, \mathbb{R}^3)$. Furthermore, we obtain

$$X_\varepsilon(w) = X(w) \quad \text{for } w \in B \setminus S_r(w_0).$$

Hence X and X_ε have the same boundary values on C. Moreover, for $\mathfrak{L}^1$-almost all $w \in I$, we have $X(w) \in S$ and therefore $|X(w) - x^*| \ge \delta(x^*)$ whence $\rho - |X(w) - x^*| < 0$. This implies $\eta(w) = 0$ for $\mathfrak{L}^1$-a.a. $w \in I$. Consequently we obtain $X_\varepsilon \in \mathcal{C}(\Gamma, S)$ for $0 \le \varepsilon < \varepsilon_0$ and for any $\varepsilon_0 > 0$. As $\eta \in H_2^1 \cap L_\infty(B, \mathbb{R}^3)$, we conclude that X_ε is an admissible variation of type II in the sense of Definition 2, Section 1.4. By Section 1.4, (3) and (7), it follows that

$$\int_{S_r(w_0)} \langle \nabla X, \nabla \eta \rangle \, du \, dv \le 0$$

(in fact, even the equality sign holds true since we are allowed to take $\varepsilon \in (-\varepsilon_0, \varepsilon_0)$), and therefore

$$\begin{aligned} \int_{S_r(w_0)} & |\nabla X|^2 \lambda(\rho - |X(w) - x^*|) \, du \, dv \\ & \le \int_{S_r(w_0)} \lambda'(\rho - |X - x^*|) |X - x^*|^{-1} \\ & \quad \cdot \{\langle X_u, X - x^* \rangle^2 + \langle X_v, X - x^* \rangle^2\} \, du \, dv. \end{aligned}$$

By virtue of the conformality relations

$$|X_u|^2 = |X_v|^2, \quad \langle X_u, X_v \rangle = 0$$

we have

$$\{\ldots\} \le \tfrac{1}{2} |\nabla X|^2 |X - x^*|^2,$$

and the factor $1/2$ will be essential for the following reasoning.

It follows that

$$\begin{aligned} (18) \qquad & \int_{S_r(w_0)} |\nabla X|^2 \lambda(\rho - |X - x^*|) \, du \, dv \\ & - \frac{1}{2} \int_{S_r(w_0)} \lambda'(\rho - |X - x^*|) |\nabla X|^2 |X - x^*| \, du \, dv \le 0. \end{aligned}$$

Since

$$\lambda'(\rho - |X - x^*|) = 0 \quad \text{if } |X - x^*| \geq \rho$$

it follows that

$$|X - x^*|\lambda'(\rho - |X - x^*|) \leq \rho\lambda'(\rho - |X - x^*|),$$

and (18) yields

$$2\varphi(\rho) - \rho\varphi'(\rho) \leq 0. \tag{18'}$$

Thus,

$$\frac{d}{d\rho}\{\rho^{-2}\varphi(\rho)\} \geq 0,$$

and it follows that

$$\rho^{-2}\varphi(\rho) \leq (\rho')^{-2}\varphi(\rho') \quad \text{for } 0 < \rho \leq \rho' < R^*, \tag{19}$$

where we have set

$$R^* := \min\{\delta(x^*), d^2 R_2\}.$$

Now we choose λ in such a way that it also satisfies

$$\lambda(t) = 1 \quad \text{for any } t \geq \varepsilon,$$

where ε denotes some positive number (in other words, we consider a family $\{\lambda_\varepsilon\}$ of cut-off functions $\lambda_\varepsilon(t)$ with the parameter ε).

Then we obtain

$$\frac{1}{2}\rho^{-2}\int_{S_r(w_0)\cap K_{\rho-\varepsilon}(x^*)} |\nabla X|^2\, du\, dv \leq \rho^{-2}\varphi(\rho),$$

where we have set

$$K_\tau(x^*) := \{w \in B\colon |X(w) - x^*| < \tau\}.$$

Letting $\varepsilon \to +0$ and then $\rho' \to R^* - 0$, we find that

$$\rho^{-2}\int_{S_r(w_0)\cap K_\rho(x^*)} |\nabla X|^2\, du\, dv \leq (R^*)^{-2}\int_{S_r(w_0)\cap K_{R^*}(x^*)} |\nabla X|^2\, du\, dv$$

taking $\lambda(t) \leq 1$ and (19) into account. Now let $\rho \to +0$. Then it follows from Lemma 1 that

$$2\pi R^{*2} \leq \int_{S_r(w_0)\cap K_{R^*}(x^*)} |\nabla X|^2\, du\, dv. \tag{20}$$

In case that $d^2 R_2 \leq \delta(x^*)$, we have by definition of R^* that $R^* = d^2 R_2$, and (20) implies

$$(21) \qquad R_2 \le \left\{ \frac{1}{2\pi d^4} \int_{S_r(w_0)} |\nabla X|^2 \, du \, dv \right\}^{1/2} \qquad \text{if } d^2 R_2 \le \delta(x^*).$$

Now we treat the opposite case $\delta(x^*) < d^2 R_2$ where we have $R^* = \delta(x^*)$. Because of (20), we have already proved that

$$(22) \qquad 2\pi\delta^2(x^*) \le \int_{S_r(w_0) \cap K_{\delta(x^*)}(x^*)} |\nabla X|^2 \, du \, dv.$$

(The still missing case $\delta(x^*) = 0$ is formally included and will be treated at the end of our discussion.)

First we choose some point $f \in S$ which satisfies

$$|f - x^*| = \operatorname{dist}(x^*, S) = \delta(x^*) < d^2 R_2 \le \tfrac{1}{4} R_2.$$

Then we choose an admissible boundary coordinate system $\{\mathcal{U}, g\}$ for S centered at $x_0 := f$ as described in Definition 2, with the diffeomorphisms g and $h = g^{-1}$. As before we define by $g_{jk}(y)$ the components of the fundamental tensor:

$$g_{jk}(y) := \frac{\partial h^l}{\partial y^j}(y) \frac{\partial h^l}{\partial y^k}(y).$$

Let us introduce the transformed surface $Y(w)$ by

$$Y(w) := g(X(w)) = (y^1(w), y^2(w), y^3(w)),$$

and set

$$\|Y(w)\| := \{g_{jk}(Y(w)) y^j(w) y^k(w)\}^{1/2}.$$

For ρ with $d^{-1}\delta(x^*) < \rho < dR_2$, we define

$$\eta(w) := \begin{cases} \lambda(\rho - \|Y(w)\|) Y(w) & \text{for } w \in \overline{S}_r(w_0), \\ 0 & \text{if } w \in \overline{B} \setminus \overline{S}_r(w_0). \end{cases}$$

Firstly we prove that $\eta \in \mathring{H}_2^1(B, \mathbb{R}^3)$. For this it suffices to show that η vanishes on $C_r(w_0)$. For this purpose, let w be an arbitrary point on $C_r(w_0)$. By assumption (16) we have

$$R_2 \le |X(w) - x^*|,$$

whence

$$R_2 \le \delta(x^*) + |X(w) - f| \le R_2/4 + |X(w) - f|,$$

and this implies

$$R_2/2 \le |X(w) - f| \quad \text{for all } w \in C_r(w_0).$$

On the other hand, since $h(0) = f$, we obtain

$$|X(w) - f| = \left| \int_0^1 h_{y^k}(tY(w))y^k(w)\,dt \right|$$
$$\leq \sqrt{m_2}|Y(w)| \leq (m_2/m_1)^{1/2}\|Y(w)\|.$$

Thus

$$\|Y(w)\| \geq (1/2)(m_1/m_2)^{1/2} R_2 = dR_2 > \rho,$$

and therefore

$$\eta(w) = 0 \quad \text{for } w \in C_r(w_0).$$

For $0 \leq \varepsilon < 1/2$ we consider the family X_ε of surfaces which are defined by

$$X_\varepsilon(w) := h(Y(w) - \varepsilon\eta(w)).$$

We want to show that X_ε is an admissible variation of X which is of type II. In fact, we have $X_\varepsilon \in H_2^1(B, \mathbb{R}^3)$ and $X_\varepsilon(w) = X(w)$ for all $w \in C$ since $\eta(w) = 0$ for $w \in C$. Now we want to show that X_ε maps $\mathfrak{L}^1$-almost all points of I into S. To this end, we pick some $w \in I$ with $X(w) \in S$. If $\eta(w) = 0$, then $X_\varepsilon(w) = X(w)$, and therefore $X_\varepsilon(w) \in S$. On the other hand, if $\eta(w) \neq 0$, we have $\|Y(w)\| < \rho$ and therefore

$$|Y(w)| < \rho/\sqrt{m_1} < dR_2/\sqrt{m_1} = \frac{dR_2}{2\sqrt{m_2}} \left(\frac{1}{2}\sqrt{m_1/m_2} \right)^{-1} = \frac{R_2}{2\sqrt{m_2}}$$
$$< \frac{R_1}{2\sqrt{m_2}} = \frac{R\sqrt{m_2}}{2\sqrt{m_2}} = R/2.$$

Since $X(w) \in S$, this estimate yields $y^3(w) = 0$ (see (2′)), whence

$$[Y(w) - \varepsilon\eta(w)]^3 = 0.$$

Taking the inequalities

$$|Y(w) - \varepsilon\eta(w)| \leq 2|Y(w)| < R$$

into account, we infer that

$$X_\varepsilon = h(Y - \varepsilon\eta) \in \mathcal{C}(\Gamma, S)$$

provided that $\partial S = \emptyset$. This inclusion holds as well if ∂S is nonvoid, since $y^1(w) \geq \sigma$ and $-1 \leq \sigma \leq 0$ implies

$$y^1(w) - \varepsilon\eta^1(w) = y^1(w)\{1 - \varepsilon\lambda(w)\} \geq \sigma\{1 - \varepsilon\lambda(w)\} \geq \sigma.$$

Now we define

$$\Psi(\varepsilon, w) := \begin{cases} \varepsilon^{-1}[h(Y(w) - \varepsilon\eta(w)) - h(Y(w))] & \text{for } \varepsilon > 0, \\ -\frac{\partial h}{\partial y^k}(Y(w))\eta^k(w) & \text{for } \varepsilon = 0. \end{cases}$$

Then we have

$$X_\varepsilon := h(Y - \varepsilon\eta) = X + \varepsilon\Psi(\cdot, \varepsilon) \quad \text{for } 0 \le \varepsilon < 1/2,$$

and Taylor's formula yields

$$\Psi(\varepsilon, w) = \Psi_0(w) + o(\varepsilon)$$

with

$$\Psi_0 := -h_{y^k}(Y)\eta^k \in H_2^1 \cap L_\infty(B, \mathbb{R}^3)$$

and

$$\Psi(\varepsilon, w) \to \Psi_0(w) \quad \text{a.e. on } B \text{ as } \varepsilon \to 0.$$

Moreover, the reader readily checks that

$$|\nabla\Psi(\varepsilon, \cdot)|_{L_2(B)} \le \text{const}$$

holds for some constant independent of $\varepsilon \in [0, 1/2)$. Hence the variations $\{X_\varepsilon\}_{0 \le \varepsilon < 1/2}$ of X are admissible, and we infer from (9) that

$$\delta E(Y, -\eta) \ge 0,$$

or

$$\delta E(Y, \eta) \le 0,$$

which implies

$$\int_{S_r(w_0)} \left[g_{jk}(Y) D_\alpha y^j D_\alpha \eta^k + \frac{1}{2} g_{jk,l}(Y) D_\alpha y^j D_\alpha y^k \eta^l \right] du\, dv \le 0,$$

where we have set

$$u^1 = u, \quad u^2 = v, \quad D_1 = \frac{\partial}{\partial u}, \quad D_2 = \frac{\partial}{\partial v}$$

(summation with respect to Greek indices from 1 to 2, and with respect to Latin indices from 1 to 3).

Then it follows that

$$\begin{aligned}
&\int_{S_r(w_0)} g_{jk}(Y) D_\alpha y^j D_\alpha y^k \lambda(\rho - \|Y\|)\, du\, dv \\
&\quad - \int_{S_r(w_0)} \lambda'(\rho - \|Y\|) g_{mn}(Y)(D_\alpha y^m) y^n \frac{1}{2} \|Y\|^{-1} \\
&\quad \cdot \{2 g_{jk}(Y)(D_\alpha y^j) y^k + g_{jk,l}(y)(D_\alpha y^l) y^j y^k\}\, du\, dv \\
&\le -\frac{1}{2} \int_{S_r(w_0)} g_{jk,l}(y) D_\alpha y^j D_\alpha y^k y^l \lambda(\rho - \|Y\|)\, du\, dv.
\end{aligned}$$

This is equivalent to

$$
\begin{aligned}
(23)\quad &\int_{S_r(w_0)} g_{jk}(Y) D_\alpha y^j D_\alpha y^k \lambda(\rho - \|Y\|)\, du\, dv \\
&\quad - \int_{S_r(w_0)} \|Y\|\lambda'(\rho - \|Y\|) \left\{ \left[g_{jk}(Y) y_u^j \frac{y^k}{\|Y\|} \right]^2 \right. \\
&\qquad\qquad\qquad \left. + \left[g_{jk}(Y) y_v^j \frac{y^k}{\|Y\|} \right]^2 \right\} du\, dv \\
&\le -\frac{1}{2} \int_{S_r(w_0)} g_{jk,l}(Y) D_\alpha y^j D_\alpha y^k y^l \lambda(\rho - \|Y\|)\, du\, dv \\
&\quad + \frac{1}{2} \int_{S_r(w_0)} \lambda'(\rho - \|Y\|) g_{jk}(Y) D_\alpha y^j \frac{y^k}{\|Y\|} g_{mn,l}(Y) y^m y^n D_\alpha y^l\, du\, dv.
\end{aligned}
$$

Now we set

$$
\psi(\rho) := \int_{S_r(w_0)} g_{jk}(Y) D_\alpha y^j D_\alpha y^k \lambda(\rho - \|Y\|)\, du\, dv.
$$

Then, by virtue of the conformality relations (15), we obtain the estimate

$$
\left[g_{jk}(Y) y_u^j \frac{y^k}{\|Y\|} \right]^2 + \left[g_{jk}(Y) y_v^j \frac{y^k}{\|Y\|} \right]^2 \le \frac{1}{2} \|\nabla Y\|^2,
$$

where we have set

$$
\|\nabla Y\|^2 := g_{jk}(Y) D_\alpha y^j D_\alpha y^k.
$$

Moreover we have

$$
\begin{aligned}
\|Y\|\lambda(\rho - \|Y\|) &\le \rho\lambda(\rho - \|Y\|), \\
\|Y\|\lambda'(\rho - \|Y\|) &\le \rho\lambda'(\rho - \|Y\|).
\end{aligned}
$$

Hence the left-hand side of (23) can be estimated from below by

$$
\psi(\rho) - \tfrac{1}{2}\rho\psi'(\rho);
$$

compare (18) and (18′) for an analogous computation.

The first term on the right-hand side of (23) can be estimated from above by

$$
\begin{aligned}
c(n)K &\int_{S_r(w_0)} |\nabla Y|^2 m_1^{-1/2} \|Y\| \lambda(\rho - \|Y\|)\, du\, dv \\
&\le c(n) K m_1^{-3/2} \int_{S_r(w_0)} \rho \|\nabla Y\|^2 \lambda(\rho - \|Y\|)\, du\, dv \\
&\le \tilde{M} \rho \psi(\rho),
\end{aligned}
$$

where we have set

$$\tilde{M} := c(n)Km_1^{-3/2},$$

and where $c(n)$ denotes a constant depending on the space dimension (in our case: $n = 3$).

Analogously, the second term is bounded from above by

$$\begin{aligned}
c(n)K & \int_{S_r(w_0)} \lambda'(\rho - \|Y\|)\|\nabla Y\||\nabla Y||Y|^2 \, du \, dv \\
& \leq c(n)Km_1^{-3/2} \int_{S_r(w_0)} \|Y\|^2 \lambda'(\rho - \|Y\|)\|\nabla Y\|^2 \, du \, dv \\
& \leq \tilde{M}\rho^2\psi'(\rho).
\end{aligned}$$

Thus we have derived the following differential inequality

$$\psi(\rho) - \tfrac{1}{2}\rho\psi'(\rho) \leq \tilde{M}[\rho\psi(\rho) + \rho^2\psi'(\rho)]$$

which is equivalent to

$$-\frac{d}{d\rho}[\rho^{-2}\psi(\rho)] \leq 2M\rho^{-2}\psi(\rho) + M\frac{d}{d\rho}[\rho^{-1}\psi(\rho)]$$

with

$$M := 2\tilde{M}.$$

Multiplying by $e^{2M\rho}$, we obtain

$$0 \leq \frac{d}{d\rho}[e^{2M\rho}\rho^{-2}\psi(\rho)] + Me^{2M\rho}\frac{d}{d\rho}[\rho^{-1}\psi(\rho)].$$

Then by integrating between the limits ρ and $\rho', \rho < \rho'$, and by applying an integration by parts, we infer that

$$\begin{aligned}
0 & \leq [e^{2M\rho}\rho^{-2}\psi(\rho)]_\rho^{\rho'} + \int_\rho^{\rho'} Me^{2M\rho}\frac{d}{d\rho}[\rho^{-1}\psi(\rho)] \, d\rho \\
& = [e^{2M\rho}\rho^{-2}\psi(\rho)]_\rho^{\rho'} + [Me^{2M\rho}\rho^{-1}\psi(\rho)]_\rho^{\rho'} - \int_\rho^{\rho'} 2M^2e^{2M\rho}\rho^{-1}\psi(\rho) \, d\rho.
\end{aligned}$$

Therefore,

$$0 \leq [e^{2M\rho}\rho^{-2}\psi(\rho) + Me^{2M\rho}\rho^{-1}\psi(\rho)]_\rho^{\rho'}$$

whence

$$\rho^{-2}\psi(\rho) \leq \frac{e^{2M\rho'} + \rho' Me^{2M\rho'}}{e^{2M\rho} + \rho Me^{2M\rho}}(\rho')^{-2}\psi(\rho').$$

Applying once again the reasoning which led to (20) (that is, choosing $\lambda = \lambda_\varepsilon$, and letting first $\varepsilon \to +0$ and then $\rho' \to dR_2 - 0$) and setting

$$C(R_2) := (1 + dR_2M)e^{2MdR_2},$$

we arrive at

$$\text{(24)} \qquad \rho^{-2} \int_{S_r(w_0) \cap \{w:\ \|Y(w)\| < \rho\}} \|\nabla Y\|^2 \, du \, dv \le C(R_2)(dR_2)^{-2} \int_{S_r(w_0)} \|\nabla Y\|^2 \, du \, dv.$$

Furthermore,

$$S_r(w_0) \cap \{w:\ |X(w) - f| < 2\delta(x^*)\} \subset S_r(w_0) \cap \{w:\ \|Y(w)\| < \rho\}$$

since

$$Y(w) = \int_0^1 g_{x^j}(tX + (1-t)f)(x^j - f^j) \, dt$$

implies

$$\|Y(w)\| \le (m_2/m_1)^{1/2} |X(w) - f| < 2(m_2/m_1)^{1/2} \delta(x^*) = d^{-1}\delta(x^*) < \rho.$$

For $\delta(x^*) > 0$ and $\rho \to d^{-1}\delta(x^*) + 0$, we then infer from (24) that

$$\frac{d^2}{\delta^2(x^*)} \int_{S_r(w_0) \cap \{w:\ |X(w) - f| < 2\delta(x^*)\}} \|\nabla Y\|^2 \, du \, dv \le C(R_2) d^{-2} R_2^{-2} \int_{S_r(w_0)} \|\nabla Y\|^2 \, du \, dv,$$

and this inequality can be rewritten in the form

$$\text{(25)} \qquad \delta(x^*)^{-2} \int_{S_r(w_0) \cap K_{2\delta(x^*)}(f)} |\nabla X|^2 \, du \, dv \le C(R_2) d^{-4} R_2^{-2} \int_{S_r(w_0)} |\nabla X|^2 \, du \, dv.$$

By virtue of

$$|X(w) - f| \le |X(w) - x^*| + |f - x^*|$$

we obtain

$$K_{\delta(x^*)}(x^*) \subset K_{2\delta(x^*)}(f),$$

and therefore

$$R_2^2 \delta(x^*)^{-2} \int_{S_r(w_0) \cap K_{\delta(x^*)}(x^*)} |\nabla X|^2 \, du \, dv \le C(R_2) d^{-4} \int_{S_r(w_0)} |\nabla X|^2 \, du \, dv.$$

By virtue of (22), the left-hand side is bounded from below by $2\pi R_2^2$. Thus we obtain

$$\text{(26)} \qquad R_2 \le \left\{ \frac{C(R_2)}{2\pi d^4} \int_{S_r(w_0)} |\nabla X|^2 \, du \, dv \right\}^{1/2} \quad \text{if } 0 < \delta(x^*) < d^2 R_2.$$

Combining (21) and (26), we obtain from $C(R_2) \le C(R_1), R_1 = R\sqrt{m_2}$ and $d^{-2} = 4(m_2/m_1)$ that

$$(27) \qquad R_2 \le c\left\{\int_{S_r(w_0)} |\nabla X|^2 \, du \, dv\right\}^{1/2} \quad \text{in Case (i): } \delta(x^*) > 0,$$

if we set

$$c := (2^3\pi^{-1}(m_2/m_1)^2 C(R_1))^{1/2} = c(R, K, m_1, m_2).$$

Case (ii): $\delta(x^*) = 0$.

Here we take $f = x^*$ as the center of an admissible boundary coordinate system $\{\mathcal{U}, g\}$ for S. Then we obtain (24) for any $\rho \in (0, dR_2)$. Setting $\rho' := \rho\sqrt{m_1/m_2}$, it follows that

$$\rho^{-2} \int_{S_r(w_0)\cap K_{\rho'}(x^*)} |\nabla X|^2 \, du \, dv \le C(R_2) d^{-2} R_2^{-2} \int_{S_r(w_0)} |\nabla X|^2 \, du \, dv.$$

Now let $\rho' \to +0$; then another application of Lemma 1 yields

$$\begin{aligned} 2\pi &\le \limsup_{\rho' \to +0} (\rho')^{-2} \int_{S_r(w_0)\cap K_{\rho'}(x^*)} |\nabla X|^2 \, du \, dv \\ &\le \frac{C(R_2)}{4d^4 R_2^2} \int_{S_r(w_0)} |\nabla X|^2 \, du \, dv \end{aligned}$$

whence we obtain

$$(27') \qquad R_2 \le c\left\{\int_{S_r(w_0)} |\nabla X|^2 \, du \, dv\right\}^{1/2} \quad \text{in Case (ii): } \delta(x^*) = 0.$$

Combining (27) and (27′), we arrive at (17). □

Now we turn to the *second step* with the aim to prove

Theorem 2. *Let S be an admissible support surface of class C^2, and suppose that $X(w)$ is a stationary point of Dirichlet's integral in the class $\mathcal{C}(\Gamma, S)$. Then there exists a constant $\alpha \in (0,1)$ such that the following holds true:*

For every $d \in (0,1)$, there exists a constant $c > 0$ such that

$$(28) \qquad \int_{S_r(w_0)} |\nabla X|^2 \, du \, dv \le c r^{2\alpha}$$

holds true for every $w_0 \in \overline{Z}_d$ and for all $r > 0$. In particular, X is of class $C^{0,\alpha}(B \cup I, \mathbb{R}^3)$.

We shall use the following simple but quite effective

Lemma 5. *Let $\varphi(r), 0 < r \le 2R$ be a nondecreasing and nonnegative function satisfying*

$$\varphi(r) \le \theta\varphi(2r) \tag{29}$$

for some $\theta \in (0,1)$ and for all $r \in (0,R]$. Then, for

$$\alpha := {}_2\log\frac{1}{\theta},$$

we have

$$\varphi(r) \le 2^\alpha\varphi(R)(r/R)^\alpha \quad \textit{for all } r \in (0,R]. \tag{30}$$

Proof. For any $r \in (0,R]$ and for any $\nu = 0,1,2,\ldots$, we have

$$\varphi(2^{-\nu}r) \le \theta\varphi(2^{-\nu+1}r).$$

Iterating these inequalities, we obtain

$$\varphi(2^{-\nu}r) \le \theta^\nu\varphi(r) \quad \text{for } 0 < r \le R.$$

Fix some $r \in (0,R]$. Then there exists some integer $\nu \ge 0$ such that

$$2^{-\nu-1} < r/R \le 2^{-\nu}.$$

Since $\theta = 2^{-\alpha}$ and $\varphi(r)$ is nondecreasing, we see that

$$\varphi(r) \le \varphi(2^{-\nu}R) \le \theta^\nu\varphi(R) \le 2^{-\nu\alpha}\varphi(R) \le 2^\alpha\varphi(R)(r/R)^\alpha. \qquad \square$$

For later use we note a generalization of Lemma 5.

Lemma 6. *Let $\varphi(r), 0 < r \le 2R$, by a nondecreasing and nonnegative function satisfying*

$$\varphi(r) \le \theta\{\varphi(2r) + r^\sigma\} \tag{31}$$

for some $\theta \in (0,1)$, $\sigma > 1$, $0 < R < 1$, and for all $r \in (0,R]$. Then, for $\varepsilon \in (0, o-1)$ and for

$$\theta^* := \max\{\theta, 2^{\varepsilon-\sigma}(\theta R^\varepsilon + 1)\}, \quad \alpha := {}_2\log\frac{1}{\theta^*}, \tag{32}$$

we have

$$\varphi(r) \le 2^\alpha\{\varphi(R) + R^{\sigma-\varepsilon}\}(r/R)^\alpha \quad \textit{for all } r \in (0,R]. \tag{33}$$

Proof. Since $R^\varepsilon\theta < 1$ and $2^{\varepsilon-\sigma} < 1/2$, we infer that

$$2^{\varepsilon-\sigma}(\theta R^\varepsilon + 1) < 1$$

and therefore $0 < \theta^* < 1$. Set

$$\varphi^*(r) := \varphi(r) + r^{\sigma-\varepsilon}.$$

Then, for any $r \in (0, R]$, it follows that

$$\begin{aligned}\varphi^*(r) &\le \theta\varphi(2r) + \theta r^\sigma + r^{\sigma-\varepsilon} = \theta\varphi(2r) + r^{\sigma-\varepsilon}(\theta r^\varepsilon + 1)\\ &\le \theta\varphi(2r) + (2r)^{\sigma-\varepsilon}2^{\varepsilon-\sigma}(\theta r^\varepsilon + 1)\\ &\le \theta^*\{\varphi(2r) + (2r)^{\sigma-\varepsilon}\} = \theta^*\varphi^*(2r).\end{aligned}$$

Applying Lemma 5, we infer

$$\varphi^*(r) \le 2^\alpha\varphi^*(R)(r/R)^\alpha \quad \text{for all } r \in (0, R] \text{ and } \alpha := {}_2\log\frac{1}{\theta^*}$$

whence

$$\varphi(r) \le \varphi(r) + r^{\sigma-\varepsilon} \le 2^\alpha\{\varphi(R) + R^{\sigma-\varepsilon}\}(r/R)^\alpha \quad \text{for } 0 < r \le R. \qquad \square$$

Proof of Theorem 2. We want to show that the growth estimate (28) is satisfied for any $w_0 \in I$. Let us first assume that ∂S is empty. We introduce an admissible boundary coordinate system $\{\mathfrak{U}, g\}$ for S centered at $x_0 := X(w_0)$ with the inverse mapping $h = g^{-1}$, and we set $Y := g(X)$. Then we have $Y \in C^0(B \cup I, \mathbb{R}^3)$ and $Y(w_0) = 0$, and we can find some number $\rho_0 \in (0, 1 - |w_0|)$ such that

$$|Y(w)| \le R/2 \quad \text{for } w \in \overline{S}_{\rho_0}(w_0), \qquad Y^3(w) = 0 \quad \text{for } w \in I \cap \overline{S}_{\rho_0}(w)$$

(cf. Definition 2 for the meaning of R, as well as the discussion following Definition 2).

Suppose that $X_\varepsilon := h(Y - \varepsilon\phi), |\varepsilon| < \varepsilon_0(\phi), \phi = (\varphi^1, \varphi^2, \varphi^3)$, is a family of admissible variations with $X_\varepsilon \in \mathcal{C}(\Gamma, S)$. Then we have

$$(34) \qquad \int_B g_{jk}(Y)D_\alpha y^j D_\alpha\varphi^k\,du\,dv \le -\frac{1}{2}\int_B g_{jk,l}(Y)D_\alpha y^j D_\alpha y^k\varphi^l\,du\,dv.$$

(In fact, equality holds true.) Now let $r \in (0, \rho_0/2]$, and choose some cut-off function $\xi \in C_c^\infty(B_{2r}(w_0))$ with $\xi(w) \equiv 1$ on $B_r(w_0)$ and $0 \le \xi \le 1, |\nabla\xi| \le 2/r$.

Set $T_{2r} := S_{2r}(w_0) \setminus S_r(w_0)$,

$$\omega^1 := \fint_{T_{2r}} y^1\,du\,dv, \quad \omega^2 := \fint_{T_{2r}} y^2\,du\,dv, \quad \omega^3 := 0,$$

where

$$\fint_\Omega \cdots \quad \text{stands for} \quad \frac{1}{\operatorname{meas}\Omega}\int_\Omega \cdots$$

$\phi = (\varphi^1, \varphi^2, \varphi^3), \varphi^k(w) := (y^k(w) - \omega^k)\xi^2(w)$ for $w \in B \cup I$. Then the test vector ϕ is admissible in (34), and we obtain

$$\int_B g_{jk}(Y) D_\alpha y^j D_\alpha y^k \xi^2 \, du \, dv + \frac{1}{2} \int_B g_{jk,l}(Y) D_\alpha y^j D_\alpha y^k (y^l - \omega^l) \xi^2 \, du \, dv$$
$$\leq -2 \int_B g_{jk}(Y) D_\alpha y^j (y^k - \omega^k) \xi D_\alpha \xi \, du \, dv.$$

Hence, for any $\varepsilon > 0$ and some constant $K_1(\varepsilon) > 0$, we find the inequality

$$(35) \quad m_1 \int_B |\nabla Y|^2 \xi^2 \, du \, dv - \frac{1}{2} \int_B |g_{jk,l}(Y)||D_\alpha y^j||D_\alpha y^k||y^l - \omega^l|\xi^2 \, du \, dv$$
$$\leq \varepsilon \int_B |\xi|^2 \|\nabla Y\|^2 \, du \, dv + K_1(\varepsilon) \int_B \|Y - \omega\|^2 |\nabla \xi|^2 \, du \, dv,$$

where $\omega = (\omega^1, \omega^2, \omega^3)$. Since

$$\|\nabla Y\|^2 \leq m_2 |\nabla Y|^2$$

we can absorb the term

$$\varepsilon \int_B \xi^2 \|\nabla Y\|^2 \, du \, dv$$

by the first term on the left-hand side, if we choose

$$\varepsilon = \frac{m_1}{2m_2}.$$

Moreover, the absolute value of the second term of the left-hand side of (35) can be bounded from above by

$$\frac{m_1}{4} \int_B |\nabla Y|^2 \xi^2 \, du \, dv,$$

if we choose $r \in (0, \rho_1)$, where $\rho_1 \in (0, \rho_0/2)$ is a sufficiently small number depending on the modulus of continuity of X. Hence there is a number $K_2 > 0$ such that

$$(36) \quad \int_{S_r(w_0)} |\nabla Y|^2 \, du \, dv \leq \int_{S_{2r}(w_0)} \xi^2 |\nabla Y|^2 \, du \, dv$$
$$\leq K_2 r^{-2} \int_{T_{2r}} |Y - \omega|^2 \, du \, dv$$

holds for all $r \in (0, \rho_1)$.

By Poincaré's inequality, there is a constant $K_3 > 0$ such that

$$(37) \quad \int_{T_{2r}} |Y - \omega|^2 \, du \, dv \leq K_3 r^2 \int_{T_{2r}} |\nabla Y|^2 \, du \, dv$$

is satisfied for $0 < r < \rho_1$. Consequently, there is a constant K_4 such that

$$\int_{S_r(w_0)} |\nabla Y|^2 \, du \, dv \le K_4 \int_{S_r(w_0) \setminus S_r(w_0)} |\nabla Y|^2 \, du \, dv$$

for all $r \in (0, \rho_1)$.

Now we fill the hole $S_r(w_0)$ by adding the term $K_4 \int_{S_r(w_0)} |\nabla Y|^2 \, du \, dv$ to both sides. Then we arrive at

$$(1 + K_4) \int_{S_r(w_0)} |\nabla Y|^2 \, du \, dv \le K_4 \int_{S_{2r}(w_0)} |\nabla Y|^2 \, du \, dv$$

whence, setting

$$\theta := \frac{K_4}{1 + K_4},$$

we attain

$$\int_{S_r(w_0)} |\nabla Y|^2 \, du \, dv \le \theta \int_{S_{2r}(w_0)} |\nabla Y|^2 \, du \, dv$$

for every $r \in (0, \rho_1)$. As $0 < \theta < 1$, we can apply Lemma 5 to $R = \rho_1$ and to $\varphi(r) := \int_{S_r(w_0)} |\nabla Y|^2 \, du \, dv$, thus obtaining

$$\int_{S_r(w_0)} |\nabla Y|^2 \, du \, dv \le 2^{2\alpha} \int_{S_r(w_0)} |\nabla Y|^2 \, du \, dv \left(\frac{r}{\rho_1} \right)^{2\alpha}$$

for $0 < r < \rho_1$, if we set $\alpha := \frac{1}{2} \, {}_2\log \theta$. For

$$K_5 := 2^{2\alpha} (m_2/m_1), \quad (K_5 > 1),$$

and by virtue of

$$\|\nabla Y\|^2 = |\nabla X|^2, \quad m_1 |\nabla Y|^2 \le \|\nabla Y\|^2 \le m_2 |\nabla Y|^2,$$

we obtain

$$\text{(38)} \qquad \int_{S_r(w_0)} |\nabla X|^2 \, du \, dv \le K_5 D(X) (r/\rho_1)^{2\alpha}$$

for all $r \in (0, \rho_1)$, and consequently for all $r > 0$.

Combining (38) in a suitable way with interior estimates for X, we arrive at (28). We can omit this reasoning since it would be a mere repetition of the arguments used in the second part of the proof of Theorem 1 in Section 2.5.

Finally, Morrey's Dirichlet growth theorem yields $X \in C^{0,\alpha}(B \cup I, \mathbb{R}^3)$. Thus we have proved Theorem 2 in the case that ∂S is empty.

The general case where ∂S is not necessarily empty can be settled by a slight modification of our previous reasoning.

First we note the that test function

$$\phi = (0, \varphi^2, \varphi^3), \quad \varphi^k = (y^k - \omega^k) \xi^2, \quad k = 2, 3$$

is admissible in (34), where ω^k and ξ are chosen as before. Then we obtain an inequality which coincides with (35) except for the term

$$m_1 \int_B |\nabla Y|^2 \xi^2 \, du \, dv,$$

which is to be replaced by

$$m_1 \int_B (|\nabla y^2|^2 + |\nabla y^3|^2)\xi^2 \, du \, dv.$$

However, this expression can be estimated from below by

$$\frac{m_1}{1+K^*} \int_B |\nabla Y|^2 \xi^2 \, du \, dv$$

since there is a constant $K^* > 0$ such that

$$|\nabla y^1|^2 \le K^*(|\nabla y^2|^2 + |\nabla y^3|^2)$$

holds true, and this inequality is an immediate consequence of the conformality relations (15), written in the complex form

$$\langle\langle Y_w, Y_w \rangle\rangle = 0,$$

where we have set

$$\langle\langle \xi, \eta \rangle\rangle := g_{jk} \xi^j \eta^k$$

(cf. Section 2.3, proof of Theorem 2, part (i)).

Thus we arrive again at an inequality of the type (36) from where we can proceed as before. This completes the proof of the theorem. □

Remark 1. A close inspection of the proof of Theorem 2 shows that we would obtain a priori estimates for the α-Hölder seminorm in the case that we had bounds on the modulus of continuity of X. Hence only the approach used in the proof of Theorem 1 is indirect.

Remark 2. Without any essential change we can replace the class $\mathcal{C}(\Gamma, S)$ in the previous reasoning by $\mathcal{C}(S)$. In other words, we have analogues to Theorems 1 and 2 for stationary points of Dirichlet's integral in the free boundary class $\mathcal{C}(S)$.

2.7 $C^{1,1/2}$-Regularity

In this section we want to prove $C^{1,1/2}$-regularity of a stationary point X of Dirichlet's integral up to its free boundary. As we have seen in Section 2.4, this regularity result is optimal, that is, we can in general not prove $X \in$

$C^{1,\alpha}(B \cup I, \mathbb{R}^3)$, I being the free boundary, for some $\alpha > 1/2$, if the boundary of the support surface S is nonvoid. On the other hand, if ∂S is empty or if $X|_I$ does not touch ∂S, then one might be able to achieve higher regularity as we shall see in the next section.

As in Section 2.6 we shall restrict our considerations to minimal surfaces with partially free boundaries or, more precisely, to stationary points of Dirichlet's integral in $\mathcal{C}(\Gamma, S)$; stationary points with completely free boundaries can be treated in exactly the same way, and perfectly analogous results hold true.

Consequently we can use the same notation as in Section 2.6. Our main result will be the following

Theorem 1. *Let S be an admissible support surface of class C^4, and suppose that $X(w)$ is a stationary point of Dirichlet's integral in the class $\mathcal{C}(\Gamma, S)$. Then X is of class $C^{1,1/2}(B \cup I, \mathbb{R}^3)$.*

The proof of this result is quite involved; it will be carried out in three steps. In the first step we prove that $X \in H_2^2(Z_d, \mathbb{R}^3)$ for any $d \in (0,1)$, using Nirenberg's difference quotient technique to derive L_2-estimates for $\nabla^2 X$. Secondly, using ideas related to those of Section 2.3, it will be shown that $X \in C^1(B \cup I, \mathbb{R}^3)$. In the third part of our investigation we shall see how the boundary regularity can be pushed up to $X \in C^{1,1/2}(B \cup I, \mathbb{R}^3)$ by applying an appropriate iteration procedure.

Let us note that, assuming $X \in C^0(B \cup I, \mathbb{R}^3)$, all regularity results will be proved directly by establishing a priori estimates. Thus the only indirect proof entering into our discussion is that of Theorem 1 of Section 2.6.

Step 1. L_2-estimates for $\nabla^2 X$ up to the free boundary. Let us begin with a few remarks on difference quotients which either are well known (cf. Nirenberg [1], Gilbarg and Trudinger [1]) or can easily be derived.

We consider some function $Y \in H_2^s(Z_{d_0}, \mathbb{R}^m)$ with $0 < d_0 < 1$ and $m \geq 1$, $s \geq 1$. For $w \in Z_d$ and t with $|t| < d_0 - d$, we define the tangential shift Y_t by

$$Y_t(u,v) := Y(u+t, v)$$

and the tangential difference quotient $\Delta_t Y$ by

$$\Delta_t Y(u,v) = \frac{1}{t}[Y(u+t,v) - Y(u,v)],$$

that is,

$$\Delta_t Y(w) = \frac{1}{t}[Y_t(w) - Y(w)], \quad w = u + iv.$$

Moreover, let $D_u = \frac{\partial}{\partial u}$ be the tangential derivative with respect to the free boundary I. Then we have:

Lemma 1. (i) *Let $Y \in H_2^s(Z_{d_0}, \mathbb{R}^m), s \geq 1, m \geq 1, d_0 \in (0,1), d \in (0, d_0)$, $|t| \leq d_0 - d$. Then $Y_t, \Delta_t Y \in H_2^s(Z_d, \mathbb{R}^m)$, and*

$$\int_{Z_d} |\Delta_t Y|^2 \, du \, dv \le \int_{Z_{d_0}} |D_u Y|^2 \, du \, dv, \quad \lim_{t\to 0} \int_{Z_d} |D_u Y - \Delta_t Y|^2 \, du \, dv = 0.$$

The operators ∇ *and* Δ_t *commute; more precisely,*

$$(\Delta_t \nabla Y)(w) = (\nabla \Delta_t Y)(w) \quad \text{for } w \in Z_d,$$

and similarly

$$(\nabla Y)_t(w) = (\nabla Y_t)(w) \quad \text{for } w \in Z_d.$$

Moreover, we have the product rule

$$\Delta_t(\varphi Y) = (\Delta_t \varphi) Y_t + \varphi \Delta_t Y = (\Delta_t \varphi) Y + \varphi_t \Delta_t Y$$

on Z_d *for scalar functions* φ, *and*

$$\int_B \varphi \Delta_{-t} \psi \, du \, dv = -\int_B (\Delta_t \varphi) \psi \, du \, dv \quad \text{for } 0 < |t| \ll 1$$

if either φ *or* ψ *has compact support in* $B \cup I$.

(ii) *Similarly, if* Y *and* $D_u Y \in L_q(Z_{d_0}, \mathbb{R}^m), q \ge 1$, *then*

$$\int_{Z_d} |\Delta_t Y|^q \, du \, dv \le \int_{Z_{d_0}} |D_u Y|^q \, du \, dv, \quad \lim_{t\to 0} \int_{Z_d} |D_u Y - \Delta_t Y|^q \, du \, dv = 0.$$

(iii) *Finally, if* $Y \in H_2^s(Z_{d_0}, \mathbb{R}^m)$, *then*

$$(\nabla^p Y)_t = \nabla^p Y_t,$$

$$\int_\Omega |\nabla^p Y_t|^2 \, du \, dv = \int_{\Omega_t} |\nabla^p Y|^2 \, du \, dv, \quad 0 < |t| \ll 1,$$

for $0 \le p \le s$ *and* $\Omega_t := \{w + t \colon w \in \Omega\}$, *for any open set* $\Omega \Subset Z_{d_0} \cup I$.

Now we turn to the derivation of L_2-estimates for the second derivatives of X. We begin by linearizing the boundary conditions on X. This will be achieved by introducing suitable new coordinates on $\mathbb{R}^3$. Thus let w_0 be an arbitrary point on I, and set $x_0 := X(w_0)$. Then we choose an admissible boundary coordinate system $\{\mathfrak{U}, g\}$, centered at x_0, as defined in Section 2.6. Let $h = g^{-1}$ and $Y = g \circ X$, i.e., $X = h \circ Y$. Then we can use the discussion at the beginning of Section 2.6; in particular we can employ the formulas (1)–(15) of Section 2.6.

By Theorem 1 of Section 2.6, we know that X and Y are continuous on $B \cup I$, and $Y(w_0) = 0$. Hence there is some number $\rho > 0$ such that

$$|Y(w)| < R \quad \text{for all } w \in \overline{S}_{2\rho}(w_0)$$

and therefore

$$y^3(w) = 0 \quad \text{for } w \in I_{2\rho}(w_0),$$

and, if ∂S is nonempty, we have

$$y^1(w) \geq \sigma \quad \text{for } w \in I_{2\rho}(w_0).$$

Let r be some number with $0 < r < \rho$ which is to be fixed later, and let $\eta(w)$ be some cut-off function of class $C_c^\infty(B_{2r}(w_0))$ with $\eta(w) \equiv 1$ on $B_r(w_0), 0 \leq \eta \leq 1, |\nabla \eta| \leq 2/r$, and $\eta(u,v) = \eta(u,-v)$.

Now we set

$$\phi := \Delta_{-t}\{\eta^2 \Delta_t Y\}. \tag{1}$$

We claim that

$$X_\varepsilon := h(Y + \varepsilon\phi), \quad 0 \leq \varepsilon < \varepsilon_0(\phi), \tag{2}$$

is an admissible variation of X in $\mathcal{C}(\Gamma, S)$ of type II (see Definition 2 of Section 1.4) for some sufficiently small $\varepsilon_0(\phi) > 0$. In fact, we have $\phi \in H_2^1 \cap L_\infty(B, \mathbb{R}^3)$, and

$$\begin{aligned} Y(w) + \varepsilon\phi(w) &= Y(w) + \varepsilon\Delta_{-t}\{\eta^2 \Delta_t Y\}(w) \\ &= \lambda_1 Y_t(w) + \lambda_2 Y_{-t}(w) + (1 - \lambda_1 - \lambda_2) Y(w), \end{aligned}$$

where

$$\lambda_1 := \varepsilon t^{-2} \eta^2(w), \quad \lambda_2 := \varepsilon t^{-2} \eta_{-t}^2(w), \quad 0 < |t| \ll 1.$$

Thus $Y(w) + \varepsilon\phi(w), 0 \leq \varepsilon \leq t^2/2$, is a convex combination of the three points $Y(w), Y_t(w)$, and $Y_{-t}(w)$.

Since $\eta(w) = 0$ for $|w - w_0| \geq 2r$, we obtain

$$\lambda_1(w) = 0, \quad \lambda_2(w) = 0 \quad \text{if } |w - w_0| \geq 2r + |t|, \ w \in \overline{B}.$$

Therefore we have

$$Y(w) + \varepsilon\phi(w) = Y(w) \quad \text{for } |w - w_0| \geq 2r + |t|.$$

On the other hand, if $|w - w_0| < 2r + |t|, w \in \overline{B}$, then we have

$$|w \pm t - w_0| \leq 2r + 2|t|$$

and therefore

$$w, w \pm t \in \overline{S}_{2\rho}(w_0), \quad \text{provided that } |t| < \rho - r.$$

Hence, for $w \in I_{2r+|t|}(w_0)$, the points $Y(w), Y_t(w), Y_{-t}(w)$ are contained in the convex set

$$C_R' := \{y \in \mathbb{R}^3 \colon y^3 = 0, |y| < R\} \quad \text{if } \partial S = \emptyset$$

or in

$$C''_R := \{y \in \mathbb{R}^3 \colon y^3 = 0, y^1 \geq \sigma, |y| < R\} \quad \text{if } \partial S \neq \emptyset$$

respectively, and we have

$$S \cap \mathcal{U} = \begin{cases} h(C'_R) & \text{if } \partial S = \emptyset, \\ h(C''_R) & \text{if } \partial S \neq \emptyset. \end{cases}$$

Thus we obtain

$$X_\varepsilon(w) = h(Y(w) + \varepsilon\phi(w)) \in S \quad \text{for all } w \in I,$$

provided that $0 \leq \varepsilon < t^2/2$ and $|t| < \rho - r$, and clearly

$$X_\varepsilon(w) = X(w) \quad \text{for } w \in C = \partial B \setminus I$$

since $\phi(w) = 0$ on C. Consequently, we have

$$X_\varepsilon = h(Y + \varepsilon\phi) \in \mathcal{C}(\Gamma, S) \quad \text{for } 0 \leq \varepsilon < t^2/2 \text{ and } |t| < \rho - r,$$

and it follows from Section 2.6, (9) that

$$\delta E(Y, \phi) \geq 0.$$

Inserting the expression (1) into this inequality, we obtain

$$\begin{aligned} &\int_B g_{jk}(Y) D_\alpha y^j D_\alpha\{\Delta_{-t}(\eta^2 \Delta_t y^k)\}\, du\, dv \\ &\qquad \geq -\frac{1}{2}\int_B \Delta_{-t}\{\eta^2 \Delta_t y^l\} g_{jk,l}(Y) D_\alpha y^j D_\alpha y^k\, du\, dv, \end{aligned}$$

where $D_1 = \frac{\partial}{\partial u}$, $D_2 = \frac{\partial}{\partial v}$, $u^1 = u$, $u^2 = v$, and an integration by parts yields

$$\begin{aligned} &\int_B \Delta_t[g_{jk}(Y) D_\alpha y^j] D_\alpha(\eta^2 \Delta_t y^k)\, du\, dv \\ &\qquad \leq -\frac{1}{2}\int_B \eta^2 \Delta_t y^l \Delta_t[g_{jk,l}(Y) D_\alpha y^j D_\alpha y^k]\, du\, dv; \end{aligned}$$

see Lemma 1. Since

$$\Delta_t[g_{jk}(Y) D_\alpha y^j] = g_{jk}(Y) D_\alpha \Delta_t y^j + D_\alpha y_t^j \Delta_t g_{jk}(Y)$$

and

$$D_\alpha(\eta^2 \Delta_t y^k) = D_\alpha \eta^2 \Delta_t y^k + \eta^2 D_\alpha \Delta_t y^k,$$

we arrive at

(3)
$$\int_B \eta^2 g_{jk}(Y) D_\alpha \Delta_t y^j D_\alpha \Delta_t y^k \, du\, dv$$
$$\leq - \int_B 2\eta D_\alpha \eta \Delta_t y^k [g_{jk}(Y) D_\alpha \Delta_t y^j + D_\alpha y_t^j \Delta_t g_{jk}(Y)] \, du\, dv$$
$$- \int_B \eta^2 D_\alpha \Delta_t y^k D_\alpha y_t^j \Delta_t g_{jk}(Y) \, du\, dv$$
$$- 1/2 \int_B \eta^2 \Delta_t y^l \Delta_t [g_{jk,l}(Y) D_\alpha y^j D_\alpha y^k] \, du\, dv.$$

The ellipticity condition for (g_{jk}) yields

(4)
$$m_1 \int_B \eta^2 |\nabla \Delta_t Y|^2 \, du\, dv \leq \int_B \eta^2 g_{jk}(Y) D_\alpha \Delta_t y^j D_\alpha \Delta_t y^k \, du\, dv.$$

Moreover, Lemma 1 implies

(5)
$$\Delta_t [g_{jk,l}(Y) D_\alpha y^j D_\alpha y^k]$$
$$= (\Delta_t g_{jk,l}(Y)) D_\alpha y^j D_\alpha y^k + g_{jk,l}(Y_t)(\Delta_t D_\alpha y^j) D_\alpha y^k$$
$$+ g_{jk,l}(Y_t) D_\alpha y_t^j \Delta_t D_\alpha y^k.$$

Furthermore, there is a constant $K^* > 0$ such that

(6)
$$|\Delta_t g_{jk}(Y)| + |\Delta_t g_{jk,l}(Y)| \leq K^* |\Delta_t Y|.$$

On account of (3)–(6), there is a number $c = c(m_2, K, K^*)$ independent of t such that

$$m_1 \int_B \eta^2 |\nabla \Delta_t Y|^2 \, du\, dv$$
$$\leq c \left\{ \int_B r^{-1} \eta |\Delta_t Y| (|\nabla \Delta_t Y| + |\nabla Y_t| |\Delta_t Y|) \, du\, dv \right.$$
$$+ \int_B \eta^2 |\nabla \Delta_t Y| |\nabla Y_t| |\Delta_t Y| \, du\, dv$$
$$\left. + \int_B \eta^2 |\Delta_t Y| (|\Delta_t Y| |\nabla Y|^2 + |\nabla \Delta_t Y| |\nabla Y| + |\nabla \Delta_t Y| |\nabla Y_t|) \, du\, dv \right\}.$$

By means of the elementary inequality

$$2ab \leq \varepsilon a^2 + \frac{1}{\varepsilon} b^2$$

for any $\varepsilon > 0$, we obtain the estimate

$$m_1 \int_B \eta^2 |\nabla \Delta_t Y|^2 \, du\, dv$$
$$\leq \varepsilon \int_B \eta^2 |\nabla \Delta_t Y|^2 \, du\, dv + \frac{c^*}{\varepsilon} \left[r^{-2} \int_{S_{2r}(w_0)} |\Delta_t Y|^2 \, du\, dv \right.$$
$$\left. + \int_B \eta^2 |\Delta_t Y|^2 |\nabla Y|^2 \, du\, dv + \int_B \eta^2 |\Delta_t Y|^2 |\nabla Y_t|^2 \, du\, dv \right].$$

Choosing $\varepsilon := m_1/2$, we can absorb the first integral on the right-hand side by the positive term on the left-hand side, and secondly, we have

$$\begin{aligned}\int_{S_{2r}(w_0)} |\Delta_t Y|^2 \, du \, dv &\le \int_B |D_u Y|^2 \, du \, dv \le \int_B |\nabla Y|^2 \, du \, dv \\ &\le m_1^{-1} \int_B \|\nabla Y\|^2 \, du \, dv = m_1^{-1} \int_B |\nabla X|^2 \, du \, dv \\ &= 2m_1^{-1} D(X).\end{aligned}$$

Thus we arrive at

$$\begin{aligned}(7) \quad &\int_{S_{2r}(w_0)} \eta^2 |\nabla \Delta_t Y|^2 \, du \, dv \\ &\quad \le c^{**} \left[r^{-2} D(X) + \int_{S_{2r}(w_0)} \eta^2 |\Delta_t Y|^2 (|\nabla Y|^2 + |\nabla Y_t|^2) \, du \, dv \right].\end{aligned}$$

Moreover, we claim that the estimate (28) in Theorem 2 of Section 2.6 implies the existence of some number c_0 independent of r and t such that

$$\begin{aligned}(8) \quad &\int_{S_{2r}(w_0)} \eta^2 |\Delta_t Y|^2 (|\nabla Y|^2 + |\nabla Y_t|^2) \, du \, dv \\ &\quad \le c_0 r^{2\alpha} \left\{ \int_{S_{2r}(w_0)} \eta^2 |\nabla \Delta_t Y|^2 \, du \, dv + r^{-2} D(X) \right\}.\end{aligned}$$

Let us defer the proof of the inequality (8) until we have finished the derivation of the L_2-estimates of $\nabla^2 X$. Then we can proceed as follows:

We choose $r \in (0, \rho)$ so small that $c^{**} c_0 r^{2\alpha} < 1/2$. Then we infer from (7) and (8) the existence of a number c_1 independent of t such that

$$(9) \quad \int_{S_{2r}(w_0)} \eta^2 |\nabla \Delta_t Y|^2 \, du \, dv \le c_1 D(X)$$

holds true for all t with $0 < |t| < \rho - r$. If we let t tend to zero, this inequality yields

$$(10) \quad \int_{S_{2r}(w_0)} \eta^2 |\nabla D_u Y|^2 \, du \, dv \le c_1 D(X)$$

since $Y = g \circ X$ is of class $C^3(B, \mathbb{R}^3)$, and from (8) and (9) we infer

$$(11) \quad \int_{S_{2r}(w_0)} \eta^2 |D_u Y|^2 |\nabla Y|^2 \, du \, dv \le c_2 D(X).$$

Moreover, the conformality relation

$$\|D_u Y\|^2 = \|D_v Y\|^2$$

implies that

$$|D_v Y|^2 \le (m_2/m_1)|D_u Y|^2,$$

whence we obtain

$$\int_{S_{2r}(w_0)} \eta^2 |\nabla Y|^4 \, du \, dv \le c_3 D(X), \tag{12}$$

taking (11) into account.

Moreover, by formula (14) of Section 2.6 we have

$$\Delta y^l + \Gamma^l_{jk}(Y)(y^j_u y^k_u + y^j_v y^k_v) = 0 \quad \text{in } B,$$

whence

$$|D^2_v Y|^2 \le c_4(|D^2_u Y|^2 + |\nabla Y|^4) \quad \text{in } B.$$

Combining the last relation with (10) and (12), we arrive at

$$\int_{S_{2r}(w_0)} \eta^2 |\nabla^2 Y|^2 \, du \, dv \le c_5 D(X) \tag{13}$$

whence

$$\int_{S_r(w_0)} |\nabla^2 Y|^2 \, du \, dv + \int_{S_r(w_0)} |\nabla Y|^4 \, du \, dv \le c_6 D(X). \tag{14}$$

Moreover, from $X = h(Y)$ we obtain

$$\nabla^2 X = h_{yy}(Y) \nabla Y \nabla Y + h_y(Y) \nabla^2 Y,$$

and therefore

$$|\nabla^2 X|^2 \le c_7(|\nabla Y|^4 + |\nabla^2 Y|^2).$$

By virtue of (14) it follows that

$$\int_{S_r(w_0)} |\nabla^2 X|^2 \, du \, dv \le c_8 D(X). \tag{15}$$

This is the desired estimate of $\nabla^2 X$.

Before we summarize the results of our investigation, we want to prove the estimate (8) which, so far, has remained open. We shall see how c_0 depends on X, and this will inform us about the dependence of the numbers $c_1, \dots, c_8$ on X.

The estimate (8) will be derived from Theorem 2 of Section 2.6 and from the following calculus inequality:

Lemma 2. *Let Ω be an open set in $\mathbb{C}$ of finite measure, and define $d \geq 0$ by the relation $\pi d^2 = \operatorname{meas} \Omega$. Suppose also that $q \in L_1(\Omega)$ is a function such that*

$$\int_{\Omega \cap B_r(w_0)} |q(w)|\, du\, dv \leq Q r^{2\alpha} \tag{16}$$

holds for some number $Q \geq 0$, for some exponent $\alpha > 0$ and for all disks $B_r(w_0)$ in $\mathbb{C}$. Then, for any $\nu \in (0, \alpha)$, there is a number $M(\alpha, \nu) > 0$, depending only on α and ν, such that

$$\int_{\Omega \cap B_r(w_0)} |q(w)||\phi(w)|^2\, du\, dv \leq MQD_\Omega(\phi) d^\nu r^{2\alpha - \nu} \tag{17}$$

holds true for all $w_0 \in \mathbb{C}$, for all $r > 0$, and for any function $\phi \in \mathring{H}_2^1(\Omega, \mathbb{R}^m)$, $m \geq 1$.

Proof. As the set $C_c^\infty(\Omega, \mathbb{R}^m)$ is dense in $\mathring{H}_2^1(\Omega, \mathbb{R}^m)$, it is sufficient to prove (17) for all $\phi \in C_c^\infty(\Omega, \mathbb{R}^m)$, taking Fatou's lemma into account.

Thus let $\phi \in C_c^\infty(\Omega, \mathbb{R}^m)$, $w = u^1 + iu^2$, $\zeta = \xi^1 + i\xi^2$, $d^2 w = du^1\, du^2$, $d^2\zeta = d\xi^1\, d\xi^2$. From Green's formula, we infer that

$$\phi(w) = -\frac{1}{2\pi} \int_\Omega |w - \zeta|^{-2} (\xi^\alpha - u^\alpha) D_\alpha \phi(\zeta)\, d^2\zeta$$

is satisfied for any $w \in \Omega$. Set $\Omega_r := \Omega \cap B_r(w_0)$; then we obtain

$$\begin{aligned} &\int_{\Omega_r} |q(w)||\phi(w)|\, d^2 w \\ &\quad \leq \frac{1}{2\pi} \int_{\Omega_r} \int_\Omega |q(w)||w - \zeta|^{-1} |\nabla \phi(\zeta)|\, d^2\zeta\, d^2 w \\ &\quad = \frac{1}{2\pi} \int_{\Omega_r} \int_\Omega |q(w)|^{1/2} |w - \zeta|^{-1+\nu} |q(w)|^{1/2} |w - \zeta|^{-\nu} |\nabla \phi(\zeta)|\, d^2\zeta\, d^2 w \\ &\quad \leq \frac{1}{2\pi} \left[\int_{\Omega_r} \int_\Omega |q(w)||w - \zeta|^{2\nu - 2}\, d^2\zeta\, d^2 w \right]^{1/2} \\ &\qquad \cdot \left[\int_\Omega \int_{\Omega_r} |q(w)||w - \zeta|^{-2\nu} |\nabla \phi(\zeta)|^2\, d^2 w\, d^2\zeta \right]^{1/2}. \end{aligned} \tag{18}$$

By an inequality of E. Schmidt, we have

$$\int_\Omega |w - \zeta|^{2\nu - 2}\, d^2\zeta \leq (\pi/\nu) d^{2\nu};$$

the simple proof of this fact is left to the reader. Then, by (16), we obtain

$$\int_{\Omega_r} \int_\Omega |q(w)||w - \zeta|^{2\nu - 2}\, d^2\zeta\, d^2 w \leq \frac{\pi}{\nu} d^{2\nu} Q r^{2\alpha}. \tag{19}$$

For $s > 0$ and $\zeta \in \mathbb{C}$ we introduce the function

$$\psi(s, \zeta) := \int_{\Omega_r \cap B_s(\zeta)} |q(w)| \, d^2 w.$$

By (16), we have

$$0 \leq \psi(s, \zeta) \leq Q s^{2\alpha} \quad \text{for all } s > 0 \text{ and } \zeta \in \mathbb{C}$$

as well as

$$0 \leq \psi(s, \zeta) \leq Q r^{2\alpha} \quad \text{for all } \zeta \in \mathbb{C}.$$

Introducing polar coordinates ρ, θ about ζ by $w = \zeta + \rho e^{i\theta}$ we have

$$\psi(s, \zeta) = \int_0^s \left(\int_{\Sigma_\rho} |q(\zeta + \rho e^{i\theta})| \, d\theta \right) \rho \, d\rho,$$

where

$$\Sigma_\rho := \{\theta \colon 0 \leq \theta \leq 2\pi, \zeta + \rho e^{i\theta} \in \Omega_r \cap B_s(\zeta)\}.$$

It follows that

$$\frac{d}{ds} \psi(s, \zeta) = s \int_{\Sigma_s} |q(\zeta + s e^{i\theta})| \, d\theta.$$

Case 1. Let $\zeta \in \overline{B}_r(w_0)$. Then we have $|w - \zeta| \leq 2r$ for any $w \in \Omega_r$. Accordingly,

$$\begin{aligned}
\int_{\Omega_r} |w - \zeta|^{-2\nu} |q(w)| \, d^2 w &\leq \int_0^{2r} \int_{\Sigma_s} s^{-2\nu} |q(\zeta + s e^{i\theta})| s \, d\theta \, ds \\
&= \int_0^{2r} s^{-2\nu} \frac{d}{ds} \psi(s, \zeta) \, ds = \lim_{\varepsilon \to +0} \int_\varepsilon^{2r} s^{-2\nu} \frac{d}{ds} \psi(s, \zeta) \, ds \\
&= \lim_{\varepsilon \to +0} [s^{-2\nu} \psi(s, \zeta)]_\varepsilon^{2r} + 2\nu \lim_{\varepsilon \to +0} \int_\varepsilon^{2r} s^{-2\nu - 1} \psi(s, \zeta) \, ds \\
&\leq Q (2r)^{2\alpha - 2\nu} + 2\nu Q \frac{1}{2(\alpha - \nu)} (2r)^{2\alpha - 2\nu} = c(\alpha, \nu) \, Q r^{2\alpha - 2\nu}.
\end{aligned}$$

Case 2. If $\zeta \in \Omega \setminus \overline{B}_r(w_0)$, then we have

$$\zeta_0 := w_0 + \frac{r}{|\zeta - w_0|} (\zeta - w_0) \in \overline{B}_r(w_0).$$

Moreover, for all $w \in \Omega_r$, it follows by a simple geometric consideration (cf. Fig. 1) that

$$|\zeta - w| \geq |\zeta_0 - w|.$$

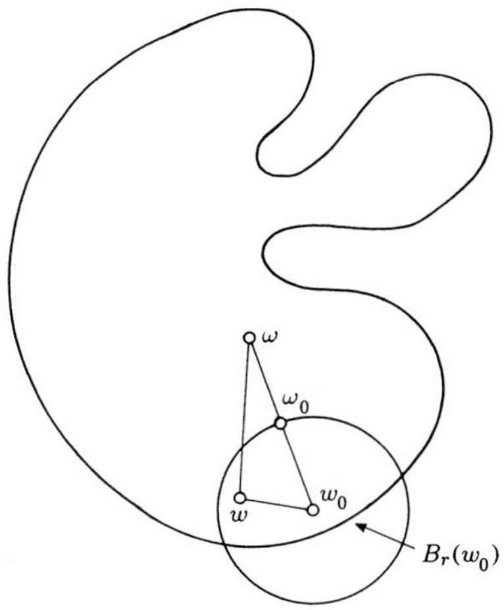

Fig. 1.

Consequently,

$$\int_{\Omega_r} |q(w)||\zeta - w|^{-2\nu}\, d^2w \le \int_{\Omega_r} |q(w)||\zeta_0 - w|^{-2\nu}\, d^2w,$$

and, by case 1,

$$\int_{\Omega_r} |q(w)||\zeta_0 - w|^{-2\nu}\, d^2w \le c(\alpha,\nu)\, Q r^{2\alpha - 2\nu}.$$

Thus we have found that

$$\int_{\Omega_r} |q(w)||\zeta - w|^{-2\nu}\, d^2w \le c(\alpha,\nu)\, Q r^{2\alpha - 2\nu} \quad \text{for all } \zeta \in \Omega.$$

Consequently,

$$\begin{aligned}(20)\qquad &\int_{\Omega}\int_{\Omega_r} |q(w)||w - \zeta|^{-2\nu} |\nabla\phi(\zeta)|^2\, d^2w\, d^2\zeta \\ &\le c(\alpha,\nu)\, Q r^{2\alpha-2\nu} \int_{\Omega} |\nabla\phi(\zeta)|^2\, d^2\zeta.\end{aligned}$$

From (18), (19), and (20) we infer that

$$(21)\qquad \int_{\Omega_r} |q(w)||\phi(w)|\, d^2w \le c^*(\alpha,\nu)\, Q d^\nu D_\Omega^{1/2}(\phi) r^{2\alpha-\nu}.$$

In other words, the function $q^* := q\phi$ satisfies

$$(21')\qquad \int_{\Omega \cap B_r(w_0)} |q^*(w)|\, d^2w \le Q^* r^{2\alpha^*} \quad \text{for all disks } B_r(w_0),$$

where

$$Q^* := c^*(\alpha,\nu) Q d^\nu D_\Omega^{1/2}(\phi), \quad 2\alpha^* := 2\alpha - \nu.$$

Now let ν and μ be two positive numbers such that $\nu + \mu < \alpha$. Then it follows that

$$\alpha^* - \mu = \alpha - \frac{\nu}{2} - \mu > \alpha - (\nu + \mu) > 0.$$

Hence we can apply the estimate (21′) to q^*, Q^*, α^*, μ instead of q, Q, α, ν, and thus we obtain

$$\int_{\Omega_r} |q^*(w)||\phi(w)|\, d^2w \le c^*(\alpha,\mu) Q^* d^\mu D_\Omega^{1/2}(\phi) r^{2\alpha^* - \mu}$$

or equivalently

$$\int_{\Omega\cap B_r(w_0)} |q(w)||\phi(w)|^2\, d^2w \le c^*(\alpha,\nu)c^*(\alpha,\mu) Q d^{\nu+\mu} D_\Omega(\phi) r^{2\alpha-(\nu+\mu)}.$$

Replacing $\nu + \mu$ by ν and $c^*(\alpha,\nu)c^*(\alpha,\mu)$ by $M(\alpha,\nu)$, we arrive at the desired inequality (17). □

Now we come to the *proof of formula* (8). From $Y = g(X)$ it follows that

$$|\nabla Y| \le \sqrt{m_2}|\nabla X|$$

whence by Section 2.6, Theorem 2 (and, in particular, Section 2.6, (28)) we obtain that

$$\int_{B\cap B_\tau(\zeta_0)} |\nabla Y|^2\, du\, dv \le Q\tau^{2\alpha} \tag{22}$$

holds for some constant $Q > 0$, some $\alpha \in (0,1)$, and for all disks $B_\tau(\zeta_0)$. Therefore,

$$\int_{S_{2r}(w_0)\cap B_\tau(\zeta_0)} (|\nabla Y|^2 + |\nabla Y_t|^2)\, du\, dv \le 2Q\tau^{2\alpha}$$

for some $Q > 0, \alpha \in (0,1)$, and for all disks $B_\tau(\zeta_0)$ and all t with $|t| < t_0$ and $0 < t_0 \ll 1$.

Let $\Omega := B_{2r}(w_0), w_0 \in I$, and set

$$q(w) := |\nabla Y(w)|^2 + |\nabla Y_t(w)|^2, \quad \phi(w) := \eta(w)\Delta_t Y(w) \quad \text{for } w \in S_{2r}(w_0)$$

and

$$q(u,v) := q(u,-v), \ \phi(u,v) := \phi(u,-v) \quad \text{for } w = u + iv \in B_{2r}(w_0) \text{ and } v < 0.$$

Applying Lemma 2, we can infer that

$$\int_{\Omega\cap B_\tau(\zeta_0)} |q(w)||\phi(w)|^2\, du\, dv \le 2MQD_\Omega(\phi)(2r)^\nu \tau^{2\alpha-\nu}.$$

In particular, for $\zeta_0 = w_0$ and $\tau = 2r$, we have $\Omega = B_\tau(\zeta_0)$ and therefore

$$\int_{B_{2r}(w_0)} |q(w)||\phi(w)|^2\, du\, dv \le 4MQD_{B_{2r}(w_0)}(\phi)(2r)^{2\alpha}$$

whence, for reasons of symmetry,

$$\begin{aligned}\int_{S_{2r}(w_0)} &\eta^2|\Delta_t Y|^2(|\nabla Y|^2 + |\nabla Y_t|^2)\, du\, dv \\ &\le 4MQ2^{2\alpha}r^{2\alpha}\int_{S_{2r}(w_0)} |\nabla(\eta\Delta_t Y)|^2\, du\, dv.\end{aligned}$$

Moreover,

$$\begin{aligned}|\nabla(\eta\Delta_t Y)|^2 &= |\nabla\eta\Delta_t Y + \eta\nabla\Delta_t Y|^2 \\ &\le 2\eta^2|\nabla\Delta_t Y|^2 + 8r^{-2}|\Delta_t Y|^2.\end{aligned}$$

Setting

(23) $$c_0 := 2^{5+2\alpha} MQ \max\{1, m_1^{-1}\},$$

we arrive at formula (8). From (38) in Section 2.6 it follows that Q is of the form

(24) $$Q = cD(X),$$

where c depends on the diffeomorphism g and on the modulus of continuity of X on $B \cup I$. Hence also the constants $c_1, \ldots, c_6$ are of the form $cD(X)$ with c depending on g and on the modulus of continuity of X.

Let us summarize the results (9)–(15), (22)–(24).

Theorem 2. *Let S be an admissible support surface of class C^3, and suppose that X is a stationary point of Dirichlet's integral in the class $\mathcal{C}(\Gamma, S)$. Then, for any $d \in (0,1)$, there is a constant $c > 0$ depending only on $d, |g|_3, D(X)$, and the modulus of continuity of X such that*

(25) $$\int_{Z_d} (|\nabla^2 X|^2 + |\nabla X|^4)\, du\, dv \le c$$

holds true.

Applying Sobolev's embedding theorem (see Gilbarg and Trudinger [1]), we derive the following result from Theorem 2:

Theorem 3. *Let S be an admissible support surface of class C^3, and suppose that X is a stationary point of Dirichlet's integral in $\mathcal{C}(\Gamma, S)$. Then, for any $d \in (0,1)$ and for any p with $2 < p < \infty$, there is a constant $c > 0$ depending only on $d, p, |g|_3, D(X)$, and the modulus of continuity of X such that*

$$\int_{Z_d} |\nabla X|^p \, du \, dv < c \tag{26}$$

holds true. Moreover both X_u and X_v have an L_2-trace on every compact subinterval of I.

In brief, we have shown that any stationary minimal surface X in $\mathcal{C}(\Gamma, S)$ is of class $H_2^2 \cap H_p^1(Z_d, \mathbb{R}^3)$ for any $d \in (0,1)$ and any p with $2 < p < \infty$, and $X_u, X_v \in L_2(I', \mathbb{R}^3)$ for every $I' \subset\subset I$.

Step 2. Continuity of the first derivatives at the free boundary. The aim of this step is the proof of the following

Theorem 4. *Let S be an admissible support surface of class C^3. Then any stationary point X of Dirichlet's integral in $\mathcal{C}(\Gamma, S)$ is of class $C^1(B \cup I, \mathbb{R}^3)$.*

Proof. We choose $w_0 \in I, x_0 = X(w_0), \rho > 0$, and a boundary coordinate system $\{\mathcal{U}, g\}$ centered at x_0 as before, and we set $Y = g \circ X = (y^1, y^2, y^3)$. Then we have

$$\Delta y^l + \Gamma^l_{jk}(Y) D_\alpha y^j D_\alpha y^k = 0. \tag{27}$$

Hence, for any $\phi = (\varphi^1, \varphi^2, \varphi^3) \in C_c^\infty(S_{2\rho}(w_0) \cup I_{2\rho}(w_0), \mathbb{R}^3)$, the equation

$$\delta E(Y, \phi) = 0 \tag{28}$$

is equivalent to

$$\int_{I_{2\rho}(w_0)} g_{jk}(Y) y_v^j \varphi^k \, du = 0. \tag{29}$$

Case 1. ∂S is empty.

Then ϕ is admissible for (28) if $\varphi^3(w) = 0$ on $I_{2\rho}(w_0)$. We conclude from (29) that

$$\begin{aligned} g_{j1}(Y) y_v^j &= 0 \quad \text{a.e. on } I_{2\rho}(w_0), \\ g_{j2}(Y) y_v^j &= 0 \quad \text{a.e. on } I_{2\rho}(w_0), \\ y^3 &= 0 \qquad\quad \text{on } I_{2\rho}(w_0) \end{aligned} \tag{30}$$

since Theorem 3 implies that both Y_u and Y_v are of class $L_2(I', \mathbb{R}^3)$ for every $I' \Subset I$.

Case 2. ∂S is nonempty.

Then ϕ is admissible for (28) if $\varphi^3(w) = 0$ on $I_{2\rho}(w_0)$ and if $\varphi^1|_{I_{2\rho}(w_0)}$ has its support in $I^+_{2\rho}(w_0) := I_{2\rho}(w_0) \cap \{y^1(w) > \sigma\}$. We conclude from (29) that

$$\begin{aligned} g_{j1}(Y) y_v^j &= 0 \quad \text{a.e. on } I^+_{2\rho}(w_0), \\ g_{j2}(Y) y_v^j &= 0 \quad \text{a.e. on } I_{2\rho}(w_0), \\ y^3 &= 0 \qquad\quad \text{on } I_{2\rho}(w_0). \end{aligned} \tag{31}$$

Now we claim that, for $S \in C^3$ and $S \in C^4$, we can find an admissible coordinate system $\{\mathfrak{U}, g\}$ for S centered at x_0 which is of class C^2 or C^3 respectively, and satisfies

$$(32) \qquad (g_{jk}(y^1, y^2, 0)) = \begin{bmatrix} g_{11}(y^1, y^2, 0) & 0 & 0 \\ 0 & g_{22}(y^1, y^2, 0) & 0 \\ 0 & 0 & 1 \end{bmatrix}$$

for all $(y^1, y^2, 0) \in B_R(0)$ if $\partial S = \emptyset$, or for all $(y^1, y^2, 0) \in B_R(0) \cap \{y^1 \geq \sigma\}$ if $\partial S \neq \emptyset$. Note, however, that we lose an order of differentiability if we pass from S to g in case of the particular coordinate system $\{\mathfrak{U}, g\}$ with property (32). Let us postpone the construction of this coordinate system; first we want to exploit (32) to derive regularity.

The special form of the metric tensor (g_{jk}) simplifies the equations (30) to

$$(33) \qquad \begin{aligned} y_v^1 &= 0 \\ y_v^2 &= 0 \quad \text{a.e. on } I_{2\rho}(w_0) \text{ if } \partial S = \emptyset, \\ y^3 &= 0 \end{aligned}$$

and (31) takes the special form

$$(34) \qquad \begin{aligned} y_v^1 &= 0 \quad \text{a.e. on } I_{2\rho}^+(w_0) \\ y_v^2 &= 0 \quad \text{a.e. on } I_{2\rho}(w_0) \quad \text{if } \partial S \neq \emptyset. \\ y^3 &= 0 \qquad\quad \text{on } I_{2\rho}(w_0). \end{aligned}$$

Furthermore, we infer from Theorem 3 that

$$(35) \qquad \Delta Y \in L_p(S_{2\rho}(w_0), \mathbb{R}^3) \quad \text{for any } p \in (1, \infty),$$

provided that S is of class C^3 which implies $h \in C^2$ and $\Gamma^l_{jk} \in C^0$. In *case* 1, we infer from (33) and (35) by means of classical results from potential theory that $Y \in H^2_p(S_{2r}(w_0), \mathbb{R}^3)$ for any $p \in (1, \infty)$, any $w_0 \in I$, and any $r \in (0, \rho)$; cf. Morrey [8], Theorem 6.3.7, or Agmon, Douglis, and Nirenberg [1, 2], for the pertinent L_p-estimates.

Then we obtain $Y \subset C^{1,\beta}(S_{2r}(w_0), \mathbb{R}^3)$ for all $\beta \in (0, 1)$, taking a Sobolev embedding theorem into account; cf. Gilbarg and Trudinger [1], Chapter 7, or Morrey [8], Theorem 3.6.6.

If $S \in C^4$, then $h \in C^3$ and $\Gamma^l_{jk} \in C^1$, and consequently $\Gamma^l_{jk}(Y) D_\alpha y^j D_\alpha y^k \in C^{0,\beta}(S_{2r}(w_0)), l = 1, 2, 3$. On account of (27) and (33), we then obtain

$$(36) \qquad \Delta Y \in C^{0,\beta}(S_{2r}(w_0), \mathbb{R}^3) \quad \text{for any } \beta \in (0, 1).$$

By classical potential-theoretic results of Korn–Lichtenstein–Schauder, we infer from (33) and (36) that $Y \in C^{2,\beta}(S_{2r}(w_0), \mathbb{R}^3)$ holds for any $\beta \in (0, 1)$ and any $r \in (0, \rho)$. A simple proof can be derived from the Korn–Privalov theorem; see Section 2.1, Lemma 6. Since $h \in C^3$ and $X = h \circ Y$, it follows that

$$X \in C^{2,\beta}(B \cup I, \mathbb{R}^3).$$

Note that the same result can be derived under the weaker assumption $S \in C^3$ if we do not work with the special coordinate system (32) where one derivative is lost. Then we have to use (36) and the more complicated boundary conditions (30). Applying Morrey's results (see [8], Chapter 6), together with a strengthening by F.P. Harth [2], we obtain the desired result.

Thus, if ∂S is empty, we have proved a result which is much stronger than Theorem 4:

Theorem 5. *Let S be an admissible support surface of class C^3, and suppose that ∂S is empty. Then any stationary point X of Dirichlet's integral in $\mathcal{C}(\Gamma, S)$ is of class $C^{2,\beta}(B \cup I, \mathbb{R}^3)$ for any $\beta \in (0,1)$.*

Remark 1. If ∂S is nonempty, the same holds true if $X|_I$ does not touch ∂S.

Remark 2. In addition to Theorem 5, the Schauder–Lichtenstein estimates together with our previous bounds (see Theorem 3) imply that there exists a number c depending only on $d \in (0,1), \beta \in (0,1), |g|_3, D(X)$, and the modulus of continuity of X such that

$$|X|_{2+\beta, Z_d} \le c \tag{37}$$

holds true for any $d \in (0,1)$ and any $\beta \in (0,1)$.

These remarks complete our discussion in the case that ∂S is empty.

Now we turn to case 2, i.e. $\partial S \neq \emptyset$. As before, we have (35), and therefore in particular

$$\Delta y^2, \Delta y^3 \in L_p(S_{2\rho}(w_0)) \quad \text{for any } p \in (1, \infty),$$

and the second and third equation of (35) yield

$$y_v^2 = 0 \quad \text{and} \quad y^3 = 0 \quad \text{a.e. on } I_{2\rho}(w_0).$$

By the same reasoning as in case 1 we first obtain $y^2, y^3 \in H_p^2(S_{2r}(w_0))$ for $p \in (1, \infty)$ and $r \in (0, \rho)$, and then

$$y^2, y^3 \in C^{1,\beta}(S_{2r}(w_0)) \quad \text{for any } \beta \in (0,1) \text{ and } r \in (0, \rho). \tag{38}$$

(For this result, we only use $S \in C^3$, whence $h \in C^2$ and $\Gamma_{jk}^l \in C^0$.)

The function $y^1(w)$ satisfies

$$\begin{gathered} \Delta y^1 \in L_p(S_{2\rho}(w_0)) \quad \text{for any } p \in (1, \infty), \\ y_v^1 = 0 \quad \text{a.e. on } I_{2\rho}^+(w_0), \quad y^1 = \sigma \quad \text{on } I_{2\rho}(w_0) \setminus I_{2\rho}^+(w_0). \end{gathered} \tag{39}$$

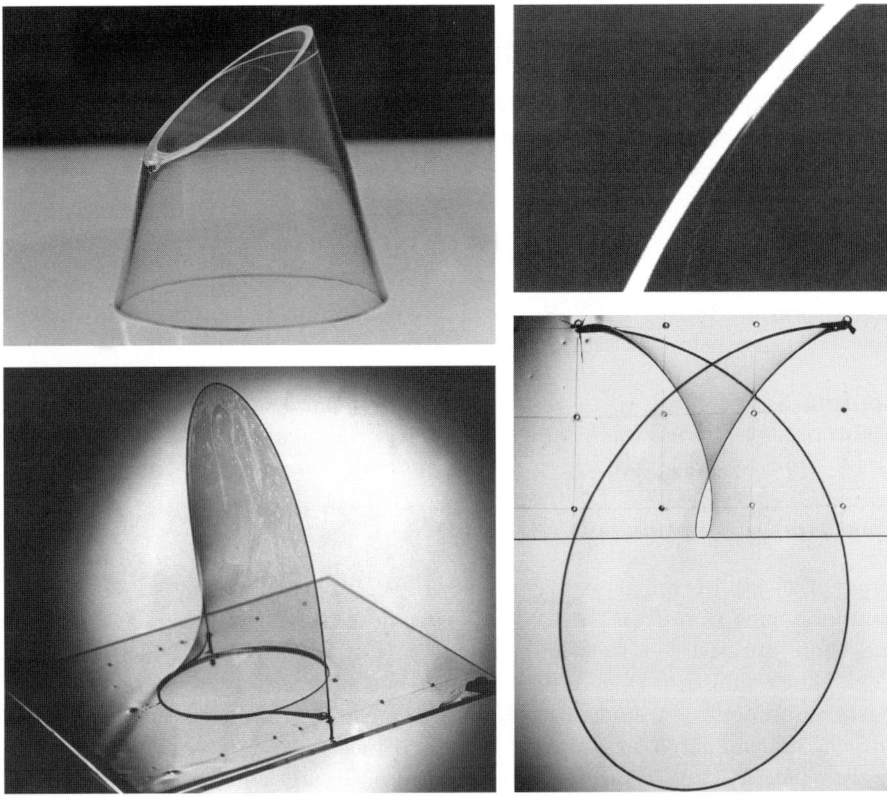

Fig. 2. Soap films attaching smoothly to the boundary of the support surface. Courtesy of E. Pitts (above) and Institut für Leichte Flächentragwerke, Stuttgart – Archive (below)

It is not at all clear how to exploit (39) (cf., however, Section 2.9). Therefore we shall instead use the conformality relations in complex notation,

$$g_{jk}(Y)y_w^j y_w^k = 0, \tag{40}$$

in order to show that $y^1 \in C^1(\overline{S}_{2r}(w_0))$ for some $r \in (0, \rho)$. Our reasoning will be similar as in the proof of Theorem 2 in Section 2.3 (see formulas (16)–(20) of Section 2.3). Since (g_{jk}) is a positive definite matrix, there is some $\gamma > 0$ such that

$$g_{11}(Y(w)) \geq \gamma \quad \text{for all } w \in \overline{S}_{2\rho}(w_0).$$

Hence we can rewrite (40) as

$$\left\{ y_w^1 + \frac{g_{1L}(Y)}{g_{11}(Y)} y_w^L \right\}^2 = \left[\frac{g_{1L}(Y)}{g_{11}(Y)} y_w^L \right]^2 - \frac{g_{LM}(Y)}{g_{11}(Y)} y_w^L y_w^M, \tag{41}$$

where repeated indices L, M are to be summed from 2 to 3. If we introduce the complex-valued function $f(w)$ by

(42) $$f(w) := y_w^1 + \frac{g_{1L}(Y)}{g_{11}(Y)} y_w^L, \quad w \in \overline{S}_{2\rho}(w_0),$$

we infer from (41) and (38) as well as from $Y \in C^{0,\beta}(B \cup I, \mathbb{R}^3)$ that

(43) $$f^2 \in C^{0,\beta}(\overline{S}_{2r}(w_0)) \quad \text{for } 0 < \beta < 1,$$

whence

$$f^2 \in C^0(\overline{S}_{2r}(w_0)).$$

In addition, we have $f \in C^0(S_{2r}(w_0))$. By the following lemma it will be seen that $f(w)$ is continuous on $\overline{S}_{2r}(w_0)$.

Lemma 3. *Let $f(w)$ be a complex-valued continuous function on an open connected set Ω in $\mathbb{C}$ such that its square $f^2(w)$ has a continuous extension to $\overline{\Omega}$. Suppose also that $\partial\Omega$ is non-degenerate in the sense that, for every $w_0 \in \partial\Omega$, there exists a $\delta > 0$ such that $\Omega_\delta(w_0) := \Omega \cap B_\delta(w_0)$ is connected. Then $f(w)$ can continuously be extended to $\overline{\Omega}$.*

Proof. Let w_0 be an arbitrary point on $\partial\Omega$. Then there exists a complex number z such that $f^2(w) \to z$ as $w \to w_0$, $w \in \Omega$. If $z = 0$, then $|f(w)|^2 \to 0$, and therefore $f(w) \to 0$ as $w \to w_0$. If $z \neq 0$, then we choose some $\zeta \neq 0$, $\zeta \in \mathbb{C}$, such that $z = \zeta^2$. We pick an $\varepsilon > 0$ such that $0 < \varepsilon < |\zeta|$. Then there exists a number $\delta > 0$ such that $\Omega_\delta(w_0)$ is connected, and that f maps $\Omega_\delta(w_0)$ into the disconnected set $B_\varepsilon(\beta) \cup B_\varepsilon(-\beta)$. Since $f\colon \Omega \to \mathbb{C}$ is continuous, the image $f(\Omega_\delta(w_0))$ is connected, and therefore already contained in one of the disks $B_\varepsilon(\beta), B_\varepsilon(-\beta)$. Thus $\lim_{w\to w_0} f(w)$ exists and is equal to β or $-\beta$. Set

$$F(w) := \begin{cases} f(w) & \text{for } w \in \Omega \\ \lim_{\tilde{w}\to w} f(\tilde{w}) & \text{for } w \in \partial\Omega \end{cases}.$$

Clearly this function is a continuous extension of f to $\overline{\Omega}$, and the lemma is proved. □

Thus we have found that the function $f(w)$, defined by (42), is continuous on $\overline{S}_{2r}(w_0)$ for some $r \in (0, \rho)$. Since

(42′) $$g(w) := \frac{g_{1L}(w)}{g_{11}(w)} y_w^L$$

is Hölder continuous on $\overline{S}_{2r}(w_0)$, we infer that

$$y_w^1 = f(w) - g(w)$$

is continuous on $\overline{S}_{2r}(w_0)$, and therefore $Y \in C^1(\overline{S}_{2r}(w_0), \mathbb{R}^3)$. This implies $X \in C^1(B \cup I, \mathbb{R}^3)$, and Theorem 4 is proved.

However, we still have to verify that we can find a coordinate system $\{\mathcal{U}, g\}$ centered at x_0 which satisfies (32). To this end, we choose a neighbourhood $\mathcal{U}$ of the point $x_0 \in \mathcal{U}$ and an orthogonal parameter representation $x = t(y^1, y^2)$, $(y^1, y^2) \in P$, of $S \cap \mathcal{U}$ with $x_0 = t(0, 0)$. In other words, we have $\mathcal{F} = 0$, where

$$\mathcal{E} := |t_{y^1}|^2, \quad \mathcal{F} := \langle t_{y^1}, t_{y^2} \rangle, \quad \mathcal{G} := |t_{y^2}|^2$$

are the coefficients of the first fundamental form of S. Moreover, set

$$\mathcal{W} := |t_{y^1} \wedge t_{y^2}| = \sqrt{\mathcal{E}\mathcal{G} - \mathcal{F}^2} = \sqrt{\mathcal{E}\mathcal{G}}$$

and let

$$n := \frac{1}{\mathcal{W}}(t_{y^1} \wedge t_{y^2})$$

be the surface normal of S. If $\partial S = \emptyset$, we can assume that the parameter domain P is given by $P = K_R$ where

$$K_R := \{(y^1, y^2) \colon |y^1|^2 + |y^2|^2 < R^2\}, \quad 0 < R < 1.$$

If $\partial S \neq \emptyset$, we can assume that S is part of a larger surface S_0 such that $S_0 \cap \mathcal{U}$ is represented on K_R in the form $x = t(y^1, y^2), (y^1, y^2) \in K_R$, and that $S \cap \mathcal{U}$ is given by $x = t(y^1, y^2), (y^1, y^2) \in P = K_R \cap \{y^1 \geq \sigma\}, \sigma \in [-1, 0]$. We can also suppose that $\partial S \cap \mathcal{U}$ is represented by t on $K_R \cap \{y^1 = \sigma\}$. Choosing $R \in (0, 1)$ sufficiently small, we can in addition assume that

$$h(y) := t(y^1, y^2) + y^3 n(y^1, y^2), \quad y = (y^1, y^2, y^3) \in B_R(0), \tag{44}$$

provides a diffeomorphism of $B_R(0) = \{y \in \mathbb{R}^3 \colon |y| < R\}$ onto some neighbourhood of x_0 which will again be denoted by $\mathcal{U}$. Then h maps C'_R or C''_R onto $S \cap \mathcal{U}$ if ∂S is void or nonvoid respectively, where

$$\begin{aligned} C'_R &= \{y \in \mathbb{R}^3 \colon y^3 = 0, |y| < R\}, \\ C''_R &= \{y \in \mathbb{R}^3 \colon y^3 = 0, y^1 \geq \sigma, |y| < R\}. \end{aligned}$$

Moreover, we may assume that h can be extended to a diffeomorphism of $\mathbb{R}^3$ onto itself; let g be its inverse. Then $\{\mathcal{U}, g\}$ is an admissible boundary coordinate system for S centered at x_0, which is of class C^2 or C^3 if S is of class C^3 or C^4, respectively (because of the special form (44) of h involving the surface normal n of S, we unfortunately lose one derivate).

The components $g_{jk} = h^l_{y^j} h^l_{y^k}$ of the metric tensor are computed as

$$\begin{aligned} g_{11} &= |h_{y^1}|^2 = \mathcal{E} - 2y^3\mathcal{L} + (y^3)^2|n_{y^1}|^2, \\ g_{22} &= |h_{y^2}|^2 = \mathcal{G} - 2y^3\mathcal{N} + (y^3)^2|n_{y^2}|^2, \\ g_{33} &= |h_{y^3}|^2 = |n|^2 = 1, \\ g_{12} &= g_{21} = \langle h_{y^1}, h_{y^2} \rangle = \mathcal{F} - 2y^3\mathcal{M} + (y^3)^2\langle n_{y^1}, n_{y^2} \rangle, \\ g_{13} &= g_{31} = \langle h_{y^1}, h_{y^3} \rangle = \langle t_{y^1}, n \rangle + y^3\langle n_{y^1}, n \rangle = 0, \\ g_{23} &= g_{32} = \langle h_{y^2}, h_{y^3} \rangle = \langle t_{y^2}, n \rangle + y^3\langle n_{y^2}, n \rangle = 0 \end{aligned}$$

for $y \in B_R(0)$. Hence we obtain

$$(g_{jk}(y^1, y^2, 0)) = \begin{bmatrix} \mathcal{E}(y^1, y^2) & 0 & 0 \\ 0 & \mathcal{G}(y^1, y^2) & 0 \\ 0 & 0 & 1 \end{bmatrix}$$

for $y \in C'_R$ or C''_R respectively, and the proof of Theorem 4 is complete. □

Step 3. Regularity of class $C^{1,1/2}$ at the free boundary.

Now we turn to the final part of our discussion. We are going to prove Theorem 1. Our main tool will be

Lemma 4. *Let $f(u)$ be a complex-valued continuous function of the real variable u on a closed subinterval I' of I, and set $a(u) = \operatorname{Re} f(u), b(u) = \operatorname{Im} f(u)$. We suppose that $a(u)b(u) \equiv 0$ on I', and that there are positive numbers α and c such that $\alpha \le 1$ and*

$$|f^2(u_1) - f^2(u_2)| \le c^2 |u_1 - u_2|^{2\alpha} \quad \textit{for all } u_1, u_2 \in I'. \tag{45}$$

Then it follows that

$$|f(u_1) - f(u_2)| \le 2c|u_1 - u_2|^{\alpha} \quad \textit{for all } u_1, u_2 \in I'. \tag{46}$$

Proof. Let $u_1, u_2 \in I', u_1 \ne u_2$, and set $f_1 := f(u_1), f_2 := f(u_2)$.

(i) Let $c|u_1 - u_2|^\alpha \le |f_1 + f_2|$. Then we obtain

$$\begin{aligned} c|u_1 - u_2|^\alpha |f_1 - f_2| &\le |f_1 + f_2||f_1 - f_2| \\ &= |f_1^2 - f_2^2| \le c^2 |u_1 - u_2|^{2\alpha} \end{aligned}$$

and consequently

$$|f_1 - f_2| \le c|u_1 - u_2|^\alpha.$$

(ii) If $|f_1 + f_2| < c|u_1 - u_2|^\alpha$ and $\operatorname{Re} f_1 = \operatorname{Im} f_2 = 0$, or else $\operatorname{Re} f_2 = \operatorname{Im} f_1 = 0$, then $|f_1 - f_2| = |f_1 + f_2|$, and consequently

$$|f_1 - f_2| < c|u_1 - u_2|^\alpha.$$

(iii) If $|f_1 + f_2| < c|u_1 - u_2|^\alpha$ and $\operatorname{Im} f_1 = \operatorname{Im} f_2 = 0$, then either $|f_1 - f_2| \le |f_1 + f_2|$, and therefore

$$|f_1 - f_2| < c|u_1 - u_2|^\alpha,$$

or else $|f_1 - f_2| > |f_1 + f_2|$, whence

$$|a_1 + a_2| < |a_1 - a_2| \quad \text{for } a_1 := \operatorname{Re} f_1, a_2 := \operatorname{Re} f_2.$$

Since $a(u)$ is continuous on I', there is a number u_0 between u_1 and u_2 such that $a(u_0) := \operatorname{Re} f_0 = 0$, where we have set $f_0 := f(u_0)$. Thus each of the pairs

$\{f_1, f_0\}$ and $\{f_2, f_0\}$ is either in case (i) or in case (ii), and by the previous conclusions we obtain

$$|f_1 - f_0| \le c|u_1 - u_0|^\alpha \quad \text{and} \quad |f_2 - f_0| \le c|u_2 - u_0|^\alpha$$

whence

$$|f_1 - f_2| \le 2c|u_1 - u_2|^\alpha.$$

(iv) If $|f_1 + f_2| < c|u_1 - u_2|^\alpha$ and $\operatorname{Re} f_1 = \operatorname{Re} f_2 = 0$, then we obtain by a reasoning analogous to (iii) that

$$|f_1 - f_2| \le 2c|u_1 - u_2|^\alpha.$$

Because of $a(u)b(u) \equiv 0$ on I', we have exhausted all possible cases and the lemma is proved. □

Proof of Theorem 1. We choose $w_0 \in I, x_0 := X(w_0), \rho > 0, r \in (0, \rho)$, and a boundary coordinate system $\{\mathfrak{U}, g\}$ with (32) as before, and set again $Y = g \circ X$. As we have now assumed that $S \in C^4$, we have $g, h \in C^3$, and therefore $\Gamma^l_{jk} \in C^1$, $\Gamma^l_{jk}(Y) \in C^1(S_{2r}(w_0))$.

Since we have already treated the case $\partial S = \emptyset$, we can concentrate our attention on the case $\partial S \ne \emptyset$ where we have the boundary conditions (34).

Let $f(w)$ be the complex-valued function defined by (42). Then, by (32), it follows that

$$f(w) = y^1_w(w) \quad \text{for all } w \in I_{2r}(w_0). \tag{47}$$

Furthermore, the equations (32) and (41) imply

$$f^2 = -\frac{g_{22}(Y)}{g_{11}(Y)}(y^2_w)^2 - \frac{1}{g_{11}(Y)}(y^3_w)^2 \quad \text{on } I_{2r}(w_0). \tag{48}$$

Since $y^1_w = \frac{1}{2}(y^1_u - iy^1_v)$ and $y^2, y^3 \in C^{1,\beta}(\overline{S}_{2r}(w_0))$ for any $\beta \in (0, 1)$ and $Y \in C^1(S_{2r}(w_0), \mathbb{R}^3)$, we infer that $f(u)$ with $u \in I' := \overline{I}_{2r}(w_0)$ satisfies the assumptions of Lemma 4 for all $\alpha \in (0, 1/2)$. Consequently y^1 is of class $C^{1,\alpha}(\overline{I}_{2r}(w_0))$ for all $\alpha \in (0, 1/2)$. Moreover, the Euler equation

$$\Delta y^1 = -\Gamma^1_{jk}(Y)(y^j_u y^k_u + y^j_v y^k_v)$$

can be written in the form

$$\Delta y^1 + a y^1_u + b y^1_v = p + q|\nabla y^1|^2$$

with functions $a, b, p, q \in C^{0,\beta}(\overline{S}_{2r}(w_0))$. Then an appropriate modification of potential-theoretic estimates (see Gilbarg and Trudinger [1], Widman [1,2]) yields $y^1 \in C^{1,\alpha}(\overline{S}_{2r}(w_0))$ for all $\alpha \in (0, 1/2)$.

Next we use the Euler equations

$$\begin{aligned} \Delta y^2 &= -\Gamma^2_{jk}(Y)(y^j_u y^k_u + y^j_v y^k_v) \\ \Delta y^3 &= -\Gamma^3_{jk}(Y)(y^j_u y^k_u + y^j_v y^k_v) \end{aligned} \quad \text{in } S_{2r}(w_0)$$

and the boundary conditions

$$y^2_v = 0, \quad y^3_v = 0 \quad \text{in } I_{2r}(w_0)$$

for any $r \in (0, \rho)$, as well as $Y \in C^{1,\alpha}(\overline{S}_{2r}(w_0), \mathbb{R}^3)$ to conclude that both y^2 and y^3 are of class $C^{2,\alpha}(\overline{S}_{2r}(w_0, \mathbb{R}^3))$ for any $r \in (0, \rho)$.

By virtue of (48), the function f^2 is Lipschitz continuous on $I' = \overline{I}_{2r}(w_0)$, whence Lemma 4 implies that y^1_w is of class $C^{0,1/2}(I')$. A repetition of the preceding argument with $\alpha = 1/2$ yields $y^1 \in C^{1,1/2}(\overline{S}_{2r}(w_0))$, and consequently $Y \in C^{1,1/2}(\overline{S}_{2r}(w_0), \mathbb{R}^3)$ for any $r \in (0, \rho)$. □

2.8 Higher Regularity in Case of Support Surfaces with Empty Boundaries. Analytic Continuation Across a Free Boundary

In this section we want to consider stationary points of Dirichlet's integral in $\mathcal{C}(\Gamma, S)$ whose support surface S has no boundary. We shall prove that any such surface X is of class $C^{m,\beta}(B \cup I, \mathbb{R}^3)$, provided that S is an admissible support surface of class $C^{m,\beta}$ with $m \geq 3$ and $\beta \in (0,1)$. Moreover, X will be seen to be real analytic on $B \cup I$ if S is real analytic, whence X can be continued analytically across its free boundary I.

Our key tool is the following

Proposition 1. *Let X be a stationary minimal surface in $\mathcal{C}(\Gamma, S)$ and suppose that S is of class $C^m, m \geq 2$. Then X is of class $C^{m-1,\alpha}(B \cup I, \mathbb{R}^3)$ for any $\alpha \in (0,1)$. Moreover, if S is of class $C^{m,\beta}$ for some $m \geq 2$ and some $\beta \in (0,1)$, then X is an element of $C^{m,\beta}(B \cup I, \mathbb{R}^3)$.*

Proof. Recall that, according to Definition 1 in Section 1.4, a stationary minimal surface in $\mathcal{C}(\Gamma, S)$ is an element of $\mathcal{C}(\Gamma, S) \cap C^1(B \cup I, \mathbb{R}^3) \cap C^2(B, \mathbb{R}^3)$ which is harmonic in B, satisfies the conformality relations

$$|X_u|^2 = |X_v|^2, \quad \langle X_u, X_v \rangle = 0,$$

and intersects S perpendicularly along its free trace Σ given by the curve $X\colon I \to \mathbb{R}^3$.

Pick some $w_0 \in I$, and set $x_0 := X(w_0)$. Without loss of generality we can assume that $x_0 = 0$, and that for some cylinder

$$C(R) := \{(x^1, x^2, x^3)\colon |x^1|^2 + |x^2|^2 \leq R^2, |x^3| \leq R\} \tag{1}$$

with $0 < R \ll 1$, the surface $S \cap C(R)$ is given by

(2) $$x^3 = f(x^1, x^2), \quad |x^1|^2 + |x^2|^2 \le R^2,$$

where f is a scalar function of class C^m or $C^{m,\beta}$ if S is of class C^m or $C^{m,\beta}$ respectively. Then S has the nonparametric representation

$$t(x^1, x^2) = (x^1, x^2, f(x^1, x^2)), \quad (x^1, x^2) \in \overline{B}_R(0),$$

with the surface normal

(3) $$n = \left(-\frac{f_1}{W}, -\frac{f_2}{W}, \frac{1}{W}\right) = (n^1, n^2, n^3),$$

where

(4) $$f_1 := f_{x^1}, \quad f_2 := f_{x^2}, \quad W := \sqrt{1 + f_1^2 + f_2^2}.$$

Now we choose some $r > 0$ such that $\overline{S}_r(w_0)$ is mapped by X into the cylinder $C(R)$. Since X_v is perpendicular to S, the vectors $X_v(w)$ and $n(X(w))$ are collinear for any $w \in I_r(w_0) := I \cap B_r(w_0)$. Consequently we have

$$X_v = \langle X_v, n(X)\rangle n(X) \quad \text{on } I_r(w_0),$$

that is,

$$x_v^j = x_v^k n^k(X) n^j(X) \quad \text{on } I_r(w_0) \text{ for } j = 1, 2, 3.$$

If we set

$$\xi^K := f_K / W^2, \quad K = 1, 2,$$

it follows that

(5) $$x_v^K = -\xi^K(x^1, x^2)\{x_v^3 - f_L(x^1, x^2) x_v^L\} \quad \text{on } I_r(w_0), K = 1, 2.$$

(Indices $K, L, M, \ldots$ run from 1 to 2; repeated indices $K, L, M, \ldots$ are to be summed from 1 to 2.) Let us introduce the function $y^3(w)$ by

(6) $$y^3(w) := x^3(w) - f(x^1(w), x^2(w)), \quad w \in \overline{S}_r(w_0).$$

Then we have the boundary condition "$X(w) \in S, w \in I$" transformed into

(7) $$y^3(w) = 0 \quad \text{for any } w \in I_r(w_0),$$

and (5) can be written as

(8) $$x_v^K = -\xi^K(x^1, x^2) y_v^3 \quad \text{on } I_r(w_0) \text{ for } K = 1, 2.$$

Moreover, from (6) and $\Delta X = 0$, we derive the equation

$$\Delta y^3 = -f_{KL}(x^1, x^2) D_\alpha x^K D_\alpha x^L \quad \text{in } S_r(w_0).$$

Thus we have the two boundary value problems

$$(*) \qquad \Delta y^3 = -f_{KL}(x^1, x^2) D_\alpha x^K D_\alpha x^L \quad \text{in } S_r(w_0), y^3 = 0 \quad \text{on } I_r(w_0)$$

with $f_{KL} := f_{x^K x^L}$, and

$$(**) \qquad \Delta x^K = 0 \quad \text{in } S_r(w_0), \quad x_v^K = -\xi^K(x^1, x^2) y_v^3 \quad \text{on } I_r(w_0), K = 1, 2.$$

Now we are going to bootstrap our regularity information by jumping back and forth from $(*)$ to $(**)$, assisted by the relation (6). To this end, we note that $f \in C^m$ or $C^{m,\beta}$; $f_K, \xi^K \in C^{m-1}$ or $C^{m-1,\beta}$; $f_{KL} \in C^{m-2}$ or $C^{m-2,\beta}$ if $S \in C^m$ of $C^{m,\beta}$, respectively.

We begin with the information $X \in C^1(\overline{S}_r(w_0), \mathbb{R}^3)$ assuming that $S \in C^2$. Then we infer from $(*)$ that

$$\Delta y^3 \in L_\infty(S_r(w_0)), \quad y^3 = 0 \quad \text{on } I_r(w_0).$$

whence $y^3 \in C^{1,\alpha}(\overline{S}_\rho(w_0))$ for any $\alpha \in (0,1)$ and $\rho \in (0,r)$. In the following, we shall always rename a number ρ with $0 < \rho < r$ in r; thus we actually obtain a sequence of decreasing numbers r.

Now we can infer from (8) that $x_v^K \in C^{0,\alpha}(I_r(w_0))$, and it follows from $(**)$ that $x^K \in C^{1,\alpha}(\overline{S}_r(w_0)), K = 1, 2$. By virtue of (6), we have

$$(9) \qquad x^3 = y^3 + f(x^1, x^2)$$

whence $X \in C^{1,\alpha}(\overline{S}_r(w_0), \mathbb{R}^3)$ for any $\alpha \in (0,1)$.

Suppose now that $S \in C^{2,\beta}$ holds for some $\beta \in (0,1)$. Then we infer from $(*)$ that

$$\Delta y^3 \in C^{0,\beta}(\overline{S}_r(w_0)), \quad y^3 = 0 \quad \text{on } I_r(w_0),$$

whence $y^3 \in C^{2,\beta}(\overline{S}_r(w_0))$. Now it follows from $(**)$ that $x_v^K \in C^{1,\beta}(I_r(w_0))$ whence $x^K \in C^{2,\beta}(\overline{S}_r(w_0))$, $K = 1, 2$. Then we obtain from (9) that $X \in C^{2,\beta}(\overline{S}_r(w_0), \mathbb{R}^3)$.

Next we assume $S \in C^3$, whence $\Delta y^3 \in C^{0,\alpha}(\overline{S}_r(w_0))$, and $(*)$ yields $y^3 \in C^{2,\alpha}(\overline{S}_r(w_0))$ for all $\alpha \in (0,1)$. Now $(**)$ implies $x_v^K \in C^{1,\alpha}(I_r(w_0))$, and therefore $x^K \in C^{2,\alpha}(\overline{S}_r(w_0))$ for any $\alpha \in (0,1)$ whence $X \in C^{2,\alpha}(\overline{S}_r(w_0))$, taking (9) into account.

In this way we can proceed to prove the proposition. □

Recall that any stationary point X of Dirichlet's integral in $\mathcal{C}(\Gamma, S)$ is a stationary minimal surface in $\mathcal{C}(\Gamma, S)$, provided that X is of class $C^1(B \cup I, \mathbb{R}^3)$ (cf. Section 1.4, Theorem 1). Hence from Proposition 1 we obtain the following result, by taking also Theorem 4 of Section 2.7 into account:

Theorem 1. *Let S be an admissible support surface of class C^m or $C^{m,\beta}$, $m \geq 3$, $\beta \in (0,1)$. Then any stationary point of Dirichlet's integral in $\mathcal{C}(\Gamma, S)$ is of class $C^{m-1,\alpha}(B \cup I, \mathbb{R}^3)$ for any $\alpha \in (0,1)$ or of class $C^{m,\beta}(B \cup I, \mathbb{R}^3)$ respectively.*

Remark 1. The result of Theorem 4 in Section 2.7 is a by-product of the general discussion of that section, the main goal of which was to deal with surfaces S having a nonempty boundary. If ∂S is void, we can use a different method that avoids both the derivation of L_2-estimates and the use of the L_p-theory. This approach is more in the spirit of Section 2.3 and uses results which are closely related to those of Sections 2.1 and 2.2. To this end we choose Cartesian coordinates $x = (x^1, x^2, x^3)$ in the neighbourhood of $0 = X(w_0) \in S$ in such a way that S is given by a nonparametric representation

$$t(x^1, x^2) = (x^1, x^2, f(x^1, x^2)).$$

Moreover, we introduce the *signed distance function*

$$d(x) := \pm \text{dist}(x, S)$$

and the *foot* $a(x)$ of the *perpendicular* line from x onto S which has the *direction* $n(x)$, $|n(x)| = 1$. Then, for all x in a sufficiently small neighbourhood of the origin 0, we have the representation

$$x = a(x) + d(x)n(x). \tag{10}$$

If $x \in S$, then clearly $x = a(x) = t(x^1, x^2)$. Note that $a(x), d(x), n(x)$ are of class C^{m-1} if $S \in C^m$, i.e., their degree of differentiability will in general drop by one. (In fact, it can be shown that $d \in C^m$.)

Let now X be the stationary point that we want to consider, and let $w_0 \in I$, $0 < r \ll 1$. Then we extend $X(w)$ from $\overline{S}_r(w_0)$ to $\overline{B}_r(w_0)$ by defining the extended surface $Z(w)$ as

$$Z(w) := \begin{cases} X(w) & \text{for } w \in \overline{S}_r(w_0), \\ a(X(\overline{w})) - d(X(\overline{w}))n(X(\overline{w})) & \text{for } \overline{w} \in \overline{S}_r(w_0). \end{cases} \tag{11}$$

It turns out that Z is a weak solution of an equation

$$\Delta Z = F(w)|\nabla Z|^2 \quad \text{in } B_r(w_0) \tag{12}$$

with some function $F \in L_\infty(B_r(w_0), \mathbb{R}^3)$, i.e. we have

$$\int_{B_r(w_0)} (\langle \nabla Z, \nabla \varphi \rangle + |\nabla Z|^2 \langle F, \varphi \rangle)\, du\, dv = 0 \tag{13}$$

for all $\varphi \in \overset{\circ}{H}{}_2^1 (B_r(w_0), \mathbb{R}^3) \cap L_\infty(B_r(w_0), \mathbb{R}^3)$. This is proved by first establishing (12) in $S_r(w_0)$ and in $S_r^*(w_0) := B_r(w_0) \setminus \overline{S}_r(w_0)$, and then multiplying (12) by φ. We integrate the resulting equation over $S_r(w_0) \cap \{\text{Im}\, w > \varepsilon\}$ and $S_r^*(w_0) \cap \{\text{Im}\, w < -\varepsilon\}, \varepsilon > 0$, and perform an integration by parts. The boundary terms on $\partial B_r(w_0)$ vanish because of $\varphi = 0$, and the remaining boundary terms cancel in the limit if we add the two equations and let $\varepsilon \to 0$; the resulting equation will be (13). The cancelling effect is derived from a *weak transversality relation* which expresses the fact that X is a stationary point of Dirichlet's integral. Concerning details of the computation, we refer the reader to Jäger [1], pp. 808–812.

Then, by a regularity theorem due to Heinz and Tomi [1] (see also the simplified version of Tomi [1]), it follows that $Z \in C^{1,\alpha}(B_{r'}(w_0), \mathbb{R}^3)$ for some $\alpha \in (0,1)$ and some $r' \in (0,r)$, whence $X \in C^{1,\alpha}(\overline{S}_r(w_0), \mathbb{R}^3)$, which was to be proved.

Remark 2. Another way to avoid L_p-estimates, $p > 2$, is the approach of Step 1 in Section 2.7. Assuming that S is of class C^4, we can estimate the L_2-norms of the third derivatives of a stationary point X up to the free boundary I, and this will imply $X \in C^1(B \cup I, \mathbb{R}^3)$. In fact, one can estimate $|D^s X|_{L_2}$ for any $s \geq 2$ thus obtaining $X \in C^{s-2,\alpha}(B \cup I, \mathbb{R}^3)$. Since one has to assume $S \in C^{s+1}$ to keep this method going, we essentially lose 2 derivatives passing from S to X. These derivatives can only be regained by potential-theoretic methods such as used in the beginning of this section. For details, we refer to Hildebrandt [3].

Analogously to Theorem 1, we obtain

Theorem 1′. *Let S be an admissible support surface of class C^m or $C^{m,\beta}$, $m \geq 3, \beta \in (0,1)$, and let B be the unit disk. Assume also that $X\colon B \to \mathbb{R}^3$ is a minimal surface of class $C^1(B \cup \gamma, \mathbb{R}^3)$ which maps some open subarc γ of ∂B into S, and which intersects S orthogonally along the trace curve $X\colon \gamma \to \mathbb{R}^3$. Then X is of class $C^{m-1,\alpha}(B \cup \gamma, \mathbb{R}^3)$ for any $\alpha \in (0,1)$, or of class $C^{m,\beta}(B \cup \gamma, \mathbb{R}^3)$ respectively.*

Now we come to the second main result of this section.

Theorem 2. *Let S be a real analytic support surface. Then any stationary point of Dirichlet's integral in $\mathcal{C}(\Gamma, S)$ is real analytic in $B \cup I$ and can be extended across I as a minimal surface.*

Note that in Theorem 2 the parameter domain B is the semidisk $\{\operatorname{Im} w > 0, |w| < 1\}$ and I is the boundary interval $\{\operatorname{Im} w = 0, |w| < 1\}$.

Analogously we have

Theorem 2′. *Let S be a real analytic support surface in $\mathbb{R}^3$, and let B be the unit disk. Assume also that X is a minimal surface of class $C^1(B \cup \gamma, \mathbb{R}^3)$ for some open subarc γ of ∂B which is mapped by X into S, and suppose that X intersects S orthogonally along the trace curve $X\colon \gamma \to \mathbb{R}^3$. Then X is real analytic in $B \cup \gamma$ and can be extended across γ as a minimal surface.*

Since both results are proved in the same way, it is sufficient to give the

Proof of Theorem 2. By Proposition 1 we already know that X is of class $C^\infty(B \cup I, \mathbb{R}^3)$. Let $X^*(w)$ be the adjoint minimal surface to $X(w)$ in B, and let

$$f(w) := X(w) + iX^*(w) = (f^1(w), f^2(w), f^3(w))$$

be the holomorphic curve in $\mathbb{C}^3$ with $X = \operatorname{Re} f$ and $X^* = \operatorname{Im} f$, satisfying

$$\langle f'(w), f'(w)\rangle = 0 \quad \text{on } B.$$

We have to show that, for any $u_0 \in I$, there is some $\delta > 0$ such that $f(w)$ can be extended across $I_\delta(u_0) = I \cap B_\delta(u_0)$ as a holomorphic mapping from $B_\delta(u_0)$ into $\mathbb{C}^3$. Without loss of generality we can assume that $u_0 = 0$. Set $B_\delta := B_\delta(0)$, $I_\delta = I_\delta(0)$ and $S_\delta := B \cap B_\delta$. We can also achieve that $f(0) = 0$ holds true. Moreover, by a suitable choice of Cartesian coordinates in $\mathbb{R}^3$, we can accomplish that S in a suitable neighbourhood $\mathfrak{U}$ of 0 is described by

$$S \cap \mathfrak{U} = \{x = (x^1, x^2, x^3) \colon x^3 = \psi(x^1, x^2), |x^1|, |x^2| < R\}$$

for some $R > 0$, where

$$\psi(0,0) = 0, \quad \psi_{x^1}(0,0) = 0, \quad \psi_{x^2}(0,0) = 0.$$

Then there is some $\delta_0 > 0$ such that

$$|x^1(u)| < R, \quad |x^2(u)| < R \quad \text{for all } u \text{ with } |u| \le \delta_0.$$

The vector fields $T_K(x)$ defined by

$$T_1 := (1, 0, \psi_{x^1}), \quad T_2 := (0, 1, \psi_{x^2})$$

are tangent to S. Moreover X_v is orthogonal to X_u, and X_u is tangent to S along I. As X_v is orthogonal to S along I, we have

$$\langle T_K(X), X_v\rangle = 0 \quad \text{on } \overline{I}_{\delta_0} \text{ for } K = 1, 2,$$

whence

$$\langle T_K(X), X_u^*\rangle = 0 \quad \text{on } \overline{I}_{\delta_0} \text{ for } K = 1, 2,$$

and consequently

$$\langle T_K(X), X_u\rangle = \langle T_K(X), f'\rangle \quad \text{on } \overline{I}_{\delta_0},\ K = 1, 2.$$

This can be written as

$$x_u^K + \psi_{x^K}(x^1, x^2) x_u^3 = \frac{d}{dw} f^K + \psi_{x^K}(x^1, x^2) \frac{d}{dw} f^3$$

on $\overline{I}_{\delta_0}$ for $K = 1, 2$, and the identity

$$x^3(u) = \psi(x^1(u), x^2(u)) \quad \text{for all } u \in \overline{I}_{\delta_0}$$

yields

$$-\psi_{x^K}(x^1, x^2) x_u^K + x_u^3 = 0 \quad \text{on } \overline{I}_{\delta_0}$$

(summation with respect to K from 1 to 2 !).

Thus we obtain

$$(14)\qquad \begin{pmatrix} 1 & 0 & \psi_{x^1}(x^1,x^2) \\ 0 & 1 & \psi_{x^2}(x^1,x^2) \\ -\psi_{x^1}(x^1,x^2) & -\psi_{x^2}(x^1,x^2) & 1 \end{pmatrix} \begin{pmatrix} x_u^1 \\ x_u^2 \\ x_u^3 \end{pmatrix} = \begin{pmatrix} f_w^1 + \psi_{x^1}(x^1,x^1) f_w^3 \\ f_w^2 + \psi_{x^2}(x^1,x^2) f_w^3 \\ 0 \end{pmatrix}$$

on $\overline{I}_{\delta_0}$. In matrix notation we may write

$$(15)\qquad A(X)X_u = l(X, f') \quad \text{on } \overline{I}_{\delta_0}$$

with a 3×3-matrix $A(X)$, the determinant of which satisfies

$$\det A(X) = 1 + \psi_{x^1}^2(x^1,x^2) + \psi_{x^2}^2(x^1,x^2) \neq 0 \quad \text{on } \overline{I}_{\delta_0}$$

for $0 < \delta_0 \ll 1$. Thus we obtain

$$(16)\qquad X_u = A^{-1}(X) l(X, f') \quad \text{on } \overline{I}_{\delta_0}.$$

Let us introduce the function $F(w,z)$ for $w \in \mathbb{C}$ and

$$z = (z^1, z^2, z^3) \in \mathbb{C}^3 \quad \text{with } |w| \le \rho_0,\ \operatorname{Im} w \ge 0, \text{ and } |z| \le \rho_1$$

(i.e., $x \in \overline{B}^3_{\rho_1}$) by setting

$$(17)\qquad F(w,z) := A^{-1}(z) l(z, f'(w)).$$

The mapping $F \colon \overline{S}_{\rho_0} \times \overline{B}^3_{\rho_1} \to \mathbb{C}^3$ is of class C^1 (differentiability meant in the "real sense" with respect to w) and holomorphic on $S_{\rho_0} \times B^3_{\rho_1}$.

Then we can write (16) in the form

$$(18)\qquad \frac{d}{du} X(u) = F(u, X(u)) \quad \text{for all } u \in \overline{I}_{\delta_0}$$

if $\delta_0 \in (0, \rho_0]$ is sufficiently small. Since $X(0) = 0$, we obtain

$$(19)\qquad X(u) = \int_0^u F(t, X(t))\, dt \quad \text{for all } u \in \overline{I}_{\delta_0}.$$

By a standard reasoning this integral equation has not more than one solution in $C^0(\overline{I}_{\delta_0}, \mathbb{R}^3)$ since $F(w,z)$ satisfies a Lipschitz condition with respect to $z \in \overline{B}^3_{\rho_1}$, uniformly for all $w \in \overline{S}_{\rho_0}$. By the same reasoning, the complex integral equation

$$(20)\qquad Z(w) = \int_0^w F(\omega, Z(\omega))\, d\omega$$

has exactly one solution $Z(w), w \in \overline{S}_\delta$, in the Banach space $\mathcal{A}(\overline{S}_\delta)$ of functions $Z\colon \overline{S}_\delta \to \mathbb{C}^3$ which are continuous on $\overline{S}_\delta$ and holomorphic in S_δ, provided that $\delta \in (0, \delta_0]$ is chosen sufficiently small (cf. the proof of Theorem 3 in Section 2.3, and in particular the footnote; one uses the standard Picard iteration). By the uniqueness principle, we have

$$Z(u) = X(u) \quad \text{for all } u \in I_\delta,$$

whence

$$\operatorname{Im} Z = 0 \quad \text{on } I_\delta.$$

Hence we can apply Schwarz's reflection principle, thus obtaining that

$$Z(w) := \overline{Z(\overline{w})} \quad \text{for } w \in B_\delta \text{ with } \operatorname{Im} w < 0$$

yields an analytic extension of Z across I_δ onto the disk B_δ centered at $u_0 = 0$.

Moreover, $f = X + iX^*$ is holomorphic in S_δ, continuous on $\overline{S}_\delta$, and

$$\operatorname{Re}(f - Z) = 0 \quad \text{on } I_\delta.$$

Thus we can extend $i(f - Z)$ analytically across I_δ by

$$i\{f(w) - Z(w)\} := -i\{\overline{f}(\overline{w}) - \overline{Z}(\overline{w})\} \quad \text{for } w \in B_\delta \text{ with } \operatorname{Im} w < 0.$$

Hence

$$f(w) := 2\overline{Z}(\overline{w}) - \overline{f}(\overline{w}) \quad \text{for } w \in B_\delta \text{ with } \operatorname{Im} w < 0$$

extends f analytically across I_δ. Now $f(w)$ is seen to be a holomorphic function on B_δ, and $X = \operatorname{Re} f$ defines the harmonic extension of X to B_δ which, by the principle of analytic continuation, has to be a minimal surface on B_δ. □

2.9 A Different Approach to Boundary Regularity

In this section we want to give a different proof of the Hölder continuity of ∇X where X is a stationary point of Dirichlet's integral in $\mathcal{C}(\Gamma, S)$. This new proof merely requires that S is of class C^3. The first step is the same as in Section 2.7 and need not be repeated: one estimates the L_2-norms of $\nabla^2 X$ up to the free boundary. The other two steps are replaced by a new argument: We insert a suitable modification of the test function $\phi = \Delta_{-t}\{\eta^2 \Delta_t Y\}$ into the variational inequality

$$\delta E(Y, \phi) \geq 0.$$

This will lead us to a Morrey condition for $\nabla^2 X$ which, in turn, implies that ∇X is of class $C^{0,\alpha}$ on $B \cup I$ for some $\alpha \in (0, \frac{1}{2}]$. The essential new feature of this approach is that we shall explicitly use the first equation of Section 2.7, (35) which states that $y_v^1 = 0$ a.e. on $I_{2\rho}^+(w_0)$, and $y^1 = 0$ on $I_{2\rho}(w_0) \setminus I_{2\rho}^+(w_0)$.

Throughout this section we shall assume that S is an admissible support surface of class C^3 in the sense of Section 2.6, *Definition* 1.

As in Steps 2 and 3 of Section 2.1, we shall use a special boundary coordinate system satisfying (32) of Section 2.7. Note that, therefore, the defining diffeomorphisms g and h of the boundary coordinates are merely of class C^2.

We also assume that we have the same situation as in Section 2.6, that is: $w_0 \in I, x_0 := X(w_0), \{\mathcal{U}, g\}$ is an admissible boundary coordinate system centered at $x_0, h = g^{-1}, Y := g(X), Y(w_0) = 0; \rho > 0$ is chosen in such a way that $|Y(w)| < R$ for all $w \in \overline{S}_{2\rho}(w_0)$; in addition, $\{\mathcal{U}, g\}$ is chosen in such a way that (32) of Section 2.7 holds true; we have

$$\begin{aligned} y^1(w) \geq \sigma \quad &\text{and} \quad y^3(w) = 0 \quad \text{for all } w \in I_{2\rho}(w_0), \\ y^1_v(w) = 0 \quad &\text{a.e. on } I^+_{2\rho}(w_0) := I_{2\rho}(w_0) \setminus \{w\colon y^1(w) = \sigma\}; \end{aligned}$$

finally, by Step 1 of Section 2.7, $Y \in H^2_2 \cap H^1_4(S_{2r}(w_0), \mathbb{R}^3)$ for any $r \in (0, \rho)$, as well as $Y \in C^{0,\alpha}(\overline{S}_{2r}(w_0), \mathbb{R}^3)$ for all $\alpha \in (0,1)$.

Lemma 1. *Let $\phi = (\varphi^1, \varphi^2, \varphi^3) \in H^1_2 \cap L_\infty(S_{2\rho}(w_0), \mathbb{R}^3)$ be a test function with $\varphi^3 = 0$ on $I_{2\rho}(w_0)$, $\operatorname{supp} \phi \Subset S_{2\rho}(w_0) \cup I_{2\rho}(w_0)$, and suppose that*

$$X_\varepsilon := h(Y + \varepsilon\phi), \quad 0 \leq \varepsilon < \varepsilon_0(\phi),$$

is an admissible type II*-variation of X in $\mathcal{C}(\Gamma, S)$. Then we have*

$$\int_B D_\alpha y^j D_\alpha \varphi^j \, du\, dv \geq \int_B \Gamma^l_{jk}(Y) D_\alpha y^j D_\alpha y^k \varphi^l \, du\, dv. \tag{1}$$

For

$$\tilde{y}^j_1 := D_1 y^j - b^j, \quad \tilde{y}^j_2 := D_2 y^j - d^j$$

with $d^1 = d^2 = 0$ and arbitrary constants b^1, b^2, b^3, d^3, we also have

$$\int_B \tilde{y}^j_\alpha D_\alpha \varphi^j \, du\, dv \geq \int_B \Gamma^l_{jk}(Y) D_\alpha y^j D_\alpha y^k \varphi^l \, du\, dv. \tag{2}$$

Proof. Because of (32) in Section 2.7, we infer that also $X^*_\varepsilon := h(Y + \varepsilon\Psi)$ with $\Psi = (\psi^1, \psi^2, \psi^3), \psi^j := g^{jk}(Y)\varphi^k, \phi = (\varphi^1, \varphi^2, \varphi^3)$ is an admissible type II-variation of X in $\mathcal{C}(\Gamma, S)$, with $\psi^3 = 0$ on $I_{2\rho}(w_0)$ and

$$\operatorname{supp} \psi \Subset S_{2\rho}(w_0) \cup I_{2\rho}(w_0).$$

Hence

$$\delta E(Y, \Psi) \geq 0,$$

and by computations similar to those in the beginning of Section 2.6, we obtain (1).

Secondly, an integration by parts yields the identities

$$\begin{aligned}\int_B \tilde{y}^j_\alpha D_\alpha \varphi^j \, du\, dv &= \int_B D_\alpha[\tilde{y}^j_\alpha \varphi^j]\, du\, dv - \int_B (\Delta y^j)\varphi^j \, du\, dv \\ &= -\int_I \tilde{y}^j_2 \varphi^j \, du - \int_B (\Delta y^j)\varphi^j \, du\, dv \\ &= -\int_I [(D_v y^1)\varphi^1 + (D_v y^2)\varphi^2]\, du - \int_B (\Delta y^j)\varphi^j \, du\, dv \\ &= -\int_I \langle D_v Y, \phi\rangle \, du - \int_B \langle \Delta Y, \phi\rangle \, du\, dv \\ &= \int_B \langle D_\alpha Y, D_\alpha \phi\rangle \, du\, dv = \int_B D_\alpha y^j D_\alpha \varphi^j \, du\, dv.\end{aligned}$$

Hence (2) is a consequence of (1). □

Next we shall prove a generalized version of Poincaré's inequality.

Lemma 2. *For any $\gamma > 0$, there is a constant $M > 0$ with the following property: If $w_0 \in \mathbb{R}, r > 0, T_{2r} := S_{2r}(w_0) \setminus S_r(w_0), \psi \in H^1_2(T_{2r})$ and*

$$\mathcal{H}^1\{w \in I_{2r}(w_0) \setminus I_r(w_0) \colon \psi(w) = 0\} \geq \gamma r,$$

then

$$\int_{T_{2r}} \psi^2 \, du\, dv \leq M r^2 \int_{T_{2r}} |\nabla \psi|^2 \, du\, dv.$$

Proof. Suppose that $r = 1, \gamma > 0$, and let $\mathcal{C}_\gamma$ be the class of functions $\psi \in H^1_2(T)$, $T := T_2$, with $\mathcal{H}^1\{w \in I_2(w_0) \setminus I_1(w_0) \colon \psi(w) = 0\} \geq \gamma$. We claim that there is some number $M > 0$ such that

$$\int_T \psi^2 \, du\, dv \leq M \int_T |\nabla \psi|^2 \, du\, dv \tag{3}$$

is satisfied for all $\psi \in \mathcal{C}_\gamma$. By a scaling argument we then obtain the assertion of the lemma.

Suppose now that there is no $M > 0$ with (3). Then there is a sequence of functions $\psi_k \in \mathcal{C}_\gamma, k \in \mathbb{N}$, such that

$$\int_T \psi_k^2 \, du\, dv > k \int_T |\nabla \psi_k|^2 \, du\, dv.$$

Without loss of generality we may assume that

$$\int_T \psi_k^2 \, du\, dv = 1, \tag{4}$$

whence

$$\int_T |\nabla \psi_k|^2 \, du\, dv < 1/k, \quad k \in \mathbb{N}. \tag{5}$$

Then, by a well-known compactness argument for bounded sequences in Hilbert spaces, there is a subsequence $\{\psi'_n\}$ of $\{\psi_k\}$ which converges weakly in $H_2^1(T)$ to some $\psi \in H_2^1(T)$. Then $\{\psi'_n\}$ converges strongly to ψ both in $L_2(T)$ and in $L_2(\partial T)$, on account of Rellich's theorem and a result by Morrey (cf. [8], pp. 75–77). As Dirichlet's integral is weakly lower semicontinuous, we infer from (5) that

$$\int_T |\nabla\psi|^2 \, du \, dv = 0.$$

Hence there is a constant c such that

$$\psi(w) = c \quad \text{a.e. on } T,$$

and, because of (4), we have $c \neq 0$.

On the other hand, we have

$$\begin{aligned}\int_{\partial T} |\psi'_n - c|^2 \, d\mathcal{H}^1 &\geq \int_{\partial T \cap \{\psi'_n = 0\}} |\psi'_n - c|^2 \, d\mathcal{H}^1 \\ &= c^2 \mathcal{H}^1(\partial T \cap \{\psi'_n = 0\}) \geq c^2\gamma \quad \text{for all } n \in \mathbb{N}.\end{aligned}$$

As

$$\lim_{n\to\infty} \int_{\partial T} |\psi'_n - c|^2 \, d\mathcal{H}^1 = 0,$$

it follows that $c^2\gamma = 0$, which is impossible since $c \neq 0$ and $\gamma > 0$. □

Now we want to use the generalized Poincaré inequality to establish the basic estimates of the stationary surface X in $\mathcal{C}(\Gamma, S)$, or rather of its transform $Y = g(X)$.

Lemma 3. *Set* $T_{2r} := S_{2r}(w_0) \setminus S_r(w_0)$, *and*

(6) $$\zeta(r) := \frac{1}{r^2} \min\left\{\int_{T_{2r}} |D_u y^1|^2 \, du \, dv, \int_{T_{2r}} |D_v y^1|^2 \, du \, dv\right\}.$$

Then, for every $\delta \in (0, 1)$, *there is a constant* $c = c(\delta) > 0$ *such that the inequality*

(7) $$\int_{S_r(w_0)} |\nabla^2 Y|^2 \, du \, dv \leq c\left\{\zeta(r) + r^{1+\delta} + \int_{T_{2r}} |\nabla^2 Y|^2 \, du \, dv\right\}$$

holds true for all $r \in (0, \rho)$.

Proof. Choose a cut-off function η as in Section 2.7, and replace the test function ϕ in Section 2.7, (1) by

(8) $$\phi := \Delta_{-t}\{\eta^2(\Delta_t Y - A)\}, \quad \phi = (\varphi^1, \varphi^2, \varphi^3),$$

where

$$A = (A^1, A^2, A^3) := (0, a, 0)$$

is a constant vector with an arbitrary constant $a \in \mathbb{R}$. We claim that ϕ is admissible for inequalities (1) and (2) in Lemma 1 provided that $|t| \ll 1$. In fact,

$$Y(w) + \varepsilon\phi(w) = \lambda_1 Y_t(w) + \lambda_2 Y_{-t}(w) + (1 - \lambda_1 - \lambda_2)Y(w) + \mu A,$$

$$\lambda_1 := \varepsilon t^{-2}\eta^2(w), \quad \lambda_2 := \varepsilon t^{-2}\eta^2_{-t}(w), \quad \mu := \varepsilon t^{-1}[\eta_t^2(w) - \eta^2(w)].$$

Thus $Y(w) + \varepsilon\phi(w)$ is a convex combination of the three points $Y(w), Y_t(w)$, $Y_{-t}(w)$ which is translated by μA, that is, in direction of the y^2-axis, provided that $0 \le \varepsilon \le \frac{1}{2}t^2$. Then, by a repetition of the reasoning used in the beginning of Section 2.7, we infer that $X_\varepsilon := h(Y + \varepsilon\phi), 0 \le \varepsilon \ll 1$, is an admissible variation of X in $\mathcal{C}(\Gamma, S)$ which is of type II. Thus we can insert ϕ in (1), whence

$$\int_B D_\alpha y^j D_\alpha \Delta_{-t}\{\eta^2(\Delta_t y^j - A^j)\}\, du\, dv$$
$$\ge \int_B \Gamma^l_{jk}(Y) D_\alpha y^j D_\alpha y^k \Delta_{-t}\{\eta^2(\Delta_t y^l - A^l)\}\, du\, dv.$$

If we multiply this inequality by -1 and perform an integration by parts (cf. Section 2.7, Lemma 1), it follows that

$$\int_B \Delta_t D_\alpha y^j \{\eta^2(\Delta_t D_\alpha y^j) + 2\eta D_\alpha\eta(\Delta_t y^j - A^j)\}\, du\, dv$$
$$\le -\int_B \Gamma^l_{jk}(Y) D_\alpha y^j D_\alpha y^k \{(\Delta_{-t}\eta^2)(\Delta_t y^l - A^l) + \eta_t^2 \Delta_{-t}\Delta_t y^l\}\, du\, dv.$$

As t tends to zero, we arrive at

$$\int_B \eta^2 |\nabla D_u Y|^2\, du\, dv$$
$$\le \int_B 2\eta|\nabla\eta||\nabla D_u Y||D_u Y - A|\, du\, dv$$
$$+ c\int_B |\nabla Y|^2 \eta|\nabla\eta||D_u Y - A|\, du\, dv + c\int_B |\nabla Y|^2\eta^2|D_u^2 Y|\, du\, dv.$$

Here and in the following, c will denote a canonical constant. Then, by means of the inequality

$$2ab \le \varepsilon a^2 + \varepsilon^{-1}b^2,$$

we obtain that

$$\int_B \eta^2|\nabla D_u Y|^2\, du\, dv \le \varepsilon\int_B \eta^2|\nabla D_u Y|^2\, du\, dv + \frac{c}{\varepsilon}\int_B \eta^2|\nabla Y|^4\, du\, dv$$
$$+ \frac{c}{\varepsilon}\int_B |\nabla\eta|^2|D_u Y - A|^2\, du\, dv.$$

By choosing $\varepsilon = 1/2$, the first term on the right can be absorbed by the left-hand side, and it follows that

$$\int_{S_{2r}(w_0)} \eta^2 |\nabla D_u Y|^2 \, du \, dv \\ \le c r^{-2} \int_{T_{2r}} |D_u Y - A|^2 \, du \, dv + c \int_{S_{2r}(w_0)} \eta^2 |\nabla Y|^4 \, du \, dv.$$

From the Euler equation

$$\Delta y^l + \Gamma^l_{jk}(Y) D_\alpha y^j D_\alpha y^k = 0$$

we infer the inequality

$$|D_v^2 Y|^2 \le |D_u^2 Y|^2 + c|\nabla Y|^4,$$

and consequently

$$\int_{S_{2r}(w_0)} \eta^2 |\nabla^2 Y|^2 \, du \, dv \\ \le c r^{-2} \int_{T_{2r}} |D_u Y - A|^2 \, du \, dv + c \int_{S_{2r}(w_0)} \eta^2 |\nabla Y|^4 \, du \, dv. \tag{9}$$

Since we have already shown that

$$Y \in H_2^2 \cap H_p^1(\overline{Z}_d, \mathbb{R}^3), \quad 0 < d < 1,$$

for any $p \in (1, \infty)$, it follows that, for any $\delta \in (0, 1)$, there is a constant $c(\delta) > 0$ such that

$$\int_{S_{2r}(w_0)} |\nabla Y|^4 \, du \, dv \le c(\delta) r^{1+\delta} \tag{10}$$

holds for all $r \in (0, \rho)$. Moreover, we have

$$\int_{T_{2r}} |D_u Y - A|^2 \, du \, dv = \int_{T_{2r}} (|D_u y^1|^2 + |D_u y^2 - a|^2 + |D_u y^3|^2) \, du \, dv.$$

Poincaré's inequality yields

$$\int_{T_{2r}} |D_u y^3|^2 \, du \, dv \le c r^2 \int_{T_{2r}} |\nabla D_u y^3|^2 \, du \, dv \tag{11}$$

for all $r \in (0, \rho)$ since y^3 vanishes on $I_{2\rho}(w_0)$. Moreover, for

$$a := \fint_{T_{2r}} D_u y^2 \, du \, dv,$$

we infer from Poincaré's inequality that

$$\int_{T_{2r}} |D_u y^2 - a|^2 \, du \, dv \le cr^2 \int_{T_{2r}} |\nabla D_u y^2|^2 \, du \, dv \tag{12}$$

is satisfied for all $r \in (0, \rho)$.

Combining inequalities (9)–(12), we find that

$$\begin{aligned} &\int_{S_r(w_0)} |\nabla^2 Y|^2 \, du \, dv \\ &\le c(\delta) \left\{ r^{-2} \int_{T_{2r}} |D_u y^1|^2 \, du \, dv + \int_{T_{2r}} |\nabla^2 Y|^2 \, du \, dv + r^{1+\delta} \right\}. \end{aligned} \tag{13}$$

By a similar reasoning, it follows that

$$\begin{aligned} &\int_{S_r(w_0)} |\nabla^2 Y|^2 \, du \, dv \\ &\le c(\delta) \left\{ r^{-2} \int_{T_{2r}} |D_v y^1|^2 \, du \, dv + \int_{T_{2r}} |\nabla^2 Y|^2 \, du \, dv + r^{1+\delta} \right\} \end{aligned} \tag{14}$$

holds true if we insert the test function

$$\phi := \eta^2 \Delta_{-t} \Delta_t Y \tag{15}$$

in inequality (2) of Lemma 1. We leave it to the reader to check that (15) is an admissible test function for (2), and to carry out the derivation of (14) in detail.

Then the desired inequality (7) is a consequence of (13) and (14). □

Now we are going to prove our main result.

Theorem 1. *Let S be an admissible support surface of class C^3, and suppose that X is a stationary point of Dirichlet's integral in $\mathcal{C}(\Gamma, S)$. Then there exists some $\alpha \in (0, 1/2)$ such that $X \in C^{1,\alpha}(B \cup I, \mathbb{R}^3)$.*

Proof. We have

$$\begin{aligned} y_v^1 &= 0 \quad \text{a.e. on } I_{2r}^+(w_0) := \{w \in I_{2r}(w_0) \colon y^1(w) > \sigma\}, \\ y_u^1 &= 0 \quad \text{a.e. on } I_{2r}^0(w_0) := \{w \in I_{2r}(w_0) \colon y^1(w) = \sigma\}. \end{aligned}$$

Hence, either

$$\mathcal{H}^1(I_{2r}^+(w_0) \setminus I_r(w_0)) \ge r$$

or

$$\mathcal{H}^1(I_{2r}^0(w_0) \setminus I_r(w_0)) \ge r$$

holds true, and we can apply Lemma 2 to $\psi = y_v^1$ or $\psi = y_u^1$ respectively, obtaining

$$\int_{T_{2r}} \psi^2 \, du \, dv \le Mr^2 \int_{T_{2r}} |\nabla \psi|^2 \, du \, dv.$$

Thus the function $\zeta(r)$, defined by (6), will satisfy

$$\zeta(r) \le M \int_{T_{2r}} |\nabla^2 y^1|^2 \, du \, dv \quad \text{for all } r \in (0, \rho),$$

and we infer from formula (7) of Lemma 3 that

$$\int_{S_r(w_0)} |\nabla^2 Y|^2 \, du \, dv \le c \left\{ r^{1+\delta} + \int_{S_{2r}(w_0)\setminus S_r(w_0)} |\nabla^2 Y|^2 \, du \, dv \right\}$$

holds true for some $\delta \in (0,1)$ and for all $r \in (0, \rho)$. Adding the term

$$c \int_{S_r(w_0)} |\nabla Y|^2 \, du \, dv$$

to both sides of the last inequality and dividing the result by $1 + c$, it follows that

$$\int_{S_r(w_0)} |\nabla^2 Y|^2 \, du \, dv \le \theta \left\{ \int_{S_{2r}(w_0)} |\nabla^2 Y|^2 \, du \, dv + r^{1+\delta} \right\}$$

holds true for some $\delta \in (0,1)$ and for all $r \in (0, \rho)$, where

$$\theta := \frac{c}{1+c};$$

that is, $0 < \theta < 1$. Hence, by Lemma 6 of Section 2.6, we infer the existence of positive numbers k and $\alpha \le 1$ such that

$$\int_{S_r(w_0)} |\nabla^2 Y|^2 \, du \, dv \le kr^{2\alpha} \quad \text{for } 0 < r < \rho, \tag{16}$$

whence by

$$|\nabla^2 X|^2 \le c\{|\nabla^2 Y|^2 + |\nabla Y|^4\}$$

and (10) we obtain

$$\int_{S_r(w_0)} |\nabla^2 X|^2 \, du \, dv \le k^* r^{2\alpha} \quad \text{for } 0 < r < \rho$$

and some constant k^* depending on ρ but not on r. By virtue of Morrey's Dirichlet growth theorem we infer that $X \in C^{1,\alpha}(\overline{Z}_d, \mathbb{R}^3)$, for any $d \in (0,1)$. □

2.10 Asymptotic Expansion of Minimal Surfaces at Boundary Branch Points and Geometric Consequences

We have seen that a minimal surface $X: B \to \mathbb{R}^3$ can be extended analytically and as a minimal surface across those parts of ∂B which are mapped by X into an analytic arc or which correspond to a free trace on an analytic support surface. Therefore, at a branch point of such a part of ∂B, the minimal surface X possesses an asymptotic expansion as described in Section 3.2 of Vol. 1. In this section we want to derive an analogous expansion of X at boundary branch points, assuming merely that Γ or S are of some appropriate class C^m. Our main tool will be a technique developed by Hartman and Wintner that is described in Chapter 3 in some detail. Presently we shall only sketch how the Hartman–Wintner technique can be used to obtain the desired expansions at boundary branch points.

Since in the preceding sections we have discussed stationary points of Dirichlet's integral in $\mathcal{C}(\Gamma, S)$, that is, stationary minimal surfaces with a partially free boundary on I, we shall begin by considering such a minimal surface X. Thus we can assume that we have the same situation as in Section 2.6:

S is assumed to be an admissible support surface of class C^3; $w_0 \in I, x_0 := X(w_0)$; $\{\mathcal{U}, g\}$ is an admissible boundary coordinate system centered at $x_0, h = g^{-1}, Y = (y^1, y^2, y^3) := g(X), Y(w_0) = 0; \rho > 0$ is chosen in such a way that $|Y(w)| < R$ for all $w \in \overline{S}_{2\rho}(w_0)$; in addition, $\{\mathcal{U}, g\}$ is chosen in such a way that (32) of Section 2.7 holds true. We have

$$y_v^2 = 0 \quad \text{and} \quad y^3 = 0 \quad \text{in } I_{2\rho}(w_0)$$

and

$$\Delta y^l + \Gamma^l_{jk}(Y) D_\alpha y^j D_\alpha y^k = 0 \quad \text{in } B.$$

Moreover, on account of

$$g_{jk}(Y) y^j_w y^k_w = 0 \quad \text{in } B,$$

it follows that

$$(1) \qquad |\nabla y^1|^2 \le c\{|\nabla y^2|^2 + |\nabla y^3|^2\}$$

and

$$(2) \qquad |\Delta y^2| + |\Delta y^3| \le c\{|\nabla y^2|^2 + |\nabla y^3|^2\}$$

holds is $S_{2\rho}(w_0)$ for some constant $c > 0$.

In Section 2.7 we have also proved that y^2 and y^3 are both of class $C^{1,\alpha}(\overline{S}_{2r}(w_0))$ and of class $H^2_p(S_{2r}(w_0))$ for any $\alpha \in (0,1), p \in (1,\infty)$, and

$r \in (0, \rho)$. Then the mapping $Z(w) = (z^1(w), z^2(w), z^3(w))$ defined by $z^1(w) := 0$ and by

$$\begin{aligned} z^2(w) &:= y^2(w), \quad z^3(w) := y^3(w) \quad \text{if } \operatorname{Im} w \geq 0, \\ z^2(w) &:= y^2(\overline{w}), \quad z^3(w) := -y^3(\overline{w}) \quad \text{if } \operatorname{Im} w < 0, \end{aligned}$$

$w = u + iv \in B_\rho(w_0), \overline{w} = u - iv$, is of class $C^{1,\alpha}(B_\rho(w_0), \mathbb{R}^3)$ and of class C^2 in $B_\rho(w_0) \setminus I_\rho(w_0)$. Furthermore, for some constant $c > 0$, we have

$$|Z_{w\overline{w}}| \leq c|Z_w| \quad \text{in } B_\rho(w_0) \setminus I_\rho(w_0). \tag{3}$$

Let Ω be an arbitrary subdomain of $B_r(w_0)$ for some $r \in (0, \rho)$ which has a piecewise smooth boundary $\partial\Omega$, and let $\phi = (\varphi^1, \varphi^2, \varphi^3)$ be an arbitrary function of class $C^1(\overline{\Omega}, \mathbb{C}^3)$. Then, by an integration by parts, we obtain that

$$\frac{1}{2i} \int_{\partial\Omega} \langle Z_w, \phi \rangle \, dw = \int_\Omega (\langle Z_w, \phi_{\overline{w}} \rangle + \langle Z_{w\overline{w}}, \phi \rangle) \, d^2 w, \tag{4}$$

where $dw = du + i\,dv, d^2 w = du\,dv$.

Combining (3) and (4), we arrive at the inequality

$$\left| \int_{\partial\Omega} \langle Z_w, \phi \rangle \, dw \right| \leq 2 \int_\Omega |Z_w|(|\phi_{\overline{w}}| + c|\phi|) \, d^2 w \tag{5}$$

which holds for all $\phi \in C^1(\overline{\Omega}, \mathbb{R}^3)$ and for all $\Omega \subset B_r(w_0)$ with a piecewise smooth boundary $\partial\Omega$. But this relation is the starting point for the Hartman–Wintner technique; cf. Chapter 3, Section 3.1.

Suppose now that $w_0 \in I$ is a branch point of X, that is,

$$|X_u(w_0)| = 0 \quad \text{and} \quad |X_v(w_0)| = 0.$$

Then we have

$$|Y_u(w_0)| = 0 \quad \text{and} \quad |Y_v(w_0)| = 0$$

and consequently

$$Y_w(w_0) = 0.$$

We claim that there is no $r \in (0, \rho)$ such that $Y_w(w) = 0$ for all $w \in S_r(w_0)$. In fact, suppose that $Y_w(w) \equiv 0$ on $S_r(w_0)$. Then we obtain $X_w(w) \equiv 0$ on $S_r(w_0)$, whence $X_w(w) \equiv 0$ on B; but this is impossible for any stationary point of the Dirichlet integral in $\mathcal{C}(\Gamma, S)$.

From $Y_w \not\equiv 0$ on $S_r(w_0)$ for any $r \in (0, \rho)$ it follows that $Z_w(w) \not\equiv 0$ on $S_r(w_0)$, on account of (1). Then by virtue of Theorem 1 of Section 3.1, there is some vector $P = (p^1, p^2, p^3) \neq 0$ in $\mathbb{C}^3$ and some number $\nu \in \mathbb{N}$ such that

$$Z_w(w) = P(w - w_0)^\nu + o(|w - w_0|^\nu) \quad \text{as } w \to w_0.$$

Because of $z^1_w(w) \equiv 0$, it follows that $p^1 = 0$:

$$P = (0, p^2, p^3).$$

Now we consider the function $e(w), w \in S_\rho(w_0) \setminus \{w_0\}$, defined by

$$e(w) := (w - w_0)^{-\nu} f(w),$$

where $f(w)$ is defined by (42) in Section 2.7. Since $Y(w_0) = 0$ and

$$(g_{jk}(0)) = \begin{bmatrix} \mathcal{E}_0 & 0 & 0 \\ 0 & \mathcal{G}_0 & 0 \\ 0 & 0 & 1 \end{bmatrix}, \quad \mathcal{E}_0 \neq 0, \quad \mathcal{G}_0 \neq 0,$$

we infer from formula (41) in Section 2.7 that $\lim_{w \to w_0} e^2(w)$ exists, and that

$$\begin{aligned} \lim_{w \to w_0} e^2(w) &= -\frac{\mathcal{G}_0}{\mathcal{E}_0} \lim_{w \to w_0} (w - w_0)^{-2\nu} (y_w^2(w))^2 \\ &\quad - \frac{1}{\mathcal{E}_0} \lim_{w \to w_0} (w - w_0)^{-2\nu} (y_w^3(w))^2. \end{aligned}$$

Then, by Lemma 3 of Section 2.7, we see that $\lim_{w \to w_0} e(w)$ does exist. Set $F := (f^1, f^2, f^3)$, where

$$f^1 := \lim_{w \to w_0} e(w) = \lim_{w \to w_0} (w - w_0)^{-\nu} y_w^1(w), \quad f^2 := p^2, \quad f^3 := p^3.$$

It follows that

$$Y_w(w) = F(w - w_0)^\nu + o(|w - w_0|^\nu) \quad \text{as } w \to w_0,$$

where $F \in \mathbb{C}^3$ satisfies $F \neq 0$ and $\langle\langle F, F \rangle\rangle = 0$, i.e.

$$g_{kl}(0) f^k f^l = 0.$$

Because of $X_w = h_y(Y) Y_w$, we obtain the following result:

Theorem 1. *Let S be an admissible support surface of class C^3 and X be a stationary point of Dirichlct's intcgral in the class $\mathcal{C}(\Gamma, S)$. Assume also that $w_0 \in I$ is a boundary branch point of X. Then there exist an integer $\nu \geq 1$ and a vector $A \in \mathbb{C}^3$ with $A \neq 0$ and*

$$\langle A, A \rangle = 0 \tag{6}$$

such that

$$X_w(w) = A(w - w_0)^\nu + o(|w - w_0|^\nu) \quad \text{as } w \to w_0. \tag{7}$$

We call ν the order of the branch point w_0.

From this expansion we can draw the same geometric conclusions as in Section 3.2 of Vol. 1. To this end we write

$$A = \tfrac{1}{2}(\alpha - i\beta) \quad \text{with } \alpha, \beta \in \mathbb{R}^3.$$

Then it follows that

$$|\alpha| = |\beta| \neq 0, \quad \langle \alpha, \beta \rangle = 0$$

and

$$\begin{aligned} X_u(w) &= \alpha \operatorname{Re}(w - w_0)^\nu + \beta \operatorname{Im}(w - w_0)^\nu + o(|w - w_0|^\nu), \\ X_v(w) &= -\alpha \operatorname{Im}(w - w_0)^\nu + \beta \operatorname{Re}(w - w_0)^\nu + o(|w - w_0|^\nu) \end{aligned} \tag{8}$$

as $w \to w_0$, whence

$$X_u(w) \wedge X_v(w) = (\alpha \wedge \beta)|w - w_0|^{2\nu} + o(|w - w_0|^{2\nu}) \quad \text{as } w \to w_0.$$

This implies that the surface normal $N(w)$, given by

$$N = |X_u \wedge X_v|^{-1}(X_u \wedge X_v),$$

tends to a limit vector N_0 as $w \to w_0$:

$$\lim_{w \to w_0} N(w) = N_0 = |\alpha \wedge \beta|^{-1}(\alpha \wedge \beta). \tag{9}$$

Consequently, the Gauss map $N(w)$ of a stationary minimal surface $X(w)$ is well-defined on all of $B \cup I$ as a continuous mapping into S^2. Therefore *the surface $X(w)$ has a well-defined tangent plane at every boundary branch point on I*, and thus at every point $w_0 \in B \cup I$.

Consider now the trace curve $X : I \to \mathbb{R}^3$ of the minimal surface X on the supporting surface S. We infer from (8) that

$$X_u(w) = \alpha(w - w_0)^\nu + o(|w - w_0|^\nu) \quad \text{as } w \to w_0, w \in I$$

and, writing $w = u, w_0 = u_0$ for $w, w_0 \in I$, we obtain for the unit tangent vector

$$t(u) := |X_u(u)|^{-1} X_u(u)$$

the expansion

$$t(u) = \frac{\alpha}{|\alpha|} \frac{(u - u_0)^\nu}{|u - u_0|^\nu} + o(1) \quad \text{as } u \to u_0. \tag{10}$$

Therefore *the nonoriented tangent moves continuously through any boundary branch point $u_0 \in I$. The oriented tangent $t(u)$ is continuous if the order ν of u_0 is even, but, for branch points of odd order, the direction of $t(u)$ jumps by 180 degrees when u passes through u_0.*

Finally, by choosing a suitable Cartesian coordinate system in $\mathbb{R}^3$, we obtain the expansion

$$(11) \qquad \begin{aligned} x(w) + iy(w) &= (x_0 + iy_0) + a(w - w_0)^{\nu+1} + o(|w - w_0|^{\nu+1}), \\ z(w) &= z_0 + o(|w - w_0|^{\nu+1}) \end{aligned}$$

as $w \to w_0$, where $X(w_0) = (x_0, y_0, z_0)$ and $a > 0$; see Section 3.2, (6), of Vol. 1.

The same reasoning can be used for the investigation of X at a boundary branch point $w_0 \in \text{int } C$. We obtain again an expansion of the kind (7) with some $\nu \geq 1$ and some $A \in \mathbb{C}^3, A \neq 0, \langle A, A \rangle = 0$. As $X : C \to \Gamma$ is a monotonic mapping, *the tangent vector*

$$t(\varphi) := |X_\varphi(e^{i\varphi})|^{-1} X_\varphi(e^{i\varphi})$$

of this mapping has to be continuous, and we infer from (7) *that* ν *is even,* provided that Γ is of class C^2.

The same result can be proved for minimal surfaces $X \in \mathcal{C}(\Gamma)$ which solve Plateau's problem for a closed Jordan curve Γ of class C^2; cf. Chapter 4 of Vol. 1 for the definition of $\mathcal{C}(\Gamma)$. Thus we obtain

Theorem 2. *Let Γ be a closed Jordan curve of class C^2 in $\mathbb{R}^3$, and suppose that $X \in \mathcal{C}(\Gamma)$ is a minimal surface spanning Γ. Then every boundary branch point $w_0 \in \partial B$ is of even order $\nu = 2p, p \geq 1$, and we have the asymptotic expansion*

$$(12) \qquad X_w(w) = A(w - w_0)^{2p} + o(|w - w_0|^{2p}) \quad \text{as } w \to w_0,$$

where $A \in \mathbb{C}^3, A \neq 0$, and $\langle A, A \rangle = 0$.

W. Jäger [3] has pointed out that $\Gamma \in C^{1,\mu}$ suffices to prove (12).

2.11 The Gauss–Bonnet Formula for Branched Minimal Surfaces

In Section 1.4 of Vol. 1 we have derived the Gauss–Bonnet formula

$$(1) \qquad \int_X K dA + \int_\Gamma \kappa_g \, ds = 2\pi$$

for regular surfaces $X \in C^2(\overline{\Omega}, \mathbb{R}^3)$ defined on a simply connected bounded domain $\Omega \subset \mathbb{C}$ which map $\partial\Omega$ onto a Jordan curve Γ. The result as well as the proof given in Section 1.4 of Vol. 1 remain correct if X does not map $\partial\Omega$ bijectively onto a Jordan curve in $\mathbb{R}^3$ provided that we replace formula (1) by

$$(2) \qquad \int_X K \, dA + \int_{\partial X} \kappa_g \, ds = 2\pi$$

or, precisely speaking, by

(3) $$\int_\Omega K|X_u \wedge X_v|\,du\,dv + \int_{\partial\Omega} \kappa_g |dX| = 2\pi.$$

Now we shall drop the assumption of regularity and, instead, admit finitely many branch points in the interior and on the boundary of the parameter domain Ω. To make our assumptions precise, we introduce the class $\mathcal{PR}(\overline{\Omega})$ of *pseudoregular surfaces* $X : \overline{\Omega} \to \mathbb{R}$ as follows:

A surface X is said to be of class $\mathcal{PR}(\overline{\Omega})$ if it satisfies the conditions
(i) $X \in C^2(\overline{\Omega}, \mathbb{R}^3)$ *and*

(4) $$|X_u| = |X_v|, \quad \langle X_u, X_v\rangle = 0.$$

(ii) *There is a continuous function $\mathcal{H}(w)$ on $\overline{\Omega}$ such that*

(5) $$\Delta X = 2\mathcal{H} X_u \wedge X_v.$$

(iii) *There is a finite set Σ_0 of points in $\overline{\Omega}$ such that $X_w(w) \neq 0$ for all $w \in \overline{\Omega} \setminus \Sigma_0$. For any point $w_0 \in \Sigma_0$ there is an integer $\nu \geq 1$ and a vector $A \in \mathbb{C}^3$ satisfying $A \neq 0$ and $\langle A, A\rangle = 0$ such that*

(6) $$X_w(w) = A(w - w_0)^\nu + o(|w - w_0|^\nu) \quad \text{as } w \to w_0.$$

We call Σ_0 the singular set *of $X \in \mathcal{PR}(\overline{\Omega})$.*

Remark 1. The set $\Omega_0 := \{w \in \Omega\colon X_w(w) \neq 0\}$ of regular points of X in Ω is open and, by Section 2.6 of Vol. 1, equations (4) yield the existence of a function $\mathcal{H} \in C^0(\Omega_0)$ such that (5) holds true on Ω_0. Moreover, the function $\mathcal{H}$ is the mean curvature of $X|_{\Omega_0}$. Thus condition (ii) is a consequence of (i) if we assume that $\mathcal{H}(w)$ can be extended from Ω_0 to $\overline{\Omega}$ as a continuous function. This extension is possible if, for some reason, we know that X is a solution of

(7) $$\Delta X = 2H(X) X_u \wedge X_v$$

in Ω, where $H \in C^0(\mathbb{R}^3)$.

If, on the other hand, $X \in C^2(\Omega, \mathbb{R}^3)$ is a solution of (4) and (7) for some $H \in C^1(\mathbb{R}^3)$, it is sometimes possible to extend X to a function of class $C^2(\overline{\Omega}, \mathbb{R}^3)$. For instance, the extendability can follow from suitable boundary conditions (e.g. from Plateau-type conditions or from free boundary conditions) as we have seen in the previous sections.

Finally if $X(w)$ is a nonconstant surface such that (4) and (5) hold for some $\mathcal{H} \in C^{0,\alpha}(\overline{\Omega}), 0 < \alpha < 1$, then the set of branch points of X defined by $\Sigma_0 := \{w \in \overline{\Omega} : X_w(w) = 0\}$ is finite (and possibly empty), and, for any $w_0 \in \Sigma_0$, the mapping X has an asymptotic expansion (6) as described in (iii). For minimal surfaces we have stated this result in Section 2.10. The general theory will be developed in Chapter 3, using Hartman–Wintner's technique.

Now we can formulate the Gauss–Bonnet theorem for pseudoregular surfaces; we shall immediately state it for multiply connected domains.

Theorem 1. *Let Ω be an m-fold connected domain in $\mathbb{C}$ bounded by m closed regular curves $\gamma_1, \dots, \gamma_m$ of class C^∞, and let $X: \overline{\Omega} \to \mathbb{R}^3$ be a pseudoregular surface with the area element $dA = |X_u \wedge X_v| \, du \, dv$, the singular set Σ_0, the Gauss curvature K in $\Omega \setminus \Sigma_0$, and the geodesic curvature κ_g of $X|_{\partial\Omega\setminus\Sigma_0}$. Suppose also that the total curvature integral $\int_X |K| \, dA$ of X exists as a Cauchy principle value. Then we obtain the generalized Gauss–Bonnet formula*

$$
(8) \qquad \int_X K \, dA = 2\pi(2-m) + 2\pi \sum_{w\in\sigma'} \nu(w) + \pi \sum_{w\in\sigma''} \nu(w) - \int_{\partial\Omega} \kappa_g |dX|,
$$

where $\sigma' := \Sigma_0 \cap \Omega$ is the set of interior branch points, $\sigma'' := \Sigma_0 \cap \partial\Omega$ the set of boundary branch points, and ν the order of a branch point $w \in \Sigma_0$.

For the proof of (8) we shall employ the reasoning of Section 1.4 of Vol. 1. To carry out these arguments in our present context, we need two auxiliary results.

Lemma 1. *Let $a > 0, I = (0, a]$, and be f a function of class $C^1(I)$ such that $|f(r)| \le m$ holds for all $r \in I$ and some constant $m \ge 0$. Then there is a sequence of numbers $r_k \in I$ satisfying $r_k \to 0$ and $r_k f'(r_k) \to 0$ as $k \to \infty$.*

Proof. Otherwise we could find two numbers $c > 0$ and $\varepsilon \in (0, a]$ such that

$$
r|f'(r)| \ge c \quad \text{for all } r \in (0, \varepsilon].
$$

Then we would either have

$$
\text{(i)} \qquad f'(r) \ge c/r \quad \text{for all } r \in (0, \varepsilon]
$$

or

$$
\text{(ii)} \qquad f'(r) \ge -c/r \quad \text{for all } r \in (0, \varepsilon].
$$

In case (i) we obtain

$$
c \log \frac{\varepsilon}{r} = c \int_r^\varepsilon \frac{dr}{r} \le \int_r^\varepsilon f'(r) \, dr = f(\varepsilon) - f(r)
$$

whence

$$
\log \frac{1}{r} \le \frac{2m}{c} - \log \varepsilon \quad \text{for all } r \in (0, \varepsilon]
$$

which yields a contradiction since $\log \frac{1}{r} \to \infty$ as $r \to +0$. Similarly case (ii) leads to a contradiction. □

Lemma 2. *Let Σ_0 be the singular set of a map $X \in \mathcal{PR}(\Omega)$. Then, for any $w_0 \in \Sigma_0$, there is a sequence of positive radii $r_k, k \in \mathbb{N}$, tending to zero such that*

$$
(9) \qquad \lim_{k\to\infty} \int_{C(w_0, r_\kappa)} \frac{\partial}{\partial r} \log \sqrt{\Lambda} \, d\sigma = \begin{cases} 2\pi\nu & \text{if } w_0 \in \Omega, \\ \pi\nu & \text{if } w_0 \in \partial\Omega, \end{cases}
$$

where $r = |w - w_0|$, $w = w_0 + re^{i\varphi}$, ν is the order of the branch point w_0 defined by the expansion (6), and $\Lambda = |X_u|^2$.

Proof. Let us write $C(w_0, r) = \{w \in \overline{\Omega}: |w - w_0| = r\}$ as

$$C(w_0, r) = \{w = w_0 + re^{i\varphi}: \varphi_1(r) \le \varphi \le \varphi_2(r)\}$$

for $0 < r \le \varepsilon \ll 1$, and set

$$f(r) := \int_{\varphi_1(r)}^{\varphi_2(r)} \log|X_w(w)||w - w_0|^{-\nu}\, d\varphi.$$

By Lemma 1, there is a sequence $r_k \to +0$ such that $r_k f'(r_k) \to 0$. Since $|X_w| = \sqrt{\Lambda}/\sqrt{2}$, we obtain

$$\log|X_w(w)||w - w_0|^{-\nu} = \log\sqrt{\Lambda(w)} - \log\sqrt{2} - \nu\log r$$

for $w \in C(w_0, r)$, whence

$$\frac{\partial}{\partial r}\log|X_w(w)||w - w_0|^{-\nu} = \frac{\partial}{\partial r}\log\sqrt{\Lambda(w)} - \frac{\nu}{r}.$$

Thus it follows that

$$\begin{aligned} r_k f'(r_k) = {} & \int_{\varphi_1(r_k)}^{\varphi_2(r_k)} \left(\frac{\partial}{\partial r}\log\sqrt{\Lambda}\right) r_k\, d\varphi - \nu \int_{\varphi_1(r_k)}^{\varphi_2(r_k)} d\varphi \\ & + r_k\varphi_2'(r_k)\log(|A| + \delta_k) - r_k\varphi_1'(r_k)\log(|A| + \delta_k^*), \end{aligned}$$

where $\{\delta_k\}$ and $\{\delta_k^*\}$ are two sequences tending to zero.

If $w_0 \in \Omega$, we can assume that $\varphi_1(r) = 0$ and $\varphi_2(r) = 2\pi$, whence

$$r_k f'(r_k) = \int_{C(w_0, r_k)} \left(\frac{\partial}{\partial r}\log\sqrt{\Lambda}\right) d\sigma - 2\pi\nu.$$

Because of $r_k f'(r_k) \to 0$, we then obtain

$$\lim_{k\to\infty}\int_{C(w_0, r_k)} \frac{\partial}{\partial r}\log\sqrt{\Lambda}\, d\sigma = 2\pi\nu.$$

If $w_0 \in \partial\Omega$, then the smoothness of $\partial\Omega$ implies

$$\varphi_1(r_k) - \varphi_1(r_k) \to \pi \quad \text{and} \quad r_k\{|\varphi_1'(r_k)| + |\varphi_2'(r_k)|\} \to 0 \quad \text{as } k \to \infty,$$

whence

$$\lim_{k\to\infty}\int_{C(w_0, r_k)} \frac{\partial}{\partial r}\log\sqrt{\Lambda}\, d\sigma = \pi\nu. \qquad \square$$

Now we turn to the

Proof of Theorem 1. Let $\sigma' = \{w_1, \ldots, w_N\}$ and $\sigma'' = \{\tilde{w}_1, \ldots, \tilde{w}_M\}$ be the sets of interior branch points and of boundary branch points respectively. We consider $N + M$ sequences $\{r_\alpha^{(j)}\}$ and $\{\tilde{r}_\beta^{(j)}\}, 1 \le \alpha \le N, 1 \le \beta \le M$, of positive numbers tending to zero as $j \to \infty$. Set

$$\Omega_j\colon = \{w \in \Omega\colon |w - w_\alpha| > r_\alpha^{(j)}, |w - \tilde{w}_\beta| > \tilde{r}_\beta^{(j)}, 1 \le \alpha \le N, 1 \le \beta \le M\}.$$

By formula (32) of Section 1.3 of Vol. 1 we have

$$-\int_{\Omega_j} K\, dA = \int_{\Omega_j} \Delta \log \sqrt{\Lambda}\, du\, dv$$

taking $|X_u \wedge X_v| = \Lambda$ into account, and an integration by parts yields

$$-\int_{\Omega_j} K\, dA = \int_{\partial\Omega_j} \left(\frac{\partial}{\partial n} \log \sqrt{\Lambda} \right) d\mathcal{H}^1, \tag{10}$$

where n denotes the exterior normal to $\partial\Omega_j$. (Actually, we should write $\int_{X_j} K\, dA$ instead of $\int_{\Omega_j} K\, dA$, with $X_j := X|_{\Omega_j}$.) According to Lemma 2, the sequences $\{r_\alpha^{(j)}\}$ and $\{\tilde{r}_\beta^{(j)}\}$ can be chosen in such a way that

$$\int_{C(w_\alpha, r_\alpha^{(j)})} \frac{\partial}{\partial n} \log \sqrt{\Lambda}\, d\mathcal{H}^1 \to 2\pi\nu(w_\alpha) \tag{11}$$

as $j \to \infty$, and that

$$\int_{C(\tilde{w}_\beta, \tilde{r}_\beta^{(j)})} \frac{\partial}{\partial n} \log \sqrt{\Lambda}\, d\mathcal{H}^1 \to \pi\nu(\tilde{w}_\beta), \tag{12}$$

where $\nu(w_\alpha)$ and $\nu(\tilde{w}_\beta)$ denote the orders of branch points w_α and $\tilde{w}_\beta$, respectively, which are defined by the corresponding expansions (6).

Moreover, let γ_k be one of the m closed curves, the union of which is $\partial\Omega$, and let $(a(\sigma), b(\sigma)), 0 \le \sigma \le L$, be a parameter representation of γ_k in terms of its parameter of arc length σ which orients $\partial\Omega$ in the positive sense with respect to Ω. Then we have $\dot{a}^2 + \dot{b}^2 = 1, a(0) = a(L), b(0) = b(L)$, and $n = (\dot{b}, -\dot{a})$ is the exterior normal to γ_k with respect to Ω. The geodesic curvature κ_g of the (oriented) curve $X \circ \gamma_k$ can, according to Vol. 1, Section 1.3, (46) be computed from the formula

$$\kappa_g \sqrt{\Lambda} = (\dot{a}\ddot{b} - \ddot{a}\dot{b}) + \frac{\partial}{\partial n} \log \sqrt{\Lambda}. \tag{13}$$

From $\dot{a}^2 + \dot{b}^2 = 1$ we infer that

$$\int_0^L (\dot{a}\ddot{b} - \ddot{a}\dot{b})\, d\sigma = \pm 2\pi, \tag{14}$$

where we have the plus-sign if γ_k is positively oriented with respect to its interior domain while otherwise the minus-sign is to be taken. As $\partial\Omega$ consists of the closed Jordan curves $\gamma_1, \gamma_2, \dots, \gamma_m$, we can assume that γ_1 forms the outer boundary curve of $\partial\Omega$ whereas $\gamma_2, \dots, \gamma_m$ lie in the interior domain of γ_1. Consequently we have the plus-sign for γ_1 and the minus-sign for $\gamma_2, \dots, \gamma_m$, and we infer from (13) and (14) that

$$(15)\qquad -\int_0^L \frac{\partial}{\partial n} \log\sqrt{\Lambda}\, d\sigma = 2\pi\varepsilon_k - \int_0^L \kappa_g \sqrt{\Lambda}\, d\sigma,$$

where $\varepsilon_k := 1$ for $k = 1$ and $\varepsilon_k := -1$ for $2 \le k \le m$. (If there are branch points on γ_k, the integrals in γ_k are to be understood as Cauchy principal values.) Adding (15) from $k = 1$ to $k = n$, we obtain

$$(16)\qquad -\int_{\partial\Omega} \left(\frac{\partial}{\partial n} \log\sqrt{\Lambda} \right) d\mathcal{H}^1 = 2\pi(2 - m) - \int_{\partial\Omega} \kappa_g |dX|.$$

Thus, letting j tend to infinity, we infer from (10) that

$$(17)\qquad \int_X K\, dA = 2\pi(2 - m) + 2\pi \sum_{\alpha=1}^{N} \nu(w_\alpha) + \pi \sum_{\beta=1}^{M} \nu(\tilde{w}_\beta) - \int_{\partial\Omega} \kappa_g |dX|$$

provided that the integral $\int_X K\, dA$ exists as principal value

$$(18)\qquad \int_X K\, dA = \lim_{j\to\infty} \int_{\Omega_j} K\, dA.$$

□

In various instances it is superfluous to assume that the principle value (18) exists. Let us consider some instructive cases.

Suppose that $X \in \mathcal{PR}(\Omega)$ is a minimal surface. Then we have $K \le 0$, and we infer that $\int_X K\, dA$ exists, but it can have the value $-\infty$. If, however, X maps $\partial\Omega$ topologically onto $\Gamma = \bigcup_{j=1}^m \Gamma_j$ where $\Gamma_1, \Gamma_2, \dots, \Gamma_m$ are mutually disjoint and regular Jordan curves of class C^2, then the geodesic curvature κ_g of $X|_{\partial\Omega}$ is bounded by $|\kappa_g| \le \kappa$ where κ denotes the curvature of κ. Hence $\int_{\partial\Omega} \kappa_g |dX|$ exists and is finite, and we infer from (10) for $j \to \infty$ that $\int_X |K|\, dA$ exists and is finite. Thus Theorem 1 implies the following result.

Theorem 2. *Let Ω be an m-fold connected, bounded domain in $\mathbb{C}$ whose boundary consists of m closed, regular, disjoint curves $\gamma_1, \dots, \gamma_m$. Secondly, let*

$$X \in C^0(\overline{\Omega}, \mathbb{R}^3) \cap C^2(\Omega, \mathbb{R}^3)$$

be a minimal surface on Ω which maps the γ_j topologically onto closed regular and disjoint Jordan curves $\Gamma_j, 1 \le j \le m$, of class $C^{2,\alpha}, 0 < \alpha < 1$, with the curvature κ. Then $\int_X |K|\, dA$ and $\int_{\partial\Omega} \kappa_g |dX| = \int_\Gamma \kappa_g\, ds$ are finite, and we have

$$(19)\qquad \int_X K\,dA + \int_\Gamma \kappa_g\,ds = 2\pi(2-m) + 2\pi\sum_{w\in\sigma'}\nu(w) + \pi\sum_{w\in\sigma''}\nu(w),$$

where $ds = |dX|$ *is the line element of* $\Gamma := \bigcup_{j=1}^m \Gamma_j, \nu(w)$ *is the order of a branch point* $w \in \Sigma, \sigma' := \Omega \cap \Sigma_0, \sigma'' := \partial\Omega \cap \Sigma_0, \Sigma_0$ *is the set of branch points of* X *in* $\overline{\Omega}$, *and* κ_g *is the geodesic curvature of* Γ *viewed as curve on the surface* X. *In particular, equation* (19) *implies that*

$$(20)\qquad 2 - m + \sum_{w\in\sigma'}\nu(w) + \frac{1}{2}\sum_{w\in\sigma''}\nu(w) \le \frac{1}{2\pi}\int_\Gamma \kappa\,ds.$$

Here we have used the fact that the assumption $\Gamma \in C^{2,\alpha}$ implies that $X \in \mathcal{PR}(\Omega)$, as we have seen in the previous sections of this chapter. We recall that *the order of boundary branch points has to be even* since X maps $\partial\Omega$ topologically onto Γ.

Remark 2. Because analogous regularity results hold for solutions $X \in C^0(\overline{\Omega}, \mathbb{R}^3) \cap C^2(\Omega, \mathbb{R}^3)$ of

$$\Delta X = 2H(X)X_u \wedge X_v,$$

$$|X_u|^2 = |X_v|^2, \quad \langle X_u, X_v\rangle = 0,$$

where $H \in C^{0,\alpha}(\mathbb{R}^3)$ (see Section 2.3), we infer from

$$K \le H^2 \le h^2, \quad h := \sup_{w\in\Omega} H(X)$$

that $\int_X |K|\,dA$ and $\int_\Gamma \kappa_g\,ds$ are finite, and that we have formula (19) as well as the estimate

$$(21)\qquad 2 - m + \sum_{w\in\sigma'}\nu(w) + \frac{1}{2}\sum_{w\in\sigma''}\nu(w) \le \frac{1}{2\pi}\int_\Gamma \kappa\,ds + h^2 A(X),$$

where $A(X) = D(X)$ denotes the area of X which in certain situations can be estimated in terms of the length of Γ by, say, by isoperimetric inequalities.

Remark 3. Let $X \in C^2 \cap H_2^1(\Omega, \mathbb{R}^3)$ be a minimal surface which is stationary with respect to a boundary configuration $\langle S_1, S_2, \dots, S_m\rangle$ consisting of m regular, sufficiently smooth surfaces S_j whose principal curvatures are bounded in absolute value by a constant $k > 0$, and suppose that Ω is an m-fold connected bounded domain. Then X is of class $\mathcal{PR}(\Omega)$ and intersects $S := \bigcup_{j=1}^m S_j$ perpendicularly. Moreover, the geodesic curvature κ_g of the free trace $\Sigma = X|_{\partial\Omega}$ can be written as

$$(22)\qquad \kappa_g = \pm\kappa_n^*,$$

where κ_n^* is the normal curvature of Σ viewed as curve(s) on S. By virtue of $|\kappa_n^*| \le k$ we then infer that

$$|\kappa_g| \le k.$$

Therefore we obtain the Gauss–Bonnet formula

$$\text{(23)} \qquad \int_X K\,dA + \int_{\partial\Omega} \kappa_g |dX| = 2\pi(2-m) + 2\pi \sum_{w\in\sigma'} \nu(w) + \pi \sum_{w\in\sigma''} \nu(w),$$

where $\sigma' = \Sigma_0 \cap \Omega, \sigma'' = \Sigma_0 \cap \partial\Omega$, and Σ_0 is the set of branch points w_0 of $X, \nu(w_0)$ is the order of $w_0 \in \Sigma_0$, and we have the estimate

$$\text{(24)} \qquad 2 - m + \sum_{w\in\sigma'} \nu(w) + \frac{1}{2} \sum_{w\in\sigma''} \nu(w) \le \frac{k}{2\pi} L(\Sigma).$$

The length $L(\Sigma) = \int_\Sigma |dX|$ of the free trace Σ can possibly be estimated by other geometric expressions (see Sections 2.12 and 4.6).

If we want to state similar formulas for minimal surfaces solving partially free boundary problems, we have to take the angles at the corners of ∂X into account (see Section 1.4 of Vol. 1, (12) and (12′)). The necessary asymptotic expansions can be found in Chapter 3.

Remark 4. For (disk-type) minimal surfaces X solving a *thread problem* (see Chapter 5), the thread Σ has a fixed length $L(\Sigma)$ and a constant geodesic curvature κ_g if we view Σ as curve on X. Hence it follows that

$$\int_\Sigma \kappa_g |dX| = \kappa_g L(\Sigma).$$

This observation can be used to draw interesting conclusions from the Gauss–Bonnet formula.

Remark 5. It is not difficult to carry over the Gauss–Bonnet formula (14) of Section 1.4 in Vol. 1 to minimal surfaces $\mathfrak{X} : M \to \mathbb{R}^3$ with branch points which are defined on a compact Riemann surface M with nonempty boundary. Suppose that ∂M consists of m disjoint, regular, smooth Jordan arcs $\gamma_1, \dots, \gamma_m$ which are topologically mapped by $\mathfrak{X}$ onto a system Γ of disjoint, regular, smooth Jordan arcs $\Gamma_1, \dots, \Gamma_m$, and let g be the genus of the orientable surfaces $\mathfrak{X}$. Then we have

$$\int_{\mathfrak{X}} K\,dA + \int_\Gamma \kappa_g\,ds + 4\pi(g-1) + 2\pi m = 2\pi \sum_{w\in\sigma'} \nu(w) + \pi \sum_{w\in\sigma''} \nu(w),$$

where σ' and σ'' denote the sets of interior and of boundary branch points, and $\nu(w)$ is the order of any $w \in \sigma' \cup \sigma''$.

2.12 Scholia

1. The first results concerning the boundary behaviour of minimal surfaces are the reflection principles of Schwarz; see Sections 3.4 and 4.8 of Vol. 1.

They insure that a minimal surface can be extended analytically across any straight part of its boundary, or across any part of its boundary where the surface meets some plane perpendicularly. Schwarz's reasoning is described in Section 3.4 of Vol. 1; cf. Schwarz [2], vol. I, p. 181. (As Schwarz mentions, he learned this reasoning from Weierstrass.) Our discussion in Section 4.8 of Vol. 1 follows the exposition in Courant [15], pp. 118–119 and pp. 218–219.

2. Another important result, found rather early, is Tsuji's theorem that a minimal surface $X \in H_2^1(B, \mathbb{R}^3)$ has boundary values $X|_{\partial B}$ of class $H_1^1(\partial B, \mathbb{R}^3)$ if their total variation is finite, i.e., if

$$\int_{\partial B} |dX| < \infty.$$

(Here we have used the parameter domain $B := \{w: |w| < 1\}$.) The importance of this result, which remained unnoticed for a long time, has been emphasized in the work of Nitsche, see [28]. Tsuji's paper [1] appeared in 1942; it is based on a classical result by F. and M. Riesz [1] from 1916 concerning the boundary values of holomorphic functions. We have presented Tsuji's result in Section 4.7 of Vol. 1.

3. The result stated as Theorem 3 in Section 2.3 is H. Lewy's celebrated regularity result from 1951; see Lewy [5]. It is the direct generalization of Schwarz's reflection principle guaranteeing that any minimal surface can be extended analytically across an analytic part of its boundary. In Courant's monograph [15], this problem was still quoted as an open question (see [15], p. 118). Lewy succeeded in proving his result without using Tsuji's theorem. Our proof essentially agrees with that of Lewy except that we use the fact that X is of class C^∞ on $B \cup \gamma$ if γ is a subarc of ∂B which is mapped by X into a real analytic arc Γ of $\mathbb{R}^3$. One can, however, avoid the use this fact (which follows from the results of Section 2.3); see Lewy [5], or Nitsche [28], pp. 297–302.

4. Hildebrandt [1] has in a first paper derived a priori estimates for minimal surfaces assuming them to be smooth up to the boundary. In conjunction with Lewy's result, one then obtains the following:

Let $X \in C^0(\overline{B}, \mathbb{R}^3) \cap H_2^1(B, \mathbb{R}^3)$ be a minimal surface which is bounded by a closed Jordan arc Γ. Suppose that $\Gamma \in C^{m,\mu}, m \geq 4, \mu \in (0,1)$, and that there is a sequence of real analytic curves Γ_n with

$$(1) \qquad |\Gamma - \Gamma_n|_{C^{m,\mu}} \to 0 \quad \text{as } n \to \infty.$$

Assume also that there is a sequence of minimal surfaces X_n bounded by Γ_n such that

$$|X - X_n|_{C^0(\overline{B})} \to 0 \quad \text{as } n \to \infty.$$

Then X is smooth up to the boundary, i.e., $X \in C^{m,\mu}(\overline{B}, \mathbb{R}^3)$.

However, it might be possible that not every solution of Plateau's problem for Γ satisfies this approximation condition; it certainly holds true for isolated local minima of Dirichlet's integral; cf. Hildebrandt [1]. By approximating a given smooth curve Γ in the sense of (1) by real analytic curves Γ_n, and by solving the Plateau problem for each of the approximating curves Γ_n, the above result yields:

Every curve $\Gamma \in C^{m,\mu}, m \geq 4, \mu \in (0,1)$, *bounds at least one minimal surface* X *of class* $C^{m,\mu}(\overline{B}, \mathbb{R}^3)$.

As a given boundary Γ may be spanning many (and, possibly, infinitely many) minimal surfaces, this regularity result by Hildebrandt [1] is considerably weaker than Theorem 1 of Section 2.3 whose global version can be formulated as follows:

Let $\Gamma \in C^0(\overline{B}, \mathbb{R}^3)$ *be a minimal surface, i.e.,*

$$\Delta X = 0, \quad |X_u|^2 = |X_v|^2, \quad \langle X_u, X_v \rangle = 0 \quad \text{in } B,$$

which is bounded by some Jordan curve Γ *of class* $C^{m,\mu}$ *with* $m \geq 1$ *and* $\mu \in (0,1)$. *Then* X *is of class* $C^{m,\mu}(\overline{B}, \mathbb{R}^3)$.

Assuming that $m \geq 4$, this result was first proved by Hildebrandt [3] in 1969. Some of the essential ideas of that paper are described in Step 1 of Section 2.7. Briefly thereafter, Heinz and Tomi [1] succeeded in establishing the result under the hypothesis $m \geq 3$, and both Nitsche [16] and Kinderlehrer [1] provided the final result for $m \geq 1$. Warschawski [6] verified that X has Dini-continuous first derivatives on $\overline{B}$, if the first derivatives of Γ with respect to arc length are Dini continuous; cf. also Lesley [1].

These results on the boundary behaviour of minimal surfaces hold for surfaces in $\mathbb{R}^n, n \geq 2$, and not only for $n = 3$; the proof requires no changes. For $n = 2$ these results include classical theorems on the boundary behaviour of conformal mappings due to Painlevé, Lichtenstein, Kellogg [2], and Warschawski [1–4]. (Concerning the older literature, we refer to Lichtenstein's article [1] in the Enzyklopädie der Mathematischen Wissenschaften; the most complete results can be found in the papers by Warschawski.)

As Nitsche has described his technique to prove boundary regularity in great detail in his monograph [28], Section 2.1, in particular pp. 283–284 and 303–312, we refer the reader to this source or to the original papers by Nitsche and Kinderlehrer quoted before. Instead we have presented a method by E. Heinz [15] which needs the slightly stronger hypothesis $m \geq 2$. By this method, Heinz could also treat H-surfaces, and Heinz and Hildebrandt [1] were able to handle minimal surfaces in Riemannian manifolds; cf. Section 2.3. The basic tools of Heinz's approach are the a priori estimates for vector-valued solutions X of differential inequalities

$$|\Delta X| \leq a |\nabla X|^2$$

which we have derived and collected in Section 2.2. They follow from classical results of potential theory which we have briefly but (more or less) completely proved in Section 2.1. The results of Section 2.2 and, in part, of Section 2.1 are taken from Heinz [2,5], and [15].

Closely related to this method is the approach of Heinz and Tomi [1] and the very useful regularity theorem of Tomi [1].

The first regularity theorem for surfaces of constant mean curvature was proved by Hildebrandt [4]; an essential improvement is due to Heinz [10]. The method of Heinz [15], described in the proof of Theorem 2 in Section 2.3, can be viewed as the optimal method. A very strong result was obtained by Jäger [3].

5. The possibility to obtain asymptotic expansions of minimal surfaces and, more generally, of H-surfaces by means of the Hartman–Wintner technique was first realized by Heinz (oral communication). A first application appeared in the paper by Heinz and Tomi [1].

6. In Theorem 2′ of Section 2.8 we proved that any minimal surface X, meeting a real-analytic support surface S perpendicularly, can be extended analytically across S. The proof basically follows ideas from H. Lewy's paper [4], published in 1951. There it was proved that *any minimizing solution X of a free boundary problem can be continued analytically across the free boundary if S is assumed to be a compact real-analytic support surface.* In fact, Lewy first had to cut off a set of hairs from the minimizer by composing it with a suitable parameter transformation before he could apply his extension technique. (Later on it was proved by Jäger [1] that the removal of these hairs is not needed since they do not exist.)

7. Combining Lewy's theorem with new a priori estimates, Hildebrandt [2] proved that the Dirichlet integral possesses at least one minimizer in $\mathcal{C}(\Gamma, S)$ which is smooth up to its free boundary provided that S is smooth and satisfies a suitable condition at infinity which enables one to prove that solutions do not escape to infinity. (A very clean condition guaranteeing this property was later formulated by Hildebrandt and Nitsche [4].)

8. The first regularity theorem for minimal surfaces with a merely smooth, but not analytic support surface S was given by Jäger [1]. He proved for instance that *any minimizer X of Dirichlet's integral in $\mathcal{C}(\Gamma, S)$ is of class $C^{m,\mu}(B \cup I, \mathbb{R}^3)$, I being the free boundary of X, provided that $S \in C^{m,\mu}$ and $m \geq 3, \mu \in (0,1)$.* Part of Jäger's method we have described or at least sketched in Section 2.8. We have not presented his main contribution, the proof of $X \in C^0(B \cup I, \mathbb{R}^3)$, which requires S to be of class C^2. Instead, in Section 2.5, we have described a method to prove continuity of minimizers up to the free boundary that needs only a *chord-arc condition for S*. Because of the Courant–Cheung example, this result is the best possible one.

The approach of Section 2.5 follows more or less the discussion in Hildebrandt [9]. The sufficiency of the chord-arc condition for proving continuity of minimizers up to the free boundary was almost simultaneously discovered by Nitsche [22] and Goldhorn and Hildebrandt [1].

Later on, Nitsche [30] showed that Jäger's regularity theorem remains valid if we relax the assumption $m \geq 3$ to $m \geq 2$. Moreover, if we assume $S \in C^2$, then every minimizer in $\mathcal{C}(\Gamma, S)$ is of class $C^{1,\alpha}(B \cup I, \mathbb{R}^3)$ for all $\alpha \in (0,1)$.

9. The regularity of stationary surfaces in $\mathcal{C}(\Gamma, S)$ up to their free boundaries was – almost simultaneously – proved by Grüter-Hildebrandt-Nitsche [1] and by Dziuk [3]. Both papers are based on the fundamental thesis of Grüter [1] (see also [2]) where interior regularity of weak H-surfaces is proved. The basic idea of Grüter's paper consists in deriving *monotonicity theorems* similar to those introduced by DeGiorgi and Almgren in geometric measure theory.

We have presented the method used in Grüter-Hildebrandt-Nitsche [1]; it has the advantage to be applicable to support surfaces with nonvoid boundary ∂S. Moreover we do not have to assume that

$$\lim_{w \to w_0} \operatorname{dist}(X(w), S) = 0 \quad \text{for any } w_0 \in I$$

as in Dziuk [5–7]. On the other hand, Dziuk's method is somewhat simpler than the other one since it reduces the boundary question to an interior regularity problem by applying Jäger's reflection method. This interior problem can be dealt with by means of the methods introduced in Grüter's thesis.

10. The results in Section 2.7 concerning the $C^{1,1/2}$-regularity of stationary minimal surfaces with a support surface S having a nonempty boundary ∂S are taken from Hildebrandt and Nitsche [1] and [2].

11. The proof of Proposition 1 in Section 2.8 is more or less that of Jäger [1], pp. 812–814.

12. The alternative method to attain the result of Step 2 in Section 2.7, given in Section 2.9, was worked out by Ye [1,4]. Ye's method is a quantitative version of the L_2-estimates of Step 2 in Section 2.7 which is based on an idea due to Kinderlehrer [6].

13. *Open questions*: (i) The regularity results for stationary minimal surfaces X with a free boundary are not yet in their final form. In particular one should prove that X is of class $C^{1,\mu}$ up to the free boundary if $S \in C^{1,\mu}_*$, and that $X \in C^{0,\alpha}$ for some $\alpha \in (0,1)$ if S satisfies a chord-arc condition (this is only known for minimizers of the Dirichlet integral). Here we say that $S \in C^{1,\mu}_*$ if $S \in C^{1,\mu}$ and if S satisfies a *uniformity condition (B) at infinity* (see Section 2.6). Dziuk [7] and Jost [8] proved that X is of class $C^{1,\mu}$ up to the free boundary, $0 < \mu < 1$, if S is of class C^2 and satisfies a suitable uniformity condition.

(ii) It would be desirable to derive a priori estimates for stationary minimal surfaces, in particular for those of higher topological type. As in general there are no estimates depending only on the geometric data of the boundary configuration (cf. the examples in Section 2.6), one could try to derive estimates depending also on certain important data of the surfaces X in consideration such as the *area* (= *Dirichlet integral*) or the *length of the free trace*.

Such estimates could be useful for approximation theorems, for results involving the deformation of the boundary configuration, for building a Morse theory, and for deriving index theorems.

Note, however, that a priori estimates depending only on boundary data can be derived in certain favourable geometric situations, for instance if the support surface is only mildly curved. Results of this kind were found by Ye [2]. Let us quote a typical result:

Suppose that S is an orientable and admissible support surface of class $C^{3,\alpha}$, $\alpha \in (0,1)$, and let n_0 be a constant unit vector and σ be a positive number, such that the surface normal $n(p)$ of S satisfies

$$\langle n(p), n_0 \rangle \geq \sigma \quad \text{for all } p \in S. \tag{2}$$

Then the length $l(\Sigma)$ of the free trace Σ of a stationary point X of Dirichlet's integral in $\mathcal{C}(\Gamma, S)$ without branch points on the free boundary I is estimated by the length of Γ via the formula

$$l(\Sigma) \leq l(\Gamma)/\sigma, \tag{3}$$

and the isoperimetric inequality yields the upper bound

$$D(X) \leq \frac{1}{4\pi}(1 + \sigma^{-2}) l^2(\Gamma) \tag{4}$$

for the Dirichlet integral of X.

Let us sketch the *proof of* (3), which is nothing but a simple variant of the reasoning used in Section 4.6.

By means of Green's formula we obtain

$$0 = \int_B \Delta X \, du \, dv = -\int_I X_v \, du + \int_C X_r \, d\varphi \tag{5}$$

with $w = u + iv = re^{i\varphi}$, where B stands for the usual semidisk. Because of (2) and of

$$X_v = |X_v| n(X) \quad \text{on } I$$

(where we possibly have to replace n by $-n$),

$$\begin{aligned} |X_r| &= |X_\varphi| \quad \text{on } C - \partial B \setminus I, \\ |X_u| &= |X_v| \quad \text{on } I, \end{aligned}$$

we then obtain

$$\begin{aligned} \sigma l(\Sigma) &= \sigma \int_I |X_u| \, du = \int_I \sigma |X_v| \, du \\ &\leq \int_I |X_v| \langle n(X), n_0 \rangle \, du = \int_I \langle X_u, n_0 \rangle \, du \\ &= \int_C \langle X_r, n_0 \rangle \, d\varphi \leq \int_C |X_r| \, d\varphi = \int_C |X_\varphi| \, d\varphi = l(\Gamma), \end{aligned}$$

i.e.,

$$\sigma l(\Sigma) \le l(\Gamma).$$

This proof of (3), (4) is not quite correct but it can easily be rectified by the reasoning in Section 4.6. We leave it to the reader to carry out the details. □

By way of an example Ye showed that the assumption $\sigma > 0$ in (2) is necessary if one wants to bound $l(\Sigma)$; see Ye [2], p. 101.

14. Now we briefly describe Courant's example of a configuration $\langle \Gamma, S\rangle$ with a continuous supporting surface S and a rectifiable Jordan arc Γ with end points on S which bounds infinitely many solutions of the corresponding free boundary problem with a discontinuous and even nonrectifiable trace curve; see Courant [15], p. 220. We firstly select a sequence of numbers $\varepsilon_n > 0, n \in \mathbb{N}$, with $\sum_{n=1}^\infty \varepsilon_n < 1/4$, and then we define set $A_n, B_n, C_n^1, C_n^2, D_n^1, D_n^2$ as follows:

$$A_n := \left\{(x,y,z)\colon z = 0, |x| < 1, \left|y - \frac{1}{n}\right| < \varepsilon_n^3\right\},$$

$$B_n := \left\{(x,y,z)\colon z = -\varepsilon_n, |x| < 1, \left|y - \frac{1}{n}\right| \le \frac{\varepsilon_n}{2}\right\},$$

$$C_n^1 := \left\{(x,y,z)\colon |x| \le 1, z = \frac{2}{2\varepsilon_n^2 - 1}\left(y - \frac{1}{n} - \varepsilon_n^3\right), -\varepsilon_n \le z \le 0\right\},$$

$$C_n^2 := \left\{(x,y,z)\colon |x| \le 1, z = \frac{2}{1 - 2\varepsilon_n^2}\left(y - \frac{1}{n} + \varepsilon_n^3\right), -\varepsilon_n \le z \le 0\right\},$$

$D_n^{1,2}$:= the compact region in $\{x = \pm 1\}$ which is bounded by $\{x = \pm 1\} \cap (\partial A_n \cup \partial B_n \cup \partial C_n^1 \cup \partial C_n^2)$,

see Fig. 1.

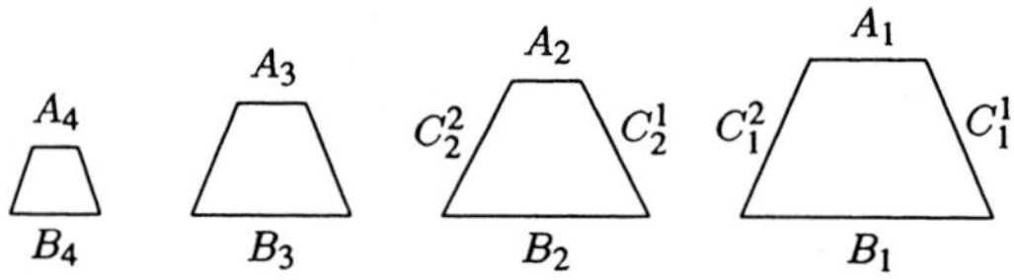

Fig. 1. Cross section of the sets A_n, B_n, C_n^1, C_n^2 at the levels $x = \pm 1$

Set

$$S_1 := \{z = 0\} \setminus \bigcup_{n=1}^\infty A_n,$$

and define S by

$$S := S_1 \cup \bigcup_{n=1}^\infty [B_n \cup C_n^1 \cup C_n^2 \cup D_n^1 \cup D_n^2]$$

(see Fig. 2).

Fig. 2. Cross sections of the surface S at the levels $x = 0, \pm 1$

Then we observe the following: If Γ_1 denotes the straight segment $\{z = 0, x = 0, |y - 1| \le \varepsilon_1^3\}$, then the associated free boundary problem $\mathcal{P}(\Gamma_1, S)$ has at least two solutions, namely the representations of the sets

$$A_1^+ := \{z = 0, 0 \le x \le 1, |y - 1| \le \varepsilon_1^3\}$$

and

$$A_1^- := \{z = 0, -1 \le x \le 0, |y - 1| \le \varepsilon_1^3\}.$$

In fact, there is still another stationary but not minimizing surface bounded by Γ_1 and S, namely the surface describing the compact region in $\{x = 0\}$ which is bounded by $\{x = 0\} \cap (\Gamma_1 \cup B_1 \cup C_1^1 \cup C_1^2)$. Similarly, if we set

$$\Gamma_n := \left\{ z = 0, x = 0, \left| y - \frac{1}{n} \right| \le \varepsilon_n^3 \right\},$$

then we obtain at leat two minimizing surfaces in $\mathcal{C}(\Gamma_n, S)$ which are determined by the sets

$$A_n^+ := \left\{ z = 0, 0 \le x \le 1, \left| y - \frac{1}{n} \right| \le \varepsilon_n^3 \right\}$$

and

$$A_n^- := \left\{ z = 0, -1 \le x \le 0, \left| y - \frac{1}{n} \right| \le \varepsilon_n^3 \right\}.$$

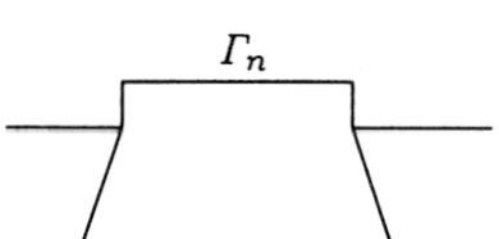

Fig. 3. The curve Γ_n lifted

Now let us lift the curves Γ_n to a height $z = \varepsilon_n^q, q \ge 4$, and connect the endpoints $P_n^1 = (0, \frac{1}{n} + \varepsilon_n^3, \varepsilon_n^q)$, $P_n^2 = (0, \frac{1}{n} - \varepsilon_n^3, \varepsilon_n^q)$ via vertical segments with S, see Fig. 3. Denoting again the lifted curves together with the vertical segments by Γ_n, it is reasonable to expect the existence of at least *two* solutions $X_n^1, X_n^2 \in \mathcal{C}(\Gamma_n, S)$ for the problem $\mathcal{P}(\Gamma_n, S)$, provided that q is large enough.

In particular, we can expect that the minimal surfaces X_n^1, X_n^2 converge to A_n^+ and A_n^- respectively, if q tends to infinity.

At a small height $z = \varepsilon < \varepsilon_{n+1}^q$, we connect Γ_n and Γ_{n+1} with a straight line segment parallel to $\{z = 0\}$ and omit the corresponding parts of the vertical segments, see Fig. 4.

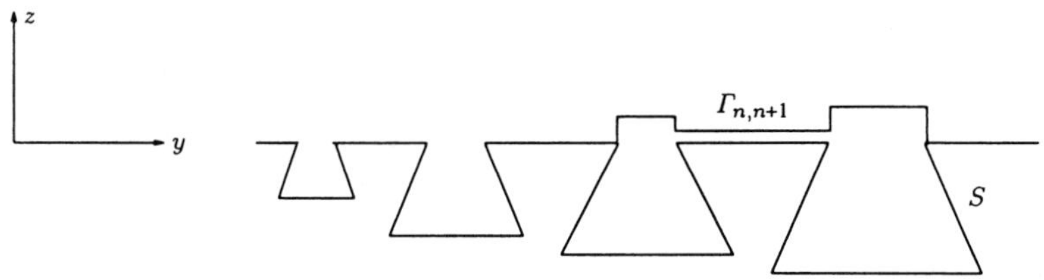

Fig. 4. Cross section of the boundary configuration $\langle \Gamma_{n,n+1}, S \rangle$ at the level $x = 0$

This way we obtain a Jordan arc $\Gamma_{n,n+1}$ with endpoints on S. Furthermore, let us assume the validity of the following bridge principle:

Given any two area-minimizing minimal surfaces $X_n \in \mathcal{C}(\Gamma_n, S)$ and $X_{n+1} \in \mathcal{C}(\Gamma_{n+1}, S)$, there exists an area-minimizing minimal surface $Y_\varepsilon \in \mathcal{C}(\Gamma_{n,n+1}, S)$ which converges (in a geometric sense) to the union of $X_n(B)$, $X_{n+1}(B)$ as ε tends to zero.

By means of this heuristic principle we obtain at least four stationary minimal surfaces in $\mathcal{C}(\Gamma_{n,n+1}, S)$ combining X_n^1 with X_{n+1}^1 or X_{n+1}^2, and X_n^2 with X_{n+1}^1 or X_{n+1}^2, respectively. Similarly we now define the Jordan arcs $\Gamma_{n,n+2}, \dots, \Gamma_{n,n+k}$ which bridge the ditches $A_n, A_{n+1}, \dots, A_{n+k}$. Then $\Gamma_{n,n+k}$ and S bound at least 2^{k+1} area-minimizing minimal surfaces which are stationary in $\mathcal{C}(\Gamma_{n,n+k}, S)$. Finally, let $\Gamma := \Gamma_{1,\infty}$ denote the rectifiable Jordan arc which bridges all the ditches and connects the points $(0,0,0)$ and $(0, 1 + \varepsilon_1, 0)$. Then it follows that there are infinitely (and even nondenumerably) many stationary minimal surfaces in $\mathcal{C}(\Gamma, S)$ each of which has a discontinuous and nonrectifiable trace curve.

Let us add that the previous reasoning is by no means rigorous; thus this example by Courant is merely of heuristic value.

15. Complementary to the existence result for the obstacle problem $\mathcal{P}(E, C)$, which we have described in the Scholia of Section 4 in Vol. 1 (see also Chapter 4 of the present volume), we want to mention some regularity properties of solutions for $\mathcal{P}(E, C)$, see also Chapter 4.

The problem $\mathcal{P} = \mathcal{P}(E, C)$ is a special case of a parametric obstacle problem which was first treated by Tomi [3,4], Hildebrandt [12,13], and Hildebrandt and Kaul [1]. Tomi's results are based on important earlier work by Lewy and Stampacchia [1,2], whereas Hildebrandt's approach uses the difference-quotient technique and some important observations due to Frehse [4].

For nonparametric obstacle problems we refer the reader to the treatise of Kinderlehrer and Stampacchia [1] and to the literature quoted there. Here we shall restrict our attention to the two-dimensional parametric case, basically following the papers by Hildebrandt and Hildebrandt-Kaul cited above. Consider the integral

$$E(X) = \int_B \{g_{ij}[X_u^i X_u^j + X_v^i X_v^j] + \langle Q(X), X_u \wedge X_v \rangle\}\, du\, dv,$$

and the class $C = C(K, \mathcal{C}^*) := \mathcal{C}^* \cap H_2^1(B, K)$, where $\mathcal{C}^*$ stands for $\mathcal{C}^*(\Gamma)$ or $\mathcal{C}^*(\Gamma, S)$ respectively and $K \subset \mathbb{R}^3$ denotes some closed set. Let us also introduce the variational problem

$$\mathcal{P}(E, C): \qquad E \to \min \text{in } C.$$

Using the abbreviation

$$e(x, p) = g_{ij}(x)\{p_1^i p_1^j + p_2^i p_2^j\} + \langle Q(x), p_1 \wedge p_2 \rangle$$

for $(x, p) \in K \times \mathbb{R}^6, p = (p_1, p_2)$, we assume that, for suitable constants $0 < m_0 \le m_1$, the coerciveness condition

$$m_0|p|^2 \le e(x, p) \le m_1|p|^2 \tag{6}$$

holds true for all $(x, p) \in K \times \mathbb{R}^6$.

Recall that the existence of a solution of $\mathcal{P}(E, C)$ can be proved under the mere assumption that K be a closed set. If we want to prove regularity, say Hölder continuity, we clearly have to add further assumptions on K. The concept of quasiregularity turns out to be of use.

Definition 1. *We call a set $K \subset \mathbb{R}^3$* quasiregular *if it is closed and if there are positive numbers δ_0, δ_1 and d such that for any point $x_0 \in K$, there exist a compact convex set K^* and a C^1-diffeomorphism g of an open neighbourhood of K^* which maps K^* onto $K \cap \overline{B}_d(x_0), B_d(x_0) = \{x \in \mathbb{R}^3 : |x - x_0| < d\}$, such that the matrix $\mathcal{H}(y) = (\frac{\partial g}{\partial y})^T \cdot (\frac{\partial g}{\partial y})$ satisfies*

$$\delta_0|\xi|^2 \le \xi \mathcal{H}(y) \xi \le \delta_1|\xi|^2 \tag{7}$$

for all $(y, \xi) \in K^* \times \mathbb{R}^3$. Here $\frac{\partial g}{\partial y}$ denotes the Jacobi matrix of g and $(\frac{\partial g}{\partial y})^T$ stands for its transpose.

Remarks. 1. Obviously, each closed convex set in $\mathbb{R}^3$ with nonvoid interior is quasiregular. Also, each compact three-dimensional submanifold of $\mathbb{R}^3$ with C^1-boundary is quasiregular.

2. The preceding Definition 1 and the following Theorem 1 extend to the case where K denotes some subset of $\mathbb{R}^N, N \ge 3$.

3. For our purposes it would be sufficient to assume that g is some bi-Lipschitz homeomorphism.

Theorem 1. *Suppose that (6) holds with functions* $g_{ij} \in C^0(K, \mathbb{R}), g_{ij} = g_{ji}$, *and* $Q \in C^0(K, \mathbb{R}^3)$. *In addition let* $K \subset \mathbb{R}^3$ *be a quasiregular set such that* $C(K, \mathcal{C}^*) = \mathcal{C}^* \cap H_2^1(B, K)$ *is nonempty. Then each solution* X *of* $\mathcal{P}(E, C)$ *satisfies a Morrey condition of the type*

$$D_{B_r(w_0)}(X) \le D_{B_R(w_0)}(X) \left(\frac{r}{R}\right)^{2\mu} \tag{8}$$

in $0 < r \le R$, *for each* $w_0 \in B_{1-R}(0)$ *and all* $R \in (0, 1)$ *and some constant* $\mu > 0$. *Hence* X *is of class* $C^{0,\mu}(B, \mathbb{R}^3)$.

Furthermore, if $\mathcal{C}^* = \mathcal{C}^*(\Gamma)$, *then* X *is also of class* $C^0(\overline{B}, \mathbb{R}^3)$, *and for* $\mathcal{C}^* = \mathcal{C}^*(\Gamma, S)$ *we infer that* $X \in C^0(\overline{B} \setminus \overline{I}, \mathbb{R}^3)$.

The idea for proving Hölder continuity is to convexify the obstacle K locally by using the definition of quasiregularity, and then to fill in harmonic functions with the right boundary values. Elementary properties of harmonic functions will yield the estimate (8). The reasoning is similar to the argument used in the proof of Theorem 1, Section 2.5; for details we refer the reader to the original paper by Hildebrandt and Kaul [1].

We now give a brief discussion of higher regularity properties of X. First we need the following

Definition 2. *A set* $K \subset \mathbb{R}^3$ *is of class* C^s *if* K *is the closure of an open set in* $\mathbb{R}^3$, *and if, for each boundary point* $x_0 \in \partial K$, *there exists a neighbourhood* U *of* x_0 *and a* C^s*-diffeomorphism* ψ *of* $\mathbb{R}^3$ *onto itself which maps* $U \cap K$ *onto*

$$B_1^+(0) = \{x \in \mathbb{R}^3 \colon |x| < 1, x^3 > 0\},$$

$U \cap \partial K$ *onto*

$$B_1^0(0) = \{x \in \mathbb{R}^3 \colon |x| < 1, x^3 = 0\},$$

and x_0 *onto* 0.

We shall also assume that the integrand $e(x, p)$ has the following **property** $(\boldsymbol{E})$:

There exist some open set $\mathcal{M} \subset \mathbb{R}^3$ *with* $K \subset \mathcal{M}$ *and functions* $Q \in C^2(\mathcal{M}, \mathbb{R}^3)$ *and* $G = (g_{jk})_{j,k=1,2,3} \in C^2(\mathcal{M}, \mathbb{R})$ *with* $g_{jk} = g_{kj}$ *such that*

$$e(x, p) = \sum_{\alpha=1}^{2} p_\alpha G(x) p_\alpha + \langle Q(x), p_1 \wedge p_2 \rangle$$

and

$$m_0 |p|^2 \le e(x, p) \le m_1 |p|^2 \quad \text{for all } (x, p) \in K \times \mathbb{R}^6, p = (p_1, p_2).$$

Theorem 2. *Suppose that* $e(x, p)$ *has property* (E), *and let* K *be quasiregular of class* C^3. *Then each solution* $X \in C(K, \mathcal{C}^*)$ *of* $\mathcal{P}(E, C)$ *is of class* $H_s^2(B', \mathbb{R}^3) \cap C^{1,\alpha}(B, \mathbb{R}^3)$ *for all* $B' \Subset B$ *and for all* $s \in [1, \infty)$ *and all* $\alpha \in (0, 1)$.

Remarks. 1. Hölder continuity of the first derivatives is still valid for solutions of the elliptic variational problem

$$\int_\Omega f(u,v,X(u,v),\nabla X(u,v))\,du\,dv \to \min \quad \text{in } C = H_2^1(\Omega,K)\cap \mathcal{C}^*,$$

with a regular Lagrangian $f\colon \Omega\times\mathcal{M}\times\mathbb{R}^{2n}\to\mathbb{R}$, where $\mathcal{M}\subset\mathbb{R}^N$ denotes some open set containing K, and $\Omega\subset\mathbb{R}^2$ denotes the domain of definition. For details we refer to Hildebrandt [12,13].

2. Assuming the conditions of Theorem 2, Gornik [1] proved that each solution $X\in C=C(K,\mathcal{C}^*)$ of $\mathcal{P}(E,C)$ is in fact of class $C^{1,1}(B,\mathbb{R}^3)$. Simple examples show that this result will in general be the best possible one. Gornik's work is based on fundamental results due to Frehse [1], Gerhardt [1], and Brézis and Kinderlehrer [1] concerning $C^{1,1}$-regularity of solutions of scalar variational inequalities.

16. The first to estimate the total order of branch points of a minimal surface via the Gauss–Bonnet formula was Nitsche [6] who reversed an idea of Sasaki [1]. R. Schneider [1] later established the formula

$$1+\sum_{w\in\sigma'}\nu(w)\le\frac{1}{2\pi}\kappa(\Gamma)$$

for all disk-type minimal surfaces $X\colon B\to\mathbb{R}^3$ which are continuous in $\overline{B}$ and map ∂B monotonically (and hence topologically) onto an arbitrary closed Jordan curve Γ which has a generalized total curvature $\kappa(\Gamma)$.

The method of Section 2.11 and the generalization of the Nitsche–Sasaki formula is taken from a paper by Heinz and Hildebrandt [2].

17. Let $\Omega\subset\mathbb{R}^2$ be an open connected domain with smooth boundary and suppose $\psi\in C^2(\overline{\Omega})$ satisfies $\max_\Omega\psi>0$ and $\psi<0$ on $\partial\Omega$. Consider the convex set of comparison functions $K_\psi:=\{v\in H_2^1(\Omega)\colon v\ge\psi \text{ in } \Omega,\ v=0 \text{ on } \partial\Omega\}$ and a solution $u\in K_\psi$ of the variational problem

$$D(u):=\frac{1}{2}\int_\Omega|\nabla u(x)|^2\,dx\to\min\quad\text{in } K_\psi.$$

One readily verifies that a solution $u\in K_\psi$ satisfies the variational inequality

$$\int_\Omega D_iuD_i(v-u)\,dx\ge 0\quad\text{for all } v\in K_\psi. \tag{9}$$

Lewy and Stampacchia [1] used the method of penalization together with suitable a priori estimates to show that u is of class $C^{1,\alpha}, \alpha<1$ (at least, if ψ is smooth and strictly concave). It is in fact true that u is of class $H^2_\infty(\Omega)$; cf. Frehse [1], Gerhardt [1], and Brézis and Kinderlehrer [1].

The set Ω may now be divided into two subsets, the coincidence set

$$\mathcal{I}=\mathcal{I}(u)=\{x\in\Omega\colon u(x)=\psi(x)\}$$

and its complement

$$\Omega \setminus \mathcal{I} = \{x \in \Omega \colon u(x) > \psi(x)\}.$$

Of particular importance is a careful analysis of the boundary $\partial\mathcal{I}$ of the set of coincidence $\mathcal{I}$. Such investigations were initiated by H. Lewy and G. Stampacchia [1] and continued by Kinderlehrer [7] and Caffarelli and Rivière [1]. It was proved that the free boundary $\partial\mathcal{I}$ is: (i) An analytic Jordan curve if ψ is strictly concave and analytic; (ii) a $C^{1,\beta}$-Jordan curve, $0 < \beta < \alpha$, if ψ is strictly concave and of class $C^{2,\alpha}$; (iii) a $C^{m-1,\alpha}$-Jordan curve if ψ is strictly concave and of class $C^{m,\alpha}$ with $m \geq 2$ and $0 < \alpha < 1$.

The investigation of $\partial\mathcal{I}$ is more difficult if we span a nonparametric surface as a graph of a function u over some obstacle graph ψ such that it minimizes area. In other words, if Ω is a strictly convex domain in $\mathbb{R}^2$ with a smooth boundary, if ψ is given as above and K_ψ is the convex set of functions $v \in \mathring{H}^1_\infty(\Omega)$ satisfying $v \geq \psi$, we consider solutions of the variational problem

$$\int_\Omega \sqrt{1 + |\nabla u|^2}\, dx \to \min \quad \text{in } K_\psi.$$

The existence of a solution $u \in K_\psi$ was proved by Lewy and Stampacchia [2] and by Giaquinta and Pepe [1]. Moreover, these authors showed that the solution u is of class $H^2_q \cap C^{1,\alpha}(\Omega)$ for every $q \in [1, \infty)$ and any $\alpha \in (0,1)$. Thus the set of coincidence $\mathcal{I} = \{x \in \Omega \colon u(x) = \psi(x)\}$ is closed, and we have

$$\operatorname{div} \frac{\nabla u}{\sqrt{1 + |\nabla u|^2}} = 0 \quad \text{in } \Omega \setminus \mathcal{I} \quad \text{as well as}$$

$$(10) \qquad \int_\Omega (1 + |\nabla u|^2)^{-1/2} \langle \nabla u, \nabla(v - u)\rangle\, dx \geq 0 \quad \text{for all } v \in K_\psi.$$

Finally, using ideas of H. Lewy (see, for instance, the proof of Theorem 2 in Section 2.8), Kinderlehrer [6] proved that the curve of separation $\Gamma := \{(x^1, x^2, x^3) \colon x^3 = u(x) = \psi(x), x \in \partial\mathcal{I}\}$ possesses a regular analytic parametrization provided that ψ is a strictly concave, analytic function.

Thin obstacle problems were treated by Lewy [6], Nitsche [19], and Giusti [2].

18. For solutions $X \in \mathcal{C}^*(\Gamma)$ of Plateau's problem satisfying a fixed three-point condition $X(w_j) = Q_j$, $j = 1, 2, 3$, $w_j \in \partial B$, $Q_j \in \Gamma$ and for $\Gamma \in C^{m,\mu}$, $m \geq 2$, $\mu \in (0,1)$, there is a number $c(m,\mu)$, independent of X such that $\|X\|_{C^{m,\mu}(\overline{B},\mathbb{R}^3)} \leq c(m,\mu)$. This a priori estimate is a quantitative version of Theorem 1 in Section 2.3, which also holds for $m = 1$ (cf. Jäger [3]).

Chapter 3

Singular Boundary Points of Minimal Surfaces

The first section of this chapter will be devoted to the study of minimal surfaces in the neighbourhood of boundary branch points. The fundamental tool for dealing with this problem is the *method of Hartman–Wintner* which yields asymptotic expansions for complex-valued solutions $f(w)$ of a differential inequality

$$(1) \qquad |f_{\overline{w}}(w)| \le c|w|^{-\lambda}|f(w)| \quad \text{on } B_R(0)$$

at the center $w = 0$ of a disk $B_R(0) = \{w \in \mathbb{C} : |w| < R\}$.

An appropriate modification of the Hartman–Wintner technique will lead to expansions for vector-valued solutions $X(w)$ of a differential inequality

$$(2) \qquad |\Delta X(w)| \le c|w|^{-\lambda}\{|X(w)| + |\nabla X(w)|\} \quad \text{on } B_R(0)$$

at $w = 0$. One of the main features of the Hartman–Wintner reasoning is that, instead of (1) and (2), one treats integral inequalities which can be considered as weak forms of the differential inequalities (1) and (2). This will enable us to deal with certain singularities at the boundary. In fact, by applying a reflection argument, it will in certain situations be possible to treat boundary singularities as interior singularities of solutions to suitably extended equations. However, to make this artifice valid, it will be indispensable to work with integral versions of (1) and (2) because they require less regularity of their solutions. We refer the reader to Section 2.10 (in particular, Theorems 1 and 2) where we have discussed the behaviour of minimal surfaces at boundary branch points in detail. We emphasize again that the Hartman–Wintner device is the essential tool in proving the asymptotic relation (7) in Section 2.10.

In Section 3.1 we shall describe an extended version of the Hartman–Wintner technique as well as some important generalizations due to Dziuk.

In Section 3.2 we shall study the asymptotic behaviour of the gradient of a minimal surface near a corner on the boundary. We shall discuss corners on a Jordan curve as well as corners between curves and supporting surfaces since

U. Dierkes, S. Hildebrandt, A.J. Tromba, *Regularity of Minimal Surfaces*,
Grundlehren der mathematischen Wissenschaften 340,
DOI 10.1007/978-3-642-11700-8_3, © Springer-Verlag Berlin Heidelberg 2010

they occur in partially free boundary problems. The results of Section 3.2 provide the *initial regularity* indispensable for the methods of Sections 3.3 and 3.4 to work. In these sections a precise discussion of the geometric behaviour of a minimal surface at corners will be given. Section 3.3 deals with the Plateau problem for piecewise smooth contours, whereas in Section 3.4 free boundary problems are investigated.

3.1 The Method of Hartman and Wintner, and Asymptotic Expansions at Boundary Branch Points

This section deals with the asymptotic behaviour of solutions to certain differential and integral inequalities at *interior singularities*. In certain situations, boundary singularities can be made into inner singularities by extending a solution, for example, by reflection.

First we shall consider complex-valued or even vector-valued solutions $f(w)$ of the differential inequality

$$|f_{\overline{w}}(w)| \le c|w|^{-\lambda}|f(w)| \tag{1}$$

in a disk $B_R(0)$, where λ and c are real constants with $0 \le \lambda < 1$ and $c > 0$, and f is of class $C^1(B_R(0), \mathbb{C}^N), N \ge 1$. As usual we write

$$g_w = \frac{\partial g}{\partial w} = \frac{1}{2}(g_u - ig_v), \quad g_{\overline{w}} = \frac{\partial g}{\partial \overline{w}} = \frac{1}{2}(g_u + ig_v).$$

Secondly, we consider vector-valued solutions $X(w) = (X^1(w), X^2(w), \dots, X^N(w))$ of

$$|\Delta X(w)| \le c|w|^{-\lambda}\{|X(w)| + |\nabla X(w)|\} \tag{2}$$

in $B_R(0), c > 0, 0 \le \lambda < 1$, which are of class C^1 of $B_R(0)$. If the right-hand side of (2) would not contain X but only ∇X, (2) could be considered as a special case of (1) by setting $f(w) := X_w(w)$.

Both (1) and (2) can be transformed into integral inequalities which require less regularity of their solutions.

For instance, let $f(w)$ be a solution of (1) in a domain $\Omega \subset \mathbb{C}$ which is of class C^1, and let $\mathcal{D} \Subset \Omega$ be an arbitrary subdomain of Ω with piecewise smooth boundary $\partial\mathcal{D}$. Choose an arbitrary function $\phi \in C^1(\Omega, \mathbb{C})$ and apply Green's formula

$$\int_{\partial\mathcal{D}} g(w)\, dw = 2i \int\int_{\mathcal{D}} \frac{\partial}{\partial \overline{w}} g(w)\, du\, dv$$

to $g(w) = \phi(w) \cdot f(w)$.

Differing from our usual notation, we denote double integrals by two integral signs.

The integral $\int_{\partial \mathcal{D}} g(w)\,dw$ stands for the complex line integral of the function g over the boundary $\partial\mathcal{D}$ which is assumed to be positively oriented with respect to $\mathcal{D}$. Then we obtain

$$\int_{\partial\mathcal{D}} \phi(w)\cdot f(w)\,dw = 2i \iint_{\mathcal{D}} [\phi_{\overline{w}}(w)\cdot f(w) + \phi(w)\cdot f_{\overline{w}}(w)]\,du\,dv,$$

and (1) yields

$$(3)\qquad \left|\int_{\partial\mathcal{D}} \phi(w)\cdot f(w)\,dw\right| \le 2\iint_{\mathcal{D}} [|\phi_{\overline{w}}(w)| + c|w|^{-\lambda}|\phi(w)|]|f(w)|\,du\,dv.$$

This is the integral inequality associated with (1).

Similarly, we have

$$\int_{\partial\mathcal{D}} \phi\cdot X_w\,dw = 2i\iint_{\mathcal{D}} [\phi_{\overline{w}}\cdot X_w + \phi\cdot X_{w\overline{w}}]\,du\,dv,$$

and we derive from (2) the inequality

$$(4)\qquad \left|\int_{\partial\mathcal{D}} \phi(w)\cdot X_w(w)\,dw\right| \le 2\iint_{\mathcal{D}} \{|\phi_{\overline{w}}(w)||X_w(w)| + c|w|^{-\lambda}|\phi(w)|[|X(w)| + |X_w(w)|]\}\,du\,dv.$$

Here c is one quarter of the constant c in (2) because of $\Delta X = 4X_{w\overline{w}}$.

In the following we shall work with inequalities (3) and (4) rather than with (1) or (2) respectively. Note that (3) makes sense even for continuous $f(w)$, and (4) can even be considered for functions X of class C^1. Hence we give the following

Definition 1. *A mapping $f(w) = (f^1(w), \ldots, f^N(w)), w \in B_R(0)$, is said to satisfy* **Assumption (A1)** *on $B_R(0)$ if it is of class $C^0(B_R(0)\setminus\{0\}, \mathbb{C}^N)$ and fulfils* (3) *for every $\phi \in C^1(B_R(0), \mathbb{C})$ and for every $\mathcal{D} \Subset B_R(0)$ with piecewise smooth boundary $\partial\mathcal{D}$.*

Similarly, $X(w) = (X^1(w), \ldots, X^N(w)), w \in B_R(0)$, is said to fulfil **Assumption (A2)** *on $B_R(0)$ if it is of class $C^1(B_R(0), \mathbb{R}^N)$ and satisfies* (4) *for every $\phi \in C^1(B_R(0), \mathbb{C})$ and for each $\mathcal{D} \Subset B_R(0)$ with piecewise smooth boundary.*

Then we are going to prove the following two theorems:

Theorem 1. *Let $f(w)$ satisfy* (A1) *on $B_R(0)$, and suppose that $f(w) \not\equiv 0$ in $B_R(0)$, and that there exists a number $\lambda' \in [0,1)$ such that*

$$f(w) = O(|w|^{-\lambda'}) \quad \text{as } w \to 0.$$

Then there is a nonnegative integer ν such that $\lim_{w\to 0} w^{-\nu} f(w)$ exists and is different from zero.

Theorem 2. *Let $X(w)$ satisfy* (A2) *on $B_R(0)$ and suppose that there is a nonnegative integer ν such that*

$$X(w) = o(|w|^\nu) \quad \text{as } w \to 0. \tag{5}$$

Then the limit $\lim_{w\to 0} X_w(w)w^{-\nu}$ exists. In addition, if $X(w) \not\equiv 0$, then there is a first nonnegative integer ν such that (5) *does not hold and, moreover, that $\lim_{w\to 0} X_w(w) \cdot w^{-\mu}$ exists for $\mu = \nu - 1$ and is different from zero.*

We shall prove both theorems simultaneously. The proof of the second theorem differs from the first one in that we have as well to estimate the additional term involving $|X(w)|$. We will return to this in detail after completing the proof of Theorem 1. Without loss of generality we may assume that f is a scalar function.

Before entering into the proofs, we first mention two interesting corollaries.

Corollary 1. *Let $f(w)$ satisfy the assumptions of Theorem 1. Then there exists a nonnegative integer ν and a complex number $a \not\equiv 0$ such that*

$$f(w) = aw^\nu + o(|w|^\nu) \quad \textit{as } w \to 0.$$

Corollary 2. *Let $X(w)$ satisfy* (A2) *on $B_R(0)$ and suppose that $X(0) = 0$ but $X(w) \not\equiv 0$ on $B_R(0)$. Then there exists a nonnegative integer μ and a nonzero complex vector A such that*

$$X_w(w) = Aw^\mu + o(|w|^\mu) \quad \textit{as } w \to 0, \tag{6}$$

and

$$X(w) = \mathrm{Re}\{Bw^{\mu+1}\} + o(|w|^{\mu+1}) \quad \textit{as } w \to 0, \tag{7}$$

where $B = 2(\mu + 1)^{-1}A$.

Proof of Corollary 2. Equation (6) is nothing but a different formulation of the second statement in Theorem 2. Relation (7) follows by a suitable integration. In fact,

$$\begin{aligned} X(w) &= \int_0^1 [uX_u(tw) + vX_v(tw)]\, dt = \int_0^1 \mathrm{Re}[2wX_w(tw)]\, dt \\ &= \int_0^1 \mathrm{Re}[2(At^\mu w^{\mu+1} + t^\mu o(|w|^{\mu+1})]\, dt \\ &= \mathrm{Re}\left[\frac{2}{\mu+1} Aw^{\mu+1}\right] + o(|w|^{\mu+1}). \end{aligned}$$

□

Now we begin with the proof of Theorem 1.

Lemma 1. *Let f satisfy* (A1) *and suppose that there exists some nonnegative integer μ such that*

$$f(w) = o(|w|^{\mu-1}) \quad \textit{as } w \to 0.$$

Then $f(w) = O(|w|^\mu|)$ as $w \to 0$.

Proof. Let $r < R, \xi \in B_r(0), \xi \neq 0, \varepsilon < \min(\frac{|\xi|}{2}, r - |\xi|)$, and put

$$\mathcal{D}_{r,\varepsilon} := B_r(0) \setminus [B_\varepsilon(0) \cup B_\varepsilon(\xi)].$$

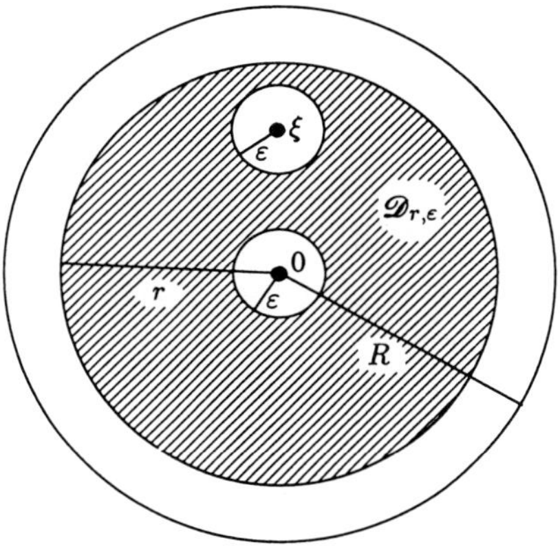

Fig. 1.

Now we test inequality (3) with the function $\phi(w) = \frac{1}{w^\mu}\frac{1}{w-\xi}$ and the domain $\mathcal{D}_{r,\varepsilon}$. This yields the estimate

$$(8) \quad \left| \int_{\partial \mathcal{D}_{r,\varepsilon}} w^{-\mu}(w-\xi)^{-1} f(w)\, dw \right| \leq 2c \iint_{\mathcal{D}_{r,\varepsilon}} |w|^{-\mu-\lambda} |w-\xi|^{-1} |f(w)|\, du\, dv.$$

The result will now follow by letting ε tend to zero. To accomplish this it will be necessary to consider the boundary integrals on the left-hand side of (6) separately. Firstly, we have

$$\int_{|w-\xi|=\varepsilon} w^{-\mu}(w-\xi)^{-1} f(w)\, dw = i \int_0^{2\pi} (\xi + \varepsilon e^{i\varphi})^{-\mu} f(\xi + \varepsilon e^{i\varphi})\, d\varphi$$

whence

$$(9) \quad \begin{aligned} \lim_{\varepsilon \to 0} \int_{|w-\xi|=\varepsilon} w^{-\mu}(w-\xi)^{-1} f(w)\, dw &= i \int_0^{2\pi} \xi^{-\mu} f(\xi)\, d\varphi \\ &= 2\pi i f(\xi) \xi^{-\mu}. \end{aligned}$$

Furthermore we obtain

$$\begin{aligned} \left| \int_{|w|=\varepsilon} w^{-\mu}(w-\xi)^{-1} f(w)\, dw \right| &\leq \int_{|w|=\varepsilon} \left| \frac{f(w)}{w^{\mu-1}} \right| |w(w-\xi)|^{-1} |dw| \\ &\leq \frac{2}{|\xi|} \int_0^{2\pi} \frac{|f(\varepsilon e^{i\varphi})|}{\varepsilon^{\mu-1}}\, d\varphi, \end{aligned}$$

where we have used that $|w - \xi| > \frac{|\xi|}{2}$ for $w \in \partial B_\varepsilon(0)$. Hence

$$\lim_{\varepsilon \to 0} \left| \int_{|w|=\varepsilon} w^{-\mu}(w-\xi)^{-1} f(w)\, dw \right| = 0, \tag{10}$$

taking $f(w) = o(|w|^{\mu-1})$ into account.

Now we conclude from (8) the inequality

$$\begin{aligned} \left| 2\pi i f(\xi)\xi^{-\mu} - \int_{|w|=r} w^{-\mu}(w-\xi)^{-1} f(w)\, dw \right| \\ \le 2c \iint_{|w|\le r} |w|^{-\mu-\lambda} |w-\xi|^{-1} |f(w)|\, du\, dv. \end{aligned} \tag{11}$$

Define J_1 and J_2 by the formulas

$$J_1(\xi) := \int_{|w|=r} |w|^{-\mu} |w-\xi|^{-1} |f(w)| |dw|$$

and

$$J_2(\xi) := \iint_{|w|\le r} |w|^{-\mu-\lambda} |w-\xi|^{-1} |f(w)|\, du\, dv;$$

then (11) implies the inequality

$$2\pi |f(\xi)\xi^{-\mu}| \le J_1(\xi) + 2c J_2(\xi). \tag{12}$$

It follows from (12) that the lemma is proved if we find uniform bounds for $J_1(\xi)$ and $J_2(\xi)$ and all $\xi \in B_{r_0}(0)$ for some $0 < r_0 < r$. The uniform boundedness of $J_1(\xi)$ is obvious since the singularities of the integrand, 0 and ξ, have positive distance from the circle $|w| = r$. Now we show that $J_1(\xi)$ provides an upper bound for $J_2(\xi)$. To this end we multiply (12) by $|\xi|^{-\lambda}|\xi - w_0|^{-1}$ where $w_0 \in B_r(0)$ and integrate over the disk $B_r(0)$; it follows that

$$\begin{aligned} & 2\pi J_2(w_0) \\ & = 2\pi \iint_{|\xi|<r} |\xi|^{-\mu-\lambda} |\xi - w_0|^{-1} |f(\xi)|\, d\xi_1\, d\xi_2 \le I_1(w_0) + I_2(w_0), \end{aligned} \tag{13}$$

where $\xi = \xi_1 + i\xi_2$ and

$$\begin{aligned} I_1(w_0) &:= \iint_{|\xi|<r} |\xi|^{-\lambda} |\xi - w_0|^{-1} J_1(\xi)\, d\xi_1\, d\xi_2, \\ I_2(w_0) &:= 2c \iint_{|\xi|<r} |\xi|^{-\lambda} |\xi - w_0|^{-1} J_2(\xi)\, d\xi_1\, d\xi_2. \end{aligned}$$

Using the identity

$$(w-\xi)^{-1}(\xi-w_0)^{-1} = (w-w_0)^{-1}[(w-\xi)^{-1}+(\xi-w_0)^{-1}]$$

and interchanging the order of integration we obtain

$$\begin{aligned} I_1(w_0) &\le \iint_{|\xi|<r} \left\{ \int_{|w|=r} [|w-\xi|^{-1}+|\xi-w_0|^{-1}]|\xi|^{-\lambda}|w-w_0|^{-1} \right. \\ &\qquad \left. \cdot|w|^{-\mu}|f(w)||dw| \right\} d\xi_1\, d\xi_2 \\ &\le Mr^{1-\lambda}\int_{|w|=r} |w|^{-\mu}|w-w_0|^{-1}|f(w)||dw| \\ &= Mr^{1-\lambda}J_1(w_0), \end{aligned}$$

where we have used the inequality

$$(*) \qquad \iint_{|\xi|<r} \{|w-\xi|^{-1}+|\xi-w_0|^{-1}\}|\xi|^{-\lambda}\, d\xi_1\, d\xi_2 \le Mr^{1-\lambda}$$

which will be proved later. Similarly we obtain the estimate

$$\begin{aligned} I_2(w_0) &\le 2c \iint_{|\xi|<r} \left\{ \iint_{|w|<r} [|w-\xi|^{-1}+|\xi-w_0|^{-1}]|\xi|^{-\lambda} \right. \\ &\qquad \left. \cdot|w-w_0|^{-1}|w|^{-\mu-\lambda}|f(w)|du\, dv \right\} d\xi_1\, d\xi_2 \\ &\le 2Mr^{1-\lambda}cJ_2(w_0) \end{aligned}$$

with the same constant M.

Finally we infer from (13) the inequality

$$2\pi J_2(w_0) \le Mr^{1-\lambda}[J_1(w_0)+2cJ_2(w_0)]$$

which implies

$$(14) \qquad 2(M^{-1}\pi r^{\lambda-1}-c)J_2(w_0) \le J_1(w_0).$$

If we now choose $r_0 < (\frac{\pi}{cM})^{1/(1-\lambda)}$, then the inequality

$$J_2(w_0) \le \tfrac{1}{2}(M^{-1}\pi r^{\lambda-1}-c)^{-1}J_1(w_0)$$

holds for all $w_0 \in B_{r_0}(0)$, and Lemma 1 is proved. □

Now we have to add a proof of inequality $(*)$. In fact we shall prove a slightly more general result known as

E. *Schmidt's inequality* (see Vekua [1], p. 39). Suppose that $w_1, w_2 \in B_r(0)$, and that $\alpha, \beta < 2$ are positive real constants. Then

$$(15)\qquad \iint_{B_r(0)} |\xi - w_1|^{-\alpha}|\xi - w_2|^{-\beta}\, d\xi_1\, d\xi_2 \leq \begin{cases} M_1|w_1 - w_2|^{2-\alpha-\beta} & \text{if } a+\beta > 2, \\ M_2 + 8\pi|\log|w_1 - w_2|| & \text{if } a+\beta = 2, \\ M_3 r^{2-\alpha-\beta} & \text{if } \alpha+\beta < 2, \end{cases}$$

where $\xi = \xi_1 + i\xi_2$, and M_1, M_2, M_3 are constants depending only on α and β.

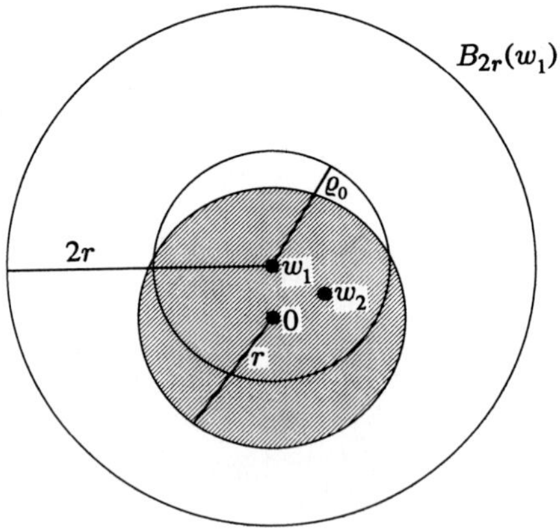

Fig. 2.

Proof of (15). We replace $B_r(0)$ by the larger domain $B_{2r}(w_1) \supset B_r(0)$. If we put $\rho_0 = 2|w_1 - w_2|$, we have for all $\xi \in B_{2r}(w_1) \setminus B_{\rho_0}(w_1)$ that $2|\xi - w_2| \geq |\xi - w_1|$ which yields

$$(16)\qquad \iint_{B_{2r}(w_1)\setminus B_{\rho_0}(w_1)} |\xi - w_1|^{-\alpha}|\xi - w_2|^{-\beta}\, d\xi_1\, d\xi_2 \leq 2^{1+\beta}\pi \int_{\rho_0}^{2r} \rho^{1-\alpha-\beta}\, d\rho$$

$$\leq \begin{cases} 2^{1+\beta}\pi \frac{|w_1-w_2|^{2-\alpha-\beta}}{\alpha+\beta-2} & \text{if } \alpha+\beta > 2, \\ 2^{1+\beta}\pi \log\frac{r}{|w_1-w_2|} & \text{if } \alpha+\beta = 2, \\ \frac{2^{3-\alpha}}{2-\alpha-\beta} r^{2-\alpha-\beta} & \text{if } \alpha+\beta < 2. \end{cases}$$

Applying the linear transformation $\xi^* = \frac{\xi - w_1}{|w_2 - w_1|}$ which maps $B_{\rho_0}(w_1)$ onto $B_2(0)$, we conclude from the change-of-variables formula that

$$(17)\qquad \iint_{B_{\rho_0}(0)} |\xi - w_1|^{-\alpha}|\xi - w_2|^{-\beta}\, d\xi_1\, d\xi_2$$

$$= |w_1 - w_2|^{2-\alpha-\beta} \iint_{B_2(0)} |\xi^*|^{-\alpha} \left|\xi^* - \frac{w_2 - w_1}{|w_2 - w_1|}\right|^{-\beta} d\xi_1^*\, d\xi_2^*.$$

By virtue of $\alpha, \beta < 2$, the integral on the right-hand side can be estimated by a *finite* constant $M(\alpha, \beta)$. Inequality (15) follows by combining the above estimates. □

Lemma 2. *Suppose that f satisfies assumption* (A1) *and that* $f(w) = o(|w|^{\mu-1})$ *as* $w \to 0$ *for some nonnegative integer* μ. *Then the limit* $\lim_{w\to 0} f(w)w^{-\mu}$ *exists.*

Proof. Let $g(w) := f(w)w^{-\mu}$ and

$$F_r(\xi) = (2\pi i)^{-1} \int_{|w|=r} g(w)(w-\xi)^{-1}\, dw, \quad r \in (0, R).$$

Then $F_r(\xi)$ is holomorphic on $B_r(0)$, and from inequality (11) we infer for all $\xi \in B_r(0) \setminus \{0\}$ the relation

$$\begin{aligned} |g(\xi) - F_r(\xi)| &\le \frac{c}{\pi} \iint_{B_r(0)} |w|^{-\mu-\lambda}|w-\xi|^{-1}|f(w)|\, du\, dv \\ &\le c_1 \iint_{B_r(0)} |w|^{-\lambda}|w-\xi|^{-1}\, du\, dv, \end{aligned}$$

where we have used Lemma 1. We infer from inequality (15) the estimate

$$|g(\xi) - F_r(\xi)| \le c_2 r^{1-\lambda}$$

for all $\xi \in B_r(0) \setminus \{0\}$. Again from the boundedness of $g(w)$ we conclude the existence of a sequence $\{w_n\}_{n\in\mathbb{N}}$ tending to zero such that

$$a := \lim_{n\to\infty} g(w_n) \in \mathbb{C},$$

whence

$$|a - F_r(0)| \le c_2 r^{1-\lambda},$$

and since $\lambda < 1$, we obtain

$$\lim_{r\to 0} F_r(0) = a.$$

Finally we conclude from

$$|g(\xi) - a| \le |g(\xi) - F_r(\xi)| + |F_r(\xi) - F_r(0)| + |F_r(0) - a| \le c_3 r^{1-\lambda}$$

the relation

$$\lim_{\xi\to 0} g(\xi) = a.$$ □

Proof of Theorem 1. The theorem will be proved if we can find an integer $\nu \ge -1$ with the properties $f(w) = o(|w|^\nu)$ but $f(w) \ne o|w|^{\nu+1}$ as $w \to 0$, taking Lemma 2 into account. Let us assume on the contrary that for all nonnegative ν the relation $f(w) = o(|w|^\nu)$ holds true. We will then show that $f \equiv 0$ on $B_R(0)$.

To accomplish this, we recall inequality (12) with $w_0 = 0$:

$$2(M^{-1}\pi r^{\lambda-1} - c)J_2(0) \le J_1(0),$$

where

$$J_1(0) = \int_{|w|=r} |w|^{-\nu-1}|f(w)||dw|,$$

and

$$J_2(0) = \int_{|w|=r} |w|^{-\nu-\lambda-1}|f(w)|\,du\,dv.$$

We select $r < (\frac{\pi}{cM})^{1/(1-\lambda)}$ and suppose that there exists some $\xi_0 \in B_r(0)$ with $f(\xi_0) \ne 0$. Clearly there exist numbers $0 < \delta_1 \le \delta_2, \varepsilon > 0$, such that $B_\varepsilon(\xi_0) \Subset B_r(0)$ and

$$2(M^{-1}\pi r^{\lambda-1} - c)\delta_1[|\xi_0| + \varepsilon]^{-\nu-\lambda-1} \le \delta_2 r^{-\nu-1}.$$

Therefore there exists some constant c_1 independent of ν such that

$$0 < c_1 \le \left(\frac{\varepsilon + |\xi_0|}{r}\right)^{\nu+1}.$$

This relation, however, cannot hold for all $\nu \in \mathbb{Z}$ since $|\xi_0| + \varepsilon < r$. In conclusion we have shown that $f = 0$ on $B_r(0)$ for some sufficiently small r, and *a* continuation argument implies $f = 0$ on $B_R(0)$. This completes the proof of Theorem 1. □

Next we are going to prove Theorem 2.

Lemma 3. *Suppose $X(w) \in C^1(B_r(0), \mathbb{R}^N)$ satisfies $X(0) = 0$ and $X_w(w) = o(|w|^{\mu-1})$ as $w \to 0$. Then $X(w) = o(|w|^\mu)$.*

Proof. Fix $w \in B_r(0)$. Then a simple integration yields

$$\begin{aligned} \frac{X(w)}{w^\mu} &= \int_0^1 \left\{\frac{u}{w^\mu}X_u(tw) + \frac{v}{w^\nu}X_v(tw)\right\} dt \\ &= \int_0^1 \frac{2}{w^\mu}\mathrm{Re}(wX_w(tw))\,dt \\ &= 2\int_0^1 t^{\mu-1}\left\{\frac{1}{(tw)^\mu}\mathrm{Re}(twX_w(tw))\right\} dt \\ &= \frac{2}{\mu}\frac{1}{(t_0w)^\mu}\mathrm{Re}(t_0wX_w(t_0w)) \end{aligned} \tag{18}$$

for some $t_0 \in (0, 1)$. Consequently,

$$\lim_{w\to 0} \frac{X(w)}{w^\mu} = 0.$$ □

The following auxiliary result provides a counterpart to Lemma 1.

Lemma 4. *Let X satisfy Assumption* (A2), *and suppose that there exists some nonnegative integer μ such that $X_w(w) = o(|w|^{\mu-1})$ as $w \to 0$. Then $X_w(w) = O(|w|^{\mu})$ as $w \to 0$.*

Proof. As in the proof of Lemma 1 we put

$$\mathcal{D}_{r,\varepsilon} = B_r(0) \setminus [B_\varepsilon(0) \cup B_\varepsilon(\xi)]$$

and

$$\phi(w) = \frac{1}{w^\mu} \cdot \frac{1}{w-\xi}.$$

Then (4) yields the inequality

$$\left| \int_{\partial \mathcal{D}_{r,\varepsilon}} w^{-\mu}(w-\xi)^{-1} X_w(w)\, dw \right| \le 2c \iint_{\mathcal{D}_{r,\varepsilon}} |w|^{-\mu-\lambda} |w-\xi|^{-1} [|X(w)| + |X_w(w)|]\, du\, dv, \tag{19}$$

taking $\phi_{\overline{w}} = 0$ on $\mathcal{D}_{r,\varepsilon}$ into account. Note that Lemma 3 and the inequality $0 \le \lambda < 1$ imply the boundedness of the integral

$$J_3(\xi) := \iint_{B_r(0)} |w|^{-\mu-\lambda} |w-\xi|^{-1} |X(w)|\, du\, dv.$$

Now we can proceed as in the proof of Lemma 1, i.e. we let $\varepsilon \to 0$ and obtain the estimate

$$2\pi |X_w(\xi)\xi^{-\mu}| \le J_1(\xi) + 2c[J_2(\xi) + J_3(\xi)], \tag{20}$$

where f has to be replaced by X_w in the formulas for J_1 and J_2 respectively, i.e.,

$$J_1(\xi) := \int_{|w|=r} |w|^{-\mu} |w-\xi|^{-1} |X_w(w)| |dw|,$$

and

$$J_2(\xi) := \iint_{|w|\le r} |w|^{-\mu-\lambda} |w-\xi|^{-1} |X_w(w)|\, du\, dv.$$

Since the boundedness of J_1 is obvious for small ξ, we only show the boundedness of J_2. To this end we multiply (18) by $|\xi|^{-\lambda}|\xi - w_0|^{-1}$, $w_0 \in B_r(0)$, and integrate over $B_r(0)$. Then we obtain

$$2\pi J_2(w_0) \le I_1(w_0) + I_2(w_0) + I_3(w_0), \tag{21}$$

where

$$I_1(w_0) := \iint_{|\xi|<r} |\xi|^{-\lambda}|\xi - w_0|^{-1} J_1(\xi)\, d\xi_1\, d\xi_2,$$

$$I_2(w_0) := 2c \iint_{|\xi|<r} |\xi|^{-\lambda}|\xi - w_0|^{-1} J_2(\xi)\, d\xi_1\, d\xi_2,$$

$$I_3(w_0) := 2c \iint_{|\xi|<r} |\xi|^{-\lambda}|\xi - w_0|^{-1} J_3(\xi)\, d\xi_1\, d\xi_2,$$

and $\xi = \xi_1 + i\xi_2$.

As in the proof of Lemma 1 we conclude

$$I_1(w_0) \le Mr^{1-\lambda} J_1(w_0),$$

$$I_2(w_0) \le 2Mr^{1-\lambda} c J_2(w_0).$$

Similarly we infer from (15) and

$$(w-\xi)^{-1}(\xi - w_0)^{-1} = (w - w_0)^{-1}[(w-\xi)^{-1} + (\xi - w_0)^{-1}]$$

the estimate

$$I_3(w_0) \le 2M_3^2 c_1 r^{2(1-\lambda)} \quad \text{for some constant } c_1.$$

Finally, the boundedness of J_2 follows from (21) and the above estimates if we choose $r > 0$ suitably small. The assertion of the lemma follows from relation (20) since the right-hand side of (20) remains bounded as $\xi \to 0$. □

Lemma 5. *Let $X(w)$ satisfy assumption* (A2) *and suppose that for some non-negative integer μ we have*

$$X_w(w) = o(|w|^{\mu-1}) \quad \text{as } w \to 0.$$

Then the limit $\lim_{w\to 0} X_w(w) w^{-\mu}$ exists.

Proof. We put $g(w) := X_w(w) w^{-\mu}$ and

$$F_r(\xi) := (2\pi i)^{-1} \int_{|w|=r} g(w)(w-\xi)^{-1}\, dw.$$

In the relation (19) we let ε tend to zero (cf. the proof of Lemma 1) and obtain the inequality

$$|g(\xi) - F_r(\xi)| \le \frac{c}{\pi} \iint_{B_r(0)} |w|^{-\mu-\lambda}|w-\xi|^{-1}[|X(w)| + |X_w(w)|]\, du\, dv,$$

holding for all $\xi \in B_r(0) \setminus \{0\}$. Now Lemmata 3 and 4 imply

$$|g(\xi) - F_r(\xi)| \le c_1 \iint_{B_r(0)} |w|^{-\lambda}|w-\xi|^{-1}\, du\, dv$$

for all $\xi \in B_r(0) \setminus \{0\}$ and some constant c_1. From here on we can proceed exactly as in the proof of Lemma 2. □

Proof of Theorem 2. Recall that $X(w)$ satisfies assumption (A2) and that for some nonnegative $\nu \in \mathbb{Z}$ we have

$$X(w) = o(|w|^\nu) \quad \text{as } w \to 0. \tag{22}$$

We first show that the limit $\lim_{w\to 0} X_w(w)w^{-\nu}$ exists. Since X is supposed to be differentiable, this clearly holds when $\nu = 0$. On the other hand, if $\nu = 1$ we infer from (22) that

$$X_w(w) = o(1) \quad \text{as } w \to 0,$$

and an application of Lemma 5 implies the existence of

$$\lim_{w\to 0} X_w(w)w^{-1}.$$

In order to prove the general case $\nu > 1$, we shall inductively show that

$$X_w(w) = o(|w|^{\mu-1}) \quad \text{as } w \to 0 \tag{23}$$

holds for all $\mu \in [1, \nu]$. (The result will then follow by a further application of Lemma 5.)

Assume the validity of (23) for some $\mu < \nu$; then there exists some number $a \in \mathbb{C}$ such that

$$\lim_{w\to 0} X_w(w)w^{-\mu} = a. \tag{24}$$

We show that $a = 0$. To this end, observe that we can write

$$\begin{aligned} \frac{X(u,0)}{u^{\mu+1}} &= \int_0^1 \frac{u}{u^{\mu+1}} X_u(tu,0)\,dt \\ &= \int_0^1 t^\mu \frac{X_u(tu,0)}{(tu)^\mu}\,dt = \frac{X_u(t_0 u,0)}{(t_0 u)^\mu} \int_0^t t^\mu\,dt \\ &= \frac{1}{\mu+1}\frac{X_u(t_0 u,0)}{(t_0 u)^\mu} \quad \text{for some } t_0 \in (0,1). \end{aligned} \tag{25}$$

On the other hand we infer from (24) that the function $g(w) := X_w(w)w^{-\mu}$ is continuous at $w = 0$, and again (24) implies

$$a = \lim_{n\to\infty} \frac{X_w(u_n,0)}{u_n^\mu} = \lim_{n\to\infty} \frac{X_u(u_n,0) - iX_v(u_n,0)}{2u_n^\mu}$$

whence

$$\operatorname{Re} a = \lim_{n\to\infty} \frac{X_u(u_n,0)}{2u_n^\mu} \quad \text{for every sequence } u_n \to 0. \tag{26}$$

Now (25), (26), the assumption $X(w) = o(|w|^\nu)$ as $w \to 0$, and the continuity of $g(w)$ yield the relation

$$\operatorname{Re} a = 0.$$

Furthermore, we infer from (24) that

$$a = \lim_{n\to\infty} \frac{X_u(0, v_n) - iX_v(0, v_n)}{2i^\mu v_n^\mu},$$

and in particular, if μ is even,

$$\pm \operatorname{Im} a = \lim_{n\to\infty} \frac{X_v(0, v_n)}{2v_n^\mu}, \quad \text{where } v_n \to 0.$$

Hence the same argument yields that $\operatorname{Im} a = 0$, provided that μ is even. If μ is odd we consider the function $Y(w) = wX(w)$. Then $Y_w(w) = X(w) + wX_w(w)$, and therefore

$$\lim_{w\to 0} \frac{Y_w(w)}{w^{\mu+1}} = \lim_{w\to 0} \frac{X(w)}{w^{\mu+1}} + \lim_{w\to 0} \frac{X_w(w)}{w^\mu} = a.$$

Also $Y(w) = o(|w|^{\mu+2})$ as $w \to 0$, whence

$$a = \lim_{n\to\infty} \frac{Y_u(0, v_n) - iY_v(0, v_n)}{2i^{\mu+1} v_n^{\mu+1}},$$

whenever $v_n \to 0$ with $n \to \infty$. Thus we obtain that $\operatorname{Im} a = 0$ if we repeat the argument above. This proves the first part of Theorem 2. To establish the second statement we assume on the contrary that, for all nonnegative $\mu \in \mathbb{Z}$, we have $X(w) = o(|w|^\mu)$. It will then be shown that $X_w \equiv 0$ in $B_R(0)$ contradicting the assumption that $X(w) \not\equiv 0$ on $B_R(0)$.

Note that, by the first part of Theorem 2, we obtain the relation

$$X_w(w) = O(|w|^\mu) \quad \text{for all } \mu,$$

and in particular

$$X_w(w) = o(|w|^{\mu-1}) \quad \text{as } w \to 0$$

and for all nonnegative $\mu \in \mathbb{Z}$. We are thus in a position to repeat the argument given in the proof of Lemma 4. Inequality (21) with $w_0 = 0$ now reads as

$$2\pi J_2(0) \le I_1(0) + I_2(0) + I_3(0), \tag{27}$$

where

$$\begin{aligned} I_1(0) &= \iint_{|\xi|<r} |\xi|^{-\lambda-1} J_1(\xi)\, d\xi_1\, d\xi_2, \\ I_2(0) &= 2c \iint_{|\xi|<r} |\xi|^{-\lambda-1} J_2(\xi)\, d\xi_1\, d\xi_2, \\ I_3(0) &= 2c \iint_{|\xi|<r} |\xi|^{-\lambda-1} J_3(\xi)\, d\xi_1\, d\xi_2, \quad \xi = \xi_1 + i\xi_2, \end{aligned}$$

and

$$\begin{aligned} J_1(\xi) &= \int_{|w|=r} |w|^{-\mu}|w-\xi|^{-1}|X_w(w)||dw|, \\ J_2(\xi) &= \iint_{|w|\le r} |w|^{-\mu-\lambda}|w-\xi|^{-1}|X_w(w)|\,du\,dv, \\ J_3(\xi) &= \iint_{|w|<r} |w|^{-\mu-\lambda}|w-\xi|^{-1}|X_w(w)|\,du\,dv, \qquad w=u+iv. \end{aligned}$$

As in the proof of Lemma 1, i.e. using the Schmidt's inequality and the identity

$$(w-\xi)^{-1}\xi^{-1} = w^{-1}[(w-\xi)^{-1}+\xi^{-1}], \tag{28}$$

we obtain the estimates

$$I_1(0) \le Mr^{1-\lambda}J_1(0), \tag{29}$$

$$I_2(0) \le 2Mr^{1-\lambda}cJ_2(0). \tag{30}$$

Again we infer from (28) and (15) that

$$\begin{aligned} I_3(0) &= 2c\iint_{|\xi|<r} d\xi_1\,d\xi_2|\xi|^{-\lambda}|\xi|^{-1}\iint_{|w|<r}|w|^{-\mu-\lambda}|w-\xi|^{-1}|X_w(w)|\,du\,dv \\ &= 2c\iint_{|w-\xi|<r} du\,dv|w|^{-\mu-\lambda}|w|^{-1}|X_w(w)| \\ &\quad\cdot\iint_{|\xi|<r}|\xi|^{-\lambda}[(w-\xi)^{-1}+\xi^{-1}]\,d\xi_1\,d\xi_2 \\ &\le 4cM_3r^{1-\lambda}\iint_{|w|<r}|w|^{-\mu-\lambda-1}\left[\int_0^1\left|\frac{d}{dt}X(tw)\right|\,dt\right]\,du\,dv \\ &\le 4cM_3r^{1-\lambda}\int_0^1 dt\left[\iint_{|w|<r}|w|^{-\mu-\lambda}|X_w(tw)|\,du\,dv\right]. \end{aligned}$$

Now we put $z := tw$, $z = z_1 + iz_2$, and employ the change-of-variables formula. This yields

$$\begin{aligned} I_3(0) &\le 4cM_3r^{1-\lambda}\int_0^1 dt\ t^{\mu+\lambda-2}\iint_{|z|<tr}|z|^{-\mu-\lambda}|X_w(z)|\,dz_1\,dz_2 \\ &\le \frac{4cM_3r^{1-\lambda}}{\mu+\lambda-1}\iint_{|z|<r}|z|^{-\mu-\lambda}|X_w(z)|\,dz_1\,dz_2, \end{aligned}$$

where we have assumed that $\mu+\lambda\ge 2$.

In the following estimates we let $r<1$. Then

$$I_3(0) \le 4cM_3r^{1-\lambda}\iint_{|w|<r}|w|^{-\mu-\lambda-1}|X_w(w)|\,du\,dv,$$

or equivalently

$$I_3(0) \le 4cM_3 r^{1-\lambda} J_2(0). \tag{31}$$

The estimates (27), (29), (30), and (31) now imply that

$$2\pi J_2(0) \le Mr^{1-\lambda} J_1(0) + 2Mr^{1-\lambda} cJ_2(0) + 4cM_3 r^{1-\lambda} J_2(0),$$

whence we obtain for small $r > 0$ and some $\delta > 0$ independent of μ that

$$\delta J_2(0) \le J_1(0),$$

or, more explicitly

$$\delta \iint_{|w|<r} |w|^{-\mu-\lambda-1} |X_w(w)|\, du\, dv \le \iint_{|w|=r} |w|^{-\mu-1} |X_w(w)| |dw|$$

for all nonnegative μ. If we now assume the existence of some w_0 such that $X_w(w_0) \ne 0$, we are led to a contradiction exactly as in the proof of Theorem 1.

We have shown that there exists some finite integer ν with

$$\begin{aligned} X(w) &= o(|w|^{\nu-1}), \\ X(w) &\ne o(|w|^{\nu}) \quad \text{as } w \to 0. \end{aligned} \tag{32}$$

By the first part of Theorem 2 we conclude the existence of the limit

$$\lim_{w\to 0} X_w(w) w^{-\nu+1} = A.$$

If $A = 0$, we could infer from Lemma 3 that $X(w) = o(|w|^\nu)$ contradicting (32), and Theorem 2 is proved. □

Now we shall consider a further generalization of Theorem 1 which will enable us to treat certain systems of differential inequalities as well. This will be of importance in Sections 3.3 and 3.4.

Definition 2. *Two complex-valued functions $F(w), G(w)$ are said to satisfy* **Assumption (A3)** *if they are of class $C^{0,1}(B'_\delta, \mathbb{C})$, $B'_\delta = \{0 < |w| < \delta\}$, and if there are numbers $\alpha, \beta, \nu \in (0,1)$, $\alpha + \beta = 1$ such that the relations*

$$\begin{cases} |F(w)| = O(|w|^{\nu-\alpha}) \\ |G(w)| = O(|w|^{\nu-\beta}) \end{cases} \quad \text{as } w \to 0, \tag{33}$$

and the inequalities

$$\begin{cases} |F_{\overline{w}}(w)| \le c[|w|^{-\beta}|F(w)|^2 + |w|^{\beta-2\alpha}|G(w)|^2], \\ |G_{\overline{w}}(w)| \le c[|w|^{\alpha-2\beta}|F(w)|^2 + |w|^{-\alpha}|G(w)|^2] \end{cases} \tag{34}$$

hold true almost everywhere on $B'_\delta = B_\delta \setminus \{0\}$ for some constant $c > 0$.

Here and in the following we shall work with the concept of generalized complex derivatives which are defined analogously to generalized real (or weak) derivatives, and we refer the interested reader to the monograph of Vekua [1,2] for more detailed background information. Note that by a theorem of Rademacher (see e.g. Federer [1]) every Lipschitz-continuous function has a weak derivative which is bounded.

Theorem 3. *Suppose that F and G satisfy assumption* (A3) *on B'_δ. Then there exists a nonnegative integer m such that the functions*

$$f^m(w) := w^{-m}F(w), \quad g^m(w) := w^{-m}G(w)$$

satisfy one of the following two conditions (i) *or* (ii)*:*

(i) $$f^m(w) \in C^{0,\mu}(B_\delta, \mathbb{C}) \quad \textit{for all } \mu < \min(1, m+\alpha),$$

$$\begin{aligned} &f^m(0) \neq 0, \\ &|f^m_{\overline{w}}(w)| = O(|w|^{m-\beta}) \quad \textit{as } w \to 0, \\ &|g^m_{\overline{w}}(w)| = O(|w|^{m+\alpha-2\beta}) \quad \textit{as } w \to 0; \end{aligned}$$

(ii) $$g^m(w) \in C^{0,\mu}(B_\delta, \mathbb{C}) \quad \textit{for all } \mu < \min(1, m+\beta),$$

$$\begin{aligned} &g^m(0) \neq 0, \\ &|f^m_{\overline{w}}(w)| = O(|w|^{m+\beta-2\alpha}) \quad \textit{as } w \to 0, \\ &|g^m_{\overline{w}}(w)| = O(|w|^{m-\alpha}) \quad \textit{as } w \to 0. \end{aligned}$$

For the proof of Theorem 3 we shall need the following auxiliary results.

Lemma 6. *Suppose that $f \in C^{0,1}(B'_\delta, \mathbb{C})$ satisfies*

(35) $$|f(w)| = o(|w|^{-1}) \quad \textit{as } w \to 0$$

and

(36) $$|f_{\overline{w}}(w)| = O(|w|^{\lambda}) \quad \textit{as } w \to 0$$

with some exponent $\lambda > -2$. Then we have:

$$f \in C^{0,\mu}(B_\delta, \mathbb{C}) \quad \textit{for all } \mu < \min(1, 1+\lambda), \ \textit{if } \lambda > -1,$$

or

$$|f(w)| = O(|w|^{-\varepsilon}), \quad (w \to 0), \quad \textit{for all } \varepsilon > 0, \ \textit{if } \lambda = -1,$$

or

$$|f(w)| = O(|w|^{1+\lambda}), \quad (w \to 0), \ \textit{if } \lambda < -1,$$

Proof. Since $f_{\overline{w}}(w) \in L_1(B_\delta(0), \mathbb{C})$, we can apply Theorem 1.16 in Vekua [1] which implies that the sum

$$f(w) + \frac{1}{\pi} \iint_{B_\delta(0)} f_{\overline{w}}(\xi)(\xi - w)^{-1}\, d\xi_1\, d\xi_2$$

is holomorphic in $B_\delta(0)$. Hence it is sufficient to prove that the above alternative holds for the function

$$g(w) = \frac{1}{\pi} \iint_{B_\delta(0)} f_{\overline{w}}(\xi)(\xi - w)^{-1}\, d\xi_1\, d\xi_2.$$

If $\lambda > -1$, we conclude for $w_1, w_2 \in B'_\delta = B_\delta \setminus \{0\}$ the inequality

$$\begin{aligned} |g(w_1) - g(w_2)| &= \frac{1}{\pi} \left| \iint_{B_\delta(0)} f_{\overline{w}}(\xi) \frac{(w_1 - w_2)}{(w_1 - \xi)(w_2 - \xi)}\, d\xi_1\, d\xi_2 \right| \\ &\le \text{const } |w_1 - w_2| \iint_{B_\delta(0)} \frac{|\xi|^\lambda}{|w_1 - \xi||w_2 - \xi|}\, d\xi_1\, d\xi_2. \end{aligned}$$

Using Hölder's inequality, we obtain for each $\mu \in (0, 1 + \lambda)$ the estimate

$$\begin{aligned} |g(w_1) - g(w_2)| \le c|w_1 - w_2| &\left[\iint_{B_\delta(0)} |\xi|^{2\lambda/(1-\mu)}\, d\xi_1\, d\xi_2 \right]^{(1-\mu)/2} \\ &\cdot \left[\iint_{B_\delta(0)} (|w_1 - \xi||w_2 - \xi|)^{-2/(1+\mu)}\, d\xi_1\, d\xi_2 \right]^{(1+\mu)/2}. \end{aligned}$$

Now inequality (15) implies

$$|g(w_1) - g(w_2)| \le c|w_1 - w_2|^{1+(2-(4/(1+\mu)))(1+\mu)/2} = c|w_1 - w_2|^\mu.$$

If $\lambda < -1$, we infer again from (15) that

$$|g(w)| \le c \iint_{B_\delta(0)} |\xi|^\lambda |\xi - w|^{-1}\, d\xi_1\, d\xi_2 \le c|w|^{1+\lambda}$$

for some suitable constant c. Finally, if $\lambda = -1$, it follows that

$$|g(w)| \le c_1 + c_2|\log|w||.$$ □

In the discussion to follow we shall always assume that $0 < \alpha \le \frac{1}{2} \le \beta < 1$. Note that this is without loss of generality since $\alpha + \beta = 1$ and because of the symmetry of the following assertions both in α and β and with respect to F and G. Observe also that we can (and will, if necessary) decrease the decay exponent ν in the relation (33).

Lemma 7. *Suppose that F and G satisfy assumption* (A3) *on $B'_\delta = B_\delta \setminus \{0\}$ with $\alpha \le \frac{1}{2}$. Then $F \in C^{0,\mu}(B_\delta, \mathbb{C})$ for all $\mu \in (0, \alpha)$ and, furthermore, the relations*

$$\begin{aligned} |F_{\overline{w}}(w)| &= O(|w|^{-\beta}) \\ & \qquad\qquad \text{as } w \to 0 \\ |G_{\overline{w}}(w)| &= O(|w|^{\alpha-2\beta}) \end{aligned}$$

hold true almost everywhere on B_δ.

Proof. The proof is based on an iteration argument where one has to use Lemma 6 in each step. To start, let us assume that $\nu < \alpha$, whence for some $k_0 \in \mathbb{N} \cup \{0\}$ we have that $2^{k_0}\nu \le \alpha < 2^{k_0+1}\nu$. Now assume that for some $k \in \mathbb{N} \cup \{0\}, k \le k_0$, we have

$$|F(w)| = O(|w|^{2^k\nu-\alpha}), \quad |G(w)| = O(|w|^{2^k\nu-\beta}).$$

Then (34) implies

$$|F_{\overline{w}}(w)| = O(|w|^{2^{k+1}\nu-\alpha-1}), \quad |G_{\overline{w}}(w)| = O(|w|^{2^{k+1}\nu-\beta-1}).$$

From Lemma 6 we infer

$$|F(w)| = O(|w|^{2^{k+1}\nu-\alpha}) \quad \text{as } w \to 0 \text{ if } k < k_0$$

or

$$F(w) \in C^{0,\mu}(B_\delta, \mathbb{C}) \quad \text{for all } \mu < 2^{k_0+1}\nu - \alpha \text{ if } k = k_0.$$

Also,

$$|G(w)| = O(1 + |w|^{2^{k+1}\nu-\beta}) \quad \text{if } k \le k_0.$$

By virtue of (33) we can start the iteration by putting $k = 0$. In conclusion we obtain that

$$F \in C^{0,\mu} \quad \text{for all } \mu < 2^{k_0+1}\nu - \alpha$$

and in particular

$$|F(w)| = O(1), \quad |G(w)| = O(1 + |w|^{2^{k_0+1}\nu-\beta}).$$

Again we infer from (34) that

$$|F_{\overline{w}}(w)| = O(|w|^{-\beta}) = O(|w|^{\alpha-1}),$$

since $2^{k_0+2}\nu > 2\alpha$, and

$$|G_{\overline{w}}(w)| = O(|w|^{\alpha-2\beta}) = O(|w|^{1-3\beta}).$$

Finally we infer from Lemma 6 that $F \in C^{0,\mu}(B_\delta, \mathbb{C})$ for all $0 < \mu < \alpha$. □

In the next lemma we improve the regularity of G provided we know that $F(0) = 0$.

Lemma 8. *Suppose F and G satisfy* (A3) *on $B'_\delta = B_\delta \setminus \{0\}$ with $\alpha \le \frac{1}{2}$, and that $F(0) = 0$. Then we have $G \in C^{0,\mu}(B_\delta, \mathbb{C})$ for all $\mu \in (0, \beta)$ and*

$$|F_{\overline{w}}(w)| = O(|w|^{1-3\alpha}), \quad |G_{\overline{w}}(w)| = O(|w|^{-\alpha}) \quad \textit{as } w \to 0.$$

Proof. Since $F(0) = 0$, we infer from Lemma 7 that $|F(w)| = O(|w|^\mu)$ for all $\mu < \alpha$. Hence the function $f(w) := \frac{F(w)}{w}$ satisfies

$$|f(w)| = O(|w|^{\mu-1}) \quad \text{as } w \to 0, \text{ for all } 0 < \mu < \alpha.$$

By Lemma 7 we have $|G_{\overline{w}}(w)| = O(|w|^{\alpha-2\beta})$ as $w \to 0$, and Lemma 6 yields

$$|G(w)| = O(1 + |w|^{\alpha-2\beta+1}) \quad \text{if } \alpha - 2\beta \ne -1,$$

that is,

$$|G(w)| = O(|w|^{-\varepsilon}) \quad \text{for all } \varepsilon > 0, \text{ if } \alpha - 2\beta = -1 \left(\text{i.e. } \alpha = \frac{1}{3} \right).$$

Using inequalities (34) we obtain the system

$$(37) \qquad \begin{cases} |f_{\overline{w}}(w)| \le c_1[|w|^\alpha |f|^2 + |w|^{-3\alpha}|G|^2], \\ |G_{\overline{w}}(w)| \le c_2[|w|^{3\alpha}|f|^2 + |w|^{-\alpha}|G|^2], \end{cases}$$

which holds true for almost all $w \in B_\delta$.

If $\alpha - 2\beta = -1$ (or equivalently $\alpha = \frac{1}{3}, \beta = \frac{2}{3}$), we infer from (37) the relations

$$(38) \qquad \begin{cases} |f_{\overline{w}}(w)| = O(|w|^{-1-\varepsilon}) & \text{as } w \to 0, \\ |G_{\overline{w}}(w)| = O(|w|^{-1/3-\varepsilon}) & \text{as } w \to 0, \end{cases} \qquad \text{for all } \varepsilon > 0$$

whence in particular

$$G(w) \in C^{0,\mu}(B_\delta, \mathbb{C}) \quad \text{for all } \mu < \frac{2}{3} = \beta$$

and

$$(39) \qquad |G(w)| = O(1), \quad |f(w)| = O(|w|^{-\varepsilon}),$$

for all $\varepsilon > 0$. Inserting (39) into (37) we obtain

$$\begin{aligned} |f_{\overline{w}}(w)| &= O(|w|^{-1}) = O(|w|^{-3\alpha}), \\ |G_{\overline{w}}(w)| &= O(1) = O(|w|^{1-3\alpha}), \end{aligned}$$

and therefore

$$|F_{\overline{w}}(w)| = O(1) = O(|w|^{1-3\alpha}),$$

because of $F_{\overline{w}} = w f_{\overline{w}}$.

Now we deal with the case $\alpha - 2\beta > -1$ (or equivalently $\beta < \frac{2}{3}, \alpha > \frac{1}{3}$): Inserting the relations

$$|G(w)| = O(1) \quad \text{and} \quad |f(w)| = O(|w|^{\mu-1}), \quad \mu < \alpha,$$

in (37), we obtain

$$\left.\begin{aligned} |f_{\overline{w}}(w)| &= O(|w|^{-3\alpha}) \\ |G_{\overline{w}}(w)| &= O(|w|^{-\alpha}) \end{aligned}\right\} \quad \text{as } w \to 0.$$

Now Lemma 6 implies that

$$|F_{\overline{w}}(w)| = O(|w|^{1-3\alpha}),$$

$$G(w) \in C^{0,\mu}(B_\delta, \mathbb{C}) \quad \text{for all } \mu < 1 - \alpha = \beta.$$

Finally, we have to treat the case $\alpha - 2\beta < -1$ (or $\beta > \frac{2}{3}$ and $\alpha < \frac{1}{3}$): To this end we fix some $\mu < \alpha$ and select some $k_0 \in \mathbb{N} \cup \{0\}$ with the property $2^{k_0}(\mu + \alpha) < 1 - \alpha < 2^{k_0+1}(\mu + \alpha)$. Assume that for some $k \le k_0$ the relations

$$(40_k) \qquad \begin{aligned} |f(w)| &= O(|w|^{2^k(\mu+\alpha)-\alpha-1}) \quad \text{as } w \to 0, \\ |G(w)| &= O(|w|^{2^k(\mu+\alpha)+\alpha-1}) \quad \text{as } w \to 0, \end{aligned}$$

hold true. Then it follows from (37) that

$$|f_{\overline{w}}(w)| = O(|w|^{2^{k+1}(\mu+\alpha)-\alpha-2})$$

and

$$|G_{\overline{w}}(w)| = O(|w|^{2^{k+1}(\mu+\alpha)+\alpha-2}).$$

If $k < k_0$, then Lemma 6 applies and we arrive at the relations

$$|f(w)| = O(|w|^{2^{k+1}(\mu+\alpha)-\alpha-1}),$$

$$|G(w)| = O(|w|^{2^{k+1}(\mu+\alpha)+\alpha-1});$$

in other words, the validity of (40_k) implies the validity of (40_{k+1}). On the other hand, for $k = k_0$ we obtain

$$(41) \qquad \begin{aligned} |f(w)| &= O(1 + |w|^{2^{k_0+1}(\mu+\alpha)-\alpha-1}), \\ |G(w)| &= O(1). \end{aligned}$$

We can start the induction because (40_k) holds with $k = 0$ taking $\mu < \alpha$ into account.

We insert (41) into (37) and get $|f_{\overline{w}}(w)| = O(|w|^{-3\alpha})$ as $w \to 0$ and $|G_{\overline{w}}(w)| = O(|w|^{-\alpha})$ whence we infer by means of Lemma 6 that $G \in C^{0,\mu}(B_\delta, \mathbb{C})$ for all $\mu < 1 - \alpha = \beta$ and also $|F_{\overline{w}}(w)| = O(|w|^{1-3\alpha})$. □

Next we suppose that both F and G vanish at zero. Then on account of the Lemmata 8 and 6 we conclude that the functions

$$f(w) := w^{-1}F(w) \quad \text{and} \quad g(w) := w^{-1}G(w)$$

fulfil the relations

$$|f(w)| = \begin{cases} O(1+|w|^{1-3\alpha}) & \text{if } \alpha \neq \frac{1}{3}, \\ O(|w|^{-\varepsilon}) & \text{for all } \varepsilon > 0, \text{ if } \alpha = \frac{1}{3} \end{cases}$$

and

$$|g(w)| = O(|w|^{-\alpha}).$$

Therefore there exists some number $\lambda' \in (0,1)$ such that the mapping

$$h(w) := (f(w), g(w))$$

satisfies the relation

$$|h(w)| = O(|w|^{-\lambda'}) \quad \text{as } w \to 0.$$

From (34) we easily infer an estimate of the type

$$|h_{\overline{w}}(w)| \leq c|w|^{-\lambda}|h(w)|$$

holding almost everywhere on B_δ with some constants c and $\lambda \in (0,1)$. Thus we are in a position to apply Corollary 1 of this section to the function h and obtain the existence of some positive integer m and of a complex vector $A \in \mathbb{C}^2 \setminus \{0\}$ such that

$$h(w) = Aw^{m-1} + o(|w|^{m-1}) \quad \text{as } w \to 0 \tag{42}$$

holds true on B_δ.

Now we come to the *proof of Theorem* 3.

Without loss of generality we only consider the case $\alpha \leq \frac{1}{2}$. We distinguish between the following alternatives (which clearly exhaust all possibilities!):

$$(\alpha)\ F(0) \neq 0, \quad G(0) \neq 0, \quad (\beta)\ F(0) \neq 0, \quad G(0) = 0,$$
$$(\gamma)\ F(0) = 0, \quad G(0) \neq 0, \quad (\delta)\ F(0) = 0, \quad G(0) = 0.$$

If (α) or (β) hold true, then Lemma 5 yields that (i) must be satisfied with $m = 0$. In view of Lemma 8 we obtain (ii) with $m = 0$ provided that (γ) holds true. Finally, let us assume that $F(0) = G(0) = 0$. Then (42) is equivalent to

$$\left.\begin{aligned} F(w) &= aw^m + o(|w|^m) \\ G(w) &= bw^m + o(|w|^m) \end{aligned}\right\} \quad \text{as } w \to 0$$

with complex numbers a, b which are not both equal to zero, and we obtain

$$\left\{\begin{array}{l} f^m(w) = a + o(1) \\ g^m(w) = b + o(1) \end{array}\right. \quad \text{as } w \to 0. \tag{43}$$

On the other hand, we easily derive from (34) the inequalities

$$\left\{\begin{array}{l} |f^m_{\overline{w}}(w)| \le c[|w|^{m-\beta}|f^m(w)|^2 + |w|^{m+\beta-2\alpha}|g^m(w)|^2], \\ |g^m_{\overline{w}}(w)| \le c[|w|^{m+\alpha-2\beta}|f^m(w)|^2 + |w|^{m-\alpha}|g^m(w)|^2], \end{array}\right. \tag{44}$$

and, together with (41), this yields

$$|f^m_{\overline{w}}(w)| = O(|w|^{m-\beta}) \quad \text{and} \quad |g^m_{\overline{w}}(w)| = O(|w|^{m+\alpha-2\beta}) \quad \text{as } w \to 0.$$

But then Lemma 6 can be applied which proves that $f^m \in C^{0,\mu}(B_\delta, \mathbb{C})$ for all $\mu < 1$. Assuming that $f^m(0) \ne 0$ we have thus shown that (i) holds true.

So let us assume that $f^m(0) = a = 0$ (whence $b = g^m(0) \ne 0$). Then clearly $|f^m(w)| = O(|w|^\mu)$ as $w \to 0$ for all $\mu < 1$, and (44) implies

$$|f^m_{\overline{w}}(w)| = O(|w|^{m+\beta-2\alpha})$$

and

$$|g^m_{\overline{w}}(w)| = O(|w|^{m-\alpha}).$$

Again, by Lemma 6 it follows that $g^m \in C^{0,\mu}(B_\delta, \mathbb{C})$ for all $\mu < 1$, and hence (ii) holds true; thus Theorem 3 is proved. □

3.2 A Gradient Estimate at Singularities Corresponding to Corners of the Boundary

In this section we consider solutions $X = X(u, v)$ of the Plateau problem $\mathcal{P}(\Gamma)$ for a Jordan curve Γ consisting of two regular pieces Γ^+ and Γ^- of class $C^{2,\mu}$ which enclose a positive angle $\beta < \pi$ at a common point $P \in \Gamma^+ \cap \Gamma^-$. We are then interested in the behaviour of X near the corner point P and, in particular, in asymptotic expansions for the gradient $\nabla X(u, v)$ near the point $w_0 \in \partial B$ which corresponds to P. More generally, let $X \in \mathcal{C}(\Gamma, S)$ be a solution to the free boundary problem $\mathcal{P}(\Gamma, S)$ and suppose that the configuration $\langle \Gamma, S\rangle$ satisfies some chord-arc condition (see Section 2.5). Then we conclude from Theorem 2 of Section 2.5 that X is globally Hölder continuous on the closure of the semi-disk $B = \{(u, v)\colon u^2 + v^2 < 1, v > 0\}$, i.e.,

$$X \in C^{0,\alpha}(\overline{B}, \mathbb{R}^3) \cap C^2(B, \mathbb{R}^3)$$

for some $\alpha > 0$. Assuming the usual three-point condition, the points (1, 0) and $(-1, 0)$ are mapped onto the corner points $P_1, P_2 \in \Gamma \cap S$ respectively. Hence our interest is concentrated on the behaviour of $\nabla X(w)$ when $w \to \pm 1$ respectively.

We first mention a (local) result concerning the Plateau problem.

Theorem 1. *Let $\Gamma^+, \Gamma^- \subset \mathbb{R}^3$ be pieces of regular Jordan arcs of class $C^{2,\mu}$ which meet at a point $P \in \mathbb{R}^3$ forming a positive angle $\beta < \pi$. Suppose that*

$$X \in C^{0,\alpha}(\overline{B}_\delta^+, \mathbb{R}^3) \cap C^2(\overline{B}_\delta^+ \setminus \{0\}, \mathbb{R}^3),$$

where $B_\delta^+ := \{w = (u,v) \colon |w| < \delta, v > 0\}$ is a minimal surface which satisfies the boundary conditions $X \colon I_\delta^\pm \to \Gamma^\pm$ with $I_\delta^\pm := \{(u,0) \colon 0 < \pm u < \delta\}$ and $X(0) = P$. Then we obtain the asymptotic relation

$$|\nabla X(w)| = O(|w|^{\alpha-1}) \quad \textit{as } w \to 0.$$

For the free boundary problem we shall prove

Theorem 2. *Let Γ be a regular Jordan curve of class $C^{2,\mu}$ which has only its two endpoints P_1, P_2 in common with a regular closed surface S of class C^3. Suppose that $X \in \mathcal{C}(\Gamma, S)$ solves the partially free minimum problem $\mathcal{P}(\Gamma, S)$ and that Γ, S satisfy some chord-arc condition. Then $X(u,v)$ is of class $C^{0,\alpha}(\overline{B}, \mathbb{R}^3) \cap C^{2,\alpha}(\overline{B} \setminus \{1,-1\})$ for some $\alpha > 0$ where $B = \{(u,v) \colon u^2 + v^2 < 1, v > 0)$, and there holds the expansion*

$$|\nabla X(w)| = O(|w \mp 1|^{\alpha-1}) \quad \text{as } w \to \pm 1. \tag{1}$$

We shall only prove Theorem 2 since the proof of the first theorem is similar. Note that we only have to show the asymptotic relation (1) since the asserted regularity properties of X were already proved in Chapter 2. Also, it will be convenient to replace the semi-disk B by the upper half-plane

$$H = \{(u,v) \in \mathbb{R}^2 \colon v > 0\}.$$

We may further assume that the point $(u,v) = (0,0)$ is mapped into the corner point $P_1 \in \Gamma \cap S$. Observe that this simplification is without loss of generality since the conformal map

$$w = w(z) = -\left[\frac{1-z}{1+z}\right]^2$$

maps the semi-disk $B = \{(u,v) \colon u^2 + v^2 < 1, v > 0\}$ conformally onto H, and the point $(1,0)$ into $(0,0)$. (Note that $w(z)$ is not conformal at the boundary point $z = 1$.) Furthermore, if X is of class $C^{0,\alpha}(\overline{B}) \cap C^2(B)$, then $Y(w) := X(z(w))$ is of class $C^{0,\alpha/2}(\overline{H})$, and if Y satisfies an asymptotic relation of the type

$$|\nabla Y(w)| = O(|w|^{\alpha/2-1}) \quad \text{as } w \to 0,$$

then also

$$\begin{aligned} |\nabla X(z)| &= O\left(|\nabla Y(w)| \left|\frac{dw}{dz}\right|\right) = O(|1-z|^{\alpha-2} \cdot |1-z|) \\ &= O(|1-z|^{\alpha-1}) \quad \text{as } z \to 1, z \in B. \end{aligned}$$

Since we only deal with local properties of X we may throughout this section require the following Assumption A to be satisfied by the minimal surface X.

Assumption A. *Let $\delta > 0$ be some positive number and put*

$$B_\delta^+ := \{w = (u, v) \in \mathbb{R}^2 \colon |w| < \delta, v > 0\},$$

$$I_\delta^+ := \{w = (u, 0) \colon 0 < u < \delta\},$$

$$I_\delta^- := \{w = (u, 0) \colon -\delta < u < 0\}.$$

Suppose that the minimal surface $X = X(u, v)$ is of class $C^{0,\alpha}(\overline{B}_\delta^+, \mathbb{R}^3) \cap C^{2,\alpha}(\overline{B}_\delta^+ \setminus \{0\})$ and satisfies the following boundary conditions:
(i) *$X \colon I_\delta^- \to \Gamma$ is weakly monotonic;*
(ii) *$X(I_\delta^+) \subset S, X(0) = 0 = P_1 \in \Gamma \cap S$;*
(iii) *$X_v|_{I_\delta^+}$ is orthogonal to S along the free trace $X|_{I_\delta^+}$.*

Then Theorem 2 follows from

Proposition 1. *Let $X \in C^{2,\alpha}(\overline{B}_\delta^+ \setminus \{0\}) \cap C^{0,\alpha}(\overline{B}_\delta^+)$ be a minimal surface which fulfills assumption* (A). *Then the gradient ∇X satisfies*

$$|\nabla X(w)| = O(|w|^{\alpha-1}) \quad \text{as } w \to 0. \tag{2}$$

The proof of Proposition 1 rests on a further investigation of solutions $\tilde{X}(w)$ of the differential inequality

$$|\Delta \tilde{X}(u, v)| \le a|\nabla \tilde{X}(u, v)|^2 \tag{3}$$

which was already considered in Section 2.2. We recall Proposition 1 of Section 2.2.

Proposition A. *There is a continuous function $\kappa(t), 0 \le t < 1$, with the following properties: For any solution $\tilde{X} \in C^2(B_R(w_0), \mathbb{R}^N)$ of the differential inequality* (3) *satisfying*

$$|\tilde{X}(w)| \le M, \quad w \in B_R(w_0) \tag{4}$$

for some M with $aM < 1$, the estimates

$$|\nabla \tilde{X}(w_0)| \le \kappa(aM) \frac{M}{R} \quad \text{and} \tag{5}$$

$$|\nabla \tilde{X}(w_0)| \le \frac{\kappa(aM)}{R} \sup_{w \in B_R(w_0)} |\tilde{X}(w) - \tilde{X}(w_0)| \tag{6}$$

hold true.

Lemma 1. *Let $D \subset B_1(0)$ be a domain such that $\overline{D}$ contains the origin. Suppose that $\tilde{X} \in C^2(D, \mathbb{R}^N) \cap C^0(\overline{D}, \mathbb{R}^N)$ satisfies inequality* (3)*. Then there exists some $\delta > 0$ such that the estimate*

$$(7) \qquad |\nabla \tilde{X}(w_0)| \le \varepsilon^{-1} \cdot \text{const} \sup_{B_\varepsilon(w_0)} |\tilde{X}(w) - \tilde{X}(w_0)|$$

holds true for all $w_0 \in D \cap B_\delta(0)$ and for all $\varepsilon > 0$ with $B_\varepsilon(w_0) \subset D \cap B_\delta(0)$.

Proof. We put $Y(w) = \frac{1}{2a}[\tilde{X}(w) - \tilde{X}(0)]$, $w \in D$, and choose $\delta > 0$ so small that $\sup_{D \cap B_\delta(0)} |Y(w)| < 1$. Then Y satisfies (3) on $D \cap B_\delta(0)$ with $a = \frac{1}{2}$. Applying Proposition A to the function $Y \in C^2(B_\varepsilon(w_0))$ and to $M = 1, a = \frac{1}{2}$, we get the estimate

$$|\nabla Y(w_0)| \le \frac{\kappa(1/2)}{\varepsilon} \sup_{B_\varepsilon(w_0)} |Y(w) - Y(w_0)|,$$

i.e.,

$$|\nabla \tilde{X}(w_0)| \le \frac{\kappa(1/2)}{\varepsilon} \sup_{B_\varepsilon(w_0)} |\tilde{X}(w) - \tilde{X}(w_0)|$$

as required. □

In order to state our results in a convenient way, we make the following

Assumption B. *For some fixed angle $\pi \ge \gamma > 0$ we denote by D_ρ the domain*

$$D_\rho := \{w = re^{i\varphi} : 0 < \varphi < \gamma, r < \rho\}$$

where r, φ denote polar coordinates about the origin. Let

$$\tilde{X}(w) = (\tilde{X}^1(w), \dots, \tilde{X}^N(w)), \quad w = (u, v) \in D_\rho,$$

be a mapping of class $C^0(\overline{D}_\rho, \mathbb{R}^N) \cap C^2(D_\rho, \mathbb{R}^N)$ which satisfies

$$(3) \qquad |\Delta \tilde{X}(w)| \le a |\nabla \tilde{X}(w)|^2 \quad \text{on } D_\rho$$

and

$$(8) \qquad |\tilde{X}(w)| \le c_1 |w|^\alpha \quad \text{on } D_\rho$$

with numbers $a, c_1 > 0$ and $0 < \alpha < 1$.

For arbitrary fixed $\theta \in (0, \gamma/2)$ we put

$$\begin{aligned} D_{\rho,\theta} &:= \{w = re^{i\varphi} : \theta < \varphi < \gamma - \theta, 0 < r < \rho\}, \\ D^1_{\rho,\theta} &:= \{w = re^{i\varphi} : 0 < \varphi < \theta, 0 < r < \rho\}, \\ D^2_{\rho,\theta} &:= \{w = re^{i\varphi} : \gamma - \theta < \varphi < \gamma, 0 < r < \rho\}. \end{aligned}$$

Then we have

Lemma 2. *Suppose $\tilde{X}$ satisfies Assumption* B *on D_ρ. Then, for every $\theta \in (0, \frac{\gamma}{2})$, there exists a constant $c_2 = c_2(\theta, a, c_1)$ such that the inequality*

$$|\nabla \tilde{X}(w_0)| \le c_2 |w_0|^{\alpha-1} \tag{9}$$

holds true for all $w_0 \in D_{\delta_1,\theta}$ and for some $\delta_1 \in (0, \rho)$.

Proof. Let $\delta > 0$ denote the number determined in Lemma 1. We take $\delta_1 := \frac{1}{2}\min(\delta, \rho)$ and put $\varepsilon := \frac{1}{2}|w_0| \sin\theta$. Then $B_\varepsilon(w_0) \subset D_\rho \cap B_\rho(0)$ for all $w_0 \in D_{\delta_1,\theta}$, and Lemma 1 implies the estimate

$$\begin{aligned} |\nabla \tilde{X}(w_0)| &\le \text{const}\, \varepsilon^{-1} \sup_{B_\varepsilon(w_0)} |\tilde{X}(w) - \tilde{X}(w_0)| \\ &\le \text{const}\, c_1 \varepsilon^{-1} [|w_0|^\alpha + (|w_0| + \varepsilon)^\alpha] \\ &\le c_2(\theta, a, c_1)|w_0|^{\alpha-1}. \end{aligned}$$

The estimate (9) controls the behaviour of the gradient on the segments $D_{\delta,\theta}$. To obtain also some information on the remaining parts $D^1_{\delta,\theta}$ or $D^2_{\delta,\theta}$, we have to make additional assumptions.

Lemma 3. *Suppose that $\tilde{X}$ satisfies Assumption* B, *and let $\theta \in (0, \min\{\frac{\pi}{16}, \frac{\gamma}{4}\})$. In addition, assume that $\tilde{X}(re^{i\varphi}) = 0$ on $0 < r < \rho$ and $\varphi = 0$ or $\varphi = \gamma$, respectively. Then for small $\delta > 0$ we obtain the estimate*

$$|\nabla \tilde{X}(w_0)| \le \text{const}\, |w_0|^{\alpha-1} \tag{10}$$

on $D^1_{\delta,\theta}$ or $D^2_{\delta,\theta}$ respectively.

Proof. It is sufficient to prove (10) for $w_0 \in D^1_{\delta,\theta}$. To this end we select some $\delta < \min(\rho, 1)$ such that

$$aM < 1, \tag{11}$$

where

$$M := \sup_{D_\delta} |\tilde{X}(w)|$$

and where a denotes the constant in (3). Applying Proposition A, we derive the gradient bound

$$|\nabla \tilde{X}(w_0)| \le c\varepsilon^{-1} \sup_{B_\varepsilon(w_0)} |\tilde{X}(w) - \tilde{X}(w_0)| \tag{12}$$

holding for some constant c independent of ε and for all $\varepsilon > 0$ satisfying $0 < \varepsilon < \text{dist}(w_0, \partial D_\delta)$.

Now we restrict w_0 further so that $|w_0| < \frac{\delta}{2}$. Put $u_0 = \operatorname{Re} w_0$, $R_\theta := 2u_0 \sin\theta$, $w_1 = (u_0, 0)$ and $B^+_{R_\theta}(w_1) := B_{R_\theta}(w_1) \cap \{(u, v) \colon v > 0\}$.

Then we find

$$B_\varepsilon(w_0) \subset B^+_{R_\theta}(w_1) \quad \text{for all } \varepsilon < \operatorname{dist}(w_0, \partial D_\delta)$$

and

$$B^+_{2R_\theta}(w_1) \subset D_\delta$$

taking the smallness of θ into account. We define harmonic functions $\varphi(w) = (\varphi^1(w), \dots, \varphi^N(w))$ and $\psi(w)$ by

$$\Delta\varphi = 0 \quad \text{on } B^+_{2R_\theta}(w_1), \quad \varphi(w) = \tilde{X}(w) \quad \text{on } \partial B^+_{2R_\theta}(w_1),$$

and

$$\Delta\psi = 0 \quad \text{on } B^+_{2R_\theta}(w_1), \quad \psi(w) = |\tilde{X}(w)|^2 \quad \text{on } \partial B^+_{2R_\theta}(w_1).$$

Consider the function

$$K(w) := \langle \tilde{X}(w) - \varphi(w), e\rangle + \frac{a}{2(1-aM)}\{\psi(w) - |\tilde{X}(w)|^2\},$$

$w \in B^+_{2R_\theta}(w_1)$, where $e \in \mathbb{R}^N$ is an arbitrary unit vector. Then

$$\begin{aligned} \Delta K(w) &= \langle \Delta\tilde{X}, e\rangle - \frac{a}{1-aM}\{|\nabla\tilde{X}|^2 + \langle \Delta\tilde{X}, \tilde{X}\rangle\} \\ &\le |\Delta\tilde{X}| - \frac{a}{1-aM}|\nabla\tilde{X}|^2 + \frac{a}{1-aM}|\Delta\tilde{X}||\tilde{X}| \\ &\le a|\nabla\tilde{X}|^2 - \frac{a}{1-aM}|\nabla\tilde{X}|^2 + \frac{a^2M}{1-aM}|\nabla\tilde{X}|^2 = 0 \end{aligned}$$

for $w \in B^+_{2R_\theta}(w_1)$. Furthermore we have $K(w) = 0$ along $\partial B^+_{2R_\theta}(w_1)$; hence we conclude from the maximum principle that $K(w) \ge 0$ on $B^+_{2R_\theta}(w_1)$. In other words,

$$\langle \varphi(w) - \tilde{X}(w), e\rangle \le \frac{a}{2(1-aM)}\psi(w) - \frac{a}{2(1-aM)}|\tilde{X}(w)|^2.$$

Since e is an arbitrary unit vector, this implies the estimate

$$|\varphi(w) - \tilde{X}(w)| \le \frac{a}{2(1-aM)}\{\psi(w) - |\tilde{X}(w)|^2\},$$

in particular

$$|\tilde{X}(w)| \le |\varphi(w)| + \frac{a}{2(1-aM)}|\psi(w)| \quad \text{for } w \in B^+_{2R_\theta}(w_1). \tag{13}$$

On the other hand, we infer from (8) the inequality

$$\begin{aligned} |\tilde{X}(w)| &\le c_1\{|w_1| + 2R_\theta\}^\alpha \\ &\le c_1\{1 + 4\sin\theta\}^\alpha |w_0|^\alpha \end{aligned}$$

for all $w \in \partial B^+_{2R_\theta}(w_1)$, whence

$$(14)\qquad \begin{aligned} &|\varphi(w)| \le c_2(\theta)|w_0|^\alpha, \quad w \in B^+_{2R_\theta}(w_1),\\ &|\psi(w)| \le c_2^2(\theta)|w_0|^{2\alpha} \le c_2^2(\theta)|w_0|^\alpha, \quad w \in B^+_{2R_\theta}(w_1), \end{aligned}$$

since $|w_0| < \delta < 1$. Employing the reflection principle for harmonic functions, it is possible to extend φ and ψ harmonically onto the disk $B_{2R_\theta}(w_1)$, taking account of the fact that φ, ψ vanish along the line $\{(u, 0)\colon u_0 - 2R_\theta < u < u_0 + 2R_\theta\}$. Denoting the reflected functions again by φ and ψ, we see that (14) continues to hold. The mean value theorem yields the relations

$$|\nabla\varphi(w)| \le \frac{1}{R_\theta} \sup_{B_{R_\theta}(w_1)} |\varphi|, \quad w \in B_{R_\theta}(w_1),$$

$$|\nabla\psi(w)| \le \frac{1}{R_\theta} \sup_{B_{R_\theta}(w_1)} |\psi|, \quad w \in B_{R_\theta}(w_1).$$

Together with (14) this implies

$$|\nabla\varphi(w)| \le c_3(\theta)|w_0|^{\alpha-1},$$

$$|\nabla\psi(w)| \le c_4(\theta)|w_0|^{\alpha-1}$$

for all $w \in B_{R_\theta}(w_1)$.

Finally we conclude from (13) and from the mean value theorem that

$$(15)\qquad \begin{aligned} |\tilde{X}(w)| &\le |\varphi(w) - \varphi(w_1)| + \frac{a}{2(1-aM)}|\psi(w) - \psi(w_1)|\\ &\le c_5(a, M, \theta)|w_0|^{\alpha-1}|w - w_1|\\ &\le c_5(a, M, \theta)|w_0|^{\alpha-1} 2\,\mathrm{dist}(w_0, \partial D_\delta), \end{aligned}$$

for all $w \in B_{\mathrm{dist}(w_0, \partial D_\delta)}(w_0)$. The desired result than follows from (15) and (12) taking $\varepsilon = \frac{1}{2}\mathrm{dist}(w_0, \partial D_\delta)$. □

Lemmata 2 and 3 imply the following

Proposition 2. *Suppose that $\tilde{X}$ satisfies Assumption* B *and that $\tilde{X}(re^{i\theta}) = 0$ for $0 < r < \rho, \varphi = 0$ or $\varphi = \gamma$. Then the asymptotic relation*

$$|\nabla\tilde{X}(w)| = O(|w|^{\alpha-1}) \quad \textit{as } w \to 0$$

holds true.

Now we turn to the

Proof of Proposition 1. (and hence of Theorem 2). Since we have assumed that $X(0) = P_1 = 0$, we infer from the Hölder continuity the estimate

$$|X(w)| \le c_1|w|^\alpha \quad \text{as } w \to 0.$$

Let us fix some $\theta \in (\theta, \frac{\pi}{16})$ and take $\gamma = \pi$ (see Assumption B). It follows that the minimal surface $X(w)$ satisfies Assumption B with $\rho = \delta, a = 0$, and from Lemma 2 we infer the estimate

$$|\nabla X(w)| \le c_2 |w|^{\alpha - 1} \quad \text{for all } w \in D_{\delta_1,\theta} \tag{16}$$

and some $\delta_1 \in (0, \rho)$.

Next we prove (16) on

$$D^2_{\delta_3,\theta} = \{w = re^{i\varphi} : 0 < r < \delta, \pi - \theta < \varphi < \pi\}$$

for some $\delta_3 \le \delta$. Recall that the segment $I^-_\delta = \{re^{i\varphi} : 0 < r < \delta, \varphi = \pi\}$ is mapped onto Γ. Employing a suitable orthogonal transformation of $\mathbb{R}^3$ we assume that Γ is locally described by two differentiable functions $x = h_1(z), y = h_2(z), z \in [0, \varepsilon)$, with the properties $h_1(0) = h_2(0) = h'_1(0) = h'_2(0) = 0$ and

$$|h'_i(z)| < \frac{1}{4}, \quad i = 1, 2, \; z \in [0, \varepsilon). \tag{17}$$

We extend the functions h_1, h_2 as even functions to the interval $(-\varepsilon, \varepsilon)$ and define

$$\begin{aligned} \tilde{x}(w) &:= x(w) - h_1(z(w)) \\ \tilde{y}(w) &:= y(w) - h_2(z(w)) \end{aligned} \qquad \text{for } w \in D_{\delta_2}, \tag{18}$$

where we have chosen δ_2 so as to satisfy $z(D_{\delta_2}) \subset (-\varepsilon, \varepsilon)$. Consider the mapping $\tilde{X}(w) := (\tilde{x}(w), \tilde{y}(w)), w \in D_{\delta_2}$, which fulfils

$$\tilde{X}(w) = (0, 0) \quad \text{on } I^-_{\delta_2}. \tag{19}$$

Furthermore, since $X(w) = (x(w), y(w), z(w))$ is harmonic we obtain

$$\Delta \tilde{x}(w) = -h''_1(z(w))|\nabla z(w)|^2, \quad w \in D_{\delta_2},$$

$$\Delta \tilde{y}(w) = -h''_2(z(w))|\nabla z(w)|^2, \quad w \in D_{\delta_2},$$

whence

$$|\Delta \tilde{X}(w)| \le c|\nabla z(w)|^2, \quad w \in D_{\delta_2}, \tag{20}$$

with a suitable constant c. Relations (17) and (18) imply the estimate

$$\begin{aligned} |\nabla x(w)| &\le \frac{1}{4}|\nabla z(w)| + |\nabla \tilde{x}(w)|, \\ |\nabla y(w)| &\le \frac{1}{4}|\nabla z(w)| + |\nabla \tilde{y}(w)|, \end{aligned} \qquad w \in D_{\delta_2}. \tag{21}$$

From the conformality condition we conclude

$$\begin{aligned}|\nabla z(w)|^2 &\le |\nabla x(w)|^2 + |\nabla y(w)|^2\\ &\le \frac{1}{16}|\nabla z|^2 + \frac{1}{2}|\nabla z||\nabla \tilde{x}| + |\nabla \tilde{x}|^2\\ &\quad + \frac{1}{16}|\nabla z|^2 + \frac{1}{2}|\nabla z||\nabla \tilde{y}| + |\nabla \tilde{y}|^2\\ &\le \frac{5}{8}|\nabla z|^2 + \frac{5}{4}\{|\nabla \tilde{x}|^2 + |\nabla \tilde{y}|^2\} \quad \text{on } D_{\delta_2},\end{aligned}$$

thus

$$|\nabla z(w)|^2 \le \frac{10}{3}|\nabla \tilde{X}(w)|^2, \quad w \in D_{\delta_2}. \tag{22}$$

Inequality (20) now yields $|\Delta \tilde{X}(w)| \le a|\nabla \tilde{X}(w)|^2, w \in D_{\delta_2}$, for some constant a. By virtue of the relation (19) we are in a position to apply Lemma 3 to the function $\tilde{X}$, and we obtain the estimate

$$|\nabla \tilde{X}(w)| \le c|w|^{\alpha-1} \quad \text{on } D^2_{\delta_3,\theta} \tag{23}$$

for some number $\delta_3 \le \delta_2$. Finally it follows from (21) and (22) that X itself satisfies (23), i.e. $|\nabla X| \le c|w|^{\alpha-1}$ on $D^2_{\delta_3,\theta}$.

Now we have to verify (23) on the set

$$D^1_{\delta_6,\theta} = \{w = re^{i\varphi} : 0 < r < \delta_6, 0 < \varphi < \theta\}$$

with $\delta_6 > 0$ chosen appropriately. Performing a suitable rotation in $\mathbb{R}^3$ we can assume that S is locally given by

$$z = f(x, y)$$

with some differentiable function f defined in a neighbourhood of zero such that

$$f(0,0) = 0, \quad \nabla f(0,0) = 0.$$

Define

$$\begin{aligned}\tilde{z}(w) &:= z(w) - f(x(w), y(w)),\\ \tilde{x}(w) &:= x(w) + \tilde{z}(w) f_x(x(w), y(w)) n(w),\\ \tilde{y}(w) &:= y(w) + \tilde{z}(w) f_y(x(w), y(w)) n(w),\end{aligned}$$

where

$$n(w) := [1 + f_x^2(x(w), y(w)) + f_y^2(x(w), y(w))]^{-1}$$

and $w \in D_{\delta_2}$ with δ_2 so small that $(x(w), y(w))$ is contained in a neighbourhood of zero where f is defined. We remark that $\tilde{z}(w) = 0$ on $I^+_{\delta_2}$ and secondly, because of

$$\begin{aligned}\tilde{x}_v(u,v) &= x_v(u,v) + \tilde{z}_v(u,v) f_x(x(u,v), y(u,v)) n(u,v)\\ &\quad + \tilde{z}(u,v)[f_x(x(u,v), y(u,v)) n(u,v)]_v,\end{aligned}$$

we have for $w \in I^+_{\delta_2}$ the equality

$$\tilde{x}_v(u,v) = x_v(u,v) + \{z_v(u,v) - f_x(x(u,v),y(u,v))x_v(u,v) \\ - f_y(x(u,v),y(u,v))y_v(u,v)\} f_x(x(u,v),y(u,v))n(u,v).$$

Equivalently, for $w \in I^+_{\delta_2}$,

$$\tilde{x}_v(u,v) = x_v(u,v) - \langle X_v(u,v), N_S(X(u,v))\rangle n^1(X(u,v)),$$

where

$$N_s(X(u,v)) = (n^1(X(u,v)), n^2(X(u,v)), n^3(X(u,v)))$$

denotes the upward unit normal of S at $X(u,v)$. However, X intersects S orthogonally along $I^+_{\delta_2}$; thus

$$\tilde{x}_v(u,v) = 0 \quad \text{on } I^+_{\delta_2}.$$

Analogously we find

$$\tilde{y}_v(u,v) = 0 \quad \text{on } I^+_{\delta_2},$$

whence the function $\tilde{X}(u,v) := (\tilde{x}(u,v), \tilde{y}(u,v))$, $(u,v) \in D_{\delta_2}$, satisfies

(24) $$\tilde{X}_v(u,v) = 0 \quad \text{on } I^+_{\delta_2}.$$

Furthermore we infer from the definition of $\tilde{x}, \tilde{y}, \tilde{z}$, from $f(0,0) = 0$, $\nabla f(0,0) = 0$, $X(0,0) = 0$, as well as from the continuity of X the relation

$$|\tilde{x}_v(w)|^2 \geq \text{const}\{|x_v(w)|^2 - \varepsilon[|y_v(w)|^2 + |z_v(w)|^2]\},$$

which holds true for $w \in D_{\delta_3}$, $\delta_3 = \delta_3(\varepsilon) \leq \delta_2$, and for arbitrary fixed $\varepsilon > 0$. We observe that similar relations hold for $\tilde{x}_u, \tilde{y}_v$ and $\tilde{y}_u$.

From the conformality condition we first obtain that $|\nabla z|^2 \leq |\nabla x|^2 + |\nabla y|^2$ and hence

(25) $$|\nabla X(u,v)|^2 \leq \text{const } |\nabla \tilde{X}(u,v)|^2 \quad \text{on } D_{\delta_3}.$$

Similar arguments show that for some $\delta_4 \leq \delta_3$ the estimate

(26) $$|\Delta \tilde{X}(u,v)| \leq \text{const } |\nabla \tilde{X}(u,v)|^2, \quad (u,v) \in D_{\delta_4},$$

holds true. We reflect $\tilde{X}(u,v)$ so as to obtain a function $\overline{X}(u,v)$ given by

$$\overline{X}(u,v) = \begin{cases} \tilde{X}(u,v), & (u,v) \in \overline{D}_{\delta_4}, \\ \tilde{X}(u,-v), & (u,-v) \in D_{\delta_4}. \end{cases}$$

By virtue of (24) we obtain for each function $\Phi \in C^1_c(B_{\delta_4}(0) \setminus \overline{I^-_{\delta_4}}, \mathbb{R}^2)$ the equalities

$$\begin{aligned}
&\int_{B_{\delta_4}(0)} \nabla \overline{X}(u,-v)\cdot\nabla\Phi(u,v)\,du\,dv \\
&\quad = \int_{D_{\delta_4}} \nabla\tilde{X}(u,v)\cdot\nabla\Phi(u,v)\,du\,dv + \int_{B_{\delta_4}\setminus D_{\delta_4}} \nabla\tilde{X}(u,-v)\cdot\nabla\Phi(u,v)\,du\,dv \\
&\quad = -\int_{D_{\delta_4}} \Delta\tilde{X}\cdot\Phi\,du\,dv - \int_{I^+_{\delta_4}} \tilde{X}_v(u,0)\cdot\Phi(u,0)\,du \\
&\qquad - \int_{B_{\delta_4}\setminus\overline{D}_{\delta_4}} \Delta\tilde{X}\cdot\Phi\,du\,dv + \int_{I^+_{\delta_4}} \tilde{X}_v(u,0)\cdot\Phi(u,0)\,du \\
&\quad = \int_{B_{\delta_4}} \overline{F}(u,v,\overline{X}(u,v),\nabla\overline{X}(u,v))\cdot\Phi(u,v)\,du\,dv
\end{aligned}$$

for some function $\overline{F}$ which grows quadratically in $|\nabla\overline{X}|$ (compare with inequality (26)). By construction, the function $\overline{X}(u,v)$ is of class $C^0(\overline{B_{\delta_4}(0)},\mathbb{R}^2)\cap C^1(B_{\delta_4}(0)\setminus\overline{I^-_{\delta_4}})$, and the preceding discussion shows that it is a weak solution of the two-dimensional system

$$\Delta\overline{X} = \overline{F}(u,v,\overline{X},\nabla\overline{X}) \quad \text{in } B_{\delta_4}(0)\setminus\overline{I^-_{\delta_4}}. \tag{27}$$

Standard regularity theory (see, for instance, Section 2.1, and Morrey [8], Gilbarg-Trudinger [1]) implies that $\overline{X}$ is in fact of class $C^2(B_{\delta_4}(0)\setminus\overline{I}_{\delta_4})$ and satisfies (27) classically on all $B_{\delta_4}(0)\setminus\overline{I^-_{\delta_4}}$. Finally we apply Lemma 1 to the domain $D = B_{\delta_4}\setminus\overline{I^-_{\delta_4}}$ and to the function $\overline{X}$; the resulting inequality is

$$|\nabla\overline{X}(w_0)| \le \text{const}\ \varepsilon^{-1} \sup_{B_\varepsilon(w_0)} |\overline{X}(w)-\overline{X}(w_0)|$$

for all w_0 and ε with the property $B_\varepsilon(w_0)\subset B_{\delta_5}\setminus\overline{I^-_{\delta_5}}$, where $\delta_5\le\delta_4$ is the constant determined in Lemma 1. If w_0 is restricted to lie in $D^1_{\delta_6,\theta}$, $\delta_6 := \frac{1}{2}\delta_5$, then a suitable choice of ε would be $\varepsilon = \frac{1}{2}|w_0|$. Hence

$$\begin{aligned}
|\nabla\overline{X}(w_0)| &\le \text{const}\ \varepsilon^{-1}[|w_0|^\alpha + (|w_0|+\varepsilon)^\alpha] \\
&\le \text{const}\ |w_0|^{\alpha-1},
\end{aligned}$$

that is, (23) holds true on $D^1_{\delta_6,\theta}$. Because of (25) we finally arrive at relation (2). □

3.3 Minimal Surfaces with Piecewise Smooth Boundary Curves and Their Asymptotic Behaviour at Corners

In the previous section we proved an asymptotic estimate for the gradient of a minimal surface X at a corner P of a given piecewise smooth boundary arc $\Gamma^+ \cup \Gamma^-$. It is the purpose of this section to obtain some more precise information on the asymptotic behaviour of X_w near the corner P. To give an idea what might happen we start with a simple but characteristic *example*:

Let $\alpha \in (0,1)$ and $k \in \mathbb{N} \cup \{0\}$ be given and define

$$X(u,v) = (x(u,v), y(u,v), z(u,v)), \quad (u,v) \in \overline{B} = \{u^2 + v^2 \le 1, v \ge 0\}$$

by

$$\begin{aligned} x(u,v) &= \text{Re}(w^{\alpha+2k}), \\ y(u,v) &= \text{Im}(w^{\alpha+2k}), \quad w = u + iv \in \overline{B}, \\ z(u,v) &\equiv 0. \end{aligned}$$

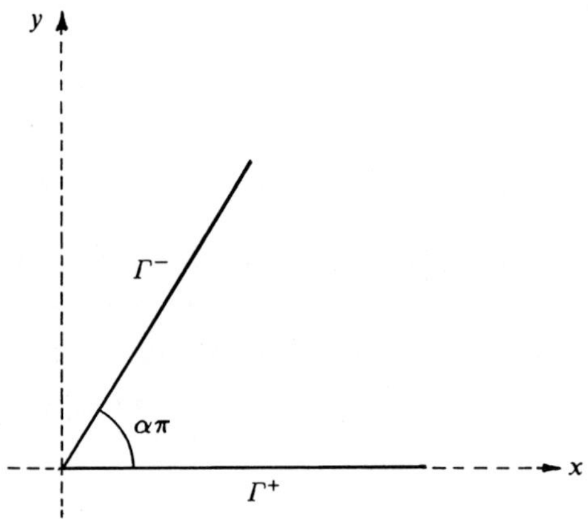

Fig. 1.

Then $X(u,v)$ is a minimal surface (i.e. $\Delta X = 0, \langle X_w, X_w \rangle = 0$) which maps the intervals $I^+ = (0,1), I^- = (-1,0)$ onto the straight arcs

$$\Gamma^+ = \{(x,y,z) \in \mathbb{R}^3 \colon z = 0, \arg(x+iy) = 0, 0 < x^2 + y^2 < 1\}$$

and

$$\Gamma^- = \{(x,y,z) \in \mathbb{R}^3 \colon z = 0, \arg(x+iy) = \pi\alpha,\ 0 < x^2 + y^2 < 1\}$$

respectively, and the point $w = 0$ into the origin of $\mathbb{R}^3$. Note that Γ^+, Γ^- form an angle $\beta = \alpha\pi$ at zero and that X has a branch point at zero if $k \ge 1$ whence X winds around zero k-times. However, there is another possible solution to the Plateau problem determined by Γ^+ and Γ^-, namely the surface

$$X_1(u,v) = (x_1(u,v), y_1(u,v), z_1(u,v))$$

the components of which are defined by

$$x_1(u,v) = \operatorname{Re}(\overline{w}^{2-\alpha+2k}), \quad y_1(u,v) = \operatorname{Im}(\overline{w}^{2-\alpha+2k}), \quad z_1(u,v) = 0,$$

with $w = u + iv \in \overline{B}$ and $\overline{w} = u - iv$. Here the semi-disk $\overline{B}$ is mapped into the great angle $(2-\alpha)\pi$ which is formed by Γ^+, Γ^- at zero. Again it is possible that branch points occur and that the surface winds about the origin. In Theorem 1 of this section we shall show that this behaviour is typical of a minimal surface X which is bounded by two Jordan arcs forming a positive angle $\alpha\pi$ at a corner P where Γ^+ and Γ^- are tangent to the x,y-plane.

Before we can formulate the main theorem of this section, we have to state the basic assumptions describing the geometric situation which is to be considered.

Assumption A. *Γ^+, Γ^- are regular arcs of class $C^{2,\mu}, \mu \in (0,1)$, which intersect at the origin, thereby enclosing an angle of $\pi\alpha, \alpha \in (0,1)$. The sets $B_\delta^+, I_\delta^+, I_\delta^-$ are defined by*

$$\begin{aligned} B_\delta^+ &:= \{w = (u,v) \in \mathbb{R}^2 \colon |w| < \delta, v > 0\}, \\ I_\delta^+ &:= \{w = (u,0) \in \mathbb{R}^2 \colon 0 < u < \delta\}, \\ I_\delta^- &:= \{w = (u,0) \in \mathbb{R}^2 \colon -\delta < u < 0\}, \end{aligned}$$

(and, as usual, we will identify $w = u + iv \in \mathbb{C}$ with $w = (u,v) \in \mathbb{R}^2$ and I_δ^+ with $(0,\delta) \subset \mathbb{R}$, etc.).

Let X be a minimal surface which is of class $C^{0,\nu}(\overline{B_{\delta+}}, \mathbb{R}^3) \cap C^2(\overline{B_\delta^+} \setminus \{0\}, \mathbb{R}^3)$ for some $\nu \in (0,1)$ and $\delta > 0$, and satisfies the boundary conditions $X \colon I_\delta^\pm \to \Gamma^\pm$ and $X(0) = 0$. Moreover, we assume that there exist functions $h_1^\pm, h_2^\pm \in C^{2,\mu}(\overline{I_\varepsilon^\pm}, \mathbb{R}), \varepsilon > 0$, such that

$$\Gamma^+ = \{(t, h_1^+(t), h_2^+(t)) \colon t \in \overline{I_\varepsilon^+}\} \quad \text{and} \quad \Gamma^- = \{(t, h_1^-(t), h_2^-(t)) \colon t \in \overline{I_\varepsilon^-}\},$$

and that furthermore

$$h_j^\pm(0) = 0, \quad j = 1,2, \quad \text{and} \quad h_1^{\pm\prime}(0) = \pm\cot\left(\frac{\alpha\pi}{2}\right), \quad h_2^{\pm\prime}(0) = 0$$

hold true.

We note that Assumption A is quite natural and not restrictive since, by performing suitable translations and rotations, we can achieve that any pair of piecewise smooth boundary curves Γ^+, Γ^- will satisfy this assumption. Also, by the results of Chapter 2, any minimal surface bounded by Γ^+, Γ^- has the desired regularity properties.

The main result of this section will be

Theorem 1. *Suppose that the Assumption* A *holds. Then there exist Hölder continuous complex valued functions Φ_1 and Φ_2 defined on the closure of some semidisk $B_\delta^+, \delta > 0$, such that the following assertions hold true:*

$$\Phi_1(0) \neq 0, \quad \Phi_2(0) \neq 0, \quad \Phi_1^2(0) + \Phi_2^2(0) = 0, \tag{1}$$

$$x_w(w) = w^\gamma \Phi_1(w), \quad y_w(w) = w^\gamma \Phi_2(w), \quad \text{and} \quad |z_w(w)| = O(|w|^\lambda), \tag{2}$$

where $\gamma = \alpha - 1 + 2k$ *or* $\gamma = 1 - \alpha + 2k$ *for some* $k \in \mathbb{N} \cup \{0\}$ *and* $\lambda > \gamma$. *Furthermore there exists some* $c \in \mathbb{C} \setminus \{0\}$ *such that*

$$x(w) + iy(w) = \begin{cases} w^{\alpha+2k}[c + o(1)] \\ or \\ \overline{w}^{2-\alpha+2k}[c + o(1)] \end{cases} \quad as\ w \to 0, \tag{3}$$

and

$$z(w) = O(|w|^{\lambda+1}) \quad as\ w \to 0.$$

Finally, the unit normal $N(w) = \frac{(X_u \wedge X_v)(w)}{|(X_u \wedge X_v)(w)|}$ *tends to a limit as* $w \to 0$:

$$\lim_{w \to 0} N(w) = \begin{pmatrix} 0 \\ 0 \\ \pm 1 \end{pmatrix}. \tag{4}$$

Remark 1. Theorem 1 extends without essential changes to conformal solutions $X(w)$ of the system $\Delta X = f(X, \nabla X)$, where the right-hand-side grows quadratically in $|\nabla X|$. Also two-dimensional surfaces in $\mathbb{R}^n$, $n \geq 3$, can be treated.

In the case of polygonal boundaries we can say more:

Theorem 2. *Suppose that* Assumption A *holds where* Γ^+, Γ^- *are straight lines. Then there exist holomorphic functions* H_j *and* $\hat{H}_j, j = 1, 2, 3$, *which are defined on a disk* B_δ *for some* $\delta > 0$, *such that the following holds true*:

$$wH_1^2(w) + 4H_2(w)H_3(w) = 0, \tag{5}$$

$$X_w(w) = w^{\alpha-1}H_2(w)\begin{pmatrix} 1 \\ -i \\ 0 \end{pmatrix} + w^{-\alpha}H_3(w)\begin{pmatrix} 1 \\ i \\ 0 \end{pmatrix} + H_1(w)\begin{pmatrix} 0 \\ 0 \\ 1 \end{pmatrix}, \tag{6}$$

$$\begin{aligned} x(w) + iy(w) &= w^\alpha \hat{H}_2(w) + \overline{w}^{1-\alpha}\hat{H}_3(\overline{w}), \\ z(w) &= \operatorname{Re}(w\hat{H}_1(w)), \end{aligned} \tag{7}$$

where $w \in \overline{B}_\delta^+ \setminus \{0\}$. *Furthermore,* (1) *holds true as well.*

The idea of the proof of Theorems 1 and 2 is to eventually apply Theorem 3 of Section 3.1 to a certain set of functions involving the gradient X_w. Here it is necessary and convenient to use first a reflection procedure followed by a smoothing argument. The new function of interest is then defined on a neighbourhood $B_\delta \setminus \{0\}$ of zero, and it turns out that Theorem 3 of Section 3.1 can be employed. A further essential ingredient is Theorem 1 of Section 3.2 which provides the starting regularity and thus makes our argument work.

Proof of Theorem 1. Because of the continuity of X it is possible to select $\delta > 0$ so small that $x(\overline{B}_\delta^+) \subset [-\varepsilon, \varepsilon]$; this will henceforth be assumed. By Assumption A we have on $I_\delta^\pm$ the equality

$$X(u,0) = (x(u,0), h_1^\pm(x(u,0)), h_2^\pm(x(u,0))),$$

whence

$$X_u(u,0) = (1, h_1^{\pm\prime}(x(u,0)), h_2^{\pm\prime}(x(u,0)))x_u(u,0).$$

The conformality conditions (which also hold on $I_\delta^\pm$) imply that

(8) $$\langle X_v(u,0), (1, h_1^{\pm\prime}(x(u,0)), h_2^{\pm\prime}(x(u,0)))\rangle = 0 \quad \text{on } I_\delta^\pm.$$

Now we put

$$a^\pm(t) := [1 + h_1^{\pm\prime}(t)^2 + h_2^{\pm\prime}(t)^2]^{-1/2}(1, h_1^{\pm\prime}(t), h_2^{\pm\prime}(t)), \quad t \in [-\varepsilon, \varepsilon],$$

and consider the linear mappings

$$S^\pm(t)y := 2\langle a^\pm(t), y\rangle a^\pm(t) - y$$

which are defined for $t \in [-\varepsilon, \varepsilon]$ and $y \in \mathbb{R}^3$. Then, using (8), we infer

$$S^\pm(x(u,0))X_u(u,0) = X_u(u,0),$$

$$S^\pm(x(u,0))X_v(u,0) = -X_v(u,0), \quad \text{where } (u,0) \in I_\delta^\pm.$$

This may be rewritten as

(9) $$S^\pm(x(w))X_w(w) = X_{\overline{w}}(w) \quad \text{for all } w \in I_\delta^\pm.$$

Since $S^\pm(t), t \in [-\varepsilon, \varepsilon]$, is a family of reflections, there exist orthogonal matrices $O^\pm(t)$ such that

$$S^\pm(t) = O^\pm(t)\,\mathrm{Diag}[-1,-1,1]O^\pm(t)^t,$$

where we have used the notation

$$\mathrm{Diag}[\alpha,\beta,\gamma] = \begin{pmatrix} \alpha & 0 & 0 \\ 0 & \beta & 0 \\ 0 & 0 & \gamma \end{pmatrix}$$

and A^t denotes the transpose of the matrix A. Furthermore we define

$$T^\pm(t) := O^\pm(0)O^\pm(t)^t$$

with

$$O^+(0) = \lim_{t\to 0+} O^+(t) \quad \text{and} \quad O^-(0) = \lim_{t\to 0-} O^-(t).$$

Now put

$$T(t) := \begin{cases} T^+(t) & \text{if } 0 \le t \le \varepsilon, \\ T^-(t) & \text{if } -\varepsilon \le t < 0. \end{cases}$$

It follows that the matrix function T is of class $C^0[(-\varepsilon, \varepsilon], \mathbb{R}^9)$ since

$$\lim_{t\to 0^+} T(t) = T^+(0) = \mathrm{Id} = T^-(0) = \lim_{t\to 0^-} T(t),$$

and, because of the assumptions on $h_1^\pm, h_2^\pm$, the matrix T is even of the class

$$C^{0,1}([-c, c], \mathbb{R}^9)$$

(although $S^\pm(t)$ is not even continuous at zero).

Next we consider the complex valued function $g(w)$ defined by

$$g(w) := T(x(w)) \cdot X_w(w) \quad \text{for all } w \in B_\delta^+. \tag{10}$$

We claim that g has the reflection property

$$S^\pm(0)g(w) = \overline{g(w)} \quad \text{for all } w \in I_\delta^\pm. \tag{11}$$

In fact, it follows from (9) that

$$\begin{aligned} S^\pm(0)g(w) &= S^\pm(0)T^\pm(x(w))X_w(w) \\ &= S^\pm(0)O^\pm(0)O^\pm(x(w))^t X_w(w) \\ &= O^\pm(0)\,\mathrm{Diag}[-1,-1,1]O^\pm(x(w))^t X_w(w) \\ &= T^\pm(x(w))S^\pm(x(w))X_w(w) \\ &= T^\pm(x(w))X_{\overline{w}}(w) = \overline{g(w)}. \end{aligned}$$

We now reflect g across the u-axis by

$$G(w) = \begin{cases} g(w) & \text{if } w \in \overline{B_\delta^+} \setminus \{0\}, \\ S^+(0)\overline{g}(\overline{w}) & \text{if } \overline{w} \in B_\delta^+. \end{cases} \tag{12}$$

Then we have

Lemma 1. *The function G is of class $C^{0,1}(B_\delta \setminus \overline{I_\delta^-}, \mathbb{C}^3)$, and there exists some constant $c > 0$ such that the estimate*

$$|G_{\overline{w}}(w)| \le c|G(w)|^2 \tag{13}$$

holds true almost everywhere on $B_\delta \setminus \{0\}$. Furthermore $G(w)$ satisfies

$$G_1^2(w) + G_2^2(w) + G_3^2(w) = 0, \quad w \in B_\delta \setminus \overline{I_\delta^-}, \tag{14}$$

$$|G(w)| = O(|w|^{\nu-1}) \quad \text{as } w \to 0, \tag{15}$$

where ν denotes the Hölder exponent of X. Finally there holds the jump relation

$$\lim_{v\to 0^+} G(u,v) = S^-(0)S^+(0) \lim_{v\to 0^-} G(u,v) \tag{16}$$

for all $u \in I_\delta^-$.

Proof. Since T is Lipschitz continuous and $X \in C^2(\overline{B}_\delta^+ \setminus \{0\}, \mathbb{R}^3)$ we also have $g \in C^{0,1}(\overline{B}_\delta^+ \setminus \{0\}, \mathbb{C}^3)$, and because of (11) we obtain $G \in C^{0,1}(B_\delta \setminus \overline{I_\delta^-}, \mathbb{C}^3)$. To establish (13), we remark that almost everywhere on B_δ we find

$$G_{\overline{w}}(w) = \begin{cases} g_{\overline{w}} = [T'(x(w))x_{\overline{w}}(w)]X_w(w) & \text{if } w \in B_\delta^+, \\ S^+(0)\overline{g}_{\overline{w}}(\overline{w}) = S^+(0)[T'(x(\overline{w}))\overline{x}_{\overline{w}}(\overline{w})]X_{\overline{w}}(\overline{w}) & \text{if } \overline{w} \in B_\delta^+ \end{cases}$$

whence

$$\begin{aligned} |G_{\overline{w}}(w)| &\le c_1|T'(x(w))||x_w||X_w| \le c_2|X_w(w)|^2 \\ &\le c_3|T^{-1}(x(w))g(w)|^2 \le c_4|g(w)|^2 \le c_5|G(w)|^2 \end{aligned}$$

for suitable constants $c_1, \dots, c_5$. From the conformality condition $\langle X_w, X_w\rangle = 0$ we easily conclude (14), taking the orthogonality of the matrices $T^\pm$ into account.

The relation (15) follows from the estimate $|G(w)| \le c_6|\nabla X|$ and from Theorem 1 of Section 3.2. Finally, to prove (16), we calculate by means of (11) that

$$\begin{aligned} \lim_{v\to 0+} G(u,v) &= g(u,0) = S^-(0)\overline{g}(u,0) \\ &= S^-(0)S^+(0)S^+(0)\overline{g}(u,0) = S^-(0)S^+(0) \lim_{v\to 0^-} G(u,v), \end{aligned}$$

where we have used that $S^+(0)S^+(0) = \text{Id}$. □

The function $G(w)$ itself is not yet accessible to the methods which were developed at the end of Section 3.1, because of the jump relation (16). To overcome this difficulty, we have to smooth the function G, which will be carried out in what follows. Recall that the jump of G at I_δ^- is given by

$$S^-(0)S^+(0) = \begin{pmatrix} \cos 2\pi\alpha & -\sin 2\pi\alpha & 0 \\ \sin 2\pi\alpha & \cos 2\pi\alpha & 0 \\ 0 & 0 & 1 \end{pmatrix}.$$

We can diagonalize $S^-(0)S^+(0)$ using the unitary matrix

$$U = \frac{1}{\sqrt{2}} \begin{pmatrix} 0 & 1 & 1 \\ 0 & -i & i \\ \sqrt{2} & 0 & 0 \end{pmatrix}$$

and obtain

$$S^-(0)S^+(0) = U\,\text{Diag}[1, e^{2\pi i(\alpha-1)}, e^{-2\pi i\alpha}]U^*,$$

where $U^* = \overline{U}^t$. We define a new function $F(w), w \in B_\delta \setminus \overline{I_\delta^+}$, by

$$F(w) := \text{Diag}[1, w^{1-\alpha}, w^\alpha]U^*G(w) \tag{17}$$

or, more explicitly,

$$F(w) = \begin{pmatrix} F_1(w) \\ F_2(w) \\ F_3(w) \end{pmatrix} = \begin{pmatrix} G_3(w) \\ \frac{1}{\sqrt{2}} w^{1-\alpha}[G_1(w) + iG_2(w)] \\ \frac{1}{\sqrt{2}} w^{\alpha}[G_1(w) - iG_2(w)] \end{pmatrix}.$$

We claim that F is continuous on the punctured disk $B_\delta(0) \setminus \{0\}$. In fact, we infer from (16) the relation

$$\begin{aligned}
\lim_{v \to 0+} F(u,v) &= \mathrm{Diag}[1, u^{1-\alpha} e^{i\pi(1-\alpha)}, u^\alpha e^{i\pi\alpha}] U^* \lim_{v \to 0+} G(u,v) \\
&= \mathrm{Diag}[1, u^{1-\alpha} e^{i\pi(1-\alpha)}, u^\alpha e^{i\pi\alpha}] U^* S^-(0) S^+(0) \lim_{v \to 0-} G(u,v) \\
&= \mathrm{Diag}[1, u^{1-\alpha} e^{i\pi(1-\alpha)}, u^\alpha e^{i\pi\alpha}] U^* U \\
&\quad \cdot \mathrm{Diag}[1, e^{2\pi i(\alpha-1)}, e^{-2\pi i\alpha}] U^* \lim_{v \to 0-} G(u,v)) \\
&= \mathrm{Diag}[1, u^{1-\alpha} e^{-i\pi(1-\alpha)}, u^\alpha e^{-i\pi\alpha}] U^* \lim_{v \to 0-} G(u,v) \\
&= \lim_{v \to 0-} F(u,v).
\end{aligned}$$

Since $G \in C^{0,1}(B_\delta \setminus \overline{I_\delta^-}, \mathbb{C}^3)$, and by Assumption A, it follows that F is even Lipschitz continuous on the punctured disk $B_\delta \setminus \{0\}$.

Lemma 2. *The function $F(w) = (F_1(w), F_2(w), F_3(w))$ defined by (17) belongs to the class $C^{0,1}(B_\delta(0) \setminus \{0\}, \mathbb{C}^3)$ and satisfies*

(18)
$$F_1^2(w)w + 2F_2(w)F_3(w) = 0 \quad \textit{for } w \in B_\delta \setminus \{0\},$$
$$|F_1(w)| = O(|w|^{\nu-1}) \quad \textit{as } w \to 0,$$

and

(19)
$$|F_2(w)| = O(|w|^{\nu-\alpha}) \quad \textit{as } w \to 0,$$
$$|F_3(w)| = O(|w|^{\nu-\beta}) \quad \textit{as } w \to 0,$$

where ν denotes the Hölder exponent of X, and $\beta = 1 - \alpha$. Furthermore, the following differential inequalities hold true:

(20)
$$\begin{aligned}
|F_{1\overline{w}}(w)| &\le c\{|w|^{-2\beta}|F_2(w)|^2 + |w|^{-2\alpha}|F_3(w)|^2\}, \\
|F_{2\overline{w}}(w)| &\le c\{|w|^{-\beta}|F_2(w)|^2 + |w|^{\beta-2\alpha}|F_3(w)|^2\}, \\
|F_{3\overline{w}}(w)| &\le c\{|w|^{\alpha-2\beta}|F_2(w)|^2 + |w|^{-\alpha}|F_3(w)|^2\}
\end{aligned}$$

almost everywhere on $B_\delta \setminus \{0\}$ for some constant $c > 0$.

Proof. We conclude from (14) and (17) that

$$\begin{aligned} 0 &= G_1^2(w) + G_2^2(w) + G_3^2(w) = \tfrac{1}{2}[w^{\alpha-1}F_2(w) + w^{-\alpha}F_3(w)]^2 \\ &\quad - \tfrac{1}{2}[w^{\alpha-1}F_2(w) - w^{-\alpha}F_3(w)]^2 + F_1^2(w) \\ &= 2w^{-1}F_2(w)F_3(w) + F_1^2(w), \end{aligned}$$

whence (18) follows. From the definition of $F(w) = (F_1(w), F_2(w), F_3(w))$ and from (15) we infer the relations (19). To prove the inequalities (20) we first note that

$$|G(w)|^2 = |F_1(w)|^2 + |w|^{-2\beta}|F_2(w)|^2 + |w|^{-2\alpha}|F_3(w)|^2,$$

whence we obtain from (13) and (17) the inequalities

$$\begin{aligned} |F_{1\overline{w}}(w)| &\le c[|F_1(w)|^2 + |w|^{-2\beta}|F_2(w)|^2 + |w|^{-2\alpha}|F_3(w)|^2], \\ |F_{2\overline{w}}(w)| &\le c[|w|^{\beta}|F_1(w)|^2 + |w|^{-\beta}|F_2(w)|^2 + |w|^{\beta-2\alpha}|F_3(w)|^2], \\ |F_{3\overline{w}}(w)| &\le c[|w|^{\alpha}|F_1(w)|^2 + |w|^{\alpha-2\beta}|F_2(w)|^2 + |w|^{-\alpha}|F_3(w)|^2]. \end{aligned}$$

On the other hand, relation (18) yields the estimate

$$\begin{aligned} |F_1(w)|^2 &\le |w|^{-1}|w|^{2\alpha-1}|F_2(w)|^2 + |w|^{-1}|w|^{1-2\alpha}|F_3(w)|^2 \\ &= |w|^{2(\alpha-1)}|F_2(w)|^2 + |w|^{-2\alpha}|F_3(w)|^2 \\ &= |w|^{-2\beta}|F_2(w)|^2 + |w|^{-2\alpha}|F_3(w)|^2. \end{aligned}$$

Together with the above inequalities we finally obtain (20). This finishes the proof of Lemma 2. □

Now we are in a position to apply Theorem 3 of Section 3.1. We can assume without loss of generality that $0 < \alpha \le \beta < 1$.

Lemma 3. *There exists a nonnegative integer m such that the functions $f_i^m(w) := w^{-m}F_i(w), i = 1, 2, 3,$ do not vanish simultaneously at zero and that one of the following conditions holds true:*

(i) $\quad f_2^m \in C^{0,\mu}(B_\delta, \mathbb{C})$ *for all* $\mu < \min(1, m+\alpha)$, $\quad f_2^m(0) \ne 0$,

$$\begin{aligned} |f_{1\overline{w}}^m(w)| &= O(|w|^{m-2\beta}) \quad \text{as } w \to 0, \\ |f_{2\overline{w}}^m(w)| &= O(|w|^{m-\beta}) \quad \text{as } w \to 0, \\ |f_{3\overline{w}}^m(w)| &= O(|w|^{m+\alpha-2\beta}) \quad \text{as } w \to 0, \end{aligned}$$

a.e. on $B_\delta \setminus \{0\}$.

(ii) $\quad f_3^m \in C^{0,\mu}(B_\delta, \mathbb{C})$ *for all* $\mu < \min(1, m+\beta)$, $\quad f_3^m(0) \ne 0$

$$|f^m_{1\overline{w}}(w)| = O(|w|^{m-2\alpha}) \quad as\ w \to 0,$$
$$|f^m_{2\overline{w}}(w)| = O(|w|^{m+\beta-2\alpha}) \quad as\ w \to 0,$$
$$|f^m_{3\overline{w}}(w)| = O(|w|^{m-\alpha}) \quad as\ w \to 0,$$

a.e. on $B_\delta \setminus \{0\}$.

In addition, if $m \geq 1$*, then in both cases*

$$f_2^m(0) f_3^m(0) = 0. \tag{21}$$

Proof of Lemma 3. From Theorem 3 of Section 3.1 we infer that (i) or (ii) has to hold, except for the assertions concerning f_1^m. We recall the cases $(\alpha), (\beta), (\gamma)$, and (δ) which occurred in the proof of Theorem 3 in Section 3.1. Let us treat these cases separately.

(α) $F_2(0) \neq 0, F_3(0) \neq 0$: Then (i) of Theorem 3 in Section 3.1 holds with $m = 0$. In particular, $|F_2(w)| = O(1)$ as $w \to 0$, and $|F_{3\overline{w}}(w)| = O(|w|^{\alpha-2\beta})$ as $w \to 0$. But then Lemma 6 of Section 3.1 implies that

$$|F_3(w)| = \begin{cases} O(1 + |w|^{\alpha-2\beta+1}) & \text{if } \alpha - 2\beta \neq -1, \\ O(|w|^{-\varepsilon}) \quad \text{for all } \varepsilon > 0 & \text{if } \alpha - 2\beta = -1. \end{cases}$$

Now relation (20_1) yields

$$|F_{1\overline{w}}(w)| = O(|w|^{-2\beta}| \quad \text{as } w \to 0,$$

which is the desired assertion.

(β) $F_2(0) \neq 0, F_3(0) = 0$: Here we obtain (i) of Section 3.1, Theorem 3 with $m = 0$. Thus we can proceed as in case (α).

(γ) $F_2(0) = 0, F_3(0) \neq 0$: In this case we obtain (ii) of Theorem 3 in Section 3.1 with $m = 0$. In particular,

$$|F_3(w)| = O(1) \quad \text{as } w \to 0,$$

$$|F_{2\overline{w}}(w)| = O(|w|^{\beta-2\alpha}) \quad \text{as } w \to 0.$$

But $0 < \alpha \leq \frac{1}{2} \leq \beta < 1$ and $\beta - 2\alpha = 1 - 3\alpha \geq -\frac{1}{2}$; therefore we conclude from Lemma 6 of Section 3.1 that $F_2 \in C^{0,\mu}(B_\delta, \mathbb{C})$ for all $\mu < \min(1, 1 + \beta - 2\alpha)$. Since $F_2(0) = 0$ we have that $|F_2(w)| = O(|w|^\mu), w \to 0, \mu < \min(1, 1+\beta-2\alpha)$, and relation (20_1) implies

$$|F_{1\overline{w}}(w)| = O(|w|^{-2\beta+2\mu} + |w|^{-2\alpha}) = O(|w|^{-2\alpha}) \quad \text{as } w \to 0,$$

if we choose μ in such a way that $2\mu - 2\beta \geq 0$.

(δ) $F_2(0) = F_3(0) = 0$: In this case we find

$$(22) \qquad \begin{aligned} F_2(w) &= aw^m + o(|w|^m) \quad \text{as } w \to 0, \\ F_3(w) &= bw^m + o(|w|^m) \quad \text{as } w \to 0, \end{aligned}$$

with $a, b \in \mathbb{C}$ not both equal to zero and $m \geq 1$. A direct consequence of (20) is the following system:

$$(23) \qquad \begin{aligned} |f^m_{1\overline{w}}(w)| &\leq c[|w|^{m-2\beta}|f^m_2(w)|^2 + |w|^{m-2\alpha}|f^m_3(w)|^2], \\ |f^m_{2\overline{w}}(w)| &\leq c[|w|^{m-\beta}|f^m_2(w)|^2 + |w|^{m+\beta-2\alpha}|f^m_3(w)|^2], \\ |f^m_{3\overline{w}}(w)| &\leq c[|w|^{m+\alpha-2\beta}|f^m_2(w)|^2 + |w|^{m-\alpha}|f^m_3(w)|^2], \end{aligned}$$

while (18) yields

$$(24) \qquad w[f^m_1(w)]^2 + 2f^m_2(w)f^m_3(w) = 0 \quad \text{in } B_\delta \setminus \{0\}.$$

The relations (22) and (24) imply that

$$|f^m_2(w)|, |f^m_3(w)| = O(1) \quad \text{as } w \to 0$$

and

$$|f^m_1(w)| = o(|w|^{-1}) \quad \text{as } w \to 0.$$

Now (23_1) yields $|f^m_{1\overline{w}}(w)| = O(|w|^{m-2\beta})$, and by Lemma 6 of Section 3.1 we find that $f^m_1 \in C^{0,\mu}(B_\delta, \mathbb{C})$ for all $\mu < \min(1, m - 2\beta + 1)$. By letting $w \to 0$ in relation (24) we conclude (21): $ab = f^m_2(0)f^m_3(0) = 0$.

First subcase: $a \neq 0, b = 0$. Then case (i) of Theorem 3, Section 3.1, holds with $m \geq 1$, and this implies (i) of Lemma 3 since we have already shown that

$$|f^m_{1\overline{w}}(w)| = O(|w|^{m-2\beta}) \quad \text{as } w \to 0.$$

Second subcase: $a = 0, b \neq 0$. Here case (ii) of Theorem 3, Section 3.1, holds with $m \geq 1$. In particular,

$$\begin{aligned} |f^m_{2\overline{w}}(w)| &= O(|w|^{m+\beta-2\alpha}) \quad \text{as } w \to 0, \\ |f^m_{3\overline{w}}(w)| &= O(|w|^{m-\alpha}) \quad \text{as } w \to 0. \end{aligned}$$

By virtue of $a = f^m_2(0) = 0$ and Lemma 6 of Section 3.1 we find

$$|f^m_2(w)| = O(|w|^\mu) \quad \text{for all } \mu < \min(1, m + \beta - 2\alpha + 1) = 1.$$

Finally we obtain from (23_1)

$$|f^m_{1\overline{w}}(w)| = O(|w|^{m-2\beta}|w|^{2\mu} + |w|^{m-2\alpha}) = O(|w|^{m-2\alpha}) \quad \text{as } w \to 0.$$

Thus Lemma 3 is proved. □

Now we finish the *proof of Theorem* 1.
From (10), (12) and (17) we infer

$$\begin{aligned} X_w(w) &= T(x(w))^* U \text{ Diag}[1, w^{\alpha-1}, w^{-\alpha}]F(w) \\ &= w^{\alpha-1} T(x(w))^* U \text{ Diag}[w^{1-\alpha}, 1, w^{1-2\alpha}]F(w). \end{aligned}$$

Let us assume that (i) of Lemma 3 holds true whence in particular

$$f_2^m \in C^{0,\mu}(B_\delta, \mathbb{C}), \quad f_2^m(0) \neq 0.$$

Now we define $\psi = (\psi_1, \psi_2, \psi_3)$ by

$$\begin{aligned} \psi(w) &:= w^{-m} U \text{ Diag}\,[w^{1-\alpha}, 1, w^{1-2\alpha}]F(w) \\ &= \frac{1}{\sqrt{2}} f_2^m(w) \begin{pmatrix} 1 \\ -i \\ 0 \end{pmatrix} + \frac{1}{\sqrt{2}} w^{1-2\alpha} f_3^m(w) \begin{pmatrix} 1 \\ i \\ 0 \end{pmatrix} \\ &\quad + w^{1-\alpha} f_1^m(w) \begin{pmatrix} 0 \\ 0 \\ 1 \end{pmatrix} \end{aligned} \tag{25}$$

and claim that ψ is Hölder continuous in $\overline{B}_\delta^+$ for $0 < \delta \ll 1$ and satisfies

$$\psi_1(0) \neq 0, \quad \psi_2(0) \neq 0, \quad \psi_3(0) = 0. \tag{+}$$

First we note that $\psi_3(w) = w^{1-\alpha} f_1^m(w)$ is Hölder continuous in $\overline{B}_\delta^+$ with $\psi_3(0) = 0$. In fact, if $m \geq 1$ then f_1^m is Hölder continuous according to part (δ) in the proof of Lemma 3. If, however, $m = 0$, then $f_1^0 = F_1$, and so we have according to Lemma 3, (i), and formula (18) that

$$|F_1(w)| = O(|w|^{\nu-1}|), |F_{1,\overline{w}}(w)| = O(|w|^{-2\beta}) \quad \text{for } w \to 0. \tag{++}$$

We distinguish two cases:

(i) $2\alpha > 1$: Then we have $-2\beta > -1$, and by (++) and Lemma 6 in Section 3.1, the function f_1^0 is Hölder continuous in $\overline{B}_\delta^+$. Thus also ψ_3 is Hölder continuous in $\overline{B}_\delta^+$, and $\psi_3(0) = 0$ since $\alpha < 1$.

(ii) $2\alpha < 1$: By (++) and Lemma 6 in Section 3.1 the function $wf_1^0(w)$ is of class $C^\mu(B_\delta)$ for all $\delta < 2\alpha$, and $wf_1^0(w) \to 0$ as $w \to 0$. Therefore,

$$|w^{1-\alpha} f_1^0(w)| = O(|w|^{\alpha-\epsilon}) \quad \text{for } w \to 0 \text{ and } 0 < \epsilon \ll 1;$$

consequently ψ_3 is continuous in $\overline{B}_\delta^+$ with $\psi_3(0) = 0$. Now we estimate $|\psi_3(w_1) - \psi_3(w_2)|$ for any $w_1, w_2 \in \overline{B}_\delta^+ \setminus \{0\}$; w.l.o.g. we assume $|w_1| \leq |w_2|$, whence $|w_2| \geq \frac{1}{2}|w_2 - w_1|$. Then

$$\begin{aligned}
&|\psi_3(w_1) - \psi_3(w_2)| \\
&\quad \le |w_1^{-\alpha} - w_2^{-\alpha}||w_1 f_1^0(w_1)| + |w_2|^{-\alpha}|w_1 f_1^0(w_1) - w_2 f_2^0(w_2)| \\
&\quad \le c\{|w_1|^{-\alpha}|w_2|^{-\alpha}|w_1 - w_2|^{\alpha}|w_1|^{2\alpha-\epsilon} + |w_2|^{-\alpha}|w_1 - w_2|^{2\alpha-\epsilon}\} \\
&\quad \le c\{|w_2|^{-\alpha}|w_1 - w_2|^{\alpha}|w_1|^{\alpha-\epsilon} + |w_1 - w_2|^{\alpha-\epsilon}\} \le c|w_1 - w_2|^{\alpha-\epsilon},
\end{aligned}$$

which shows that ψ_3 is Hölder continuous in $\overline{B_\delta^+}$ also in case (ii).

Now we infer from (18), (24) the identity

$$w^{1-2\alpha} f_3^m(w) = -\frac{[w^{1-\alpha} f_1^m(w)]^2}{2 f_2^m(w)}.$$

Since the right-hand side is Hölder continuous in B_δ^+ (note that $f_2^m(0) \neq 0$), the same holds for $w^{1-2\alpha} f_2^m(w)$; thus we arrive at $w^{1-2\alpha} f_3^m(w) \to 0$ for $w \to 0$. Therefore, ψ is Hölder continuous and satisfies (+). Because of $T(0) = \mathrm{Id}$, also the function $\Phi(w) := T(x(w))^* \psi(w)$ satisfies $\Phi_1(0) \neq 0, \Phi_2(0) \neq 0$ and $\Phi_3(0) = 0$. Since T is Lipschitz continuous and ψ and X are Hölder continuous, also Φ is of class $C^{0,\nu_1}(\overline{B_\delta^+}, \mathbb{C}^3)$ where $\nu_1 := \min(\nu, \mu)$. Because of (25) we have

$$X_w(w) = w^{\alpha-1+m}\Phi(w) = w^{\alpha-1+m} T(x(w))^* \psi(w),$$

that is,

$$\begin{aligned}
x_w(w) &= w^{\alpha-1+m}\Phi_1(w), \quad \Phi_1(0) \neq 0, \\
y_w(w) &= w^{\alpha-1+m}\Phi_2(w), \quad \Phi_2(0) \neq 0, \qquad (26)\\
|z_w(w)| &= |w^{\alpha-1+m}||\Phi_3(w)| = O(|w|^{\lambda}),
\end{aligned}$$

as $w \to 0$, with $\lambda > \alpha - 1 + m$. This proves relation (2).

On the other hand, let us assume that (ii) of Lemma 3 occurs; then we argue with the function

$$\tilde{\psi}(w) := \frac{1}{\sqrt{2}} f_3^m(w) \begin{pmatrix} 1 \\ i \\ 0 \end{pmatrix} + \frac{1}{\sqrt{2}} w^{2\alpha-1} f_2^m(w) \begin{pmatrix} 1 \\ -i \\ 0 \end{pmatrix} + w^{\alpha} f_1^m(w) \begin{pmatrix} 0 \\ 0 \\ 1 \end{pmatrix}$$

instead of ψ, and similar arguments show that $\tilde{\psi}(w)$ is Hölder continuous on $\overline{B_\delta^+}$ and that $\tilde{\psi}_1(0) \neq 0, \tilde{\psi}_2(0) \neq 0, \tilde{\psi}_3(0) = 0$. In this case we have $\gamma = m - \alpha$ because of

$$w^{m-\alpha}\tilde{\psi}(w) = U \,\mathrm{Diag}[1, w^{\alpha-1}, w^{-\alpha}] F(w)$$

and

$$X_w(w) = w^{m-\alpha} T(x(w))^* \tilde{\psi}(w);$$

thus (2) holds with $\gamma = m - \alpha$.

From the conformality condition

$$x_w^2(w) + y_w^2(w) + z_w^2(w) = 0, \quad w \in I_\delta^- \cup I_\delta^+,$$

we infer, using (2), that

$$0 = w^{2\gamma}[\Phi_1^2(w) + \Phi_2^2(w)] + O(|w|^{2\lambda}) \quad \text{as } w \to 0,$$

where $\gamma = \alpha - 1 + m$ or $m - \alpha$. Letting $w \to 0$, we obtain

$$0 = \Phi_1^2(0) + \Phi_2^2(0)$$

which proves (1).

Relation (3) follows by integrating formula (2), using the fact that

$$X(w) = 2\ \mathrm{Re}\left[\int_0^r X_w(\underline{r}e^{i\varphi})e^{i\varphi}\, d\underline{r}\right].$$

Thus

$$\text{(27)} \qquad x(w) + iy(w) = \begin{cases} w^{\alpha+m}[c + o(1)] \\ \overline{w}^{m-\alpha+1}[c + o(1)] \end{cases} \quad \text{as } w \to 0$$

in the two cases respectively. Relation (27) and the boundary conditions imply that $m = 2k$ in the first and $m = 2k + 1$ in the second case.

Finally we have to consider the normal

$$N(w) = (N_1, N_2, N_3) = \frac{X_u \wedge X_v}{|X_u \wedge X_v|}.$$

Since

$$|X_u \wedge X_v| = 2|X_w|^2 = 2|w|^{2\gamma}[|\Phi_1(w)|^2 + |\Phi_2(w)|^2] + O(|w|^{2\lambda}), \quad \lambda > \gamma.$$

and

$$X_u \wedge X_v = 2(\mathrm{Im}(y_w z_{\overline{w}}), -\mathrm{Im}(x_w z_{\overline{w}}), \mathrm{Im}(x_w y_{\overline{w}}))$$

we find by means of (2) that

$$\lim_{w\to 0} N_1(w) = \lim_{w\to 0}\left[\mathrm{const}\frac{|w|^{\gamma+\lambda}}{|w|^{2\gamma}}\right] = 0 \quad \text{since } \lambda > \gamma,$$

and

$$\lim_{w\to 0} N_2(w) = 0.$$

Finally

$$\lim_{w\to 0} N_3(w) = 2\lim_{w\to 0}\frac{\mathrm{Im}(x_w(w)y_{\overline{w}}(w))}{|X_w(w)|^2} = 2\frac{\mathrm{Im}(\Phi_1(0)\overline{\Phi_2(0)})}{|\Phi_1(0)|^2 + |\Phi_2(0)|^2} = \pm 1$$

since $\Phi_1(0) = \pm i\Phi_2(0)$, and Theorem 1 is proved. □

Proof of Theorem 2. If Γ^+ and Γ^- are straight lines, then the matrix T is the identity $\mathrm{Id}|^3_{\mathbb{R}}$, whence $g(w) = X_w(w)$ and $\frac{dG}{d\overline{w}}(w) = X_{w\overline{w}}(w) = 0$ almost everywhere in $B_\delta - \{0\}$. According to Theorem 1.15 in Vekua [1] (or Satz 1.17 in Vekua [2]) we see that G and hence F are holomorphic on $B_\delta(0)$. By the definition of F we obtain

$$\begin{aligned} X_w(w) &= G(w) = U \ \mathrm{Diag}[1, w^{\alpha-1}, w^{-\alpha}]F(w) \\ &= \frac{w^{\alpha-1}F_2(w)}{\sqrt{2}}\begin{pmatrix} 1 \\ -i \\ 0 \end{pmatrix} + \frac{w^{-\alpha}F_3(w)}{\sqrt{2}}\begin{pmatrix} 1 \\ i \\ 0 \end{pmatrix} + F_1(w)\begin{pmatrix} 0 \\ 0 \\ 1 \end{pmatrix}. \end{aligned}$$

Putting $H_1 := F_1, H_2 := (\sqrt{2})^{-1}F_2$ and $H_3 := (\sqrt{2})^{-1}F_3$ we obtain representation (6). Finally (7) follows by integration, and (5) is a consequence of (18). Thus Theorem 2 is proved. □

3.4 An Asymptotic Expansion for Solutions of the Partially Free Boundary Problem

The aim of this section is to prove an analogue of Theorem 1 in Section 3.3 for minimal surfaces with partially free boundaries. Here the point of interest is the intersection point of the boundary arc Γ with the supporting surface S. Let us again start with an instructive *example*:

Let S be the coordinate plane $\{z = 0\}$ and

$$\Gamma = \{(x, y, z) \colon z = x \tan(\alpha\pi), y = 0, 0 \le x \le 1\},$$

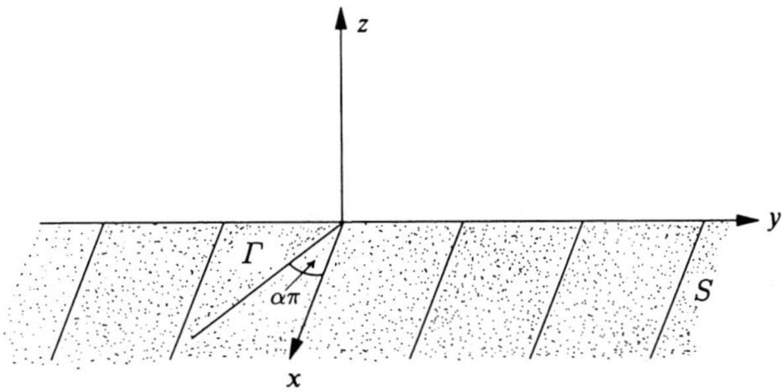

Fig. 1.

where $\alpha \in (0, \frac{1}{2})$. For each $k \in \mathbb{N} \cup \{0\}$ we consider the functions

$$f_1(w) = w^{\alpha+2k}, \quad f_2(w) = \overline{w}^{2-\alpha+2k},$$

$$f_3(w) = -w^{\alpha+1+2k}, \quad f_4(w) = -\overline{w}^{1-\alpha+2k},$$

and the associated minimal surfaces

$$X_j(u,v) = (x_j(u,v), y_j(u,v), z_j(u,v)), \quad j \in \{1,2,3,4\};$$

given by

$$x_j(u,v) = \operatorname{Re} f_j(w), \quad y_j(u,v) = 0, \quad z_j(u,v) = \operatorname{Im} f_j(w),$$
$$w \in B = \{(u,v) \in \mathbb{R}^2 \colon u^2 + v^2 < 1, v > 0\}, \quad w = u + iv.$$

Then each $X_j, j = 1,2,3,4$, is a minimal surface which maps the interval $[-1,0]$ onto Γ and $[0,1]$ into S while $X(0,0) = 0$. Also X_j meets the surface S orthogonally along its trace $X_j|_{[0,1]}$, and hence it is a stationary solution of a free boundary problem determined by Γ and S. We shall prove that any minimal surface with a free boundary behaves near the corner point like one of the four solutions constructed above. More precisely, it will be shown that

(1) $$X_w(w) = w^\gamma \Phi(w) \quad \text{as } w \to 0,$$

where $\gamma > -1$, and $\Phi(w) = (\Phi_1(w), \Phi_2(w), \Phi_3(w))$ denotes some Hölder continuous complex valued function with $\Phi_1(0) \neq 0, \Phi_3(0) \neq 0$, and $\Phi_2(0) = 0$ if $\alpha \neq \frac{1}{2}$. From the representation (1) we deduce that the surface normal tends to a limiting position as $w \to 0$. If in particular $\alpha \neq \frac{1}{2}$, then the tangent space of X at the corner $P \in \Gamma \cap S$ is spanned by the normal to S at P and the tangent to Γ at P. Thus the solution surface X must meet the point P at one of the angles $\alpha\pi, (2-\alpha)\pi, (1-\alpha)\pi$ and $(\alpha+1)\pi$ depending on whether X behaves like f_1, f_2, f_3, or f_4, respectively. In each of these cases X may penetrate S and can wrap P k-times.

Let us recall some notation. We define the sets I_δ^-, I_δ^+ as in Sections 3.2 and 3.3, and we formulate Assumption A similar as in Section 3.2:

Assumption A. *Let S be a regular surface of class C^3, and Γ be a regular arc of class $C^{2,\mu}$ which meets S in a common point P at an angle $\alpha\pi$ with $0 < \alpha \leq \frac{1}{2}$. We assume that P is the origin O, that the x,y-plane is tangent to S at O, and that the tangent vector to Γ at O lies in the x,z-plane. Moreover, let $X(u,v)$ be a minimal surface of class $C^{0,v}(\overline{B_\delta^+}, \mathbb{R}^3) \cap C^2(\overline{B_\delta^+} \setminus \{0\}, \delta > 0,$ which satisfies the boundary conditions*

(2) $$X \colon I_\delta^- \to \Gamma, \quad X \colon I_\delta^+ \to S, \quad X(0) = P.$$

We also suppose that X intersects S orthogonally along its free trace $X|_{I_\delta^+}$.

The main result of this section is

Theorem 1. *Suppose that Assumption* A *holds. Then there exists an* $R > 0$ *and a Hölder continuous function* $\Phi(w) = (\Phi_1(w), \Phi_2(w), \Phi_3(w))$ *defined on* $\overline{B_R^+}$ *such that*

(3) $$X_w(w) = w^\gamma \Phi(w)$$

holds true on $\overline{B_R^+} \setminus \{0\}$ *with either* $\gamma = \alpha - 1 + m$ *or* $\gamma = -\alpha + m$ *for some integer* $m \geq 0$. *Moreover, we have* $\Phi_1(0), \Phi_2(0), i\Phi_3(0) \in \mathbb{R}$ *and*

(4) $$\Phi_1(0) = \pm i\Phi_3(0) \neq 0, \quad \Phi_2(0) = 0 \quad \text{if } \alpha \neq \tfrac{1}{2},$$

that is,

(5) $$\Phi_1^2(0) + \Phi_2^2(0) + \Phi_3^2(0) = 0$$

and at least two $\Phi_j(0) \neq 0$ *if* $\alpha = \frac{1}{2}$. *The unit normal vector*

$$N(w) = (N_1(w), N_2(w), N_3(w)) = \frac{X_u \wedge X_v}{|X_u \wedge X_v|}(w)$$

satisfies

(6)
$$\begin{cases} \lim_{w \to 0} N(w) = \begin{pmatrix} 0 \\ \pm 1 \\ 0 \end{pmatrix} & \text{if } \alpha \neq \tfrac{1}{2}, \\ \lim_{w \to 0} N(w) = \begin{pmatrix} c_1 \\ c_2 \\ 0 \end{pmatrix}, & \text{if } \alpha = \tfrac{1}{2}, \text{ where } c_1, c_2 \in \mathbb{R} \text{ and } c_1^2 + c_2^2 = 1. \end{cases}$$

For the trace $X(u, 0), u \in \overline{I_R^+}$, *we find*

(7) $$X(u, 0) = u^{\gamma+1}\psi(u)$$

with some Hölder continuous function ψ *such that* $\psi(0) = (\Phi_1(0), \Phi_2(0), 0)$. *Furthermore, the oriented tangent vector* $t(u) = \frac{X_u(u,0)}{|X_u(u,0)|}$, $u \in I_R^+$, *satisfies*

(8) $$\lim_{w \to 0+} t(u) = \begin{pmatrix} \pm 1 \\ 0 \\ 0 \end{pmatrix} \quad \text{if } \alpha \neq \frac{1}{2},$$

and

(9) $$\lim_{u \to 0+} t(u) = \begin{pmatrix} d_1 \\ d_2 \\ 0 \end{pmatrix} \quad \text{if } \alpha = \frac{1}{2}, \text{ where } d_1^2 + d_2^2 = 1.$$

If, in addition, S *is a plane and if* Γ *is a straight line segment, then there exist functions* H_1, H_2, H_3, *holomorphic on* $B_R(0)$, *such that*

$$(10)\quad X_w(w) = w^{\alpha-1}H_1(w)\begin{pmatrix} i \\ 0 \\ 1 \end{pmatrix} + w^{-\alpha}H_3(w)\begin{pmatrix} -i \\ 0 \\ 1 \end{pmatrix} + w^{-1/2}H_2(w)\begin{pmatrix} 0 \\ 1 \\ 0 \end{pmatrix}$$

holds true on $\overline{B_R^+} \setminus \{0\}$ *and*

$$(11)\qquad H_2^2(w) + 4H_1(w)H_3(w) = 0 \quad \text{on } B_R(0).$$

Corollary 1. *If* $\alpha \neq \frac{1}{2}$, *then there exist some* $c \in \mathbb{C} \setminus \{0\}$ *and some integer* $k \geq 0$ *such that one of the following four expansions holds true:*

$$(12)\qquad (x+iz)(w) = \begin{cases} w^{\alpha+2k}[c+o(1)], & w \to 0, \\ \overline{w}^{2-\alpha+2k}[c+o(1)], & w \to 0, \\ w^{\alpha+1+2k}[c+o(1)], & w \to 0, \\ \overline{w}^{1-\alpha+2k}[c+o(1)], & w \to 0. \end{cases}$$

Moreover

$$|y(w)| = O(|w|^{\lambda+1}) \quad \text{as } w \to 0, \text{ for some } \lambda > \gamma$$

where γ *is the exponent in the expansion* $(x+iz)(w) = w^\gamma[c+o(1)]$ *stated in* (12).

The *proof of Theorem* 1 consists in an adaptation of the method which was developed in Section 3.3 for the proof of the corresponding result, see Theorem 1 in Section 3.3. So from time to time our presentation will be sketchy and leave the details to the reader as an instructive exercise. We begin the proof of Theorem 1 with a description of a *reflection* and a *smoothing procedure*. To this end let us henceforth assume that S is locally described by

$$z = f(x,y), (x,y) \in B_\varepsilon(0) = \{(x,y) \in \mathbb{R}^2 \colon x^2 + y^2 < \varepsilon\},$$

where $f \in C^3(B_\varepsilon(0), \mathbb{R})$, and $f(0,0) = 0, \nabla f(0,0) = 0$. Also, Γ may locally be described by two functions $h_1(t)$ and $h_2(t)$ of class $C^{2,\mu}([0,\varepsilon], \mathbb{R})$ such that $(h_1(t), h_2(t), t) \in \Gamma$ for $t \in [0,\varepsilon]$, and $h_1(0) = h_2(0) = h_2'(0) = 0$ while $h_1'(0) = \cot \alpha\pi$. Thus it follows that the unit tangent vector of Γ at zero is then given by $(\cos\alpha\pi, 0, \sin\alpha\pi)$. Because of the continuity of X we can select a number $R > 0$ such that

$$X(\overline{B_\delta^+}) \subset \mathcal{K}_\varepsilon(0) = \{(x,y,z) \in \mathbb{R}^3 \colon x^2 + y^2 + z^2 < \varepsilon\}.$$

We define the unit vector $a(t), t \in [0,\varepsilon]$, by

$$a(t) := [h_1'(t)^2 + h_2'(t)^2 + 1]^{-1/2}\begin{pmatrix} h_1'(t) \\ h_2'(t) \\ 1 \end{pmatrix}$$

and the reflection across Γ by

$$R_\Gamma(t)Q := 2\langle a(t), Q\rangle a(t) - Q$$

for $Q \in \mathbb{R}^3, t \in [0, \varepsilon]$. Similarly, we define reflections across S by

$$R_S(x,y)Q := Q - 2\langle N_S(x,y), Q\rangle N_S(x,y),$$

for all $Q \in \mathbb{R}^3$ and $(x,y) \in B_\varepsilon(0) \subset \mathbb{R}^2$, where

$$N_S(x,y) = [1 + f_x^2(x,y) + f_y^2(x,y)]^{-1/2} \begin{pmatrix} -f_x(x,y) \\ -f_y(x,y) \\ 1 \end{pmatrix}$$

is the unit normal of S at the point $(x, y, f(x,y))$. Identifying the reflections R_Γ and R_S with their respective matrices $R_\Gamma(t)$ and $R_S(x,y)$, we may construct orthogonal matrices $O_\Gamma(t)$ and $O_S(x,y)$ with the properties[1]

$$R_\Gamma(t) = O_\Gamma(t)\,\mathrm{Diag}[-1,-1,1]O_\Gamma^*(t),$$

$$R_S(x,y) = O_S(x,y)\,\mathrm{Diag}[1,1,-1]O_S^*(x,y).$$

We put

$$T_\Gamma(t) := O_\Gamma(0)O_\Gamma^*(t)$$

and

$$T_S(x,y) := O_S(0,0)O_S^*(x,y).$$

Thus we have obtained matrices R_S and T_S which are of class $C^2(B_\varepsilon(0), \mathbb{R}^9)$, $B_\varepsilon(0) \subset \mathbb{R}^2$, while R_Γ and T_Γ are of class $C^{1,\mu}([0,\varepsilon], \mathbb{R}^9)$. If we extend $a(t), t \in [0,\varepsilon]$ by $\tilde{a}(t) = a(-t)$ for $t \in [-\varepsilon, 0]$ and call the extended functions again a, R_Γ and T_Γ, then also $a, R_\Gamma, T_\Gamma \in C^{1,\mu}([-\varepsilon,\varepsilon])$. Now let K_τ denote the cone with vertex 0 and opening angle τ whose axis is given by $x = z \cot \alpha\pi, z \geq 0, y = 0$. We assume that τ is so small that the vertex 0 is the only point of $K_{2\tau} \cap S$ in the ball $\mathcal{K}_\varepsilon(0)$. Next we choose a real valued differentiable function η defined on the punctured ball $\mathcal{K}_\varepsilon(0) \setminus \{0\} = \{0 < x^2 + y^2 + z^2 < \varepsilon\}$ which satisfies

$$\eta(x,y,z) = \begin{cases} 1 & \text{on } K_\tau \cap [\mathcal{K}_\varepsilon(0) \setminus \{0\}], \\ 0 & \text{on } \mathcal{K}_\varepsilon(0) \setminus \{0\} \setminus K_{2\tau}, \end{cases}$$

and

$$|\nabla\eta(x,y,z)| \leq \mathrm{const}[x^2 + y^2 + z^2]^{-1/2} \quad \text{on } \mathcal{K}_\varepsilon \setminus \{0\}.$$

We extend η (noncontinuously) by defining $\eta(0,0,0) = 0$, and denote by $T = T(x,y,z), (x,y,z) \in \mathcal{K}_\varepsilon(0)$, the matrix-valued function

$$T(x,y,z) := \eta(x,y,z)[T_\Gamma(z) - T_S(x,y)] + T_S(x,y).$$

[1] As the symbols t and T are used otherwise, we presently denote the transpose of a matrix A by A^*.

Then T is continuous at zero because $\lim_{(x,y,z)\to 0} T(x,y,z)$ exists and is equal to $\mathrm{Id}_{\mathbb{R}^3}$. In fact, T is even Lipschitz continuous on $\mathcal{K}_\varepsilon(0) \subset \mathbb{R}^3$ because of

$$|T_\Gamma(z) - T_S(x,y)| \le \text{const}[x^2 + y^2 + z^2]^{1/2}$$

and hence $|\nabla T(x,y,z)|$ stays bounded as $(x,y,z) \to 0$. Defining

$$g(w) := T(X(w))X_w(w) \quad \text{for } w \in \overline{B_R^+} \setminus \{0\}$$

we then obtain

Lemma 1. *The function $g(w)$ is of class $C^{0,1}(\overline{B_R^+} \setminus \{0\}, \mathbb{C}^3)$ and has the following properties:*

$$(13) \qquad R_\Gamma(0)g(w) = \overline{g(w)} \quad \textit{for all } w \in I_R^-,$$

and

$$(14) \qquad R_S(0)g(w) = \overline{g(w)} \quad \textit{for all } w \in I_R^+,$$

where $R_S(0) := R_S(0,0)$.

Proof. The Lipschitz continuity of $g(w)$ is an immediate consequence of the Lipschitz continuity of T and of the regularity properties of X. Relation (13) follows similarly as equation (11) in Section 3.3 using the fact that $T(X(w)) = T_\Gamma(z(w))$ if $w \in I_R^-$. To prove (14), we let $w \in I_R^+$; then

$$X_u(w) = (x_u(w), y_u(w), f_x(x,y)x_u(w) + f_y(x,y)y_u(w))$$

and

$$\langle X_u(w), N_S(x(w), y(w))\rangle = 0.$$

From the transversality condition we infer that

$$X_v(w) = \langle X_v(w), N_S(x(w), y(w))\rangle N_S(x(w), y(w)),$$

for all $w \in I_R^+$ whence

$$R_S(x(w), y(w))X_u(w) = X_u(w),$$

and

$$R_S(x(w), y(w))X_v(w) = -X_v(w)$$

or equivalently

$$(15) \qquad R_S(x(w), y(w))X_w(w) = X_{\overline{w}}(w), \quad w \in I_R^+.$$

Now, using (15) and the definition of T, we obtain

$$
\begin{aligned}
\overline{g(w)} &= T(X(w))X_{\overline{w}}(w) = T_S(x(w), y(w))X_{\overline{w}}(w) \\
&= T_S(x,y)R_S(x,y)X_w \\
&= O_S(0)O_S^*(x,y)O_S(x,y)\,\mathrm{Diag}[1,1,-1]O_S^*(x,y)X_w \\
&= O_S(0)\,\mathrm{Diag}[1,1,-1]O_S^*(0)O_S(0)O_S^*(x,y)X_w \\
&= R_S(0)T_S(x,y)X_w = R_S(0)T(X(w))X_w(w) \\
&= R_S(0)g(w),
\end{aligned}
$$

where the argument of X, x, y is always $w \in I_R^+$. □

We now reflect $g(w)$ so as to obtain a function $G(w)$,

$$
G(w) := \begin{cases} g(w) & \text{if } w \in \overline{B_R^+} \setminus \{0\}, \\ R_S(0)\overline{g(\overline{w})} & \text{if } \overline{w} \in B_R^+; \end{cases} \tag{16}
$$

then $G \in C^{0,1}(B_R(0) \setminus \overline{I_R^-}, \mathbb{C}^3)$ and $\lim_{v\to 0^+} G(w) = R_\Gamma(0)R_S(0) \times \lim_{v\to 0^-} G(w)$ for all $w = (u,v)$ with $u \in I_R^-$. Furthermore, G satisfies

$$
|G_{\overline{w}}(w)| \le c|G(w)|^2 \tag{17}
$$

almost everywhere in B_R, and we infer from Proposition 1 in Section 3.2 that

$$
|G(w)| \le c|w|^{\nu-1}, \quad w \in B_R \setminus \{0\}, \tag{18}
$$

with some constant c, where ν denotes the Hölder exponent of X.

Next we are going to smoothen the jump of G on the interval I_R^- by multiplication with a singular matrix function which is related to the eigenvalues of the matrix $R_\Gamma(0)R_S(0)$. It follows easily that

$$
R_\Gamma(0)R_S(0) = \begin{pmatrix} \cos 2\pi\alpha & 0 & -\sin 2\pi\alpha \\ 0 & -1 & 0 \\ \sin 2\pi\alpha & 0 & \cos 2\pi\alpha \end{pmatrix}
$$

and

$$
R_\Gamma(0)R_S(0) = U\ \mathrm{Diag}[e^{i2\pi(\alpha-1)}, e^{-i\pi}, e^{-i2\pi\alpha}]U^*,
$$

where U^* is the unitary matrix

$$
\frac{1}{\sqrt{2}}\begin{pmatrix} i & 0 & -i \\ 0 & \sqrt{2} & 0 \\ 1 & 0 & 1 \end{pmatrix}.
$$

The smoothed function $F(w) = (F_1(w), F_2(w), F_3(w))$ is now defined by

$$
F(w) := \mathrm{Diag}[w^{1-\alpha}, w^{1/2}, w^{\alpha}]U^*G(w), \quad \text{for all } w \in B_R(0) \setminus \{0\}. \tag{19}
$$

Equation (19) is equivalent to

$$X_w(w) = T(X(w))^{-1} U \,\mathrm{Diag}[w^{\alpha-1}, w^{-1/2}, w^{-\alpha}] F(w) \quad \text{for all } w \in B_R(0) \setminus \{0\}. \tag{20}$$

It is easily seen that F is continuous; in particular, we have

$$\lim_{v\to+0} F(u,v) = \lim_{v\to-0} F(u,v) \quad \text{for all } u \in I_R^-.$$

In fact we find

Lemma 2. *The function $F(w)$ is of class $C^{0,1}(B_R(0)) \setminus \{0\}, \mathbb{C}^3)$ and satisfies the relations*

$$\begin{aligned} |F_1(w)| &= O(|w|^{\nu-\alpha}) \\ |F_2(w)| &= O(|w|^{\nu-1/2}) \quad \textit{as } w \to 0 \\ |F_3(w)| &= O(|w|^{\nu-\beta}) \end{aligned} \tag{21}$$

and $\beta = 1 - \alpha$. Furthermore the following differential system holds almost everywhere on $B_R(0)$:

$$\begin{aligned} |F_{1\overline{w}}| &\le c[|w|^{\alpha-1}|F_1|^2 + |w|^{1-3\alpha}|F_3|^2], \\ |F_{2\overline{w}}| &\le c[|w|^{(1/2)-2\beta}|F_1|^2 + |w|^{(1/2)-2\alpha}|F_3|^2], \\ |F_{3\overline{w}}| &\le c[|w|^{\alpha-2\beta}|F_1|^2 + |w|^{-\alpha}|F_3|^2], \end{aligned} \tag{22}$$

where we have dropped the argument w. Moreover, there exist complex-valued functions χ_1, χ_2, χ_3 which are Hölder continuous on $B_R(0)$ such that

$$\begin{aligned} &F_2^2(w)\chi_1(w) + 2F_1(w)F_3(w)\chi_2(w) \\ &\quad = [w^{2\alpha-1}F_1^2(w) + w^{1-2\alpha}F_3^2(w)](1 - \chi_3(w)), \\ &\quad \textit{and} \quad \chi_j(0) = 1 \quad \textit{for } j = 1, 2, 3. \end{aligned} \tag{23}$$

Proof. Relations (21) follow from the definition of F and from (18). The Lipschitz continuity of F on the punctured disk is a consequence of the Lipschitz continuity of G and of the continuity of F at I_R^-. The conformality condition $\langle X_w, X_w \rangle = 0$, the definition of G and the relation $T(0) = \mathrm{Id}$ imply the existence of Hölder continuous functions $a_1(w), a_2(w), a_3(w)$ such that

$$a_1(w)G_1^2(w) + a_2(w)G_2^2(w) + a_3(w)G_3^2(w) = 0 \quad \text{in } B_R(0) \setminus \{0\},$$

and

$$a_1(0) = a_2(0) = a_3(0) = 1.$$

Then (23) follows with

$$\chi_1(w) = a_2(w), \quad \chi_2(w) = \tfrac{1}{2}(a_1(w) + a_3(w)),$$

$$\chi_3(w) = 1 + \tfrac{1}{2}(a_1(w) - a_2(w)).$$

From the definition of G we derive

$$|G(w)|^2 = |w|^{-2\beta}|F_1(w)|^2 + |w|^{-1}|F_2(w)|^2 + |w|^{-2\alpha}|F_3(w)|^2,$$

and inequality (17) together with (19) yields

$$\begin{aligned} |F_{1\overline{w}}(w)| &\le c|w|^{1-\alpha}|G(w)|^2, \\ |F_{2\overline{w}}(w)| &\le c|w|^{1/2}|G(w)|^2, \\ |F_{3\overline{w}}(w)| &\le c|w|^{\alpha}|G(w)|^2, \end{aligned}$$

whence

$$\begin{aligned} |F_{1\overline{w}}(w)| &\le c[|w|^{-\beta}|F_1|^2 + |w|^{-\alpha}|F_2|^2 + |w|^{\beta-2\alpha}|F_3|^2], \\ |F_{2\overline{w}}(w)| &\le c[|w|^{(1/2)-2\beta}|F_1|^2 + |w|^{-1/2}|F_2|^2 + |w|^{(1/2)-2\alpha}|F_3|^2], \\ |F_{3\overline{w}}(w)| &\le c[|w|^{\alpha-2\beta}|F_1|^2 + |w|^{-\beta}|F_2|^2 + |w|^{-\alpha}|F_3|^2]. \end{aligned}$$

On the other hand, we deduce from (23) the inequality

$$\begin{aligned} |F_2|^2 &\le c[|F_1||F_3| + |w|^{2\alpha-1}|F_1|^2 + |w|^{1-2\alpha}|F_3|^2] \\ &\le c[|w|^{2\alpha-1}|F_1|^2 + |w|^{1-2\alpha}|F_3|^2]. \end{aligned}$$

These inequalities imply system (22). □

Relations (21_1), (21_3) and (22_1), (22_3) are equivalent to (33) and (34) respectively stated in Section 3.1. Hence we infer from Theorem 3 in Section 3.1, similarly as in Lemma 3 of Section 3.3, the following

Lemma 3. *There exists a nonnegative integer m such that the functions $f_j^m(w) := w^{-m}F_j(w), j = 1, 2, 3$ either satisfy*
(i) $f_1^m(0) \neq 0, f_1^m \in C^{0,\mu}(B_R, \mathbb{C})$ for all $\mu < \min(1, m + \alpha)$, and

$$\begin{aligned} |f_{1\overline{w}}^m(w)| &= O(|w|^{m-\beta}) \\ |f_{2\overline{w}}^m(w)| &= O(|w|^{m+2\alpha-3/2}) \quad \text{as } w \to 0, \\ |f_{3\overline{w}}^m(w)| &= O(|w|^{m+3\alpha-2}) \end{aligned}$$

or
(ii) $f_1^m(0) = 0$ and $f_3^m(0) \neq 0, f_3^m \in C^{0,\mu}(B_R, \mathbb{C})$ for every $\mu < \min(1, m+\beta)$, and

$$\begin{aligned} |f_{1\overline{w}}^m(w)| &= O(|w|^{m+\beta-2\alpha}) \\ |f_{2\overline{w}}^m(w)| &= O(|w|^{m+1/2-2\alpha}) \quad \text{as } w \to 0, \\ |f_{3\overline{w}}^m(w)| &= O(|w|^{m-\alpha}) \end{aligned}$$

almost everywhere on B_R.

If $m \ge 1$, then in both cases

$$[f_2^m(0)]^2 + 2f_1^m(0)f_3^m(0) = 0. \tag{24}$$

Proof. This can be proved like the corresponding result, Lemma 3, in Section 3.3. □

Now we can continue with the *proof of Theorem* 1. Assume that case (i) of Lemma 3 holds true; then we put

$$\psi(w) := \frac{1}{\sqrt{2}} f_1^m(w) \begin{pmatrix} i \\ 0 \\ 1 \end{pmatrix} + \frac{1}{\sqrt{2}} w^{1-2\alpha} f_3^m(w) \begin{pmatrix} -i \\ 0 \\ 1 \end{pmatrix} + w^{1/2-\alpha} f_2^m(w) \begin{pmatrix} 0 \\ 1 \\ 0 \end{pmatrix}$$

and

$$\Phi(w) := T^{-1}(X(w))\psi(w), \quad w \in \overline{B_R^+}.$$

Now we claim that ψ is Hölder continuous in $\overline{B_R^+}$. On account of Lemma 3, (i) we first have $f_1^m \in C^{0,\mu}(B_R, \mathbb{C})$ for all $\mu < \min\{1, m+\alpha\}$. Then we distinguish two cases:

1.) $m \geq 1$. Then the functions $w^{1-2\alpha} f_3^m(w)$ and $w^{1/2-\alpha} f_2^m(w)$ are Hölder continuous. Indeed, we have for $w \to 0$:

$$(+) \quad \begin{aligned} &|f_3^m(w)| = O(1), \quad \text{by construction;} \\ &|f_2^m(w)| \leq c\{|w|^{\alpha-1/2}|f_1^m(w)| + |w|^{1/2-\alpha}|f_3(w)|\} = O(|w|^{\alpha-1/2}). \end{aligned}$$

Here we have employed (23) and $\alpha \leq 1$. On account of Lemma 3, (i), we see that Lemma 6 in Section 3.1 yields the Hölder continuity of f_2^m and f_3^m, and therefore of ψ, in $\overline{B_R^+}$.

2.) $m = 0$. Now we use (21) instead of (+). By Lemma 6 in Section 3.1 and Lemma 3, (i), we see that $f_2^0 = F_2$ is Hölder continuous for $\alpha > 1/4$, and so is $f_3^0 = F_3$ for $\alpha > 1/3$.

If $\alpha \leq 1/3$, we consider the function $wF_3(w)$, which satisfies

$$|wF_3(w)| = O(|w|^{\nu+\alpha}), \quad |[wF_3(w)]_{\overline{w}}| = O(|w|^{3\alpha-1}) \quad \text{for } w \to 0.$$

Hence $wF_3(w)$ is Hölder continuous for any exponent $< 3\alpha$, and it vanishes for $w = 0$.

For arbitrary $w_1, w_2 \in \overline{B_R^+} \setminus \{0\}$ and $0 < \epsilon \ll 1$ we estimate the expression $|w_1^{1-2\alpha} F_3(w_1) - w_2^{1-2\alpha} F_3(w_2)|$ as follows, using w.l.o.g. that $|w_1| \leq |w_2|$ whence $|w_2| \geq (1/2)|w_1 - w_2|$:

$$\begin{aligned} &|w_1^{1-2\alpha} F_3(w_1) - w_2^{1-2\alpha} F_3(w_2)| \\ &\quad \leq |w_1^{-2\alpha} - w_2^{-2\alpha}||w_1 F_3(w_1)| + |w_2|^{-2\alpha}|w_1 F_3(w_1) - w_2 F_3(w_2)| \\ &\quad \leq c\{|w_1|^{-2\alpha}|w_2|^{-2\alpha}|w_1 - w_2|^{2\alpha}|w_1|^{3\alpha-\epsilon} + |w_2|^{-2\alpha}|w_1 - w_2|^{3\alpha-\epsilon}\} \\ &\quad \leq c|w_1 - w_2|^{\alpha-\epsilon}. \end{aligned}$$

Thus $w^{1-2\alpha}F_3(w)$ is Hölder continuous in $\overline{B}_R^+$, and similarly one shows the Hölder continuity of $w^{1/2-\alpha}F_2(w)$. Since T^{-1} is Lipschitz continuous, and $X(w)$ is Hölder continuous in $\overline{B}_R^+$, we see that $T^{-1}(X(w))$ is Hölder continuous in $\overline{B}_R^+$. This yields the Hölder continuity of $\Phi(w) = T^{-1}(X(w))\psi(w)$. On the other hand, it follows from definition (20) that

$$X_w(w) = w^{\alpha-1+m}\Phi(w), \quad w \in \overline{B_R^+} \setminus \{0\}. \tag{25}$$

Because of $T(0) = \mathrm{Id}$, we obtain for $\alpha < \frac{1}{2}$ the relations

$$\begin{aligned} \Phi_1(0) = \frac{1}{\sqrt{2}} i f_1^m(0), \quad \Phi_3(0) = \frac{1}{\sqrt{2}} f_1^m(0) \neq 0 \\ |\Phi_2(w)| = O(|w|^{\mu'}) \quad \text{for some } \mu' > 0. \end{aligned} \tag{26}$$

Then (25) yields

$$|y_w(w)| = O(|w|^{\lambda'}) \quad \text{for some } \lambda' > \alpha - 1 + m,$$

and we also have $\Psi_2(0) = 0$, which is sufficient for the proof of the theorem. If the second alternative of Lemma 3 holds true, we consider instead of ψ the function $\tilde{\psi}$ given by

$$\tilde{\psi}(w) := \frac{1}{\sqrt{2}} f_3^m(w) \begin{pmatrix} -i \\ 0 \\ 1 \end{pmatrix} + \frac{1}{\sqrt{2}} w^{2\alpha-1} f_1^m(w) \begin{pmatrix} i \\ 0 \\ 1 \end{pmatrix} + w^{\alpha-1/2} f_2^m(w) \begin{pmatrix} 0 \\ 1 \\ 0 \end{pmatrix}$$

and

$$\tilde{\Phi}(w) := T^{-1}(X(w))\tilde{\psi}(w).$$

Then $\tilde{\Phi}(w)$ is Hölder continuous and we have

$$X_w(w) = w^{-\alpha+m}\tilde{\Phi}(w), \quad w \in \overline{B_R^+}(0) \setminus \{0\}, \tag{27}$$

which together with (25) proves (3) of Theorem 1. Also we find for $\alpha < \frac{1}{2}$ that

$$\Phi_1(0) = \frac{-i}{\sqrt{2}} f_3^m(0) \neq 0, \quad \Phi_3(0) = \frac{1}{\sqrt{2}} f_3^m(0) \neq 0, \quad \Phi_2(0) = 0,$$

since $(w^{2\alpha-1}f_1^m)(0) = 0$ and $(w^{\alpha-1/2}f_2^m)(0) = 0$. The last relation follows because of $f_1^m(0) = 0$, relation (24) if $m \geq 1$ or (23) for $m = 0$, Lemma 3 (ii) and Lemma 6 of Section 3.1.

If $\alpha = \frac{1}{2}$, we obtain relation (3) with

$$\Phi(w) = T^{-1}(X(w)) \left[\frac{1}{\sqrt{2}} f_3^m(w) \begin{pmatrix} -i \\ 0 \\ 1 \end{pmatrix} + \frac{1}{\sqrt{2}} f_1^m(w) \begin{pmatrix} i \\ 0 \\ 1 \end{pmatrix} + f_2^m(w) \begin{pmatrix} 0 \\ 1 \\ 0 \end{pmatrix} \right]$$

and

$$\begin{aligned}\Phi_1^2(0)+\Phi_2^2(0)+\Phi_3^2(0) &= -\tfrac{1}{2}[f_1^2(0)-2f_1(0)f_3(0)+f_3^2(0)] \\ &\quad +f_2^2(0)+\tfrac{1}{2}[f_1^2(0)+2f_1(0)f_3(0)+f_3^2(0)] \\ &= 2f_1(0)f_3(0)+f_2^2(0)=0\end{aligned}$$

by (23) and (24).

Then the unit normal of $X(w)$ given by $N(w)=\frac{X_u\wedge X_v}{|X_u\wedge X_v|}(w)$ satisfies by virtue of (25) or (27) and because of

$$(X_u\wedge X_v)(w)=2(\mathrm{Im}(y_w z_{\overline{w}}),-\mathrm{Im}(x_w z_{\overline{w}}),\mathrm{Im}(x_w y_{\overline{w}}))$$

the relation

$$\lim_{w\to 0} N(w)=2[|\Phi_1(0)|^2+|\Phi_2(0)|^2+|\Phi_3(0)|^2]^{-1}\begin{pmatrix}\mathrm{Im}(\Phi_2(0)\overline{\Phi_3}(0))\\ \mathrm{Im}(\Phi_3(0)\overline{\Phi_1}(0))\\ \mathrm{Im}(\Phi_1(0)\overline{\Phi_2}(0))\end{pmatrix}.$$

But now relation (15) implies $R_S(0)\Phi(0)=\overline{\Phi(0)}$, and this means that

$$\mathrm{Im}\,\Phi_1(0)=0,\quad \mathrm{Im}\,\Phi_2(0)=0,\quad\text{and}\quad \mathrm{Re}\,\Phi_3(0)=0.$$

Also, if $\alpha<\frac{1}{2}$, then $\Phi_2(0)=0$, and we arrive at

$$N_1(0)=0,\quad N_3(0)=0,\quad N_2(0)=\pm 1,$$

whereas, if $\alpha=\frac{1}{2}$, we conclude that

$$N_j(0)=\pm\,\mathrm{Re}\,\Phi_j(0)[(\mathrm{Re}\,\Phi_1(0))^2+(\mathrm{Re}\,\Phi_2(0))^2]^{-1/2},\quad j=1,2,$$

and

$$N_3(0)=0.$$

Finally, we obtain for the tangent vector $t(u)=\frac{X_u(u,0)}{|X_u(u,0)|}$, $u>0$, the asymptotic behaviour

$$\lim_{u\to 0+} t(u)=[(\mathrm{Re}\,\Phi_1(0))^2+(\mathrm{Re}\,\Phi_2(0))^2]^{-1/2}\begin{pmatrix}\mathrm{Re}\,\Phi_1(0)\\ \mathrm{Re}\,\Phi_2(0)\\ 0\end{pmatrix},$$

which proves the relations (8) and (9).

If S is a plane and Γ is a straight line, then $T=\mathrm{Id}_{\mathbb{R}^3}$ and $g=X_w$. Hence G is holomorphic on $B_R\setminus\{0\}$ and F is holomorphic on B_R. Finally (10) and (11) follows from (20) if we take

$$H_1:=\frac{1}{\sqrt{2}}F_1,\quad H_2:=F_2,\quad H_3:=\frac{1}{\sqrt{2}}F_3,$$

and Theorem 1 is proved. □

3.5 Scholia

3.5.1 References

The basic idea of this chapter, the Hartman–Wintner method, was described and developed in the paper [1] of Hartman and Wintner in 1953. Its relevance for the theory of nonlinear elliptic systems with two independent variables was emphasized by E. Heinz. In particular, he discovered the use of this method for obtaining asymptotic expansions of minimal surfaces at boundary branch points, and of H-surfaces at branch points in the interior and at the boundary.

The results of Sections 3.2–3.4 concerning minimal surfaces with non-smooth boundaries are due to Dziuk (cf. his papers [1–4]). His work is based on methods by Vekua [1,2], Heinz [5], and Jäger [1–3].

Earlier results on the behaviour of minimal surfaces at a corner were derived by H.A. Schwarz [3] and Beeson [1]. The boundary behaviour of conformal mappings at corners was first treated by Lichtenstein, and then by Warschawski [4]. The continuity of minimal surfaces in Riemannian manifolds at piecewise smooth boundaries was investigated by Jost [12].

The proofs in the paper [1] of Marx based on joint work of Marx and Shiffman concerning minimal surfaces with polygonal boundaries are somewhat sketchy and contain several large gaps. Heinz [19–24] was able to fill these gaps and to develop an interesting theory of quasi-minimal surfaces bounded by polygons, thereby generalizing classical work of Fuchs and Schlesinger on linear differential equations in complex domains that have singularities (see Schlesinger [1]). A survey of Heinz's work can be found in the Scholia of Chapter 6 of Vol. 1.

In this context we also mention the work of Sauvigny [3–6]. The papers of Garnier are also essentially concerned with minimal surfaces having polygonal boundaries, but apparently these results were rarely studied in detail and did not have much influence on the further progress. This might be both unjustified and unfortunate, see the recent thesis by L. Desideri.

3.5.2 Hölder Continuity at Intersection Points

In Theorem 1 of Section 3.4 we have derived asymptotic expansions for $X_w(w)$ and $N(w)$ at the points $w_0 = \pm 1$ if $X : B \to \mathbb{R}^3$ is a minimal surface of class $\mathcal{C}(\Gamma, S)$ with the parameter domain $B = \{w = u + iv : |w| \le 1, v > 0\}$ that is bounded by $I = \{(u, 0) : |u| < 1\}$ and $C = \{w : |w| = 1, v \ge 0\}$, and $w_0 = \pm 1$ are mapped onto the two points P_1, P_2 where the arc Γ meets the surface S. The basic assumption (cf. Assumption A) was that X is Hölder continuous on $\overline{B}$. Recently, F. Müller [4] has proved that Hölder continuity of X on $\overline{B}$ follows from the much weaker assumption that X merely be continuous on $\overline{B}$. His reasoning even applies to continuous solutions X of

$$(1) \qquad |\Delta X| \le a |\nabla X|^2,$$

satisfying also

$$(2) \qquad |X_u|^2 = |X_v|^2, \quad \langle X_u \cdot X_v \rangle = 0,$$

i.e. to H-surfaces with $\sup |H| \le \text{const}$.

Let us choose the corner $w_0 = 1$ of B, and consider the 3-gon $\Omega_\delta := B \cap B_\delta(1)$ as well as the arcs $I_\delta := I \cap \partial\Omega_\delta$ and $C_\delta := C \cap \partial\Omega_\delta$. We assume that both Γ and S are of class C^3, and that Γ meets S in P_1, P_2 only. We fix $P := P_2$ which is assumed to correspond to the corner $w_0 = 1$, i.e. $X(1) = P$. Then F. Müller's result reads as follows:

Theorem 1. *Suppose that*

$$X \in C^0(\overline{\Omega}_\delta, \mathbb{R}^3) \cap H_2^1(\Omega_\delta, \mathbb{R}^3) \cap C^2(\overline{\Omega}_\delta \setminus \{1\}, \mathbb{R}^3)$$

satisfies (1) *and* (2) *in* Ω_δ *as well as*

$$(3) \qquad X(w) \in \Gamma \quad \text{for } w \in C_\delta, \quad X(1) = P,$$

$$(4) \qquad X(w) \in S \quad \text{and} \quad X_v(w) \perp T_{X(w)} S \quad \text{for } w \in I_\delta.$$

Then we obtain $X \in C^{0,\mu}(\overline{\Omega}_{\delta'})$ *for some* $\mu \in (0,1)$ *and some* $\delta' \in (0, \delta)$.

Sketch of the Proof. 1. Let us introduce local coordinates $y = (y^1, y^2, y^3)$ about P in the same way as in Section 2.7 such that 0 corresponds to P. Suppose that x and y are related by a C^2-diffeomorphism $y \mapsto x = h(y)$ from the ball $\mathcal{K}_r(0) := \{y \in \mathbb{R}^3 : |y| < r\}$ onto a neighbourhood U of P such that $h^{-1}(S \cap U) = \mathcal{K}_r(0) \cap \{y^3 = 0\} = B_r(0) \times \{0\}$ and

$$(5) \quad g_{jk}(y^1, y^2, 0) = \operatorname{diag}(\mathcal{E}(y^1, y^2), \mathcal{E}(y^1, y^2), 1) \quad \text{for } (y^1, y^2) \in B_r(0),$$

as well as $g_{13} = g_{31} = g_{23} = g_{32} = 0$ and $g_{33} = 1$ in $\mathcal{K}_r(0)$,

$$(6) \qquad m|\xi|^2 \le g_{jk}(y)\xi^j\xi^k \le m^{-1}|\xi|^2 \quad \text{for } y \in \mathcal{K}_r(0), \ \xi \in \mathbb{R}^3,$$

$$(7) \qquad \left| \frac{\partial g_{jk}}{\partial y^\ell}(y) \right| \le M \quad \text{for } y \in \mathcal{K}_r(0).$$

2. Then there is an $\epsilon \in [0, \delta]$ such that $X(\overline{\Omega}_\epsilon) \subset U$. We may assume that $\epsilon = \delta$. Then $Y := h^{-1}(X)$ lies in the same class as X and satisfies

$$(8) \qquad \begin{aligned} &|\Delta Y| \le b|\nabla Y|^2 \quad \text{in } \Omega_\epsilon \text{ for some } b \in \mathbb{R}, b < 0, \\ &g_{jk}(y) y_w^j y_w^k = 0 \quad \text{in } \Omega_\epsilon, \\ &y(w) \in \Gamma^* := h^{-1}(\Gamma \cap U) \quad \text{for } w \in C_\epsilon, \\ &y_v^1(w) = 0, \quad y_v^2(w) = 0, \quad y^3(w) = 0 \quad \text{for } w \in I_\epsilon. \end{aligned}$$

We can assume that $\Gamma^* \setminus \{0\} \subset \mathcal{H}^+ := \{y^3 > 0\}$. Set

$$\tilde{Y}(w) = (\tilde{y}^1(w), \tilde{y}^2(w), \tilde{y}^3(w)) := \begin{cases} Y(w) & \\ & \text{for} \quad \begin{matrix} w \in \overline{\Omega}_\epsilon, \\ \\ w \in \overline{\Omega}^*_\epsilon, \end{matrix} \\ (y^1(w), y^2(w), -y^3(w)) & \end{cases}$$

where $\Omega^*_\epsilon := \{w \in \mathbb{C} : \overline{w} \in \Omega_\epsilon\}$.

Let $\tau : B \to \tilde{\Omega}_\epsilon := \Omega_\epsilon \cup I_\epsilon \cup \Omega^*_\epsilon$ be a conformal mapping of the unit disk B onto $\tilde{\Omega}_\epsilon$, and set $Z := \tilde{Y} \circ \tau$ and

$$\gamma_{jk} := \tilde{g}_{jk} \circ \tau \quad \text{with } \tilde{g}_{jk}(w) := \begin{cases} g_{jk}(Y(w)) & w \in \overline{\Omega}_\epsilon, \\ & \text{for} \\ g_{jk}(Y(\overline{w})) & w \in \overline{\Omega}^*_\epsilon. \end{cases}$$

Furthermore, let

$$\Gamma^+ := \Gamma^*, \quad \Gamma^- := \{(z^1, z^2, z^3) \in \mathbb{R}^3 : (z^1, z^2, -z^3) \in \Gamma^+\}.$$

Then for some $\rho \in (0,1)$, $S_\rho(0) := B \cap B_\rho(0)$, and $I^+_\rho := [0, \rho)$, $I^-_\rho := (-\rho, 0]$ and a proper choice of τ we obtain for $Z|_{S_\rho(0)}$, which is again denoted by Z, the following relations, by employing the special form of the g_{jk}:

$$|\Delta Z| \le b|\nabla Z|^2 \quad \text{in } S_\rho(0),$$

(9) $$\gamma_{jk} z^j_w z^k_w = 0 \quad \text{in } S_\rho(0),$$

$$Z(w) \in \Gamma^+ \quad \text{for } w \in I^+_\rho, \quad Z(w) \in \Gamma^- \quad \text{for } w \in I^-_\rho.$$

By a suitable change of the z-coordinates we can arrange for

$$\Gamma^\pm = \{(z^1, z^2, z^3) \in \mathbb{R}^3 : z^j = h^j_\pm(z^1), 0 \le \pm z^3 \le \epsilon_0, j = 1, 2\}$$

with some $\epsilon_0 > 0$ and $h^j_- \in C^2([-\epsilon_0, 0]), h^j_+ \in C^2([0, \epsilon_0])$,

(10) $$(h^1_\pm)'(0) = \pm\text{cotg}\left(\frac{\alpha\pi}{2}\right), \quad (h^2_\pm)'(0) = 0, \quad \alpha \in [0, 1].$$

Set

$$h^j(t) := \left\{ \begin{matrix} h^j_-(t) & & -\epsilon_0 \le t_0 \le 0 \\ & \text{for} & \\ h^j_+(t) & & 0 \le t_0 \le \epsilon_0 \end{matrix} \right\}, \quad j = 1, 2.$$

For $0 < \rho \ll 1$ we define $\zeta = (\zeta^1, \zeta^2)$ by

(11) $$\zeta^j = z^j - h^j(z^3), \quad j = 1, 2.$$

Then $\zeta \in C^{0,1}(\overline{S}_\rho(0) \setminus \{0\}, \mathbb{R}^2) \cap C^0(\overline{S}_\rho(0), \mathbb{R}^2) \cap H^1_2(S_\rho(0), \mathbb{R}^2)$ satisfies

$$\int_{S_\rho(0)} \sum_{j=1}^{2} \nabla\zeta^j \nabla\varphi^j \, du\, dv = \int_{S_\rho(0)} \sum_{j=1}^{2} [g^j \nabla\varphi^j + f^j \varphi^j] \, du\, dv \tag{12}$$
$$\text{for all } \varphi = (\varphi^1, \varphi^2) \in C_c^\infty(S_\rho(0), \mathbb{R}^2),$$

where we have set for $j = 1, 2$:

$$f^j := -\Delta z^j \in L_1(S_\rho(0)), \quad g^j := -(h^j)'(z^2)\nabla z^3 \in L_2(S_\rho(0), \mathbb{R}^2). \tag{13}$$

Claim. *For $0 < \rho \ll 1$ we have*

$$|g^1 \nabla\zeta^1| \le a_1 |\nabla\zeta^1|^2 + b_1 |\nabla\zeta^2|^2, \tag{14}$$

$$|g^1| \le a_2 |\nabla\zeta^1| + b_2 |\nabla\zeta^2|, \tag{15}$$

$$|g^2| \le a_3(\rho) |\nabla\zeta|, \tag{16}$$

$$|f^j| \le b_3 |\nabla\zeta|^2 \quad \textit{for } j = 1, 2, \tag{17}$$

with positive constants $a_1, a_2 \in [0, 1), b_1, b_2, b_3$, and a function $a_3(t) \to +0$ as $t \to +0$.

Suppose that the claim is proved. Using the boundary condition

$$\zeta(w) = 0 \quad \text{for } w \in I_\rho := \{w = u \in \mathbb{R} : |u| \le \rho\} \tag{18}$$

we extend ζ to a continuous function $\tilde{\zeta}$ on $\overline{B}_\rho(0)$by setting

$$\tilde{\zeta}(w) := \begin{cases} \zeta(w) & \text{for } w \in S_\rho(0), \\ -\zeta(\overline{w}) & \text{for } \overline{w} \in S_\rho(0). \end{cases}$$

Furthermore, we have $\tilde{\zeta} \in H_2^1(S_\rho(0), \mathbb{R}^2)$. In addition, we reflect $g_1^j = -(h^j)'(z^3)z_u^3$ and f^j in an odd way and $g_2^j = -(h^j)'(z^3)z_v^3$ evenly across I_ρ, obtaining $\tilde{g}_1^j, \tilde{f}^j, \tilde{g}_2^j$. Then it follows

$$\int_{B_\rho(0)} \sum_{j=1}^{2} \nabla\tilde{\zeta}^j \nabla\varphi^j \, du\, dv = \int_{B_\rho(0)} \sum_{j=1}^{2} [\tilde{g}^j \nabla\varphi^j + \tilde{f}^j \varphi^j] \, du\, dv \tag{19}$$
$$\text{for all } \varphi \in \overset{\circ}{H}{}_2^1 (B_\rho(0), \mathbb{R}^2) \cap L_\infty(B_\rho(0), \mathbb{R}^2).$$

One checks that $\tilde{f}^j$ and $\tilde{g}^j$ satisfy growth conditions analogous to (14)–(17) where the ζ^j are to be replaced by $\tilde{\zeta}^j$, whereas $a_1, a_2, a_3(\rho), b_1, b_2, b_3$ remain the same. Now one can apply a procedure due to Dziuk [1] (cf. the proof of Satz 1 in [1]) to show that $\tilde{\zeta}$ satisfies a "Dirichlet growth condition" on some disk $\overline{B}_{\rho'}(0)$ with $0 < \rho' \ll 1$, and the same holds for ζ on $\overline{S}_{\rho'}(0)$. From (11) one infers that also z^1 and z^2 satisfy such a condition on $\overline{S}_{\rho'}(0)$, using also (15), (16), and

$$|z_w^j| \overset{(11)}{\leq} |\zeta_w^j| + |(h^j)' z_w^3| \overset{(15),(16)}{\leq} c|\zeta_w|, \quad j = 1, 2,$$

and

$$\gamma_{jk} z_w^j z_w^k = 0 \tag{20}$$

implies that

$$|\nabla z^3|^2 \leq \text{const}(|\nabla z^1|^2 + |\nabla z^2|^2) \quad \text{on } \overline{S}_{\rho'}(0) \quad \text{for } 0 < \rho' \ll 1.$$

Consequently, $Z = (z^1, z^2, z^3)$ satisfies a Dirichlet growth condition on $\overline{S}_{\rho'}(0)$, and therefore Z is Hölder continuous on $\overline{S}_{\rho'}(0)$. Since $\tilde{Y} = Z \circ \tau^{-1}$, it follows that $\tilde{Y}$ is Hölder continuous on the closure of $\tilde{\Omega}_{\epsilon'}$ for $0 < \epsilon' \ll 1$, and Y is Hölder continuous on $\overline{\Omega}_{\epsilon'}$ for $0 < \epsilon' \ll 1$. Since $X = h(Y)$, we finally conclude that $X \in C^{0,\mu}(\overline{\Omega}_{\delta'})$ for some $\mu \in (0,1)$ and some $\delta' \in (0, \delta)$.

It remains to prove the Claim. We begin with (15). From (20) and the special structure of the g_{jk}, and therefore of the γ_{jk}, it follows that

$$-(z_w^3)^2 - \gamma_{11}(z_w^1)^2 = 2\gamma_{12} z_w^1 z_w^2 + \gamma_{22}(z_w^2)^2 \quad \text{in } S_\rho(0).$$

Inserting $z_w^1 = \zeta_w^1 + (h^1)'(z^3) z_w^3$ into the left-hand side we find

$$-\gamma(z_w^3 - \xi^1 \zeta_w^1)(z_w^3 - \xi_2 \zeta_w^1) = 2\gamma_{12} z_w^1 z_w^2 + \gamma_{22}(z_w^2)^2 \quad \text{in } S_\rho(0) \tag{21}$$

with

$$\xi_{1,2} := -\gamma^{-1}[\gamma_{11}(h^1)'(z^3) \pm i\sqrt{\gamma_{11}}],$$

$$\gamma := 1 + \gamma_{11}[(h^1)'(z^3)]^2.$$

We have

$$|\xi_1| = |\xi_2| = \left\{ \frac{\tilde{\gamma_{11}}}{1 + \tilde{\gamma}_{11}[(h^1)'(z^3)]^2} \right\}^{\frac{1}{2}} \quad \text{in } S_\rho(0). \tag{22}$$

If $|z_w^3| \leq |\xi_1||\zeta_w^1|$, we find

$$|g^1| \leq 2|(h^1)'(z^3)||\xi_1||\zeta_w^1| \leq a_2|\nabla \zeta^1|,$$

and this is (15) with $a_2 < 1$ and $b_2 = 0$.

Otherwise we infer from (21)

$$\gamma(|z_w^3| - |\xi_1||\zeta_w^1|) \leq |2\gamma_{12} z_w^1 z_w^2 + \gamma_{22}(z_w^2)^2|;$$

thus,

$$\sqrt{\gamma}|z_w^3| \le \left\{\sqrt{|\gamma_{12}|}|(h^1)'(z^3)| + \sqrt{\gamma_{22}+|\gamma_{12}|}(h^2)'(z^3)\right\}|z_w^3|$$
$$+\left\{\sqrt{|\gamma_{12}|}+\sqrt{\gamma}|\xi_1|\right\}|\zeta_w^1| + \sqrt{\gamma_{22}+|\gamma_{12}|}|\zeta_w^2|.$$

Furthermore there is a function $c(t)$ with $c(t) \to +0$ as $t \to 0$ such that

$$|\tilde{\gamma}_{12}| + |(h^2)'(z^3)| \le c(\rho) \quad \text{on } S_\rho(0), \tag{23}$$

due to (5) and (10). Thus we have for $0 < \rho \ll 1$ that

$$|z_w^3| \le [1-\tilde{c}(\rho)]^{-1}\left\{\left[\left(\frac{|\gamma_{12}|}{\gamma}\right)^{\frac{1}{2}} + |\xi_1|\right]|\zeta_w^1| + \left[\left(\frac{\gamma_{22}+|\gamma_{12}|}{\gamma}\right)^{\frac{1}{2}}|\zeta_w^2|\right]\right\} \quad \text{in } S_\rho(0) \tag{24}$$

with $\tilde{c}(\rho) \to +0$ as $\rho \to 0$.

Using (22) and again (23), we obtain for $0 < \rho \ll 1$ that

$$|g^1| \le 2|(h^1)'(z^3)||z_w^3| \le a_2|\nabla\zeta^1| + b_2|\nabla\zeta^2| \quad \text{in } S_\rho(0)$$

with $a_2 \in (0,1)$ and $b_2 > 0$, as claimed in (15).

The estimate (14) follows easily from (15), and (16) and (17) are derived from (10) and (24). Thus we have verified the "Dziuk estimates" of the Claim, and the proof of the theorem is complete. □

3.5.3

We also note that Dziuk [1] has proved Hölder continuity of a minimal surface $X \in \mathcal{C}(\Gamma)$ at a corner of the boundary contour Γ, assuming only $X \in C^0(\overline{B}, \mathbb{R}^3)$. This is relevant for Theorem 1 in Section 3.3 where we have assumed that $X \in C^{0,\mu}(\overline{B}_\delta(0), \mathbb{R}^3)$, which in Chapter 2 was only proved for minimizers.

Part II

Geometric Properties of Minimal Surfaces and H-Surfaces

Chapter 4

Enclosure and Existence Theorems for Minimal Surfaces and H-Surfaces. Isoperimetric Inequalities

In this chapter we shall discuss certain quantitative geometric properties of minimal surfaces and surfaces of prescribed mean curvature.

We begin by deriving *enclosure theorems*. Such results give statements about the confinement of minimal surfaces to certain "enclosing sets" on the basis that one knows something about the position of their boundaries. For example, any minimal surface is contained in the convex hull of its boundary values. All of our results will in one way or another be founded on some version of the maximum principle for subharmonic functions.

Closely related to these theorems are *nonexistence theorems for multiply connected surfaces*. Everyone who has played with wires and soap films will have noticed that a soap film catenoid between two coaxial parallel circles will be torn up if one moves the two wires too far apart. Section 4.1 supplies a very simple proof of the corresponding mathematical assertion which again relies on the maximum principle for subharmonic functions.

A comparison principle for solutions of the equation of prescribed mean curvature is employed in the study of points where two (parametric) surfaces of continuous mean curvature H ("H-surfaces" for short) touch without crossing each other. The resulting *touching point theorem* (Section 4.2) implies further enclosure and nonexistence theorems. Since the proofs are nearly identical for minimal surfaces (where $H \equiv 0$) and for surfaces of continuous mean curvature H, we shall deal with the latter.

We have chosen to extend these principles to submanifolds of arbitrary dimension and, if possible, of arbitrary codimension as well (Section 4.3). In Section 4.4 we discuss a *"barrier principle"* for submanifolds of $\mathbb{R}^{n+k}$ with bounded mean curvature and arbitrary codimension k. Furthermore, a similar argument is used to prove a *"geometric inclusion principle"* for strong (possibly branched) subsolutions of a variational inequality, which is later used (Section 4.7) in a crucial way to solve the Plateau problem for H-surfaces in Euclidean space. Additionally we present some existence theorems for sur-

U. Dierkes, S. Hildebrandt, A.J. Tromba, *Regularity of Minimal Surfaces*,
Grundlehren der mathematischen Wissenschaften 340,
DOI 10.1007/978-3-642-11700-8_4, © Springer-Verlag Berlin Heidelberg 2010

faces of prescribed mean curvature with a given boundary in a Riemannian manifold (Section 4.8).

The enclosure theorems of this chapter also serve to find conditions ensuring that the solutions of the free (Chapter 1) or semifree (Chapter 4 of Vol. 1) variational problems for minimal surfaces remain on one side of their supporting surface. Only such solutions describe the soap films produced in experiments because these can evidently never pass through a supporting surface made of e.g. plexiglas, whereas in general we cannot exclude this phenomenon for the solutions of the corresponding variational problems (unless we consider problems with obstructions; see Vol. 1, Section 4.10, no. 5).

Moreover, if the minimal surface remains on one side of the supporting surface, then there are no branch points on the free boundary, as follows from the asymptotic expansions in Chapter 3 (see also Section 2.10). This will be of importance for some of the trace estimates proved in Section 4.6.

The two Sections 4.5 and 4.6 deal with the relationship between the area of a minimal surface and the length of its boundary. In particular, *isoperimetric inequalities* bound the area in terms of the length of the boundary and, possibly, of other geometric quantities. It is a surprising fact that minimal surfaces satisfy the same isoperimetric inequalities as a planar domain Ω for which the relation

$$4\pi A \leq L^2$$

holds true, A being the area of Ω and L the length of $\partial\Omega$.

In Section 4.6 we shall derive upper and lower bounds for the length $L(\Sigma)$ of the free trace Σ of a stationary minimal surface X in a semifree or a free boundary configuration $\langle \Gamma, S\rangle$ or $\langle S\rangle$ respectively. These bounds will depend on geometric quantities such as the area of X, the length of the fixed part $\Gamma$of its boundary, and of parameters bounding the curvature of the supporting surface S. We shall close this section by discussing analogous questions for solutions of a *partition problem* which turn out to be stationary surfaces X of constant mean curvature with a free boundary on the surface S of a body U which is partitioned by X.

4.1 Applications of the Maximum Principle and Nonexistence of Multiply Connected Minimal Surfaces with Prescribed Boundaries

Our first result is the prototype of an *enclosure theorem*; it will be obtained by a straight-forward application of the maximum principle for harmonic functions.

Theorem 1 (Convex hull theorem). *Suppose that* $X \in C^0(\overline{\Omega}, \mathbb{R}^3) \cap C^2(\Omega, \mathbb{R}^3)$ *is harmonic in a bounded and connected open set* $\Omega \subset \mathbb{R}^2$*. Then* $X(\overline{\Omega})$ *is contained in the convex hull of its boundary values* $X(\partial\Omega)$*.*

Proof. Let A be a constant vector in $\mathbb{R}^3$. Then $h(w) := \langle A, X(w)\rangle$ is harmonic in Ω, and we apply the maximum principle to h. Hence, if for some number $d \in \mathbb{R}$, the inequality

$$\langle A, X(w)\rangle \le d$$

holds true for all $w \in \partial\Omega$, it is also satisfied for all $w \in \overline{\Omega}$. As any closed convex set is the intersection of its supporting half-spaces, the assertion is proved. □

Throughout this section, let us agree upon the following *terminology*:

A finite connected minimal surface is a nonconstant mapping

$$X \in C^0(\overline{\Omega}, \mathbb{R}^3) \cap C^2(\Omega, \mathbb{R}^3)$$

which is defined on the closure of a bounded, open, connected set $\Omega \subset \mathbb{R}^2$ and satisfies

$$(1) \qquad \Delta X = 0$$

and

$$(2) \qquad |X_u|^2 = |X_v|^2, \quad \langle X_u, X_v\rangle = 0$$

in Ω. We call Ω the parameter domain of X.

Then, on account of Theorem 1, we obtain

Corollary 1. *Any finite connected minimal surface X with the parameter domain Ω is contained in the convex hull of its boundary values $X|_{\partial\Omega}$, that is,*

$$(3) \qquad X(\overline{\Omega}) \subset \textit{convex hull } X(\partial\Omega).$$

In fact, we can sharpen this statement by inspecting the proof of Theorem 1. Suppose that

$$h(w_0) := \langle A, X(w_0)\rangle = d$$

holds for some $w_0 \in \Omega$, in addition to

$$h(w) \le d \quad \text{for all } w \in \partial\Omega.$$

Then the maximum principle implies

$$h(w) = d \quad \text{for all } w \in \overline{\Omega}.$$

Thus we obtain

Corollary 2. *If a finite connected minimal surface X with the parameter domain Ω touches the convex hull $\mathcal{K}$ of its boundary values $X(\partial\Omega)$ at some "interior point" $X(w_0), w_0 \in \Omega$, then X is a planar surface. In particular, X cannot touch any corner of $\partial\mathcal{K}$ nor any other nonplanar point of $\partial\mathcal{K}$.*

The reader will have noticed that, so far, we have nowhere used the conformality relations (2). In other words, all the previous results are even true for harmonic mappings. Thus we may expect that by using (2) we shall obtain stronger enclosure theorems which will better reflect the saddle-surface character of nonplanar minimal surfaces. In fact, we have

Theorem 2 (Hyperboloid theorem). *If $X(w) = (x(w), y(w), z(w))$ is a finite connected minimal surface with the parameter domain Ω, whose boundary $X(\partial\Omega)$ is contained in the hyperboloid*

$$\mathcal{K}_\varepsilon := \left\{(x, y, z) \in \mathbb{R}^3 : x^2 + y^2 - z^2 \le \varepsilon^2\right\},$$

$\varepsilon > 0$, then $X(\overline{\Omega})$ lies in $\mathcal{K}_\varepsilon$. Moreover, we even have $X(\Omega) \subset \operatorname{int} \mathcal{K}_\varepsilon$.

Proof. Note that $\mathcal{K}_\varepsilon$ is the sublevel set

$$\mathcal{K}_\varepsilon = \left\{(x, y, z) : f(x, y, z) \le \varepsilon^2\right\} \tag{4}$$

of the quadratic form

$$f(x, y, z) := x^2 + y^2 - z^2.$$

Let us therefore compute the Laplacian of the composed map $h := f \circ X = f(X)$. We obtain

$$\Delta h = \langle \nabla X, D^2 f(X) \nabla X\rangle + \langle Df(x), \Delta X\rangle. \tag{5}$$

Because of (1) and

$$D^2 f = \begin{pmatrix} 2 & 0 & 0 \\ 0 & 2 & 0 \\ 0 & 0 & -2 \end{pmatrix},$$

it follows that

$$\Delta h = 2(|\nabla x|^2 + |\nabla y|^2 - |\nabla z|^2) \quad \text{in } \Omega. \tag{6}$$

Moreover, we can write (2) in the complex form

$$\langle X_w, X_w\rangle = 0, \tag{7}$$

that is,

$$x_w^2 + y_w^2 + z_w^2 = 0,$$

whence we obtain

$$|\nabla z|^2 \le |\nabla x|^2 + |\nabla y|^2 \quad \text{in } \Omega. \tag{8}$$

From (6) and (8) we infer that

$$\Delta h \ge 0 \quad \text{in } \Omega,$$

i.e., h is subharmonic, and the assumption yields $h(w) \le \varepsilon^2$ for all $w \in \partial\Omega$, taking (4) into account. Then the maximum principle implies $h(w) \le \varepsilon^2$ for all $w \in \overline{\Omega}$ whence $X(\overline{\Omega}) \subset \mathcal{K}_\varepsilon$.

Suppose that $X(w_0) \in \partial\mathcal{K}_\varepsilon$ for some $w_0 \in \Omega$. Then we would have $h(w_0) = \varepsilon^2$, and the maximum principle would imply $h(w) \equiv \varepsilon^2$, i.e., $X(w) \in \partial\mathcal{K}_\varepsilon$ for all $w \in \overline{\Omega}$. As $X(w) \not\equiv$ const, we know that $X(w)$ has zero mean curvature (except for the isolated branch points) which contradicts the relation $X(\Omega) \subset \partial\mathcal{K}_\varepsilon$, since no open part of $\partial\mathcal{K}_\varepsilon$ is a minimal surface. □

Let us take one step further and assume that the boundary of the minimal surface X is even contained in the cone

$$\mathcal{K}_0 = \{(x,y,z) \in \mathbb{R}^3 : f(x,y,z) \le 0\} = \bigcap_{\varepsilon>0} \mathcal{K}_\varepsilon.$$

Then in view of the hyperboloid theorem the whole surface $X(\overline{\Omega})$ is contained in the cone $\mathcal{K}_0$.

Can it be true that, in addition, the boundary $X(\partial\Omega)$ intersects both cones

$$\mathcal{K}_0^{\pm} := \mathcal{K}_0 \cap \{z \lesseqgtr 0\}?$$

If so, then there is some $w \in \Omega$ such that the point $X(w)$ of the minimal surface lies in the vertex of the cone $\mathcal{K}_0$, that is, $X(w_0) = 0$ for some $w_0 \in \Omega$.

On the other hand, as $X(w) \not\equiv$ const, the minimal surface X has a (possibly generalized) tangent plane T at $X(w_0) = 0$; cf. Section 3.2 of Vol. 1. Clearly, there is no neighbourhood U of 0 in $\mathbb{R}^3$ such that $T \cap U \subset \mathcal{K}_0$. Then one infers that the relation $X(w_0) = 0$ is impossible, taking the asymptotic expansion

$$X_w(w) = A(w - w_0)^m + O(|w - w_0|^{m+1}) \quad \text{as } w \to w_0$$

with $A \in \mathbb{C}^3$, $A \ne 0$, $m \ge 0$, into account.

Hence, except for a suitable congruence mapping, we have shown the following result:

Theorem 3 (Cone theorem). *Let $\mathcal{K}$ be a cone congruent to $\mathcal{K}_0$ which consists of the two half-cones $\mathcal{K}^+$ and $\mathcal{K}^-$ corresponding to $\mathcal{K}_0^+$ and $\mathcal{K}_0^-$. Then there is no finite connected minimal surface the boundary of which lies in $\mathcal{K}$ and intersects both $\mathcal{K}^+$ and $\mathcal{K}^-$.*

The cone theorem can be used to prove nonexistence results for Plateau problems, or for free (or partially free) boundary value problems. Instead of formulating a general theorem, we shall merely consider a special case that illustrates the situation. The reader can easily set up other – and possibly more interesting – examples, or he may himself formulate a general *necessary* criterion for the existence of stationary minimal surfaces within a given boundary configuration $\langle \Gamma_1, \dots, \Gamma_l, S_1, \dots, S_m \rangle$.

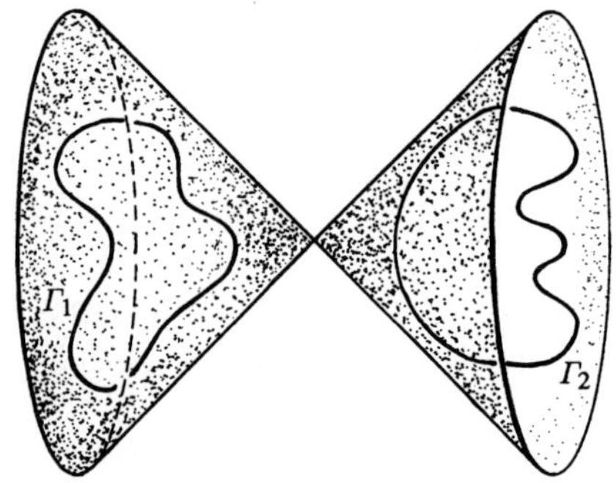

Fig. 1. Two suitable cones give a nonexistence result

Consider two closed Jordan curves Γ_1 and Γ_2 which can be separated by some cone $\mathcal{K}$ as described in Theorem 3. That is, we can move the *test cone* $\mathcal{K}_0$ into such a position $\mathcal{K}$ that Γ_1 lies in the half-cone $\mathcal{K}^+$ and Γ_2 is contained in $\mathcal{K}^-$. *Then there is no connected solution of the general Plateau (or Douglas) Problem for the boundary configuration* $\langle \Gamma_1, \Gamma_2 \rangle$. This corresponds to the experimental fact mentioned in the introduction to this chapter: A soap film spanned into two closed (non-linked) wires Γ_1 and Γ_2 will decompose into two parts separately spanning Γ_1 and Γ_2 if Γ_1 and Γ_2 are moved sufficiently far apart.

We shall show at the end of the next section that the "test cone $\mathcal{K}_0$ for non-existence" may even be replaced by a slightly larger set.

Further results about enclosure and nonexistence of minimal surfaces can be obtained by an elaboration and extension of the ideas used in the proof of the Theorems 1–3, some of which will be worked out in the next three sections. Note, however, that the use of the maximum principle was by no means the first way to obtain information about the extension of minimal surfaces and about nonexistence of solutions to boundary value problems, though the maximum principle is certainly the simplest tool to obtain such results. Concerning other methods we refer to Nitsche's monograph [28], Kap. VI, 3.1, pp. 474–498, and pp. 707–708 of the Appendix (=Anhang).

4.2 Touching H-Surfaces and Enclosure Theorems. Further Nonexistence Results

In the sequel we shall look for other sets $\mathcal{K}$ enclosing any finite connected minimal surface whose boundary is confined to $\mathcal{K}$. Since nothing is gained if we restrict our attention to minimal surfaces, we shall more generally study surfaces of continuous mean curvature H (or "H-surfaces").

To avoid confusion we recall our notation from Chapter 1 of Vol. 1: $\mathcal{E}$, $\mathcal{F}$, $\mathcal{G}$ and $\mathcal{L}$, $\mathcal{M}$, $\mathcal{N}$ denote the coefficients of the first and second fundamental form of a surface X; H and K stand for its mean curvature and Gauss curvature respectively.

Assumption. *Throughout this section we will assume that H is a continuous real-valued function on $\mathbb{R}^3$.*

Definition 1. *An H-surface X is a nonconstant map $X \in C^2(\Omega, \mathbb{R}^3)$ defined on an open set Ω satisfying*

$$\Delta X = 2H(X) X_u \wedge X_v \tag{1}$$

and

$$|X_u|^2 = |X_v|^2, \quad \langle X_u, X_v \rangle = 0. \tag{2}$$

We call Ω the parameter domain of the H-surface X. An H-surface X is said to be finite and connected if its parameter domain Ω is a bounded, open, connected set in $\mathbb{R}^2$, and if $X \in C^0(\overline{\Omega}, \mathbb{R}^3)$.

Clearly, minimal surfaces are H-surfaces with $H \equiv 0$.

In order to study touching H-surfaces, we need the following

Lemma 1. *Suppose that $\Phi : B_r(0) \to \mathbb{C}$ is a function of class C^1 which can be written in the form*

$$\Phi(w) = a(w - w_0)^m + \Psi(w), \quad w \in B_r(0), \tag{3}$$

for some $w_0 \in B_r(0)$, some real number $a > 0$, some integer $m \geq 1$, and some mapping $\Psi : B_r(0) \to \mathbb{C}$ with $\Psi(w_0) = 0$ and

$$\nabla \Psi(w) = o(|w - w_0|^{m-1}) \quad \text{as } w \to w_0. \tag{4}$$

Then there is some neighbourhood U of w_0 and some C^1-diffeomorphism φ from U onto $\varphi(U)$ such that

$$\Phi(w) = [\varphi(w)]^m \quad \text{for all } w \in U \tag{5}$$

holds true.

Proof. Clearly, if there exists some function φ satisfying (5), it has to be the function

$$\varphi(w) := (w - w_0) \sqrt[m]{\chi(w)}, \tag{6}$$

where

$$\chi(w) = a + (w - w_0)^{-m} \Psi(w). \tag{7}$$

We shall have to prove that φ is well defined and has the desired properties.

First of all, the relation (4) implies

$$\Psi(w) = o(|w - w_0|^m) \quad as\ w \to w_0$$

since $\Psi(w_0) = 0$. Therefore $\chi(w)$ tends to a as $w \to w_0$, and we set $\chi(w_0) := a$. Hence there is a neighbourhood U_0 of w_0 where a single-valued branch $\sqrt[m]{\ }$ of the m-th root can be defined. Thus the function φ defined by (6) and (7) is a well-defined function near w_0.

Now (4) implies for the derivatives of χ in $U_0 - \{w_0\}$ that

$$\begin{aligned}
\chi_u(w) &= -m(w - w_0)^{-m-1}\Psi(w) + (w - w_0)^{-m}\Psi_u(w) = o(|w - w_0|^{-1}),\\
\chi_v(w) &= -mi(w - w_0)^{-m-1}\Psi(w) + (w - w_0)^{-m}\Psi_v(w) = o(|w - w_0|^{-1}),
\end{aligned}$$

whence

$$\begin{aligned}
\varphi_u(w) &= \sqrt[m]{\chi(w)} + \frac{1}{m}(w - w_0)\chi(w)^{(1-m)/m}\chi_u(w)\\
&= \sqrt[m]{\chi(w)} + o(1),\\
\varphi_v(w) &= i\sqrt[m]{\chi(w)} + \frac{1}{m}(w - w_0)\chi(w)^{(1-m)/m}\chi_v(w)\\
&= i\sqrt[m]{\chi(w)} + o(1),
\end{aligned}$$

and therefore

$$\lim_{w \to w_0} D\varphi(w) = \begin{pmatrix} \sqrt[m]{a} & 0 \\ 0 & i\sqrt[m]{a} \end{pmatrix}.$$

On the other hand, we have

$$\lim_{w \to w_0} \frac{\varphi(w)}{w - w_0} = \lim_{w \to w_0} \sqrt[m]{\chi(w)} = \sqrt[m]{a}.$$

Thus φ is a C^1-function, and the lemma follows from the inverse mapping theorem. □

Let us now describe what we can say about touching points of two H-surfaces, one of which is assumed to be regular.

Theorem 1. *Suppose that G is a domain in $\mathbb{R}^3$ and that $\partial_0 G$ is an open part of the boundary of G with $\partial_0 G \in C^2$. Secondly let X be a finite connected H-surface with the parameter domain Ω whose image $X(\Omega)$ lies in $G \cup \partial_0 G$. Finally, denoting the mean curvature of $\partial_0 G$ at P with respect to the interior normal by $\Lambda(P)$, we assume that*

$$\sup_{\overline{G}} |H| \le \inf_{\partial_0 G} \Lambda \tag{8}$$

holds true. Then $X(\Omega)$ is completely contained in $\partial_0 G$ if $X(\Omega) \cap \partial_0 G$ is nonempty (that is, if $X(\Omega)$ "touches" $\partial_0 G$).

Remark 1. This is, in fact, a local result. Instead of (8), it suffices to assume that every point $P \in \partial_0 G$ has a neighbourhood U in $\overline{G}$ such that

$$(8') \qquad \sup_U |H| \le \inf_{U \cap \partial_0 G} \Lambda.$$

This remark implies the following

Enclosure Theorem I. *Let G be a domain in $\mathbb{R}^3$ with $\partial G \in C^2$, and let H be a continuous function on $\mathbb{R}^3$ satisfying*

$$|H(P)| < \Lambda(P) \quad \textit{for all } P \in \partial G,$$

where Λ denotes again the mean curvature of ∂G with respect to the inward normal. Then every finite connected H-surface X with the parameter domain Ω whose image $X(\Omega)$ is confined to the closure $\overline{G}$ lies in G, i.e. $X(\Omega) \subset G$.

Remark 2. Note that the condition $|H(P)| \le \Lambda(P)$ for all $P \in \partial G$ is not sufficient to conclude the assertion of the theorem. Indeed this follows easily by considering a plane with $\Lambda \equiv 0$ and a paraboloid of fourth order lying on one side of the plane and touching it in a single point.

Proof of Theorem 1. Clearly we have $\Omega = \Omega_1 \cup \Omega_2$ where

$$\Omega_1 := X^{-1}(G), \quad \Omega_2 := X^{-1}(\partial_0 G).$$

Since X is continuous, the set Ω_1 is open. Suppose that $X(\Omega)$ touches $\partial_0 G$; then $\Omega_2 = \Omega \setminus \Omega_1$ is not empty. We show that the assumption "$\Omega_1 \ne \emptyset$" will lead to a contradiction.

In fact, suppose $\Omega_1 \ne \emptyset$. Then also $\partial\Omega_1 \cap \Omega$ is nonempty and we can select a point $z_0 \in \Omega_1$ which is closer to $\partial\Omega_1 \cap \Omega$ than to $\partial\Omega$. Since Ω_1 is open, there is a maximal open disc $B_r(z_0) \subset \Omega_1$ with the property $w_0 \in \partial B_r(z_0) \cap \partial\Omega_1 \cap \Omega$ for (at least) one point $w_0 \in \Omega_2$, i.e. $X(w_0) = P_0 \in \partial_0 G$. Without loss of generality we may suppose that $w_0 = 0$. By the reasoning of Section 2.10, we may assume after a suitable shift and rotation of the coordinate system that, close to $w_0 = 0$, the surface $X(w) = (x(w), y(w), z(w))$ has the asymptotic expansion

$$x(w) + iy(w) = aw^m + o(|w|^m),$$
$$z(w) = o(|w|^m),$$

for some integer $m > 0$ and some $a > 0$. According to the preceding Lemma 1, there is a neighbourhood $U \subset \Omega$ of 0 and a C^1-diffeomorphism $\varphi : U \to \varphi(U)$ such that for $w \in U$

$$x(w) + iy(w) = [\varphi(w)]^m.$$

Next we choose an $\varepsilon > 0$ so small that the disk $B_\varepsilon(0)$ is contained in $\varphi(U)$, whence $B_{\varepsilon^m}(0)$ lies in $\varphi^m(U)$. Therefore all the disks

$$\Omega_\varepsilon(\xi, \eta) = B_{\varepsilon^m/2}(\xi + i\eta) \quad \text{with } \xi^2 + \eta^2 = \left(\frac{\varepsilon^m}{2}\right)^2,$$

which cover $B_{\varepsilon^m}(0) \setminus \{0\}$, are subsets of $\varphi^m(U)$, and their preimages under the mapping φ^m cover a punctured neighbourhood of 0.

Now let $\sqrt[m]{\ }$ denote an arbitrary single-valued branch of the m-th root defined on $\Omega_\varepsilon(\xi,\eta)$. Then

$$z'(x,y) := z\left(\varphi^{-1}(\sqrt[m]{x+iy})\right)$$

defines a C^1-non-parametric representation of a part of the H-surface X, namely the one defined on $\varphi^{-1}(\sqrt[m]{\Omega_\varepsilon(\xi,\eta)})$. For the construction to follow it is convenient and necessary to choose $\Omega_\varepsilon(\xi,\eta) \subset B_r(z_0)$ such that $w_0 = 0 \in \partial\Omega_\varepsilon(\xi,\eta)$.

The plane $\{z = 0\}$ is the (possibly "generalized") tangent plane of X at $P_0 = X(w_0)$. Thus

(9) $$\lim_{(x,y)\to 0} \nabla z'(x,y) = 0.$$

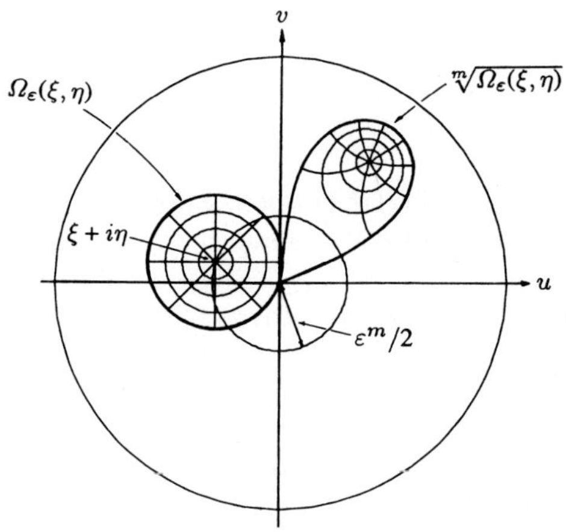

Fig. 1. The domains used in the proof of Theorem 1

Since $X(\Omega)$ lies on one side of $\partial_0 G$ and since $X(w_0)$ belongs to $\partial_0 G$, the set $\{z = 0\}$ is also the tangent plane of $\partial_0 G$ at $X(w_0)$. Therefore (after decreasing ε if necessary) we obtain also a local non-parametric representation of $\partial_0 G$ by means of a function

$$z'' = z''(x,y) \quad \text{for } (x,y) \in \overline{\Omega}_\varepsilon(\xi,\eta).$$

By assumption, we have $z'' \in C^2(\overline{\Omega}_\varepsilon(\xi,\eta))$. If the interior normal of $\partial_0 G$ at $X(w_0)$ points in the direction of the positive z-axis, (the other case is handled similarly), we have by assumption

(10) $$z'' < z' \quad \text{on } \Omega_\varepsilon(\xi,\eta), \quad \text{and also} \quad z''(0) = z'(0).$$

Since $\{z \equiv 0\}$ is also the tangent plane of $\partial_0 G$ at $X(w_0)$ we have

$$\lim_{(x,y)\to 0} \nabla z''(x,y) = \nabla z''(0,0) = 0. \tag{11}$$

Moreover, z' and z'' are solutions of the corresponding equations of prescribed mean curvature (cf. Section 2.7 of Vol. 1), i.e.,

$$Q(z') := \operatorname{div} \frac{\nabla z'}{\sqrt{1+|\nabla z'|^2}} = \pm 2H(x,y,z'(x,y)),$$
$$Q(z'') := \operatorname{div} \frac{\nabla z''}{\sqrt{1+|\nabla z''|^2}} = 2\Lambda(x,y,z''(x,y))$$

for all $(x,y) \in \Omega_\varepsilon(\xi,\eta)$. By assumption, it follows that

$$Q(z') \le Q(z'') \quad \text{in } \Omega_\varepsilon(\xi,\eta).$$

It now readily follows from the theorem of the mean, that the difference $\hat{z} := z'' - z'$ satisfies a linear differential inequality of the type

$$L(\hat{z}) = a_{ij}(x) D_{ij}\hat{z} + b_i(x) D_i \hat{z} \ge 0 \quad \text{in } \Omega_\varepsilon(\xi,\eta),$$

where the coefficients b_i are locally bounded and the a_{ij}'s are elliptic (for a similar argument see e.g. the proof of Theorem 10.1 in Gilbarg and Trudinger [1]).

Now (9) and (11) yield that

$$\lim_{(x,y)\to 0} \nabla \hat{z}(x,y) = 0,$$

and hence also the normal derivative $\frac{\partial \hat{z}}{\partial n}(0,0) = 0$.

However, because of $\hat{z}(0) = 0 > \hat{z}(x,y)$ for all $(x,y) \in \Omega_\varepsilon(\xi,\eta)$, the point $w_0 = 0 \in \partial\Omega_\varepsilon(\xi,\eta)$ is a strict maximum, which contradicts Hopf's boundary point lemma (Lemma 3.4 in Gilbarg and Trudinger [1]). Consequently Ω_1 has to be empty and hence $\Omega = \Omega_2$ or $X(\Omega) \subset \partial_0 G$. This completes the proof of Theorem 1. □

Proof of Enclosure Theorem I. The condition $|H(P)| < \Lambda(P)$ for all $P \in \partial G$ clearly implies that every point $P \in \partial G$ has a neighbourhood U in $\overline{G}$, such that

$$\sup_U |H| \le \inf_{U \cap \partial G} \Lambda$$

holds true. Therefore a local version of Theorem 1 is applicable and we assume, contradictory to the assertion, that some interior point $w_0 \in \Omega$ is mapped onto ∂G, i.e. X touches ∂G at $X(w_0)$. It then follows from Theorem 1 that $X(\Omega) \subset \partial G$. On the other hand X is an H-surface, which in particular means that X has mean curvature H, except possibly at isolated singular points, compare the derivation of the asymptotic expansion near branch points in Section 2.10. Whence, by continuity, it follows that $|H(P)| = \Lambda(P)$ for all $P \in \partial G$, a contradiction to the assumption of the theorem. Enclosure Theorem I is proved. □

The reasoning used to prove Theorems 2 and 3 in Section 4.1, may be generalized to H-surfaces X. In fact consider the quadratic function

$$f(x,y,z) = x^2 + y^2 - bz^2,$$

with $0 \le b < 1$, and compute the Laplacian of the composed map $h := f \circ X$.

We obtain similarly as in Theorem 2 of Section 6.1

$$\Delta h = \langle \nabla X, D^2 f(X) \nabla X \rangle + \langle Df(X), \Delta X \rangle.$$

Because of (1) and

$$D^2 f = \begin{pmatrix} 2 & 0 & 0 \\ 0 & 2 & 0 \\ 0 & 0 & -2b \end{pmatrix},$$

it follows that

$$\begin{aligned} \Delta h &= 2|\nabla x|^2 + 2|\nabla y|^2 - 2b|\nabla z|^2 + 4H(X) \cdot \langle (x,y,-bz), X_u \wedge X_v \rangle \\ &\ge 2|\nabla x|^2 + 2|\nabla y|^2 - 2b|\nabla z|^2 - 4|H(X)||X_u \wedge X_v| \cdot \sqrt{x^2+y^2+b^2z^2}. \end{aligned}$$

From the conformality condition (2) we obtain

$$|\nabla z|^2 \le |\nabla x|^2 + |\nabla y|^2,$$

whence

$$|X_u \wedge X_v| \le |\nabla x|^2 + |\nabla y|^2.$$

Concluding we find

$$\begin{aligned} \Delta h &\ge 2|\nabla x|^2 + 2|\nabla y|^2 - 2b|\nabla z|^2 - 4|H(X)|(|\nabla x|^2 + |\nabla y|^2)\sqrt{x^2+y^2+b^2z^2} \\ &\ge 2(|\nabla x|^2 + |\nabla y|^2)\left[1 - b - 2|H(X)| \cdot \sqrt{x^2+y^2+b^2z^2}\right]. \end{aligned}$$

Thus we have proved

Theorem 2. *Let X be an H-surface on Ω and $f(x,y,z) = x^2 + y^2 - bz^2$, $0 \le b < 1$. Then the function $h = h(u,v) = f \circ X(u,v)$, $(u,v) \in \Omega$ is subharmonic on Ω, provided that*

$$b + 2|H(X)| \cdot \sqrt{x^2+y^2+b^2z^2} \le 1 \quad \text{on } \Omega.$$

A consequence of this result and the asymptotic expansion for H-surfaces in singular points is the following

Theorem 3 (Cone Theorem). *Suppose that $X \in C^2(\Omega) \cap C^0(\overline{\Omega})$ is an H-surface on Ω which satisfies*

$$\sup_{w \in \Omega} |X(w)| \cdot |H(X(w))| = q < \frac{1}{2}.$$

Then, for $b = 1-2q \in (0,1]$ *the function* $h(u,v) = x^2(u,v)+y^2(u,v)-bz^2(u,v)$ *is subharmonic on* Ω *and therefore by the maximum principle*

$$\sup_{\Omega} h \le \sup_{\partial\Omega} h.$$

Moreover let $\mathcal{K} = \mathcal{K}^+ \cup \mathcal{K}^- \cup \{0\}$ *where*

$$\mathcal{K}^\pm = \left\{(x,y,z) : x^2 + y^2 - bz^2 \le 0,\ \pm z > 0\right\}.$$

Suppose that $X(\partial\Omega)$ *is contained in* $\mathcal{K}$*, such that both intersections* $X(\partial\Omega) \cap \mathcal{K}^+$ *and* $X(\partial\Omega) \cap \mathcal{K}^-$ *are not empty; then* Ω *cannot be connected.*

Proof. The asymptotic expansion for H-surfaces, cp. Section 2.10 and Chapter 3, or the discussion in the proof of Theorem 1, imply the existence of a tangent plane for X at every point $w \in \Omega$. Hence the H-surface cannot pass through the vertex of the cone, cp. the discussion in Section 4.1. □

For our next enclosure theorem we need some further terminology which will allow us to give a lucid formulation of the result.

Definition 2. *Let* $\mathcal{J}$ *be an interval in* $\mathbb{R}$*. We shall say that a family of domains in* $\mathbb{R}^3$*,* $(G_\alpha)_{\alpha\varepsilon\mathcal{J}}$*, depends continuously on the parameter* α*, if for all* $\alpha_0 \in \mathcal{J}$ *the symmetric difference*

$$G_\alpha \Delta G_{\alpha_0} := (G_\alpha \cup G_{\alpha_0}) \setminus (G_\alpha \cap G_{\alpha_0})$$

tends to ∂G_{α_0} *as* α *tends to* α_0*, i.e., if for all* $\alpha_0 \in \mathcal{J}$ *and all* $\varepsilon > 0$ *there is a* $\delta > 0$ *such that* $|\alpha - \alpha_0| < \delta$ *implies that*

$$G_\alpha \Delta G_{\alpha_0} \subset T_\varepsilon(\partial G_{\alpha_0}) := \{P : \operatorname{dist}(P, \partial G_{\alpha_0}) < \varepsilon\}.$$

Definition 3. *If* M *is a simply connected subset of an open set* G *in* $\mathbb{R}^3$*, then a family* $(G_\alpha)_{\alpha\varepsilon\mathcal{J}}$ *of domains depending continuously on its parameter* α *is called an* **enclosure** *of* M **with respect to** G *(or it is said:* $(G_\alpha)_{\alpha\varepsilon\mathcal{J}}$ **encloses** M with respect to G*) if*

(i) $M \subset G_\alpha$ *for all* $\alpha \in \mathcal{J}$*;*
(ii) *every* $P \in G \setminus M$ *does not belong to at least one of the* G_α*;*
(iii) *every compact subset* $\mathcal{K}$ *of* G *lies in at least one of the* G_α*;*

Here are two examples:

1 Let M be a star-shaped domain in $\mathbb{R}^3$ whose boundary may be considered as a graph of a positive real-valued function $f : S^2 \to (0,\infty)$ of class C^2, i.e. we assume that

$$M = \left\{\lambda P : P \in S^2 \text{ and } 0 \le \lambda < f(P)\right\}.$$

Then ∂M is the level set

$$\partial M = \{P \in \mathbb{R}^3 \setminus \{0\} : F(P) = 1\}$$

of the function

$$F(P) := |P|/f\left(\frac{P}{|P|}\right) \quad \text{for } P \in \mathbb{R}^3 \setminus \{0\}\,,\ F(0) := 0$$

satisfying $F(cP) = cF(P)$ for $c > 0$. In particular, we have for $c > 0$ that

$$F(P) = 1 \quad \text{if and only if} \quad F(cP) = c,$$

i.e., the level sets of F are homothetic, hence the mean curvature of $\{F = 1\}$ at P is equal to c-times the mean curvature of $\{F = c\}$ at the point cP.

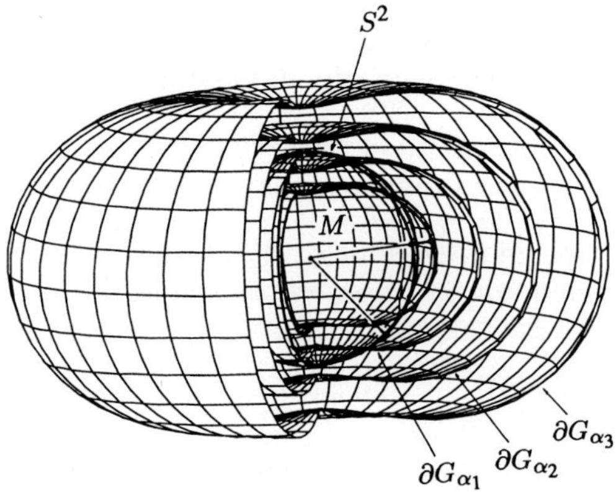

Fig. 2. A star-shaped domain M, whose boundary is the graph of a smooth function $f : S^2 \to \mathbb{R}$ defined on the unit sphere S^2, is enclosed with respect to $\mathbb{R}^3$ by the family of domains $G_\alpha = \{\alpha P : P \in M\}$, $\alpha > 1$, which are homothetic to M

Moreover, the family

$$G_\alpha := \{F < \alpha\} \quad \text{for } \alpha > 1$$

defines an enclosure of M with respect to $\mathbb{R}^3$.

$\boxed{2}$ Let $\tau = 1.199678640257\ldots$ be the solution of the equation $\tau \sinh \tau = \cosh \tau$. Then, for any $c > 0$, the cone

$$\mathcal{K}^c := \left(\mathcal{K}^+ \cup \{0\} \cup \mathcal{K}^-\right) \cap \{|z| < c\}$$

with

$$\mathcal{K}^\pm := \left\{(x, y, z) \in \mathbb{R}^3 : x^2 + y^2 < (\sinh^2 \tau) z^2,\ z \lessgtr 0\right\}$$

is enclosed by the domains

$$\mathcal{K}^c_\alpha := \mathcal{K}_\alpha \cap \{|z| < c\}\,,$$

$$\mathcal{K}_\alpha := \left\{(x,y,z) \in \mathbb{R}^3 : x^2 + y^2 < \alpha^2 \cosh^2 \frac{z}{\alpha}\right\}, \quad \alpha > 0.$$

Note that the $\mathcal{K}_\alpha$ have catenoids, i.e. minimal surfaces, as their boundaries $\partial\mathcal{K}_\alpha$, cf. Osserman and Schiffer [1].

By the way, the angle of aperture of the cone $\mathcal{K}^+$ is $\alpha = \arctan(\sinh^2 \tau) \mathrel{\widehat{=}} 56.4658\ldots$ degrees whereas the angle of the cone $\mathcal{K}^+$ appearing in the *cone theorem* of Section 6.1 is 45°.

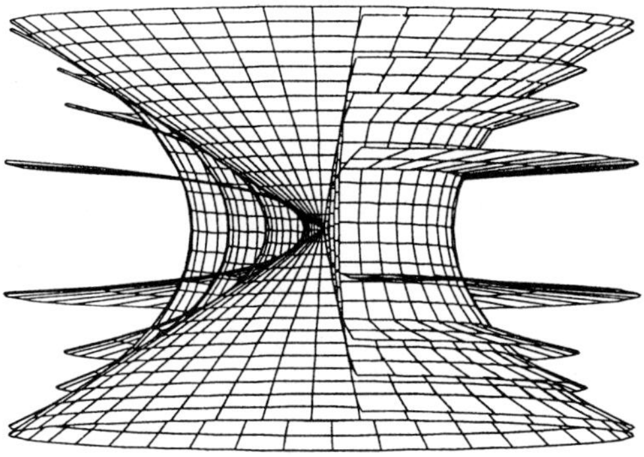

Fig. 3. Let τ be the solution of the equation $\tau \sinh(\tau) = \cosh(\tau)$. Then the cone $\{x^2+y^2 < \sinh^2(\tau)z^2, |z| < c\}$ is enclosed by the family of domains $\{x^2 + y^2 < \alpha^2 \cosh^2(z/\alpha), |z| < c\}$ having catenoids as parts of their boundaries

Assumption. *In the sequel let M be a simply connected subset of a domain G in $\mathbb{R}^3$ which possesses an enclosure $(G_\alpha)_{\alpha\in\mathcal{J}}$ with respect to G such that each subset $\partial_0 G_\alpha := G \cap \partial G_\alpha$ of ∂G_α is of class C^2.*

Denote by Λ_α the mean curvature of ∂G_α with respect to the inward normal of ∂G_α.

Recall that $H \in C^0(\mathbb{R}^3)$, and *suppose that we have*

(12) $$\sup_{\overline{G}_\alpha} |H| \le \inf_{\partial_0 G_\alpha} \Lambda_\alpha$$

for every $\alpha \in \mathcal{J}$.

Under this assumption we can formulate the

Enclosure Theorem II. *Let $X \in C^2(\Omega) \cap C^0(\overline{\Omega})$ be a finite connected H-surface with the parameter domain Ω whose image $X(\Omega)$ lies in G, and whose boundary $X(\partial\Omega)$ is contained in M. Then the image $X(\Omega)$ must, in fact, lie in M.*

Proof. If $X(\Omega)$ is not contained in M, then, according to the definition of an enclosure $(G_\alpha)_{\alpha\in\mathcal{J}}$, there is an α_1 such that $X(\Omega)$ does *not* lie in G_{α_1}, and an α_2 (without loss of generality greater than α_1) such that $X(\Omega)$ remains in G_{α_2}. Therefore the number

$$\alpha_0 := \sup\{\alpha \in \mathfrak{J} : \alpha < \alpha_2 \text{ and } X(\Omega) \not\subseteq G_\alpha\}$$

is well defined and finite. We shall presently show that

(I) $$X(\Omega) \subset G_{\alpha_0} \cup \partial_0 G_{\alpha_0},$$

(II) $$X(\Omega) \cap \partial_0 G_{\alpha_0} \neq \emptyset.$$

Then, on account of Theorem 1, we obtain that $X(\Omega)$ lies in $\partial_0 G_{\alpha_0}$; in particular, $X(\partial\Omega)$ is confined to ∂G_{α_0}. This contradicts the assumption that

$$X(\partial\Omega) \subset M \subset G_{\alpha_0}.$$

Now, as for (I), let us assume that for some $w \in \Omega$, the point $X(w)$ lies at a distance $d > 0$ from $\overline{G}_{\alpha_0}$. Then the continuity of the family G_α with respect to α implies that, for some small $\varepsilon > 0$, the point $X(w)$ is not contained in $G_{\alpha_0+\varepsilon}$ either. This, however, contradicts the definition of α_0.

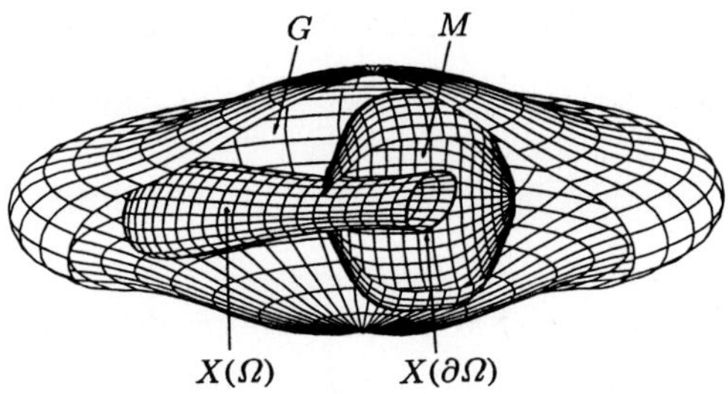

Fig. 4. A simply connected set M which has an enclosure G_α as shown before with respect to an open set G, and an H-surface X whose image $X(\Omega)$ is confined to G and whose boundary even lies in the smaller set M. If the H-surface would satisfy the curvature condition of the enclosure theorem II, then all of $X(\Omega)$ would remain in M

As for (II), since $X(\Omega)$ is contained in G, it will suffice to show that $X(\Omega)$ does not lie in G_{α_0}. Otherwise, as follows from the compactness of $X(\overline{\Omega})$, we have

$$d' := \operatorname{dist}(X(\overline{\Omega}), \partial_0 G_{\alpha_0}) > 0,$$

which also implies that α_0 is not the supremum since, once again, in view of the continuity of G_α with respect to α, the set $X(\Omega)$ lies in $G_{\alpha_0-\varepsilon}$ for some small $\varepsilon > 0$. □

As an illustrative application of the last enclosure theorem, we have the following

Enclosure Theorem III. *Let $f : S^2 \to (0,\infty)$ be some C^2-function on S^2, and let $F : \mathbb{R}^3 \to (0,\infty)$ be its homogeneous extension to $\mathbb{R}^3$ defined by*

$F(0) := 0$ and $F(P) := |P|/f(\frac{P}{|P|})$ for $P \neq 0$. Denote by M the star-shaped domain $\{F < 1\}$ and assume that the mean curvature of ∂M with respect to the inward normal is everywhere nonnegative. Then every connected finite minimal surface X with the parameter domain Ω satisfies $X(\Omega) \subset M$ if we assume that $X(\partial\Omega) \subset \overline{M}$ and if the intersection of $X(\partial\Omega)$ with M is nonvoid.

This result follows from Theorem 1 and from the remarks about Example $\boxed{1}$ in connection with the Enclosure Theorem II. Instead of going into the details we shall state a nonexistence result that follows from the Enclosure Theorem III; it can be proved like the nonexistence result in Section 4.1.

Nonexistence Theorem. *Assume that M, G, G_α satisfy the assumptions stated above, and suppose in addition that there are finitely many points $P_1, \dots, P_m$ in M such that $M \setminus \{P_1, \dots, P_m\}$ decomposes into $n \geq 2$ simply connected components $M_1, \dots, M_n$. Then there is no finite connected H-surface with a parameter domain Ω which has the following properties:*

(i) $X(\Omega) \subset G$;
(ii) $X(\partial\Omega) \subset \overline{M}$;
(iii) $X(\partial\Omega)$ *intersects at least two of the components* $M_1, \dots, M_n$.

Applying the last theorem to Example $\boxed{2}$, we obtain the following improvement of the cone theorem of Section 4.1:

Corollary 1. *Set*

$$\mathcal{K}^{\pm} := \{(x, y, z) \in \mathbb{R}^3 : z \gtrless 0 \text{ and } x^2 + y^2 < z^2 \sinh^2 \tau\},$$

where $\tau = 1.199678640257\ldots$ *is a solution of the equation*

$$\tau \sinh \tau = \cosh \tau,$$

and define $\mathcal{K}$ *by*

$$\mathcal{K} := \mathcal{K}^{+} \cup \{0\} \cup \mathcal{K}^{-}.$$

Then there is no connected finite minimal surface with boundary which intersects both $\mathcal{K}^+$ and $\mathcal{K}^-$.

This "nonexistence test-cone" $\mathcal{K}$ cannot be further increased as one can see by means of catenoids between suitable circles as boundary curves, see Fig. 1 in the introduction of this chapter.

4.3 Minimal Submanifolds and Submanifolds of Bounded Mean Curvature. An Optimal Nonexistence Result

It is the aim of this section to generalize the results of Sections 4.1 and 4.2 to higher dimensions and codimensions. To accomplish this, we first define a concept of n-dimensional surfaces or submanifolds in $\mathbb{R}^{n+k}$. It turns out that, for the present purpose, it is not necessary to develop the complete differential geometric notion of submanifolds in arbitrary ambient manifolds, as e.g. described in Gromoll, Klingenberg, and Meyer [1], do Carmo [3], Jost [18] and Kühnel [2], but rather the more elementary concepts of submanifolds in $\mathbb{R}^{n+k}$ (although later in Section 4.8 we shall also treat surfaces of prescribed mean curvature in Riemannian manifolds). We start with the following

Definition 1. *A subset $M \subset \mathbb{R}^{n+k}$ is called an n-dimensional submanifold of class C^s, if for each $x \in M$ there are open neighbourhoods U, $V \subset \mathbb{R}^{n+k}$ of x and 0 in $\mathbb{R}^{n+k}$ respectively, and a C^s-diffeomorphism $\varphi : V \to U$, such that $\varphi(0) = x$ and $\varphi(V \cap \mathbb{R}^n \times \{0\}) = U \cap M$. Here $\varphi_{|V\cap\mathbb{R}^n\times\{0\}}$ is a local parametrization and φ^{-1} is called a local chart for M. In case that $k = 1$, $M \subset \mathbb{R}^{n+1}$ is also called a* **hypersurface** *(of class C^s).*

Given M, x and φ as in Definition 1 we have

Definition 2. *The tangent space T_xM of M at x is the n-dimensional linear subspace of R^{n+k} which is spanned by the independent vectors $\varphi_{x^1}(0), \dots, \varphi_{x^n}(0)$.*

One easily convinces oneself that the tangent space T_xM is given by all vectors $\xi = \dot\alpha(0)$, where $\alpha : (-\varepsilon, \varepsilon) \to M$ is a regular curve in M with $\alpha(0) = x$. That is we have

Proposition 1. *The tangent space of M at x is given by*

$$T_xM = \Big\{\dot\alpha(0) : \alpha : (-\varepsilon,\varepsilon) \to M \text{ is a regular curve with } \alpha(0) = x\Big\}.$$

Now consider a function $f : M \to \mathbb{R}^m$. One way of defining differentiability of f is to consider all possible compositions of f with parametrizations φ and to requiring the composition $f \circ \varphi : V \cap \mathbb{R}^n \times \{0\} \to \mathbb{R}^m$ to be differentiable, see e.g. Chapter 1. Here we define differentiability somewhat different (but equivalently)

Definition 3. *Let $M \subset \mathbb{R}^{n+k}$ be a submanifold of class C^s and $f : M \to \mathbb{R}^m$. f is differentiable of class C^r, $r \le s$, if there exists an open subset $U \subset \mathbb{R}^{n+k}$ with $M \subset U$ and a C^r-function $F : U \to \mathbb{R}^m$ such that $f = F|_M$.*

In other words, $f : M \to \mathbb{R}^m$ is differentiable, if it is the restriction of a differentiable map from an open set $U \subset \mathbb{R}^{n+k}$. Of particular interest are the cases $m = 1$ *(scalar functions)* and $m = n + k$ (vector fields). If $f : M \to \mathbb{R}$ is differentiable we define the *(intrinsic)* gradient of f as follows

Definition 4. *The gradient of f on M, in symbols $\nabla_M f$, is defined by $\nabla_M f = (Df)^\top$, where $Df = (f_{x^1}, \dots, f_{x^{n+k}})$ denotes the usual (Euclidean) gradient and $(\xi)^\top$ stands for the orthogonal projection of the vector $\xi \in \mathbb{R}^{n+k}$ onto the tangent space of M at x. (Note that here and in the discussion to follow we tacitly assume, that f coincides with its differentiable extension F, cp. Definition 3).*

Definition 5. *The normal space of M at x is given by*

$$T_xM^\perp := \{n \in R^{n+k} : \langle n, t\rangle = 0 \text{ for all } t \in T_xM\}.$$

Here $\langle\cdot,\cdot\rangle$ denotes the Euclidean scalar product in $\mathbb{R}^{n+k}$.

Let $N_1, \ldots, N_k$ be an orthonormal basis of $T_xM^\perp$. Then we obtain $\nabla_M f = Df - \langle Df, N_1\rangle N_1 - \cdots - \langle Df, N_k\rangle N_k$, for the intrinsic gradient of a function $f : M \to \mathbb{R}$.

Equivalently we now consider an orthonormal basis $t_1, \ldots, t_n$ of the tangent space $T_xM \subset \mathbb{R}^{n+k}$ and an arbitrary vector $t \in T_xM$. Recall that the directional derivative D_tf of $f : M \to \mathbb{R}^m$ at x in the direction of t is given by

$$D_tf(x) := \frac{d}{d\varepsilon} f(\alpha(\varepsilon))_{\varepsilon=0},$$

where

$$\alpha : (-\delta, \delta) \to M$$

is a regular curve in M with $\alpha(0) = x$ and $\alpha'(0) = t$. It is easily seen, that this definition is meaningful (i.e. independent of the particular curve α), and furthermore we have by the chain rule

$$D_tf(x) = Df(x) \cdot t.$$

Definition 6. *Let $f : M \to \mathbb{R}^m$ be differentiable. The differential $df(x)$ of f at x is the linear map $df(x) : T_xM \to \mathbb{R}^m$*

$$t \mapsto df(x)(t) := D_tf(x).$$

In fact, it follows immediately from the definition that $df(x)$ is linear. Observe now that the gradient of $f : M \to \mathbb{R}$ is equivalently given by

$$\nabla_M f = (D_{t_1}f)t_1 + \cdots + (D_{t_n}f)t_n = \sum_{i=1}^{n}(D_{t_i}f)t_i, \tag{1}$$

for any orthonormal basis $t_1, \ldots, t_n$ of T_xM. Then equation (1) easily follows from the previous relation by multiplication with the basis vectors $t_1, \ldots, t_n$ respectively.

Note that (1) is already meaningful for functions $f : M \to \mathbb{R}$ which are merely defined on M, whereas Definition 4 assumes f to be defined (locally) on an open neighbourhood of M, however we shall not dwell on this.

The next important notion is that of the divergence on M.

Definition 7. *Let $X : M \to \mathbb{R}^{n+k}$*

$$X(x) = (X^1(x), \ldots, X^{n+k}(x))$$

be a differentiable function on a differentiable submanifold $M \subset \mathbb{R}^{n+k}$, i.e. a—not necessarily tangential—vector field on M. The divergence $\mathrm{div}_M X$ of X on M is given by

$$\mathrm{div}_M X = \sum_{i=1}^{n} \langle t_i, D_{t_i} X \rangle,$$

where $t_1, \dots, t_n \in T_x M$ is an orthonormal basis of the tangent space $T_x M$.

We observe here that the definition of div_M is independent of the particular orthonormal basis $t_1, \dots, t_n$ of the tangent space $T_x M$. To see this we compute

$$\begin{aligned}
\sum_{i=1}^{n} \langle t_i, dX(t_i) \rangle &= \sum_{i=1}^{n} \langle t_i, D_{t_i} X \rangle = \sum_{i=1}^{n} \left\langle t_i, D_{t_i} \left(\sum_{j=1}^{n+k} e_j \, X^j \right) \right\rangle \\
&= \sum_{i=1}^{n} \left\langle t_i, \sum_{j=1}^{n+k} e_j \, D_{t_i} X^j \right\rangle = \sum_{i=1}^{n} \sum_{j=1}^{n+k} \langle t_i, e_j \, D_{t_i} X^j \rangle \\
&= \sum_{j=1}^{n+k} \left\langle e_j, \sum_{i=1}^{n} (D_{t_i} X^j) t_i \right\rangle = \sum_{j=1}^{n+k} \langle e_j, \nabla_M X^j \rangle
\end{aligned}$$

by equation (1), where $e_1, \dots, e_{n+k}$ denotes the canonical basis of $\mathbb{R}^{n+k}$ and $\nabla_M X^j$ is the gradient of the j-th component X^j of the vector field X on M.

For later computations we note here

Proposition 2. *Let $X(x) = (X^1(x), \dots, X^{n+k}(x))$ be a differentiable vector field on M. Then the divergence of X on M is given by the relation*

$$\mathrm{div}_M X = \sum_{j=1}^{n+k} \langle e_j, \nabla_M X^j \rangle,$$

where $e_1, \dots, e_{n+k}$ stands for the canonical basis of $\mathbb{R}^{n+k}$.

The next important operator is the *Laplace–Beltrami* operator.

Definition 8. *For $f : M \to \mathbb{R}$ of class C^2 we put $\Delta_M f := \mathrm{div}_M(\nabla_M f)$. Then Δ_M is called the* **Laplacian** *on M or* **Laplace–Beltrami operator**.

Note that Δ_M coincides with the Laplace–Beltrami operator on a surface X given in Chapter 1.5 of Vol. 1, equations (15) and (16). Observe also that Δ_M is an elliptic operator on M; this will be used later in this section when we compute the Laplacian of a certain quadratic form.

Finally we have to introduce some curvature quantities for the submanifold M. To this end we choose an orthonormal basis $t_1, \dots, t_n$ of $T_x M$, which together with an orthonormal basis $N_1, \dots, N_k$ of the normal space $T_x M^\perp$ forms an orthonormal basis of $\mathbb{R}^{n+k}$.

Let us initially assume that the codimension k is equal to 1, so that (up to a sign) there is only one unit normal $N = N_1$. Consider $N = N(x)$ as a function of $x \in M$ and assume that $N(\cdot)$ is differentiable, which is true if $M \in C^2$. We then define the **Weingarten map** (cp. Section 1.2 of Vol. 1) of M at $x \in M$ to be the linear map

$$-dN(x) : T_x M \to \mathbb{R}^{n+1} \text{ defined by } t \mapsto -dN(x)(t) = -D_t N(x),$$

where D_t denotes the derivative in the direction of t. Because of $|N|^2 = 1$ it easily follows that $-dN(x)$ is a linear map from $T_x M$ into itself.

The **second fundamental form** $\mathrm{II} = \mathrm{II}_x(\cdot, \cdot)$ of M at x with respect to N is defined to be the bilinear form

$$\begin{aligned} &\mathrm{II} : T_x M \times T_x M \to \mathbb{R} \quad \text{with} \\ &(t, \tau) \longmapsto \mathrm{II}_x(t, \tau) := -\langle dN(t), \tau \rangle = -\langle D_t N, \tau \rangle, \end{aligned}$$

where $\langle \cdot, \cdot \rangle$ denotes the scalar product in $\mathbb{R}^{n+1}$ and $N = N(x)$. It is convenient to consider also the bilinear map $A_x(t, \tau) := \mathrm{II}_x(t, \tau) \cdot N$, which—by a slight abuse of notation—is again called the second fundamental form of M. Observe that for every $x \in M$ the bilinear maps $A_x : T_x M \times T_x M \to T_x M^\perp$ and $\mathrm{II}_x : T_x M \times T_x M \to \mathbb{R}$ are *symmetric*, and that $-dN(x) : T_x M \to T_x M$ is a symmetric endomorphism field. To see this, consider a mapping $\Phi : B_\varepsilon(0) \subset \mathbb{R}^2 \to M \subset \mathbb{R}^{n+1}$ such that $\Phi(0,0) = x$, $\Phi_{x^1}(0,0) = t$, $\Phi_{x^2}(0,0) = \tau$. Differentiating the identities

$$\langle \Phi_{x^1}, N \rangle = 0 = \langle \Phi_{x^2}, N \rangle$$

and putting $x_1 = x_2 = 0$, we infer $\langle \Phi_{x^1 x^2}(0,0), N \rangle + \langle t, D_\tau N \rangle = 0$ and $\langle \Phi_{x^1 x^2}(0,0), N \rangle + \langle \tau, D_t N \rangle = 0$, whence

$$\begin{aligned} \mathrm{II}_x(t, \tau) &= -\langle D_t N, \tau \rangle = \langle \Phi_{x^1 x^2}(0,0), N \rangle \\ &= -\langle D_\tau N, t \rangle = \mathrm{II}_x(\tau, t). \end{aligned} \tag{2}$$

Similarly

$$A_x(t, \tau) = A_x(\tau, t) = \langle \Phi_{x^1 x^2}(0,0), N \rangle \cdot N = [\Phi_{x^1 x^2}(0,0)]^\perp,$$

where $\xi^\perp$ stands for the orthogonal projection of the vector $\xi \in \mathbb{R}^{n+1}$ onto the normal space $T_x M^\perp$.

As in the case of surfaces in $\mathbb{R}^3$ we define the *principal directions* of M at x to be the unit eigenvectors of the Weingarten map

$$-dN = -dN(x) : T_x M \to T_x M$$

and the *principal curvatures* $\lambda_1, \ldots, \lambda_n$ to be the corresponding eigenvalues. Note that there is an orthonormal basis of $T_x M$ consisting of principal directions. Also, if $t_1, \ldots, t_n \in T_x M$ are orthonormal principal directions, then obviously the matrix

$$b_{ij} := \mathrm{II}_x(t_i, t_j) = \mathrm{diag}(\lambda_1, \ldots, \lambda_n).$$

More generally, we conclude from a discussion similar to the one in Section 1.2 of Vol. 1, that the principal curvatures are the eigenvalues of the matrix $G^{-1}B$, where

$$\begin{aligned} B &= (b_{ij})_{i,j=1,\ldots,n}, \quad b_{ij} := \mathrm{II}_x(\xi_i, \xi_j), \\ G &= (g_{ij})_{i,j=1,\ldots,n}, \quad g_{ij} := \langle \xi_i, \xi_j \rangle, \end{aligned}$$

and $\xi_1, \ldots, \xi_n \in T_xM$ denotes an arbitrary basis of the tangent space T_xM. In particular, the principal curvatures are eigenvalues of the symmetric matrix $b_{ij} = \mathrm{II}_x(t_i, t_j)$ for any *orthonormal* basis $t_1, \ldots, t_n$ of T_xM.

Another description of the principal curvatures might also be of interest: Suppose that near a point $x \in M$, the manifold M is locally defined by a smooth function $\varphi : B_\varepsilon(0) \subset \mathbb{R}^n \to \mathbb{R}$, $x^{n+1} = \varphi(x^1, \ldots, x^n)$ and that $e_1, \ldots, e_n$ are principal directions corresponding to the curvatures $\lambda_1, \ldots, \lambda_n$. Such a coordinate system is called a *principal coordinate system*. Without loss of generality assume that $x = 0$, i.e. $\varphi(0) = 0$, $D\varphi(0) = 0$ or $N(0) = e_{n+1}$. It is not difficult to see that M can locally be represented in this way. Now consider the mapping

$$\Phi : B_\varepsilon(0) \subset \mathbb{R}^n \to M \subset \mathbb{R}^{n+1}$$

given by $\Phi(x^1, \ldots, x^n) := (x^1, \ldots, x^n, \varphi(x^1, \ldots, x^n))$, i.e. Φ is a *local parametrization* of M. By arguments similar to those leading to equation (2) we infer

$$D^2\varphi(0) = (\varphi_{x^i x^j}(0))_{i,j=1,\ldots,n} = \mathrm{II}_x(e_i, e_j) = b_{ij} = \mathrm{diag}(\lambda_1, \ldots, \lambda_n),$$

since $e_1, \ldots, e_n$ are principal directions at $x = 0$.

Using the *elementary symmetric functions* of n variables $\sigma_1, \ldots, \sigma_n$, it is now possible to define corresponding curvature quantities K_j by putting

$$K_j(x) := \frac{1}{\binom{n}{j}} \sigma_j(\lambda_1, \ldots, \lambda_n).$$

The cases $j = 1$ and $j = n$ deserve special attention: The *mean curvature* H and the *Gauß(–Kronecker)* curvature K are defined by

$$\begin{aligned} H(x) &:= K_1(x) = \frac{1}{n}(\lambda_1 + \cdots + \lambda_n), \quad \textit{and} \\ K(x) &:= K_n(x) = \lambda_1 \cdot \cdots \cdot \lambda_n \end{aligned}$$

corresponding to the elementary symmetric functions σ_1 and σ_n.

In other words we have

$$\begin{aligned} H(x) &= \frac{1}{n}\mathrm{trace}(G^{-1}B) = \frac{1}{n} \sum_{j,k=1}^{n} g^{jk} b_{jk} \quad \text{and} \\ K(x) &= \det(G^{-1}B) = \frac{\det B}{\det G}, \end{aligned} \tag{3}$$

where

$$B = (b_{ij})_{i,j=1,\dots,n}, \quad b_{ij} = \mathrm{II}_x(\xi_i, \xi_j),$$
$$G = (g_{ij})_{i,j=1,\dots,n}, \quad g_{ij} = \langle \xi_i, \xi_j \rangle, \quad G^{-1} = (g^{ij})_{i,j=1,\dots,n}$$

and $\xi_1, \dots, \xi_n$ stand for a basis of the tangent space $T_x M$. Therefore the mean curvature is (up to the factor $\frac{1}{n}$) just the trace of the Weingarten map $-dN(x)$, or—equivalently—of the second fundamental form II_x.

For arbitrary codimension $k > 1$ it is not possible to define principal directions and curvatures. However we can define principal curvatures and directions with respect to a given normal N_j, $j = 1, \dots, k$, and a corresponding second fundamental form, but we shall not dwell on this here (for a further discussion see e.g. Spivak [1]).

Instead we define for arbitrary $k \geq 1$ and $M \subset \mathbb{R}^{n+k}$ the second fundamental form of M at x as the bilinear form $A_x : T_x M \times T_x M \to T_x M^\perp$ given by $A_x(t, \tau) = -\sum_{j=1}^k \langle dN_j(t), \tau \rangle N_j(x)$.

Arguments similar to those mentioned above prove that $A_x(\cdot, \cdot)$ is a symmetric bilinear form.

Motivated by the foregoing discussion, in particular relation (3), we define the *mean curvature vector* $\vec{H}$ of M at x to be $\frac{1}{n}$trace A_x, i.e.

$$\vec{H}(x) := \frac{1}{n} \sum_{i=1}^n A_x(t_i, t_i), \tag{4}$$

where $t_1, \dots, t_n \in T_x M$ is some orthonormal basis.

In the codimension one case we obtain for the mean curvature vector

$$\begin{aligned} \vec{H}(x) &= \frac{1}{n} \sum_{i=1}^n A_x(t_i, t_i) = -\sum_{i=1}^n \langle dN(t_i), t_i \rangle N \\ &= \frac{1}{n} \left(\sum_{i=1}^n \mathrm{II}_x(t_i, t_i) \right) N(x) \underset{(\text{by } (3))}{=} H(x) N(x), \end{aligned} \tag{5}$$

where $H(x)$ is the mean curvature of M at x with respect to the normal $N(= N_1)$.

We are thus led to

Definition 9. *An n-dimensional C^2-submanifold $M \subset \mathbb{R}^{n+k}$ is called minimal submanifold, if and only if $\vec{H} = 0$ on M.*

A different expression for $\vec{H}$ is obtained as follows:

$$\begin{aligned} \vec{H}(x) &= \frac{1}{n} \sum_{i=1}^n A_x(t_i, t_i) = -\frac{1}{n} \sum_{i=1}^n \sum_{j=1}^k \langle dN_j(t_i), t_i \rangle N_j \\ &= -\frac{1}{n} \sum_{j=1}^k \sum_{i=1}^n \langle D_{t_i} N_j, t_i \rangle N_j = -\frac{1}{n} \sum_{j=1}^k (\mathrm{div}_M N_j) N_j, \end{aligned}$$

taking Definition 7 into account. Thus we obtain

Proposition 3. *Let $M \subset \mathbb{R}^{n+k}$ be an n-dimensional C^2-submanifold of $\mathbb{R}^{n+k}$ and $N_1, \ldots, N_k$ be an orthonormal basis of the normal space $T_xM^\perp$. Then the mean curvature vector $\vec{H} = \vec{H}(x)$ of M at x is given by*

$$\vec{H}(x) = -\frac{1}{n}\sum_{j=1}^{k}(\operatorname{div}_M N_j)N_j. \tag{6}$$

Remark 1. The mean curvature vector $\vec{H}$ is independent of the particular choice of the (local) orthonormal fields $t_1, \ldots, t_n$ and $N_1, \ldots, N_k$; in particular independent of the orientation of M.

Remark 2. Using equations (5) and (6) we infer for hypersurfaces the relation

$$H(x) = -\frac{1}{n}\operatorname{div}_M N, \tag{7}$$

where the mean curvature H corresponds to the unit normal N of M.

We should point out here, that (7) also leads to an alternative proof of the Theorem in Section 2.7 of Vol. 1. In fact, suppose that M is the level surface of some regular function

$$S : G \subset \mathbb{R}^{n+1} \to \mathbb{R},$$

say $M = \{x \in G : S(x) = c\}$, $c \in \mathbb{R}$, and $N(x) = \frac{\nabla S(x)}{|\nabla S(x)|}$ denotes a unit normal field along M. Then we claim that

$$H(x) = -\frac{1}{n}\operatorname{div} N(x), \tag{8}$$

where ∇ and div denote the Euclidean gradient and divergence respectively.

Proof of (8). With Definition 4 and Proposition 2 we find for the divergence of $N(x)$ on M the expression

$$\operatorname{div}_M N(x) = \sum_{j=1}^{n+1}\langle e_j, \nabla_M N^j\rangle = \sum_{j=1}^{n+1}\langle e_j, \nabla N^j - \langle \nabla N^j, N\rangle \cdot N\rangle, \tag{9}$$

where we have put $N = (N^1, \ldots, N^{n+1})$. On the other hand by taking partial derivatives $\frac{\partial}{\partial x^i}$ we infer from $|N|^2 = 1$, the relation $\langle N, \frac{\partial N}{\partial x^i}\rangle = 0$ for any $i = 1, \ldots, n+1$, or

$$\sum_{j=1}^{n+1} N^j(x)\frac{\partial N^j}{\partial x^i} = 0 \quad \text{for } i = 1, \ldots, n+1. \tag{10}$$

Now we get by (10)

$$\sum_{j=1}^{n+1} \langle e_j, \langle \nabla N^j, N \rangle \cdot N \rangle = \sum_{j=1}^{n+1} \langle \nabla N^j, N \rangle N^j = \sum_{j=1}^{n+1} \sum_{i=1}^{n+1} \left(\frac{\partial N^j}{\partial x^i} \cdot N^i \right) N^j$$
$$= \sum_{i=1}^{n+1} \sum_{j=1}^{n+1} \left(\frac{\partial N^j}{\partial x^i} N^j \right) N^i = 0.$$

Therefore (9) yields

$$\mathrm{div}_M N(x) = \sum_{j=1}^{n+1} \langle e_j, \nabla N^j \rangle = \sum_{j=1}^{n+1} \frac{\partial N^j}{\partial x^j} = \mathrm{div}\, N(x).$$

This proves (cf. Vol. 1, Section 2.7, Theorem)

Proposition 4. *If G is a domain in $\mathbb{R}^{n+1}$, and if S is a function of class $C^2(G)$ such that $\nabla S(x) \neq 0$ on G, then the mean curvature $H(x)$ of the level hypersurface $\mathcal{F}_c = \{x \in G; S(x) = c\}$ passing through $x \in G$ with respect to the unit normal field $N(x) = |\nabla S(x)|^{-1} \nabla S(x)$ of $\mathcal{F}_c$ is given by the equation*

$$H(x) = -\frac{1}{n} \mathrm{div}\, N(x).$$

Proposition 4 also permits to carry over the Schwarz–Weierstraß field theory for two-dimensional minimal surfaces to $\mathbb{R}^{n+1}$; compare the discussion in Section 2.8 of Vol. 1. By essentially the same arguments, using Gauss's theorem, we derive

Theorem 1. *A C^2-family of embedded hypersurfaces $\mathcal{F}_c$ covering a domain G in $\mathbb{R}^{n+1}$ is a* Mayer family *of minimal submanifolds if and only if its normal field is divergence free. Such a foliation by minimal submanifolds is area minimizing in the following sense:*

(i) *Let $\mathcal{F}$ be a piece of some of the minimal leaves $\mathcal{F}_c$ with $\mathcal{F} \Subset G$. Then we have*

$$\mathrm{Area}(\mathcal{F}) = \int_{\mathcal{F}} dA \leq \int_{\mathcal{S}} dA = \mathrm{Area}(\mathcal{S})$$

for each C^1-hypersurface $\mathcal{S}$ contained in G with $\partial \mathcal{F} = \partial \mathcal{S}$.

(ii) *("Kneser's transversality Theorem"): Let T be a hypersurface in G which, in all of its points, is tangent to the normal field of the minimal foliation, and suppose that T cuts out of each leaf $\mathcal{F}_c$ some piece $\mathcal{F}_c^*$ whose boundary $\partial \mathcal{F}_c$ lies on T. Then we have*

$$\int_{\mathcal{F}_{c_1}^*} dA = \int_{\mathcal{F}_{c_2}^*} dA$$

for all admissible parameter values c_1 and c_2, and secondly

$$\int_{\mathcal{F}_c} dA \leq \int_{\mathcal{S}} dA$$

for all C^1-hypersurfaces $\mathcal{S}$ contained in G whose boundary $\partial\mathcal{S}$ is homologous to $\partial\mathcal{F}$ on T.

We remark here that a result similar to—but more general than—Theorem 1 has been used by Bombieri, De Giorgi and Giusti [1] to show that the seven-dimensional "Simons-cone"

$$C = \{(x, y) \in \mathbb{R}^4 \times \mathbb{R}^4 : |x|^2 = |y|^2\}$$

is area minimizing in a very general sense. Indeed they were able to construct a foliation of $\mathbb{R}^8$ consisting of smooth minimal hypersurfaces and the singular minimal cone C. This was also the first example of an area-minimizing boundary in $\mathbb{R}^{n+1}$ with an **interior** singularity, namely the origin, which dashed the hope to prove interior regularity of area minimizing boundaries in arbitrary dimensions.

The *divergence theorem* for a C^2-compact manifold $M \subset \mathbb{R}^{n+k}$ with smooth boundary $\partial M = \overline{M} \setminus M$ states that for any C^1-vector field $X : \overline{M} \to \mathbb{R}^{n+k}$ the identity

$$\int_M \operatorname{div}_M X \, dA = -n \int_M X \cdot \vec{H} \, dA + \int_{\partial M} X \cdot \nu \, dA$$

holds where ν denotes the exterior unit normal field to ∂M which is tangent to M along ∂M. Here $\vec{H} = -\frac{1}{n} \sum_{i=1}^k (\operatorname{div}_M N_i) N_i$ denotes the mean curvature vector and integration over ∂M is with respect to the standard $(n-1)$-dimensional area measure (or, equivalently, $(n-1)$-dimensional Hausdorff measure $\mathcal{H}^{n-1}$).

In particular, if X is a tangential vector field, i.e. $X(x) \in T_x M$ for each $x \in M$ or if M is minimal, then we have the formula

$$\int_M \operatorname{div}_M X \, dA = \int_{\partial M} X \cdot \nu \, dA.$$

Similarly, if X has compact support, or if $\partial M = \emptyset$, then the divergence theorem yields

$$\int_M \operatorname{div}_M X \, dA = -n \int_M X \cdot \vec{H} \, dA,$$

and finally

$$\int_M \operatorname{div}_M X \, dA = 0,$$

if X is a compactly supported, tangential vector field on M.

Remark 3. It can be shown that $M \subset \mathbb{R}^{n+k}$ is stationary for the n-dimensional area functional, if and only if $\vec{H} \equiv 0$; see Vol. 3, Section 3.2, for details.

Remark 4. Some authors use trace A_x – instead of $\frac{1}{n}$trace A_x – as a definition of the mean curvature vector. This clearly is irrelevant when working with minimal submanifolds; but it is of importance when $\vec{H} \not\equiv 0$.

Next we shall derive a generalization of Theorem 1 in Section 2.5 of Vol. 1 (compare also Theorem 1 in Vol. 1, Section 2.6). To accomplish this we simply compute the Laplace (–Beltrami) operator of the vector field $X(x) = x$. Assuming that $M \subset \mathbb{R}^{n+k}$ is an n-dimensional submanifold of class C^2, we find for the gradient of X on M the expression

$$\nabla_M X^i = \nabla_M x^i = e_i - \langle N_1, e_i \rangle N_1 - \cdots - \langle N_k, e_i \rangle N_k,$$

$i = 1, \ldots, n+k$, where $e_1, \ldots, e_{n+k}$ stands for the canonical basis of $\mathbb{R}^{n+k}$. Applying div_M to this relation we obtain the identity

$$\begin{aligned} \Delta_M x^i &= \operatorname{div}_M(\nabla_M x^i) = -\langle N_1, e_i \rangle \operatorname{div}_M N_1 - \cdots - \langle N_k, e_i \rangle \operatorname{div}_M N_k \\ &= -\sum_{j=1}^{k} \langle N_j, e_i \rangle \operatorname{div}_M N_j, \end{aligned}$$

$$\text{since} \quad \big\langle \nabla_M \langle N_j, e_i \rangle, N_j \big\rangle = 0, \quad \forall i, j.$$

Thus we have for $i = 1, \ldots, n+k$

$$\Delta_M(\langle x, e_i \rangle) = -\sum_{j=1}^{k} \langle N_j, e_i \rangle \operatorname{div}_M N_j = -e_i \left(\sum_{j=1}^{k} N_j \cdot \operatorname{div}_M N_j \right).$$

By Proposition 3 this implies $\Delta_M x^i = n \cdot H^i$, where $\vec{H} = (H^1, \ldots, H^{n+k})$ is the mean curvature vector of M. Thus we have proved

Theorem 2. *Let $M \subset \mathbb{R}^{n+k}$ be an n-dimensional C^2-submanifold. Then the position vector x fulfills the identity*

$$\Delta_M x = n \vec{H}.$$

Corollary 1. *$M \subset \mathbb{R}^{n+k}$ is a minimal submanifold, if and only if $\Delta_M x = 0$ holds on M.*

A straight-forward application of the maximum principle for harmonic functions yields the following enclosure results (cp. Theorem 1 in Section 4.1 for the case $n = 2$, $k = 1$ and its proof).

Corollary 2 (Convex hull theorem). *Let $M \subset \mathbb{R}^{n+k}$ be a compact n-dimensional minimal submanifold. Then M is contained in the convex hull $\mathcal{K}$ of its boundary ∂M. Moreover if M touches the convex hull $\mathcal{K}$ at some interior point, then M is part of a plane. In particular there is no compact minimal submanifold M without boundary.*

We now consider the possibility of obtaining polynomials p which are subharmonic functions on M, i.e. which satisfy

$$\Delta_M p \geq 0 \quad \text{on } M \text{ if } H = 0.$$

To achieve this, we define for any $j = 1, \dots, n-1$ a quadratic function $p_j = p_j(x^1, \dots, x^{n+k})$ by

$$p_j(x) := \sum_{i=1}^{n+k-j} |x^i|^2 - \frac{(n-j)}{j} \sum_{i=n+k-j+1}^{n+k} |x^i|^2.$$

Note that for $n = 2$, $j = k = 1$, we recover the polynomial considered in Theorem 2 of Section 4.1.

We have the following

Theorem 3. *Let $M \subset \mathbb{R}^{n+k}$ be an n-dimensional minimal submanifold of class C^2. Then for each $j = 1, \dots, n-1$ the quadratic form $p_j(\cdot)$ is a subharmonic function on M.*

Proof. Fixing $j \in \{1, \dots, n-1\}$ we set $P := p_j$ and compute the Laplace–Beltrami expression $\Delta_M P$ as follows:

$$\begin{aligned}
\frac{1}{2}\Delta_M P &= \frac{1}{2}\operatorname{div}_M(\nabla_M P) \\
&= \frac{1}{2}\operatorname{div}_M\Big\{2x^1\nabla_M x^1 + \cdots + 2x^{n+k-j}\nabla_M x^{n+k-j} \\
&\quad - \frac{(n-j)}{j}\Big[2x^{n+k-j+1}\nabla_M x^{n+k-j+1} + \cdots + 2x^{n+k}\nabla_M x^{n+k}\Big]\Big\} \\
&= |\nabla_M x^1|^2 + \cdots + |\nabla_M x^{n+k-j}|^2 + x^1\Delta_M x^1 + \cdots + x^{n+k-j}\Delta_M x^{n+k-j} \\
&\quad - \frac{(n-j)}{j}\Big[|\nabla_M x^{n+k-j+1}|^2 + \cdots + |\nabla x^{n+k}|^2 \\
&\quad + x^{n+k-j+1}\Delta_M x^{n+k-j+1} + \cdots + x^{n+k}\Delta_M x^{n+k}\Big].
\end{aligned}$$

Since M is minimal this gives

$$\frac{1}{2}\Delta_M P = \sum_{s=1}^{n+k-j} |\nabla_M x^s|^2 - \frac{n-j}{j}\sum_{s=1}^{j} |\nabla_M x^{n+k-j+s}|^2.$$

To compute the terms $|\nabla_M x^i|^2$ we denote by $\mathcal{P} : \mathbb{R}^{n+k} \to T_x M$ the orthogonal projection of $\mathbb{R}^{n+k}$ onto the tangent space $T_x M$. Let $(p_{ij})_{i,j=1,\dots,n+k}$ stand for the matrix of $\mathcal{P}$ with respect to the canonical basis $e_1, \dots, e_{n+k}$ of $\mathbb{R}^{n+k}$. Then we have (by Definition 4)

$$\nabla x^i = \mathcal{P}(e_i) = \sum_{l=1}^{n+k} p_{li} e_l \quad \text{and}$$

$$\begin{aligned} |\nabla_M x^i|^2 &= \left(\sum_{l=1}^{n+k} p_{li} e_l\right)\left(\sum_{j=1}^{n+k} p_{ji} e_j\right) \\ &= \sum_{l,j=1}^{n+k} p_{li} p_{ji} e_l e_j = \sum_{j=1}^{n+k} p_{ji}^2 . \end{aligned}$$

Since $\mathcal{P}$ is a projection we clearly have $p_{ij} = p_{ji}$ and $\mathcal{P} = \mathcal{P}^2$, whence

$$p_{ij} = \sum_{l=1}^{n+k} p_{il} p_{lj},$$

in particular

$$p_{ii} = \sum_{j=1}^{n+k} p_{ij}^2 = |\nabla_M x^i|^2 .$$

Again, since $\mathcal{P}$ is a projection, all eigenvalues are either equal to one or zero and the sum of the eigenvalues is equal to n:

$$\operatorname{trace} \mathcal{P} = \sum_{i=1}^{n+k} p_{ii} = \sum_{i=1}^{n+k} |\nabla_M x^i|^2 = n .$$

Concluding we find for $\frac{1}{2}\,\Delta_M P$ the estimate

$$\begin{aligned} \frac{1}{2}\Delta_M P &= \sum_{s=1}^{n+k-j} |\nabla_M x^s|^2 - \frac{n-j}{j} \sum_{s=1}^{j} |\nabla_M x^{n+k-j+s}|^2 \\ &= \sum_{s=1}^{n+k-j} p_{ss} - \frac{n-j}{j} \sum_{s=1}^{j} p_{n+k-j+s,n+k-j+s} \\ &= \sum_{s=1}^{n+k} p_{ss} - \sum_{s=n+k-j+1}^{n+k} p_{ss} - \frac{n-j}{j} \sum_{s=1}^{j} p_{n+k-j+s,n+k-j+s} \\ &\geq \operatorname{trace} \mathcal{P} - j - (n-j) \\ &\geq n - j - (n-j) = 0 . \end{aligned}$$

□

Remark 5. Clearly, for any $j \geq n$ and $n + k - j \geq 1$ the polynomials p_j are trivially subharmonic on M, since $-\frac{(n-j)}{j} \geq 0$ in this case.

Again, by a straight-forward application of maximum principle we obtain

Corollary 3. *Suppose $M \subset \mathbb{R}^{n+k}$ is a minimal submanifold with boundary ∂M contained in a body congruent to*

$$\mathcal{H}_j(\varepsilon) := \Big\{(x^1, \dots, x^{n+k}) \in \mathbb{R}^{n+k} : p_j(x^1, \dots, x^{n+k}) \le \varepsilon\Big\},$$

for any $\varepsilon \in \mathbb{R}$. *Then* $M \subset \mathcal{H}_j(\varepsilon)$, $j = 1, \dots, n-1$.

In this Corollary one can take $j = 1$ obtaining "*nonexistence cones*" for any dimension n and any codimension k. In other words we consider the cones $C_{n+k} = C^+_{n+k} \cup C^-_{n+k} \cup \{0\}$ defined by

$$\begin{aligned} C^{\pm}_{n+k} := & \Bigg\{(x^1, \dots, x^{n+k}) \in \mathbb{R}^{n+k} : \pm x^{n+k} > 0, \text{ and} \\ & \quad \sum_{i=1}^{n+k-1} |x^i|^2 \le (n-1)|x^{n+k}|^2\Bigg\} \\ = & \Big\{x \in \mathbb{R}^{n+k} : \pm x^{n+k} > 0 \text{ and } p_1(x) \le 0\Big\}. \end{aligned}$$

Theorem 4. *Let* $C \subset \mathbb{R}^{n+k}$ *be a cone with vertex* P_0 *which is congruent to* C_{n+k} *and let* $C^{\pm}$ *denote the two disjoint parts which correspond to* $C^{\pm}_{n+k}$. *Then there is no connected, compact, n-dimensional minimal submanifold* $M \subset \mathbb{R}^{n+k}$ *with* $\partial M \subset C$ *such that both* $\partial M \cap C^+$ *and* $\partial M \cap C^-$ *are nonempty.*

Proof. By performing a rotation and translation we may assume without loss of generality that $C = C_{n+k}$. Suppose on the contrary that there is a minimal M satisfying the assumptions of Theorem 4. By Theorem 3 we obtain the inequality

$$\Delta_M \left[\sum_{i=1}^{n+k-1} |x^i|^2 - (n-1)|x^{n+k}|^2\right] \ge 0$$

and by the hypothesis of Theorem 4 we have

$$\left[\sum_{i=1}^{n+k-1} |x^i|^2 - (n-1)|x^{n+k}|^2\right]\Bigg|_{\partial M} \le 0.$$

The maximum principle yields

$$\left[\sum_{i=1}^{n+k-1} |x^i|^2 - (n-1)|x^{n+k}|^2\right]\Bigg|_{M} \le 0,$$

or equivalently, $M \subset C_{n+k}$. Since M is connected and $\partial M \cap C^+_{n+k} \ne \emptyset$, $\partial M \cap C^-_{n+k} \ne \emptyset$, M must contain the vertex 0 of the cone, which clearly contradicts the manifold property of M. □

We remark that Theorem 4 may be used to derive necessary conditions for the existence of compact, connected minimal submanifolds with several boundary components.

Corollary 4 (Necessary Condition). *Let $B_1, B_2 \subset \mathbb{R}^{n+k}$ be closed sets and suppose there exists an n-dimensional compact, connected minimal submanifold $M \subset \mathbb{R}^{n+k}$ with $\partial M \subset B_1 \cup B_2$ and that both $\partial M \cap B_1 \neq \phi$ and $\partial M \cap B_2 \neq \phi$. Then we have:*

(i) *If B_i, $i = 1, 2$ are closed balls with centers x_i and radii δ_i and $R := |x_1 - x_2|$, then*

$$R \leq \left(\frac{n}{n-1}\right)^{\frac{1}{2}} (\delta_1 + \delta_2).$$

(ii) *If B_1 and B_2 are arbitrary compact sets of diameters d_1 and d_2 which are separated by a slab of width $r > 0$, then*

$$r \leq \frac{1}{2}\left(\frac{2n(n+k)}{(n-1)(n+k+1)}\right)^{\frac{1}{2}} (d_1 + d_2). \qquad \square$$

Next we consider arbitrary n-dimensional submanifolds $M \subset \mathbb{R}^{n+k}$ with mean curvature vector $\vec{H}$. According to Theorem 2 we have the identity

$$\Delta_M x = n \vec{H},$$

and by Proposition 3,

$$\vec{H}(x) = -\frac{1}{n}\sum_{j=1}^{k} (\operatorname{div}_M N_j) N_j$$

for an arbitrary orthonormal basis $N_1, \dots, N_k \in \mathbb{R}^{n+k}$ of the normal space $T_x M^\perp$.

Let $H^1, \dots, H^k$ be the components of $\vec{H}$ with respect to that basis $N_1, \dots, N_k$ i.e.

$$\begin{aligned} H &= H^1 N_1 + \cdots + H^k N_k, \quad \text{or} \\ H^i &= -\frac{1}{n}\operatorname{div}_M N_i \quad \text{for } i = 1, \dots, k, \end{aligned}$$

and put

$$p(x) := \sum_{i=1}^{n+k-j} |x^i|^2 - \frac{(n-j)}{j} b \sum_{i=n+k-j+1}^{n+k} |x^i|^2,$$

where $b \in \mathbb{R}$ and $j = 1, \dots, n-1$. Defining r_j and s_j by

$$r_j(x) := \sum_{i=1}^{n+k-j} |x^i|^2 \quad \text{and} \quad s_j(x) := \sum_{i=n+k-j+1}^{n+k} |x^i|^2$$

we obtain

$$p(x) = r_j(x) - \frac{(n-j)}{j} b s_j(x).$$

By the same arguments as in the proof of Theorem 3 we conclude

$$\begin{aligned}
\frac{1}{2}\Delta_M p = & \sum_{i=1}^{n+k-j} x^i \Delta_M x^i - b\frac{(n-j)}{j} \sum_{i=n+k-j+1}^{n+k} x^i \Delta_M x^i \\
& + \sum_{i=1}^{n+k-j} |\nabla_M x^i|^2 - b\frac{(n-j)}{j} \sum_{i=n+k-j+1}^{n+k} |\nabla_M x^i|^2 \\
& \geq n \left\langle \vec{H}, \left(x^1, \dots, x^{n+k-1}, -b\frac{(n-j)}{j} x^{n+k-j+1}, \dots, -b\frac{(n-j)}{j} x^{n+k} \right) \right\rangle \\
& + (n-j)(1-b) \\
& \geq -n|\vec{H}| \left[r_j + \frac{b^2(n-j)^2}{j^2} s_j \right]^{\frac{1}{2}} \\
& + (n-j)(1-b), \quad \text{by Schwarz's inequality.}
\end{aligned}$$

Finally we obtain the estimate

$$\frac{1}{2}\Delta_M p \geq (n-j) \left\{ (1-b) - n|\vec{H}| \left[\frac{r_j}{(n-j)^2} + \frac{b^2}{j^2} s_j \right]^{\frac{1}{2}} \right\}.$$

Thus we have proved:

Theorem 5. *Let $M \subset \mathbb{R}^{n+k}$ be an n-dimensional submanifold with mean curvature vector $\vec{H} = H^1 N_1 + \dots + H^k N_k$, $0 \leq b \leq 1$, $1 \leq j \leq n-1$ and $p(x) = \sum_{i=1}^{n+k-j} |x^i|^2 - \frac{(n-j)}{j} b \sum_{i=n+k-j+1}^{n+k} |x^i|^2 = r_j(x) - \frac{(n-j)}{j} b s_j(x)$. Then $p(x)$ is subharmonic on M, if*

$$\text{(11)} \qquad b + n|\vec{H}| \left[\frac{r_j(x)}{(n-j)^2} + \frac{b^2}{j^2} s_j(x) \right]^{1/2} \leq 1$$

holds true, where

$$|\vec{H}| = (|H^1|^2 + \dots + |H^k|^2)^{1/2}.$$

□

Observe that (11) is satisfied for example if

$$\text{(12)} \qquad q := \sup_{x \in M} |x||\vec{H}(x)| < \frac{1}{n} \quad \text{and} \quad b := 1 - nq.$$

Corollary 5. *Suppose that condition (12) holds true. Then for any $j = 1, \dots, n-1$ the quadratic polynomial $p(x) = r_j(x) - (\frac{(n-j)}{j}) b s_j(x)$ is subharmonic on M. Therefore, if M is compact the estimate $\sup_M p \leq \sup_{\partial M} p$ is fulfilled. In particular, if $K := K^+ \cup \{0\} \cup K^-$, where $K^\pm := \{x \in \mathbb{R}^{n+k} : \sum_{i=1}^{n+k-1} |x^i|^2 - (n-1)(1-nq)|x^{n+k}|^2 \leq 0, \pm x^{n+k} > 0\}$ and $\partial M \subset K$ such that both $\partial M \cap K^+$ and $\partial M \cap K^-$ are nonempty, then M cannot be connected.*

Alternatively, (11) is fulfilled provided

(13) $$q := \sup_M |x||\vec{H}(x)| < \frac{n-j}{n},$$

and $b := \min(\frac{1}{n-1}, 1 - \frac{n\,q}{n-j})$.

Corollary 6. *Suppose that* (13) *holds for some* $j = 1, \ldots, n-1$. *Then* $p(x) = r_j(x) - \frac{b(n-j)}{j} \cdot s_j(x)$ *is subharmonic on* M. *In particular, if this holds with* $j = 1$ *then there is no connected compact submanifold with mean curvature* $\vec{H}$ *which satisfies* $\partial M \subset K$ *and* $\partial M \cap K^+ \neq \emptyset$, *and* $\partial M \cap K^- \neq \emptyset$, *where*

$$K = K^+ \cup \{0\} \cup K^- \quad \textit{and}$$
$$K^\pm := \left\{ x \in \mathbb{R}^{n+k} : \sum_{i=1}^{n+k-1} |x^i|^2 - (n-1)b|x^{n+k}|^2 \leq 0, \pm x^{n+k} > 0 \right\}. \quad \square$$

4.3.1 An Optimal Nonexistence Result for Minimal Submanifolds of Codimension One

Now we address the question whether the *"nonexistence cones"* C_{n+k} considered in Theorem 4 can still be enlarged. In Section 6.2, Corollary, we have considered the cone

$$\mathcal{K} := \mathcal{K}^+ \cup \{0\} \cup \mathcal{K}^-,$$

where

$$\mathcal{K}^\pm := \left\{ (x, y, z) \in \mathbb{R}^3 : z \gtrless 0 \text{ and } x^2 + y^2 < z^2 \sinh^2 \tau \right\}$$

and $\tau = 1.1996\ldots$ is a solution of the equation

$$\tau \sinh \tau = \cosh \tau.$$

This cone $\mathcal{K}$ is in fact a "nonexistence cone" for $n = 2$, $k = 1$ which cannot be enlarged further, since it is the envelope of a field of suitable catenoids; in other words $\mathcal{K}$ is "enclosed" by the "catenoidal domains"

$$\mathcal{K}_\alpha = \left\{ (x, y, z) \in \mathbb{R}^3 : x^2 + y^2 < \alpha^2 \cosh^2 \frac{z}{\alpha} \right\},$$

cp. the discussion in Section 4.2. We generalize this argument as follows: Consider a curve $(x, y(x))$ in the Euclidean plane and its rotational symmetric graph (of dimension $n+1$)

$$\mathcal{M}_{\text{rot}} := \{ (x, y(x) \cdot w) \in \mathbb{R} \times \mathbb{R}^{n+1} : x \in [a, b], w \in S^n \},$$

where $S^n = \{z \in \mathbb{R}^{n+1} : |z| = 1\}$ denotes the unit n-sphere. One readily convinces oneself that the $(n+1)$-dimensional area of $\mathcal{M}_{\text{rot}}$ is proportional to the one-dimensional variational integral

$$\mathcal{J} = \mathcal{J}(y) = \int_a^b y^n(x)\sqrt{1 + y'^2(x)}\, dx.$$

In other words, extremals of $\mathcal{J}$ correspond to $(n+1)$-dimensional minimal submanifolds in $\mathbb{R}^{n+2}$, which are rotationally symmetric, the so-called "*n-catenoids*" (or, to be more precise, "$(n+1)$-*catenoids*"). The Euler equation of the integral $\mathcal{J}$ is simply

$$\frac{d}{dx}\left(\frac{y'y^n}{\sqrt{1+y'^2}}\right) = ny^{n-1}\sqrt{1+y'^2}. \tag{14}$$

Since the integrand f of $\mathcal{J}(\cdot)$ does not explicitly depend on the variable x we immediately obtain a first integral of (14), namely

$$y^n = \lambda\sqrt{1+y'^2}$$

for any $\lambda > 0$. A further integration gives the inverse of a solution $y = y(x)$ of the Euler equation (14) as follows:

$$x = x(y) = \lambda \int_{\sqrt[n]{\lambda}}^{y} \frac{d\xi}{\sqrt{\xi^{2n} - \lambda^2}} + c. \tag{15}$$

These inverse functions are defined for any $\lambda > 0$, $c \in \mathbb{R}$ and all $y \geq \sqrt[n]{\lambda}$. Note that (15) with $n = 1$ leads to the classical catenaries, which—upon rotation into $\mathbb{R}^3$—determine the well known catenoids. Of importance in our following construction here, is the one parameter family of "n-catenaries" (or rather of their inverses)

$$x = g(y, \lambda) := \lambda \int_{\sqrt[n]{\lambda}}^{y} \frac{d\xi}{\sqrt{\xi^{2n} - \lambda^2}}, \quad y \geq \sqrt[n]{\lambda}.$$

Claim. *The envelope of the family $g(y, \lambda)$, $\lambda > 0$, is the straight line $y = \tau_0 x$, $x > 0$, where $\tau_0 := \sqrt{z_0^{2n} - 1}$, and z_0 is the unique solution of the equation*

$$\frac{z}{\sqrt{z^{2n} - 1}} = \int_1^z \frac{d\xi}{\sqrt{\xi^{2n} - 1}}. \tag{16}$$

Proof. First note that (15) implies that $\frac{d^2x}{dy^2} < 0$, whence the solutions $x = g(y, \lambda)$, $\lambda > 0$, are strictly convex functions when considered as graphs over x. Hence for each $\lambda > 0$ there exist unique numbers $\tau = \tau(\lambda)$, $x = x(\lambda) > 0$ and $y = y(\lambda) > 0$ with the properties

$$x(\lambda) = g(y(\lambda), \lambda) \quad \text{and} \quad \tau(\lambda) = \frac{y(\lambda)}{x(\lambda)} = y'(x(\lambda)),$$

where $y'(x(\lambda))$ denotes the slope of the curve $x = g(y, \lambda)$ considered as a function $y(x)$ at the particular point $x(\lambda)$. Since $y^n = \lambda\sqrt{1+y'^2}$, this last requirement can be written as

$$\tau(\lambda) = \frac{y(\lambda)}{x(\lambda)} = \frac{\sqrt{y^{2n}(\lambda) - \lambda^2}}{\lambda} = \left\{ \left[\frac{y(\lambda)}{\sqrt[n]{\lambda}} \right]^{2n} - 1 \right\}^{\frac{1}{2}}.$$

We now claim that the quotient $q(\lambda) := \frac{y(\lambda)}{\sqrt[n]{\lambda}}$ is independent of λ, i.e. $q(\lambda) =$ const. Indeed we find successively $x(\lambda) = \frac{y(\lambda)}{\tau(\lambda)} = \frac{y(\lambda)}{[q^{2n}(\lambda)-1]^{\frac{1}{2}}}$; on the other hand,

$$x(\lambda) = \lambda \int_{\sqrt[n]{\lambda}}^{y} \frac{d\xi}{\sqrt{\xi^{2n} - \lambda^2}} = \int_{\sqrt[n]{\lambda}}^{y} \frac{d\xi}{\{(\frac{\xi}{\sqrt[n]{\lambda}})^{2n} - 1\}^{\frac{1}{2}}}.$$

Thus $q(\lambda) = \frac{y(\lambda)}{\sqrt[n]{\lambda}}$ satisfies (16). However, there is only one solution of (16), since the left hand side of (16) is monotonically decreasing, while the right hand side monotonically increases, and both sides are continuous. Concluding we have shown that each member of the family $g(y, \lambda)$, $\lambda > 0$, $y \geq \sqrt[n]{\lambda}$, touches the half line $y = \tau_0 x$, $\tau_0 = \sqrt{z_0^{2n} - 1}$ precisely at one point, namely at $x_0(\lambda) = \frac{z_0}{\tau_0} \sqrt[n]{\lambda}$, $y_0(\lambda) = z_0 \sqrt[n]{\lambda}$. Also, each point of the half line $y = \tau_0 x$, $x > 0$, is the point of contact for precisely one member of the family $g(\cdot, \lambda)$, $\lambda > 0$. This proves the claim. □

Let $f(\cdot, \lambda)$ denote the family of inverse functions, that is we have

$$f(g(y,\lambda),\lambda) = y, \quad \text{for } y \geq \sqrt[n]{\lambda} \quad \text{and} \quad g(f(x,\lambda),\lambda) = x \quad \text{for } x \geq 0.$$

We extend f by an even reflection i.e. $f(x, \lambda) = f(-x, \lambda)$ for $x \leq 0$, so as to obtain a smooth function defined on the real axis. Observe that for $n = 1$, these are precisely the catenaries $f(x, \lambda) = \lambda \cosh(\frac{x}{\lambda})$. Put $r := \{\sum_{i=1}^{n+1} |x^i|^2\}^{\frac{1}{2}}$; then for each $\lambda > 0$ the hypersurfaces

$$\mathcal{M}_\lambda = \left\{ x \in \mathbb{R}^{n+2} : r = f(x^{n+2}, \lambda) \right\}$$

are smooth $(n+1)$-dimensional minimal submanifolds of $\mathbb{R}^{n+2}$. Furthermore the foregoing construction shows that the sets

$$\mathcal{G}_\lambda := \left\{ x \in \mathbb{R}^{n+2} : r < f(x^{n+2}, \lambda) \right\}$$

for $\lambda > 0$ *enclose* the cone

$$\mathcal{K}_{\tau_0} := \left\{ x \in \mathbb{R}^{n+2} : \pm \tau_0 x^{n+2} > r \right\} \cup \{0\}$$

in the sense of Section 4.2.

By a straightforward modification of Theorem 1, Section 4.2, i.e. by Hopf's maximum principle and the arguments in the proof of the Enclosure Theorem II, Section 4.2 we conclude the following "*Nonexistence Theorem*".

Theorem 6. *The family of domains* $\{\mathcal{G}_\lambda\}_{\lambda>0}$ *enclose the cone* $\mathcal{K}_{\tau_0}$*, where* $\tau_0 := \sqrt{z_0^{2n} - 1}$ *and* z_0 *is a solution of the equation* (16). *Furthermore, if* $\mathcal{C} = \mathcal{C}^+ \cup \{0\} \cup \mathcal{C}^- \subset \mathbb{R}^{n+2}$ *is a cone with vertex* p_0 *which is congruent to* $\mathcal{K}_{\tau_0}$*, then there is no connected, compact* $(n+1)$*-dimensional minimal submanifold* $M \subset \mathbb{R}^{n+2}$ *with* $\partial M \subset \mathcal{C}$ *such that both* $\partial M \cap \mathcal{C}^+$ *and* $\partial M \cap \mathcal{C}^-$ *are nonempty.*

Remark 1. By construction, the hypersurfaces $r = f(x^{n+2}, \lambda)$, $\lambda > 0$ are minimal in $\mathbb{R}^{n+2}$ and intersect the boundary of the cone $\mathcal{K}_{\tau_0}$ in an n-dimensional sphere. Thus there is no "larger" cone with the nonexistence property described in Theorem 6. In particular the corresponding nonexistence cones introduced in Theorem 4 are "smaller" than $\mathcal{K}_{\tau_0}$. This is illustrated in the following table. Observe that the cones $\mathcal{K}_{\tau_0}$ become larger when the dimension increases.

Dimension of the surface: $n+1$	τ_0	Angle of aperture	$\sqrt{n}$
2	1.51	56.46	1
3	2.37	67.15	1.414
4	3.15	72.40	1.732
5	3.89	75.60	2
6	4.63	77.81	2.236
7	5.44	79.59	2.449
8	6.02	80.58	2.645

4.4 Geometric Maximum Principles

4.4.1 The Barrier Principle for Submanifolds of Arbitrary Codimension

Let $\mathcal{S}$ (the "barrier") be a C^2 hypersurface of $\mathbb{R}^{n+1}$ with mean curvature Λ with respect to the local normal field ν. Assume that $M \subset \mathbb{R}^{n+1}$ is another hypersurface with mean curvature H which lies locally on that side of $\mathcal{S}$ to which the normal ν points, and that the inequality

$$\sup_{U \cap M} |H| \le \inf_{U \cap \mathcal{S}} \Lambda \tag{1}$$

holds in a neighbourhood $U = U(p_0) \subset \mathbb{R}^{n+1}$ of any point $p_0 \in \mathcal{S} \cap M$. If the intersection $\mathcal{S} \cap M$ is nonempty (in other words, if M touches $\mathcal{S}$ in some interior point p_0) then, using Hopf's lemma and an argument similar as in the proof of Theorem 1 in Section 4.2, it follows that M must be locally contained in $\mathcal{S}$. In Section 4.2 we have admitted one of the surfaces to be singular in possible points of intersection.

Now we discuss a version of this barrier principle for n-dimensional submanifolds $M \subset \mathbb{R}^{n+k}$ with bounded mean curvature vector $\vec{H}$. The crucial requirement is again a condition of type (1); however, the mean curvature Λ has to be replaced by the "*n-mean curvature*" Λ_n, which is the arithmetic mean of the sum of the n smallest principal curvatures of $\mathcal{S}$, while $|H|$ has to be replaced by the length of the mean curvature vector $\vec{H}$ of the submanifold M.

Let us recall some notations: $\mathcal{S} \subset \mathbb{R}^{n+k}$ denotes a C^2-hypersurface with (local) normal field ν and $\lambda_1 \leq \cdots \leq \lambda_{n+k-1}$ stand for the principal curvatures of $\mathcal{S}$ with respect to that normal ν. We define the "*n-mean curvature*" Λ_n with respect to the normal ν as

$$\Lambda_n := \frac{1}{n}(\lambda_1 + \cdots + \lambda_n), \quad \text{where } \lambda_1 \leq \lambda_2 \leq \cdots \leq \lambda_n \leq \cdots \lambda_{n+k-1}.$$

Furthermore let $M \subset \mathbb{R}^{n+k}$ be an n-dimensional C^2-submanifold with mean curvature vector

$$\vec{H} = -\frac{1}{n}\sum_{j=1}^{k} (\operatorname{div}_M N_j)N_j,$$

where $N_1, \ldots, N_k$ denotes an orthonormal basis of the normal space $T_xM^\perp$, cp. Section 4.3 for definition and properties of the mean curvature vector.

Theorem 1. *Let $M \subset \mathbb{R}^{n+k}$ be an n-dimensional C^2-submanifold with mean curvature vector $\vec{H}$, and $\mathcal{S} \subset \mathbb{R}^{n+k}$ be a C^2-hypersurface. Suppose that M lies locally on that side of $\mathcal{S}$ into which the normal ν is pointing. Finally assume that M touches $\mathcal{S}$ at an interior point $p_0 \in M \cap \mathcal{S}$ and that in some neighbourhood $U(p_0) \subset \mathbb{R}^{n+k}$ the inequality*

$$\sup_{U\cap M} |\vec{H}| \leq \inf_{U\cap\mathcal{S}} \Lambda_n \tag{2}$$

holds true. Then, near p_0, M is contained in $\mathcal{S}$, i.e. we have $M \cap U \subset \mathcal{S} \cap U$.

Corollary 1. *Suppose that M lies locally on that side of $\mathcal{S}$ into which the normal ν is pointing. Then M and $\mathcal{S}$ cannot touch at an interior point $p_0 \in M \cap \mathcal{S}$ if $|\vec{H}(p_0)| < \Lambda_n(p_0)$ holds.*

This theorem implies the following

Enclosure Theorem 1. *Let $G \subset \mathbb{R}^{n+k}$ be a domain with boundary $\mathcal{S} = \partial G \in C^2$ and M be an n-dimensional C^2-submanifold with mean-curvature vector $\vec{H}$ which is confined to the closure $\overline{G}$. Also, let Λ_n denote the n-mean curvature of $\mathcal{S} = \partial G$ with respect to the inward unit normal ν. Finally assume that, if M touches $\mathcal{S}$ at some interior point p_0, then the inequality*

$$\sup_{U\cap M} |\vec{H}| \leq \inf_{U\cap\mathcal{S}} \Lambda_n$$

holds true for some neighbourhood $U = U(p_0) \subset \mathbb{R}^{n+k}$. Then M lies in the interior of G, if at least one of its points lies in G.

Remark 1. Clearly, the hypothesis

$$|\vec{H}(p_0)| < \Lambda_n(p_0),$$

implies (2), but excludes e.g. the case $\vec{H} \equiv 0$ and $\Lambda_n \equiv 0$.

Remark 2. Let us consider an example which shows that Theorem 1 is optimal. To see that let $\mathcal{S} \subset \mathbb{R}^3$ be the cylinder $\{x^2 + y^2 = R^2\}$; then the principal curvatures with respect to the inward unit normal ν are given by $\lambda_1 = 0 \leq \lambda_2 = \frac{1}{R}$ and the n-mean curvature ($n = 1$ or 2 is possible) are $\Lambda_1 = 0$ and $\Lambda_2 = \Lambda = \frac{1}{2R}$. Take $n = 1$; then Theorem 1 requires $\vec{H} \equiv 0$, and this implies that M is a straight line. This is indeed necessary for the conclusion of Theorem 1 to hold since there are circles of arbitrary small "mean curvature" $|\vec{H}| = \frac{1}{r}$, $r > 0$, which locally are on the interior side of the cylinder $\mathcal{S}$ and touch $\mathcal{S}$ in exactly one point; yet these circles are not locally contained in $\mathcal{S}$.

For the proof of Theorem 1 we need to recall some important facts about the distance function, a proof of which can be found in Gilbarg and Trudinger [1], Chapter 14.6, or Hildebrandt [19], Section 4.6.

Let $\mathcal{S} \subset \mathbb{R}^{n+k}$ be a hypersurface with orientation ν. The distance function $d = d(x)$ is defined by

$$d(x) = \operatorname{dist}(x, \mathcal{S}) = \inf_{y \in \mathcal{S}} |x - y|.$$

Locally we can orient d so as to obtain the *signed* or *oriented* distance function ρ as follows: Choose a point $p_0 \in \mathcal{S}$. Then there is an open ball $B_\varepsilon(p_0) \subset \mathbb{R}^{n+k}$ which is partitioned by $\mathcal{S}$ into two open sets B_ε^+ and B_ε^-. Let B_ε^+ denote the set into which the normal ν points. The oriented distance ρ is then given by

$$\rho(x) = \begin{cases} d(x), & \text{for } x \in B_\varepsilon^+, \\ -d(x), & \text{for } x \in B_\varepsilon^-. \end{cases}$$

It follows easily that d and ρ are Lipschitz-continuous functions with Lipschitz constant equal to one. In fact, let $y \in \mathbb{R}^{n+k}$ and choose $z \in \mathcal{S}$ such that $d(y) = |z - y|$. Then for any $x \in \mathbb{R}^{n+k}$ we have

$$d(x) \leq |x - z| \leq |x - y| + |y - z| = |x - y| + d(y)$$

and the some inequality holds with x replaced by y, whence we obtain $|d(x) - d(y)| \leq |x - y|$. Observe that this holds without any assumption on the set $\mathcal{S}$. Similarly, for $x \in B_\varepsilon^+$, $y \in B_\varepsilon^-$ there exists $t_0 \in [0, 1]$ with $z_{t_0} = t_0 y + (1 - t_0)x \in \mathcal{S}$ and $\rho(x) - \rho(y) = d(x) + d(y) = d(x) - d(z_{t_0}) + d(y) - d(z_{t_0}) \leq |x - z_{t_0}| + |y - z_{t_0}| = |x - y|$, whence also ρ is Lipschitz continuous.

Much more is true, if $\mathcal{S}$ is of class $C^j, j \geq 2$.

Lemma 1. *Let $\mathcal{S} \subset \mathbb{R}^{n+k}$ be a hypersurface of class $C^j, j \geq 2$, and $p_0 \in \mathcal{S}$ be arbitrary. Then there is a constant $\varepsilon > 0$ (depending on p_0 in general) such that $d \in C^j(\overline{B}_\varepsilon^+)$, $d \in C^j(\overline{B}_\varepsilon^-)$ and the oriented distance $\rho \in C^j(B_\varepsilon(p_0))$.*

For a proof – which consists in an application of the implicit function theorem – we refer the reader to Gilbarg and Trudinger [1], Section 14.6, or Hildebrandt [19], Section 4.6.

Remark I. Obviously $d \notin C^1(B_\varepsilon(p_0))$ for $p_0 \in \mathcal{S}$, $\varepsilon > 0$.

Remark II. In Gilbarg and Trudinger [1] only the unoriented distance d is considered; however the proofs can be easily modified with almost no alterations.

Remark III. If $\mathcal{S} \subset \mathbb{R}^{n+k}$ is a compact closed hypersurface of class C^j, $j \geq 2$, then it satisfies a uniform interior (as well as exterior) *sphere condition*; that is at each point $p_0 \in \mathcal{S}$ there exists a ball B_{ε_0} of uniform radius $\varepsilon_0 > 0$ which lies in the interior (or exterior) side of $\mathcal{S}$ respectively and such that the closure $\overline{B}_{\varepsilon_0}$ has just one point in common with the surface $\mathcal{S}$, namely p_0. In this case the distance function is of class C^j on a tube T_{ε_0} of uniform width ε_0 where $T_{\varepsilon_0} := T_{\varepsilon_0}^+ \cup T_{\varepsilon_0}^-$ with

$$T_{\varepsilon_0}^+ := \{x \in \mathbb{R}^{n+k};\ 0 \leq \rho(x) < \varepsilon_0\}, \quad T_{\varepsilon_0}^- := \{x \in \mathbb{R}^{n+k};\ -\varepsilon_0 < \rho(x) \leq 0\}.$$

Then we have $d \in C^j(T_{\varepsilon_0}^+)$, $d \in C^j(T_{\varepsilon_0}^-)$ and $\rho \in C^j(T_{\varepsilon_0})$.

Choose $p_0 \in \mathcal{S}$ and $\varepsilon > 0$ such that $\rho \in C^j(B_\varepsilon(p_0))$; consider the *parallel surface*

$$\mathcal{S}_\tau := \{x \in \mathbb{R}^{n+k} \cap B_\varepsilon(p_0) : \rho(x) = \tau\},$$

$-\varepsilon < \tau < \varepsilon$, which is again of class C^j, if $\mathcal{S} \in C^j$, $j \geq 2$. The unit normal of $\mathcal{S}_\tau$ at $x \in \mathcal{S}_\tau$ directed towards increasing ρ is given by $\nu(x) = D\rho(x) = (\rho_{x^1}(x), \ldots, \rho_{x^{n+k}}(x))$. (Note that here – for simplicity of notation – we refrain from writing ν_τ instead of ν, so as to obtain a function $\nu \in C^{j-1}(B_\varepsilon(p_0))$ which, on $\mathcal{S} \cap B_\varepsilon(p_0)$ coincides with the unit normal on $\mathcal{S}$.)

For every point $x_0 \in \overline{B}_\varepsilon^+(p_0)$ or $B_\varepsilon(p_0)$ there exists a unique point $y_0 = y(x_0) \in \mathcal{S}$ such that $d(x_0) = |x_0 - y_0|$ or $\rho(x_0) = \pm|x_0 - y_0|$ respectively, in particular $x_0 = y_0 + \nu(y_0) \cdot \rho(x_0)$. We need to compare the principal curvatures $\lambda_1(y_0), \ldots, \lambda_{n+k-1}(y_0)$ of $\mathcal{S}$ at y_0 with the principal curvatures of $\mathcal{S}_{\rho(x_0)}$ at x_0. We recall the following

Lemma 2. *Let $x_0 \in \mathcal{S}_\tau$, $y_0 \in \mathcal{S}$ be such that $\rho(x_0) = \pm|x_0 - y_0|$. If y is a principal coordinate system at y_0 and $x = y + \nu(y)\rho$ then we have*

$$D^2\rho(x_0) = \rho_{x^i x^j}(x_0) = \operatorname{diag}\left(\frac{-\lambda_1(y_0)}{1 - \lambda_1(y_0)\rho(x_0)}, \ldots, \frac{-\lambda_{n+k-1}(y_0)}{1 - \lambda_{n+k-1}(y_0)\rho(x_0)}, 0\right).$$

For a *proof of Lemma 2* we refer to Gilbarg and Trudinger [1], Lemma 14.17. □

We now claim that the Hessian matrix $(\rho_{x^i x^j}(x_0))$, $i,j = 1,\dots,n+k-1$ is also given by the diagonal matrix $(-\lambda_i(x_0)\delta_{ij})$, $i,j = 1,\dots,n+k-1$, where $\lambda_1(x_0),\dots,\lambda_{n+k-1}(x_0)$ stand for the principal curvatures of $\mathcal{S}_{\rho(x_0)}$ at x_0. To see this consider $\nu(x_0) = D\rho(x_0) = (\rho_{x^1}(x_0),\dots,\rho_{x^{n+k}}(x_0))$ and suppose without loss of generality that $D\rho(x_0) = (0,\dots,0,1)$. Then there is some C^j-function $x^{n+k} = \varphi(x^1,\dots,x^{n+k-1})$ such that $\rho(x^1,\dots,x^{n+k-1},\varphi(x^1,\dots,x^{n+k-1})) = \rho(x_0)$. Differentiating this relation with respect to x^i, $i = 1,\dots,n+k-1$ yields $\varphi_{x^i} = -\frac{\rho_{x^i}}{\rho_{x^{n+1}}}$, for $i = 1,\dots n+k-1$, and

$$\varphi_{x^i x^j} = -\left(\frac{\rho_{x^i x^j}\rho_{x^{n+1}} - \rho_{x^i}\rho_{x^{n+1}x^j}}{\rho^2_{x^{n+1}}}\right).$$

Hence we get

$$\varphi_{x^i x^j}(\hat{x}_0) = -\rho_{x^i x^j}(x_0), \quad i,j = 1,\dots,n+k-1,$$

where $x_0 = (\hat{x}_0, x^{n+k})$.

On the other hand we have seen in the beginning of Section 4.3 that the eigenvalues of $D^2\varphi(\hat{x}_0)$ are precisely the principal curvatures $\lambda_i(x_0)$ of the graph of φ, i.e. of the distance surface $\mathcal{S}_\tau$, $\tau = \rho(x_0)$, at x_0. We have shown

Lemma 3. *Let $\mathcal{S}$ and $\mathcal{S}_\tau$ be as above, and $x_0 \in \mathcal{S}_\tau$, $y_0 \in \mathcal{S}$ be such that $\rho(x_0) = \pm|x_0 - y_0|$, i.e. $\tau = \rho(x_0)$. Denote by $\lambda_1(y_0),\dots,\lambda_{n+k-1}(y_0)$ the principal curvatures of $\mathcal{S}$ at y_0 and by $\lambda_1(x_0),\dots,\lambda_{n+k-1}(x_0)$ the principal curvatures of the parallel surface $\mathcal{S}_\tau$ at x_0. Then we have*

$$\lambda_i(x_0) = \frac{\lambda_i(y_0)}{1 - \lambda_i(y_0)\rho(x_0)} \qquad \textit{for } i = 1,\dots,n+k-1.$$

We continue with further preparatory results for the proof of Theorem 1 and select an orthonormal basis $t_1,\dots,t_n$ of the tangent space T_xM of M at x, assuming that x is close to $\mathcal{S}$. Introducing the orthogonal projection

$$t_i^\top := t_i - \langle t_i, \nu\rangle \cdot \nu$$

of t_i onto the tangent space $T_x\mathcal{S}_{\rho(x)}$ of the parallel surface $\mathcal{S}_{\rho(x)}$ at the point x. Also let $T_xM^\top$ stand for the orthogonal projection of the n-dimensional tangent space T_xM onto the $(n+k-1)$-dimensional tangent space $T_x\mathcal{S}_{\rho(x)}$.

Finally $\mathrm{II} = \mathrm{II}_x(\cdot,\cdot)$ denotes the second fundamental form of the distance hypersurface $\mathcal{S}_{\rho(x)}$ with respect to the normal $\nu = D\rho$ at the particular point $x \in \mathcal{S}_{\rho(x)}$, i.e. (cp. Section 4.3) $\mathrm{II}_x(t,\tau) = \langle -D_t\nu, \tau\rangle$, for $t,\tau \in T_x\mathcal{S}_{\rho(x)}$.

Lemma 4. *Let M and $\mathcal{S}$ be as in Theorem 1 and $p_0 \in M \cap \mathcal{S}$. Then the distance function $\rho = \rho(x)$ satisfies the equation*

$$\Delta_M \rho + b_i(\nabla_M \rho)_i - n\langle \vec{H}, D\rho\rangle + \operatorname{trace} \mathrm{II}|_{T_x M^\top} = 0$$

in a neighbourhood $V \subset \mathbb{R}^{n+k}$ *of* p_0. *Here trace* $\mathrm{II}|_{T_x M^\top}$ *denotes the trace of the second fundamental form* II *of* $\mathcal{S}_{\rho(x)}$ *at* x *restricted to the subspace* $T_x M^\top$ *of* $T_x \mathcal{S}_{\rho(x)}$, $b_i = b_i(x) := \frac{-\mathrm{II}(t_i^\top, t_j^\top)(\nabla_M \rho)_j}{1-|\nabla_M \rho|^2}$, *for* $i = 1, \dots, n$, $(\nabla_M \rho)_i = D_{t_i}\rho$ *and* $D\rho = (\rho_{x_1}, \dots, \rho_{n+k})$.

Proof of Lemma 4. We have $\nabla_M \rho = D\rho - \langle D\rho, N_1\rangle N_1 - \cdots - \langle D\rho, N_k\rangle N_k$, where $N_1, \dots, N_k$ is an orthonormal basis of the normal space $T_x M^\perp$. Therefore

$$\begin{aligned}(3)\qquad \Delta_M \rho &= \operatorname{div}_M \nabla_M \rho \\ &= \operatorname{div}_M D\rho - \langle D\rho, N_1\rangle \operatorname{div}_M N_1 - \cdots - \langle D_M\rho, N_k\rangle \operatorname{div}_M N_k \\ &= \operatorname{div}_M D\rho + n\langle \vec{H}, (D\rho)^\perp\rangle,\end{aligned}$$

where $(D\rho)^\perp = \langle D\rho, N_1\rangle N_1 + \cdots + \langle D\rho, N_k\rangle N_k$ is the normal part of $\nu = D\rho$ relative to M, and $\vec{H} = -\frac{1}{n}\sum_{j=1}^k (\operatorname{div} N_j)N_j$ is the mean curvature vector of M (see Proposition 3 of Section 4.3).

Now equation (3) obviously is equivalent to $\Delta_M \rho = \operatorname{div}_M D\rho + n\langle \vec{H}, D\rho\rangle$ and since the divergence on M is the operator $\sum_{i=1}^n t_i D_{t_i}$ we find, because of $D\rho(x) = \nu(x)$

$$(4)\qquad \Delta_M \rho = \sum_{i=1}^n t_i D_{t_i}\nu(x) + n\langle \vec{H}, D\rho\rangle.$$

To relate the expression $t_i D_{t_i}\nu$ to the second fundamental form of $\mathcal{S}_{\rho(x)}$ we put $t_i = t_i^\top + \langle t_i, \nu\rangle \nu$ and obtain

$$t_i D_{t_i}\nu = (t_i^\top + \langle t_i, \nu\rangle\nu) D_{t_i^\top + \langle t_i,\nu\rangle\nu}\nu = t_i^\top D_{t_i^\top}\nu = -\mathrm{II}_x(t_i^\top, t_i^\top),$$

where we have used that $\langle \nu, D_{t_i^\top}\nu\rangle = 0$ and $D_\nu \nu(x) = 0$ which is a consequence of the relations $|\nu(x)|^2 = 1$ and $\nu(x + t\nu(x)) = \nu(x)$ for $|t| \ll 1$.

Thus (4) implies

$$(5)\qquad \Delta\rho + \sum_{i=1}^n \mathrm{II}_x(t_i^\top, t_i^\top) - n\langle \vec{H}, D\rho\rangle = 0$$

in a neighbourhood of $p_0 \in \mathcal{S} \cap M$.

In general the projections $t_i^\top$, $i = 1, \dots, n$ are neither of unit length nor pairwise perpendicular. Therefore, in order to compute the trace of II_x on $T_x M^\top \subset T_x \mathcal{S}_{\rho(x)}$, we put

$$\begin{aligned} g_{ij}(x) &= g_{ij} := \langle t_i^\top, t_j^\top\rangle = \langle t_i, -\langle t_i, \nu\rangle\nu, t_j - \langle t_j, \nu\rangle\nu\rangle \\ &= \delta_{ij} - \langle t_i, \nu\rangle\langle t_j, \nu\rangle.\end{aligned}$$

If $x = p_0 \in M \cap \mathcal{S}$ we have $g_{ij} = g_{ij}(x) = \delta_{ij}$; hence in some neighbourhood V of p_0 we can assume that $\sum_{i=1}^n \langle t_i, \nu \rangle^2 < 1$ and that the inverse matrix $g^{ij} = g^{ij}(x)$ is simply

$$g^{ij} = \delta_{ij} + \frac{\langle t_i, \nu \rangle \langle t_j, \nu \rangle}{1 - \sum\limits_{i=1}^n \langle t_i, \nu \rangle^2} =: \delta_{ij} + \varepsilon_{ij}, \quad \text{for } i, j = 1, \dots, n.$$

Therefore we get for the trace of II_x on the subspace $T_x M^\top \subset T_x \mathcal{S}_{\rho(x)}$

$$\begin{aligned}\operatorname{trace} \mathrm{II}|_{T_x M^\top} &= \sum_{i,j=1}^n g^{ij} \mathrm{II}(t_i^\top, t_j^\top) \\ &= \sum_{i,j=1}^n \mathrm{II}(t_i^\top, t_j^\top) + \sum_{i,j=1}^n \varepsilon_{ij} \mathrm{II}(t_i^\top, t_j^\top).\end{aligned}$$

By virtue of (5) this yields

$$\Delta \rho - \sum_{i,j=1}^n \varepsilon_{ij} \mathrm{II}(t_i^\top, t_j^\top) + \operatorname{trace} \mathrm{II}|_{T_x M^\top} - n \langle \vec{H}, D\rho \rangle = 0 \quad \text{in } V \subset \mathbb{R}^{n+k}.$$

Lemma 4 follows by noting that

$$\varepsilon_{ij} = \frac{\langle t_i, \nu \rangle \langle t_j, \nu \rangle}{1 - \sum\limits_{i=1}^n \langle t_i, \nu \rangle^2} = \frac{(\nabla_M \rho)_i (\nabla_M \rho)_j}{1 - |\nabla_M \rho|^2}$$

and taking $b_i = -(1 - |\nabla_M \rho|^2)^{-1} \sum_{j=1}^n \mathrm{II}(t_i^\top, t_j^\top)(\nabla_M \rho)_j$. □

Lemma 5. *Let* II *be a quadratic form on an* n*-dimensional Euclidean space* V *with eigenvalues* $\lambda_1 \le \cdots \le \lambda_n$*. Then for any* k*-dimensional subspace* $W \subset V$ *we have the estimate*

$$\operatorname{trace} \mathrm{II}|_W \ge \lambda_1 + \cdots + \lambda_k.$$

The *proof of Lemma 5* is carried out by induction on $k + n$. The case $k + n = 2$ is trivial. By the induction hypothesis we may assume that the assertion holds for all quadratic forms II and linear spaces $V, W \subset V$ of dimension n and k respectively, $k \le n$, such that $k + n \le N$, $N \ge 2$. For given II, V and W we hence assume that $k + n = N + 1$. By $v_1 \in V$ we denote an eigenvector of II corresponding to the smallest eigenvalue λ_1 and put $V_1 := (\operatorname{span} v_1)^\perp$ to denote the $(n-1)$-dimensional orthogonal complement of v_1. We distinguish between the following two cases:
First case: $W \subset V_1$, then by induction hypotheses we have

$$\operatorname{trace} \mathrm{II}|_W \ge \lambda_2 + \cdots + \lambda_{k+1} \ge \lambda_1 + \cdots + \lambda_k.$$

Second case: $W \not\subset V_1$, then there is a nonzero vector $w_1 \in W$ such that

$$(w_1 - v_1) \perp W \quad \text{or, equivalently,}$$
$$\langle w_1, w\rangle = \langle v_1, w\rangle \quad \forall w \in W. \tag{6}$$

Select on orthonormal basis $\frac{w_1}{|w_1|}, w_2, \dots, w_k$ of W, then by (6) we find $w_2, \dots, w_k$ perpendicular to v_1, in other words, $w_2, \dots, w_k \in V_1$. Applying the induction hypothesis to the triple $\mathrm{II}|_{V_1}, V_1$ and $W_1 := \mathrm{span}(w_2, \dots, w_k)$ yields the estimate

$$\operatorname{trace} \mathrm{II}|_{W_1} = \sum_{j=2}^{k} \mathrm{II}(w_j, w_j) \geq \lambda_2 + \dots + \lambda_k$$

and therefore

$$\begin{aligned} \operatorname{trace} \mathrm{II}|_W &= \sum_{j=2}^{k} \mathrm{II}(w_j, w_j) + \mathrm{II}\left(\frac{w_1}{|w_1|}, \frac{w_1}{|w_1|}\right) \\ &\geq \lambda_2 + \dots + \lambda_k + \mathrm{II}\left(\frac{w_1}{|w_1|}, \frac{w_1}{|w_1|}\right) \geq \lambda_1 + \dots + \lambda_k. \end{aligned}$$ □

Proof of Theorem 1. We claim that, under the assumptions of the theorem, the inequality

$$-n\langle \vec{H}, D\rho\rangle + \operatorname{trace} \mathrm{II}|_{T_x M^\top} \geq 0 \tag{7}$$

holds true in a neighbourhood of any point $p_0 \in M \cap \mathcal{S}$. To prove this let $y_0 \in \mathcal{S}$, $x_0 \in M$ close to $\mathcal{S}$ and $\lambda_1(y_0) \leq \lambda_2(y_0) \leq \dots \leq \lambda_{n+k-1}(y_0)$ denote the principal curvatures of $\mathcal{S}$ with respect to the unit normal ν. By Lemma 3 we infer for the principal curvatures of $\mathcal{S}_\tau$, $\tau = \rho(x_0)$, at x_0:

$$\lambda_1(x_0) = \frac{\lambda_1(y_0)}{1 - \lambda_1(y_0)\rho(x_0)} \leq \dots \leq \lambda_{n+k-1}(x_0) = \frac{\lambda_{n+k-1}(y_0)}{1 - \lambda_{n+k-1}(y_0)\rho(x_0)}.$$

Lemma 5 now implies the estimate

$$\begin{aligned} \frac{1}{n} \operatorname{trace} \mathrm{II}|_{T_x M^\top} &\geq \frac{1}{n}\left(\frac{\lambda_1(y)}{1 - \lambda_1(y)\rho(x)} + \dots + \frac{\lambda_n(y)}{1 - \lambda_n(y)\rho(x)}\right) \\ &\geq \frac{1}{n}(\lambda_1(y) + \dots + \lambda_n(y)) = \Lambda_n(y), \end{aligned} \tag{8}$$

where $y \in \mathcal{S}$ is such that $\rho(x) = |x - y|$. By assumption (2) of Theorem 1,

$$\inf_{U \cap \mathcal{S}} \Lambda_n \geq \sup_{U \cap M} |\vec{H}|,$$

we infer from (8)

$$\frac{1}{n}\operatorname{trace}\mathrm{II}|_{T_xM^\top} \geq |\vec{H}(x)|$$

for every $x \in M$ close to $\mathcal{S}$. Inequality (7) then follows immediately by Schwarz's inequality. Now Theorem 1 is a consequence of Lemma 4. Indeed by relation (7) and Lemma 4 we conclude the inequality

$$\Delta_M \rho + b_i(\nabla_M \rho)_i \leq 0$$

in a neighbourhood of every point $p_0 \in M \cap \mathcal{S}$. E. Hopf's maximum principle (see e.g. Gilbarg and Trudinger [1], Theorem 3.5) finally proves that $\rho = 0$ in a neighbourhood of any point $p_0 \in M \cap \mathcal{S}$. Theorem 1 is proved. □

Proof of Corollary 1. Assuming the contrary we conclude from Theorem 1 the inclusion $M \cap U \subset \mathcal{S} \cap U$ for some neighbourhood U of $p_0 \in M \cap \mathcal{S}$. Therefore we had $\rho \equiv 0$ on $M \cap U$ and Lemma 4 implied the relation

$$\operatorname{trace}\mathrm{II}_{|T_xM} = n\langle \vec{H}, D\rho\rangle = n\langle \vec{H}, \nu\rangle \quad \text{on } M \cap U,$$

since $T_xM^T = T_xM$ and also $\nabla_M \rho = 0 = \triangle_M \rho$ on $M \cap U$. In particular we obtain the estimate

$$\frac{1}{n}\operatorname{trace}\mathrm{II}_{|T_xM} \leq |\vec{H}(x)| \quad \text{on } M \cap U.$$

On the other hand, by Lemma 5, this leads to the inequality

$$\Lambda_n(x) \leq |\vec{H}(x)| \quad \text{for all } x \in M \cap U,$$

which obviously contradicts the assumption

$$|\vec{H}(p_0)| < \Lambda_n(p_0). \qquad \square$$

Remark 3. We observe here that the estimate (7) is an immediate consequence of an hypothesis of the type $|\vec{H}(p_0)| < \Lambda_n(p_0)$, $p_0 \in \mathcal{S} \cap M$, the continuity of the involved functions, and Lemma 5, without using the explicit estimates in Lemma 3.

4.4.2 A Geometric Inclusion Principle for Strong Subsolutions

We now present a version of Theorem 1 for strong (but not necessarily classical) subsolutions of the parametric mean curvature equation, since they arise naturally as solutions of suitable obstacle problems to be considered later. This will be of importance for the existence proof for surfaces of prescribed mean curvature that will be carried out in Section 4.7.

If $\mathcal{S} \subset \mathbb{R}^3$ is a regular surface of class C^2 with unit normal $\nu = D\rho$ and mean curvature Λ with respect to that normal, we let $\mathcal{S}_\tau, |\tau| \ll 1$, denote the local parallel surface at (small) distance τ and $\Lambda_\tau(x)$ denote the mean curvature of $\mathcal{S}_\tau$ with respect to the normal $\nu(x) = D\rho(x)$ at the point $x \in \mathcal{S}_\tau$. Clearly $\tau = \rho(x)$ and $\Lambda_\tau(x) = \Lambda_{\rho(x)}(x)$.

Theorem 2. *Let $\mathcal{S} \subset \mathbb{R}^3$ be a regular surface of class C^2 with unit normal ν and mean curvature Λ (with respect to this normal). Furthermore let H denote some bounded continuous function on $\mathbb{R}^3$ and $\Omega \subset \mathbb{R}^2$ be a bounded, open and connected set. Suppose $X \in C^1(\Omega, \mathbb{R}^3) \cap H^2_{2,\mathrm{loc}}(\Omega, \mathbb{R}^3)$ lies locally on that side of $\mathcal{S}$ into which the normal ν is directed, and is a conformal solution of the variational inequality*

$$\delta\mathcal{F}(X, \varphi) = \int_\Omega \{\langle \nabla X, \nabla \varphi\rangle + 2H(X)\langle X_u \wedge X_v, \varphi\rangle\}\, du\, dv \geq 0 \tag{9}$$

for all functions $\varphi \in \mathring{H}^1_2(\Omega, \mathbb{R}^3) \cap L_\infty(\Omega, \mathbb{R}^3)$ with $X + \varepsilon\varphi$ locally on the same side of $\mathcal{S}$ for $0 < \varepsilon \ll 1$. Then the following conclusions hold:

(a) *Assume that $X_0 = X(w_0) \in \mathcal{S}$ and that for some neighbourhood $U = U(X_0) \subset \mathbb{R}^3$ one has*

$$|H(x)| \leq \Lambda_{\rho(x)}(x) \quad \textit{for all } x \in U. \tag{10}$$

Then there exists a disk $B_\varepsilon(w_0) \subset \Omega$ such that $X(B_\varepsilon(w_0)) \subset \mathcal{S}$.

(b) *Suppose that (10) holds for every point $X_0 \in \mathcal{S}$. Then $X(\Omega)$ is completely contained in $\mathcal{S}$, if $X(\Omega) \cap \mathcal{S}$ is nonempty.*

Corollary 2. *The conclusion of the Theorem holds if* (10) *is replaced by the (stronger) assumption*

$$\sup_U |H| \leq \inf_{U \cap \mathcal{S}} \Lambda. \tag{10'}$$

Corollary 3. *Suppose* (10) *is replaced by the (stronger) hypotheses*

$$|H(P_0)| < \Lambda(P_0) \tag{10''}$$

for some $P_0 \in \mathcal{S}$. Then there is no $w_0 \in \Omega$ such that $X(w_0) = P_0$. Clearly, this conclusion holds for a whole neighbourhood U of P_0 in $\mathcal{S}$. In particular if (10″) *is fulfilled for all points $P_0 \in \mathcal{S}$ then the intersection $X(\Omega) \cap \mathcal{S}$ is empty.*

As a further consequence of Theorem 2 we have the following

Enclosure Theorem 2. *Let $G \subset \mathbb{R}^3$ be a domain with $\partial G \in C^2$ and H be a bounded continuous function on $\mathbb{R}^3$. Assume that every point $P \in \partial G$ has a neighbourhood $U \subset \mathbb{R}^3 \cap \overline{G}$ such that*

$$|H(x)| \leq \Lambda_{\rho(x)}(x) \quad \textit{for all } x \in U, \tag{11}$$

where $\Lambda_{\rho(x)}$ stands for the mean curvatures of $\partial G_{\rho(x)}$ with respect to the inward unit normal $\nu = D\rho(x)$. Suppose $X \in C^1(\Omega, \mathbb{R}^3) \cap H^2_{2,\mathrm{loc}}(\Omega, \mathbb{R}^3)$ is a strong subsolution of the H-surface equation, whose image $X(\Omega)$ is confined to the closure $\overline{G}$, i.e. X is a conformal solution of the variational inequality

$$\delta \mathcal{F}(X,\varphi) = \int_\Omega \{\langle \nabla X, \nabla \varphi\rangle + 2H(X)\langle X_u \wedge X_v, \varphi\rangle\}\, du\, dv \geq 0$$

for all functions $\varphi \in \mathring{H}_2^1(\Omega, \mathbb{R}^3) \cap L_\infty(\Omega, \mathbb{R}^3)$ *with* $X + \varepsilon\varphi \in H_2^1(\Omega, \overline{G})$ *for* $0 \leq \varepsilon < \varepsilon_0(\varphi)$. *Then* $X(\Omega) \subset G$ *if at least one of the points* $X(w)$ *lies in* G.

Corollary 4. *The strong inclusion* $X(\Omega) \subset G$ *holds for example, if, in addition to the assumption of Enclosure Theorem 2,* X *is of class* $C^0(\overline{\Omega}, \mathbb{R}^3) \cap C^1(\Omega, \mathbb{R}^3) \cap H^2_{2,\mathrm{loc}}(\Omega, \mathbb{R}^3)$ *and maps one point* $w_0 \in \partial\Omega$ *into the interior of* G.

Corollary 5. *Enclosure Theorem 2 is valid if (11) is replaced by the (stronger) assumption*

$$(11') \qquad |H(P)| < \Lambda(P) \quad \text{for all } P \in \partial G.$$

Remark 4. Suppose $X \in C^1(\Omega, \mathbb{R}^3) \cap H_2^2(\Omega, \mathbb{R}^3)$ satisfies the assumptions of the Enclosure Theorem and that $X(\Omega) \subset G$. Then X is a *strong* (and of course also weak) H-surface in the sense that

$$\int_\Omega \Big\{\langle \nabla X, \nabla \varphi\rangle + 2H(X)\langle X_u \wedge X_v\rangle \varphi\Big\}\, du\, dv = 0$$

for all $\varphi \in \mathring{H}_2^1(\Omega, \mathbb{R}^3) \cap L_\infty(\Omega, \mathbb{R}^3)$. Furthermore, by elliptic regularity results it follows that X is a *classical* $C^{2,\alpha}$-solution of the H-surface equation if H is Hölder continuous. This means that X is an H-surface, i.e.

$$\begin{aligned} \Delta X &= 2H(X) X_u \wedge X_v, \\ |X_u|^2 &= |X_v|^2, \langle X_u, X_v\rangle = 0 \quad \text{in } \Omega. \end{aligned}$$

In Section 4.7 we will see how to find subsolutions X of the kind needed in Enclosure Theorem 2 by solving suitable obstacle problems.

While condition (11) is sufficient to show strong inclusion $X(\Omega) \subset G$ relative to the hypotheses in Enclosure Theorem 2 this is not true under the weaker assumption $|H(P)| \leq \Lambda(P)$ for all $P \in \partial G$, Λ the inward mean curvature of ∂G, see Remark 2 following Enclosure Theorem I in Section 4.2. However this still leaves open the possibility that X might satisfy the H-surface system a.e. in $\overline{G}$. The next result shows that this in indeed the case:

Theorem 3 (Variational equality). *Suppose that* $G \subset \mathbb{R}^3$ *is a domain of class* C^2, H *is bounded and continuous with* $H(P) \leq \Lambda(P)$ *for all* $P \in \partial G$. *Let* $X \in C^1(\Omega, \mathbb{R}^3) \cap H^2_{2,\mathrm{loc}}(\Omega, \mathbb{R}^3)$ *satisfy the assumptions of Enclosure Theorem 2. Then* X *is a strong* H*-surface in* $\overline{G}$, *i.e. we have* $\triangle X = 2H(X) X_u \wedge X_v$ *a.e. in* Ω, *and* $|X_u|^2 = |X_v|^2, \langle X_u, X_v\rangle = 0$ *in* Ω.

Proof of Theorem 2. Define the sets Ω_1, Ω_2 and Ω_3 by $\Omega_1 := X^{-1}(\mathcal{S})$, $\Omega_2 := \Omega \setminus \Omega_1$, and $\Omega_3 := \{w \in \Omega : |X_u(w)| = |X_v(w)| = 0\}$. Observe that Ω_1 and Ω_3 are closed, while Ω_2 is an open set. Then the function

$$\mathcal{H}^*(w) := \begin{cases} \pm\Lambda(X(w)), & \text{for } w \in \Omega_1, \\ H(X(w)), & \text{for } w \in \Omega_2 \end{cases}$$

is of class $L_{\infty,\text{loc}}(\Omega)$ and we claim that

$$\Delta X = 2\mathcal{H}^*(w)(X_u \wedge X_v) \tag{12}$$

holds a.e. on Ω. In fact, on Ω_3, the (possibly empty) set of branch points of X, we have $X_u(w) = X_v(w) = 0$. Since $X \in H^2_{2,\text{loc}}(\Omega, \mathbb{R}^3)$ this implies that also $X_{uu} = X_{vv} = X_{uv} = 0$ a.e. on Ω_3 (compare e.g. Gilbarg and Trudinger [1], Lemma 7.7); in particular (12) holds a.e. on Ω_3. Again, because of $X \in H^2_{2,\text{loc}}(\Omega, \mathbb{R}^3)$, we infer from (9) using an integration by parts and the fundamental lemma of the calculus of variations that (12) is satisfied a.e. on Ω_2. Finally, to verify equation (12) a.e. on $\Omega_1 \setminus \Omega_3$ we use the same argument as in the proof of Theorem 1 in Chapter 2.6 of Vol. 1, observing that X is a conformal and regular parametrization of $\mathcal{S}$ on $\Omega_1 \setminus \Omega_3$ and that $\mathcal{S}$ has mean curvature Λ.

Now the reasoning of Hartman and Wintner as outlined in Section 2.10 and Chapter 3 yields the asymptotic expansion

$$X_u - iX_v = (a - ib)(w - w_0)^l + o(|w - w_0|^l) \tag{13}$$

in a sufficiently small neighbourhood of an arbitrary point $w_0 \in \Omega$, where $l \geq 0$ is an integer and $a, b \in \mathbb{R}^3$ satisfy the relations $|a| = |b| \neq 0$ and $\langle a, b\rangle = 0$. In particular $\lambda(w) := |X_u(w)| = |X_v(w)| > 0$ on a punctured neighbourhood of w_0 and $\lambda(w_0) = 0$, if and only if w_0 is a branch point of X. Introducing polar coordinates $w = re^{i\varphi}$ around w_0 we infer from (13) the asymptotic relations

$$\begin{aligned} X_u(re^{i\varphi}) &= ar^l \cos(l\varphi) + br^l \sin(l\varphi) + o(r^l), \\ X_v(re^{i\varphi}) &= br^l \cos(l\varphi) - ar^l \sin(l\varphi) + o(r^l), \\ |X_u|^2 = |X_v|^2 &= |a|^2 r^{2l} + o(r^{2l}), \end{aligned}$$

all holding for $r \to 0$. Therefore $\lambda(w) = |a|r^l + o(r^l)$, for $r \to 0$, and consequently the unit normal has the asymptotic expansion

$$\frac{X_u \wedge X_v}{|X_u \wedge X_v|}(w) = \frac{a \wedge b}{|a \wedge b|} + o(1) \quad \text{as } w \to w_0.$$

In particular, the normal $N(w) = \frac{X_u \wedge X_v}{|X_u \wedge X_v|}(w)$ is continuous in Ω and

$$\lim_{w \to w_0} N(w) = \frac{a \wedge b}{|a \wedge b|} = \frac{a \wedge b}{|a|^2} = \frac{a \wedge b}{|b|^2}. \tag{14}$$

In other words, the surface X has a tangent plane at any point $w_0 \in \Omega$.

Suppose now that $w_0 \in \Omega_1$, i.e. $X(w_0) \in \mathcal{S}$, then since $X \in C^1$ lies locally on one side of $\mathcal{S}$ and because of (14) we obtain

$$\frac{a \wedge b}{|a \wedge b|} = \pm\, \nu(X(w_0)). \tag{15}$$

In fact, (15) can be proved rigorously by the same argument as used in the proof of Enclosure Theorem 1 of Chapter 4.2, namely by invoking a local non-parametric representation of the surfaces X and $\mathcal{S}$ near a punctured neighbourhood of the point $X(w_0)$.

Consider now the oriented distance function $\rho(x) = \text{dist}(x, \mathcal{S})$ which is of class C^2 near $\mathcal{S}$ and put $\nu(x) = D\rho(x) = (\rho_{x_1}, \rho_{x_2}, \rho_{x_3})$, cp. the discussion in the beginning of this section. Recall that $\rho(X(w)) \geq 0$ and "=" if and only if $w \in \Omega_1$ and that $\nu(x)$ is the unit normal of $\mathcal{S}$ at x. For the computations to follow it is convenient to put $u = u_1$ and $v = u_2$, and define for a – sufficiently small – neighbourhood $B_\rho(w_0)$ of an arbitrary point $w_0 \in \Omega_1$,

$$X^t_{u^\alpha}(w) := \frac{X_{u^\alpha}(w)}{|X_{u^\alpha}|} - \left\langle \frac{X_{u^\alpha}(w)}{|X_{u^\alpha}|}, \nu(X(w)) \right\rangle \nu(X(w)),$$

for $w \in B_\rho(w_0) \setminus \{w_0\}$ and $\alpha = 1, 2$, to denote the orthogonal projection of the unit tangent vector $\frac{X_{u^\alpha}(w)}{|X_{u^\alpha}|}$ of X onto the tangent space of the parallel surface $\mathcal{S}_{\rho(X(w))} := \{y \in \mathbb{R}^3 : \rho(y) = \rho(X(w))\}$ to $\mathcal{S}$ at distance $\rho(X(w))$.

The vectors $X^t_{u^\alpha}(w)$ are continuous in $B_\delta(w_0) \setminus \{w_0\}$ but merely bounded on $B_\delta(w_0)$. Define the metric

$$\begin{aligned} g_{\alpha\beta} &= g_{\alpha\beta}(w) := \langle X^t_{u^\alpha}(w), X^t_{u^\beta}(w) \rangle \\ &= \delta_{\alpha\beta} - \left\langle \frac{X_{u^\alpha}}{|X_{u^\alpha}|}, \nu \right\rangle \left\langle \frac{X_{u^\beta}}{|X_{u^\beta}|}, \nu \right\rangle \quad \text{for } w \in B_\delta(w_0) \setminus \{w_0\} \end{aligned}$$

and $\alpha, \beta = 1, 2$, where $\nu = \nu(X(w)) = D\rho(X(w))$.

We assert that

$$\lim_{w \to w_0} \left\langle \frac{X_{u^\alpha}(w)}{|X_{u^\alpha}|}, \nu(X(w)) \right\rangle = 0 \quad \text{holds true.} \tag{16}$$

To see this note that (15) yields the relation

$$\nu(X(w)) = \pm \frac{a \wedge b}{|a \wedge b|} + o(1) \quad \text{as } w \to w_0,$$

which implies (16) by virtue of the asymptotic expansions

$$\begin{aligned} \tfrac{X_u(w)}{|X_u|} &= \tfrac{ar^l \cos(l\varphi) + br^l \sin(l\varphi)}{|a| r^l} + o(1), \quad w \to w_0, \\ \tfrac{X_v(w)}{|X_v|} &= \tfrac{br^l \cos(l\varphi) - ar^l \sin(l\varphi)}{|b| r^l} + o(1), \quad w \to w_0. \end{aligned}$$

On the other hand relation (16) shows that the metric $g_{\alpha\beta}$ is continuous on $B_\delta(w_0)$ and

$$\lim_{w \to w_0} g_{\alpha\beta}(w) = \delta_{\alpha\beta}, \quad \text{for } \alpha, \beta = 1, 2.$$

Hence, by possibly decreasing δ, we can consider the inverse metric $g^{\alpha\beta} \in C^0(B_\delta(w_0))$, $g^{\alpha\beta} = g^{\alpha\beta}(w) = \delta_{\alpha\beta} + \varepsilon_{\alpha\beta}$, where

$$\varepsilon_{\alpha\beta} = \varepsilon_{\alpha\beta}(w) = \frac{\left\langle \frac{X_{u^\alpha}(w)}{|X_{u^\alpha}|}, \nu(X(w))\right\rangle \left\langle \frac{X_{u^\beta}(w)}{|X_{u^\beta}|}, \nu(X(w))\right\rangle}{1 - \left\langle \frac{X_u}{|X_u|}, \nu\right\rangle^2 - \left\langle \frac{X_v}{|X_v|}, \nu\right\rangle^2}$$

and $\varepsilon_{\alpha\beta} \in C^0(B_\delta(w_0))$ with $\varepsilon_{\alpha\beta}(w_0) = 0$ (by (16)). For $|\tau| \ll 1$ we denote by $\Lambda_\tau(x)$ the mean curvature of the distance surface $\mathcal{S}_\tau$ with respect to the unit normal $\nu(x)$ at x. Also let II_x stand for the second fundamental form of $\mathcal{S}_{\rho(x)}$ with respect to $\nu(x)$ at the point x, cp. Chapter 1 of Vol. 1 or Section 4.3, in particular we have $2\Lambda_{\rho(x)}(x) = \operatorname{trace} \mathrm{II}_x$, and since

$$\mathrm{II}_x(t,\tau) = \langle -D_t\nu, \tau\rangle = -\langle D\nu(x)t, \tau\rangle$$

for $t, \tau \in T_x\mathcal{S}_{\rho(x)}$ this implies for every $w \in B_\delta(w_0) \setminus \{w_0\}$,

$$\begin{aligned} -2\Lambda_{\rho(X)}(X(w)) &= g^{\alpha\beta}(X(w))\langle X^t_{u^\alpha}(w), D\nu(X(w))X^t_{u^\alpha}(w)\rangle \\ &= \langle X^t_u, D\nu(X)X^t_u\rangle + \langle X^t_v, D\nu(X)X^t_v\rangle + \varepsilon_{\alpha\beta}\langle X^t_{u^\alpha}(w), D\nu(X)X^t_{u^\beta}\rangle. \end{aligned}$$

Equivalently,

$$\begin{aligned} -2|X_u|^2\Lambda_{\rho(X)}(X(w)) &= -(|X_u|^2 + |X_v|^2)\Lambda_{\rho(X)}(X) \\ &= |X_u|^2\langle X^t_u, D\nu(X)X^t_u\rangle + |X_v|^2\langle X^t_v, D\nu(X)X^t_v\rangle \\ &\quad + \varepsilon_{\alpha\beta}|X_u||X_v|\langle X^t_{u^\alpha}, D\nu(X)X^t_{u^\beta}\rangle. \end{aligned} \tag{17}$$

Now, we look at the term

$$\begin{aligned} &|X_u|^2\langle X^t_u, D\nu(X)X^t_u\rangle \\ &= \Big\langle X_u - \langle X_u, \nu\rangle\nu, D\nu(X)\big[X_u - \langle X_u, \nu\rangle\nu\big]\Big\rangle \\ &= \langle X_u, D\nu(X)X_u\rangle - \langle X_u, \nu\rangle\langle \nu, D\nu(X)X_u\rangle \\ &\quad - \langle X_u, \nu\rangle\langle X_u, D\nu(X)\nu\rangle + \langle X_u, \nu\rangle^2\langle \nu, D\nu(X)\nu\rangle \\ &= \langle X_u, D\nu(X)X_u\rangle, \end{aligned}$$

since $D\nu(X)\nu = 0$ and II_x is symmetric. Similarly, we find

$$|X_v|^2\langle X^t_v, D\nu(X)X^t_v\rangle = \langle X_v, D\nu(X)X_v\rangle$$

and

$$|X_u||X_v|\langle X^t_u, D\nu(X)X^t_v\rangle = \langle X_u, D\nu(X)X_v\rangle.$$

This combined with (17) yields

$$\begin{aligned} &\langle X_u, D\nu(X)X_u\rangle + \langle X_v, D\nu(X)X_v\rangle \\ &\quad = -(|X_u|^2 + |X_v|^2)\Lambda_{\rho(X)}(X) - \varepsilon_{ij}\langle X_{u_i}, D\nu(X)X_{u_j}\rangle, \end{aligned} \tag{18}$$

for all $w \in B_\delta(w_0) \setminus \{w_0\}$. However, for continuity reasons (18) clearly holds on $B_\delta(w_0) \subset \Omega$.

So far we have not exploited the variational inequality

$$(19) \qquad \int_\Omega \{\langle \nabla X, \nabla\varphi\rangle + 2H(X)\langle X_u \wedge X_v, \varphi\rangle\}\, du\, dv \ge 0$$

holding for all $\varphi \in \mathring{H}_2^1(\Omega, \mathbb{R}^3) \cap L_\infty$, such that $X + \varepsilon\varphi$ lies locally on the same side of $\mathcal{S}$ as X for all $\varepsilon \in [0, \varepsilon(\varphi)]$. We choose as a test function $\varphi(w) = \eta(w) \cdot \nu(X(w))$, where $0 \le \eta \in C_c^\infty(B_\delta(w_0))$ is arbitrary and $\nu(x) = D\rho(x)$.

Clearly $\varphi \in \mathring{H}_2^1(\Omega, \mathbb{R}^3) \cap L_\infty$ is an admissible function in (19) and we compute

$$\nabla\varphi = (\varphi_u, \varphi_v) = \nabla\eta \cdot \nu(X) + \eta[D\nu(X)\nabla X]$$

where $\nabla X = (X_u, X_v)$, $\nabla\eta = (\eta_u, \eta_v)$. Plugging this relation into the variational inequality (19) we obtain

$$\int_{B_\delta(w_0)} \{\langle \nabla X, \nabla\eta \cdot \nu + \eta[D\nu(X)\nabla X]\rangle + 2\eta H(X)\langle X_u \wedge X_v, \nu(X)\rangle\}\, du\, dv \ge 0$$

from which we infer, by virtue of

$$\begin{aligned} \langle \nabla X, \nabla\eta \cdot \nu\rangle &= \eta_u \langle X_u, \nu(X)\rangle + \eta_v \langle X_v, \nu(X)\rangle \\ &= \eta_u \frac{\partial}{\partial u}\rho(X(w)) + \eta_v \frac{\partial}{\partial v}\rho(X(w)) = \langle \nabla\eta, \nabla\rho(X)\rangle \end{aligned}$$

and

$$\eta\langle \nabla X, D\nu(X)\nabla X\rangle = \eta\langle X_u, D\nu(X)X_u\rangle + \eta\langle X_v, D\nu(X)X_v\rangle$$

the inequality

$$\begin{aligned} \int_{B_\delta(w_0)} \Big\{ &\langle \nabla\eta, \nabla\rho(X)\rangle + \eta\langle X_u, D\nu(X)X_u\rangle \\ &+ \eta\langle X_v, D\nu(X)X_v\rangle + 2\eta H(X)\langle X_u \wedge X_v, \nu\rangle \Big\}\, du\, dv \ge 0. \end{aligned}$$

In this inequality we replace the expression $\langle X_u, D\nu(X)X_u\rangle + \langle X_v, D\nu(X)X_v\rangle$ using (18) and get

$$\begin{aligned} 0 \le \int_{B_\delta(w_0)} \Big\{ &\langle \nabla\eta, \nabla\rho(X)\rangle - \eta(|X_u|^2 + |X_v|^2)\Lambda_{\rho(X)}(X) \\ &- \eta\varepsilon_{\alpha\beta}\langle X_{u^\alpha}, D\nu(X)X_{u^\beta}\rangle + 2\eta H(X)\langle X_u \wedge X_v, \nu\rangle \Big\}\, du\, dv, \end{aligned}$$

or equivalently,

$$(20) \quad \begin{aligned} 0 \le \int_{B_\delta(w_0)} \{ &\langle \nabla\rho(X), \nabla\eta\rangle + \eta(|X_u|^2 + |X_v|^2)\big[|H(X)| - \Lambda_{\rho(X)}(X)\big] \\ &- \eta\,\varepsilon_{\alpha\beta}\langle X_{u^\alpha}, D\nu(X)X_{u^\beta}\rangle\}\, du\, dv. \end{aligned}$$

To see that the function $\rho(X(w))$ satisfies a differential equation of second order we compute the term

$$
\begin{aligned}
(21)\qquad & \varepsilon_{\alpha\beta}\langle X_{u^\alpha}, D\nu(X)X_{u^\beta}\rangle \\
&= \frac{\left\langle \frac{X_{u^\alpha}}{|X_{u^\alpha}|}, \nu\right\rangle \left\langle \frac{X_{u^\beta}}{|X_{u^\beta}|}, \nu\right\rangle}{1 - \left\langle \frac{X_u}{|X_u|}, \nu\right\rangle^2 - \left\langle \frac{X_v}{|X_v|}, \nu\right\rangle^2} \cdot \langle X_{u^\alpha}, D\nu(X)X_{u^\beta}\rangle \\
&= \frac{\frac{1}{|X_{u^\alpha}|}\frac{\partial}{\partial u^\alpha}\rho(X)\left\langle \frac{X_{u^\beta}}{|X_{u^\beta}|}, \nu\right\rangle}{1 - \langle\cdot,\cdot\rangle^2 - \langle\cdot,\cdot\rangle^2} \cdot \langle X_{u^\alpha}, D\nu(X)X_{u^\beta}\rangle \\
&=: b_\alpha(w)\frac{\partial}{\partial u^\alpha}\rho(X),
\end{aligned}
$$

where we have put

$$
b_\alpha(w) := \frac{\left\langle \frac{X_{u^\alpha}}{|X_{u^\alpha}|}, D\nu(X)X_{u^\beta}\right\rangle \left\langle \frac{X_{u^\beta}}{|X_{u^\beta}|}, \nu\right\rangle}{1 - \left\langle \frac{X_u}{|X_u|}, \nu\right\rangle^2 - \left\langle \frac{X_v}{|X_v|}, \nu\right\rangle^2}.
$$

Note that by (16), we have $\lim_{w\to w_0}\langle \frac{X_{u_i}}{|X_{u_i}|}, \nu\rangle = 0$. Thus $b_\alpha(\cdot)$ is continuous in $B_\delta(w_0)$ with $b_\alpha(w_0) = 0$. By assumption (10) we have

$$
(22)\qquad |H(X(w))| - \Lambda_{\rho(X)}(X(w)) \le 0 \quad \text{for all } w \in B_\varepsilon(w_0)
$$

and some positive $\varepsilon \le \delta$.

From (20), (21) and (22) we finally infer the inequality

$$
0 \le \int_{B_\varepsilon(w_0)} \left\{ \langle \nabla\rho(X), \nabla\eta\rangle - \eta \cdot b_i(w)\frac{\partial}{\partial u_i}\rho(X) \right\} du\, dv
$$

which holds for all nonnegative $\eta \in C_c^\infty(B_\varepsilon(w_0))$. Thus the function $f(w) := \rho(X(w))$ is an $H_2^2 \cap C^1(B_\varepsilon(w_0))$ strong (and therefore an almost everywhere) solution of the inequality

$$
\Delta f(w) + b_i\frac{\partial f}{\partial u_i}(w) \le 0 \quad \text{in } B_\varepsilon(w_0).
$$

By the strong maximum principle (see e.g. Gilbarg and Trudinger [1], Theorem 9.6) it follows that $f(w) = \rho(X(w)) \equiv 0$ in $B_\varepsilon(w_0)$. This clearly means that $X(w) \in \mathcal{S}$ for all $w \in B_\varepsilon(w_0)$. Theorem 2 is proved. □

Proof of Corollary 2. We use Lemma 3 to control the mean curvature $\Lambda_{\rho(X)}(X)$ of the parallel surface at X as follows:

$$
\begin{aligned}
\Lambda_{\rho(X)}(X) &= \frac{1}{2}\left(\frac{\lambda_1(y)}{1-\lambda_1(y)\rho(X)} + \frac{\lambda_2(y)}{1-\lambda_2(y)\rho(X)}\right) \\
&\ge \frac{1}{2}(\lambda_1(y) + \lambda_2(y)) = \Lambda_0(y) = \Lambda(y),
\end{aligned}
$$

where $y \in \mathcal{S}$ is such that $\rho(X) = |X - y|$ and $\lambda_1(y) \leq \lambda_2(y)$ are the principal curvatures of $\mathcal{S}$ at y. By assumption (10′) we have

$$\inf_{U \cap \mathcal{S}} \Lambda \geq \sup_U |H|$$

for some neighbourhood $U = U(X(w_0)) \subset \mathbb{R}^3$. Hence for some $\varepsilon > 0$ there holds

$$\Lambda_{\rho(X)}(X) \geq \inf_{U \cap \mathcal{S}} \Lambda \geq \sup_U |H| \geq |H(X)| \quad \text{on } B_\varepsilon(w_0),$$

that is $|H(X(w))| - \Lambda_{\rho(X)}(X(w)) \leq 0$ for all $w \in B_\varepsilon(w_0)$. The proof of Corollary 2 can now be completed in the same way as in Theorem 2. □

Proof of Corollary 3. Assume on the contrary the existence of some $w_0 \in \Omega$ such that $X(w_0) = P_0$. By Theorem 2 there exists a disk $B_\epsilon(w_0) \subset \Omega$ such that X maps $B_\epsilon(w_0)$ into $\mathcal{S}$. Therefore we obtain on $B_\epsilon(w_0)$ the identities

$$\nabla \rho(X(w)) = 0 \quad \text{and} \quad \epsilon_{\alpha\beta} \langle X_{u^\alpha}(w), D\nu(X) X_{u^\beta}(w) \rangle = 0.$$

The variational inequality (20) then yields the estimate

$$0 \leq \int_{B_\epsilon(w_0)} \eta(|X_u|^2 + |X_v|^2)[|H(X)| - \Lambda_{\rho(X)}(X)]\, du\, dv$$

for all $\eta \in C_c^\infty(B_\epsilon(w_0)), \eta \geq 0$. However, this contradicts the assumption $|H(P_0)| < \Lambda(P_0)$, since X cannot be constant on $B_\epsilon(w_0)$ and H and Λ are continuous. □

Remark 5. We point out here that the stronger assumption $|H(X(w_0))| < \Lambda(X(w_0))$ (replacing (10) in Theorem 2) leads to a somewhat more straightforward proof of the fact $X(B_\varepsilon(w_0)) \subset \mathcal{S}$, starting from inequality (20); namely we have

$$\begin{aligned} &(|X_u|^2 + |X_v|^2)\Big[|H(X)| - \Lambda_\rho(X)\Big] - \varepsilon_{\alpha\beta} \langle X_{u^\alpha}, D\nu(X) X_{u^\beta} \rangle \\ &= 2|X_u|^2 \left\{ [|H| - \Lambda_\rho] - \varepsilon_{\alpha\beta} \left\langle \frac{X_{u^\alpha}}{|X_{u^\alpha}|}, D\nu(X) \frac{X_{u^\beta}}{|X_{u^\beta}|} \right\rangle \right\}. \end{aligned}$$

Put $\sigma(w) := \varepsilon_{\alpha\beta} \langle \frac{X_{u^\alpha}}{|X_{u^\alpha}|}, D\nu(X) \frac{X_{u^\beta}}{|X_{u^\beta}|} \rangle$; then $\sigma \in C^0(B_\delta(w_0))$ with $\sigma(w_0) = 0$, since $\varepsilon_{\alpha\beta}(w_0) = 0$. Thus the assumption $|H(X(w_0))| < \Lambda(X(w_0))$ implies the inequality

$$2|X_u|^2 \Big\{ [|H(X(w))| - \Lambda_\rho(X(w))] - \sigma(w) \Big\} \leq 0$$

on a suitable disc $B_\varepsilon(w_0) \subset B_\delta(w_0)$, whence (20) yields

$$0 \leq \int_{B_\varepsilon(w_0)} \langle \nabla \rho(X(w)), \nabla \eta(w) \rangle\, du\, dv,$$

i.e. $f(w) = \rho(X(w))$ is strongly superharmonic on $B_\varepsilon(w_0)$, or $\Delta f(w) \leq 0$ a.e. in $B_\varepsilon(w_0)$.

Remark 6. Recall that assumption (10) cannot be replaced by $|H(P)| \le \Lambda(P)$ for all $P \in \mathcal{S}$, see Remark 2 following Enclosure Theorem I in Section 4.2.

Proof of Enclosure Theorem 2. The coincidence set $\Omega_1 = X^{-1}(\partial G)$ is a closed set in Ω. By Theorem 2 Ω_1 is also open, whence either $\Omega_1 = \emptyset$ or $\Omega_1 = \Omega$. However, by assumption there exists a $w_0 \in \Omega$ with $X(w_0) \in G$ and therefore only the alternative $\Omega_1 = \emptyset$ can hold true, i.e. $X(\Omega) \subset G$. □

Proof of Theorem 3. We let $\mathcal{T} := \{w \in \Omega : X(w) \in \partial G\}$ denote the (closed) coincidence set and put, for $\varepsilon > 0$, $\mathcal{T}_\varepsilon := \{w \in \Omega : \operatorname{dist}(w, \mathcal{T}) < \varepsilon\}$. Then $\mathcal{T}_\varepsilon$ is open and $\bigcap_{\varepsilon>0} \mathcal{T}_\varepsilon = \mathcal{T}$. Extend $\nu(x)$ to a bounded C^1-vector field $\tilde{\nu}$ on $\mathbb{R}^3$ which coincides with $D\rho(x) = \nu(x)$ on a neighbourhood of ∂G. Then we take nonnegative functions $\eta \in C_c^\infty(\Omega)$ and $\eta_\varepsilon \in C_c^\infty(\mathcal{T}_\varepsilon)$ with $\eta = \eta_\varepsilon$ on $\mathcal{T}_{\varepsilon/2}$ and put

$$\varphi(w) := \eta(w)\tilde{\nu}(X(w)), \quad \varphi_\varepsilon(w) := \eta_\varepsilon(w)\tilde{\nu}(X(w)).$$

Since both $(\varphi - \varphi_\varepsilon)$ and $(\varphi_\varepsilon - \varphi) \in C_c^1(\Omega \setminus \overline{\mathcal{T}}_{\varepsilon/4}, \mathbb{R}^3)$ are admissible in the variational inequality (19) we have $\delta\mathcal{F}(X, \varphi - \varphi_\varepsilon) = 0$; whence, since also φ and φ_ε are admissible functions we find as in the proof of Theorem 2, cp. (20),

$$\begin{aligned}(23)\qquad 0 \le \delta\mathcal{F}(X,\varphi) &= \delta\mathcal{F}(X,\varphi_\varepsilon)\\ &= \int_{\mathcal{T}_\varepsilon} [\langle \nabla\rho(X), \nabla\eta_\varepsilon\rangle + \eta_\varepsilon(|X_u|^2 + |X_v|^2)(|H(X)| - \Lambda_{\rho(X)}(X))\\ &\quad - \eta_\varepsilon \varepsilon_{\alpha\beta}\langle X_{u^\alpha}, D\nu(X)X_{u^\beta}\rangle]\,du\,dv,\end{aligned}$$

assuming that $\varepsilon > 0$ is choosen suitably small. We infer from $\rho(X(\cdot)) \in H^2_{2,\mathrm{loc}}(\mathcal{T}_\varepsilon)$ and an integration by parts

$$\int_{\mathcal{T}_\varepsilon} \langle \nabla\rho(X), \nabla\eta_\varepsilon\rangle\,du\,dv = -\int_{\mathcal{T}_\varepsilon} \Delta\rho(X)\cdot\eta_\varepsilon\,du\,dv.$$

Taking the relations $\Delta\rho(X(w)) = 0$ a.e. on $\mathcal{T}$ and $\varepsilon_{\alpha\beta}(w) = 0$ a.e. on $\mathcal{T}$ for $\alpha, \beta = 1, 2$ and Lebesgue's dominated convergence theorem into account we arrive at

$$\begin{aligned}\lim_{\varepsilon\to 0} \int_{\mathcal{T}_\varepsilon} &[\langle \nabla\rho(X), \nabla\eta_\varepsilon\rangle + \eta_\varepsilon(|X_u|^2 + |X_v|^2)(|H(X)| - \Lambda_\rho(X))\\ &\quad - \eta_\varepsilon \varepsilon_{\alpha\beta}\langle X_{u^\alpha}, D\nu(X)X_{u^\beta}\rangle]\,du\,dv\\ &= \int_{\mathcal{T}} \eta(|X_u|^2 + |X_v|^2)(|H(X)|) - \Lambda(X))\,du\,dv.\end{aligned}$$

By (23) and the assumption $|H| \le \Lambda$ along ∂G we obtain the variational equality

$$(24)\qquad \delta\mathcal{F}(X,\varphi) = 0$$

for all functions φ of the type $\varphi = \eta(w)\tilde{\nu}(X(w))$, $\eta \in C_c^\infty(\Omega), \eta \geq 0$. Clearly, this also holds for all normal variation $\varphi = \eta\tilde{\nu}(X)$ without assuming the sign restriction on η.

We can now exploit the variational inequality $\delta\mathcal{F}(X,\varphi) \geq 0$ which holds for all $\varphi \in \overset{\circ}{H}{}_2^1(\Omega,\mathbb{R}^3) \cap L_\infty(\Omega,\mathbb{R}^3)$ with $X + \varepsilon\varphi \in H_2^1(\Omega,\overline{G})$. Note that we may hence admit variations $\varphi(w) = \eta(w)\zeta(X(w))$ where ζ denotes a C^1-vectorfield defined on a neighbourhood of ∂G with $\zeta(P) = 0$ for all $P \in \partial G$ or $\langle\zeta(P),\nu(P)\rangle > 0$ along ∂G and $\eta \in C_c^1(\mathcal{T}_\epsilon)$, $\eta \geq 0$, $\varepsilon > 0$ suitably small. By an approximation argument this also follows for C^1-vectorfields as above which are directed weakly into the interior of ∂G, i.e. $\langle\zeta(P),\nu(P)\rangle \geq 0$ along ∂G. In particular we have $\delta\mathcal{F}(X,\varphi) = 0$, if ζ as above is tangential to ∂G along ∂G, since in this case $\langle\pm\zeta,\nu\rangle \geq 0$ on ∂G.

Suppose $\varphi \in C_c^1(\mathcal{T}_\varepsilon,\mathbb{R}^3)$ is arbitrary, then we decompose $\varphi = \varphi^\perp + \varphi^T$ where $\varphi^\perp = \eta(w)\tilde{\nu}(X(w))$ with $\eta(w) := \langle\varphi(w),\tilde{\nu}(X(w))\rangle$ denotes the "normal component" and $\langle\varphi^T(w),\tilde{\nu}(X(w))\rangle = 0$ for all $w \in \mathcal{T}_\varepsilon$. Concluding we find $\delta\mathcal{F}(X,\varphi^T) = 0$ and because of (24) also $\delta\mathcal{F}(X,\varphi^\perp) = 0$, whence $\delta\mathcal{F}(X,\varphi) = 0$ for all $\varphi \in C_c^1(\mathcal{T}_\epsilon,\mathbb{R}^3)$. Since, on the other hand $\delta\mathcal{F}(X,\varphi) = 0$ whenever φ is supported in $\Omega\setminus\mathcal{T}$ we finally conclude the result by virtue of the fundamental lemma of the calculus of variations. □

4.5 Isoperimetric Inequalities

For the sake of completeness we first repeat the proof of the isoperimetric inequality for disk-type minimal surfaces $X : B \to \mathbb{R}^3 \in H_2^1(B,\mathbb{R}^3)$ with the parameter domain $B = \{w \in \mathbb{C} : |w| < 1\}$, the boundary of which is given by $C = \partial B = \{w \in \mathbb{C} : |w| = 1\}$. Recall that any $X \in H_2^1(B,\mathbb{R}^3)$ has boundary values $X|_C$ of class $L_2(C,\mathbb{R}^3)$. Denote by $L(X)$ the length of the boundary trace $X|_C$, i.e.,

$$L(X) = L(X|_C) := \int_C |dX|.$$

We recall a result that, essentially, has been proved in Section 4.7 of Vol. 1.

Lemma 1. (i) *Let $X : B \to \mathbb{R}^3$ be a minimal surface with a finite Dirichlet integral $D(X)$ and with boundary values $X|_C$ of finite total variation*

$$L(X) = \int_C |dX|.$$

Then X is of class $H_2^1(B,\mathbb{R}^3)$ and has a continuous extension to $\overline{B}$, i.e., $X \in C^0(\overline{B},\mathbb{R}^3)$. Moreover, the boundary values $X|_C$ are of class $H_1^1(C,\mathbb{R}^3)$. Setting $X(r,\theta) := X(re^{i\theta})$, we obtain that, for any $r \in (0,1]$, the function $X_\theta(r,\theta)$ vanishes at most on a set of θ-values of one-dimensional Hausdorff measure zero, and that the limits

$$\lim_{r\to 1-0} X_r(r,\theta) \quad \textit{and} \quad \lim_{r\to 1-0} X_\theta(r,\theta)$$

exist, and that

$$\frac{\partial}{\partial\theta} X(1,\theta) = \lim_{r\to 1-0} X_\theta(r,\theta) \quad \text{a.e. } \textit{on } [0,2\pi]$$

holds true. Finally, setting $X_r(1,\theta) := \lim_{r\to 1-0} X_r(r,\theta)$*, it follows that*

$$\int_B \langle \nabla X, \nabla \phi \rangle \, du\, dv = \int_C \langle X_r, \phi \rangle \, d\theta \tag{1}$$

is satisfied for all $\phi \in H_2^1 \cap L_\infty(B, \mathbb{R}^3)$*. Moreover, we have*

$$\lim_{r\to 1-0} \int_0^{2\pi} |X_\theta(r,\theta)| r \, d\theta = \int_0^{2\pi} |dX(1,\theta)|. \tag{2}$$

(ii) *If* $X : B \to \mathbb{R}^3$ *is a minimal surface with a continuous extension to* $\overline{B}$ *such that* $L(X) := \int_C |dX| < \infty$*, then we still have* (2).

Proof. Since $L(X) < \infty$, the finiteness of $D(X)$ is equivalent to the relation $X \in H_2^1(B, \mathbb{R}^3)$, on account of Poincaré's inequality. Hence X has an $L_2(C)$-trace on the boundary C of ∂B which, by assumption, has a finite total variation $\int_C |dX|$. Consequently, the two one-sided limits

$$\lim_{\theta\to\theta_0-0} X(1,\theta) \quad \text{and} \quad \lim_{\theta\to\theta_0+0} X(1,\theta)$$

exists for every $\theta_0 \in \mathbb{R}$. In conjunction with the Courant–Lebesgue lemma, we obtain that $X(1,\theta)$ is a continuous function of $\theta \in \mathbb{R}$ whence $X \in C^0(\overline{B}, \mathbb{R}^3)$. The rest of the proof follows from Theorems 1 and 2 Vol. 1, in Section 4.7. □

Lemma 2 (Wirtinger's inequality). *Let* $Z : \mathbb{R} \to \mathbb{R}^3$ *be an absolutely continuous function that is periodic with the period* $L > 0$ *and has the mean value*

$$P := \frac{1}{L} \int_0^L Z(t)\, dt. \tag{3}$$

Then we obtain

$$\int_0^L |Z(t) - P|^2 \, dt \le \left(\frac{L}{2\pi}\right)^2 \int_0^L |\dot{Z}(t)|^2 \, dt, \tag{4}$$

and the equality sign holds if and only if there are constant vectors A_1 *and* B_1 *in* $\mathbb{R}^3$ *such that*

$$Z(t) = P + A_1 \cos\left(\frac{2\pi}{L} t\right) + B_1 \sin\left(\frac{2\pi}{L} t\right) \tag{5}$$

holds for all $t \in \mathbb{R}$.

Proof. We first assume that $L = 2\pi$ and $\int_0^{2\pi} |\dot{Z}|^2\, dt < \infty$. Then we have the expansions

$$Z(t) = P + \sum_{n=1}^{\infty} (A_n \cos nt + B_n \sin nt), \quad \dot{Z}(t) = \sum_{n=1}^{\infty} n(B_n \cos nt - A_n \sin nt)$$

of Z and $\dot{Z}$ into Fourier series with A_n, $B_n \in \mathbb{R}^3$, and

$$\begin{aligned} \int_0^{2\pi} |Z - P|^2\, dt &= \pi \sum_{n=1}^{\infty} (|A_n|^2 + |B_n|^2), \\ \int_0^{2\pi} |\dot{Z}|^2\, dt &= \pi \sum_{n=1}^{\infty} n^2 (|A_n|^2 + |B_n|^2). \end{aligned} \tag{6}$$

Consequently it follows that

$$\int_0^{2\pi} |Z - P|^2\, dt \le \int_0^{2\pi} |\dot{Z}|^2\, dt, \tag{7}$$

and the equality sign holds if and only if all coefficients A_n and B_n vanish for $n > 1$. Thus we have verified the assertion under the two additional hypotheses. If $\int_0^{2\pi} |\dot{Z}|^2\, dt = \infty$, the statement of the lemma is trivially satisfied, and the general case $L > 0$ can be reduced to the case $L = 2\pi$ by the scaling transformation $t \mapsto (2\pi/L)t$. $\square$

Now we state the isoperimetric inequality for minimal surfaces in its simplest form.

Theorem 1. *Let $X \in C^2(B, \mathbb{R}^3)$ with $B = \{w : |w| < 1\}$ be a nonconstant minimal surface, i.e., $\Delta X = 0, |X_u|^2 = |X_v|^2, \langle X_u, X_v \rangle = 0$. Assume also that X is either of class $H_2^1(B, \mathbb{R}^3)$ or of class $C^0(\overline{B}, \mathbb{R}^3)$, and that $L(X) = \int_C |dX| < \infty$. Then $D(X)$ is finite, and we have*

$$D(X) \le \frac{1}{4\pi} L^2(X). \tag{8}$$

Moreover, the equality sign holds if and only if $X : B \to \mathbb{R}^3$ represents a (simply covered) disk.

Remark 1. Note that for every minimal surface $X : B \to \mathbb{R}^3$ the area functional $A(X)$ coincides with the Dirichlet integral $D(X)$. Thus (8) can equivalently be written as

$$A(X) \le \frac{1}{4\pi} L^2(X). \tag{8'}$$

Proof of Theorem 1. (i) Assume first that X is of class $H_2^1(B, \mathbb{R}^3)$, and that P is a constant vector in $\mathbb{R}^3$. Because of $L(X) < \infty$, the boundary values $X|_C$ are bounded whence X is of class $L_\infty(B, \mathbb{R}^3)$ (this follows from the maximum principle in conjunction with a suitable approximation device). Thus we can apply formula (1) to $\phi = X - P$, obtaining

(9)

$$\begin{aligned}\int_B \langle \nabla X, \nabla X \rangle \, du \, dv &= \int_B \langle \nabla X, \nabla (X - P) \rangle \, du \, dv \\ &= \int_C \langle X_r, X - P \rangle \, d\theta \le \int_C |X_r||X - P| \, d\theta \\ &= \int_C |X_\theta||X - P| \, d\theta = \int_0^{2\pi} |X_\theta(1,\theta)||X(1,\theta) - P| \, d\theta.\end{aligned}$$

Introducing $s = \sigma(\theta)$ by

$$\sigma(\theta) := \int_0^\theta |X_\theta(1,\theta)| \, d\theta,$$

we obtain that $\sigma(\theta)$ is a strictly increasing and absolutely continuous function of θ, and $\dot\sigma(\theta) = |X_\theta(1,\theta)| > 0$ a.e. on $\mathbb{R}$. Hence $\sigma : \mathbb{R} \to \mathbb{R}$ has a continuous inverse $\tau : \mathbb{R} \to \mathbb{R}$. Let us introduce the reparametrization

$$Z(s) := X(1, \tau(s)), \quad s \in \mathbb{R},$$

of the curve $X(1,\theta)$, $\theta \in \mathbb{R}$. Then, for any $s_1, s_2 \in \mathbb{R}$ with $s_1 < s_2$, the numbers $\theta_1 := \tau(s_1)$, $\theta_2 := \tau(s_2)$ satisfy $\theta_1 < \theta_2$ and

(10)
$$\int_{s_1}^{s_2} |dZ| = \int_{\theta_1}^{\theta_2} |dX| = \sigma(\theta_2) - \sigma(\theta_1) = s_2 - s_1,$$

whence

$$|Z(s_2) - Z(s_1)| \le s_2 - s_1.$$

Consequently, the mapping $Z : \mathbb{R} \to \mathbb{R}^3$ is Lipschitz continuous and therefore also absolutely continuous, and we obtain from (10) that

(11)
$$\int_{s_1}^{s_2} |Z'(s)| \, ds = s_2 - s_1$$

whence

(12)
$$|Z'(s)| = 1 \quad \text{a.e. on } \mathbb{R}.$$

In other words, the curve $Z(s)$ is the reparametrization of $X(1,\theta)$ with respect to the parameter s of its arc length.

As the mapping $\sigma : \mathbb{R} \to \mathbb{R}$ is absolutely continuous, it maps null sets onto null sets, and we derive from

$$\frac{\tau(s_2) - \tau(s_1)}{s_2 - s_1} = \frac{1}{\frac{\sigma(\theta_2)-\sigma(\theta_1)}{\theta_2-\theta_1}}$$

and from $\dot{\sigma}(\theta) > 0$ a.e. on $\mathbb{R}$ that

$$\tau'(s) = \frac{1}{\dot{\sigma}(\tau(s))} > 0 \quad \text{a.e. on } \mathbb{R}. \tag{13}$$

On account of

$$\dot{\sigma}(\theta) = |X_\theta(1, \theta)| \quad \text{a.e. on } \mathbb{R}$$

it then follows that

$$|X_\theta(1, \tau(s))| \frac{d\tau}{ds}(s) = 1 \quad \text{a.e. on } \mathbb{R}, \tag{14}$$

and thus we obtain

$$\int_0^{2\pi} |X_\theta(1, \theta)||X(1, \theta) - P| \, d\theta = \int_0^L |Z(s) - P| \, ds. \tag{15}$$

We now infer from (9) and (15) that

$$\int_B \langle \nabla X, \nabla X \rangle \, du \, dv \le \int_0^L |Z(s) - P| \, ds. \tag{16}$$

By Schwarz's inequality, we have

$$\int_0^L |Z(s) - P| \, ds \le \sqrt{L} \left\{ \int_0^L |Z(s) - P|^2 \, ds \right\}^{1/2}, \tag{17}$$

and Wirtinger's inequality (4) together with (12) implies that

$$\left\{ \int_0^L |Z(s) - P|^2 \, ds \right\}^{1/2} \le L^{3/2}/(2\pi) \tag{18}$$

if we choose P as the barycenter of the closed curve $Z : [0, L] \to \mathbb{R}^3$, i.e., if

$$P := \frac{1}{L} \int_0^L Z(s) \, ds.$$

By virtue of (16)–(18), we arrive at

$$\int_B |\nabla X|^2 \, du \, dv \le \frac{1}{2\pi} L^2 \tag{19}$$

which is equivalent to the desired inequality (8).

Suppose that equality holds true in (8) or, equivalently, in (19). Then equality must also hold in Wirtinger's inequality (18), and by Lemma 2 we infer

$$Z(s) = P + A_1 \cos\left(\frac{2\pi}{L}s\right) + B_1 \sin\left(\frac{2\pi}{L}s\right).$$

Set $R := L/(2\pi)$ and $\varphi = s/R$. Because of $|Z'(s)| \equiv 1$, we obtain

$$R^2 = |A_1|^2 \sin^2\varphi + |B_1|^2 \cos^2\varphi - 2\langle A_1, B_1\rangle \sin\varphi\cos\varphi.$$

Choosing $\varphi = 0$ or $\frac{\pi}{2}$, respectively, it follows that $|A_1| = |B_1| = R$, and therefore $\langle A_1, B_1\rangle = 0$. Then the pair of vectors $E_1, E_2 \in \mathbb{R}^3$, defined by

$$E_1 := \frac{1}{R}A_1, \quad E_2 := \frac{1}{R}B_1,$$

is orthonormal, and we have

$$Z(R\varphi) = P + R\{E_1\cos\varphi + E_2\sin\varphi\}.$$

Consequently $Z(R\varphi), 0 \le \varphi \le 2\pi$, describes a simply covered circle of radius R, centered at P, and the same holds true for the curve $X(1,\theta)$ with $0 \le \theta \le 2\pi$. Hence $X : \overline{B} \to \mathbb{R}^3$ represents a (simply covered) disk of radius R, centered at P, as we infer from the "convex hull theorem" of Section 4.2 and a standard reasoning.

Conversely, if $X : \overline{B} \to \mathbb{R}^3$ represents a simply covered disk, then the equality sign holds true in (8′) and, therefore also in (8).

Thus the assertion of the theorem is proved under the assumption that $X \in H_2^1(B, \mathbb{R}^3)$.

(ii) Suppose now that X is of class $C^0(\overline{B}, \mathbb{R}^3)$. Then we introduce nonconstant minimal surfaces $X_k : B \to \mathbb{R}^3$ of class $C^\infty(\overline{B}, \mathbb{R}^3)$ by defining

$$X_k(w) := X(r_k w) \quad \text{for } |w| < 1, \quad r_k := \frac{k}{k+1}.$$

We can apply (i) to each of the surfaces X_k, thus obtaining

$$4\pi D(X_k) \le \left\{\int_0^{2\pi} |dX_k(1,\theta)|\right\}^2. \tag{20}$$

For $k \to \infty$, we have $r_k \to 1-0$, $D(X_k) \to D(X)$, and part (ii) of Lemma 1 yields

$$\lim_{k\to\infty} \int_0^{2\pi} |dX_k(1,\theta)| = \int_0^{2\pi} |dX(1,\theta)|.$$

Thus we infer from (20) that $4\pi D(X) \le L^2(X)$ which implies in particular that X is of class $H_2^1(B, \mathbb{R}^3)$. For the rest of the proof, we can now proceed as in part (i). □

If the boundary of a minimal surface X is very long in comparison to its "diameter", then another estimate of $A(X) = D(X)$ might be better which depends only linearly on the length $L(X)$ of the boundary of X. We call this estimate *the linear isoperimetric inequality.* It reads as follows:

Theorem 2. *Let X be a nonconstant minimal surface with the parameter domain $B = \{w : |w| < 1\}$, and assume that X is either continuous on $\overline{B}$ or of class $H_2^1(B, \mathbb{R}^3)$. Moreover, suppose that the length $L(X) = \int_C |dX|$ of its boundary is finite, and let $\mathcal{K}_R(P)$ be the smallest ball in $\mathbb{R}^3$ containing $X(\partial B)$ and therefore also $X(\overline{B})$. Then we have*

$$D(X) \le \frac{1}{2} RL(X). \tag{21}$$

Equality holds in (21) *if and only if $X(B)$ is a plane disk.*

Proof. By Theorem 1 it follows that $D(X) < \infty$ and $X \in H_2^1(B, \mathbb{R}^3)$, and formula (9) implies

$$2D(X) \le \int_C |X_\theta||X - P| \, d\theta \le RL(X) \tag{22}$$

whence we obtain (21). Suppose now that

$$D(X) = \frac{1}{2} RL(X). \tag{23}$$

Then we infer from (9) and (22) that

$$\int_C \langle X_r, X - P \rangle \, d\theta = \int_C |X_r||X - P| \, d\theta$$

is satisfied; consequently we have

$$\langle X_r, X - P \rangle = |X_r||X - P|$$

a.e. on C, that is, the two vectors X_r and $X - P$ are collinear a.e. on C.

Secondly we infer from (22) and (23) that

$$|X - P| = R \quad \text{a.e. on } C.$$

Hence the H_1^1-curve Σ defined by $X : C \to \mathbb{R}^3$ lies on the sphere $S_R(P)$ of radius R centered at P, and the side normal X_r of the minimal surface X at Σ is proportional to the radius vector $X - P$. Thus $X_r(1, \theta)$ is perpendicular to $S_R(P)$ for almost all $\theta \in [0, 2\pi]$. Hence the surface X meets the sphere $S_R(P)$ orthogonally a.e. along Σ. As in the proof of Theorem 1 in Section 1.4 we can show that X is a stationary surface with a free boundary on $S_R(P)$ and that X can be viewed as a stationary point of Dirichlet's integral in the class $\mathcal{C}(S_R(P))$. By Theorems 1 and 2 of Section 2.8, the surface X is real analytic on the closure $\overline{B}$ of B. Then it follows from the Theorem in Section 1.7 that $X(\overline{B})$ is a plane disk.

Conversely, if $X : B \to \mathbb{R}^3$ represents a plane disk, then (23) is fulfilled. □

Now we want to state a more general version of the isoperimetric inequality, valid for global minimal surfaces with boundaries.

Definition 1. *A* global minimal surface *(in $\mathbb{R}^3$) is a nonconstant map*

$$\mathfrak{X} \in C^0(M, \mathbb{R}^3) \cap C^2(\overset{\circ}{M}, \mathbb{R}^3)$$

from a two-dimensional manifold M of class C^k, $k \geq 2$, with the boundary ∂M and the interior $\overset{\circ}{M} = \operatorname{int} M$ into the three-dimensional Euclidean space $\mathbb{R}^3$ which has the following properties:

(i) *M possesses an atlas $\mathcal{C}$ which defines a conformal structure on the interior $\overset{\circ}{M}$ of M;*

(ii) *for every chart φ belonging to the conformal structure $\mathcal{C}$ the local map*

$$X = \mathfrak{X} \circ \varphi^{-1} : \ \operatorname{int} \varphi(G) \to \mathbb{R}^3, \quad G \subset M,$$

is harmonic and conformal, i.e. a minimal surface as defined in Section 2.6.

In other words, a global minimal surface is defined on a Riemann surface M with a smooth boundary ∂M (which might be empty).

Note that $\mathfrak{X}$ may have branch points and selfintersections. Moreover we know that, away from the branch points, the map $\mathfrak{X} : M \to \mathbb{R}^3$ induces a Riemannian metric on $\overset{\circ}{M}$. With respect to the local coordinates determined by the charts φ of the atlas $\mathcal{C}$ this metric is given by

$$g_{\alpha\beta}(u, v) = \lambda(u, v)\delta_{\alpha\beta},$$

where $\lambda = |X_u|^2 = |X_v|^2$, so that the gradient ∇_M and the Laplace–Beltrami operator Δ_M are proportional to the corresponding Euclidean operators ∇ and Δ with respect to the local coordinates u and v,

$$\nabla_M = \frac{1}{\lambda}\nabla, \quad \Delta_M = \frac{1}{\lambda}\Delta.$$

In particular, the function $|\mathfrak{X}|^2 = \sum_{j=1}^{3} |\mathfrak{X}^j|^2$ satisfies

$$\Delta_M |\mathfrak{X}|^2 = 4. \tag{24}$$

Moreover, if M is compact, $\mathfrak{X}$ is of class C^1 up to its boundary, and if $\mathfrak{X}$ has only finitely many branch points in M, then $M \setminus \{\text{branch points}\}$ is a Riemannian manifold, and Green's formulas (in the sense of the Riemannian metric) are meaningful and true for smooth functions defined on M; for example, we obtain from (24) the formula

$$\begin{aligned} 4 \operatorname{area} \mathfrak{X} &= 4 \int_M d\operatorname{vol}_M = \int_M \Delta_M |\mathfrak{X}|^2 \, d\operatorname{vol}_M \\ &= 2 \int_{\partial M} |\mathfrak{X}| \frac{\partial}{\partial \nu} |\mathfrak{X}| \, d\operatorname{vol}_{\partial M}, \end{aligned} \tag{25}$$

where ν is the exterior unit normal to ∂M in the tangent bundle $TM|_{\partial M}$.

In Chapter 2 we have seen that boundary branch points of $\mathfrak{X}$ on ∂M are isolated. Hence, for reasonably regular surfaces $\mathfrak{X}$, there exist only finitely many branch points in the interior and on the boundary.

Definition 2. *Let $\mathfrak{X} : M \to \mathbb{R}^3$ be a global minimal surface defined on a compact manifold M. Then the boundary $\partial\mathfrak{X} := \mathfrak{X}(\partial M)$ of $\mathfrak{X}$ is called* weakly connected *if there is a system of Cartesian coordinates (x^1, x^2, x^3) in $\mathbb{R}^3$ such that no hyperplane $H := \{x^j = \text{const}\}$, $j = 1, 2, 3$, separates $\partial\mathfrak{X}$, that is, if H is any hyperplane orthogonal to one of the coordinate axes and if $H \cap \partial\mathfrak{X}$ is empty, then $\partial\mathfrak{X}$ lies on one side of H. Moreover $\mathfrak{X} : M \to \mathbb{R}^3$ is called* compact *if M is compact.*

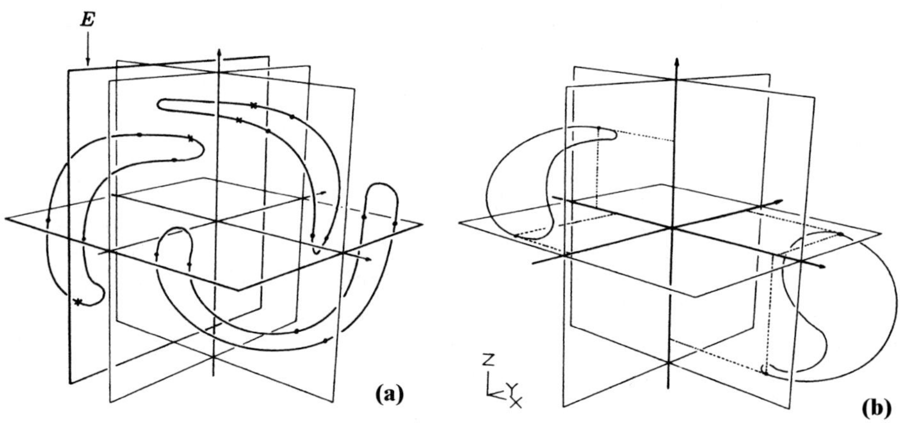

Fig. 1. (**a**) Three weekly connected curves. No plane E parallel to any of the coordinate planes shown separates them. (**b**) Two curves in $\mathbb{R}^3$ which are not weakly connected. It is shown in the text that they lie in opposite quadrants of a suitable coordinate system

Now we can formulate a *general version of the isoperimetric inequality.*

Theorem 3. *Let $\mathfrak{X} : M \to \mathbb{R}^3$ be a global compact minimal surface of class C^1 having at most finitely many branch points defined on a compact Riemann surface M. Suppose also that the boundary $\partial\mathfrak{X}$ is weakly connected. Then the area $A(\mathfrak{X})$ of $\mathfrak{X}$ is bounded from above in terms of the length $L(\mathfrak{X})$ of $\partial\mathfrak{X}$ by the inequality*

$$A(\mathfrak{X}) \le \frac{1}{4\pi} L^2(\mathfrak{X}). \tag{26}$$

Moreover, equality holds if and only if $\mathfrak{X}$ is a plane disk in $\mathbb{R}^3$.

Proof. Let (x^1, x^2, x^3) be the coordinates appearing in the definition of the weakly connected boundary $\partial\mathfrak{X}$. By means of a suitable shift we may even assume that the center of mass of the boundary $\partial \mathcal{X}$ lies at the origin, i.e. that for $j = 1, 2, 3$

$$\int_{\partial M} \mathcal{X}^j \, d\,\mathrm{vol}_{\partial M} = 0, \tag{27}$$

where $\mathcal{X}^j$ is, of course, the j-th coordinate function of the surface $\mathcal{X}$.

On account of (25), it follows that

$$2A(\mathcal{X}) = \int_{\partial M} |\mathcal{X}| \frac{\partial}{\partial \nu} |\mathcal{X}| \, d\,\mathrm{vol}_{\partial M},$$

and it is easily seen that $\frac{\partial}{\partial \nu}|\mathcal{X}| \le 1$. Therefore Schwarz's inequality implies that

$$\begin{aligned} 2A(\mathcal{X}) &= \int_{\partial M} |\mathcal{X}| \frac{\partial}{\partial \nu} |\mathcal{X}| \, d\,\mathrm{vol}_{\partial M} \le \int_{\partial M} |\mathcal{X}| \, d\,\mathrm{vol}_{\partial M} \\ &\le \left(\int_{\partial M} d\,\mathrm{vol}_{\partial M} \int_{\partial M} |\mathcal{X}|^2 \, d\,\mathrm{vol}_{\partial M} \right)^{1/2} \\ &= L^{1/2}(\mathcal{X}) \left(\int_{\partial M} |\mathcal{X}|^2 \, d\,\mathrm{vol}_{\partial M} \right)^{1/2}. \end{aligned} \tag{28}$$

Case (i): Suppose that $\partial \mathcal{X} = \mathcal{X}(\partial M)$ is connected, i.e., $\partial \mathcal{X}$ is a closed curve. Then the proof is essentially that of Theorem 1. In fact, let s be the parameter of arc length of $\partial \mathcal{X}$, and assume that $\partial \mathcal{X}$ is parametrized by s, we write $\mathcal{X}(s)$ for the parameter representation of $\partial \mathcal{X}$ with respect to s. Because of (27) we have $\int_0^L \mathcal{X}(s)\, ds = 0$, where $L := L(\mathcal{X})$, and Wirtinger's inequality yields

$$\begin{aligned} \int_{\partial M} |\mathcal{X}|^2 \, d\,\mathrm{vol}_{\partial M} &= \int_0^L |\mathcal{X}(s)|^2 \, ds \\ &\le \frac{L^2}{4\pi^2} \int_0^L \left| \frac{\partial \mathcal{X}}{ds}(s) \right|^2 ds = \frac{L^3}{4\pi^2}. \end{aligned} \tag{29}$$

From (28) and (29) we derive the desired inequality (26).

Case (ii): $\partial \mathcal{X}$ is weakly connected, but not connected. Hence we are not allowed to apply Wirtinger's inequality, and we have to look for some substitute. Again, we introduce $L = L(\mathcal{X})$ as length of $\mathcal{X}(\partial M)$.

Since M is compact and regular, its boundary ∂M consists of finitely many, say, p closed curves $\partial^1 M, \dots, \partial^p M$. Denote their images under $\mathcal{X}$ by $\sigma_1, \sigma_2, \dots, \sigma_p$, and fix some index $j \in \{1, 2, 3\}$. By assumption, no hyperplane $\{x^k = \text{const}\}$ separates σ_1 from $\sigma_2, \dots, \sigma_p$. Hence, for at least one of these curves, say, for σ_2, we have following property:

There are two points P_1 and Q_1 on σ_1 and σ_2, respectively, whose j-th components P_1^j and Q_1^j coincide. The translation $A_2 : \mathbb{R}^3 \to \mathbb{R}^3$ defined by $P \mapsto P + (P_1 - Q_1)$ leaves the j-th component of every point of $\mathbb{R}^3$ unchanged. Thus $\sigma_1 \cup A_2 \sigma_2$ is connected. In a second step we find points

$$P_2 \in \sigma_2 \quad \text{and} \quad Q_2 \in \sigma_3 \cup \cdots \cup \sigma_p, \quad \text{say,} \quad Q_2 \in \sigma_3,$$

such that $P_2^j = Q_2^j$, and a translation $A_3 : \mathbb{R}^3 \to \mathbb{R}^3$ defined by

$$P \mapsto P + (P_1 - Q_1) + (P_2 - Q_2).$$

Again, A_3 leaves the j-th component of every point in $\mathbb{R}^3$ unchanged, and $\sigma_1 \cup A_2\sigma_2 \cup A_3\sigma_3$ is connected. Proceeding by induction, we find translations $A_4, \ldots, A_p$ such that $c_j := \sigma_1 \cup A_2\sigma_2 \cup \cdots \cup A_p\sigma_p$ is a connected curve.

Now let $\mathfrak{X}_1(s), \ldots, \mathfrak{X}_p(s)$ be the parametrizations of $\sigma_1, \ldots, \sigma_p$ with respect to their arc lengths, and

$$x_1(s), \quad 0 \le s \le L_1, \ldots, x_p(s), \quad 0 \le s \le L_p,$$

be their j-th components. We can assume that $\mathfrak{X}_1(0) = P_1$ and $\mathfrak{X}_2(0) = Q_1$ whence $x_1(0) = x_1(L_1) = x_2(0)$. Define

$$y_1(s) := \begin{cases} x_1(s) & \text{for } 0 \le s \le L_1, \\ x_2(s - L_1) & \text{for } L_1 \le s \le L_1 + L_2 \end{cases}$$

and

$$z_2(s) := y_1(s + s_2),$$

where s_2 is chosen in such a way that $z_2(0) = y_1(s_2) = P_2^j = Q_2^j$. Then both $y_1(s)$ and $z_2(s)$ are continuous and periodic with the period $L_1 + L_2$, and we have a.e. that $|\dot{y}_1(s)| = 1$ and $|\dot{z}_2(s)| = 1$.

In the second step we define

$$y_2(s) := \begin{cases} z_2(s) & \text{for } 0 \le s \le L_1 + L_2, \\ x_3(s - L_1 - L_2) & \text{for } L_1 + L_2 \le s \le L_1 + L_2 + L_3 \end{cases}$$

and

$$z_3(s) := y_2(s + s_3),$$

where s_3 is chosen in such a way that $z_3(0) = y_2(s_3) = P_3^j = Q_3^j$. Finally, after $p-1$ steps, we obtain a continuous function $y_p(s), 0 \le s \le L := L_1 + \cdots + L_p$, which is periodic with the period L, and $|\dot{y}_p(s)| = 1$ a.e. on $[0, L]$.

By Wirtinger's inequality we obtain

$$\int_0^L |y_{p-1}(s)|^2 \, ds \le \left(\frac{L}{2\pi}\right)^2 \int_0^L |\dot{y}_{p-1}(s)|^2 \, ds, \tag{30}$$

as the mean value of the function y_{p-1} is zero. By construction it follows that

$$\int_0^L |y_{p-1}(s)|^2 \, ds = \int_{\partial M} |\mathfrak{X}^j(s)|^2 \, d\,\mathrm{vol}_{\partial M}$$

and that

$$\int_0^L |\dot{y}_{p-1}(s)|^2 \, ds = \int_{\partial M} \left| \frac{d}{ds} \mathcal{X}^j \right|^2 \, d\,\mathrm{vol}_{\partial M}$$

whence

$$\int_{\partial M} |\mathcal{X}^j|^2 \, d\,\mathrm{vol}_{\partial M} \leq \left(\frac{L}{2\pi} \right)^2 \int_M \left| \frac{d}{ds} \mathcal{X}^j \right|^2 \, d\,\mathrm{vol}_{\partial M}. \tag{31}$$

As j is an arbitrary index in $\{1, 2, 3\}$, we may sum up the equations (31) for $j = 1, 2, 3$, thus obtaining

$$\int_{\partial M} |\mathcal{X}^2| \, d\,\mathrm{vol}_{\partial M} \leq \left(\frac{L}{2\pi} \right)^2 \int_{\partial M} \left| \frac{d}{ds} \mathcal{X} \right|^2 \, d\,\mathrm{vol}_{\partial M}. \tag{32}$$

Thus Wirtinger's inequality can be generalized to weakly connected boundaries $\mathcal{X} : \partial M \to \mathbb{R}^3$ in the form (32). Now we can proceed as in case (i) to obtain the isoperimetric inequality (26).

Let us now suppose that equality holds in the isoperimetric inequality, i.e.,

$$4\pi A(\mathcal{X}) = L^2(\mathcal{X}).$$

Then, in particular, equality holds in (28) implying that

$$|\mathcal{X}| \equiv \text{const} =: R \quad \text{on } \partial M,$$

i.e., $\partial\mathcal{X}$ lies on a sphere of radius R, and $R > 0$ since $X(w) \not\equiv 0$.

Now let P be some point on the curve σ_1 which is not the image of a branch point of $\mathcal{X}$. The parametrization of the curves c_j introduced above with respect to the arc length s can now be chosen such that

$$c_j(0) = P \quad \text{for all } j$$

and, if P is suitably selected, that for some neighbourhood $(-\varepsilon, \varepsilon)$ of 0 the curve $c_j(s)$ parametrizes a part of σ_1. Now equality in the isoperimetric inequality implies equality in Wirtinger's inequality for the j-th component c_j^j of the curve c_j, thus

$$c_j^j(s) = a^j \cos\left(\frac{2\pi}{L} s \right) + b^j \sin\left(\frac{2\pi}{L} s \right)$$

for two constants a^j and b^j, $L = L(\mathcal{X})$; in particular, we have for all j that

$$c_j^j(0) = p^j = a^j,$$
$$\left(\frac{d}{ds} c_j^j \right)(0) = \frac{d\sigma_1^j}{ds}(0) = b^j \frac{2\pi}{L}.$$

Since $\partial\mathcal{X}$ lies on a sphere, the vectors

$$a = (a^1, a^2, a^3), \quad b = (b^1, b^2, b^3)$$

are mutually perpendicular, and they satisfy

$$R = |a| = |b| = \frac{L}{2\pi}.$$

Since $R > 0$, at least one of the components of a, say a^{j_0}, does not vanish; consequently the function $c_{j_0}^{j_0}(s)$ has exactly two critical points in the interval $[0, L)$. This implies that the boundary $\partial\mathfrak{X}$ of the minimal surface $\mathfrak{X}$ under consideration has only one component. In fact, $c_{j_0}^{j_0}$ is the j_0-th component of the curve obtained by shifting the boundary components $\sigma_1, \ldots, \sigma_p$ of $\partial\mathfrak{X}$ together in a plane perpendicular to the j_0-axis, and every curve σ_j contributes at least two critical points to the function $c_{j_0}^{j_0}$, so that $c_{j_0}^{j_0}$ has at least four critical points if p is greater than one.

This proves that the functions c_j^j are simply the j-th components of the one and only boundary curve σ_1. The preceding identities show that σ_1 is a circle of radius $R = \frac{L}{2\pi}$, the boundary of a plane disk containing $\mathfrak{X}$; see the convex hull theorem in Section 4.1. □

We shall now study a minimal surface $\mathfrak{X} : M \to \mathbb{R}^3$ in the three-dimensional space defined on a compact manifold M whose boundary ∂M has exactly two components $\partial^+ M$ and $\partial^- M$. Let us see what happens if $\partial\mathfrak{X}$ is *not* weakly connected.

Denote by $\partial^+\mathfrak{X} = \mathfrak{X}(\partial^+ M)$ and $\partial^-\mathfrak{X} = \mathfrak{X}(\partial^- M)$ the components of $\partial\mathfrak{X}$. They lie in some ball $\overline{B}_R(0) \subset \mathbb{R}^3$. We claim that there is a hyperplane E_*^1 with normal $N_* \in S^2$ through a point P_* such that the components $\partial^\pm\mathfrak{X}$ of $\partial\mathfrak{X}$ lie in the two closed half spaces $H_1^\pm$ defined by E_*^1 respectively and such that $\partial^+\mathfrak{X}$ and $\partial^-\mathfrak{X}$ touch E_*^1.

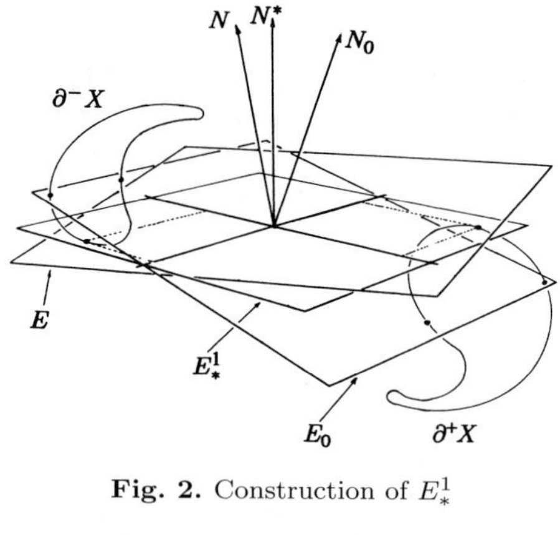

Fig. 2. Construction of E_*^1

Such a plane E_*^1 can be constructed as follows: First of all, there is a plane E_0 with normal N_0 which intersects $\partial^+\mathfrak{X}$ and $\partial^-\mathfrak{X}$. Then consider the open

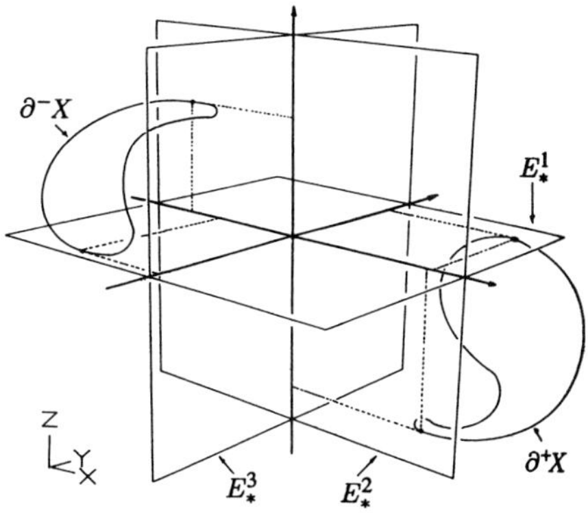

Fig. 3. The planes E^1_*, E^2_*, E^3_*

set $U \subset S^2$ of all unit vectors N for which there is a $P \in \overline{B}_R(0)$ such that the oriented plane $E(P, N)$ through P with normal N separates $\partial\mathfrak{X}$, (i.e., $\partial^+\mathfrak{X}$ and $\partial^-\mathfrak{X}$ lie in the *open* half spaces defined by $E(P, N)$). U is not empty since by assumption $\partial\mathfrak{X}$ is not weakly connected (see Fig. 2).

Now take a sequence of planes $E_n = E(P_n, N_n)$ such that $P_n \in \overline{B}_r(0)$, $N_n \in U$, and such that

$$\lim_{n\to\infty} \langle N_n, N_0\rangle = \sup\{\langle N, N_0\rangle : N \in U\};$$

this expression is positive since U is open. Passing to a subsequence we may assume that P_n converges to P_* and N_n to N_*. The plane $E^1_* = E(P_*, N_*)$ then has the desired property (cf. Fig. 2).

No plane parallel to E^1_* separates $\partial\mathfrak{X}$, therefore some plane E^2 orthogonal to E^1_* separates $\partial\mathfrak{X}$ since it is not weakly connected. Proceeding as above we can now construct a plane E^2_* perpendicular to E^1_* such that $\partial^+\mathfrak{X}$ and $\partial^-\mathfrak{X}$ lie again in the two closed half spaces defined by E^2_* and such that both components of $\partial\mathfrak{X}$ touch E^2_*. Once again none of the planes parallel to E^2_* separates $\partial\mathfrak{X}$, hence there has to be a third plane E^3_* orthogonal to E^1_* as well as E^2_* which separates ∂X (Fig. 3). Thus we can choose x, y, z-coordinate axes such that E^1_*, E^2_* and E^3_* correspond to the x, y-, x, z- and y, z-planes respectively and such that $\partial^\pm\mathfrak{X}$ lies in the octant

$$\{(x, y, z) : x, y, z \geq 0(\leq 0)\};$$

in particular, putting $V = \frac{1}{\sqrt{3}}(1, 1, 1)$, the components $\partial^\pm\mathfrak{X}$ lie in the cones

$$C^\pm = \left\{P \in \mathbb{R}^3 : \pm\langle P, V\rangle \geq \frac{|P|}{\sqrt{3}}\right\}$$

respectively. The opening angle of this cone is 54.7356103... degrees.

As we have proved in Section 4.2, this implies that the minimal surface $\mathfrak{X}$ is not connected. According to Section 3.6 of Vol. 1, there are not compact global minimal surfaces without boundary. Therefore M has exactly two components M^+ and M^- with boundaries $\partial^+ M$ and $\partial^- M$ respectively. Applying the isoperimetric inequality to both of them we obtain

$$\begin{aligned} 4\pi A(\mathfrak{X}) &= 4\pi \left\{ A(\mathfrak{X}^+) + A(\mathfrak{X}^-) \right\} \\ &\le L^2(\partial^+ \mathfrak{X}) + L^2(\partial^- \mathfrak{X}) \\ &< L^2(\mathfrak{X}). \end{aligned}$$

(Here $L(\partial^\pm \mathfrak{X})$ denotes the length of $\partial^\pm \mathfrak{X}$, and $L(\mathfrak{X})$ is the length of $\partial \mathfrak{X}$, i.e., $L(\mathfrak{X}) = L(\partial^+ \mathfrak{X} \cup \partial^- \mathfrak{X})$.) Thus we have proved the following

Corollary 1. *If $\mathfrak{X}$ is a global compact minimal surface of class $C^1(M, \mathbb{R}^3)$ having at most finitely many branch points and whose boundary has no more than two connected components, then we have the isoperimetric inequality*

$$4\pi A(\mathfrak{X}) \le L^2(\mathfrak{X}),$$

and equality holds if and only if $\mathfrak{X}$ is a plane disk.

One undesirable feature of our isoperimetric inequality is that the minimal surface $\mathfrak{X}$ has to be of class C^1 up to the boundary. For a minimal surface $X : B \to \mathbb{R}^3$ defined on the disk $B = \{w : |w| < 1\}$, it follows that the lengths of the boundaries of the surfaces $Z^{(r)}(w) := X(rw)$, $0 < r < 1$, and $w \in B$, tend to the length of the boundary of X, if $X \in C^0(\overline{B}, \mathbb{R}^3)$ and $X|_C$ is rectifiable.

Such a continuity property is also known for doubly connected minimal surfaces defined on annuli; cf. Feinberg [1]. Thus we obtain also

Corollary 2. *If $X : \Omega \to \mathbb{R}^3$ is a minimal surface with $X \in C^0(\overline{\Omega}, \mathbb{R}^3)$, which has a rectifiable boundary and whose parameter domain Ω is either a disk or an annulus, then we have*

$$A(X) \le \frac{1}{4\pi} L^2(X).$$

It can be seen that equality holds if and only if $X(\Omega)$ is a plane disk. Note that Corollary 2 is a generalization of Theorem 1.

4.6 Estimates for the Length of the Free Trace

In this section we want to estimate the length of the free trace of a minimal surface $X : B \to \mathbb{R}^3$ in two situations. First we assume that the image $X(I)$, $I \subset \partial B$, is contained in some part S_0 of the support surface S which can be viewed as the graph of some function $\psi : \Omega \to \mathbb{R}$, $\Omega \subset \mathbb{R}^2$, having a bounded gradient (that is, the Gauss image of S is compactly contained in some open hemisphere of S^2). Secondly, we shall study the case that S satisfies a (two-sided) sphere condition.

To begin with the first situation, we assume that S is an embedded regular surface of class C^1 in $\mathbb{R}^3$, and that Γ is a rectifiable Jordan arc of length $L(\Gamma)$ with endpoints P_1 and P_2 on S. We shall not exclude that Γ and S have also other points in common. Nevertheless, we can define the class $\mathcal{C}(\Gamma, S)$ of admissible surfaces $X : B \to \mathbb{R}^3$ for the semifree problem with respect to the boundary configuration $\langle \Gamma, S \rangle$ as in 4.6 of Vol. 1. For technical reasons we imagine such surfaces to be parametrized on the semidisk $B = \{w : |w| < 1, \operatorname{Im} w > 0\}$, the boundary of which consists of the interval I and the circular arc C. For any $X \in \mathcal{C}(\Gamma, S)$, the Jordan arc Γ is the weakly monotonic image of C under X, by Σ we want to denote the free trace $X : I \to \mathbb{R}^3$ of the mapping X on the support surface S. The total variation

$$L(\Sigma) := \int_I |dX|$$

will be called the *length of the free trace* Σ.

Definition 1. *We say that some orientable part S_0 of S fulfils a λ-graph condition, $\lambda > 0$, if there is a unit vector $N_0 \in \mathbb{R}^3$ such that the (suitably chosen) field $N_S(P)$ of unit normals on S satisfies the condition*

(1) $$\langle N_0, N_S(P) \rangle \geq \lambda \quad \text{for all } P \in S_0.$$

Proposition 1. *Let X be a stationary minimal surface in $\mathcal{C}(\Gamma, S)$ (see Section 1.4, Definition 1) which satisfies the following two conditions:*

(i) The free boundary curve $X(I)$ is contained in an open, orientable part S_0 of S which fulfils a λ-graph condition, $\lambda > 0$.

(ii) The scalar product $\langle X_v, N_S(X) \rangle$ does not change its sign on I.

Then the length $L(\Sigma)$ of the free trace Σ, given by $X : I \to \mathbb{R}^3$, is estimated from above by

(2) $$L(\Sigma) \leq \lambda^{-1} L(\Gamma),$$

and the area $A(X) = D(X)$ is bounded by

(3) $$A(X) \leq \frac{(1+\lambda)^2}{4\pi\lambda^2} L^2(\Gamma).$$

Moreover, the surface X is continuous on $\overline{B}$.

Supplement. *If we drop the assumption that X* maps *C monotonically onto Γ, we obtain the estimates*

$$L(\Sigma) \leq \lambda^{-1} \int_C |dX|, \quad A(X) \leq \frac{(1+\lambda)^2}{4\pi\lambda^2} \left(\int_C |dX| \right)^2$$

instead of (2) *and* (3).

Proof of Proposition 1. We can assume that both

$$\langle X_v, N_S(X)\rangle \geq 0$$

and

(4) $$\langle N_0, N_S(X)\rangle \geq \lambda > 0$$

hold on I (we possibly have to replace N_S and N_0 by $-N_S$ and $-N_0$ respectively). As X is assumed to be stationary in $\mathcal{C}(\Gamma, S)$, we have by definition that X is of class $C^1(B \cup I, \mathbb{R}^3)$ and meets S_0 perpendicularly. Consequently we have

$$X_v = |X_v| N_S(X) \quad \text{on } I,$$

and the conformality relation $|X_u| = |X_v|$ yields

(5) $$X_v = |X_u| N_S(X) \quad \text{on } I.$$

Integration by parts implies

$$0 = \int_B \Delta X \, du \, dv = \int_{\partial B} \frac{\partial}{\partial \nu} X \, d\mathcal{H}^1,$$

where ν is the exterior normal to ∂B. Introducing polar coordinates r, φ by $u + iv = re^{i\varphi}$, we arrive at

$$\int_I X_v \, du = \int_C X_r \, d\varphi,$$

and (5) yields

$$\int_I N_S(X)|X_u| \, du = \int_C X_r \, d\varphi.$$

Multiplying this identity by N_0, we arrive at

(6) $$\lambda L(\Sigma) \leq \int_I \langle N_0, N_S(X)\rangle |dX| = \int_C \langle N_0, X_r\rangle \, d\varphi,$$

taking (4) into account, and the conformality relation

$$|X_r| = |X_\varphi| \quad \mathcal{H}^1\text{-a.e. on } C$$

yields

(7) $$\lambda L(\Sigma) \leq \int_C \cos\alpha(\varphi) |X_\varphi| \, d\varphi \leq \int_C |dX|,$$

where $\alpha(\varphi)$ denotes the angle between N_0 and the side normal $X_r(1, \varphi)$ to Γ on X at the point $X(1, \varphi)$. This implies (2), and (3) follows from the isoperimetric inequality

$$A(X) \leq \frac{1}{4\pi} \left(\int_{\partial B} |dX| \right)^2 .$$

Finally, a by now standard reasoning yields $X \in C^0(\overline{B}, \mathbb{R}^3)$, taking the relations $D(X) < \infty$ and $L(\Gamma) < \infty$ into account. The "Supplement" is proved by the same reasoning. □

Remark 1. As X intersects S_0 perpendicularly along Σ, the assumption (ii) is certainly satisfied if X possesses no boundary branch points on the free boundary I. Taking the asymptotic expansion of X at boundary branch points into account (see Section 2.10), we see that there are no branch points on I if, for any $r \in (0,1)$, there is a $\delta \in (0, \sqrt{1-r^2})$ such that the part $X : \{w = u + iv : |u| < r, 0 < |v| < \delta\} \to \mathbb{R}^3$ of the minimal surface X lies "on one side of S_0". The last assumption means that, close to I, the minimal surface X does not penetrate the supporting surface S_0.

Moreover, we read off from the asymptotic expansion that $\langle X_v, N_S(X)\rangle$ does not change its sign on I close to branch points of even order. Thus condition (ii) is even fulfilled if branch points of odd order are excluded on I.

Remark 2. By exploiting (6) somewhat more carefully, we can derive an improvement of estimate (2). To this end, we introduce the representation $\{\xi(s) : 0 < s \le l\}$, $l = L(\Gamma)$, of the Jordan arc Γ with respect to its parameter s of the arc length. Then $\xi'(s)$ is defined a.e. on $[0, l]$, and $|\xi'(s)| = 1$. Let $\beta(s) \in [0, \frac{\pi}{2}]$ be the angle between N_0 and the unoriented tangent $T(s)$ of Γ at $\xi(s)$, given by $\pm\xi'(s)$. Then we obtain

$$\langle X_r, N_0\rangle \le |X_r| \cos\left(\frac{\pi}{2} - \beta\right) = |X_r| \sin\beta = |X_\varphi| \sin\beta$$

and, because of $ds = |X_\varphi|\, d\varphi$ and of the monotonicity of the mapping $X : C \to \Gamma$, we infer from (6) the following variant of (7):

$$\lambda L(\Sigma) \le \int_0^l \sin\beta(s)\, ds.$$

This yields the following sharpened version of (2):

$$L(\Sigma) \le \frac{1}{\lambda} \int_\Gamma \sin\beta(s)\, ds. \tag{8}$$

Remark 3. The estimate (2) *is* optimal in the sense that the number λ^{-1} cannot be replaced by a smaller constant. In order to see this, we consider for $0 < \gamma < \frac{\pi}{2}$ the surface

$$S := \{(x, y, z) : y = (\tan\gamma)(x + 1) \text{ for } x \le 0, y = (\tan\gamma)(1 - x) \text{ for } x \ge 0\}$$

and the arc

$$\Gamma := \{(x, 0, 0) : |x| \le 1\}$$

(cf. Fig. 1). Let

$$N_0 := (0, 1, 0),$$

and consider the minimal surface

$$X(w) := (\operatorname{Re}\tau(w), \operatorname{Im}\tau(w), 0),$$

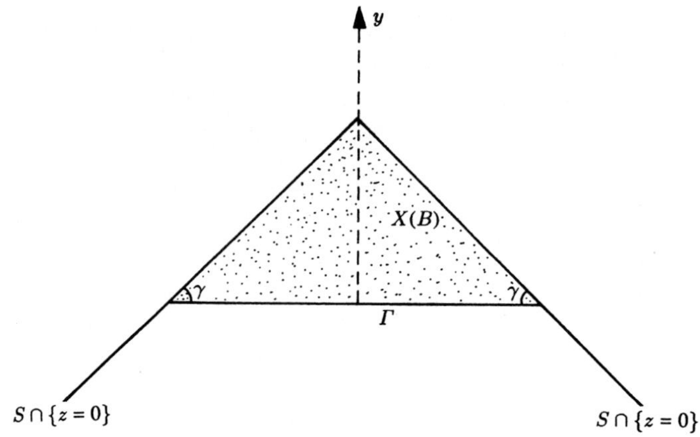

Fig. 1. Remark 3: The estimate (2) is sharp

where $\tau(\omega)$ denotes the conformal mapping of B onto the triangle $\Delta \subset \mathbb{C}$ with the vertices $-1, 1, i \tan \gamma$ keeping ± 1 fixed and mapping i onto 0. Here we have

$$\langle N_0, N_S(P) \rangle = \cos \gamma > 0$$

and

$$L(\Sigma) = \frac{1}{\cos \gamma} L(\Gamma),$$

which shows that the estimate (2) is sharp. However the support surface S of our example does not quite match with the assumptions of Proposition 1 as it is only a Lipschitz surface. By smoothing the surface S at the edge $E := \{(0, \tan \gamma, z)\}$, we can construct a sequence of support surfaces $S_n \in C^\infty$ and a sequence of minimal surfaces $X_n \in \mathcal{C}(\Gamma, S_n)$ whose free traces Σ_n are estimated by

$$L(\Sigma_n) \le \lambda_n^{-1} L(\Gamma)$$

with numbers λ_n tending to $\lambda := \cos \gamma$. As we have

$$\inf_{P \in S_n} \langle N_0, N_{S_n}(P) \rangle = \cos \gamma$$

for all $n = 1, 2, \ldots$ if we construct S_n from S by smoothing around the edge E, it follows that (2) is also sharp in the class of C^∞-support surfaces.

Remark 4. The λ-graph condition (i) in Proposition 1 is crucial. By way of example we shall, in fact, show that one cannot bound the length $L(\Sigma)$ of the free trace Σ in terms of $L(\Gamma)$ and S alone if the λ-graph condition is dropped.

To this end we construct a regular support surface S of class C^∞ which is perpendicularly intersected by the planes

$$\Pi_n := \{(x, y, z) : x = n\}, \quad n = 1, 2, \ldots .$$

We can arrange matters in such a way that the Gauss image of S is contained in the northern hemisphere $S^2 \cap \{z \geq 0\}$ of S^2 and that every intersection curve $S \cap \Pi_n$ consists of an semi-ellipse

$$E_n := \{(x, y, z) : x = n, y^2 + n^{-2}z^2 = 1, z \geq 0\}$$

and of two rays $\{(n, \pm 1, z) : z \leq 0\}$; cf. Fig. 2. Moreover, we choose Γ_n as straight segments in Π_n connecting the endpoints of E_n,

$$\Gamma_n := \{(n, y, 0) : |y| \leq 1\}.$$

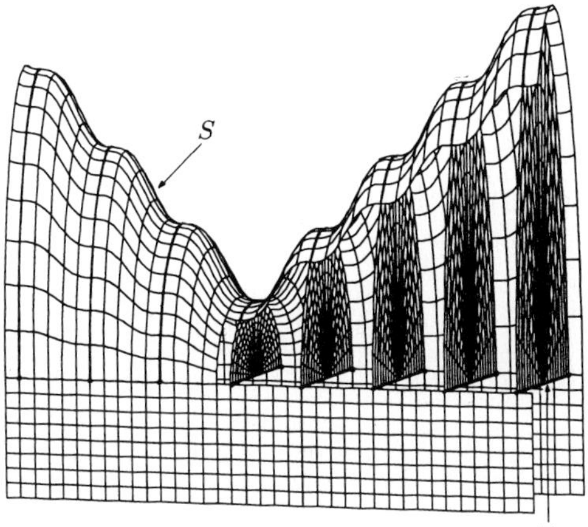

Fig. 2. A supporting surface S, Jordan curves Γ_n of length 2 with endpoints on S, and a sequence of stationary minimal surfaces for these boundary configurations whose surface areas and the lengths of whose free boundaries are unbounded, cf. Remark 4

Finally we choose conformal maps $\tau_n(w) = y_n(w) + iz_n(w)$ of B onto the solid semi-ellipse E_n^* in the y, z-plane, given by

$$E_n^* := \{(y, z) : y^2 + n^{-2}z^2 < 1, z > 0\},$$

which map C onto Γ_n. Then the minimal surfaces

$$X_n(w) := (n, \operatorname{Re} \tau_n(w), \operatorname{Im} \tau_n(w))$$

are stationary in $\mathcal{C}(\Gamma_n, S)$. Their areas $A(X_n)$ and the lengths $L(\Sigma_n)$ of their free traces tend to infinity as $n \to \infty$ whereas $L(\Gamma_n)$ is always equal to 2.

Note that the support surface S of our example satisfies a λ-graph condition with the forbidden value $\lambda = 0$ if we choose N_0 as $(0, 0, 1)$, but it does not fulfil a λ-graph condition for any $\lambda > 0$, no matter what we choose N_0 to be.

By a slight change of the previous reasoning, the reader may construct a similar example of a support surface S with only one Jordan arc Γ such that $\langle \Gamma, S\rangle$ bounds infinitely many stationary minimal surfaces $X_n \in \mathcal{C}(\Gamma, S)$, $n \in \mathbb{N}$, having the property that $A(X_n) \to \infty$ and $L(\Sigma_n) \to \infty$ as $n \to \infty$.

Remark 5. We can use Proposition 1 to derive a priori estimates for the derivatives of X up to the free boundary I. The key step is the following: Suppose that the assumptions of Proposition 1 are satisfied. Let w_0 be some point on I, $d := 1 - |w_0|$, and let r, θ be polar coordinates around w_0, that is, $w = w_0 + re^{i\theta}$. Set $S_r(w_0) := B \cap B_r(w_0)$ and

$$\varphi(r) := \int_{S_r(w_0)} |\nabla X|^2 \, du \, dv = 2 \int_0^r \int_0^\pi |X_\theta|^2 \rho^{-1} \, d\rho \, d\theta.$$

Then we have

$$\varphi'(r) = 2r^{-1} \int_0^\pi |X_\theta(r, \theta)|^2 \, d\theta.$$

By an obvious modification of the proof of Proposition 1 we obtain

$$\varphi(r) \le 2\lambda_1 \left\{ \int_0^\pi |X_\theta(r, \theta)| \, d\theta \right\}^2, \quad \lambda_1 := \frac{(1+\lambda)^2}{4\pi\lambda^2},$$

and Schwarz's inequality yields

$$\varphi(r) \le \pi\lambda_1 r \varphi(r) \quad \text{for } 0 < r < d$$

whence

(9) $$\varphi(r) \le \varphi(d)(r/d)^{2\mu} \quad \text{for } 0 \le r \le d$$

with

$$\mu := \frac{1}{2\pi\lambda_1} = \frac{2\lambda^2}{(1+\lambda)^2}.$$

Because of

(10) $$\varphi(d) \le 2D(X) \le 2\lambda_1 L^2(\Gamma)$$

we arrive at the following result:

If the assumptions of Proposition 1 *are satisfied, then, for any* $w_0 \in I_d := \{w \in I : |w| < 1 - d\}$, $0 < d < 1$, *we have*

(11) $$\int_{S_r(w_0)} |\nabla X|^2 \, du dv \le K(r/d)^{2\mu} \quad \text{for } r \in [0, d],$$

where

(12) $$\mu := \frac{2\lambda^2}{(1+\lambda)^2}, \quad K := \frac{(1+\lambda)^2}{2\pi\lambda^2} L^2(\Gamma).$$

By a reasoning used in the proofs of the Theorems 1 and 4 of Section 2.5 we obtain:

There is a constant K^* depending only on λ and $L(\Gamma)$ such that, for any $w_0 \in \overline{B}$ satisfying $|w_0| \le 1 - d$ and any $r \in [0, d], 0 < d < 1$, we have

$$\int_{S_r(w_0)} |\nabla X|^2 \, du \, dv \le K^*(r/d)^{2\mu}, \tag{13}$$

and Morrey's Dirichlet growth theorem yields

$$[X]_{\mu, \overline{Z}_d} \le c(\mu) \, d^{-\mu} \sqrt{K^*} \tag{14}$$

(cf. Section 2.5, Theorem 1).

Remark 6. In consideration of Remark 4 and of the observation stated at the beginning of Section 2.6 it cannot be expected that estimates of the type (13) and (14) hold with some constant K^* depending only on $L(\Gamma)$ and S if we drop assumption (i) in Proposition 1. Nevertheless one could expect such estimates to be true with numbers K^* depending solely on $L(\Gamma), S$ and $D(X)$.

This seems to be unknown in general except for the following particular case which we want to formulate as

Proposition 2. *Let X be a stationary minimal surface in $\mathcal{C}(\Gamma, S)$ which lies in the exterior of an open, convex subset $\mathcal{K}$ of $\mathbb{R}^3$ that is bounded by S. Suppose also that $S = \partial\mathcal{K}$ is a regular surface of class C^3, and suppose that the unit normal N_S of S pointing into the set $\mathcal{K}$ satisfies the following condition:*

(iii) *There exist two constants $\rho > 0$ and $\lambda > 0$ such that $\langle N_S(P), N_2(Q)\rangle \ge \lambda$ is fulfilled for any two points $P, Q \in S$ whose S-intrinsic distance is at most ρ.*

Then there is a constant K^* depending only on $L(\Gamma)$, S and $D(X)$ such that the inequalities (13) *and* (14) *hold true.*

Let us sketch the proof. We begin with the following

Lemma 1. *Suppose that the assumptions of Proposition 2 are satisfied. Let r, θ be polar coordinates about some points $w_0 \in I$, defined by $w = w_0 + re^{i\theta}$, and set*

$$\psi(r) := \int_0^\pi |X_\theta(r, \theta)|^2 \, d\theta, \quad 0 \le r \le 1 - |w_0|.$$

Then $\psi(r)$ is a monotonically increasing function of r in $[0, 1 - |w_0|]$.

Proof. By Proposition 1 of Section 2.8 it follows that $X \in C^2(B \cup I, \mathbb{R}^3)$. Set $I_r(w_0) := \{w \in I : |w - w_0| < r\}$. Then, by partial integration we obtain

$$\begin{aligned} r\psi'(r) &= \int_0^\pi \frac{\partial}{\partial r} |X_\theta(r, \theta)|^2 r \, d\theta \\ &= \int_{S_r(w_0)} \Delta |X_\theta|^2 \, du \, dv + \int_{I_r(w_0)} \frac{\partial}{\partial v} |X_\theta|^2 \, du. \end{aligned} \tag{15}$$

Let X^* be the adjoint minimal surface to X. Then the mapping $f : B \to \mathbb{C}^3$ defined by $f(w) = X(w) + iX^*(w)$ is holomorphic. Consequently also $wf'(w)$ is holomorphic, and

$$|w|^2|f'(w)|^2 = r^2|\nabla X|^2 = 2|X_\theta|^2$$

is subharmonic. Thus we arrive at

$$\Delta|X_\theta|^2 \geq 0 \quad \text{on } B. \tag{16}$$

Moreover, the conformality relations imply

$$|X_\theta|^2 = r^2|X_u|^2,$$

where $r^2 = |w - w_0| = (u - u_0)^2 + v^2$ for $w_0 = u_0 \in I$, and therefore

$$\frac{\partial}{\partial v}|X_\theta|^2 = 2v|X_u|^2 + 2r^2\langle X_u, X_{uv}\rangle.$$

Thus we obtain

$$\frac{\partial}{\partial v}|X_\theta|^2 = 2(u - u_0)^2\langle X_u, X_{uv}\rangle \quad \text{on } I.$$

Differentiating $\langle X_u, X_v = 0\rangle$ with respect to u it follows that

$$\langle X_u, X_{uv}\rangle = -\langle X_{uu}, X_v\rangle \quad \text{on } B \cup I,$$

and consequently

$$\frac{\partial}{\partial v}|X_\theta|^2 = -2(u - u_0)^2\langle X_{uu}, X_v\rangle \quad \text{on } I.$$

Note that X_v points in the direction of the exterior normal of $\mathcal{K}$ whereas X_{uu} points into the interior of $\mathcal{K}$ since $X : I \to \mathbb{R}^3$ maps I into the boundary S of the open convex set $\mathcal{K}$. Thus we have

$$\langle X_{uu}, X_v\rangle \leq 0 \quad \text{on } I$$

and therefore

$$\frac{\partial}{\partial v}|X_\theta|^2 \geq 0 \quad \text{on } I. \tag{17}$$

On account of (15)–(17) we infer that $\psi'(r) \geq 0$. □

In the same way, the following result can be established, we leave its proof to the reader.

Lemma 2. *Suppose that the assumptions of Proposition 2 are satisfied, and let $w = w_0 + re^{i\theta}$ for some point $w_0 \in I$. Then, for any $p \in [1, \infty)$, the function*

$$\psi_p(r) := \int_0^\pi |X_\theta(r, \theta)|^p \, d\theta$$

is a monotonically increasing function of $r \in [0, 1 - |w_0|]$.

Now we turn to the

Proof of Proposition 2. Fix some $d \in (0,1)$, and let w_0 be an arbitrary point on I with $|w_0| < 1-d$. By $X(r,\theta)$ we denote the representation of X in polar coordinates r, θ about w_0 (i.e. $w = w_0 + re^{i\theta}$). Set

$$\chi(r) := \frac{\sqrt{\pi}}{\lambda}\left\{\int_0^\pi |X_\theta(r,\theta)|^2\,d\theta\right\}^{1/2},$$

$$\chi^*(r) := \left\{\frac{2\pi D(X)}{\lambda^2 \log 1/r}\right\}^{1/2}.$$

By the reasoning of the Courant–Lebesgue lemma (see Section 4.4) we infer that, for any $r \in (0, d^2)$, there exists some $r' \in (r, \sqrt{r})$ such that $\chi(r') \le \chi^*(r)$ holds true. On account of Lemma 1, the function χ is increasing whence

(18) $$\chi(r) \le \chi^*(r) \quad \text{for all } r \in (0, d^2).$$

Since χ^* is strictly increasing, we have

(19) $$\chi^*(r) < \rho \quad \text{if and only if} \quad r < \lambda_2,$$

where the number λ_2 is defined by

$$\lambda_2 := \exp\left(-\frac{2\pi D(X)}{\lambda^2 \rho^2}\right).$$

Let us now introduce the increasing function

$$l(r) := \int_{I_r(w_0)} |dX|,$$

and set

$$\mathcal{I}(w_0) := \left\{r \in (0, d^2) : l(r) < \rho\right\}.$$

Clearly, $\mathcal{I}(w_0)$ is an open and non-empty interval contained in $(0, d^2)$, and therefore

$$m := \sup \mathcal{I}(w_0)$$

is a positive number which is not contained in $\mathcal{I}(w_0)$. Set

$$\delta := \min\{d^2, \lambda^2\}.$$

We claim that the interval $(0, \delta)$ is contained in $\mathcal{I}(w_0)$, independently of the choice of w_0. Otherwise we had $m < \delta$, whence

$$m < d^2 \quad \text{and} \quad m < \lambda^2,$$

and therefore also

(20) $$\chi^*(m) < \rho,$$

on account of (19). For any $r \in (0, m)$ we have $r \in \mathcal{J}(w_0)$, and therefore $l(r) < \rho$. Applying assumption (iii), we infer as in Proposition 1 that

(21) $$l(r) \le \lambda^{-1} \int_0^\pi |X_\theta(r,\theta)|\, d\theta$$

and Schwarz's inequality yields

(22) $$l(r) \le \chi(r) \quad \text{for all } r \in (0, m).$$

From (18)–(21) we infer that

$$l(r) \le \chi(r) \le \chi^*(r) < \chi^*(m) < \rho \quad \text{for all } r \in (0, m)$$

whence, by $r \to m - 0$, we deduce that

$$l(m) < \rho.$$

This implies $m \in \mathcal{J}(w_0)$ which is impossible. Thus we have proved:

The interval $(0, \delta)$ *lies in* $\mathcal{J}(w_0)$, *for any* $w \in I$ *with* $|w_0| < 1 - d$.

Thus we obtain (21) for all $r \in (0, \delta)$, and the isoperimetric inequality yields

(23) $$\int_{S_r(w_0)} |\nabla X|^2\, du\, dv \le 2\lambda_1 \left\{ \int_0^\pi |X_\theta(r,\theta)|\, d\theta \right\}^2$$

with $\lambda_1 = (1+\lambda)^2/(4\pi\lambda^2)$, for all $r \in (0, \delta)$ and for all $w_0 \in I$ with $|w_0| < 1-d$.

Now we can proceed as in Remark 5 in order to prove the assertion of Proposition 2.

Theorem 1. *Let S be an admissible*[1] *surface of class C^3, and assume that X is a critical point of Dirichlet's integral which has the following properties:*

(i) *The free trace $X(I)$ is contained in an open, orientable part S_0 of S that fulfils a λ-graph condition, $\lambda > 0$.*

(ii) *There exist no branch points of X on I which are of odd order.*

Then the length $L(\Sigma) = \int_I |dX|$ of the free trace Σ given by $X : I \to \mathbb{R}^3$ is estimated by (2):

$$L(\Sigma) \le \lambda^{-1} L(\Gamma),$$

and the area $A(X)$ of X is estimated by (3):

$$A(X) \le \lambda_1 L^2(\Gamma), \quad \lambda_1 := \frac{(1+\lambda)^2}{4\pi\lambda^2}.$$

[1] The condition of "admissibility" is essentially a uniformity condition at infinity which is formulated in Section 2.6.

Proof. By Theorem 4 of Section 2.7, the surface X is a stationary minimal surface in $\mathcal{C}(\Gamma, S)$ which is of class $C^1(B \cup I, \mathbb{R}^3)$. Then the assertion follows from Proposition 1 and from Remark 1. □

Note that the λ-graph condition imposes no bounds on the principal curvature of S_0. Thus S_0 was allowed to have arbitrarily sharp wrinkles.

The following assumption is in a sense complementary to the λ-graph condition; it implies a bound on the principal curvatures of S but does not restrict the position of the Gauss image N_S of S.

Definition 2. *We say that a surface S in $\mathbb{R}^3$ satisfies a* (two-sided) R-sphere condition, *if S is a C^2-submanifold of $\mathbb{R}^3$ which is the boundary of an open set U of $\mathbb{R}^3$, and if for every $P \in S$ the tangent balls*

(24) $$B^{\pm}(P, R) := \{Q \in \mathbb{R}^3 : |P \pm RN_S(P) - Q| < R\}$$

do not contain any points of S. Here N_S denotes the exterior unit normal of S with respect to U (see Fig. 3).

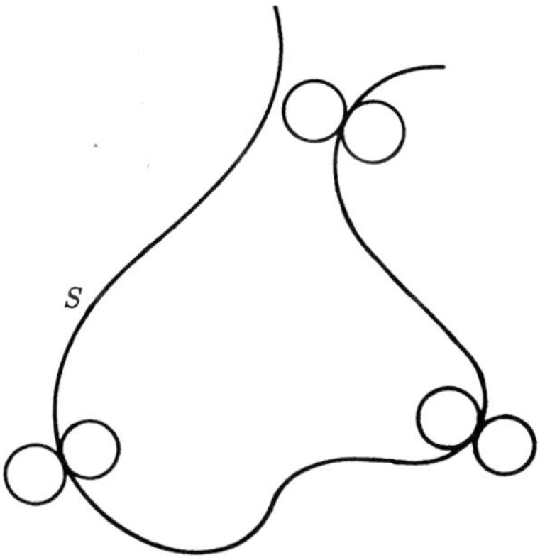

Fig. 3. The R-sphere condition

Theorem 2. *Let S be a support surface satisfying an R-sphere condition, and let Γ be a rectifiable Jordan arc with its endpoints on S. Let X be a stationary minimal surface in $\mathcal{C}(\Gamma, S)$ with the free trace Σ given by $X : I \to \mathbb{R}^3$. Then the length $L(\Sigma)$ of Σ can be estimated by*

(25) $$L(\Sigma) \le L(\Gamma) + \frac{2}{R} D(X).$$

This estimate is optimal in the sense that 2 cannot be replaced by any smaller number.

For the proof we need the following

Lemma 3. *Suppose that the surface S satisfies an R-sphere condition. Then its principal curvatures are bounded from above by $1/R$. Moreover, any point P in the tubular neighbourhood*

$$T_R := \{Q \in \mathbb{R}^3 : \text{dist}(Q, S) < R\} \tag{26}$$

has a unique representation of the form

$$P = F(P) + \rho(P) N_S(F(P)), \tag{27}$$

where $F(P) \in S$ is the unique foot of P on S, $\rho(P)$ is the oriented distance from S to P, and $N_S(Q)$ denotes the exterior normal to S at $Q \in S$ (i.e., $\rho(P) < 0$ if $P \in U$, and $\rho(P) \geq 0$ if $P \in \mathbb{R}^3 - U$). The distance function ρ is of class C^2 (and of class C^m or $C^{m,\alpha}$ if $S \in C^m$ or $C^{m,\alpha}$ respectively, $m \geq 2, 0 < \alpha < 1$), and we have

$$D\rho(P) = N_S(F(P)) \quad \textit{for all } P \in T_R. \tag{28}$$

Finally the eigenvalues of the Hessian matrix $H(P) = D^2\rho(P) = (\rho_{x^i x^k}(P))$ at any $P \in T_R$ are bounded from above by $[R - |\rho(P)|]^{-1}$, and the Hessian annihilates normal vectors, i.e.,

$$H(P) N_S(F(P)) = 0 \quad \textit{for all } P \in T_R. \tag{29}$$

(Here D denotes the three-dimensional gradient in $\mathbb{R}^3$.)

Proof. The representation formula (27) in the tubular neighbourhood T_R is fairly obvious. The other results follow from (27) by means of the implicit function theorem using the fact that the principal curvature of S at P are precisely the eigenvalues of the Hessian of a nonparametric representation of S close to P whose x, y-plane is parallel to the tangent plane of S at P. We omit the details and refer the reader to Gilbarg and Trudinger [1], Appendix (pp. 383–384), for the pertinent estimates.

Proof of Theorem 2 in the special case that X has no branch point of odd order on I. For any $\delta > 0$ we can choose a function $\varphi(t), t \in \mathbb{R}$, of class $C_C^\infty((-R, R))$ having the following properties:

$$\begin{aligned} 0 \leq \varphi \leq 1, \quad \varphi(0) = 1, \quad \varphi(t) = \varphi(-t), \\ \varphi(t) \leq (1 - R^{-1}|t|)(1 + \delta) \quad \text{for } |t| \leq R, \\ |\varphi'(t)| \leq R^{-1}(1 + \delta). \end{aligned} \tag{30}$$

Then we define a C^1-vector field Z on $\mathbb{R}^3$ by

$$Z(P) = \begin{cases} \varphi(\rho(P)) N_S(F(P)) & \text{for } P \in T_R, \\ 0 & \text{otherwise.} \end{cases} \tag{31}$$

We clearly have

$$|Z(P)| \le 1 \quad \text{for all } P \in \mathbb{R}^3, \qquad Z(Q) = N_S(Q) \quad \text{for } Q \in S, \tag{32}$$

and we claim that also

$$|\nabla Z(P) \le R^{-1}(1+\delta) \quad \text{for all } P \in \mathbb{R}^3 \tag{33}$$

holds true. As ∇Z vanishes in the exterior of T_R, we have to prove (33) only for $P \in T_R$. Thus we fix some $P \in T_R$ and some unit vector $\nu \in \mathbb{R}^3$. Then the directional derivative $\frac{\partial Z}{\partial \nu}(P)$ is given by

$$\frac{\partial Z}{\partial \nu}(P) = \varphi'(\rho(P))\frac{\partial \rho}{\partial \nu}(P)N_S(F(P)) + \varphi(\rho(P))\frac{\partial}{\partial \nu}N_S(F(P)).$$

If $\nu = \pm N_S(F(P))$, then

$$\frac{\partial Z}{\partial \nu}(P) = \varphi'(\rho(P))\frac{\partial \rho}{\partial \nu}(P)N_S(F(P)),$$

and by (28) and (30_3) it follows that

$$\left|\frac{\partial Z}{\partial \nu}(P)\right| \le \frac{1+\delta}{R}. \tag{34}$$

If ν is orthogonal to $N_S(F(P))$, then $\frac{\partial \rho}{\partial \nu}(P) = 0$, and Lemma 3 yields

$$\frac{\partial}{\partial \nu}N_S(F(P)) = \frac{\partial}{\partial \nu}\nabla\rho(P) = \nabla^2\rho(P)\nu$$

and

$$|\nabla^2\rho(P)\nu| \le |\nabla^2\rho(P)||\nu| \le (R - |\rho(P)|)^{-1}.$$

In conjunction with (30_2) it follows that

$$\left|\frac{\partial Z}{\partial \nu}(P)\right| \le (1 - R^{-1}|\rho(P)|)(1+\delta)(R - |\rho(P)|)^{-1} = \frac{1+\delta}{R}$$

and thus (34) holds true if $\nu \perp N_S(F(P))$. Hence (34) is satisfied for all unit vectors ν, and we have established property (33).

By means of the vector field Z on $\mathbb{R}^3$ we define a surface $Y(w), w \in B$, of class $L_\infty \cap H_2^1(B, \mathbb{R}^3)$, setting $Y(w) = Z(X(w))$.

Given any $\varepsilon \in (0,1)$, we can find two numbers $\varepsilon_1, \varepsilon_2 \in (\varepsilon^2, \varepsilon)$ such that

$$\int_{\gamma_1(\varepsilon)} |dX| + \int_{\gamma_2(\varepsilon)} |dX| \le 2\left\{\frac{\pi D(X)}{\log(1/\varepsilon)}\right\}^{1/2}, \tag{35}$$

where $\gamma_1(\varepsilon)$ and $\gamma_2(\varepsilon)$ denote the circular arcs

$$\gamma_1(\varepsilon) := \{w \in B : |w-1| = \varepsilon_1, \operatorname{Im} w > 0\},$$

$$\gamma_2(\varepsilon) := \{w \in B : |w+1| = \varepsilon_2, \operatorname{Im} w > 0\};$$

see Section 4.4 of Vol. 1.

Now we apply Green's formula to the functions X, Y and to the domain $\Omega(\varepsilon)$ which is obtained from the semidisk B by removing the parts which are contained in the disks $B_{\varepsilon_1}(1)$ or $B_{\varepsilon_2}(-1)$, respectively:

$$\Omega(\varepsilon) := B \setminus [B_{\varepsilon_1}(1) \cup B_{\varepsilon_2}(-1)].$$

Thus we obtain

$$(36) \quad \int_{\Omega(\varepsilon)} \langle \nabla X, \nabla Y \rangle \, du \, dv = -\int_{\Omega(\varepsilon)} \langle \Delta X, Y \rangle \, du \, dv + \int_{\partial\Omega(\varepsilon)} \left\langle \frac{\partial X}{\partial \nu}, Y \right\rangle d\mathcal{H}^1,$$

where ν denotes the exterior normal on $\partial\Omega(\varepsilon)$. Set

$$I(\varepsilon) := I \cap \partial\Omega(\varepsilon) \quad \text{and} \quad C(\varepsilon) := C \cap \partial\Omega(\varepsilon).$$

Then

$$\partial\Omega(\varepsilon) = I(\varepsilon) \cup C(\varepsilon) \cup \gamma_1(\varepsilon) \cup \gamma_2(\varepsilon).$$

On the interval $I(\varepsilon)$, we have $d\mathcal{H}^1 = du$, $\frac{\partial X}{\partial \nu} = -X_v$, $Y = N_S(X)$, and $X_v = \pm|X_v|N_S(X)$. As there exist no branch points of odd order on I, the vector X_v always points in the direction of $N_S(X)$ or in the direction of $-N_S(X)$. Thus we can assume that

$$X_v = |X_v| N_s(X) \quad \text{on } I(\varepsilon),$$

and we arrive at

$$\left\langle \frac{\partial X}{\partial \nu}, Y \right\rangle = -|X_v| = -|X_u| \quad \text{on } I(\varepsilon).$$

This implies

$$(37) \qquad \int_{I(\varepsilon)} \left\langle \frac{\partial X}{\partial \nu}, Y \right\rangle d\mathcal{H}^1 = \int_{I(\varepsilon)} |dX|.$$

Let $\frac{\partial}{\partial \tau} X$ be the tangential derivative of X along $\partial\Omega(\varepsilon)$. The conformality relations yield

$$\left| \frac{\partial X}{\partial \nu} \right| = \left| \frac{\partial X}{\partial \tau} \right|$$

and therefore

$$\left| \left\langle \frac{\partial X}{\partial \nu}, Y \right\rangle \right| \le \left| \frac{\partial X}{\partial \nu} \right| |Y| \le \left| \frac{\partial X}{\partial \tau} \right| \quad \text{along } \partial\Omega(\varepsilon).$$

Consequently, we have

$$(38)\qquad \left| \int_{C(\varepsilon)+\gamma_1(\varepsilon)+\gamma_2(\varepsilon)} \left\langle \frac{\partial X}{\partial \nu}, Y \right\rangle d\mathcal{H}^1 \right|$$
$$\le \int_{C(\varepsilon)} |dX| + \int_{\gamma_1(\varepsilon)} |dX| + \int_{\gamma_2(\varepsilon)} |dX| \le L(\Gamma) + f(\varepsilon),$$

where the remainder term $f(\varepsilon)$ tends to zero as $\varepsilon \to +0$, by virtue of (35).

Finally we infer from $Y = Z \circ X$ and from (33) that

$$|\nabla Y| \le \frac{1+\delta}{R} |\nabla X|$$

whence

$$(39)\qquad \left| \int_{\Omega(\varepsilon)} \langle \nabla X, \nabla Y \rangle \, du\, dv \right| \le \frac{1+\delta}{R} \int_{\Omega(\varepsilon)} |\nabla X|^2 \, du\, dv.$$

Because of $\Delta X = 0$ we infer from (36) in conjunction with (37)–(39) that

$$(40)\qquad \int_{I(\varepsilon)} |dX| \le L(\Gamma) + \frac{1+\delta}{R} \int_{\Omega(\varepsilon)} |\nabla X|^2 \, du\, dv + f(\varepsilon).$$

Letting $\varepsilon \to +0$, it follows that

$$L(\Sigma) \le L(\Gamma) + \frac{1+\delta}{R} \int_B |\nabla X|^2 \, du\, dv.$$

Since we can choose $\delta > 0$ as small as we please, we arrive at the desired in equality

$$L(\Sigma) \le L(\Gamma) + 2R^{-1} D(X).$$

In order to show that the estimate (25) is optimal we consider the following *examples*. Let S be the circular cylinder of radius R given by

$$S := \left\{ (x, y, z) : x^2 + y^2 = R^2 \right\},$$

and let Γ be the straight arc

$$\Gamma := \left\{ (x, y, z) : x = a, y^2 \le R^2 - a^2, z = 0 \right\},$$

where a denotes some number with $0 < a < R$. Then it is easy to define a planar minimal surface $X : B \to \mathbb{R}^3$ which is stationary in $\mathcal{C}(\Gamma, S)$ and maps B conformally onto the planar domain $\Omega = \{(x, y, z) : z = 0, x^2 + y^2 < R^2, x < a\}$.

If a tends to R then $L(\Sigma)$ converges to $2\pi R$ and $cR^{-1}D(X)$ to $c\pi R$ whereas $L(\Gamma)$ shrinks to zero. This shows that $c = 2$ is the optimal value in the estimate

$$L(\Sigma) \le L(\Gamma) + cR^{-1}D(X),$$

and the proof of Theorem 2 is complete in the special case that there are no branch points of odd order on I. □

The proof of Theorem 2 in the general case will be based on the relation

$$|X_v| = |D_v\rho(X)| \quad \text{along } I. \tag{41}$$

This follows by differentiating the relation

$$X = F(X) + \rho(X)N_S(F(X)) \quad \text{on } B \cup I$$

which holds on $B \cup I$ close to I (cf. (27)). Hence we can express the length of the free trace Σ as

$$L(\Sigma) = \int_I |D_v\rho(X)|\, du. \tag{42}$$

If $X(B)$ were contained in the tubular neighbourhood T_R of S, we could write

$$\int_I D_v\rho(X)\, du = -\int_B \Delta\rho(X)\, du\, dv + \int_C \frac{\partial}{\partial\nu}\rho(x)\, d\mathcal{H}^1.$$

If $D_v\rho(X)$ has a uniform sign on I, we could use this identity to derive an estimate for $L(\Sigma)$. However, since both facts are not guaranteed, we shall instead construct some function $\eta(w)$ of which we can prove that

$$\eta \ge |D_\nu\rho(X)| \quad \text{on } I \tag{43}$$

holds true. Then we can estimate $L(\Sigma)$ from above by the integral $\int_I \eta_v\, du$ which is transformed into

$$-\int_B \Delta\eta\, du\, dv + \int_C \frac{\partial}{\partial\nu}\eta\, d\mathcal{H}^1,$$

and this integral will be estimated in terms of X.

In order to define η we first introduce

$$\Psi(t) := \int_0^t \varphi(s)\, ds,$$

where φ is a function of class $C_c^\infty((-R,R))$ satisfying (30). Then Ψ satisfies

$$\begin{aligned} \Psi(t) &= \Psi(R) \quad \text{for } t \ge R, \quad \Psi(t) = -\Psi(R) \quad \text{for } t \ge -R,\\ \Psi(0) &= 0, \quad \Psi'(0) = 1, \quad 0 \le \Psi' \le 1,\\ \Psi(t) &\le (1 - R^{-1}|t|)(1+\delta) \quad \text{for } |t| \le R,\\ |\Psi''(t)| &\le R^{-1}(1+\delta). \end{aligned} \tag{44}$$

Secondly we define

$$\zeta(P) := \Psi^2(\rho(P)) \quad \text{if } P \in \mathbb{R}^3,$$

$$\alpha(w) := \delta v, \quad w = u + iv.$$

Then $\eta(w)$ will be defined as

$$\eta(w) := \left\{\alpha^2(w) + \zeta(X(w))\right\}^{1/2} \quad \text{for } w \in \overline{B}.$$

The function η is of class $C^2(B) \cap C^1(B \cup I)$, and its boundary values on C are absolutely continuous. Moreover, we have

$$\nabla\eta = \left\{\alpha^2 + \zeta(X)\right\}^{1/2} \left[\alpha\nabla\alpha + \frac{1}{2}\nabla\zeta(X)\right]$$

whence it follows that

$$\eta_v = \left\{\alpha^2 + \zeta(X)\right\}^{-1/2} \left[\delta\alpha + \Psi'(\rho(X))\Psi(\rho(X))D_v\rho(X)\right].$$

Here and in the sequel we use the notation $\nabla\zeta(X)$ for $\nabla(\zeta \circ X)$, $D_v\rho(X)$ for $D_v(\rho \circ X)$, etc.

Set $w = u_0 + iv$, $v > 0$, and let $v \to +0$. Then $\Psi(\rho(X)) \to 0$, $\Psi'(\rho(X)) \to 1$, and l'Hospital's rule yields

$$\frac{\Psi(\rho(X(u_0 + iv)))}{v} \to D_v\rho(X)\Big|_{w=w_0}.$$

Hence η_v tends to

$$\frac{\delta^2 + |D_v\rho(X)|^2}{\{\delta^2 + |D_v\rho(X)|^2\}^{1/2}} = \left\{\delta^2 + |D_v\rho(X)|^2\right\}^{1/2} \geq |D_v\rho(X)|$$

whence we have established (43).

Next we want to estimate $-\Delta\eta$ from above. We have

$$\begin{aligned} \nabla\zeta(X) &= 2\Psi'(\rho(X))\Psi(\rho(X))\nabla\rho(X), \\ \Delta\zeta(X) &= 2\gamma'(\rho(X))|\nabla\rho(X)|^2 + 2\gamma(\rho(X))\Delta\rho(X), \end{aligned}$$

where we have set

$$\gamma := \Psi\Psi',$$

and

$$\begin{aligned} -\Delta\eta = {} & \left\{\alpha^2 + \zeta(X)\right\}^{-3/2} \left|\alpha\nabla\alpha + \frac{1}{2}\nabla\zeta(X)\right|^2 \\ & - \left\{\alpha^2 + \zeta(X)\right\}^{-1/2} \left[|\nabla\alpha|^2 + \frac{1}{2}\Delta\zeta(X)\right]. \end{aligned}$$

This implies (with $\zeta = \zeta(X)$, $\rho = \rho(X)$, $\gamma = \gamma(\rho(X))$, etc.) that

$$\begin{aligned}
-\Delta\eta &= \left\{\alpha^2+\zeta\right\}^{-3/2}|\alpha\nabla\alpha+\gamma\nabla\rho|^2 \\
&\quad -\left\{\alpha^2+\zeta\right\}^{-1/2}(|\nabla\alpha|^2+\Psi'^2|\nabla\rho|^2+\Psi\Psi''|\nabla\rho|^2+\gamma\Delta\rho) \\
&= -\left\{\alpha^2+\zeta\right\}^{-3/2}|\alpha\Psi'\nabla\rho-\Psi\nabla\alpha|^2 \\
&\quad -\left\{\alpha^2+\zeta\right\}^{-1/2}(\Psi\Psi''|\nabla\rho|^2+\gamma\Delta\rho) \\
&\le -\left\{\alpha^2+\zeta\right\}^{-1/2}\Psi(\Psi''|\nabla\rho|^2+\Psi'\Delta\rho) \\
&\le \left\{\alpha^2+\Psi^2\right\}^{-1/2}\Psi(|\Psi''||\nabla\rho|^2+|\Psi'||\Delta\rho|) \\
&\le |\Psi''||\nabla\rho|^2+|\Psi'||\Delta\rho|.
\end{aligned}$$

Thus we have

$$-\Delta\eta \le |\Psi''(\rho(X))||\nabla\rho(X)|^2+|\Psi'(\rho(X))||\Delta\rho(X)|. \tag{45}$$

We can restrict our attention to the set

$$B' := \{w\in B : \rho(X(w)) < R\}$$

since $\Psi'(\rho(X))$ and $\Psi''(\rho(X))$ vanish in $B\setminus B'$ whence also $\Delta\eta=0$ in $B\setminus B'$. In B' we have

$$|\Psi''(\rho(X))||\nabla\rho(X)|^2 \le \frac{1+\delta}{R}|\nabla\rho(X)|^2 \tag{46}$$

and

$$|\Psi'(\rho(X))| \le (1+\delta)(1-R^{-1}|\rho(X)|), \tag{47}$$

taking (44) into account.

Furthermore we have

$$\Delta\rho(X) = X_u H(X) X_u + X_v H(X) X_v \tag{48}$$

with $H(X)=(\rho_{x^i x^k}(X))$ = Hessian matrix of ρ composed with X. By means of Lemma 3 we infer that

$$|X_u H(X) X_u + X_v H(X) X_v| \le (R-|\rho(X)|)^{-1}\left\{|\nabla X|^2-|\nabla\rho(X)|^2\right\} \tag{49}$$

since $|X_u|^2 - D_u\rho(X)|^2$ is the square of the norm of the tangential component of X_u, and an analogous statement holds for $|X_v|^2 - D_v\rho(X)|^2$. Combining (47), (48) and (49), we arrive at

$$|\Psi'(\rho(X))||\Delta\rho(X)| \le \frac{1+\delta}{R}[|\nabla X|^2-|\nabla\rho(X)|^2]. \tag{50}$$

Then (45), (46) and (50) yield

$$-\Delta\eta \le \frac{1+\delta}{R}|\nabla X|^2 \tag{51}$$

on B', and therefore also on B.

Moreover, a straight-forward estimation yields

$$\left|\frac{\partial \eta}{\partial \nu}\right| \le \left\{\left|\frac{\partial \alpha}{\partial \nu}\right|^2 + \left|\frac{\partial}{\partial \nu}\Psi(X)\right|^2\right\}^{1/2} \le \sqrt{\delta^2 + |X_r|^2} = \sqrt{\delta^2 + |X_\theta|^2} \quad \text{on } C$$

and therefore

$$\left|\frac{\partial \eta}{\partial \nu}\right| \le \delta + |X_\theta| \quad \text{on } C. \tag{52}$$

Now choose $\Omega(\varepsilon)$ as in the proof of the special case as

$$\Omega(\varepsilon) = B \setminus [B_{\varepsilon_1}(1) \cup B_{\varepsilon_2}(-1)]$$

with

$$\partial\Omega(\varepsilon) = I(\varepsilon) \cup C(\varepsilon) \cup \gamma_1(\varepsilon) \cup \gamma_2(\varepsilon).$$

Then we obtain

$$\int_{I(\varepsilon)} D_v \eta \, du = -\int_{\Omega(\varepsilon)} \Delta\eta \, du \, dv + \int_{C(\varepsilon)+\gamma_1(\varepsilon)+\gamma_2(\varepsilon)} \frac{\partial}{\partial \nu}\eta \, d\mathcal{H}^1. \tag{53}$$

By (41) and (43) it follows that

$$\begin{aligned} \int_{I(\varepsilon)} |dX| &= \int_{I(\varepsilon)} |X_u| \, du = \int_{I(\varepsilon)} |X_v| \, dv \\ &= \int_{I(\varepsilon)} |D_v \rho(X)| \, du \le \int_{I(\varepsilon)} D_v \eta \, du, \end{aligned} \tag{54}$$

taking $|X_u| = |X_v|$ into account. Thus, by virtue of (51)–(54), we obtain that

$$\begin{aligned} \int_{I(\varepsilon)} |dX| &\le \int_{\Omega(\varepsilon)} (\Delta\eta) \, du \, dv + \int_{C(\varepsilon)+\gamma_1(\varepsilon)+\gamma_2(\varepsilon)} \left|\frac{\partial \eta}{\partial \nu}\right| d\mathcal{H}^1 \\ &\le \frac{1+\delta}{R} \int_{\Omega(\varepsilon)} |\nabla X|^2 \, du \, dv + \int_{C(\varepsilon)} \{\delta + |X_\theta|\} \, d\mathcal{H}^1 \\ &\quad + \int_{\gamma_1(\varepsilon)+\gamma_2(\varepsilon)} \frac{\partial \eta}{\partial \nu} \, d\mathcal{H}^1. \end{aligned}$$

Letting first δ and then ε tend to zero, we arrive at

$$\int_I |dX| \le \frac{1}{R} \int_R |\nabla X|^2 \, du \, dv + \int_C |dX|,$$

where the integral over $\gamma_1(\varepsilon)$ is dealt with in the same way as in the previous proof for the special case. □

Remark 7. There is no estimate of the form

$$L(\Sigma) \le c_1 L(\Gamma) + c_2 H_0 D(X)$$

or of the form

$$L(\Sigma) \le c_1 L(\Gamma) + c_2 K_0 D(X)$$

with absolute constants c_1 and c_2, where H_0 and K_0 denote upper bounds for $|H|$ and $|K|^{1/2}$, respectively, H and K being the mean curvature and Gauss curvature of S. In fact, the second inequality is ruled out by the cylinder example discussed before, and the first is disproved by a similar example where one replaces the cylinder surface by a suitable catenoid as supporting surface S (see Fig. 4). In other words, it is quite natural that in (25) an upper bound for the two principal curvatures κ_1 and κ_2 of S enters and not an upper bound for the mean curvature H or for the Gauss curvature K.

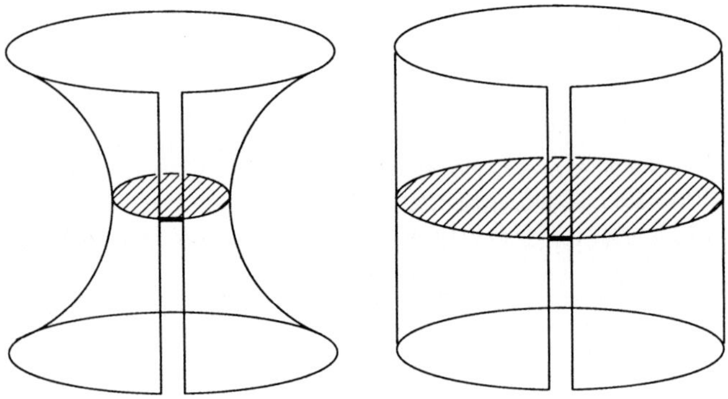

Fig. 4. The examples of Remark 7

Remark 8. Suppose that not all of Γ lies in S. Then, by choosing $\varphi(t)$ in such a way that $\varphi(t) < 1$ for all $t \neq 0$, a close inspection of the proof of Theorem 2 shows that we have in fact the strict inequality

$$L(\Sigma) < L(\Gamma) + \frac{2}{R} D(X) \tag{55}$$

instead of (25).

Remark 9. In addition to the assumptions of Theorem 2 we now assume that $X(B)$ is contained in a ball $\mathcal{K}_{R_0}(P) = \{Q : |P - Q| \le R_0\}$ of $\mathbb{R}^3$. Then the linear isoperimetric inequality of Section 6.3 implies that

$$D(X) \le \frac{R_0}{2} \{L(\Gamma) + L(\Sigma)\}.$$

If $R > R_0$, we infer in conjunction with (25)

$$L(\Sigma) \le \frac{R + R_0}{R - R_0} L(\Gamma). \tag{56}$$

This is an analogue to the inequality (2) in Proposition 1. From the example $S = \partial\mathcal{K}_{R_0}(P)$ we infer that $L(\Sigma)$ can in general not be bounded from above by $L(\Gamma)$. In this case the inequality (56) fails since we have $R = R_0$.

Remark 10. An estimate similar to (25) can be given for stationary minimal surfaces with completely free boundaries. In fact, *suppose that* $X : B \to \mathbb{R}3$ *is a stationary minimal surface in* $\mathcal{C}(S)$ *and assume that* S *satisfies an* R*-sphere condition. Then it follows that the free trace* Σ *of* X *satisfies*

$$L(\Sigma) \le 2R^{-1} D(X). \tag{57}$$

Note that we cannot prove strict inequality as equality holds for the cylinder

$$S = \left\{(x, y, z) : x^2 + y^2 = 1\right\},$$

where $R = 1$ and for

$$X(w) = (\mathrm{Re}(w^n), \mathrm{Im}(w^n), 0), \quad w = u + iv,\ n \in \mathbb{N}.$$

Let us conclude this section by a brief discussion of surfaces $X : B \to \mathbb{R}^3$, parametrized over the unit disk which are the class $H_2^1 \cap C^2(B, \mathbb{R}^3)$ and satisfy both

$$\Delta X = 2H X_u \wedge X_v$$

and

$$|X_u|^2 = |X_v|^2, \quad \langle X_u, X_v\rangle = 0.$$

That is, the surface X has constant mean curvature H at all points w where $\nabla X(w) \ne 0$. We shall in the following assume that $X(w) \not\equiv$ const. Then branch points w_0 of X are isolated, and $X_w(w)$ possesses an asymptotic expansion

$$X_w(w) = A(w - w_0)^m + o(|w - w_0|^m) \quad \text{as } w \to w_0,$$

$$A \in \mathbb{C}^3, \quad A \ne 0, \quad \langle A, A\rangle = 0, \quad m \in \mathbb{N},$$

which is completely analogous to asymptotic expansions of minimal surfaces at branch points w_0 derived in Vol. 1, Section 3.2 (see Section 3.1).

Moreover we assume that X is of class $\mathcal{C}(S)$ where the support surface S satisfies an R-sphere condition (cf. Definition 2), and that $S = \partial U$ where U is an open (nonempty) set in $\mathbb{R}^3$.

Finally we suppose that X is of class $C^1(\overline{B}, \mathbb{R}^3)$ and intersects S perpendicularly along its free trace Σ given by $X : \partial B \to \mathbb{R}^3$.

We shall call such surfaces *stationary H-surfaces in* $\mathcal{C}(S)$. Then, by the same computations as in the proof of Theorem 2, we obtain the following *analogue of* (57) *for "stationary H-surfaces in the class* $\mathcal{C}(S)$*"*:

$$(58) \qquad L(\Sigma) \le 2(|H|) + R^{-1})D(X).$$

Whenever X satisfies an isoperimetric inequality of the kind

$$(59) \qquad D(X) \le cL^2(\Sigma),$$

it follows that

$$L(\Sigma) \le 2(|H| + R^{-1})cL^2(\Sigma)$$

whence

$$(60) \qquad L(\Sigma) \ge \frac{1}{2c(|H| + R^{-1})}.$$

In particular, for stationary minimal surfaces in $\mathcal{C}(S)$ we have $H = 0$ and $c = \frac{1}{4\pi}$, whence

$$(61) \qquad L(\Sigma) \ge 2\pi R.$$

This is a remarkable *lower bound* for the length of the free trace of a stationary minimal surface in $\mathcal{C}(S)$.

One encounters stationary H-surfaces as solutions of the so-called *partition problem*. Given an open set U in $\mathbb{R}^3$ of finite volume V and with $S = \partial U$, this is the following task:

Among all surfaces Z of prescribed topological type which are contained in $\overline{U}$, have their boundaries on S, and divide U in two disconnected parts U_1 and U_2 of prescribed ratio of volumes, one is to find a surface X which assigns a minimal value or at least a stationary value to its surface area (Dirichlet integral).

One can show[2] that any solution $X : B \to \mathbb{R}^3$ of the partition problem is a surface of constant mean curvature H which is regular up to its free boundary and intersects $S = \partial U$ perpendicularly along $\Sigma = X|_{\partial B}$. That is, *any solution of the partition problem for U is a stationary H-surface in $\mathcal{C}(S)$, $S := \partial U$.*

If $\overline{U}$ is a closed convex body $\mathcal{K}$ whose boundary $S = \partial\mathcal{K}$ satisfies an R-sphere condition, and if R_* is the inradius of $\mathcal{K}$ (i.e., the radius of the largest ball contained in $\mathcal{K}$), then one can also prove the following lower bound for the length $L(\Sigma)$ of the free trace Σ of any stationary H-surface $X : B \to \mathbb{R}^3$ in $\mathcal{C}(S)$ that is parametrized on the unit disk and satisfies $X(B) \subset \mathcal{K}$:

$$(62) \qquad L(\Sigma) \ge \frac{2\pi R_*}{1 + (\operatorname{diam}\mathcal{K} - R_*)|H|}.$$

For $H = 0$ this reduces to

$$(63) \qquad L(\Sigma) \ge 2\pi R_*.$$

As we have $R_* \ge R$, this inequality is an improvement of (61).

[2] Cf. Grüter-Hildebrandt-Nitsche [2].

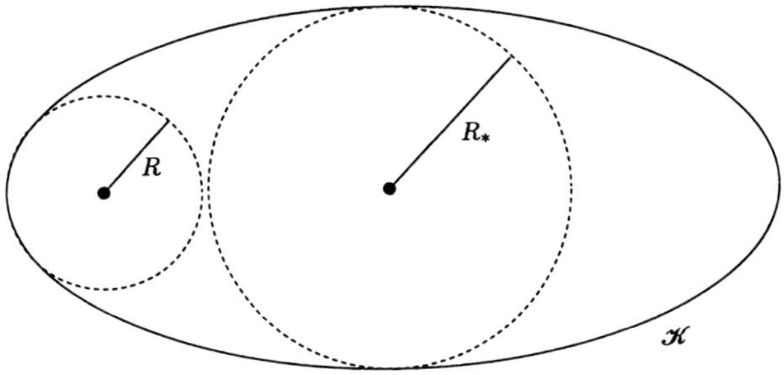

Fig. 5. The inradius R_*, and the smallest curvature radius R

Proof of estimate (62). Set $L := L(\Sigma)$, and define the parameter of the arc length of Σ by

$$s(\theta) := \int_0^\theta |X_\theta(e^{i\theta})|\, d\theta = \int_0^\theta |X_r(e^{i\theta})|\, d\theta$$

($r, \theta =$ polar coordinates about the origin $w = 0$).

Let $\theta(s)$ be the inverse function, $0 \le s \le L$, and introduce the representation

$$Z(s) := X(e^{i\theta(s)}), \quad 0 \le s \le L,$$

of Σ with respect to the parameter s. Moreover let $N_S(P)$ be the exterior unit normal of S at the point $P \in S$. As the H-surface X meets S perpendicularly along Σ, we have

$$X_r(e^{i\theta}) = |X_r(e^{i\theta})| N_S(X(e^{i\theta}))$$

and therefore

$$\int_{\partial B} X_r\, d\theta = \int_\Sigma N_S(Z)\, ds := \int_0^L N_S(Z(s))\, ds. \tag{64}$$

Secondly, a partial integration yields

$$2\int_B X_u \wedge X_v\, du\, dv = \int_{\partial B} X \wedge dX = \int_\Sigma Z \wedge dZ, \tag{65}$$

and another partial integration implies

$$\int_{\partial B} X_r\, d\theta = \int_B \Delta X\, du\, dv.$$

On account of $\Delta X = 2H X_u \wedge X_v$ we thus obtain

$$\int_{\partial B} X_r\, d\theta = 2H \int_B X_u \wedge X_v\, du\, dv. \tag{66}$$

Now we infer from (64)–(66) that

(67)
$$\int_\Sigma \{N_S(Z)\, ds - HZ \wedge dZ\} = 0.$$

Set
$$\overline{Z} := \fint_0^L Z(s)\, ds.$$

Then Wirtinger's inequality (Section 6.3, Lemma 2) yields

(68)
$$\int_0^L |Z - \overline{Z}|^2\, ds \le \frac{L^3}{4\pi^2}.$$

Let us now introduce the support function $\sigma(P)$ of the convex surface S by
$$\sigma(P) := \langle P, N_S(P)\rangle,$$
where we have identified P with the radius vector $\overrightarrow{OP}$ from the origin 0 to the point P. We can assume that 0 is the center of the in-ball $B_{R_*}(0)$ of $\mathcal{K}$. Then we obtain
$$\sigma(P) \ge R_* \quad \text{for all } P \in S.$$

Consequently we have
$$\begin{aligned} R_* L - \int_0^L \langle \overline{Z}, N_S(Z)\rangle\, ds &\le \int_0^L \langle Z, N_S(Z)\rangle\, ds - \int_0^L \langle \overline{Z}, N_S(Z)\rangle\, ds \\ &= \int_0^L \langle Z - \overline{Z}, N_S(Z)\rangle\, ds \\ &\le L^{1/2}\left\{\int_0^L |Z - \overline{Z}|^2\, ds\right\}^{1/2} \le \frac{L^2}{2\pi}, \end{aligned}$$
taking also (68) into account.

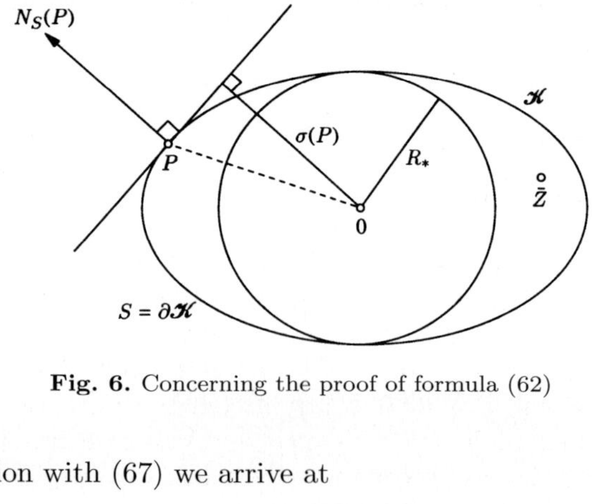

Fig. 6. Concerning the proof of formula (62)

In conjunction with (67) we arrive at

$$
\begin{aligned}
R_* L &= \left\{ R_* L - \int_0^L \langle \overline{Z}, N_S(Z) \rangle \, ds \right\} + \int_0^L \langle \overline{Z}, N_S(Z) \rangle \, ds \\
&\le \frac{L^2}{2\pi} + H \int_0^L [\overline{Z}, Z, Z'] \, ds.
\end{aligned}
$$

Here $[A_1, A_2, A_3]$ denotes the volume form $\langle A_1, A_2 \wedge A_3 \rangle = \det(A_1, A_2, A_3)$ of three vectors A_1, A_2, A_3 of $\mathbb{R}^3$. Because of the identity

$$[\overline{Z}, Z, Z'] = [\overline{Z}, Z - \overline{Z}, Z']$$

we arrive at

$$
\begin{aligned}
R_* L &\le \frac{L^2}{2\pi} + H \int_0^L [\overline{Z}, Z - \overline{Z}, Z'] \, ds \\
&\le \frac{L^2}{2\pi} + |H| \int_0^L |\overline{Z}||Z - \overline{Z}||Z'| \, ds \\
&\le \frac{L^2}{2\pi} + |H||\overline{Z}|\sqrt{L} \left\{ \int_0^L |Z - \overline{Z}|^2 \, ds \right\}^{1/2} \\
&\le \frac{L^2}{2\pi}(1 + |H\overline{Z}|).
\end{aligned}
$$

Moreover, an elementary estimation yields

$$|\overline{Z}| \le \operatorname{diam} \mathcal{K} - R_*,$$

and therefore

$$R_* L \le \frac{1}{2\pi} \{1 + (\operatorname{diam} \mathcal{K} - R_*)|H|\} \, L^2.$$

Now (62) is an obvious consequence of this inequality. □

Let us conclude this section with the remark that equality in (63) implies that X is a disk.

4.7 Obstacle Problems and Existence Results for Surfaces of Prescribed Mean Curvature

In this section we treat obstacle problems, that is, we look for surfaces of minimal area (or minimal Dirichlet integral) which are spanning a prescribed closed boundary curve Γ and avoid certain open sets (the "obstacles"). This means that the competing surfaces of the variational problem are confined to some closed set $\mathcal{K}$ which is a subset of $\mathbb{R}^3$ or, more generally a subset of a three dimensional manifold M. In Chapter 4 of Vol. 1 we have very thoroughly described the minimization procedure which leads to a solution of Plateau's problem for minimal surfaces. In addition we have outlined the extension of this argument to a more general variational integral, see Theorem in No. 6 of the Scholia to that chapter. Therefore we refrain from repeating the procedure here and refer to Chapter 4 of Vol. 1 as well as to the pertinent literature cited therein. Instead we focus on higher regularity results for obstacle problems. Note that the optimal regularity which can be expected is $C^{1,1}$-regularity of a solution. Indeed, this can already be seen by considering a thread of minimal length which is spanned between two fixed points and touches an (analytic) obstacle in a whole interval.

In a first step we prove Hölder continuity of any solution, and later in Theorem 6 we use a difference quotient technique to show $H^2_{s,\mathrm{loc}}$-regularity for any solution of the variational problem. By standard Sobolev imbedding results this implies the Hölder continuity of the first derivatives.

We also study the Plateau problem for surfaces of prescribed mean curvature in Euclidean space $\mathbb{R}^3$. Here one prescribes a real valued function H on $\mathbb{R}^3$ and asks for a surface X which is bounded by a given closed Jordan curve Γ and has prescribed mean curvature $H(X(u,v))$ at a particular point $X(u,v)$. Clearly, if $H \equiv 0$, we recover the classical Plateau problem for minimal surfaces. In this section we discuss some classical existence and also non-existence results for the general Plateau problem described above.

Set

$$B = \{w \in \mathbb{C} : |w| < 1\} \quad \text{and} \quad C := \{w \in \mathbb{C} : |w| = 1\} = \partial B$$

and let Γ denote a closed Jordan curve in $\mathbb{R}^3$ i.e. a topological image of C. Let $H : \mathbb{R}^3 \to \mathbb{R}$ be a given function which is bounded and continuous.

Definition 1. *Given a closed Jordan curve Γ in $\mathbb{R}^3$ and a bounded continuous function $H : \mathbb{R}^3 \to \mathbb{R}$. We say that $X : \overline{B} \to \mathbb{R}^3$ is a solution of Plateau's problem determined by Γ and H (in short: an "H-surface spanned by Γ") if it fulfills the following three conditions:*

(i) $X \in C^0(\overline{B}, \mathbb{R}^3) \cap C^2(B, \mathbb{R}^3)$.
(ii) *X satisfies in B the equations*

$$\Delta X = 2H(X(u,v)) X_u \wedge X_v \tag{1}$$

and

$$|X_u|^2 = |X_v|^2, \quad \langle X_u, X_v \rangle = 0. \tag{2}$$

(iii) *The restriction $X_{|C}$ of X to the boundary C of the parameter domain B is a homeomorphism of C onto Γ.*

It follows from Chapter 2.5 of Vol. 1 that every H-surface X spanned by Γ has mean curvature $H = H(X(u,v))$ at each *regular* point $(u,v) \in B$.

Since, for $H \equiv 0$, each H-surface with boundary Γ provides a solution to the classical Plateau problem for minimal surfaces, it is conceivable that a similar variational approach using a more general energy functional instead of Dirichlet's integral might be successful.

Before we define a suitable energy functional we remark at the outset that Plateau's problem can certainly not be solvable for arbitrary Γ and H, in other words there are necessary conditions for existence.

To see this let us suppose that $X \in C^2(\overline{B}, \mathbb{R}^3)$ is a solution of (ii) and (iii) with $H \equiv$ const.

Then, by integrating (ii) we obtain

$$\int_B \triangle X\, du\, dv = \int_B \operatorname{div} \nabla X\, du\, dv = 2H \int_B (X_u \wedge X_v)\, du\, dv$$

and Gauß' and Green's theorem yield

$$\begin{aligned}
&\int_{\partial B} \nabla X \cdot n\, ds \\
&= 2H \int_B \begin{bmatrix} y_u z_v - z_u y_v \\ -x_u z_v + x_v z_u \\ x_u y_v - x_v y_u \end{bmatrix} du\, dv \\
&= H \int_B \begin{bmatrix} (y z_v)_u - (z_u y)_v \\ (z x_v)_u - (x_u z)_v \\ (x y_v)_u - (x y_u)_v \end{bmatrix} du\, dv + H \int_B \begin{bmatrix} (z y_u)_v - (z y_v)_u \\ (x z_u)_v - (x z_v)_u \\ (y x_u)_v - (y x_v)_u \end{bmatrix} du\, dv \\
&= H \int_{\partial B} \begin{bmatrix} y z_u\, du + y z_v\, dv \\ x_u z\, du + z x_v\, dv \\ x y_u\, du + x y_v\, dv \end{bmatrix} + H \int_{\partial B} \begin{bmatrix} -z y_u\, du - z y_v\, dv \\ -x z_u\, du - x z_v\, dv \\ -y x_u\, du - y x_v\, dv \end{bmatrix}.
\end{aligned}$$

On the other hand we have

$$X \wedge X_u = \begin{pmatrix} y z_u - z y_u \\ -x z_u + z x_u \\ x y_u - y x_u \end{pmatrix} \quad \text{and} \quad X \wedge X_v = \begin{pmatrix} y z_v - z y_v \\ -x z_v + z x_v \\ x y_v - y x_v \end{pmatrix}$$

and therefore

$$\begin{aligned}
\int_{\partial B} \frac{\partial X}{\partial r}\, ds &= H \int_{\partial B} (X \wedge X_u)\, du + H \int_{\partial B} (X \wedge X_v)\, dv \\
&= H \int_{\partial B} X \wedge dX.
\end{aligned}$$

In particular this implies the relation

$$|H| \left| \int_{\partial B} X \wedge dX \right| = \left| \int_{\partial B} \frac{\partial X}{\partial r}\, ds \right|$$

from which we conclude the necessary condition

$$\begin{aligned}
|H| \left| \int_{\partial B} X \wedge dX \right| &\le \int_{\partial B} \left| \frac{\partial X}{\partial r} \right| ds = \int_{\partial B} |X_\theta(1,\theta)|\, d\theta = L(\Gamma) \\
&= \text{length of the curve } \Gamma.
\end{aligned}$$

(Note that here we have used the conformality relation (ii).)

Putting $k(\Gamma) := |\int_{\partial B} X \wedge dX|$ we obtain the following *necessary condition of Heinz [12]* which we formulate as a nonexistence result.

Theorem 1. *Suppose $k(\Gamma) > 0$. Then there is no solution $X \in C^2(\overline{B}, \mathbb{R}^3)$ of Plateau's problem determined by Γ and $H \equiv$ const, if*

$$|H| > \frac{L(\Gamma)}{k(\Gamma)}.$$

This theorem also holds for solutions $X \in C^2(B, \mathbb{R}^3) \cap C^0(\overline{B}, \mathbb{R}^3)$ as was proved by Heinz [12] using an appropriate approximation procedure.

Example. Let Γ be a circle of radius R,

$$\Gamma = \{(R\cos\theta, R\sin\theta, 0) \in \mathbb{R}^3 : \theta \in [0, 2\pi)\}.$$

Then

$$k(\Gamma) = \left| \int_{\partial B} X \wedge dX \right| = \left| \int_{\partial B} X \wedge X_\theta d\theta \right| = \left| \int_{\partial B} R^2 \begin{pmatrix} 0 \\ 0 \\ 1 \end{pmatrix} d\theta \right| = 2\pi R^2.$$

Hence there is no solution of Plateau's problem for a circle of radius R and constant mean curvature H if

$$|H| > \frac{2\pi R}{2\pi R^2} = \frac{1}{R}.$$

Also, if Γ is "close to" a circle of radius R, we cannot expect the existence of an H-surface bounded by Γ and constant H bigger than $\frac{1}{R}$. We will see later on in this section, that this conditions is sharp.

Recall now that every minimizer X of the Dirichlet integral within the class $\mathcal{C}(\Gamma)$ is harmonic in B.

Furthermore we have seen in Theorem 1 of Section 4.5 in Vol. 1 that the conformality conditions (2) hold if the first inner variation $\partial D(X, \lambda)$ vanishes for all vector fields $\lambda \in C^1(\overline{B}, \mathbb{R}^3)$ (which is the case for a minimizer of $D(\cdot)$). As a suitable energy functional to be considered one might therefore try an integral $\mathcal{F}$ of the type

$$\mathcal{F}(X) = D(X) + V(X)$$

consisting of the Dirichlet integral and a "volume" term

$$V(X) := \int_B \langle Q(X), X_u \wedge X_v \rangle \, du \, dv,$$

where $Q = (Q^1, Q^2, Q^3)$ denotes a C^1-vector field defined on $\mathbb{R}^3$ or a subset $\mathcal{K}$ of $\mathbb{R}^3$. Since $V(\cdot)$ is invariant with respect to all orientation preserving C^1-diffeomorphisms of $\overline{B}$ this term would not alter the conformality of minimizers.

Note also that $V = V(X)$ equals the algebraic volume enclosed by the surface X and the cone over the boundary Γ weighted by the factor div Q, as follows easily by applying Gauß's theorem.

We observe that the Euler equation for the functional $\mathcal{F}$ is given by the system

$$\triangle X^\ell = \operatorname{div} Q(X)(X_u \wedge X_v)_\ell \tag{3}$$

for $\ell = 1, 2, 3$. (Compare Vol. 1, Section 4.5; here we have put $g_{ij} = \delta_{ij}$ and $\Gamma^k_{ij} = 0$).

If in addition to the first outer variation also the first inner variation $\partial\mathcal{F}(X, \lambda) = \partial D(X, \lambda)$ vanishes for all C^1-vector fields $\lambda = (\mu, \nu)$, then it follows that the conformality condition

$$|X_u|^2 = |X_v|^2, \quad \langle X_u, X_v\rangle = 0 \tag{4}$$

hold true (almost every where) in B.

Theorem 1 of Vol. 1, Section 2.6 now states that a solution X of (3) which satisfies (4) has mean curvature

$$H(X) = \frac{1}{2}\operatorname{div} Q(X)$$

at each regular point $(u, v) \in B$ of X.

We are thus led to consider the "energy" functional

$$\mathcal{F}(X) = \frac{1}{2}\int_B |\nabla X|^2\, du\, dv + \int_B \langle Q(X), X_u \wedge X_v\rangle\, du\, dv,$$

where the vector field Q is of class $C^1(\mathbb{R}^3, \mathbb{R}^3)$ or $C^1(\mathcal{K}, \mathbb{R}^3), \mathcal{K} \subset \mathbb{R}^3$, and has to be determined such that

$$\operatorname{div} Q(x) = 2H(x) \tag{5}$$

for all $x \in \mathbb{R}^3$ or $\mathcal{K}$ respectively.

In addition $\mathcal{F}(\cdot)$ has to be coercive on the set of admissible functions, i.e. there are positive numbers $m_0 \le m_1$ so that

$$m_0 D(X) \le \mathcal{F}(X) \le m_1 D(X) \tag{6}$$

holds for every admissible X.

The Lagrangian $e \equiv e(x, p_1, p_2)$ of $\mathcal{F}$ is given by

$$e(x, p_1, p_2) = \frac{1}{2}(|p_1|^2 + |p_2|^2) + \langle Q(x), p_1 \wedge p_2\rangle,$$

where $x \in \mathbb{R}^3$ or $\mathcal{K}$ and $p_1, p_2 \in \mathbb{R}^3$.

Assuming that

$$\sup_{\mathcal{K}} |Q| = |Q|_{0,\mathcal{K}} < 1 \tag{7}$$

we immediately conclude coerciveness of $\mathcal{F}(\cdot)$ since we obtain from Schwarz's inequality

$$\frac{1}{2}(1-|Q|_{0,\mathcal{K}})(|p_1|^2+|p_2|^2) \le e(x,p_1,p_2) \le \frac{1}{2}(1+|Q|_{0,\mathcal{K}})(|p_1|^2+|p_2|^2),$$

that is (6) follows with constants

$$m_0 := (1-|Q|_{0,\mathcal{K}}) > 0 \quad \text{and} \quad m_1 := (1+|Q|_{0,\mathcal{K}}).$$

In order to avoid additional difficulties which arise from the discussion of an obstacle problem it would be desirable to construct a vector field Q of class C^1 which is defined on $\mathcal{K} = \mathbb{R}^3$ and is subject to (5) and (7). However, even in the case $H = \text{const}$, a quick inspection of equation (5), using Gauß's theorem, shows that the quantity $|Q|_{0,\partial B_R}$ has to grow linearly in the radius R; in other words (7) can not hold for $\mathcal{K} = \mathbb{R}^3$, even if $H = \text{const}$.

Hence we consider the following strategy:

I) **The vector field Q:**
For given Γ and H satisfying conditions to be determined later, find a closed set $\mathcal{K} \subset \mathbb{R}^3$ such that $\Gamma \subset \mathcal{K}$ together with a vector field $Q \in C^1(\mathcal{K}, \mathbb{R}^3)$ which fulfills the conditions (5) and (7).

II) **The obstacle problem:**
Define the set of admissible functions $\mathcal{C} = \mathcal{C}(\Gamma, \mathcal{K}) := \mathcal{C}^*(\Gamma) \cap H_2^1(B, \mathcal{K})$, where $\mathcal{C}^*(\Gamma)$ denotes the class of H_2^1-surfaces spanning Γ which are normalized by a three point condition, and $H_2^1(B, \mathcal{K})$ denotes the subset of all Sobolev functions $f \in H_2^1(B, \mathbb{R}^3)$ which map almost all of B into $\mathcal{K}$. Solve the obstacle problem

$$\mathcal{P}(\Gamma, \mathcal{K}): \qquad \mathcal{F}(\cdot) \to \min \quad \text{in } \mathcal{C}(\Gamma, \mathcal{K})$$

and establish some initial regularity of the solutions assuming appropriate regularity hypotheses on $\mathcal{K}$. Instead of a variational equality $\delta\mathcal{F} = 0$, a solution X of $\mathcal{P}(\Gamma, \mathcal{K})$ in general merely satisfies a variational inequality $\delta\mathcal{F} \ge 0$. Therefore we have to apply a suitable inclusion principle.

III) **Geometric maximum principle:**
Determine conditions on H and $\mathcal{K}$ (or $\partial\mathcal{K}$ respectively) which guarantee that the "coincidence" set

$$\mathcal{T} := \{w \in B : X(w) \in \partial\mathcal{K}\}$$

is empty for a minimizer or a stationary point X of $\mathcal{F}$ in $\mathcal{C}$. In this case X maps B into the interior of $\mathcal{K}$ and hence satisfies the Euler-equation $\delta\mathcal{F} = 0$ in a weak sense. We refer to the Enclosure Theorems 2 and 3 in Section 4.4 for the pertinent results; however note that more elementary arguments suffice, when $\mathcal{K}$ is a ball or a cylinder.

IV) **Regularity:**
Show that under natural assumption on H (and Γ) a minimizer of $\mathcal{F}$ in $\mathcal{C}$ is a classical $C^{2,\alpha}$ solutions of the H-surface system (1) and (2). Note that the conformality conditions (2) are automatically satisfied, compare the discussion in Vol. 1, Section 4.5, and in No. 6 of the Scholia to Chapter 4 of Vol. 1.

Ad I) **Construction of the vector field Q**

The construction device requires $Q \in C^1(\mathcal{K}, \mathbb{R}^3)$ with the properties

$$\operatorname{div} Q(x) = 2H(x) \quad \text{for all } x \in \mathcal{K}$$

and some given $H \in C^0(\mathcal{K}, \mathbb{R})$ and, in addition,

$$|Q|_{0,\mathcal{K}} < 1, \quad \text{see (5) and (7).}$$

The simplest situation occurs, when $\mathcal{K} = \overline{B}_R(0) \subset \mathbb{R}^3$ and $H \in C^1(\mathbb{R}^3, \mathbb{R}^3)$. The vectorfield

$$\text{(8)}\quad Q(x) := \frac{2}{3}\left(\int_0^{x^1} H(\tau, x^2, x^3)\, d\tau, \int_0^{x^2} H(x^1, \tau, x^3)\, d\tau, \int_0^{x^3} H(x^1, x^2, \tau)\, d\tau\right)$$

clearly is of class $C^1(\mathbb{R}^3, \mathbb{R}^3)$ and satisfies (5) on $\mathbb{R}^3$ (and in particular on $\mathcal{K}$). Also

$$|Q(x)| \le \frac{2}{3}|x|\,|H|_{0,\mathcal{K}} \quad \text{for all } x \in \mathcal{K},$$

whence $|Q|_{0,\mathcal{K}} \le \frac{2}{3}R|H|_{0,\mathcal{K}}$; therefore $\mathcal{F}(\cdot)$ is coercive, if we take $\mathcal{K} = B_R(0)$ and

$$\text{(9)}\qquad |H|_{0,B_R(0)} < \frac{3}{2}R^{-1}.$$

Now let $\mathcal{K} = Z_R(0)$ be the cylinder

$$Z_R := \{(x^1, x^2, x^3) \in \mathbb{R}^3 : (x^1)^2 + (x^2)^2 \le R^2\}$$

and $H \in C^1(\mathbb{R}^3)$. Instead of (8) we put

$$\text{(10)}\qquad Q(x) := \left(\int_0^{x^1} H(\tau, x^2, x^3)\, d\tau, \int_0^{x^2} H(x^1, \tau, x^3)\, d\tau, 0\right),$$

which is again of class $C^1(\mathbb{R}^3, \mathbb{R}^3)$ and fulfills relation (5) for all $x \in \mathbb{R}^3$. Furthermore

$$|Q(x)| \le |H|_{0,\mathcal{K}}((x^1)^2 + (x^2)^2)^{1/2} \quad \text{for } x \in \mathbb{R}^3,$$

that is

$$|Q|_{0,\mathcal{K}} \le R \cdot |H|_{0,\mathcal{K}}.$$

In particular $\mathcal{F}(\cdot)$ is coercive if $\mathcal{K} = Z_R(0)$ and

$$|H|_{0,Z_R} < \frac{1}{R}. \tag{11}$$

Finally suppose $\mathcal{K} \subset S_R$ is a slab of width $2R$,

$$S_R = \{(x^1, x^2, x^3) \in \mathbb{R}^3 : -R \le x_3 \le R\}.$$

Putting

$$Q(x) := 2\left(0, 0, \int_0^{x^3} H(x^1, x^2, \tau)\, d\tau\right)$$

we then have

$$\operatorname{div} Q(x) = 2H(x) \quad \text{in } S_R$$

and

$$|Q(x)| \le 2|H(x)| \cdot |x_3|.$$

Therefore $\mathcal{F}(\cdot)$ is coercive in this case if $\mathcal{K} \subset S_R$ and

$$|H|_{0,S_R} < \frac{1}{2R}.$$

The situation for general $\mathcal{K} \subset \mathbb{R}^3$ is more involved, although the essential idea is fairly simply, namely to consider a Dirichlet problem for the nonparametric mean curvature equation in $\mathcal{K}$. To this end suppose that $u = u(x^1, x^2, x^3) \in C^1(\mathcal{K}, \mathbb{R})$ solves the mean curvature equation

$$\operatorname{div}\left(\frac{\nabla u}{\sqrt{1 + |\nabla u|^2}}\right) = 2H \quad \text{in } \mathcal{K} \tag{12}$$

then the vector field

$$Q(x) := \frac{\nabla u(x)}{\sqrt{1 + |\nabla u(x)|^2}}$$

certainly satisfies (5) and also (7) $|Q|_{0,\mathcal{K}} < 1$ holds, provided u has globally bounded gradient on $\mathcal{K}$.

For a bounded set $\mathcal{K} \subset \mathbb{R}^3$ with boundary $\partial\mathcal{K} \in C^2$ and for constant H the equation (12) with boundary condition $u = 0$ on $\partial\mathcal{K}$ is uniquely solvable with $u \in C^{2,\alpha}(\overline{\mathcal{K}})$ if and only if the inward mean curvature Λ of $\partial\mathcal{K}$ satisfies

$$(13) \qquad |H| \le \Lambda \quad \text{along } \partial\mathcal{K},$$

for a proof of this result see e.g. Gilbarg and Trudinger [1] Theorem 16.11, or Serrin [4].

To describe the condition on $\mathcal{K}$ and Λ in the case of variable H we let $\rho(x) := \text{dist}\,(x, \partial\mathcal{K})$ denote the distance of $x \in \mathcal{K}$ to the boundary $\partial\mathcal{K}$ of $\mathcal{K}$, cp. the discussion of the distance function in Section 4.4. Furthermore we extend the mean curvature function Λ from $\partial\mathcal{K}$ to $\mathcal{K}$ by putting

$$\Lambda(x) = \Lambda_{\rho(x)}(x)$$

to equal the mean curvature at x of the local surface $\mathcal{S}_{\rho(x)}$ through x which is parallel to $\partial\mathcal{K}$ at distance $\rho(x)$ in case this surface exists and is of class C^2. Otherwise we let $\Lambda_{\rho(x)}(x) = +\infty$. Condition (13) may now be replaced by

$$(14) \qquad |H(x)| \le (1 - a\rho(x))\Lambda_{\rho(x)}(x) + \frac{a}{2}$$

for $x \in \mathcal{K}$, where a denotes some number with $0 \le a \le \inf_{x\in\mathcal{K}} \rho^{-1}(x)$.

Theorem 2. *Suppose $\mathcal{K} \subset \mathbb{R}^3$ is the closure of a C^2 domain whose boundary $\partial\mathcal{K}$ has uniformly bounded principal curvatures and a global inward parallel surface at distance $\varepsilon > 0$. In addition assume that $\sup_{\mathcal{K}} \rho(x) < \infty$ and let $H \in C^1(\mathcal{K}, \mathbb{R})$ have uniformly bounded C^1-norm on $\mathcal{K}$ with* (13) *and* (14) *being fulfilled for some a, $0 \le a \le \inf_{\mathcal{K}} \rho^{-1}(x)$. Then there exists a solution $u \in C^2(\mathcal{K})$ of equation* (12) *with uniformly bounded gradient on $\mathcal{K}$. In particular there exists a C^1-vector field Q satisfying* (5) *and* (7).

The proof of Theorem 2 in case of bounded domains is due to Serrin [4]; the generalization to unbounded $\mathcal{K}$ can be found in Gulliver and Spruck [2].

Ad II) The obstacle problem

Let $\Gamma \in \mathbb{R}^3$ be a closed Jordan curve and $\mathcal{K} \subset \mathbb{R}^3$ a closed set which contains Γ. Also put

$$\mathcal{C} = \mathcal{C}(\Gamma, \mathcal{K}) = \mathcal{C}^*(\Gamma) \cap H_2^1(B, \mathcal{K})$$

to denote the class of $H_2^1(B, \mathbb{R}^3) \cap C^0(\partial B, \mathbb{R}^3)$-surfaces which map ∂B weakly monotonic onto Γ, satisfy a three point condition and have an image almost everywhere in $\mathcal{K}$.

Since, in Section 4.8, we study surfaces of prescribed mean curvature in a Riemannian three-manifold we consider now somewhat more generally functionals $\mathcal{F}(\cdot)$ which are the sum of a Riemannian Dirichlet integral and a suitable volume term.

Put $\mathcal{F}(X) = E(X) + V(X)$, where

$$E(X) := \frac{1}{2}\int_B g_{ij}(X)(X_u^i X_u^j + X_v^i X_v^j)\, du\, dv$$

and

$$V(X) := \int_B \langle Q(X), X_u \wedge X_v \rangle\, du\, dv,$$

that is

$$\mathcal{F}(X) = \int_B e(X, \nabla X)\, du\, dv$$

with the Lagrangian

$$e(x,p) = \frac{1}{2} g_{ij}(x)(p_1^i p_1^j + p_2^i p_2^j) + \langle Q(x), p_1 \wedge p_2 \rangle,$$

where $x = (x^1, x^2, x^3) \in \mathbb{R}^3$ and $p = (p_1, p_2) \in \mathbb{R}^3 \times \mathbb{R}^3$. In No. 6 of the Scholia to Vol. 1, Chapter 4, we have outlined the proof of the following

Theorem 3. *Suppose $Q \in C^0(\mathcal{K}, \mathbb{R}^3), g_{ij} \in C^0(\mathcal{K})$ $g_{ij} = g_{ji}$ for all $i,j = 1,2,3$, and let $0 < m_0 \le m_1$ be constants with the property $m_0(|p_1|^2 + |p_2|^2) \le e(x,p) \le m_1(|p_1|^2 + |p_2|^2)$ for all $(x, p_1, p_2) \in \mathcal{K} \times \mathbb{R}^3 \times \mathbb{R}^3$. Moreover assume that $\mathcal{K}$ is a closed set in $\mathbb{R}^3$ such that $\mathcal{C} = \mathcal{C}(\Gamma, \mathcal{K})$ is nonempty. Then the variational problem*

$\mathcal{P} = \mathcal{P}(\Gamma, \mathcal{K})$: $\qquad \mathcal{F} \to \min$ *in* $\mathcal{C}$

has a solution. Every solution $X \in \mathcal{C}$ satisfies the conformality relations

$$g_{ij} X_u^i X_u^j = g_{ij} X_v^i X_v^j \quad \text{and} \quad g_{ij} X_u^i X_v^j = 0 \tag{15}$$

almost everywhere in B. □

In order to obtain continuity for solutions of $\mathcal{P}$ we have to assume more regularity of $\mathcal{K}$ or $\partial\mathcal{K}$ respectively. A reasonable quantitative notion is the "quasiregularity" of $\mathcal{K}$.

Definition 2. *A closed set $\mathcal{K} \subset \mathbb{R}^3$ is called "quasiregular", if*

(a) *$\mathcal{K}$ is equal to the closure of its interior $\mathring{\mathcal{K}}$;*
(b) *there are positive numbers d and M such that for each point $x_0 \in \mathcal{K}$ there exists a compact convex set $K^*(\mathring{K}^* \ne \emptyset)$ and a C^1-diffeomorphism g defined on some open neighbourhood of K^* with $g : K^* \to \mathcal{K} \cap \overline{B_d(x_0)}$ with*

$$|Dg|^2_{0,K^*} \le M \quad \text{and} \quad |Dg^{-1}|^2_{0,\mathcal{K}\cap\overline{B_d}(x_0)} \le M.$$

Remarks. (i) Closed convex sets $\mathcal{K}$ with $\mathcal{K} = \overline{\mathring{\mathcal{K}}}$ are quasiregular.

(ii) If $\mathcal{K} = \overline{\mathring{\mathcal{K}}} \subset \mathbb{R}^3$ is compact with $\partial\mathcal{K} \in C^1$, then $\mathcal{K}$ is quasiregular.

(iii) Suppose $\mathcal{K} = \overline{\mathring{\mathcal{K}}}$, $\partial\mathcal{K} \in C^2$ and $\partial\mathcal{K}$ has uniformly bounded principal curvatures and a global parallel surface in $\mathring{\mathcal{K}}$, then $\mathcal{K}$ is quasiregular; for a proof see Gulliver and Spruck [2].

Theorem 4. *Let the assumption of Theorem 3 be satisfied and suppose that $\mathcal{K} \subset \mathbb{R}^3$ is quasiregular. Furthermore let X be a solution of the problem*

$$\mathcal{P}(\Gamma, \mathcal{K}) : \qquad \mathcal{F} \to \min \ \text{in } \mathcal{C}(\Gamma, \mathcal{K}).$$

Then there is a number $\mu > 0$ such that

$$(16) \qquad \int_{B_r(w_0)} |\nabla X|^2 \, du \, dv \le \left(\frac{r}{R}\right)^{2\mu} \int_{B_R(w_0)} |\nabla X|^2 \, du \, dv$$

for all $r \in (0, R]$ and $w_0 \in B$ with $0 < R \le \operatorname{dist}(w_0, \partial B)$. It follows that X is of class $C^{0,\mu}(B, \mathbb{R}^3)$. Furthermore X is continuous up to the boundary.

Proof. Let X be a minimizer of the functional $\mathcal{F}$ in $\mathcal{C}$. For an arbitrary point $w_0 \in B$ we define

$$\phi(r) = \phi(r, w_0) = \int_{B_r(w_0)} |\nabla X|^2 \, du \, dv,$$

where $0 < r \le R = \operatorname{dist}(w_0, \partial B)$.

Introducing polar coordinates (ρ, θ) around w_0 by $w = w_0 + \rho e^{i\theta}$ and writing (with a slight but convenient abuse of notation) $X(w) = X(w_0 + \rho e^{i\theta}) = X(\rho, \theta)$, we get

$$\phi(r) = \int_0^r \int_0^{2\pi} \left\{ |X_\rho|^2 + \frac{1}{\rho^2} |X_\theta|^2 \right\} \rho \, d\rho \, d\theta.$$

Furthermore, by selecting an ACM-representative of X again denoted by X, we can assume that for almost all $\theta \in [0, 2\pi]$ the restriction $X(\cdot, \theta)$ is absolutely continuous in $\rho \in [\varepsilon, R]$, $\varepsilon > 0$, and $X(\rho, \cdot)$ is absolutely continuous in $\theta \in [0, 2\pi]$ for almost all $\rho \in [0, R]$.

There is a Lebesgue null set $\mathcal{N} \subset [0, R]$ such that for $r \in [0, R] \setminus \mathcal{N}$ we have

(i) $X(r, \cdot)$ is absolutely continuous on $[0, 2\pi]$,

(ii) $\int_0^{2\pi} |X_\theta(r, \theta)|^2 \, d\theta < \infty$,

(iii) $\phi(r)$ is differentiable with

$$\phi'(r) = \int_0^{2\pi} \left\{ |X_\rho(r, \theta)|^2 + \frac{1}{r^2} |X_\theta(r, \theta)|^2 \right\} r \, d\theta \geq \frac{1}{r} \int_0^{2\pi} |X_\theta(r, \theta)|^2 \, d\theta,$$

i.e.

(17) $$\int_0^{2\pi} |X_\theta(r, \theta)|^2 \, d\theta \leq r \cdot \phi'(r) \quad \text{for all } r \in [0, R].$$

Take a radius $r \in [0, R] \setminus \mathcal{N}$ for which

(18) $$\int_0^{2\pi} |X_\theta(r, \theta)|^2 \, d\theta < \frac{\pi^{-1}}{2} d^2,$$

where d denotes the constant in the definition of quasiregularity. Then for any $\theta_0, \theta_1 \in [0, 2\pi]$ we infer the estimate

$$|X(r, \theta_1) - X(r, \theta_0)| \leq \left| \int_{\theta_1}^{\theta_2} |X_\theta(r, \theta)| \, d\theta \right| \leq \sqrt{2\pi} \left\{ \int_0^{2\pi} |X_\theta(r, \theta)|^2 \, d\theta \right\}^{\frac{1}{2}} < d$$

and hence the image of the curve $X(r, \cdot)$ is contained in $\mathcal{K} \cap \overline{B_d(x_0)}$, where $x_0 = X(r, \theta_0)$ is an arbitrary point on that curve. According to the definition of quasiregularity there is a C^1-diffeomorphism $h = g^{-1} : \mathcal{K} \cap \overline{B_d(x_0)} \to K^*$, where K^* is a compact and convex set. Hence the curve $\zeta(\theta) := h(X(r, \theta))$ is of class $H_2^1([0, 2\pi], \mathbb{R}^3)$ with values in the convex set K^*. Now let $H = H(w)$ denote the harmonic vector function defined in $B_r(w_0)$ whose boundary values are given by $\zeta(\theta)$, i.e.

$$H(w_0 + re^{i\theta}) = \zeta(\theta) = h(X(r, \theta))$$

for $0 \leq \theta \leq 2\pi$. By the maximum principle and the convexity of K^* it follows that the image $H(B_r(w_0)) \subset K^*$ and therefore the function $g \circ H \in H_2^1(B_r(w_0), \mathcal{K})$ with boundary trace $X(r, \theta)$. Setting

$$Y(w) := \begin{cases} g \circ H(w) & \text{for } w \in B_r(w_0), \\ X(w) & \text{for } w \in B \setminus B_r(w_0) \end{cases}$$

we therefore obtain a function $Y \in \mathcal{C}(\Gamma, \mathcal{K})$. Since X is a minimizer of $\mathcal{F}$ in $\mathcal{C} = \mathcal{C}(\Gamma, \mathcal{K})$ we have

$$\mathcal{F}(X) \leq \mathcal{F}(Y)$$

and by the coercivity assumption and the quasiregularity of $\mathcal{K}$ it follows

$$\begin{aligned} m_0 \int_{B_r(w_0)} |\nabla X|^2 \, du \, dv &\leq m_1 \int_{B_r(w_0)} |\nabla Y|^2 \, du \, dv \\ &= m_1 \int_{B_r(w_0)} |\nabla (g \circ H)|^2 \, du \, dv \leq m_1 M \int_{B_r(w_0)} |\nabla H|^2 \, du \, dv, \end{aligned}$$

that is

$$\phi(r) \leq \frac{m_1}{m_0} M \int_{B_r(w_0)} |\nabla H|^2 \, du \, dv. \tag{19}$$

On the other hand an expansion of ζ and H in Fourier series yields

$$\zeta(\theta) = A_0 + \sum_{n=1}^{\infty} (A_n \cos(n\theta) + B_n \sin(n\theta)),$$

and

$$H(w) = A_0 + \sum_{n=1}^{\infty} \left(\frac{\rho}{r}\right)^n [A_n \cos(n\theta) + B_n \sin(n\theta)],$$

which yields

$$\int_{B_r(w_0)} |\nabla H|^2 \, du \, dv = \pi \sum_{n=1}^{\infty} n(|A_n|^2 + |B_n|^2),$$

and

$$\int_0^{2\pi} |\zeta_\theta|^2 \, d\theta = \pi \sum_{n=1}^{\infty} n^2(|A_n|^2 + |B_n|^2).$$

In particular we have

$$\int_{B_r(w_0)} |\nabla H|^2 \, du \, dv \leq \int_0^{2\pi} |\zeta_\theta|^2 \, d\theta. \tag{20}$$

But from $\zeta(\theta) = h(X(r, \theta))$ we obtain, using the quasiregularity of $\mathcal{K}$ again,

$$\int_0^{2\pi} |\zeta_\theta|^2 \, d\theta \leq M \int_0^{2\pi} |X_\theta|^2 \, d\theta. \tag{21}$$

Relations (19), (20), (21) and (17) now yield the estimate

$$\phi(r) \leq \frac{m_1}{m_0} M^2 \int_0^{2\pi} |X_\theta|^2 \, d\theta \leq \frac{m_1}{m_0} M^2 r \phi'(r)$$

for almost every $r \in [0, R]$.

On the other hand, if (18) does not hold, then we trivially have

$$\phi(r) \le \phi(R) \le \phi(R) \cdot \frac{2\pi}{d^2} \int_0^{2\pi} |X_\theta|^2 \, d\theta \le \frac{2\pi}{d^2} D(X) r \cdot \phi'(r),$$

again by using (17). Concluding we obtain in both cases the inequality $\phi(r) \le C \cdot r\phi'(r)$, where we have put $C := \max(2\pi d^{-2} D(X), \frac{m_1}{m_0} M^2)$. From this inequality we finally obtain by a simple integration

$$\phi(r) \le \left(\frac{r}{R}\right)^{2\mu} \phi(R)$$

for all $r \in [0, R]$ and $\mu := \frac{1}{2C}$.

Now $X \in C^{0,\mu}(B)$ follows from Dirichlet's growth theorem, see e.g. Gilbarg and Trudinger [1], Theorem 7.19.

To prove continuity of X up to the boundary, we apply a conformal mapping τ which maps the unit disk onto the upper half plane and the unit circle onto the real axis. Since τ maps circles onto circles, leaves the Dirichlet integral invariant and is locally bi-Lipschitz, it follows that $X \circ \tau^{-1}$ satisfies again condition (16) in a neighbourhood of any boundary point of the half plane, possibly with an additional constant factor K on the right hand side. In addition we may choose τ in such a way that an arbitrary but fixed point $e^{i\Theta}$ is mapped onto the origin. We are thus led to consider the following situation: Let Ω be the rectangle $\{w = u + iv \in \mathbb{C} : |u| < 2, 0 < v < 2\}$ and suppose $X \in H_2^1(\Omega, \mathbb{R}^3)$ possesses continuous boundary trace $\xi(u) = X(u, 0), u \in (-2, 2)$. Then we have to show that $X(w) \to \xi(0)$ as $w \to 0$. To this end we introduce the entities

$$\varepsilon(X, u, h) := \left(\int_{u-2h}^{u+2h} \int_0^{2h} |\nabla X|^2 \, du \, dv \right)^{\frac{1}{2}},$$

$$\omega(\xi, h) := \sup_{|u'-u''| \le h} |\xi(u') - \xi(u'')|$$

and let $w = u + ih$ be an arbitrary point with $|u| < 1, 0 < h < \frac{1}{2}$. Recalling Morrey's proof of Dirichlet's growth theorem (see Morrey [8], Theorem 3.5.2) we obtain by virtue of condition (16) the estimate

$$|X(u, h) - X(u', h)| \le C_0 k \varepsilon(X, u, h) |u - u'|^\mu h^{-\mu}$$

for all u' with $|u - u'| \le h < \frac{1}{2}$ with some constant c_0 depending only on μ.

Next we select a $u_1 \in [-1, 1]$, $|u - u_1| < h$ with the properties

(i) $X(u_1, \cdot) \in H_2^1([0, 2], \mathbb{R}^3)$,

(ii) $X(u_1, v) \to \xi(u_1)$ as $v \to 0^+$,

(iii) $\varepsilon^2(X, u, h) = \int_{u-2h}^{u+2h} \int_0^{2h} |\nabla X|^2 \, du \, dv \ge h \int_0^h |X_v(u_v, u)|^2 \, dv.$

Consequently

$$
\begin{aligned}
|X(u_1,h)-\xi(u_1)| &\le \int_0^h |X_v(u_1,v)|\,dv \\
&\le \sqrt{h}\left(\int_0^h |X_v(u_1,v)|^2\,dv\right)^{\frac{1}{2}} \le \varepsilon(X,u,h)
\end{aligned}
$$

by (iii). Finally we obtain for all $u \in \mathbb{R}$ with $|u| < h' < \frac{1}{2}$,

$$
\begin{aligned}
|X(u,h)-\xi(0)| \le\ & |X(u,h)-X(u_1,h)| + |X(u_1,h)-\xi(u_1)| \\
&+ |\xi(u_1)-\xi(u)| + |\xi(u)-\xi(0)| \le (c_0 k+1)\varepsilon(X,u,h) \\
&+ \omega(\xi,h)+\omega(\xi,h'),
\end{aligned}
$$

whence $X(u,h) \to \xi(0)$ as $(u,h) \to (0,0)$. This proves that $X \in C^0(\overline{B},\mathbb{R}^3)$.□

By the same reasoning we can show

Proposition 1. *Let $\mathbb{F}$ be a family of functions $X \in H_2^1(B,\mathbb{R}^3)$ whose boundary values are equicontinuous on ∂B. Suppose that*

$$
\int_{B_r(w_0)} |\nabla X|^2\,du\,dv \le k^2\left(\frac{r}{R}\right)^{2\mu}\int_{B_R(w_0)} |\nabla X|^2\,du\,dv
$$

holds for all $r \in (0,R]$ and $w_0 \in B$ with $0 < R \le \operatorname{dist}(w_0,\partial B)$ and uniform constants k and μ for all $X \in \mathbb{F}$. Furthermore, assume that there exist a number $A > 0$ and a function $\eta(r)$ on $0 < r < \infty$ with $\lim_{r\to 0}\eta(r) = 0$, all independent of $X \in \mathbb{F}$, such that $D_B(X) = \int_B |\nabla X|^2\,du\,dv \le A$, $D_{B\cap B_r(w^*)}(X) \le \eta(r)$ for $w^* \in \partial\Omega$ and $0 < r < \infty$, for all $X \in \mathbb{F}$. Then the family $\mathbb{F}$ is equicontinuous on $\overline{B}$. □

In Section 4.5 we have derived a formula for the *inner variation* of a functional $\mathcal{F}$, see the formulae in Section 4.5 of Vol. 1, (15) and (20). In particular the conformality relations (15) hold if the first inner variation $\partial\mathcal{F}$ vanishes for all vector fields λ.

Now we have to consider "*outer variations*", that is variations of the type $X_\varepsilon = X + \varepsilon\varphi$.

Assumption A. *Let $\mathcal{K} \in \mathbb{R}^3$ be a closed set and $Q \in C^1(S,\mathbb{R}^3)$, $g_{ij} \in C^1(S,\mathbb{R})$ for all $i,j = 1,2,3$ and some open set S containing $\mathcal{K}$. In addition suppose that Q and g_{ij} satisfy*

$$
\left|\frac{\partial Q^j}{\partial x^i}\right|_{0,\mathcal{K}} < \infty, \qquad \left|\frac{\partial g_{ij}}{\partial x^k}\right|_{0,\mathcal{K}} < \infty \tag{22}
$$

for all $i,j,k = 1,2,3$ and suppose

$$e(x,p_1,p_2) = \frac{1}{2} g_{ij}(x) p^i_\alpha p^j_\alpha + \langle Q(x), p_1 \wedge p_2 \rangle$$

is coercive, i.e.

$$m_0\{|p_1|^2 + |p_2|^2\} \le e(x,p_1,p_2) \le m_1\{|p_1|^2 + |p_2|^2\}$$

for all $(x,p_1,p_2) \in \mathcal{K} \times \mathbb{R}^3 \times \mathbb{R}^3$ *and suitable constants* $0 < m_0 \le m_1$.

Theorem 5 (First variation formula). *Assume* $Q, g_{ij}, \mathcal{K}$ *and* $e(x,p_1,p_2)$ *fulfill Assumption A. Let* $X \in H^1_2(B,\mathcal{K})$ *and* $\varphi \in L_\infty(B,\mathbb{R}^3)$ *be functions such that* $X + \varepsilon\varphi \in H^1_2(B,\mathcal{K})$ *for all* $\varepsilon \in [0,\varepsilon_0)$ *and some* $\varepsilon_0 > 0$. *Then the first (outer) variation* $\delta\mathcal{F}(X,\varphi) = \lim_{\epsilon\to 0^+} \frac{\mathcal{F}(X+\varepsilon\varphi)-\mathcal{F}(X)}{\varepsilon}$ *exists and is given by*

$$\begin{aligned}
&\delta\mathcal{F}(X,\varphi)\\
&= \int_B \Big\{ g_{ij}(X) X^i_{u^\alpha} \varphi^j_{u^\alpha} + \frac{1}{2}\frac{\partial g_{ij}(X)}{\partial x^e} X^i_{u^\alpha} X^j_{u^\alpha} \varphi^\ell \\
&\quad + \Big\langle \frac{\partial Q}{\partial x^j}(X), X_u \wedge X_v \Big\rangle \varphi^j + \langle Q(X), X_u \wedge \varphi_v + \varphi_u \wedge X_v \rangle \Big\}\, du_1\, du_2.
\end{aligned}$$

Furthermore, if $\varphi \in \overset{\circ}{H}{}^1_2(B,\mathbb{R}^3) \cap L_\infty(B,\mathbb{R}^3)$ *then*

$$\begin{aligned}
\delta\mathcal{F}(X,\varphi) = \int_B \Big\{ & g_{ij}(X) X^i_{u^\alpha} \varphi^j_{u^\alpha} + \frac{1}{2}\frac{\partial g_{ij}}{\partial x^\ell} X^i_{u^\alpha} X^j_{u^\alpha} \varphi^\ell \\
& + \operatorname{div} Q(X) \langle X_u \wedge X_v, \varphi \rangle \Big\}\, du\, dv,
\end{aligned} \tag{23}$$

where

$$\operatorname{div} Q(X) = \frac{\partial Q^1}{\partial x^2}(X) + \frac{\partial Q^2}{\partial x^2}(X) + \frac{\partial Q^3}{\partial x^3}(X).$$

Remark. $\delta\mathcal{F}(X,\varphi)$ is called the *first (outer)* variation of $\mathcal{F}$ at X in direction φ.

We have adopted the summation convention that Latin indices have to be summed from 1 to 3 and Greek indices from 1 to 2. Also we have replaced (u,v) by (u_1,u_2).

Proof of Theorem 5. We compute

$$\begin{aligned}
&\frac{1}{\varepsilon}[\mathcal{F}(X+\varepsilon\varphi)-\mathcal{F}(X)]-\delta\mathcal{F}(X,\varphi)\\
&\quad=\frac{1}{\varepsilon}\int_B\Big\{\frac{1}{2}[g_{ij}(X+\varepsilon\varphi)(X^i+\varepsilon\varphi^i)_{u^\alpha}(X^j+\varepsilon\varphi^j)_{u^\alpha}-g_{ij}(X)X^i_{u^\alpha}X^j_{u^\alpha}]\\
&\qquad+\langle Q(X+\varepsilon\varphi),(X_u+\varepsilon\varphi_u)\wedge(X_v+\varepsilon\varphi_v)\rangle-\langle Q(X),X_u\wedge X_v\rangle\Big\}\,du\,dv\\
&\qquad-\int_B\Big\{g_{ij}(X)X^i_{u^\alpha}\varphi^j_{u^\alpha}+\frac{1}{2}\frac{\partial g_{ij}}{\partial x^e}X^i_{u^\alpha}X^j_{u^\alpha}\varphi^e\\
&\qquad+\Big\langle\frac{\partial Q}{\partial x^j},X_u\wedge X_v\Big\rangle\varphi^j+\langle Q(X),X_u\wedge\varphi_v+\varphi_u\wedge X_v\rangle\Big\}\,du\,dv\\
&\quad=\int_B\Big\{\frac{1}{2}\Big[\frac{1}{\varepsilon}(g_{ij}(X+\varepsilon\varphi)-g_{ij}(X))-\frac{\partial g_{ij}}{\partial x^e}\varphi^e\Big]X^i_{u^\alpha}X^j_{u^\alpha}\\
&\qquad+\Big\langle\frac{1}{\varepsilon}(Q(X+\varepsilon\varphi)-Q(X))-\frac{\partial Q}{\partial x^j}(X)\varphi^j,X_u\wedge X_v\Big\rangle\\
&\qquad+[g_{ij}(X+\varepsilon\varphi)-g_{ij}(X)]X^i_{u^\alpha}\varphi^j_{u^\alpha}\\
&\qquad+\langle Q(X+\varepsilon\varphi)-Q(X),X_u\wedge\varphi_v+\varphi_u\wedge X_v\rangle\\
&\qquad+\frac{\varepsilon}{2}g_{ij}(X+\varepsilon\varphi)\varphi^i_{u^\alpha}\varphi^j_{u^\alpha}+\varepsilon Q(X+\varepsilon\varphi)(\varphi_u\wedge\varphi_v)\Big\}\,du\,dv\\
&\quad=\int_B a^\varepsilon_{ij}(w)X^i_{u^\alpha}X^j_{u^\alpha}\,du\,dv+\int_B b^\varepsilon_i(w)(X_u\wedge X_v)_i\,du\,dv\\
&\qquad+\int_B c^\varepsilon_{ij}(w)X^i_{u^\alpha}\varphi^j_{u^\alpha}\,du\,dv+\int_B d^\varepsilon_i(w)[(X_u\wedge\varphi_v)_i+(\varphi_u\wedge X_v)_i]\,du\,dv\\
&\qquad+\varepsilon\int_B[h^\varepsilon_{ij}(w)\varphi^i_{u^\alpha}\varphi^j_{u^\alpha}+f^\varepsilon_i(w)(\varphi_u\wedge\varphi_v)_i]\,du\,dv
\end{aligned}$$

with obvious choices of bounded and measurable functions $a^\varepsilon_{ij},\dots,f^\varepsilon_i$ on B whose $L_\infty(B)$-norms are uniformly bounded with respect to ε. Furthermore

$$a^\varepsilon_{ij}(\cdot),\ b^\varepsilon_i(\cdot),\ c^\varepsilon_{ij}(\cdot),\ d^\varepsilon_i\to 0$$

a.e. on B as $\varepsilon\to 0$.

For any measurable set $\Omega\subset B$ we have

$$\begin{aligned}
&\left|\int_\Omega a^\varepsilon_{ij}X^i_{u^\alpha}X^j_{u^\alpha}\,du\,dv\right|\le cD_\Omega(X)=c\int_\Omega|\nabla X|^2\,du\,dv,\\
&\left|\int_\Omega b^\varepsilon_i(X_u\wedge X_v)_i\,du\,dv\right|\le cD_\Omega(X),\\
&\left|\int_\Omega c^\varepsilon_{ij}X^i_{u^\alpha}\varphi^j_{u^\alpha}\,du\,dv\right|\le c(D_\Omega(X))^{\frac{1}{2}}(D_\Omega(\varphi))^{\frac{1}{2}},\\
&\left|\int_\Omega d^\varepsilon_i[(X_u\wedge\varphi_v)_i+(\varphi_u\wedge X_v)_i]\,du\,dv\right|\le c(D_\Omega(X))^{\frac{1}{2}}(D_\Omega(\varphi))^{\frac{1}{2}}
\end{aligned}$$

and

$$\left| \int_\Omega b^\varepsilon_{ij} \varphi^i_{u^\alpha} \varphi^j_{u^\alpha}\, du\, dv \right| \le c D_\Omega(\varphi),$$

$$\left| \int_\Omega f^\varepsilon_i (\varphi_u \wedge \varphi_v)_i \, du\, dv \right| \le c D_\Omega(\varphi)$$

for a constant c independent of ε. This implies the uniform absolute continuity of the integrals under consideration. By virtue of Vitali's convergence theorem the first part of Theorem 5 follows. Finally formula (23) can be derived by an integration by parts using an appropriate approximation argument. □

Remarks. (i) The statements of Theorem 5 hold true without the hypotheses (22), if $X \in C^0(\overline{B}, \mathbb{R}^3)$ or even $X \in L_{\infty,\mathrm{loc}}(B, \mathbb{R}^3)$, which is – by Theorem 4 – true for solutions X of $\mathcal{P}(\Gamma, \mathcal{K})$.

(ii) The first variation formula (23) continues to hold if Q is not necessarily C^1 but $\operatorname{div} Q$ is defined (possibly in a weak sense!). For a proof and an application of this remark see the proof of Theorem 8, in particular relation (37).

A consequence of Theorem 5 is the Euler equation for the functional $\mathcal{F} = E + V$ (see also Theorem 7), namely

$$\text{(24)} \quad \triangle X^\ell + \Gamma^\ell_{jk}(X^j_u X^k_u + X_v X^k_v) = \operatorname{div} Q(X) g^{\ell m} (X_u \wedge X_v)_m, \quad \ell = 1, 2, 3,$$

where the Christoffel symbols $\Gamma_{jk\ell}$ and Γ^ℓ_{jk} are given by (cp. Vol. 1, Chapter 1)

$$\Gamma_{jk\ell} = \frac{1}{2}\left(\frac{\partial g_{jk}}{\partial x^\ell} - \frac{\partial g_{j\ell}}{\partial x^k} + \frac{\partial g_{k\ell}}{\partial x^j} \right), \quad \Gamma^\ell_{jk} = g^{\ell m} \Gamma_{jmk}.$$

Indeed, (24) follows from the first variation formula (23) on testing with $\varphi = (\varphi^1, \varphi^2, \varphi^3)$, where $\varphi^j = g^{jk}(X)\psi^k$ with $\psi = (\psi^1, \psi^2, \psi^3) \in C^\infty_0(B, \mathbb{R}^3)$, and the fundamental lemma of the calculus of variations.

A major step in the regularity theory for obstacle problems is the following

Theorem 6. *Suppose $Q \in C^2(S, \mathbb{R}^3)$, $g_{ij} \in C^2(S, \mathbb{R})$, $i, j = 1, 2, 3$ and $e(x, p_1, p_2)$ satisfy Assumption A (possibly without relation (22)), where $\mathcal{K}$ is quasiregular and of class C^3 and $S \subset \mathbb{R}^3$ is open with $\mathcal{K} \subset S$. Then each solution $X \in \mathcal{C}(\Gamma, \mathcal{K})$ of the obstacle problem*

$$\mathcal{P}(\Gamma, \mathcal{K}): \qquad \mathcal{F} \to \min \ \text{in } \mathcal{C}(\Gamma, \mathcal{K})$$

is of class $H^2_s(B', \mathbb{R}^3) \cap C^{1,\alpha}(B, \mathbb{R}^3) \cap C^0(\overline{B}, \mathbb{R}^3)$ for all $B' \subset\subset B$ and all $s, \alpha \in \mathbb{R}$ with $0 \le s < \infty$ and $1 < \alpha < 1$.

This result holds under somewhat weaker regularity hypothesis on Q, see the Remark at the end of the proof of Theorem 6.

The key argument of the proof of Theorem 6 is given in the following Lemma 1 where the L_2-estimates of the second derivatives are established.

Definition 3. *Let $\Omega' \subset \mathbb{R}^2$ be a bounded open set, $K \subset \mathbb{R}^3$ a closed set and $S \subset \mathbb{R}^3$ some open set containing K. Consider functions $A = A(w,z,p) = (A_j^\alpha)(w,z,p), j = 1,2,3, \alpha = 1,2$ and $B = B(w,z,p) = B_j(w,z,p), j = 1,2,3$ of class C^1 on $\Omega' \times S \times \mathbb{R}^6$ such that the inequalities*

$$m_2|\eta|^2 \le A^\alpha_{jp^k_\beta}(\xi)\eta^j_\alpha\eta^k_\beta,$$

$$\left|A^\alpha_{jp^k_\beta}(\xi)\right| \le m_3$$

and

$$|A(\xi)|^2 + |A_w(\xi)|^2 + |A_z(\xi)|^2 + |B(\xi)| + |B_w(\xi)| + |B_z(\xi)| + |B_p(\xi)|^2 \\ \le m_4(1+|p|^2)$$

hold for all $\xi = (w,z,p) \in \Omega' \times K \times \mathbb{R}^6$ and for all $\eta = (\eta_1,\eta_2) \in \mathbb{R}^3 \times \mathbb{R}^3$ with positive constants $m_2, m_3, m_4 \in \mathbb{R}$ independent of ξ.

Lemma 1. *Suppose A, B and Ω' satisfy Definition 3 with $K = B_1^+(0) := \{(x,y,z) \in \mathbb{R}^3 : x^2+y^2+z^2 \le 1, z \ge 0\}$. Moreover let $z = z(w) \in H_2^1(\Omega', B_1^+)$ have the following properties*

(a) *There are positive numbers M_0 and μ such that*

$$\text{(25)} \qquad \int_{\Omega' \cap B_\rho(\zeta)} |\nabla z|^2\, du\, dv \le M_0 \rho^{2\mu} \quad \textit{for all disks } B_\rho(\zeta) \subset \mathbb{R}^2,$$

(b) *For all $\varphi \in C_c^0(\Omega', \mathbb{R}^3) \cap H_2^1(\Omega', \mathbb{R}^3)$ with $z^3 - \varepsilon\varphi^3 \ge 0$ for $\varepsilon \in [0, \varepsilon_0]$, $\varepsilon_0(\varphi) > 0$, the variational inequality*

$$\text{(26)} \qquad \int_{\Omega'} \{A_j^\alpha(w,z,\nabla z)\varphi^j_{u^\alpha} + B_j(w,z,\nabla z)\varphi^j\}\, du\, dv \le 0$$

is satisfied.

Then we have $z \in H_2^2(\Omega'', \mathbb{R}^3) \cap H_s^1(\Omega'', \mathbb{R}^3)$ for $\Omega'' \Subset \Omega'$ and all $s \in [1, \infty)$.

Proof. Pick any $\zeta_0 \in \Omega'$ and consider a disk $B_{3R_0}(\zeta_0) \Subset \Omega'$, $0 < R_0 < 1$ and choose $R \in (0, R_0)$. Then there exists a function $\eta \in C_c^\infty(B_{2R}(\zeta_0))$ satisfying $0 \le \eta \le 1$, $|\nabla\eta| \le \frac{2}{R}$ and $\eta(w) = 1$ for $w \in B_R(\zeta_0)$. Moreover let us denote by $\triangle_h z$ the difference quotient

$$\triangle_h z = \frac{1}{h}[z(w+h\zeta) - z(w)], \quad h \ne 0,$$

in the direction of a unit vector $\zeta \in \mathbb{R}^2$. Then we have the relation

$$z(w) + \epsilon\triangle_{-h}[\eta^2(w)\triangle_h z(w)] = \frac{\epsilon}{h^2}\eta^2(w)z(w+h\zeta) \\ + \left\{1 - \frac{\epsilon}{h^2}[\eta^2(w) + \eta^2(w-h\zeta)]\right\} z(w) + \frac{\epsilon}{h^2}\eta^2(w-h\zeta)z(w-h\zeta).$$

Therefore $\varphi = -\triangle_{-h}[\eta^2\triangle_h z]$ is of class $C_c^{0,\mu}(\Omega', \mathbb{R}^3) \cap H_2^1(\Omega', \mathbb{R}^3)$ for $0 < |h| < R$ and satisfies $z^3 - \epsilon\varphi^3 \geq 0$ provided $0 \leq \epsilon < \epsilon_0 = \frac{h^2}{2}$. Thus φ is admissible in (26) and we obtain

$$(27) \quad \int \{\triangle_h A(w,z,\nabla z)\nabla(\eta^2\triangle_h z) + \triangle_h B(w,z,\nabla z)(\eta^2\triangle_h z)\}\, du\, dv \leq 0,$$

where we have for simplicity omitted the domain of integration Ω'. Now we use the identity

$$(28) \quad \triangle_h A(w, z(w), \nabla z(w)) = \int_0^1 A_w(\xi(t))\, dt \cdot \zeta + \int_0^1 A_z(\xi(t))\, dt \cdot \triangle_h z(w) + \int_0^1 A_p(\xi(t))\, dt \cdot \nabla\triangle_h z(w),$$

where $\xi(t) = (w + th\zeta, z(w) + th\triangle_h z(w), \nabla z(w) + th\nabla\triangle_h z(w))$ and analogous expressions holding for $\triangle_h B(w, z(w), \nabla z(w))$. Observe that the set $B_{3R}(\zeta_0) \times B_1^+ \times \mathbb{R}^6$ is convex and $z : \Omega' \to B_1^+$; hence $\xi(t) \in B_{3R}(\zeta_0) \times B_1^+ \times \mathbb{R}^6$ for all $t \in [0,1]$, $|h| < R$ and $w \in B_{2R}(\zeta_0) \Supset \operatorname{supp} \eta$.

By virtue of Definition 3

$$|\triangle_h A(w,z,\nabla z)| \leq m_5\{(1 + |\nabla z| + |\nabla z_h|) \cdot (1 + |\triangle_h z|) + |\nabla\triangle_h z|\},$$

$$(29) \quad |\triangle_h A(w,z,\nabla z) - \int_0^1 A_p(\xi(t))\, dt \nabla\triangle_h z(w)| \leq m_5(1 + |\nabla z| + |\nabla z_h|)(1 + |\triangle_h z|),$$

$$|\triangle_h B(w,z,\nabla z| \leq m_6\{(1 + |\nabla z|^2 + |\nabla z_h|^2)(1 + |\triangle_h z|) + (1 + |\nabla z| + |\nabla z_h|)|\nabla\triangle_h z|\}$$

with suitable constants m_5, m_6 and $z_h(w) := z(w + h\zeta)$. Again from Definition 3 we infer

$$(30) \quad m_2 \int_{\Omega'} |\eta\nabla\triangle_h z|^2\, du\, dv \leq \int_{\Omega'} \eta^2 \int_0^1 A_p(\xi(t))\, dt \nabla\triangle_h z \nabla\triangle_h z\, du\, dv.$$

Now we use the variational inequality (27) and relation (28) together with $\nabla(\eta^2\triangle_h z) = 2\eta\nabla\eta\triangle_h z + \eta^2\nabla\triangle_h z$ and infer

$$\begin{aligned}
&\int_{\Omega'} \int_0^1 A_p\, dt \nabla\triangle_h z \nabla\triangle_h z \eta^2\, du\, dv \\
&\leq -\int_{\Omega'} \int_0^1 A_p\, dt \nabla\triangle_h z \nabla\eta\triangle_h z \cdot 2\eta\, du\, dv \\
&\quad - \int_{\Omega'} \int_0^1 A_w\, dt \zeta[2\eta\nabla\eta\triangle_h z + \eta^2\nabla\triangle_h z]\, du\, dv \\
&\quad - \int_{\Omega'} \int_0^1 A_z\, dt \triangle_h z[2\eta\nabla\eta\triangle_h z + \eta^2\nabla\triangle_h z]\, du\, dv \\
&\quad - \int_{\Omega'} \triangle_h B \eta^2 \triangle_h z\, du\, dv.
\end{aligned}$$

Inequality (30) implies the estimate

$$
\begin{aligned}
m_2 \int_{\Omega'} |\mu \nabla \triangle_h z|^2 \, du\, dv &\le c \int_{\Omega'} \eta |\nabla \eta| |\triangle_h z| |\nabla \triangle_h z| \, du\, dv \\
&+ c \int_{\Omega'} \int_0^1 |A_w| \, dt \; \eta |\nabla \eta| |\triangle_h z| \, du\, dv + c \int_{\Omega'} \int_0^1 |A_w| \, dt \; \eta^2 |\nabla \triangle_h z| \, du\, dv \\
&+ c \int_{\Omega'} \int_0^1 |A_z| \, dt |\triangle_h z|^2 \eta |\nabla \eta| \, du\, dv \\
&+ c \int_{\Omega'} \int_0^1 |A_z| \, dt \; \eta^2 |\triangle_h z| |\nabla \triangle_h z| \, du\, dv \\
&+ c \int_{\Omega'} |\triangle_h B| \eta^2 |\triangle_h z| \, du\, dv,
\end{aligned}
$$

where here and in the following c denotes some constant independent of h and R (and only depending on $m_2, \ldots, m_6$).

Definition 3 yields the estimates

$$
\begin{aligned}
|A_w| &\le c(1 + |\nabla z| + |\nabla z_h|), \\
|A_z| &\le c(1 + |\nabla z| + |\nabla z_h|)
\end{aligned}
$$

and together with (29) and the previous inequality we get

$$
\begin{aligned}
\int_{\Omega'} |\eta \nabla \triangle_h z|^2 \, du\, dv &\le c \int_{\Omega'} \eta |\nabla \eta| |\triangle_h z| |\nabla \triangle_h z| \, du\, dv \\
&+ c \int_{\Omega'} \eta |\nabla \eta| |\triangle_h z| \{1 + |\nabla z| + |\nabla z_h|\} \, du\, dv \\
&+ c \int_{\Omega'} \eta^2 |\nabla \triangle_h z| \{1 + |\nabla z| + |\nabla z_h|\} \, du\, dv \\
&+ c \int_{\Omega'} \eta |\nabla \eta| |\triangle_h z|^2 \{1 + |\nabla z| + |\nabla z_h|\} \, du\, dv \\
&+ c \int_{\Omega'} \eta^2 |\triangle_h z| |\nabla \triangle_h z| \{1 + |\nabla z| + |\nabla z_h|\} \, du\, dv \\
&+ c \int_{\Omega'} \eta^2 |\triangle_h z| \{(1 + |\nabla z|^2 + |\nabla z_h|^2)(1 + |\triangle_h z|) \\
&+ (1 + |\nabla z| + |\nabla z_h|) |\nabla \triangle_h z|\} \, du\, dv.
\end{aligned}
$$

Taking the elementary inequality $2ab \le \epsilon a^2 + \frac{1}{\epsilon} b^2$ for $\epsilon > 0$, into account we can estimate the different integrands as follows

$$\eta|\nabla\eta||\triangle_h z||\nabla\triangle_h z| \le \epsilon\eta^2|\nabla\triangle_h z|^2 + \frac{1}{\epsilon}|\nabla\eta|^2|\triangle_h z|^2,$$
$$\eta|\nabla\eta||\triangle_h z|^2 \le \eta^2|\triangle_h z|^2 + |\nabla\eta|^2|\triangle_h z|^2,$$
$$\eta|\nabla\eta||\triangle_h z||\nabla z| \le \eta^2|\nabla z|^2 + |\nabla\eta|^2|\triangle_h z|^2,$$
$$\eta|\nabla\eta||\triangle_h z|^2|\nabla z| \le \eta^2|\triangle_h z|^2|\nabla z|^2 + |\nabla\eta|^2|\triangle_h z|^2,$$
$$\eta^2|\triangle_h z||\nabla z||\nabla\triangle_h z| \le \epsilon\eta^2|\nabla\triangle_h z|^2 + \frac{1}{\epsilon}\eta^2|\nabla z|^2|\triangle_h z|^2,$$

and the other terms are treated similarly. In this way we get for $\epsilon > 0$ arbitrary

$$(31)\quad \int_{\Omega'} |\eta\nabla\triangle_h z|^2\,du\,dv \le \epsilon\int_{\Omega'} |\eta\nabla\triangle_h z|^2\,du\,dv$$
$$+ c\left(1+\frac{1}{\epsilon}\right)\int_{\Omega'} \eta^2(|\nabla z|^2 + |\nabla z_h|^2)|\triangle_h z|^2\,du\,dv$$
$$+ c\left(1+\frac{1}{\epsilon}\right)$$
$$\times \int_{\Omega'} \{\eta^2(1+|\triangle_h z|^2 + |\nabla z|^2 + |\nabla z_h|^2 + |\nabla\eta|^2|\triangle_h z|^2\}\,du\,dv.$$

For some constant c depending on $m_2, \ldots, m_6$ but not on h, R or ϵ. We observe that for $|h| < R$ we have (see e.g. Lemma 7.23 in Gilbarg and Trudinger [1])

$$\int_{B_{2R}(\zeta_0)} |\triangle_h z|^2\,du\,dv \le \int_{B_{3R}(\zeta_0)} |\nabla z|^2\,du\,dv,$$

and therefore

$$(32)\quad \int_{\Omega'} \{\eta^2(1+|\triangle_h z|^2 + |\nabla z|^2 + |\nabla z_h|^2) + |\nabla\eta|^2|\triangle_h z|^2\}\,du\,dv$$
$$\le \frac{c}{R^2}\int_{\Omega'} |\nabla z|^2\,du\,dv + cR^2.$$

Next we apply the Dirichlet-growth condition (25) which yields

$$\int_{B_{2R}(\zeta_0)\cap B_\rho(\xi)} (|\nabla z|^2 + |\nabla z_h|^2)\,du\,dv \le 2M_0\rho^{2\mu}$$

for all disks $B_\rho(\xi) \subset \mathbb{R}^2$. Now Lemma 2 in Section 2.7 applied to the functions $q(w) := |\nabla z(w)|^2 + |\nabla z_h(w)|^2 \in L_1(B_{2R}(\zeta_0))$ and to $\phi(w) := \eta\triangle_h z \in \overset{\circ}{H}{}^1_2(B_{2R}(\zeta_0), \mathbb{R}^3)$ gives the estimate

(33)
$$\begin{cases} \int_{B_{2R}(\zeta_0)} \eta^2(|\nabla z|^2 + |\nabla z_h|^2)|\triangle_h z|^2 \, du \, dv \\ \leq \overline{C}(M_0, \mu) R^{2\mu} \int_{B_{2R}(\zeta_0)} |\nabla(\eta \triangle_h z)^2| \, du \, dv \\ \leq C(M_0, \mu) R^{2\mu} \left\{ \int_{B_{2R}(\zeta_0)} \eta^2 |\nabla \triangle_h z|^2 \, du \, dv + R^{-2} \int_{\Omega'} |\nabla z|^2 \, du \, dv \right\} \end{cases}$$

for $|h| < R$, since

$$|\nabla(\eta \triangle_h z)|^2 \leq (|\nabla \eta||\triangle_h z| + \eta |\nabla \triangle_h z|)^2 \leq 2\eta^2 |\nabla \triangle_h z|^2 + \frac{8}{R^2} |\triangle_h z|^2$$

and with constants $C(M_0, \mu)$ indepent of h and R. The formulae (31), (32) and (33) yield

$$\begin{aligned} &\int_{\Omega'} |\eta \nabla \triangle_h z|^2 \, du \, dv \\ &\leq \left[\epsilon + c\left(1 + \frac{1}{\epsilon}\right) C(M_0, \mu) R^{2\mu} \right] \int_{B_{2R}} \eta^2 |\nabla \triangle_h z|^2 \, du \, dv \\ &\quad + c\left(1 + \frac{1}{\epsilon}\right) C(M_0, \mu) R^{2\mu - 2} \int_{\Omega'} |\nabla z|^2 \, du \, dv \\ &\quad + c\left(1 + \frac{1}{\epsilon}\right) \left\{ \frac{c}{R^2} \int_{\Omega'} |\nabla z|^2 \, du \, dv + cR^2 \right\}. \end{aligned}$$

By an appropriate choice of $\epsilon > 0$ and $R \in (0, R_0)$ the coefficient $[\ldots]$ can be made arbitrary small, for instance $[\ldots] < \frac{1}{2}$.

Hence the term $[\ldots] \int_{B_{2R}} \eta^2 |\nabla \triangle_h z|^2 \, du \, dv$ can be absorbed by the left hand side and we obtain an estimate of the type

$$\int_{B_{2R}(\zeta_0)} \eta^2 |\nabla \triangle_h z|^2 \, du \, dv \leq \text{const} \quad \text{for all } |h| < R$$

and some constant depending on $M_0, \mu, m_2, \ldots, m_6$ and the Dirichlet integral of z, but not on h. We conclude that the weak derivatives $D_i D_j z, \quad i, j = 1, 2$ exist and that $z \in H_2^2(B_R(\xi_0), \mathbb{R}^3)$, since $\eta \equiv 1$ on $B_R(\xi_0)$ (see e.g. Lemma 7.24 in Gilbarg and Trudinger [1]). Then a covering argument yields that $z \in H_2^2(\Omega'', \mathbb{R}^3)$ for all $\Omega'' \Subset \Omega'$, and by the Sobolev imbedding theorem we finally obtain $z \in H_s^1(\Omega'', \mathbb{R}^3)$ for any subset $\Omega'' \Subset \Omega'$ and all $s \in [1, \infty)$. □

Now we turn to the

Proof of Theorem 6. **Step I: L_2-estimates of the second derivatives.** By virtue of Theorem 4 we have $X \in C^{0,\mu}(B, \mathbb{R}^3) \cap C^0(\overline{B}, \mathbb{R}^3)$ for some $\mu > 0$. For some arbitrary point $\zeta_0 \in B$ either $X(\zeta_0) \in \partial\mathcal{K}$ or $X(\zeta_0) \in \operatorname{int} \mathcal{K}$. We treat the first case by reducing it to Lemma 1; the second case can be handled similarly. Since $\mathcal{K}$ is of class C^3 there exists a neighbourhood U of $X(\zeta_0)$ and a C^3-diffeomorphim $\psi : \mathbb{R}^3 \to \mathbb{R}^3$ with inverse χ which maps $U \cap \mathcal{K}$ onto $B_1^+(0)$, $U \cap \partial\mathcal{K}$ onto $B_1^+(0) \cap \{x_3 = 0\}$ and $X(\zeta_0)$ onto 0. For sufficiently small $\rho_0 > 0$ and $\Omega' := B_{\rho_0}(\zeta_0)$ we have $\Omega' \Subset B$ and $z := \psi \circ X \in H_2^1(\Omega', B_{1/2}^+(0)) \cap C^{0,\mu}(\Omega', \mathbb{R}^3)$. Consider any $\varphi \in C_c^0(\Omega', \mathbb{R}^3) \cap H_2^1(\Omega', \mathbb{R}^3)$ with the property $z^3(w) - \epsilon\varphi^3(w) \geq 0$ for all $w \in \Omega'$ and sufficiently small $\epsilon \geq 0$. Then the mapping $X_\epsilon := \chi(z - \epsilon\varphi) \in \mathcal{C}(\Gamma, \mathcal{K})$ for $\epsilon \in [0, \epsilon_0)$, $\epsilon_0 = \epsilon_0(\varphi)$, while clearly $X_0 = X$. By the minimum property of X we have $\mathcal{F}(X) \leq \mathcal{F}(X_\epsilon)$ for all $\epsilon \in [0, \epsilon_0)$. Introduce the integral $\tilde{\mathcal{F}}(Y) := \int_B \tilde{e}(Y, \nabla Y)\, du\, dv$, whose integrand is defined by

$$\tilde{e}(y, q) := e(\chi(y), \chi_y(y)q),$$

for $(y, q) \in \mathcal{K}^* \times \mathbb{R}^6$, $\mathcal{K}^* := \psi(\mathcal{K})$ and where $\chi_y = D\chi : \mathbb{R}^3 \to \mathbb{R}^3$ denotes the Jacobian of χ while $e(x, p) = \frac{1}{2} g_{ij}(x)[p_1^i p_1^j + p_2^i p_2^j] + \langle Q(x), p_1 \wedge p_2 \rangle$, $p = (p_1, p_2) \in \mathbb{R}^3 \times \mathbb{R}^3$.

Since $\chi : \mathbb{R}^3 \to \mathbb{R}^3$ is a C^3-diffeomorphism it is not difficult (but somewhat tedious) to prove that the functions defined by

$$A_j^\alpha(y, q) := \tilde{e}_{q_\alpha^j}(y, q)$$

and

$$B_j(y, q) := \tilde{e}_{y^j}(y, q) \quad \text{for } \alpha = 1, 2 \text{ and } j = 1, 2, 3,$$

satisfy the growth and coercivity conditions of Definition 3.

Furthermore, arguments similar to those used in the proof of the variation formula Theorem 5 show that the first variation

$$\delta\tilde{\mathcal{F}}(z, \varphi) = \lim_{\epsilon \to 0+} \frac{1}{\epsilon}(\tilde{\mathcal{F}}(z + \epsilon\varphi) - \tilde{\mathcal{F}}(z))$$

exists for functions φ considered above and is given by

$$\delta\tilde{\mathcal{F}}(z, \varphi) = \int_{\Omega'} \{A_j^\alpha(z, \nabla z)\varphi_{u^\alpha}^j + B_j(z, \nabla z)\varphi^j\}\, du\, dv.$$

By the minimality of X we infer that $\delta\tilde{\mathcal{F}}(z, \varphi) \leq 0$ is satisfied for all $\varphi \in C_c^0(\Omega', \mathbb{R}^3) \cap H_2^1(\Omega', \mathbb{R}^3)$ with $z^3 - \epsilon\varphi^3 \geq 0$ on Ω' and $0 \leq \epsilon < \epsilon_0(\varphi)$.

By Theorem 4 X satisfies a Dirichlet growth condition of the type (16), whence also $z = \psi \circ X$ fulfills the estimate

$$\int_{\Omega' \cap B_\rho(\xi)} |\nabla z|^2\, du\, dv \leq M_o \rho^{2\mu}$$

for every ball $B_\rho(\xi) \subset \mathbb{R}^2$ with constant $M_0 = \lambda(R-\rho_0)^{-2\mu} D_B(X)$ where $R = \operatorname{dist}(\xi_0, \partial B)$ and $\lambda := |\mathcal{G} \circ \chi|_{0,B_1^+}$, $\mathcal{G} := \psi_x^t \psi_x$; here we have used $\nabla z = \psi_x \circ X$ and $|\nabla z(w)|^2 = \nabla X(w) \mathcal{G}(X(w)) \nabla X(w) \le \lambda |\nabla X(w)|^2$ for all $w \in \Omega' = B_{\rho_0}(\zeta_0)$.

Now we can apply Lemma 1 and obtain $z \in H_2^2(\Omega'', \mathbb{R}^3) \cap H_s^1(\Omega'', \mathbb{R}^3)$ for all $1 \le s < \infty$ and domains $\Omega'' \Subset \Omega'$. Taking $X = \chi \circ z$ on $B_{\rho_0}(\zeta_0)$ into account, we see by a covering argument that $X \in H_2^2(\Omega', \mathbb{R}^3) \cap H_s^1(\Omega', \mathbb{R}^3)$ for all subsets $\Omega' \Subset B$ and all numbers $s \in [1, \infty)$.

Step II. **L_s-estimates of the second derivatives.**

Case 1. $X(\zeta_0) \in \operatorname{int} \mathcal{K}$.

Since X is continuous also $X(B_{R_0}(\zeta_0)) \subset \operatorname{int} \mathcal{K}$ for some $0 < R_0 \ll 1$, whence we obtain $\delta \mathcal{F}(X, \varphi) = 0$, i.e. by Theorem 5

$$\int_B \{g_{jk}(X) X_{u^\alpha}^j \varphi_{u^\alpha}^k + \frac{1}{2} \frac{\partial g_{jk}}{\partial x^e} X_{u^\alpha}^j X_{u^\alpha}^k \varphi^\ell + \operatorname{div} Q(X) \langle X_u \wedge X_v, \varphi \rangle \} \, du \, dv = 0,$$

for every φ of class $\mathring{H}_2^1(B_{R_0}(\zeta_0), \mathbb{R}^3) \cap L_\infty(B_{R_0}(\zeta_0), \mathbb{R}^3)$. Therefore, since $X \in H_{2,\text{loc}}^2(B, \mathbb{R}^3)$, the Euler equations

$$\triangle X^\ell + \Gamma_{jk}^\ell X_{u^\alpha}^j X_{u^\alpha}^k = \operatorname{div} Q(X) g^{\ell m} (X_u \wedge X_v)_m$$

hold almost everywhere on $B_{R_0}(\zeta_0)$, whence we have the estimate

$$|\triangle X(w)| \le C |\nabla X(w)|^2$$

a.e. on $B_{R_0}(\zeta_0)$ for some constant $c > 0$. Since $\nabla X \in L_{2s,\text{loc}}$ on B for all $s \in [1, \infty)$ we get $\triangle X \in L_s$ on $B_{R_0}(\zeta_0)$ and therefore conclude by standard L_p-theory (e.g. Gilbarg and Trudinger [1]) that $X \in H_s^2(B_{R_0}(\zeta_0), \mathbb{R}^3)$.

Case 2. $X(\zeta_0) \in \partial \mathcal{K}$.

Since $\mathcal{K}$ is of class C^3 there exists a neighbourhood U of $X(\zeta_0)$ and a C^3-diffeomorphism ψ of $\mathbb{R}^3$ onto itself which maps $U \cap \mathcal{K}$ onto B_1^+ and $U \cap \partial \mathcal{K}$ onto $B_1^0 := B_1^+ \cap \{x^3 = 0\}$ and $\psi(X(\zeta_0)) = 0$, $\det \psi_x > 0$.

For sufficiently small $R_0 > 0$ and $\Omega' = B_{R_0}(\zeta_0) \Subset B$ we have $z := \psi \circ X \in H_2^1(\Omega', B_{1/2}^+)$. Pick any $\varphi \in C_c^0(\Omega', \mathbb{R}^3) \cap H_2^1(\Omega', \mathbb{R}^3)$ with the property that $z^3(w) - \epsilon \varphi^3(w) \ge 0$ for all $w \in \Omega'$, provided that $\epsilon > 0$ is sufficiently small. As in Step I consider the functional

$$\tilde{\mathcal{F}}(Y) = \int_B \tilde{e}(Y, \nabla Y) \, du \, dv,$$

where $\tilde{e}(y, q) := e(\chi(y), \chi_y(y) q)$ and $q = (q_1, q_2) \in \mathbb{R}^3 \times \mathbb{R}^3$. A simple calculation shows that we have

$$\tilde{e}(y, q) = \tilde{g}_{\ell m}(y) q_\alpha^\ell q_\alpha^m + \langle \tilde{Q}(y), q_1 \wedge q_2 \rangle,$$

where

$$\begin{aligned}\tilde{g}_{\ell m} &= (g_{jk} \circ \chi)\chi^j_{y^\ell}\chi^k_{y^m} \quad \text{and}\\ \tilde{Q} &= (\det \chi_y)\chi_y^{-1}(Q \circ \chi).\end{aligned}$$

In other words, $\tilde{\mathfrak{F}}$ is of the same structure as $\mathfrak{F}$ and we can apply Theorem 5. Also we have $\delta\tilde{\mathfrak{F}}(z,\varphi) \le 0$ for all $\varphi \in C^0_c(\Omega', \mathbb{R}^3) \cap H^1_2(\Omega', \mathbb{R}^3)$ with the property that

$$z^3(w) - \epsilon\varphi^3(w) \ge 0 \quad \text{for all } w \in \Omega'$$

and $0 \le \epsilon \le \epsilon_0 = \epsilon_0(\varphi)$. In particular we are free to make arbitrary "tangential" variations, in other words

$$\delta\tilde{\mathfrak{F}}(z,\varphi) = 0 \quad \text{for all } \varphi = (\varphi^1, \varphi^2, 0) \in C^0_c(\Omega', \mathbb{R}^3) \cap H^1_2(\Omega', \mathbb{R}^3).$$

Theorem 5 now implies

$$\begin{aligned}\tilde{g}_{1j}(z)\triangle z^j + \tilde{\Gamma}_{j1k} z^j_{u^\alpha} z^k_{u^\alpha} &= \operatorname{div} \tilde{Q}(z)(z_u \wedge z_v)_1,\\ \tilde{g}_{2j}(z)\triangle z^j + \tilde{\Gamma}_{j2k} z^j_{u^\alpha} z^k_{u^\alpha} &= \operatorname{div} \tilde{Q}(z)(z_u \wedge z_v)_2\end{aligned} \tag{34}$$

a.e. on Ω', where $\tilde{\Gamma}_{j\ell k}$ are the Christoffel symbols of the first kind corresponding to $\tilde{g}_{ij}$. Introduce the "coincidence" set

$$\begin{aligned}\mathfrak{T}_z &:= \{w \in \Omega' = B_{R_0}(\zeta_0) : z^3(w) = 0\}\\ &= \{w \in \Omega' : X(w) \in \partial\mathcal{K}\}.\end{aligned}$$

By a well known property of Sobolev functions we get $\nabla z^3(w) = 0, \nabla^2 z^3(w) = 0$ a.e. on $\mathfrak{T}_z$. Hence, on account of (34)

$$\begin{aligned}\tilde{g}_{11}(z)\triangle z^1 + \tilde{g}_{12}\triangle z^2 &= \ell_1(z, \nabla z),\\ \tilde{g}_{21}(z)\triangle z^1 + \tilde{g}_{22}\triangle z^2 &= \ell_2(z, \nabla z),\\ \triangle z^3 &= 0\end{aligned} \tag{35}$$

a.e. on $\mathfrak{T}_z$, where the right hand side grows quadratically in $|\nabla z|$, i.e.

$$|\ell_1(z,\nabla z)| + |\ell_2(z,\nabla z)| \le c|\nabla z|^2 \quad \text{on } \Omega'$$

for some constant c.

The coercivity of $e(x,p)$ (cf. Assumption A) implies that

$$\tilde{m}_0|\xi|^2 \le \tilde{g}_{jk}(z)\xi^j\xi^k \le \tilde{m}_1|\xi|^2$$

for all $(z,\xi) \in \mathcal{K}^{**} \times \mathbb{R}^3, \mathcal{K}^{**} \subset \mathcal{K}^* = \psi(\mathcal{K})$, where $\tilde{m}_0 \le \tilde{m}_1$ are positive numbers. Therefore we infer from equation (35)

$$|\triangle z| \le c^*|\nabla z|^2 \quad \text{a.e. on } \mathfrak{T}_z$$

for some number c^*.

On the other hand we get as in case 1

$$\triangle z^\ell + \tilde{\Gamma}^\ell_{jk}(z) z^j_{u^\alpha} z^k_{u^\alpha} = \operatorname{div} \tilde{Q}(z) \tilde{g}^{\ell m}(z)(z_u \wedge z_v)_m$$

for $\ell = 1, 2, 3$ a.e. on the (open) set $\Omega' \setminus \mathcal{T}_z$; whence also

$$|\triangle z| \le c^{**} |\nabla z|^2 \quad \text{a.e. on } \Omega' \setminus \mathcal{T}_z.$$

Concluding we have

$$|\triangle z| \le c |\nabla z|^2 \quad \text{a.e. on } \Omega'$$

with $c := \max(c^*, c^{**})$. Now we can proceed as in case 1 and obtain $z \in H^2_s(B_R(\zeta_0), \mathbb{R}^3)$ for any $R \in (0, R_0)$ and any $s \in [1, \infty)$. This implies that $X \in H^2_s(\Omega', \mathbb{R}^3)$ for all $\Omega' \Subset B$ and all $s \in [1, \infty)$. Finally, by Sobolev imbedding theorem we infer that also $X \in C^{1,\alpha}(\Omega', \mathbb{R}^3)$ for all $\Omega' \Subset B$ and all $\alpha \in [0, 1)$. This completes the proof of Theorem 6. □

Remark. The assertion of Theorem 6 still holds true if the condition $Q \in C^2(S, \mathbb{R}^3)$ is replaced by the weaker assumption $Q \in C^1(\mathcal{K})$ and $\operatorname{div} Q \in C^1(\mathcal{K})$. This observation is of importance for the solution of Plateau's problem for H-surfaces in the set $\mathcal{K}$.

Proof of the Remark. A careful scrutinizing of the steps in the proof of Theorem 6 shows that Step II (L_p-estimates of second derivatives) only requires $Q \in C^1(\mathcal{K})$. Returning to Step I we consider the functional $\tilde{\mathcal{F}}(Y) = \int_B \tilde{e}(Y, \nabla Y)\, du\, dv$, where $\tilde{e}(y, q_1, q_2) = \tilde{g}_{\ell m}(y) q^\ell_\alpha q^m_\alpha + \langle \tilde{Q}(y), q_1 \wedge q_2 \rangle$ with $\tilde{g}_{\ell m}(y) = g_{jk}(\chi(y)) \chi^j_{y^\ell} \chi^k_{y^m}$ and $\tilde{Q}(y) = (\det \chi_y(y))[\chi_y(y)]^{-1} Q(\chi(y))$. By Theorem 5 the first variation $\delta\tilde{\mathcal{F}}(z, \varphi)$ for

$$z \in C^0_c(\Omega', B^+_{\frac{1}{2}}(0)) \cap H^1_2(\Omega', \mathbb{R}^3) \quad \text{and} \quad \varphi \in C^0_c(\Omega', \mathbb{R}^3) \cap H^1_2(\Omega', \mathbb{R}^3)$$

is given by

$$\delta\tilde{\mathcal{F}}(z, \varphi) = \int_{\Omega'} \left\{ \tilde{g}_{ij}(z) z^i_{u^\alpha} z^j_{u^\alpha} + \frac{1}{2} \frac{\partial \tilde{g}_{ij}}{\partial y^\ell} z^i_{u^\alpha} z^j_{u^\alpha} \varphi^\ell + \operatorname{div} \tilde{Q}(z) \langle z_u \wedge z_v, \varphi \rangle \right\} du\, dv,$$

where

$$\operatorname{div} \tilde{Q}(z) = \frac{\partial \tilde{Q}^1}{\partial y^1}(z) + \frac{\partial \tilde{Q}^2}{\partial y^2}(z) + \frac{\partial \tilde{Q}^3}{\partial y^3}(z).$$

Therefore, in order to apply Lemma 1 and the same arguments as in Step I in the proof of Theorem 6, it is sufficient to show that still we have $\operatorname{div} \tilde{Q}(y) \in C^1(K)$ under the weaker assumption Q, $\operatorname{div} Q \in C^1$. To this end we put

$\tilde{Q}(y) = (\det \chi_y(y))Q^*(y)$ with $Q^*(y) := [\chi_y(y)]^{-1}Q(\chi(y))$ and observe that it remains to show $\operatorname{div} Q^* \in C^1$, since χ is of class C^3. Let

$$\chi_y(y) = \begin{bmatrix} \frac{\partial \chi^1}{\partial y^1} & \frac{\partial \chi^1}{\partial y^2} & \frac{\partial \chi^1}{\partial y^3} \\ \vdots & \vdots & \vdots \\ \frac{\partial \chi^3}{\partial y^1} & \frac{\partial \chi^3}{\partial y^2} & \frac{\partial \chi^3}{\partial y^3} \end{bmatrix},$$

then, since $\psi(\chi(y)) = y$ we have $\psi_x(\chi(y)) \cdot \chi_y(y) = \mathrm{Id}$ and

$$\chi_y^{-1}(y) = \begin{bmatrix} \frac{\partial \psi^1}{\partial x^1} & \cdots & \frac{\partial \psi^1}{\partial x^3} \\ \vdots & & \vdots \\ \frac{\partial \psi^3}{\partial x^1} & \cdots & \frac{\partial \psi^3}{\partial x^3} \end{bmatrix} (\chi(y)).$$

In particular we have $\frac{\partial \psi^k}{\partial x^i}(\chi(y)) \cdot \frac{\partial \chi^j}{\partial y^k}(y) = \delta_i^j$ for $i, j = 1, 2, 3$. Next we compute

$$\begin{aligned} \frac{\partial Q^{*k}}{\partial y^i} &= \frac{\partial}{\partial y^i}\left[\frac{\partial \psi^k}{\partial x^j}(\chi(y)) Q^j(\chi(y))\right] \\ &= \frac{\partial}{\partial y^i}\left[\frac{\partial \psi^k}{\partial x^j}(\chi(y))\right] Q^j(\chi(y)) + \frac{\partial \psi^k}{\partial x^j}(\chi(y)) \frac{\partial Q^j}{\partial y^i}(\chi(y)) \\ &= \frac{\partial}{\partial y^i}\left[\frac{\partial \psi^k}{\partial x^j}(\chi(y))\right] Q^j(\chi(y)) + \frac{\partial \psi^k}{\partial x^j}(\chi(y)) \frac{\partial Q^j}{\partial x^\ell}(\chi(y)) \frac{\partial \chi^\ell}{\partial y^i}, \end{aligned}$$

i.e.

$$\begin{aligned} \operatorname{div} Q^*(y) &= \frac{\partial}{\partial y^k}\left[\frac{\partial \psi^k}{\partial x^j}(\chi(y))\right] Q^j(\chi(y)) + \frac{\partial \psi^k}{\partial x^j}(\chi(y)) \frac{\partial Q^j}{\partial x^\ell}(\chi(y)) \frac{\partial \chi^\ell}{\partial y^k} \\ &= \frac{\partial}{\partial y^k}\left[\frac{\partial \psi^k}{\partial x^j}(\chi(y))\right] Q^j(\chi(y)) + \operatorname{div} Q(\chi(y)) \end{aligned}$$

which is of class $C^1(K)$. Now Lemma 1 can be applied and the proof can be completed as in Theorem 6. □

Theorem 7 (Regularity off the coincidence set). *Suppose that Assumption A is satisfied (possibly without condition (22)), $\mathcal{K}$ is quasiregular and $g_{ij} \in C^{1,\beta}(\mathcal{K})$, $\operatorname{div} Q \in C^{0,\beta}(\mathcal{K})$ for $0 < \beta < 1$ and $i, j = 1, 2, 3$. Let X be a solution for $\mathcal{P}(\Gamma, \mathcal{K})$ in $\mathcal{C}(\Gamma, \mathcal{K})$ and put $\Omega := \{w \in B : X(w) \in \partial\mathcal{K}\}$ to denote the coincidence set. Then $X \in C^{2,\beta}(B \setminus \Omega, \mathbb{R}^3)$ and satisfies the Euler equation (24) classically on $B \setminus \Omega$.*

Proof. By Theorem 4, $X \in C^0(\overline{B}, \mathbb{R}^3)$; therefore $B \setminus \Omega$ is an open set and for each $w_0 \in B \setminus \Omega$ there is a disk $B_\rho(w_0)$ which is contained in $B \setminus \Omega$. Consequently for any testfunction $\varphi \in C_c^\infty(B_\rho(w_0), \mathbb{R}^3)$ we have $X + \epsilon\varphi \in \mathcal{C}(\Gamma, \mathcal{K})$ for all $\epsilon \in (-\epsilon_0, \epsilon_0)$, $\epsilon_0 = \epsilon_0(\varphi) > 0$ sufficiently small, and the minimizing property of X implies

$$\mathcal{F}(X) \le \mathcal{F}(X + \epsilon\varphi) \quad \text{for all } \epsilon \in (-\epsilon_0, \epsilon_0).$$

Whence $\delta\mathcal{F}(X, \varphi) = 0$ and by Theorem 5 we obtain

$$\int_B \left\{ g_{jk}(X) X^j_{u^\alpha} \varphi^k_{u^\alpha} + \frac{1}{2} \frac{\partial g_{jk}}{\partial x^\ell} X^j_{u^\alpha} X^k_{u^\alpha} \varphi^\ell + \operatorname{div} Q(X) \langle X_u \wedge X_v, \varphi \rangle \right\} du\, dv = 0.$$

Put $\varphi^j := g^{jk}(X)\psi^k$, where (g^{ij}) denotes the inverse of the matrix (g_{ij}) and $\psi = (\psi^1, \psi^2, \psi^3) \in C_c^\infty(B_\rho(w_0), \mathbb{R}^3)$ is arbitrary. A simple calculation yields

$$\int_B \left\{ X^\ell_{u^\alpha} \psi^\ell_{u^\alpha} - \Gamma^\ell_{jk}(X) X^j_{u^\alpha} X^k_{u^\alpha} \psi^\ell + \operatorname{div} Q(X) g^{\ell m}(X) (X_u \wedge X_v)_m \psi^\ell \right\} du\, dv$$
$$= 0$$

for all $\psi \in C_c^\infty(B \setminus \Omega, \mathbb{R}^3)$ applying appropriate partitions of unity. The fundamental lemma in the calculus of variations shows that (24) is the Euler equation of $\mathcal{F}$. A regularity theorem of Tomi [1] (for a similar reasoning due to Heinz see also Section 2.1 and 2.2) now implies that $X \in C^{1,\mu}(B \setminus \Omega, \mathbb{R}^3)$ for all $\mu \in (0, 1)$. Alternatively, we might also apply Theorem 6 assuming the somewhat stronger hypotheses $g_{ij} \in C^2(S)$ and $Q \in C^2(S, \mathbb{R}^3)$, where S denotes an open set containing $\mathcal{K}$. Finally classical results from potential theory yields that $X \in C^{2,\beta}(B \setminus \Omega, \mathbb{R}^3)$. □

Now we solve the Plateau problem for surfaces of prescribed mean curvature H. We start with Jordan curves Γ which are contained in a closed ball $\overline{B_R(P_0)} \subset \mathbb{R}^3$.

Theorem 8. *Let $\mathcal{K}$ be the closed ball $\overline{B_R(P_0)}$ of radius R and center P_0 and denote by H a function of class $C^{0,\beta}(\mathcal{K}), 0 < \beta < 1$, satisfying*

$$|H|_{0,\mathcal{K}} < \frac{3}{2} R^{-1} \quad \textit{and} \quad |H|_{0,\partial\mathcal{K}} \le R^{-1}.$$

Suppose $\Gamma \subset \mathcal{K}$ is a closed Jordan curve such that $\mathcal{C}(\Gamma, \mathcal{K})$ is nonempty. Then there exists a surface X of class $\mathcal{C}(\Gamma, \mathcal{K}) \cap C^{2,\beta}(B, \mathbb{R}^3) \cap C^0(\overline{B}, \mathbb{R}^3)$, which maps ∂B homeomorphically onto Γ and satisfies

$$\triangle X = 2H(X) X_u \wedge X_v \quad \textit{in } B,$$

and

$$|X_u|^2 = |X_v|^2, \quad \langle X_u, X_v \rangle = 0 \quad \textit{in } B.$$

Proof. Without loss of generality we take $P_0 = 0 \in \mathbb{R}^3$ and extend H to some ball $\overline{B}_{R+r_0}(0)$ such that $|H|_{0,B_{R+r_0}} < \frac{3}{2}(R+r_0)^{-1}$ and $|H(x)|\,|x| \le 1$ for all $x \in \overline{B}_{R+r_0} - B_R$ and some $r_0 > 0$. We remark here that the first variation formula (23) of Theorem 5 extends to cases where Q is not necessarily of class C^1 but $\operatorname{div} Q$ is defined (possibly in a weak sense). Here we define the vectorfield

$$Q(x) = \frac{2}{3}\left(\int_0^{x^1} H(\tau, x^2, x^3)\,d\tau, \int_0^{x^2} H(x^1, \tau, x^3)\,d\tau, \int_0^{x^3} H(x^1, x^2, \tau)\,d\tau\right)$$

which, although not necessarily of class $C^1(B_{R+r_0}, \mathbb{R}^3)$, satisfies $\operatorname{div} Q = 2H$. We claim that $\delta\mathcal{F}_Q(X,\varphi)$ exists for all $X \in H_2^1(B, \overline{B}_{R+r_0})$, $\varphi \in \overset{\circ}{H}{}_2^1(B,\mathbb{R}^3) \cap L_\infty(B,\mathbb{R}^3)$ and is given by (23) i.e.

$$\delta\mathcal{F}_Q(X,\varphi) = \int_B \{\langle \nabla X, \nabla\varphi\rangle + 2H(X)\langle X_u \wedge X_v, \varphi\rangle\}\,du\,dv.$$

Note that here we have written $\mathcal{F}_Q$ to indicate the dependence of $\mathcal{F}$ on Q. Now, to see that (23) holds in this case we take a sequence $H_n \in C^1(\overline{B}_{R+r_0})$ s.t. $|H_n - H|_{0,B_{R+r_0}} \to 0$, $n \to \infty$ and define $Q_n \in C^1(\overline{B}_{R+r_0}, \mathbb{R}^3)$ by

$$Q_n(x) = \frac{2}{3}\left(\int_0^{x^1} H_n(\tau, x^2, x^3)\,d\tau, \int_0^{x^2} H_n(x^1, \tau, x^3)\,d\tau, \int_0^{x^3} H_n(x^1, x^2, \tau)\,d\tau\right)$$

and

$$\mathcal{F}_{Q_n}(X) = \frac{1}{2}\int_B |\nabla X|^2\,du\,dv + \int_B \langle Q_n(X), X_u \wedge X_v\rangle\,du\,dv.$$

Relation (23) of Theorem 5 implies

$$\begin{aligned}\delta\mathcal{F}_{Q_n}(X,\varphi) &= \int_B \{\langle \nabla X, \nabla\varphi\rangle + \operatorname{div} Q_n(X)\langle X_u \wedge X_v, \varphi\rangle\}\,du\,dv \\ &= \int_B \{\langle \nabla X, \nabla\varphi\rangle + 2H_n(X)\langle X_u \wedge X_v, \varphi\rangle\}\,du\,dv,\end{aligned}$$

whence, as $n \to \infty$

$$\delta\mathcal{F}_{Q_n}(X,\varphi) \to \int_B \{\langle \nabla X, \nabla\varphi\rangle + 2H(X)\langle X_u \wedge X_v, \varphi\rangle\}\,du\,dv. \tag{36}$$

On the other hand we have

$$\frac{\mathcal{F}_Q(X+\epsilon\varphi)-\mathcal{F}_Q(X)}{\epsilon} = \frac{\mathcal{F}_{Q_n}(X+\epsilon\varphi)-\mathcal{F}_{Q_n}(X)}{\epsilon}$$
$$+\frac{1}{\epsilon}\{\mathcal{F}_Q(X+\epsilon\varphi)-\mathcal{F}_Q(X)-\mathcal{F}_{Q_n}(X+\epsilon\varphi)+\mathcal{F}_{Q_n}(X)\}$$
$$=\frac{\mathcal{F}_{Q_n}(X+\epsilon\varphi)-\mathcal{F}_{Q_n}(X)}{\epsilon}$$
$$+\frac{1}{\epsilon}\left\{\int_B \langle Q-Q_n, (X_u+\epsilon\varphi_u)\wedge(X_v+\epsilon\varphi_v)\rangle\,du\,dv\right.$$
$$\left.+\int_B \langle Q-Q_n, X_u\wedge X_v\rangle\,du\,dv\right\} = \frac{\mathcal{F}_{Q_n}(X+\epsilon\varphi)-\mathcal{F}_{Q_n}(X)}{\epsilon}$$
$$+\int_B \langle Q-Q_n, X_u\wedge\varphi_v+\varphi_u\wedge X_v\rangle\,du\,dv$$
$$+\epsilon\int_B \langle Q-Q_n, \varphi_u\wedge\varphi_v\rangle\,du\,dv.$$

Letting $\epsilon\to 0$ we find that $\delta\mathcal{F}_Q(X,\varphi)$ exists and is given by

$$\delta\mathcal{F}_Q(X,\varphi)=\delta\mathcal{F}_{Q_n}(X,\varphi)+\int_B \langle Q-Q_n, X_u\wedge\varphi_v+\varphi_u\wedge X_v\rangle\,du\,dv.$$

Since $|Q-Q_n|_{0,B_{R+r_0}}\le \text{const}|H-H_n|_{0,B_{R+r_0}}\to 0$ as $n\to\infty$ we conclude, by letting $n\to\infty$ and using (36), the first variation formula

$$\delta\mathcal{F}_Q(X,\varphi)=\int_B\{\langle\nabla X,\nabla\varphi\rangle+2H(X)\langle X_u\wedge X_v,\varphi\rangle\}\,du\,dv. \tag{37}$$

Next we observe that for every $x\in\overline{B_{R+r_0}}(0)$ we have $|Q(x)|\le\frac{2}{3}|x|\,|H|_{0,B_{R+r_0}}$, whence $|Q|_{0,B_{R+r_0}}<1$. By the discussion following Theorem 1 and by virtue of Theorems 3 and 4 we can find a solution $X\in\mathcal{C}(\Gamma,\overline{B_{R+r_0}(0)})$ of the variational problem $\mathcal{F}_Q(X)\to\min$ in the class $\mathcal{C}(\Gamma,\overline{B_{R+r_0}}(0))$, which in addition belongs to the spaces $C^{0,\alpha}(B,\mathbb{R}^3)\cap C^0(\overline{B},\mathbb{R}^3)$. Consider the function $\varphi(w):=\max(|X(w)|^2-R^2,0)\cdot X$ which is of class $\mathring{H}_2^1(B,\mathbb{R}^3)\cap L_\infty(B,\mathbb{R}^3)$ and satisfies $X-\epsilon\varphi\in\mathcal{C}(\Gamma,\overline{B_{R+r_0}(0)})$ for all $\epsilon\in[0,\epsilon_0)$, provided ϵ_0 is sufficiently small. Since X is a minimizer in that class we have $\mathcal{F}_Q(x)\le\mathcal{F}_Q(x-\epsilon\varphi)$ for all $\epsilon\in[0,\epsilon_0)$ and therefore $\delta\mathcal{F}_Q(X,\varphi)\le 0$. On the other hand we compute, using well known properties to Sobolev functions

$$\nabla\varphi=(\varphi_u,\varphi_v)=\begin{cases}2\langle X,\nabla X\rangle X+(|X|^2-R^2)\nabla X, & \text{on } \{w:|X(w)|>R\},\\ 0, & \text{on } \{w:|X(w)|\le R\}.\end{cases}$$

From the first variation formula (37) and the variational inequality $\delta\mathcal{F}_Q(X,\varphi)\le 0$ we derive

$$\int_{B\cap\{|X(w)|>R\}}\{2\langle X,X_u\rangle^2+2\langle X,X_v\rangle^2+(|X|^2-R^2)(|X_u|^2+|X_v|^2)$$
$$+2H(X)\langle X,X_u\wedge X_v\rangle(|X|^2-R^2)\}\,du\,dv\le 0. \tag{38}$$

But on the set $\{w : |X(w)| > R\}$ we have

$$|2H(X)(|X|^2 - R^2)\langle X, X_u \wedge X_v\rangle| \le (|X|^2 - R^2)|H(X)|\,|X|(|X_u|^2 + |X_v|^2) \\ \le (|X|^2 - R^2)(|X_u|^2 + |X_v|^2),$$

whence by (38) it follows that $\langle X, X_u\rangle = \langle X, X_v\rangle = 0$ a.e. on $\{w : |X(w)| > R\}$. This implies that the function $\eta(w) := \max(|X(w)|^2 - R^2, 0)$ belongs to $H_2^1(B) \cap C^0(\overline{B})$ whose derivative is

$$\nabla\eta = \begin{cases} \langle X, \nabla X\rangle & \text{on } \{|X(w)|^2 > R^2\}, \\ 0 & \text{on } \{|X(w)|^2 \le R^2\} \end{cases}$$

must vanish identically on B, since $\eta = 0$ on ∂B. Therefore $|X(w)| \le R$ on B and the coincidence set $\Omega = \{w \in B : X(w) \in \partial B_{R+r_0}\}$ is empty. Now observe that Theorem 7 is applicable here, since we have already proved the variational formula (37) to also hold in this case; furthermore we have by assumption $\operatorname{div} Q = 2H \in C^{0,\beta}(\mathcal{K})$. By Theorem 7 we get $X \in C^{2,\beta}(B, \mathbb{R}^3)$ and the system

$$\triangle X = 2H(X) X_u \wedge X_v, \\ |X_u|^2 = |X_v|^2, \quad \langle X_u, X_v\rangle = 0 \quad \text{in } B$$

is satisfied in a classical sense.

The topological character of the boundary mapping $X_{|\partial B} : \partial B \to \Gamma$ is proved similarly as in Theorem 3 of Chapter 4.5 in Vol. 1. Indeed in some neighbourhood of a boundary branch point $w_0 \in \partial B$ we have the asymptotic expansion $X_w(w) = a(w - w_0)^\nu + o(|w - w_0|^\nu)$ for some integer $\nu \ge 1$ and some $a \in \mathbb{C}^3 \setminus \{0\}$, provided X is of class C^1 in a neighbourhood $U_0 \subset \overline{B}$ of w_0 (cf. Section 2.10). Therefore $|\nabla X(w)| > 0$ for $w \in \partial B$ with $0 < |w - w_0| < \epsilon$. We conclude that $X(w)$ cannot be constant on any open arc $\Gamma_0 \subset \partial B$, because this would imply $X \in C^1(B \cup \Gamma_0, \mathbb{R}^3)$ and, because of the conformality relations, $\nabla X = 0$ on Γ_0, an obvious contradiction. □

Remark. The proof of Theorem 8 also shows the existence of a conformal weak solution $X \in C^0(\overline{B}, \mathbb{R}^3) \cap C^{1,\alpha}(B, \mathbb{R}^3)$ of the system $\triangle X = 2H(X) X_u \wedge X_v$, if H is only of class $C^0(\mathcal{K})$; also X maps ∂B homeomorphically onto Γ.

By Theorem 1 the sharpness of the existence result Theorem 8 follows if all closed curves $\Gamma \subset B_R(p_0)$ are considered. However, for certain shapes one expects better results for geometric reasons. Consider for instance a long and "thin" Jordan curve Γ, say a slightly perturbed rectangle of sidelengths ϵ and ϵ^{-1} respectively where $\epsilon > 0$ is small. Then Theorem 8 asserts the existence of a solution if $|H| < \epsilon$. However, a much better result holds in this situation.

Theorem 9. *Suppose $\mathcal{K} \subset \mathbb{R}^3$ is a closed circular cylinder $\overline{C}_R$ of radius $R > 0$ and $\Gamma \subset \overline{C}_R$ is a closed Jordan curve such that $\mathcal{C}(\Gamma, \mathcal{K})$ is nonempty. Denote* *by H a function of class $C^{0,\beta}(\mathcal{K}), 0 < \beta < 1$, satisfying $|H|_{0,\partial\mathcal{K}} \le \frac{1}{2R}$ and $|H|_{0,\mathcal{K}} < \frac{1}{R}$. Then the Plateau problem determined by H and Γ is solvable, i.e. there exists a surface $X \in \mathcal{C}(\Gamma, \mathcal{K}) \cap C^{2,\beta}(B, \mathbb{R}^3) \cap C^0(\overline{B}, \mathbb{R}^3)$ with*

$$\triangle X = 2H(X) X_u \wedge X_v \quad \text{in } B, \quad \text{and} \quad |X_u|^2 = |X_v|^2, \langle X_u, X_v \rangle = 0 \quad \text{in } B,$$

which maps ∂B homeomorphically onto Γ.

Proof. The proof is similar to the one of Theorem 8. Without loss of generality, we assume at the outset that

$$\mathcal{K} = \overline{C}_R = \{(a,b,c) \in \mathbb{R}^3 : a^2 + b^2 \le R^2\},$$

and $H \in C^0(\overline{C}_{R_0})$ for some $R_0 > R$, satisfies

$$|H|_{0,R_0} < \frac{1}{R_0}, \quad |y|\ |H(x)| \le \frac{1}{2} \tag{39}$$

for all $x = (x^1, x^2, x^3) \in \overline{C}_{R_0} \setminus C_R$ and $y := (x^1, x^2, 0)$. As vector field Q we choose

$$Q(x) := \left(\int_0^{x^1} H(\tau, x^2, x^3)\, d\tau, \int_0^{x^2} H(x^1, \tau, x^3)\, d\tau, 0 \right),$$

which again satisfies

$$\operatorname{div} Q(x) = 2H(X) \quad \text{in } C_{R_0}$$

and

$$|Q(x)| = \{(Q^1(x))^2 + (Q^2(x))^2\}^{\frac{1}{2}} \le |H|_{0,C_{R_0}} \{(x^1)^2 + (x^2)^2\}^{\frac{1}{2}} = |H|_{0,C_{R_0}} |y|.$$

Whence, by (39) it follows that $|Q|_{0,C_{R_0}} < 1$. Therefore the variational problem

$$(\mathcal{P}): \qquad \mathcal{F}(X) = \frac{1}{2} \int_B |\nabla X|^2\, du\, dv + \int_B \langle Q(X), X_u \wedge X_v \rangle\, du\, dv \to \min$$

in $\mathcal{C}(\Gamma, \overline{C}_{R_0})$ is solvable; let $X \in \mathcal{C}(\Gamma, \overline{C}_{R_0}) \cap C^{0,\alpha}(B, \mathbb{R}^3) \cap C^0(\overline{B}, \mathbb{R}^3)$ be a conformally parametrized solution (cf. Theorems 3 and 4). Denote by $Y(w) := (x^1(w), x^2(w), 0)$ the projection of $X(w)$ onto the plane $x^3 = 0$ and consider the $\overset{\circ}{H}{}_2^1(B, \mathbb{R}^3) \cap L_\infty(B, \mathbb{R}^3)$ function $\varphi(w) := \max(|Y(w)|^2 - R^2, 0) \cdot Y(w)$. We have $X - \epsilon\varphi \in \mathcal{C}(\Gamma, \overline{C}_{R_0}) \cap H_2^1(B, \overline{C}_{R_0})$ for all $\epsilon \in [0, \epsilon_0)$, provided $\epsilon_0 > 0$ is sufficiently small. Whence, by the minimality of X,

$$\mathcal{F}(X) \le \mathcal{F}(X - \epsilon\varphi) \quad \text{for all } \epsilon \in [0, \epsilon_0). \tag{40}$$

By the same reasoning as in the proof of Theorem 8 we see that the first variation $\delta\mathcal{F}(X, \varphi)$ exists and is given by (see relation (37))

$$\delta\mathcal{F}(X,\varphi) = \int_B \{\langle \nabla X, \nabla\varphi\rangle + 2H(X)\langle X_u \wedge X_v, \varphi\rangle\}\, du\, dv,$$

whence by (40) we arrive at the variational inequality

$$\int_B \{\langle \nabla X, \nabla\varphi\rangle + 2H(X)\langle X_u \wedge X_v, \varphi\rangle\}\, du\, dv \le 0.$$

Now, since

$$\nabla\varphi = (\varphi_u, \varphi_v) = \begin{cases} 2\langle Y, \nabla Y\rangle Y + (|Y|^2 - R^2)\nabla Y & \text{on } \{|Y(w)| > R\}, \\ 0 & \text{on } \{|Y(w)| \le R\} \end{cases}$$

we infer

$$(41) \qquad \begin{aligned} \delta\mathcal{F}(X,\varphi) = \int_{B\cap\{|Y|>R\}} &\{2\langle Y, Y_u\rangle^2 + 2\langle Y, Y_v\rangle^2 \\ &+ (|Y|^2 - R^2)(|Y_u|^2 + |Y_v|^2) \\ &+ 2H(X)\langle X_u \wedge X_v, Y\rangle(|Y|^2 - R^2)\}\, du\, dv \le 0. \end{aligned}$$

By virtue of the conformality relation $|X_u|^2 = |X_v|^2, \langle X_u, X_v\rangle = 0$ a.e. on B, we obtain as in the proof of Theorem 2 in Section 4.1 the inequality

$$(42) \qquad |\nabla x^3|^2 \le |\nabla x^1|^2 + |\nabla x^2|^2 = |\nabla Y|^2.$$

Whence

$$\begin{aligned} 2|H(X)\langle X_u \wedge X_v, Y\rangle| &\le 2|H(X)| \cdot |Y| \cdot \{|x_u^2 x_v^3 - x_u^3 x_v^2|^2 + |x_u^3 x_v^1 - x_u^1 x_v^3|^2\}^{\frac{1}{2}} \\ &\le 2|H(X)|\, |Y| \{|\nabla x^2|^2 |\nabla x^3|^2 + |\nabla x^1|^2 |\nabla x^3|^2\}^{\frac{1}{2}} = 2|H(X)||Y|\, |\nabla x^3|\, |\nabla Y| \\ &\le 2|H(X)|\, |Y|\, |\nabla Y|^2 \le |\nabla Y|^2 = |Y_u|^2 + |Y_v|^2 \quad \text{a.e. on } \{w : |Y(w)| > R\}, \end{aligned}$$

where we have used (42) and (39). By virtue of (41) this now implies that $\langle Y, Y_u\rangle = \langle Y, Y_v\rangle = 0$ a.e. on $\{w : |Y(w)| > R\}$. In other words, the H_2^1-function $\eta(w) := \max(|Y(w)|^2 - R^2, 0)$ has vanishing derivative a.e. in B and hence vanishes identically. This means that the coincidence set $\Omega := \{w \in B : X(w) \in \partial C_{R_0}\}$ is empty and by Theorem 7 we conclude that $X \in C^{2,\beta}(B, \mathbb{R}^3) \cap C^0(\overline{B}, \mathbb{R}^3)$ satisfies the Euler equation

$$\triangle X = 2H(X) X_u \wedge X_v \quad \text{in } B$$

and

$$|X_u|^2 = |X_v|^2, \quad \langle X_u, X_v\rangle = 0$$

in the classical sense. The rest of the proof is the same as in Theorem 8. □

Now we consider Plateau's problem for surfaces of prescribed mean curvature H and boundary Γ which are confined to arbitrary sets $\mathcal{K}$. In particular it is desirable to describe geometric conditions on H and $\mathcal{K}$ or $\partial\mathcal{K}$ respectively, which guarantee the existence of a solution to this problem. In this respect Theorem 2 and Enclosure Theorems 2 and 3 of Section 4.4 are of crucial importance. We recall the definition of the "mean curvature" function $\Lambda_\rho(x)$ for $x \in \mathcal{K}$ to denote the mean curvature at x of the surface $\mathcal{S}_{\rho(x)}$ through x which is parallel to $\partial\mathcal{K}$ at distance $\rho = \rho(x)$, if this is defined and is equal to infinity otherwise.

Theorem 10. *Suppose $\mathcal{K} \subset \mathbb{R}^3$ is the closure of a C^3 domain whose boundary $\partial\mathcal{K}$ has uniformly bounded principal curvatures and a global inward parallel surface at distance $\epsilon > 0$. Assume also that $\sup_{\mathcal{K}} \rho(x) < \infty$ and $H \in C^1(\mathcal{K})$ has uniformly bounded C^1-norm on $\mathcal{K}$ with*

$$|H(x)| \le \Lambda(x) \quad \textit{for all } x \in \partial\mathcal{K}, \tag{43}$$

and

$$|H(x)| \le (1 - a\rho(x))\Lambda_\rho(x) + \frac{a}{2} \quad \textit{for all } x \in \mathcal{K} \tag{44}$$

and some number a, $0 \le a \le \inf_{\mathcal{K}} \rho^{-1}(x)$. Finally let $\Gamma \subset \mathcal{K}$ denote a closed Jordan curve such that $\mathcal{C}(\Gamma, \mathcal{K}) \ne \emptyset$. Then there exists a solution $X \in C^{2,\alpha}(B, \mathbb{R}^3) \cap C^0(\overline{B}, \mathcal{K})$ of the Plateau problem which is determined by H and Γ. Furthermore X satisfies the H-surface system 1) *and* 2) *classically in B and maps the boundary of B homeomorphically onto Γ. Moreover, if in addition*

$$|H(x)| \le \Lambda_\rho(x) \tag{45}$$

holds for all x in a small strip in $\mathcal{K}$ near $\partial\mathcal{K}$ and $\Gamma \cap$ int $\mathcal{K} \ne \emptyset$, then every solution X maps B into the interior of $\mathcal{K}$. Finally, if for some point $x_0 \in \partial\mathcal{K}$ we have

$$|H(x_0)| < \Lambda(x_0), \tag{46}$$

then there is a neighbourhood $U(x_0) \subset \mathbb{R}^3$ such that no $w_0 \in B$ is mapped into $U(x_0)$. In particular if (46) *holds true for all $x_0 \in \mathcal{K}$, then $X(B) \subset \operatorname{int}\mathcal{K}$. (Clearly,* (45) *follows from* (44)*, if $a = 0$.)*

Proof. First we remark that $\mathcal{K}$ is quasiregular; for a proof see Lemma 2.4 in Gulliver and Spruck [2]. Furthermore by Theorem 2 there is a vector field $Q \in C^1(\mathcal{K}, \mathbb{R}^3)$ which satisfies

$$\operatorname{div} Q(x) = 2H(x) \quad \text{for all } x \in \mathcal{K}$$

and $|Q|_{0,\mathcal{K}} < 1$. Now Theorems 3, 4 and 6, in particular the Remark at the end of the proof of Theorem 6 imply the existence of a conformally parametrized solution $X \in \mathcal{C}(\Gamma, \mathcal{K}) \cap H^2_{s,\mathrm{loc}}(B, \mathbb{R}^3) \cap C^{1,\alpha}(B, \mathbb{R}^3) \cap C^0(\overline{B}, \mathbb{R}^3)$, for all $s < \infty$, and $0 < \alpha < 1$, of the variational problem

$$\mathcal{P}(\Gamma, \mathcal{K}): \quad \mathcal{F}(X) = \frac{1}{2} \int_B |\nabla X|^2 \, du \, dv + \int_B \langle Q(X), X_u \wedge X_v \rangle \, du \, dv \to \min$$

$$\text{in } \mathcal{C}(\Gamma, \mathcal{K}).$$

By Theorem 5 the first variation $\delta\mathcal{F}(X, \varphi)$ exists, is given by

$$\delta\mathcal{F}(X, \varphi) = \int_B \{\langle \nabla X, \nabla \varphi \rangle + 2H(X) \langle X_u \wedge X_v, \varphi \rangle\} \, du \, dv$$

and satisfies – since X is a minimum of $\mathcal{F}$ in $\mathcal{C}(\Gamma, \mathcal{K})$ – the relation

$$\delta\mathcal{F}(X, \varphi) \geq 0$$

for all $\varphi \in \overset{\circ}{H}{}^1_2 (B, \mathbb{R}^3) \cap L_\infty(B)$ such that $(X + \epsilon\varphi) \in \mathcal{C}(\Gamma, \mathcal{K})$. Assumption (43) together with Enclosure Theorem 3 of Section 4.4 yield that $X \in H^2_{s,\mathrm{loc}}(B, \mathbb{R}^3) \cap C^{1,\alpha}(B, \mathbb{R}^3)$ satisfies the system

$$\triangle X = 2H(X) X_u \wedge X_v$$

almost everywhere in B. Since the right hand side is Hölder continuous it follows from Schauder theory that X is of class $C^{2,\alpha}(B, \mathbb{R}^3)$ and satisfies the H-surface system in a classical sense.

By Enclosure Theorem 2 of Section 4.4 and since $X \in C^0(\overline{B}, \mathbb{R}^3)$, we see that $X(B) \subset \operatorname{int} \mathcal{K}$, if (45) holds and $\Gamma \cap \operatorname{int} \mathcal{K} \neq \emptyset$. The rest of the assertion is a consequence of Corollary 3 in Section 4.4. That the boundary mapping $X_{|\partial B} : \partial B \to \Gamma$ is a homeomorphism follows in a standard manner. Theorem 10 is completely proved. □

Let us close this section with a simple example when $\mathcal{K} = \{\xi \in \mathbb{R}^3 : |\xi| \leq R\}$ is the closed ball of radius R and center zero. Formula (44) then is equivalent to

$$|H(x)| \leq (1 - a(R - |x|)) \frac{1}{|x|} + \frac{a}{2},$$

where $0 \leq a \leq R^{-1}$; or

$$|H(x)| \leq \frac{1}{|x|}(1 - aR) + \frac{3a}{2}$$

for all $x \in \mathcal{K}$. For $a = R^{-1}$ we recover the result of Theorem 8, while a new existence result is obtained when $a = 0$. In this case the condition requires $|H(x)| \leq \frac{1}{|x|}$ for all $x \in \mathcal{K}$, whence we obtain the existence of an H-surface in $\mathcal{K}$ which lies strictly interior to $\mathcal{K}$ if $\Gamma \cap \operatorname{int} \mathcal{K} \neq \emptyset$.

4.8 Surfaces of Prescribed Mean Curvature in a Riemannian Manifold

In this section we shall extend the methods which we have introduced in Section 4.7 to surfaces of prescribed mean curvature in a three-dimensional Riemannian manifold. We assume the reader's acquaintance with basic Riemannian geometry; however we repeat some of the underlying concepts and calculations when assumed necessary. In particular we discuss in this section estimates for Jacobi fields. As standard reference on differential geometry we refer to the monographs by Gromoll, Klingenberg, and Meyer [1], do Carmo [3], Jost [18], and Kühnel [2], and we also refer to Chapter 1 of Vol. 1, where most of the formulas needed later can also be found. In what follows we shall assume, unless stated otherwise, that M is a three-dimensional, connected, orientable, and complete Riemannian manifold of class C^4 with scalar product $\langle X, Y\rangle$ and norm $\|X\| = \langle X, X\rangle^{\frac{1}{2}}$ for $X, Y \in T_pM$, $p \in M$, where T_pM denotes the tangent space of M at p. Observe that this notation contrasts with the one in the last section, where $\langle\cdot,\cdot\rangle$ has denoted the Euclidean scalar product, which in this chapter will simply be written as $X \cdot Y$.

If $\varphi : U \to \mathbb{R}^3$, $U \subset M$ an open set, denotes a chart we let $x = (x^1, x^2, x^3) = \varphi(p)$ stand for the local coordinates and $\partial_k = \frac{\partial}{\partial x^k} = X_k$ denote their basis fields. We put

$$
\begin{aligned}
g_{ij}(x) &= \langle \partial_i, \partial_j\rangle = \langle X_i, X_j\rangle, \quad g(x) = \det(g_{ij}(x)),\\
(g^{ij})_{i,j} &= (g_{ij})^{-1}_{i,j}, \quad D_{\partial_i}\partial_j = \Gamma^\ell_{ij}\partial_\ell = D_{X_i}X_j = \Gamma^\ell_{ij}X_\ell
\end{aligned}
$$

and $\Gamma_{ijk} = \langle D_{\partial_i}\partial_k, \partial_j\rangle$, compare the formulas in Vol. 1, Section 1.5. Here D denotes covariant differentiation on M, g_{ij} is the metric and Γ_{ijk}, Γ^k_{ij} stand for the Christoffel symbols. From Chapter 1 we recall the relation

$$
\Gamma^k_{ij} = g^{km}\Gamma_{imj} \quad \text{and} \quad \Gamma_{ijk} = \frac{1}{2}\left\{\frac{\partial g_{jk}}{\partial x^i} - \frac{\partial g_{ik}}{\partial x^j} + \frac{\partial g_{ij}}{\partial x^k}\right\}.
$$

A mapping $f : B \to M$ of the unit disk B into M represents a surface of (prescribed) mean curvature H in M, if it is of class C^2 and any local representation $X(w) = \varphi \circ f(w)$ satisfies in B (or a suitable subset of B) the system

$$
\triangle X^\ell + \Gamma^\ell_{jk}X^j_{u^\alpha}X^k_{u^\alpha} = 2H(X)\sqrt{g(x)}g^{\ell m}(X)(X_u \wedge X_v)_m
$$

for $\ell = 1, 2, 3$, and the conformality condition

$$
g_{ij}X^i_uX^j_u = g_{ij}X^i_vX^j_v, \quad g_{ij}X^i_uX^j_v = 0.
$$

We shall confine ourselves to surfaces which are contained in a "Riemann normal chart" (φ, U) with center $p \in M$. Here (φ, U) is called a Riemann normal chart with center p, if $U \subset M$ is an open set with $p \in U$ and $\varphi : U \to$

$\mathbb{R}^3$ is of the form $\varphi = j \circ \exp_p^{-1}$, where $\exp_p : T_pM \to M$ is the exponential map with center p and $j : T_pM \to \mathbb{R}^3$ is a linear isometry. Recall that the map $\exp_p : T_pM \to M$ is defined by $\exp_p(v) = c(1)$ for $v \in T_pM$, where $c = c(t)$ is the geodesic in M with $c(0) = p$ and $\dot{c}(0) = v$. Hence every point $q \in U$ can be connected with p by exactly one shortest geodesic which is the image of a straight line through 0 in T_pM under the exponential map $\exp_p$.

Since we want to solve the Plateau problem for surfaces of prescribed mean curvature in M via a minimization procedure of the functional $\mathcal{F}(X)$ which we have investigated in Section 4.7, it is of crucial importance to have a quantitative control of the metric tensor and the Christoffel symbols in terms of the curvature of the underlying manifold M. This will be established by invoking estimates for Jacobi fields along geodesics. These estimates are of independent interest and will be of importance later in Subsection 4.8.3.

4.8.1 Estimates for Jacobi Fields

Throughout this subsection we assume that M is a complete m-dimensional Riemannian manifold of class C^4 with covariant derivative D and Riemann curvature tensor $R(X,Y)Z$ (for a definition and properties of R, see e.g. Vol. 1, Sections 1.3 and 1.5). A geodesic $c(t)$ starting for $t = 0$ at $p \in M$ is then defined for all times $t \geq 0$.

A vector field J along a geodesic $c : [0, \infty) \to M$ with $\dot{c}(0) \neq 0$ is said to be a *Jacobi field* along c if it satisfies

$$(1) \qquad \frac{D}{dt}\frac{D}{dt} J + R(J, \dot{c})\dot{c} = 0.$$

If no misunderstanding is possible, we shall abbreviate both the ordinary derivation $\frac{d}{dt}$ and the covariant derivation $\frac{D}{dt}$ with a superscript dot. Then (1) takes the form

$$(1') \qquad \ddot{J} + R(J, \dot{c})\dot{c} = 0.$$

Here $R(X,Y)Z$ denotes the Riemann curvature tensor of M. The linear equation (1), the so-called *Jacobi equation* of the geodesic c, is nothing but the Euler equation of the second variation of the Dirichlet integral $\int \langle \dot{c}, \dot{c} \rangle \, dt$ at c. In local coordinates, the Jacobi equation is equivalent to the system of m linear ordinary differential equations of second order

$$\ddot{\eta}^k + R^k_{\ell r s}(c)\eta^\ell \dot{c}^r \dot{c}^s = 0$$

for the unknown functions $\eta^k(t)$, $k = 1, \ldots, m$. Thus the Jacobi fields along a geodesic c span a $2m$-dimensional linear space over $\mathbb{R}$ which we denote by J_c. In particular, the tangent vector $\dot{c}$ of a geodesic c is a Jacobi field of constant length $\|\dot{c}(0)\|$ along c, since

$$\frac{D}{dt}\dot{c} = 0, \quad R(\dot{c}, \dot{c})\dot{c} = 0$$

and

$$\frac{d}{dt}\|\dot{c}\|^2 = 2\left\langle \dot{c}, \frac{D}{dt}\dot{c}\right\rangle = 0.$$

Moreover, if J and $J^* \in J_c$, then

$$\begin{aligned}\frac{d}{dt}\left\{\langle \dot{J}, J^*\rangle - \langle J, \dot{J}^*\rangle\right\} &= \langle \ddot{J}, J^*\rangle - \langle J, \ddot{J}^*\rangle \\ &= -\langle R(J,\dot{c})\dot{c}, J^*\rangle + \langle R(J^*,\dot{c})\dot{c}, J\rangle = 0.\end{aligned}$$

We therefore obtain

$$\langle \dot{J}, J^*\rangle - \langle J, J^*\rangle = \text{const} \quad \text{for all } J, J^* \in J_c$$

and in particular, for $J^* = \dot{c}$, we arrive at

$$(2) \qquad \langle \dot{J}, \dot{c}\rangle = \text{const} \quad \text{for all } J \in J_c.$$

Suppose now that $c : [0,\infty) \to M$ is a geodesic normalized by the condition $\|\dot{c}\| = 1$. Then, by setting

$$J^T = \alpha\dot{c}, \quad \alpha = \langle J, \dot{c}\rangle, \quad J^\perp = J - J^T,$$

we can decompose each Jacobi field $J \in J_c$ into a tangential component J^T and a normal component $J^\perp$:

$$J = J^T + J^\perp.$$

We claim that both J^T and $J^\perp$ are Jacobi fields. In fact, equation (2) implies $\ddot{\alpha} = 0$, and therefore $(J^T)^{\cdot\cdot} + R(J^T,\dot{c})\dot{c} = (J^T)^{\cdot\cdot} = (\alpha\dot{c})^{\cdot\cdot} = (\dot{\alpha}\dot{c})^{\cdot} = \ddot{\alpha}\dot{c} = 0$ if we take $\ddot{c} = 0$ into account.

The tangential part J^T is of the form

$$(3) \qquad J^T(t) = \{at + b\}\dot{c}(t),$$

where

$$(3') \qquad a = \langle \dot{J}(0), \dot{c}(0)\rangle, \quad b = \langle J(0), \dot{c}(0)\rangle.$$

Thus the growth of the tangential part $J^T(t)$ can easily be determined from the initial values $J(0)$ and $\dot{J}(0)$.

Hence we can control the growth of all Jacobi fields if we can estimate the normal Jacobi fields. These are the elements of J_c orthogonal to $\dot{c}$ which, by (3), span a $(2m-2)$-dimensional subspace of J_c that is denoted by $J_c^\perp$.

Unfortunately, there is no simple way to compute the normal Jacobi fields, yet they can fairly well be estimated in terms of upper and lower bounds on the sectional curvature of M. To see this, we consider the solutions of the scalar differential equation

$$\ddot{f} + \kappa f = 0, \quad \kappa \in \mathbb{R},$$

which also satisfy

$$-\left(\frac{\dot{f}}{f}\right)^{\cdot} = \kappa + \left(\frac{\dot{f}}{f}\right)^2,$$

wherever f does not vanish. In particular the solutions s_κ and c_κ of the initial value problems

$$\boxed{\begin{aligned} &\ddot{s}_\kappa + \kappa s_\kappa = 0 \\ &s_\kappa(0) = 0, \ \dot{s}_\kappa(0) = 1 \end{aligned}} \quad \text{and} \quad \boxed{\begin{aligned} &\ddot{c}_\kappa + \kappa c_\kappa = 0 \\ &c_\kappa(0) = 1, \ \dot{c}_\kappa(0) = 0 \end{aligned}}$$

We have

$$\begin{aligned} &s_\kappa(t) = t, && c_\kappa(t) = 1 && \text{if } \kappa = 0, \\ &s_\kappa(t) = \frac{1}{\sqrt{\kappa}} \sin \sqrt{\kappa} t, && c_\kappa(t) = \cos \sqrt{\kappa} t && \text{if } \kappa > 0, \\ &s_\kappa(t) = \frac{1}{\sqrt{-\kappa}} \sinh \sqrt{-\kappa} t, && c_\kappa(t) = \cosh \sqrt{-\kappa} t && \text{if } \kappa < 0. \end{aligned}$$

Put

$$t_\kappa = \begin{cases} \dfrac{\pi}{\sqrt{\kappa}} & \text{if } \kappa > 0, \\ +\infty & \text{if } \kappa \le 0 \end{cases}$$

that is, t_κ is the first positive zero of $s_\kappa(t)$.

Lemma 1. *Let $c : [0, \infty) \to M$ be a geodesic with $\|\dot{c}\| = 1$, and suppose that some $J \in J_c^\perp$ satisfies $\|J\| > 0$ on $(0, t^*)$. Finally we assume that, for some number κ, the sectional curvature K of M is bounded on $\Gamma_{t^*} = \{c(t) : 0 \le t \le t^*\}$ by the inequality $K \le \kappa$. Then $\|J\|$ satisfies the differential inequality*

(4) $$\frac{d^2}{dt^2}\|J\| + \kappa \|J\| \ge 0 \quad \text{on } (0, t^*).$$

Proof. We first obtain

(5) $$\frac{d}{dt}\|J\| = \|J\|^{-1}\langle J, \dot{J}\rangle,$$

whence

$$\begin{aligned} \frac{d^2}{dt^2}\|J\| &= \|J\|^{-1}\langle J, \ddot{J}\rangle + \|J\|^{-1}\|\dot{J}\|^2 - \|J\|^{-3}\langle J, \dot{J}\rangle^2 \\ &= \|J\|^{-1}\langle J, \ddot{J}\rangle + \|J\|^{-3}\left\{\|J\|^2\|\dot{J}\|^2 - \langle J, \dot{J}\rangle^2\right\}, \end{aligned}$$

and, by Schwarz's inequality, we arrive at

$$\frac{d^2}{dt^2}\|J\| \geq \|J\|^{-1}\langle J, \ddot{J}\rangle. \tag{6}$$

The Jacobi equation (1′), on the other hand, implies

$$\langle J, \ddot{J}\rangle = -\langle R(J, \dot{c})\dot{c}, J\rangle.$$

The term on the right hand side is nothing but $-K\|J\|^2$, where $K = K(t)$ denotes the sectional curvature of M at $c(t)$ with respect to the two-plane spanned by $J(t)$ and $\dot{c}(t)$. Thus we find

$$\langle J, \ddot{J}\rangle\|J\|^{-1} = -K\|J\| \geq -\kappa\|J\|. \tag{7}$$

Finally, (4) follows from (6) and (7).

Lemma 2. *Let the assumption of Lemma 1 be satisfied. If, moreover, we assume that $J(0) = 0$ and $t^* \leq t_\kappa$, then*

$$\frac{d}{dt}\left\{\frac{\|J\|}{s_\kappa}\right\} \geq 0 \quad \text{on } (0, t^*). \tag{8}$$

Proof. Set

$$Z = \|J\|^{\cdot} s_\kappa - \|J\|\dot{s}_\kappa.$$

Then, for $0 < t < t^*$, we obtain $Z(t) \geq 0$ since

$$\dot{Z} = \|J\|^{\cdot\cdot} s_\kappa - \|J\|\ddot{s}_\kappa = s_\kappa\left\{\|J\|^{\cdot\cdot} + \kappa\|J\|\right\} \geq 0,$$

if we take (4) into account. Hence, for any $t_0 \in (0, t^*)$, we infer that

$$Z(t) \geq Z(t_0) \quad \text{for all } t \in (t_0, t^*).$$

Moreover, (5) yields

$$\|J\|^{\cdot} \leq \|\dot{J}\|,$$

and therefore

$$|Z| \leq \|\dot{J}\|s_\kappa + \|J\|\ |\dot{s}_\kappa| \quad \text{on } (0, t^*).$$

As $t_0 \to +0$, we have $s_\kappa(t_0) \to 0$ and $\|J(t_0)\| \to 0$, whence $Z(t_0) \to 0$ and $Z \geq 0$ on $(0, t^*)$. Then the desired inequality (8) follows from

$$\frac{d}{dt}\left\{\frac{\|J\|}{s_\kappa}\right\} = \frac{Z}{s_\kappa^2}.$$

Theorem 1. *Let $c : [0,\infty) \to M$ be a geodesic with $\|\dot{c}\| = 1$, and let J be a normal Jacobi field along c which satisfies $J(0) = 0$. We moreover suppose that the sectional curvature K of M has an upper bound κ on $\Gamma_{t_\kappa} = \{c(t) : 0 \le t \le t_\kappa\}$. Then*

$$\|\dot{J}(0)\| s_\kappa(t) \le \|J(t)\| \quad \text{for all } t \in [0, t_\kappa). \tag{9}$$

Proof. If $\dot{J}(0) = 0$, (9) obviously is correct. We therefore may assume that $\|\dot{J}(0)\| > 0$, whereas $J(0) = 0$. Then there is a number $t^* \in (0, t_\kappa)$ such that $\|J\| > 0$ on $(0, t^*)$, and Lemma 2 implies

$$\frac{\|J\|}{s_\kappa}(t_0) \le \frac{\|J\|}{s_\kappa}(t) \quad \text{for } 0 < t_0 \le t < t^*.$$

As t_0 tends to $+0$, the quotient on the left hand side is an expression of the kind $\frac{0}{0}$ which, according to L'Hospital's rule, is determined by

$$\lim_{t_0 \to +0} \frac{\|J\|^2}{s_\kappa^2} = \lim_{t_0 \to +0} \frac{\frac{d}{dt}\|J\|^2}{\frac{d}{dt} s_\kappa^2} = \lim_{t_0 \to +0} \frac{\frac{d^2}{dt^2}\|J\|^2}{\frac{d^2}{dt^2} s_\kappa^2} = \|\dot{J}(0)\|^2,$$

since

$$\begin{aligned}
&\frac{d}{dt} s_\kappa^2(t_0) \to 0, \quad \frac{d^2}{dt^2} s_\kappa^2(t_0) \to 2,\\
&\frac{d}{dt}\|J\|^2(t_0) = 2\langle J, \dot{J}\rangle(t_0) \to 0,\\
&\frac{d^2}{dt^2}\|J\|^2(t_0) = 2\left\{\|\dot{J}\|^2 + \langle \ddot{J}, J\rangle\right\}(t_0)\\
&\qquad\qquad = 2\left\{\|\dot{J}\|^2 - \langle R(J,\dot{c})\dot{c}, J\rangle\right\}(t_0) \to 2\|\dot{J}(0)\|^2,
\end{aligned}$$

and (9) is proved for $0 \le t \le t^*$. We then conclude that $J(t)$ cannot vanish before t_κ, and thus (9) must hold for all $t \in [0, t_\kappa)$.

By the same reasoning, we can prove

Theorem 1′. *Let $c : [0,\infty) \to M$ be a geodesic with $\|\dot{c}\| = 1$, and let $J \in J_c^\perp$. Suppose also that the sectional curvature K satisfies $K \le \kappa$ on $\Gamma_{\tau_\kappa} = \{c(t) : 0 \le t \le \tau_\kappa\}$ where τ_κ is the first positive zero of*

$$\varphi(t) = \|J(0)\| c_\kappa(t) + \|J\|^{\cdot}(0) s_\kappa(t),$$

and $\|J\|^{\cdot}(0) = \|J\|^{-1}\langle J, \dot{J}\rangle(0)$. We then obtain

$$\varphi(t) \le \|J(t)\| \quad \text{for } 0 \le t < \tau_\kappa \tag{10}$$

and

$$\|J(t)\| \le \frac{\|J(t^*)\|}{\varphi(t^*)}\varphi(t) \quad \text{for all } t \in [0, t^*], \tag{11}$$

where $0 < t^* < \tau_\kappa$.

Remark. Here we have assumed that $J(0) \neq 0$; the case $J(0) = 0$ is handled by a limit consideration.

We now turn to another class of Jacobi field estimates derived from a lower bound on the sectional curvature of M.

To this end, let $c : [0, \infty) \to M$ again be a unit speed geodesic, and let $X_1, X_2, \dots, X_m$ be m parallel vector fields along c which, at every point $c(t)$ of the geodesic, yield on orthogonal frame of the tangent space $T_{c(t)}M$. In other words, we have

$$\dot{X}_k = 0 \quad \text{and} \quad \langle X_k, X_\ell \rangle = \delta_{k\ell}.$$

Then every vector field U along c can be written as

$$U(t) = u^k(t) X_k(t).$$

If we identify $\mathbb{R}^m$ with $T_{c(0)}M$ and introduce the vector function $u : [0, \infty) \to \mathbb{R}^m$ by

$$u(t) = \left(u^1(t), \dots, u^m(t)\right)^T$$

we obtain a 1-1-correspondence between the vector functions $u : [0, \infty) \to \mathbb{R}^m$ and the vector fields U along c given by parallel translation.

To any $m \times m$-matrix function $B(t) = (b^\ell_k(t))$ which acts on vector functions $u(t)$ according to $(B(t)u(t))^\ell = b^\ell_k(t) u^k(t)$, we can associate an operator, again called B, acting on vector fields $U = u^k X_k$ by the rule

$$(BU)(t) = (B(t)u(t))^\ell X_\ell(t)$$

if the vectorfield U is identified with the function u.

We, in particular, can associate with every Jacobi field $J = J^k X_k$ a vector function $I = (J^1, \dots, J^m)$ which satisfies

$$\ddot{I} + R_c I = 0, \tag{12}$$

where the matrix function $R_c(t) = (R^s_k(t))$ is defined by

$$R^s_k = R^s_{k\ell r} \dot{c}^\ell \dot{c}^r,$$

where

$$\dot{c} = \dot{c}^k X_k \quad \text{and} \quad R(J, \dot{c})\dot{c} = R^s_{k\ell r} J^k \dot{c}^\ell \dot{c}^r X_s.$$

The well known symmetry relation

$$\langle R(U, \dot{c})\dot{c}, V \rangle = \langle R(V, \dot{c})\dot{c}, U \rangle$$

implies the symmetry of R_c.

Next we choose a basis $J_1, \dots, J_m$ of the m-dimensional subspace $\mathring{J}_c := \{J \in J_c : J(0) = 0\}$ of J_c with $\dot{J}_k(0) = X_k(0)$. By Theorem 1 and by (3), the tangent vectors $J_1(t), \dots, J_m(t)$ are linearly independent for all $t \in (0, t_\kappa)$ if we assume $K \leq \kappa$. Let now I_k be the vector functions corresponding to the Jacobi vectors J_k. Then the matrix $A(t)$, defined by

$$A = (I_1, I_2, \dots, I_m),$$

is invertible and satisfies

$$\ddot{A} + R_c A = 0, \quad A(0) = 0, \quad \dot{A}(0) = 1, \tag{13}$$

where 1 denotes the unit matrix (δ_k^ℓ). We therefore can define the matrix function

$$S(t) = -\dot{A}(t)A^{-1}(t) \quad \text{for } t \in (0, t_\kappa),$$

which satisfies the Riccati equation

$$\dot{S} = R_c + S^2, \tag{14}$$

since the differentiation of $AA^{-1} = 1$ and $S = -\dot{A}A^{-1}$ yields $(A^{-1})^\cdot = -A^{-1}\dot{A}A^{-1}$ and $\dot{S} = -\ddot{A}A^{-1} - \dot{A}(A^{-1})^\cdot = -\ddot{A}A^{-1} + (\dot{A}A^{-1})^2$, and from (13) we infer $\ddot{A}A^{-1} = -R_c$. Moreover,

$$S(t) = -t^{-1} \cdot 1 + 0(1) \quad \text{as } t \to +0 \tag{15}$$

since $A(t) = t \cdot 1 + \cdots$ and $\dot{A}(t) = 1 + \cdots$.

We also claim that $S(t)$ is a symmetric operator on $T_{c(t)}M$, i.e. we must prove that

$$\langle S(t_0)U_0, V_0\rangle = \langle U_0, S(t_0)V_0\rangle$$

holds for every $t_0 \in (0, t_\kappa)$ and for each pair of tangent vectors $U_0 = u_0^k X_k(t_0)$, $V_0 = v_0^k X_k(t_0) \in T_{c(t_0)}M$.

But, if we introduce the two parallel vector fields $U(t) = u^k X_k(t)$ and $V(t) = v^k X_k(t)$ with $u = A^{-1}(t_0)u_0$ and $v = A^{-1}(t_0)v_0$, this is equivalent to saying that the function

$$\phi = \langle \dot{A}U, AV\rangle - \langle AU, \dot{A}V\rangle$$

vanishes for $t = t_0$, which is proved by showing that ϕ identically vanishes on $(0, t_\kappa)$. In fact, we infer from the definition of ϕ that

$$\lim_{t \to +0} \phi(t) = 0,$$

and, on the other hand, ϕ is constant because of

$$\begin{aligned}\dot{\phi} &= \langle \ddot{A}U, AV\rangle - \langle AU, \dot{A}V\rangle \\ &= -\langle R_c AU, AV\rangle + \langle AU, R_c AV\rangle = 0.\end{aligned}$$

Let now J be an arbitrary normal Jacobi field in $\mathring{J}_c$, and let I be the associated vector function. Then we infer from $\dot{A} = -SA$ that

$$\dot{I} = -SI \quad \text{or} \quad \dot{J} = -SJ \tag{16}$$

holds on $(0, t_\kappa)$. We fix some $t_0 \in (0, t_\kappa)$ and set $U_0 = u_0^k X_k(t_0) = \|J(t_0)\|^{-1} J(t_0)$. Moreover, we define a parallel vector field U along c with $U(t_0) = U_0$ by setting $U(t) = u_0^k X_k(t)$. Then we claim that the function

$$k(t) = \langle SU, U\rangle(t)$$

satisfies

$$-k \le \frac{\dot{s}_\omega}{s_\omega} \quad \text{on } (0, t_\kappa) \tag{16$'$}$$

provided that $\omega \le K \le \kappa$ is assumed. We also note that $\omega \le \kappa$ implies $t_\kappa \le t_\omega$.

From (16′) we infer that

$$-\frac{\langle SJ, J\rangle}{\|J\|^2}(t) \le \frac{\dot{s}_\omega}{s_\omega}(t)$$

holds for $t = t_0$. Since t_0 was arbitrary, this inequality is true for all $t \in (0, t_\kappa)$, and, together with (16), we arrive at

$$\frac{\|J\|^{\cdot}}{\|J\|} = \frac{\langle J, \dot{J}\rangle}{\|J\|^2} = -\frac{\langle J, SJ\rangle}{\|J\|^2} \le \frac{\dot{s}_\omega}{s_\omega}$$

which is to hold on $(0, t_\kappa)$.

On the other hand, by repeating the proof of Lemma 2 and by taking Theorem 1 into account, we obtain

$$Z = \|J\|^{\cdot} s_\kappa - \|J\| \dot{s}_\kappa \ge 0 \quad \text{on } (0, t_\kappa).$$

Hence we have

Theorem 2. *Let J be a normal Jacobi field with $J(0) = 0$ along a unit speed geodesic $c : [0, \infty) \to M$, and suppose that $\omega \le K \le \kappa$ holds on the set $\{c(t) : t \in (0, t_\kappa)\}$. Then we may conclude that*

$$\frac{\dot{s}_\kappa}{s_\kappa} \le \frac{\langle J, \dot{J}\rangle}{\|J\|^2} \le \frac{\dot{s}_\omega}{s_\omega} \quad \textit{on } (0, t_\kappa). \tag{17}$$

It remains to prove (16′). We first note that $\|U\| = 1$ and $\langle U, \dot{c}\rangle = 0$ hold on $[0, \infty)$, since these relations are true for $t = t_0$, and $U, \dot{c}$ are parallel.

Thus we get

$$\omega \le \langle R(U, \dot{c})\dot{c}, U\rangle,$$

and

$$\langle SU, U\rangle^2 \le \|SU\|^2 = \langle S^2 U, U\rangle.$$

Furthermore, (14) yields

$$\begin{aligned} \frac{d}{dt}\langle SU, U\rangle &= \langle R_c U, U\rangle + \langle S^2 U, U\rangle \\ &= \langle R(U, \dot{c})\dot{c}, U\rangle + \langle S^2 U, U\rangle, \end{aligned}$$

and therefore

$$\dot{k} \ge \omega + k^2 \quad \text{on } (0, t_\kappa). \tag{18}$$

Consider the function

$$h = s_\omega k + \dot{s}_\omega$$

which then satisfies

$$\dot{h} \ge hk, \tag{19}$$

as we see from

$$\dot{h} = \dot{s}_\omega k + s_\omega \dot{k} + \ddot{s}_\omega \ge \dot{s}_\omega k + s_\omega k^2 + (\ddot{s}_\omega + \omega s_\omega)$$

if we take (18) and $\ddot{s}_\omega + \omega s_\omega = 0$ into account. By differentiating, one checks the identity

$$h(t) \exp\left(-\int_\epsilon^t k(s)\,ds\right) = h(\epsilon) + \int_\epsilon^t (\dot{h} - hk)(s) \exp\left(-\int_\epsilon^s k(\tau)\,d\tau\right)\,ds,$$

$0 < \epsilon < t < t_\kappa$, and thus by (19):

$$h(t) \ge h(\epsilon) \exp\left(\int_\epsilon^t k(s)\,ds\right).$$

As ϵ tends to $+0$, (15) yields $k(\epsilon) = -\frac{1}{\epsilon} + 0(1)$, whence $h(\epsilon) \to 0$ and $k(\epsilon) \to -\infty$. We infer

$$h(t) \ge 0 \quad \text{for } t \in (0, t_\kappa),$$

which is equivalent to (16′), and thus Theorem 2 is proved.

From Theorem 2 we infer that

$$\frac{d}{dt}\left\{\frac{\|J\|}{s_\omega}\right\} = \frac{\|J\|^{\cdot} s_\omega - \|J\|\dot{s}_\omega}{s_\omega^2} = \frac{\|J\|}{s_\omega}\left\{\frac{\|J\|^{\cdot}}{\|J\|} - \frac{\dot{s}_\omega}{s_\omega}\right\} = \frac{\|J\|}{s_\omega}\left\{\frac{\langle J, \dot{J}\rangle}{\|J\|^2} - \frac{\dot{s}_\omega}{s_\omega}\right\}$$
$$\leq 0,$$

i.e., the function $\|J\|/s_\omega$ is decreasing on $(0, t_\omega)$ and then the same reasoning as in the proof of Theorem 1 yields $\|\dot{J}(0)\| s_\omega(t) \geq \|J(t)\|$ for $t \in (0, t_\kappa)$.

Thus we have proved

Theorem 3. *Let J be a normal Jacobi field with $J(0) = 0$ along a unit speed geodesic $c : [0, \infty) \to M$, and suppose that the sectional curvature K of M satisfies $\omega \leq K \leq \kappa$ on the set $\{c(t) : t \in (0, t_\kappa)\}$. Then the function $\frac{\|J\|}{s_\omega}$ is decreasing in $(0, t_\kappa)$, and we have*

$$\|J(t)\| \leq \|\dot{J}(0)\| s_\omega(t) \quad \textit{for all } t \in (0, t_\kappa). \tag{20}$$

Remarks. 1. We first note that the completeness of M was not really needed. It was only used to insure the existence of $c(t)$ for all $t \in (0, t_\kappa)$. If we instead assume that $c(t)$ is defined for $0 \leq t \leq R$, the estimates (9), (17) and (20) will hold for $0 < t < \min(t_\kappa, R)$.

2. From $\omega \leq K \leq \kappa$ and $\langle R(J, \dot{c})\dot{c}, J\rangle = K(t)\|J^\perp\|^2$ we conclude that

$$\omega\|J^\perp\|^2 \leq \langle R(J, \dot{c})\dot{c}, J\rangle \leq \kappa\|J^\perp\|^2,$$

and therefore

$$\omega\|J\|^2 \leq \langle R(J, \dot{c})\dot{c}, J\rangle \leq \kappa\|J\|^2, \tag{21}$$

if we also assume that $\omega \leq 0 \leq \kappa$. The inequality (21) was all we needed to derive the statements of the Theorems 1–3, and the assumption $\langle J, \dot{c}\rangle = 0$ was nowhere else used. Thus these statements remain true for all Jacobi fields J along c with $J(0) = 0$.

3. Let us once again assume that $\omega \leq K \leq \kappa$ and $\omega \leq 0 \leq \kappa$, and suppose that $J \in \mathring{J}_c$, but not necessarily $\|\dot{c}\| = 1$. Then we define $r = \|\dot{c}\|$, $\underline{c}(\tau) = c(\tau/r)$, $\underline{J}(\tau) = J(\tau/r)$, and note that $\underline{J} \in \mathring{J}_{\underline{c}}$ and $\|\dot{\underline{c}}\| = 1$, whence, by (17),

$$\sqrt{\kappa}\,\mathrm{ctg}\,\sqrt{\kappa}\tau \leq \frac{\langle \underline{J}, \dot{\underline{J}}\rangle}{\|\underline{J}\|^2}(\tau) \leq \sqrt{-\omega}\,\mathrm{ctgh}\,\sqrt{-\omega}\tau,$$

and therefore

$$r\sqrt{\kappa}\,\mathrm{ctg}\,\sqrt{\kappa}rt \leq \frac{\langle J, \dot{J}\rangle}{\|J\|^2}(t) \leq r\sqrt{-\omega}\,\mathrm{ctgh}\,\sqrt{-\omega}rt.$$

If we introduce the functions

$$a_\kappa(t) = t\sqrt{\kappa}\,\mathrm{ctg}\,\sqrt{\kappa}t \qquad \text{for } 0 \le t < \pi/\sqrt{\kappa},$$
$$a_\omega(t) = t\sqrt{-\omega}\,\mathrm{ctgh}\,\sqrt{-\omega}t \quad \text{for } 0 \le t < \infty,$$

we arrive at

$$a_\kappa(r)\|J(1)\|^2 \le \langle J(1), \dot{J}(1)\rangle \le a_\omega(r)\|J(1)\|^2 \tag{22}$$

and

$$\{a_\kappa(r) - 1\}\|J(1)\|^2 \le \langle \dot{J} - J, J\rangle(1) \le \{a_\omega(r) - 1\}\|J(1)\|^2 \tag{23}$$

provided that $\sqrt{\kappa}r < \pi$.
By the same scaling argument, we derive from (9) and (20) the inequalities

$$\|\dot{J}(0)\|^2 r^{-2} s_\kappa^2(rt) \le \|J(t)\|^2 \le \|\dot{J}(0)\|^2 r^{-2} s_\omega^2(rt) \quad \text{if } 0 < rt < \pi/\sqrt{\kappa}.$$

By setting

$$b_\kappa(t) = \frac{\sin\sqrt{\kappa}t}{\sqrt{\kappa}t} \quad \text{and} \quad b_\omega(t) = \frac{\sinh\sqrt{-\omega}t}{\sqrt{-\omega}t},$$

we arrive at

$$\|\dot{J}(0)\|^2 b_\kappa^2(r) \le \|J(1)\|^2 \le \|\dot{J}(0)\|^2 b_\omega^2(r) \tag{24}$$

provided that $\sqrt{\kappa}r < \pi$.

Let us collect these results in the following

Theorem 4. *Let J be a Jacobi field with $J(0) = 0$ along a geodesic $c : [0,1] \to M$ with $r = \|\dot{c}(0)\|$, and suppose that the sectional curvature K of M satisfies $\omega \le K \le \kappa$ on the arc c. Then, if $\omega \le 0 \le \kappa$ and $r\sqrt{\kappa} < \pi$, the estimates (22)–(24) hold.*

Remark. We observe that $a_\omega, b_\omega \ge 1$ and $a_\kappa, b_\kappa \le 1$, in particular $a_\omega(0) = a_\kappa(0) = b_\omega(0) = b_\kappa(0) = 1$.

4.8.2 Riemann Normal Coordinates

Let $\psi(t,\alpha)$ be a mapping $\psi : [0,R] \times [-\alpha_0, \alpha_0] \to M$ such that, for every $\alpha \in [-\alpha_0, \alpha_0]$, $\alpha_0 > 0$, the curve $c(t) = \psi(t,\alpha)$ is a geodesic in M. Then $J(t) = \frac{\partial\psi}{\partial\alpha}(t,\alpha)$ is a Jacobi field along c. This follows from the identities

$$\frac{D}{\partial t}\frac{\partial\psi}{\partial\alpha} - \frac{D}{\partial\alpha}\frac{\partial\psi}{\partial t} = 0$$

and

$$\frac{D}{\partial t}\frac{D}{\partial\alpha}Z - \frac{D}{\partial\alpha}\frac{D}{\partial t}Z = R\left(\frac{\partial\psi}{\partial t}, \frac{\partial\psi}{\partial\alpha}\right)Z,$$

where Z denotes an arbitrary vector field along ψ. In fact, we have

$$\begin{aligned}\frac{D}{\partial t}\frac{D}{\partial t}\frac{\partial\psi}{\partial\alpha} &= \frac{D}{\partial t}\frac{D}{\partial\alpha}\frac{\partial\psi}{\partial t} = \frac{D}{\partial\alpha}\frac{D}{\partial t}\frac{\partial\psi}{\partial t} + R\left(\frac{\partial\psi}{\partial t},\frac{\partial\psi}{\partial\alpha}\right)\frac{\partial\psi}{\partial t}\\ &= 0 - R\left(\frac{\partial\psi}{\partial\alpha},\frac{\partial\psi}{\partial t}\right)\frac{\partial\psi}{\partial t}\end{aligned}$$

or

$$\ddot{J} + R(J,\dot{c})\dot{c} = 0.$$

This idea to construct Jacobi fields will be used in the following.

In what follows we identify the tangent space of T_pM at $v \in T_pM$ with T_pM itself and write $T_v(T_pM) \equiv T_pM$. The exponential map $\exp_p\colon T_pM \to M$ with center p is defined by $\exp_p(v) = c(1)$ for $v \in T_pM$, where c is the geodesic with $c(0) = p$, $\dot{c}(0) = v$.

Let $q = \exp_p v$. Then, by *Gauss's lemma*, the differential $(d\exp_p)_v$: $T_v(T_pM) = T_pM \to T_qM$ satisfies

$$(25) \qquad \langle \xi, \eta\rangle_p = \langle \tilde{\xi}, \tilde{\eta}\rangle_q,$$

where $\eta \in T_v(T_pM) \tilde{=} T_pM$ is the radial vector parallel to v (i.e. $\eta = v$ after identification of T_pM and $T_v(T_pM)$) and $\tilde{\xi}, \tilde{\eta}$ are defined by

$$(25') \qquad \tilde{\xi} = (d\exp_p)_v(\xi), \quad \tilde{\eta} = (d\exp_p)_v(\eta).$$

A "normal chart" (φ, U) with center $p \in M$ is given by an open set $U \subset M$ with $p \in U$, and by a mapping $\varphi : U \to \mathbb{R}^m$ of the form $\varphi = j \cdot \exp_p^{-1}$, where $j : T_pM \to \mathbb{R}^m$ is a linear isometry, and $\exp_p^{-1}$ is supposed to be existing on U.

Let $e_1, \dots, e_m$ be the orthogonal base of T_pM which under j corresponds to the standard base $(1, 0, \dots, 0), \dots, (0, \dots, 0, 1)$ of the Euclidean space $\mathbb{R}^m$. Since T_pM is identified with $T_v(T_pM)$ for all $v \in T_pM$, we may consider $e_1, \dots, e_m$ as m orthogonal vector fields on T_pM, and the base vector fields $X_1, \dots, X_m$ of the normal chart (φ, U) are given by

$$X_i(q) = (d\exp_p)e_i,$$

where $q = \exp_p v$.

Let c be the geodesic with $c(0) = p$ and $\dot{c}(0) = v$ for some $v \in T_pM$, and let $\xi = \xi^k e_k$ be an arbitrary vector in T_pM. Then $c(t) = \exp_p(tv)$, and, for each α, $\psi(t,\alpha) = \exp_p\{t(v+\alpha\xi)\}$ defines a geodesic $\psi(\cdot,\alpha) : [0,\infty) \to M$ with $\psi(0,\alpha) = p$. By our previous remarks, $J(t) = \frac{\partial\psi}{\partial\alpha}(t,0)$ therefore is a Jacobi field along c and, moreover,

$$\frac{\partial\psi}{\partial\alpha}(t,0) = (d\exp_p)_{tv}(t\xi) = t\xi^k(d\exp_p)_{tv}e_k = t\xi^k X_k(c(t)).$$

Thus we have proved:

Lemma 3. *If* $c : [0, \infty) \to M$ *is a geodesic with* $c(0) = p$ *and* $\dot{c}(0) = v \in T_pM$, *then, for every* $\xi = \xi^k e_k$, $J(t) = t\xi^k X_k(c(t))$ *defines a Jacobi field* J *along* c *with* $J(0) = 0, \dot{J}(0) = \xi^k X_k(p)$ *and, if* $q = c(1)$, *with*

$$(26) \qquad J(1) = \xi^k X_k(q), \quad \dot{J}(1) = \left\{\xi^\ell + \Gamma^\ell_{ik}(q)\xi^i \dot{c}^k(1)\right\} X_\ell(q).$$

For each normal chart (ψ, U) with center p, we may introduce Riemann normal coordinates by

$$x = \varphi(q)$$

for all $q \in U$. Let $X_1, \ldots, X_m$ be the base vector fields on U corresponding to the chart (φ, U). Then $q_{k\ell}(q) = \langle X_k(q), X_\ell(q)\rangle_q$ are the components of the fundamental tensor on U, and $\Gamma_{ik\ell}(q)$ and $\Gamma^\ell_{ik}(q)$ denote the Christoffel symbols of the first and second kind. For the sake of brevity, we set

$$g_{k\ell}(x) := g_{k\ell}\left(\varphi^{-1}(x)\right), \quad \Gamma_{ik\ell}(x) := \Gamma_{ik\ell}\left(\varphi^{-1}(x)\right), \quad \text{etc.}$$

without using different notation.

We obviously have

$$\varphi(p) = 0.$$

Moreover, $(d\exp_p)_0$ is the identical map, whence $X_i(p) = e_i$, and therefore

$$g_{k\ell}(p) = \delta_{k\ell} \quad \text{or} \quad g_{k\ell}(0) = \delta_{k\ell}.$$

Let $c(t) = \exp_p tv$, where $v = x^k e_k$ and $j(v) = x = (x^1, \ldots, x^m)$. Then $\eta(t) := \varphi(c(t))$ satisfies

$$\ddot{\eta}^\ell + \Gamma^\ell_{ik}(\eta)\dot{\eta}^i\dot{\eta}^k = 0.$$

On the other hand, the definition of φ implies $\eta(t) = tx$ and therefore $c(t) = \varphi^{-1}(tx)$ and $\Gamma^\ell_{ik}(tx)x^i x^k = 0$, in particular, $\Gamma^\ell_{ik}(0)x^i x^k = 0$ for all $x \in \mathbb{R}^m$. Therefore,

$$\Gamma^\ell_{ik}(0) = \Gamma_{ik\ell}(0) = 0 \quad \text{or} \quad \Gamma^\ell_{ik}(p) = \Gamma_{ik\ell}(p) = 0$$

since $\Gamma^\ell_{ik} = \Gamma^\ell_{ki}$.

Let $\xi = e_k$, and $\eta = x^\ell e_\ell$ be a radial vector that coincides with $v = \dot{c}(0)$. Then

$$\langle \xi, \eta\rangle_v = \langle e_k, x^\ell e_\ell\rangle_v = x^\ell \delta_{k\ell} = x^k.$$

Since $X_i(q) = (d\exp_p)_v e_i$, we infer from Gauss's lemma (25), (25′) that

$$\begin{aligned} x^k = \langle \xi, \eta\rangle_v = \langle (d\exp_p)_v \xi, (d\exp_p)_v \eta\rangle &= \langle X_k(q), x^\ell X_\ell(q)\rangle_q = x^\ell g_{k\ell}(q) \\ &= x^\ell g_{k\ell}(x). \end{aligned}$$

Thus we have

$$x^k = x^\ell g_{k\ell}(x) \quad \text{and also} \quad x^k = x^\ell g^{k\ell}(x).$$

Moreover, one also infers from Gauss's lemma that the distance $d(p,q)$ of the two points $p, q \in U$ with $p = c(0)$, $q = c(1) = \exp_p v$ is given by

$$d(p,q) = \|\dot{c}\| = \|v\| = |x|,$$

where $|x| = \sqrt{\delta_{k\ell} x^k x^\ell}$ denotes the Euclidian length of the vector $x \in \mathbb{R}^m$. Hence we have proved:

Lemma 4. *If $x = \varphi(q)$ are Riemann normal coordinates with center p on the set $U \subset M$, then*

(27) $$g_{ik}(0) = \delta_{ik}, \quad \Gamma_{ik\ell}(0) = 0, \quad \Gamma^\ell_{ik}(0) = 0,$$

(28) $$x^k = g_{k\ell}(x) x^\ell, \quad x^k = g^{k\ell}(x) x^\ell,$$

(29) $$d(p,q) = |x|.$$

Moreover, if $v = x^m e_m \in T_pM, x = (x^1, \dots, x^m) \in \mathbb{R}^m$, and if $c(t)$ denotes the geodesic $\exp_p tv$ with $c(0) = p$ and $\dot{c}(0) = v$, then $\varphi(c(t)) = tx$. □

For some real-valued function $f(x)$, we write

$$f_\ell(x) = \frac{\partial f}{\partial x^\ell}(x).$$

Then the following holds:

Lemma 5. *If $x = \varphi(q)$ are Riemann normal coordinates, then*

(30) $$x^k g_{ik,\ell}(x) = \delta_{i\ell} - g_{i\ell}(x), \quad x^k g^{ik}_\ell(x) = \delta^{i\ell} - g^{i\ell}(x),$$

(31) $$x^i x^k g_{ik,\ell}(x) = x^i x^\ell g_{ik,\ell}(x) = x^k x^\ell g_{ik,\ell}(x) = 0,$$

$$x^i x^k g^{ik}_{,\ell}(x) = x^i x^\ell g^{ik}_{,\ell}(x) = x^k x^\ell g^{ik}_{,\ell}(x) = 0,$$

(32) $$x^\ell \left\{ \Gamma_{i\ell k}(x) + \Gamma_{ik\ell}(x) \right\} = \delta_{ik} - g_{ik}(x),$$

(33) $$x^\ell \Gamma^\ell_{ik}(x) = x^\ell \Gamma_{i\ell k}(x),$$

(34) $$x^i x^k \Gamma_{ik\ell}(x) = x^i x^\ell \Gamma_{ik\ell}(x) = x^i x^k \Gamma^k_{i\ell}(x) = x^i x^\ell \Gamma^k_{i\ell}(x) = 0.$$

Proof. By differentiating the formulas (28), we obtain (30), and (31) is a consequence of (28) and (30). The identity $\Gamma_{ik\ell}+\Gamma_{i\ell k} = g_{k\ell,i}$ together with (30) yields (32). Finally, if we take $\Gamma_{i\ell k} = g_{\ell j}\Gamma^j_{ik}$ into account, (28) implies (33), and (34) follows from (31). □

Let us now return to the formulas (26) of Lemma 3. If $c(t) = \exp_p tv$ and $v = x^k e_k$, then we infer from Lemma 4 that $x = \varphi(q)$ with $q = c(1)$, and $\dot{c}^k(1) = x^k$ if $\dot{c}(t) = \dot{c}^k(t)X_k(c(t))$. Hence, the Jacobi field $J(t) = t\xi^k X_k(c(t))$ fulfills

$$\dot{J}(1) = \{\xi^\ell + \Gamma^\ell_{ik}(x)\xi^i x^k\}\, X_\ell(q). \tag{35}$$

Thus we obtain the relations

$$\|\dot{J}(0)\|^2 = \delta_{k\ell}\xi^k\xi^\ell, \quad \|J(1)\|^2 = g_{k\ell}(x)\xi^k\xi^\ell, \tag{36}$$

and

$$\begin{aligned} \langle \dot{J}(1) - J(1), J(1)\rangle &= \Gamma^\ell_{ik}(x)\xi^i x^k g_{\ell j}(x)\xi^j \\ &= \Gamma_{ijk}(x)\xi^i\xi^j x^k. \end{aligned} \tag{37}$$

We also note that $r := d(p,q) = |x| = \|\dot{c}\|$.

For any $p_0 \in M$, the interior $\mathring{S}$ of the set $\{V \in T_{p_0}M : \|V\| = d(p_0, \exp_{p_0} V)\}$ is an open, starshaped neighbourhood of 0 in $T_{p_0}M$. If we denote the cut locus of p_0 in M by $C(p_0) = \exp_{p_0}(\partial\mathring{S}) \subset M$ then the exponential map $\exp_{p_0} : \mathring{S} \to M$ is a C^2-diffeomorphism onto $S(p_0) := \exp_{p_0}(\mathring{S})$ and we can define Riemann normal coordinates $x = \varphi(q)$ for $q \in U = S(p_0)$. In addition, if K denotes the sectional curvature of M we define the numbers

$$\kappa(A) := \max\Big\{0, \sup_A K\Big\},$$

$$\omega(A) := \min\Big\{0, \inf_A K\Big\} \quad \text{for } A \subset M,$$

and

$$\begin{aligned} \kappa(x) &:= \kappa([0,x]) = \kappa([p_0,p]), \\ \omega(x) &:= \omega([0,x]) = \omega([p_0,p]), \end{aligned}$$

where $[p_0,p]$ is the geodesic segment between p_0 and p which in normal coordinates is just the segment $[0,x]$ on the ray from the origin 0 through x. Recall that we also use the notation

$$a_\kappa(t) = t\sqrt{\kappa}\,\mathrm{ctg}\,\sqrt{\kappa}t \quad \text{for } 0 \le t < \pi/\sqrt{\kappa},$$

$$a_\omega(t) = t\sqrt{-\omega}\mathrm{ctgh}\sqrt{-\omega}t \quad \text{for } 0 \le t < \infty$$

and

$$b_\kappa(t) = \frac{\sin\sqrt{\kappa}t}{\sqrt{\kappa}t}, \quad b_\omega(t) = \frac{\sinh\sqrt{-\omega}t}{\sqrt{-\omega}t}.$$

Theorem 5. *Let M be a complete Riemannian manifold and $x = \varphi(q)$ denote Riemann normal coordinates for $q \in S(p_0)$. With respect to those coordinates the following estimates are true:*

$$(38) \quad \{a_{\kappa(x)}(|x|) - 1\}g_{i\kappa}(x)\xi^i\xi^k \le \Gamma_{ik\ell}(x)x^i\xi^k\xi^\ell \le \{a_{\omega(x)}(|x|) - 1\}g_{ik}\xi^i\xi^k,$$

$$(39) \quad b^2_{\kappa(x)}(|x|)\xi^i\xi^i \le g_{ik}\xi^i\xi^k \le b^2_{\omega(x)}(|x|)\xi^i\xi^i,$$

$$(40) \quad b^2_{\kappa(x)}(|x|) \le \sqrt{g}(x) \le b^2_{\omega(x)}(|x|)$$

for all $\xi \in \mathbb{R}^m$ and all $x \in \varphi(S(p_0))$ with $|x| \cdot \kappa(x) < \pi$.

Proof. The inequalities (38) and (39) readily follow from the estimates (23) and (24) of Theorem 4 and from (36) and (37). Finally relation (28) implies that $\lambda = 1$ is one of the eigenvalues $\lambda_1, \lambda_2, \lambda_3$ of the matrix $g_{k\ell}(x)$ and by virtue of (39) we have $b^2_{\kappa(x)}(|x|) \le \lambda_k \le b^2_{\omega(x)}(|x|)$ for $k = 1, 2, 3$. This yields estimate (40). □

Theorem 6. *Let the assumptions of Theorem 5 be satisfied, and set $f(q) = \frac{1}{2}d^2(p, q)$. Then we have*

$$(41) \quad a_{\kappa(q)}(r)\|\xi\|^2 \le (D^2 f)_q(\xi, \xi) \le a_{\omega(q)}(r)\|\xi\|^2$$

for all $q \in M$ with $r = d(p, q) \le R$ and for $\xi \in T_qM$, where $(D^2f)_q(\xi, \xi)$ denotes the Hessian form of f at q (cp. Section 1.5 of Vol. 1, equation (28)).

Proof. Let $c(t) = \exp_p tv$, $q = c(1)$, and $\xi = \xi^k X_k(q) \in T_qM$. Then $J(t) = t\xi^k X_k(c(t))$ forms a Jacobi field J along c with $J(1) = \xi$ and $\|J(1)\|^2 = \|\xi\|^2$. Consider normal coordinates $x = \varphi(q)$ with center at p, and set $F(x) = f(q)$. Then

$$(D^2 f)_q(\xi, \xi) = F_{,ik}(x)\xi^i\xi^k - \Gamma^\ell_{ik}(x)F_{,\ell}(x)\xi^i\xi^k.$$

Since $F(x) = \frac{1}{2}|x|^2$, we get

$$(D^2 f)_q(\xi, \xi) = \delta_{ik}\xi^i\xi^k - \Gamma^\ell_{ik}(x)x^\ell\xi^i\xi^k = \Gamma_{ik\ell}(x)x^\ell\xi^i\xi^k + g_{ik}(x)\xi^i\xi^k$$

by virtue of (32) and (33). We then derive from (26) that

$$\langle \dot{J}(1), J(1)\rangle = (D^2 f)_q(\xi, \xi)$$

and thus (41) follows from (22). □

Theorem 7. *Let M be a complete Riemannian manifold, the sectional curvature K of which is bounded from above by*

$$K \le \kappa, \quad \kappa \ge 0,$$

on some ball $B_R(p)$ that does not meet the cut locus of its center p. Moreover, let $R\sqrt{\kappa} < \pi/2$. Then any two points q_1, q_2 of $B_R(p)$ can be connected by a geodesic arc contained in $B_R(p)$. This arc does not contain any pairs of conjugate points, and it is shortest among all arcs in $B_R(p)$ that join q_1 and q_2.

This result was proved by Jost [19].

4.8.3 Surfaces of Prescribed Mean Curvature in a Riemannian Manifold

In the following we consider a complete three-dimensional Riemannian manifold of class C^4. Since we restrict our considerations to surfaces in a normal chart (φ, U) with center p_0, we shall identify any point $q \in U \subset M$ with its normal coordinates $x = \varphi(q) \in \mathbb{R}^3$. Correspondingly any subset $\mathcal{K} \subset U$ is identified with $\varphi(\mathcal{K})$ and any surface $f : B \to U$ is identified with $X = X(w) = \varphi \circ f(w)$. In this way we obtain a natural definition of the Sobolev classes $H_s^1(B, U)$ as subsets of $H_s^1(B, \mathbb{R}^3)$. We recall the definition of a normal neighbourhood $U = S(p_0) = \exp_{p_0} \mathring{S}$, where $\mathring{S}$ is equal to the interior of $\{V \in T_{p_0}M : \|V\| = d(p_0, \exp_{p_0} V)\}$. Define the (Riemannian) cross product of two vector fields $Y = Y^k(x)X_k$, $Z = Z^\ell(x)X_\ell$ with respect to a chart x by $Y \times Z := \sqrt{g}g^{jk}(Y \wedge Z)_k$, where $(Y \wedge Z)_1 = Y^2Z^3 - Y^3Z^2$ etc. We then obtain the relation $\langle Y_1, Y_2 \times Y_3 \rangle = \sqrt{g}Y_1 \cdot (Y_2 \wedge Y_3)$, where the dot denotes the Euclidean scalar product.

Lemma 6. *For any $H \in C^1(U, \mathbb{R})$ we define the vector potentials*

$$Q(x) = \mu(x)x \quad \text{and} \quad Q^*(x) = \frac{1}{\sqrt{g(x)}}Q(x), \tag{42}$$

where $x \in U$ and

$$\mu(x) = 2\int_0^1 t^2\sqrt{g(tx)}H(tx)\,dt.$$

(i) *Q and Q^* are of class $C^1(U, \mathbb{R}^3)$ and*

$$\operatorname{div} Q = 2\sqrt{g}H, \quad \operatorname{Div} Q^* = 2H, \tag{43}$$

where div Q *denotes the (noninvariantly defined) expression $\sum_{k=1}^3 \frac{\partial Q^k}{\partial x^k}$, while* Div Q^* *stands for the divergence on M, i.e. we have*

$$\operatorname{div} Q^* + \Gamma^j_{jk}Q^{*k} = \frac{1}{\sqrt{g}}\operatorname{div}(\sqrt{g}Q^*) \quad \textit{(see Chapter 1.5 of Vol. 1).}$$

(ii) *Suppose $\mathcal{K} \subset S(p_0)$ is starshaped with respect to p_0 and let*

$$\text{(44)} \qquad \begin{cases} \rho_{\mathcal{K}}^{+} := \sup_{x \in \mathcal{K}} (|x|\sqrt{\kappa(x)}) < \pi, \\ \\ \rho_{\mathcal{K}}^{-} := \sup_{x \in \mathcal{K}} (|x|\sqrt{-\omega(x)}) < \infty, \end{cases}$$

and

$$\text{(45)} \qquad \begin{cases} b_{+}(\tau) := \dfrac{\sin \tau}{\tau} & \text{for } 0 \le \tau \le \pi, \\ \\ b_{-}(\tau) := \dfrac{\sinh \tau}{\tau} & \text{for } \tau \ge 0. \end{cases}$$

Then we have

$$\text{(46)} \qquad q_{\mathcal{K}}^{*} := \sup_{\mathcal{K}} \|Q^{*}(x)\| \le \frac{2}{3} \frac{b_{-}^{2}(\rho_{\kappa}^{-})}{b_{+}^{2}(\rho_{\kappa}^{+})} |H|_{0,\mathcal{K}} \cdot \sup_{\mathcal{K}} |x|.$$

Moreover, if in addition $\mathcal{K}$ is compact and $q_{\mathcal{K}}^{*} < 1$, then for this $\mathcal{K}$

$$\text{(47)} \qquad \begin{aligned} e(x,\eta) &:= \frac{1}{2} g_{jk}(x) \eta_{\alpha}^{j} \eta_{\alpha}^{k} + Q(x) \cdot (\eta_1 \wedge \eta_2) \\ &= \|\eta_1\|^2 + \|\eta_2\|^2 + \langle Q^{*}(x), \eta_1 \times \eta_2 \rangle \end{aligned}$$

satisfies Assumption A of Section 4.7 with

$$m_0 := (1 - q_{\mathcal{K}}^{*}) b_{+}^{2}(\rho_{\mathcal{K}}^{+}) \quad \textit{and} \quad m_1 := (1 + q_{\mathcal{K}}^{*}) b^{2}(\rho_{\mathcal{K}}^{-}).$$

Proof. (i) The function $\mu(x)$ is well defined for $x \in S(p_0)$ because this set is starshaped with respect to p_0 and the differentiability of Q and Q^* is obvious. Equation (43) follows by using an integration by parts.

(ii) The estimate (46) is obtained from (40), the definition of Q^* and the monotonicity properties of the functions b_- and b_+^{-1}. Furthermore, if $\eta = (\eta_1, \eta_2) \in \mathbb{R}^3 \times \mathbb{R}^3$ then

$$\begin{aligned} Q \cdot (\eta_1 \wedge \eta_2) = \langle Q^{*}, \eta_1 \times \eta_2 \rangle &\le \|Q^{*}\| \, \|\eta_1 \times \eta_2\| \le \|Q^{*}\| \, \|\eta_1\| \, \|\eta_2\| \\ &\le \frac{1}{2} \|Q^{*}\| \{\|\eta_1\|^2 + \|\eta_2\|^2\} \end{aligned}$$

and in view of (29) we have

$$b_{+}^{2}(\rho_{\mathcal{K}}^{+})|\xi|^2 \le \|\xi\|^2 \le b_{-}^{2}(\rho_{\mathcal{K}}^{-})|\xi|^2.$$

Combining these estimates we obtain (47),

$$\frac{1}{2}(1 - q_{\mathcal{K}}^{*}) b_{+}^{2}(\rho_{\mathcal{K}}^{+})|\xi|^2 \le e(x,\eta) \le \frac{1}{2}(1 + q_{\mathcal{K}}^{*}) b_{-}^{2}(\rho_{\mathcal{K}}^{-})|\xi|^2. \qquad \square$$

Definition 1. *A subset $\mathcal{K}$ of $S(p_0)$ is called a "gauge ball" in M with center p_0 if there exists an open neighbourhood $U \subset S(p_0)$ of p_0 which is starshaped with respect to p_0, a function $k \in C^2(U, \mathbb{R})$ and a real number $R > 0$ such that*

(i) $\mathcal{K} = \mathcal{K}_R(p_0) := \{x \in U : k(x) \le R^2\}$,

(ii) $k(0) = 0,\ Dk(0) = 0$,

(iii) $\gamma := \inf_{x\in\mathcal{K}} \gamma_k(x) > 0$, *where for $x \in U$,*

$$\gamma_k(x) := \inf\{D^2k(x;\xi,\xi) : \xi \in T_xM, \|\xi\| = 1\}$$

and $D^2k(x;\xi,\eta) = D^2k_q(\xi,\eta)$, $q = \varphi(x)$, stands for the Hessian form $\langle D_\xi Dk, \eta\rangle_q$. A function k with these properties is called a gauge function.

Remark 1. In local coordinates the coefficients of the Hessian form $D^2k(x;\xi,\eta)$ are given by

$$D^2k(x;X_j,X_\ell) = \frac{\partial^2 k(x)}{\partial x^j \partial x^\ell} - \Gamma^m_{j\ell}(x)\frac{\partial k(x)}{\partial x^m}. \tag{48}$$

Remark 2. Since we are dealing with Riemann normal coordinates, Lemma 5, (34) is applicable; in particular it follows that $x^i x^\ell \Gamma^j_{i\ell}(x) = 0$ for all $x \in S(p_0)$, $j = 1, 2, 3$. This yields by virtue of (48), (ii), (iii) and Taylor's formula the following estimates

$$x^j \frac{\partial k}{\partial x^j}(x) \ge \gamma|x|^2 \quad \text{for all } x \in \mathcal{K}, \quad \text{and} \quad k(x) \ge \frac{1}{2}\gamma|x|^2 \quad \text{for all } x \in \mathcal{K}. \tag{49}$$

Therefore, each gauge ball $\mathcal{K}_R(p_0)$ in M is bounded and hence also relatively compact in M, according to the Theorem of Hopf and Rinow. Also, every gauge ball is starshaped with respect to $p_0 = 0$. Indeed, (49) implies that the function $g(t) = k(tx)$ is strictly increasing in $t \in [0,1]$ for any $x \in \mathcal{K}$, $x \ne 0$, which yields the assertion.

The most important example of a gauge function on M is furnished by the square of the distance function (cp. Lemma 4)

$$k_0(x) := |x|^2 = d^2(p_0, p) \quad \text{on } U = S(p_0),$$

where x denotes normal coordinates around $p_0 = 0$. Using relation (28) in Lemma 4 we find

$$\frac{\partial k_0}{\partial x^j} = 2x^j = 2g_{j\ell}x^\ell$$

and

$$\begin{aligned}
\frac{\partial^2 k_0(x)}{\partial x^j \partial x^\ell} - \Gamma^m_{j\ell}(x)\frac{\partial k_0(x)}{\partial x^m} &= 2\frac{\partial}{\partial x^\ell}[g_{jk}x^k] - 2\Gamma^m_{j\ell} g_{mk} x^k \\
&= 2g_{jk,\ell}x^k + 2g_{jk}\delta^k_\ell - 2g^{mn}\Gamma_{jn\ell}g_{mk}x^k \\
&= 2g_{jk,\ell}x^k + 2g_{j\ell} - 2\Gamma_{jk\ell}x^k.
\end{aligned}$$

By virtue of (30) and (32) in Lemma 5 we can compute the coefficients of the Hessian form

$$(50)\quad D^2 k_0(x; X_j, X_\ell) = \frac{\partial^2 k_0}{\partial x^j \partial x^\ell} - \Gamma^m_{j\ell}\frac{\partial k_0}{\partial x^m} = 2\delta_{j\ell} - 2g_{j\ell} + 2g_{j\ell} + 2\Gamma_{j\ell k}x^k - 2\delta_{j\ell} + 2g_{j\ell} = 2[g_{j\ell} + \Gamma_{j\ell k}]x^k,$$

and

$$(51)\qquad \|Dk_0(x)\| = 2|x|.$$

Lemma 7. (i) *Suppose that the sectional curvature of M is bounded from above, i.e. $\kappa(M) < \infty$. Then for $\mathcal{K} = \{x \in S(p_0) : |x| \le R\}$ and $R < \frac{\pi}{2\sqrt{\kappa(M)}}$ we have*

$$\inf_{\mathcal{K}} \gamma_{k_0}(x) \ge 2a_{\kappa(M)}(R) > 0,$$

and

$$\frac{\gamma_{k_0}(x)}{\|Dk_0(x)\|} \ge \frac{a_{\kappa(M)}}{R} > 0 \quad \text{for } x \in \mathcal{K}\setminus\{0\}.$$

(ii) *If only $\rho^+_{\mathcal{K}} = \sup_{x\in\mathcal{K}}(|x|\sqrt{\kappa(x)}) < \frac{\pi}{2}$ holds, then we obtain instead $\inf_{x\in\mathcal{K}} \gamma_{k_0}(x) \ge 2a_+(\rho^+_{\mathcal{K}}) > 0$, and*

$$\frac{\gamma_{k_0}(x)}{\|Dk_0(x)\|} \ge \frac{a_+(\rho^+_{\mathcal{K}})}{R} > 0 \quad \text{for } x \in \mathcal{K}\setminus\{0\};$$

here we have put $a_+(t) := t\,\mathrm{ctg}(t)$.

Proof. (i) and (ii) follow from the definition of γ_{k_0}, relation (50), (51) and (38) of Theorem 5 and the monotonicity of the functions a_κ and a_+ respectively. □

Lemma 8 (Inclusion Principle). *Let $\mathcal{K} = \mathcal{K}_R(p_0)$ be a compact gauge ball and consider the Lagrangian (47) and the corresponding variational integral*

$$\mathcal{F}(X) = \int_B e(X, \nabla X)\,du\,dv = \int_B \left\{\frac{1}{2}g_{ij}(X)X^i_{u^\alpha}X^j_{u^\alpha} + Q(X)\cdot(X_u \wedge X_v)\right\} du\,dv.$$

Suppose that $Q \in C^1(S(p_0), \mathbb{R}^3)$ satisfies

$$(52)\qquad \operatorname{div} Q = 2\sqrt{g}H \quad \text{on } S(p_0)$$

and

(53) $$|H(x)| \le \frac{\gamma_k(x)}{\|Dk(x)\|} \quad \textit{for all } x \in \mathcal{K} \setminus \{p_0\}.$$

Moreover, denote by X *a function of class* $H_2^1(B, \mathcal{K}) \cap C^0(\overline{B}, \mathbb{R}^3)$ *satisfying*

(54) $$\delta\mathcal{F}(X, \phi) \ge 0 \quad \textit{for every } \phi \in L_{\infty,c}(B, \mathbb{R}^3)$$

such that $X + \epsilon\phi \in H_2^1(B, \mathcal{K})$ *for sufficiently small* $\epsilon > 0$. *Then* $X(\overline{B}) \subset \mathcal{K}_r$ *provided* $X(\partial B) \subset \mathcal{K}_r$ *for some* $r \le R$.

Proof. Define $\phi = (\phi^1, \phi^2, \phi^3)$ by

$$\phi^\ell(w) = \eta(w) g^{\ell m}(X(w)) \frac{\partial k}{\partial x^m}(X(w)),$$

where $\eta \in C_c^1(B, \mathbb{R})$ satisfies $0 \le \eta \le 1$ and X is a solution of (54). Since $\mathcal{K} \Subset U, X \in C^0(\overline{B}, \mathcal{K})$ and $\phi \in C_c^0(B, \mathbb{R}^3)$ there is a $\mathcal{K}' \Subset U$ such that $X - \epsilon\phi \in H_2^1(B, \mathcal{K}')$ for sufficiently small $|\epsilon|$. Hence $k(X(w) - \epsilon\phi(w))$ is defined for all $w \in B$, provided $|\epsilon|$ is small. Furthermore we have

(55) $$\begin{aligned} &k(X - \epsilon\phi) \\ &\quad = k(X) - \epsilon k_{x^j}(X)\phi^j + \epsilon^2 \int_0^1 (1-t) k_{x^j x^\ell}(X - \epsilon t\phi)\phi^j \phi^\ell \, dt \\ &\quad = k(X) - \epsilon\eta g^{\ell m}(X) k_{x^\ell}(X) k_{x^m}(X) \\ &\qquad + \epsilon^2\eta^2 \int_0^1 (1-t) k_{x^j x^\ell}(X - \epsilon t\phi) g^{jm}(X) g^{\ell n}(X) k_{x^m}(X) k_{x^n}(X)\, dt. \end{aligned}$$

Since (g^{jk}) is a positive definite matrix and $\mathcal{K}$ is compact, there is a constant $c > 0$ such that everywhere on B

$$g^{jk}(X(w))\xi^j\xi^k \ge c|\xi|^2 \quad \text{for all } \xi \in \mathbb{R}^3.$$

Also since $(X - \epsilon\phi)(w) \in \mathcal{K}' \Subset U$ for every $w \in B$ there is a constant $c' > 0$ such that the integral in (55) can be estimated in absolute value by $c' k_{x^j}(X) k_{x^j}(X)$ for all $w \in B$.

Thus we obtain

$$k(X - \epsilon\phi) \le k(X) - \epsilon\eta c k_{x^j}(X) k_{x^j}(X) + \epsilon^2\eta^2 c' k_{x^j}(X) k_{x^j}(X)$$

which implies that

$$k(X - \epsilon\phi) \le R \quad \text{for all } w \in B \text{ and } 0 < \epsilon < \epsilon_0 := \frac{c}{c'}.$$

Therefore the function $-\phi = (-\phi^1, -\phi^2, -\phi^3)$ is admissible in (54) and by Theorem 5 in Section 4.7 in particular (23), we have

$$\begin{aligned} \delta\mathcal{F}(X, \phi) = \int_B &\{g_{j\ell}(X) X^j_{u^\alpha} [\eta g^{\ell m}(X) k_{x^m}(X)]_{u^\alpha} \\ &+ \frac{1}{2}\frac{\partial g_{j\ell}}{\partial x^n} X^j_{u^\alpha} X^\ell_{u^\alpha} \eta g^{mn} k_{x^m}(X) \\ &+ \eta \operatorname{div} Q(X)(X_u \wedge X_v)^j \cdot g^{jm} k_{x^m}(X)\}\, du\, dv \le 0. \end{aligned}$$

Using the expression

$$D^2k(X; X_{u^\alpha}, X_{u^\alpha}) = \frac{\partial^2 k(X)}{\partial x^j \partial x^\ell} X^j_{u^\alpha} X^\ell_{u^\alpha} - \Gamma^m_{j\ell} \frac{\partial k(X)}{\partial x^m} X^j_{u^\alpha} X^\ell_{u^\alpha}$$

for the Hessian form of k and

$$\begin{aligned} 2H(X)\langle X_u \times X_v, Dk(X)\rangle &= 2H(X)\sqrt{g} Dk \cdot (X_u \wedge X_v) \\ &= \operatorname{div} Q g^{j\ell} k_{x^\ell}(X)(X_u \wedge X_v)^j \end{aligned}$$

we obtain the inequality

$$\begin{aligned} 0 &\geq \int_B \eta_{u^\alpha} [k(X)]_{u^\alpha} \, du \, dv \\ &\quad + \int_B \{\eta D^2 k(X; X_{u^\alpha}, X_{u^\alpha}) + 2H(X)\langle X_u \times X_v, Dk(X)\rangle\} \, du \, dv \\ &\geq \int_B \eta_{u^\alpha} [k(X)]_{u^\alpha} \, du \, dv \\ &\quad + \int_B \eta\{\gamma_k(X)(\|X_u\|^2 + \|X_v\|^2) \\ &\quad - 2|H(X)| \, \|X_u\| \, \|X_v\| \, \|Dk(X)\|\} \, du \, dv. \end{aligned}$$

By assumption (53) $|H(X)| \leq \frac{\gamma_k(X)}{\|Dk(X)\|}$ and because of $\|X_u\| \ \|X_v\| \leq \frac{1}{2}(\|X_u\|^2 + \|X_v\|^2)$ it follows that $\{\dots\} \geq 0$ on B, whence

$$\int_B \eta_{u^\alpha} [k(X)]_{u^\alpha} \, du \, dv \leq 0$$

for all $\eta \in C^1_c(B)$ with $0 \leq \eta \leq 1$. Therefore $k(X(u,v)) \in C^0(\overline{B}) \cap H^1_2(B)$ is subharmonic in B and the assertion follows from the maximum principle. □

Note that by the strong maximum principle (cp. Gilbarg and Trudinger, Theorem 8.19) we may even conclude $X(B) \subset \operatorname{int} \mathcal{K}_r$ or $X(B) \subset \partial \mathcal{K}_r$.

Now we can prove the main result of this section.

Theorem 8. *Let $\mathcal{K} = \mathcal{K}_R(p_0)$ be a compact gauge ball in M. Suppose that the restriction $\rho^+_{\mathcal{K}} < \pi$ on the sectional curvature of M is satisfied and that Γ is a closed Jordan curve in $\mathcal{K}$ such that $\mathcal{C}(\Gamma, \mathcal{K}) \neq \emptyset$. Finally let H be a function of class $C^{0,\beta}(\mathcal{K}), 0 < \beta < 1$, satisfying the conditions*

$$|H|_{0,\mathcal{K}} < \frac{3}{2} \frac{1}{\sup_{p \in \mathcal{K}} d(p_0, p)} \frac{b^2_+(\rho^+_{\mathcal{K}})}{b^2_-(\rho^-_{\mathcal{K}})}, \tag{56}$$

and

$$|H(x)| \leq \frac{\gamma_k(x)}{\|Dk(x)\|} \quad \textit{for all } x \in \mathcal{K} \setminus \{p_0\}. \tag{57}$$

Then there exists an X *of class* $\mathcal{C}(\Gamma, \mathcal{K}) \cap C^{2,\beta}(B, \mathbb{R}^3) \cap C^0(\overline{B}, \mathbb{R}^3)$ *with*

$$\triangle X^\ell + \Gamma^\ell_{ij} X^i_{u^\alpha} X^j_{u^\alpha} = 2H(X)\sqrt{g(X)} g^{\ell m}(X_u \wedge X_v)_m \tag{58}$$

in B *for* $\ell = 1, 2, 3$, *and such that the conformality relations*

$$g_{ij} X^i_u X^j_u = g_{ij} X^i_v X^j_v, \quad g_{ij} X^i_u X^j_v = 0$$

hold everywhere in B. *Furthermore* X *maps* ∂B *homeomorphically onto* Γ.

In other words, we have determined a surface X in the Riemannian manifold M which has mean curvature $H(X)$ in B (except, possibly at isolated branch points) and which is spanned by the Jordan arc Γ.

Proof of Theorem 8. We extend H continuously to some compact gauge ball $\mathcal{K}_{R+\epsilon}(p_0), \epsilon > 0$, such that (56) and (57) continue to hold for $\mathcal{K} = \mathcal{K}_{R+\epsilon}(p_0)$. Consider the variational problem

$$\mathcal{F}(X) = \int_B \left\{ \frac{1}{2} g_{ij}(X) X^i_{u^\alpha} X^j_{u^\alpha} + Q(X) \cdot (X_u \wedge X_v) \right\} du\, dv \to \min$$

in $\mathcal{C}(\Gamma, \mathcal{K}_{R+\epsilon})$ where $Q(x) = \mu(x) \cdot x$, $\mu(x) = 2\int_0^1 t^2 \sqrt{g(tx)} H(tx)\, dt$ as in Lemma 6. Relation (46) and assumption (56) imply that $\mathcal{F}(\cdot)$ is coercive; also $\mathcal{K}_{R+\epsilon}$ is quasiregular. Hence we may apply Theorems 3 and 4 in Section 4.7 and obtain the existence of a conformally parametrized solution $X \in \mathcal{C}(\Gamma, \mathcal{K}_{R+\epsilon}) \cap C^0(\overline{B}, \mathcal{K}_{R+\epsilon}) \cap C^{0,\alpha}(B, \mathbb{R}^3)$. By a reasoning analogous to the one in the proof of Theorem 8 in Section 4.7 one can see that the first variation formula

$$\delta\mathcal{F}(X, \phi) = \int_B \left\{ g_{j\ell} X^j_{u^\alpha} X^\ell_{u^\alpha} + \frac{1}{2} \frac{\partial g_{j\ell}}{\partial x^n} X^j_{u^\alpha} X^\ell_{u^\alpha} \phi^n + 2H\sqrt{g}(X_u \wedge X_v)^j \phi^j \right\} du\, dv$$

holds for all $\phi \in \mathring{H}^1_2(B, \mathbb{R}^3) \cap L_\infty(B, \mathbb{R}^3)$, cp. Theorem 5 in Section 4.7. Moreover, it follows from the minimum property of X that the variational inequality

$$\delta\mathcal{F}(X, \phi) \geq 0$$

holds for all $\phi \subset \mathring{H}^1_2(B, \mathbb{R}^3) \cap L_\infty(B, \mathbb{R}^3)$ with $X + \epsilon\phi \in H^1_2(B, \mathcal{K}_{R+\epsilon})$. The inclusion principle Lemma 8 now implies that the coincidence set $\Omega = \{w \in B : X(w) \in \partial\mathcal{K}_{R+\epsilon}\}$ must be empty. Finally Theorem 7 in Section 4.7 shows that $X \in C^{2,\beta}(B, \mathcal{K}_R) \cap C^0(\overline{B}, \mathbb{R}^3)$ is a conformal solution of the system (58).

The topological character of the boundary mapping follows in a standard way. Theorem 8 is completely proved. □

We finally consider the special case $k = k_0$.

Theorem 9. *Let $\mathcal{K}_R = \{x \in M : |x| \leq R\} \cap S(p_0)$ be a compact gauge ball, where $k_0(x) = |x| = d(p_0, p)$. Suppose that*

$$R < \frac{\pi}{2\sqrt{\kappa(M)}} \quad \textit{and} \quad \omega(M) > -\infty.$$

Let $\Gamma \subset \mathcal{K}$ be a closed Jordan curve such that $\mathcal{C}(\Gamma, \mathcal{K})$ is nonempty and suppose that H is a function of class $C^{0,\beta}(\mathcal{K}, \mathbb{R}), 0 < \beta < 1$, for which

$$|H|_{0,\mathcal{K}} < \min\left\{\frac{a_{\kappa(M)}(R)}{R}, \frac{3b_+^2(R\sqrt{\kappa(M)})}{2b_-^2(R\sqrt{-\omega(M)})}\right\}.$$

Then the assertion of Theorem 8 holds. □

4.9 Scholia

4.9.1 Enclosure Theorems and Nonexistence

The observation that a connected minimal surface lies in the convex hull of its boundary (cf. Theorem 1 of Section 4.1) has been made a long time ago and was, for instance, known to T. Radó (see e.g. [21]). Apparently S. Hildebrandt [11] was the first to observe that also certain nonconvex sets can be used for enclosing minimal surfaces and H-surfaces, and to apply this fact for proving nonexistence of connected minimal surfaces whose boundaries are "too far apart", cf. Theorem 2.3 of Section 4.1. Earlier, J.C.C. Nitsche [13,15] had proved various results about the "extension" of minimal surfaces with two boundary curves, thereby obtaining nonexistence results; cf. also Nitsche [28], pp. 474–498. The results by Hildebrandt [11] were improved and generalized in several directions; a survey of this work is presented in Sections 4.1–4.4, based on papers by Osserman and Schiffer [1], Böhme, Hildebrandt, and Tausch [1], Gulliver and Spruck [1,2], Hildebrandt [8,11], Hildebrandt and Kaul [1], U. Dierkes [1–4,6,11], Dierkes and Huisken [1,2], and Dierkes and Schwab [1]. We particularly mention the geometric maximum principle in Dierkes [6] which is based on a pull–back version of the standard monotonicity formula from geometric measure theory due to M. Grüter [2] (see also Section 2.6 of this volume as well as Böhme, Hildebrandt, and Tausch [1] for a related technique). Furthermore we refer to the maximum principles proved in Gulliver, Osserman, and Royden [1], R. Gulliver [7], and particularly we mention the work of K. Steffen [6] and of Duzaar and Steffen [5–7] where geometric maximum principles of an optimal form are derived. Theorems 3–6 in Section 4.3 are due to Dierkes [11] and Dierkes and Schwab [1]. It is interesting to note that – despite its simplicity – the argument used here is of considerable generality and is applicable to a number of important situations. For example, K. Ecker [2,3] could give a very simple proof of the "neck-pinching" phenomenon for mean curvature flow by using a parabolic version of the polynomial

$$p_j = \sum_{i=1}^{n+k-j} |x^i|^2 - \frac{(n-j)}{j} \sum_{i=n+k-j+1}^{n+k} |x^i|^2 \quad (\text{for } k=1).$$

Furthermore, U. Clarenz [1,2] applied the same argument to *$\mathcal{F}$-minimal immersions* which arise as extremals of parametric integrals of the type

$$\int_M \mathcal{F}(X, N)\, dA$$

for suitable homogenous integrands $\mathcal{F}$ depending on the position X and the normal of an immersion. Again, general necessary conditions for $\mathcal{F}$-minimal surfaces are obtained, and the method can be generalized to corresponding parabolic flow problems as well. For details in this direction see Winklmann [1].

Apparently the first *barrier-principle* for minimal immersions with arbitrary codimension is due to Jorge and Tomi [1]; however, see also the geometric inclusion principle for energy minimizers obtained earlier by R. Gulliver [1]. The barrier principle for submanifolds with arbitrary codimension and bounded mean curvature, formulated in Theorem 1 of Section 4.4, is due to Dierkes and Schwab [1].

Geometric inclusion principles valid for conformal H_2^1-solutions of the variational inequality (9), Section 4.4, were found by Steffen [6], cp. also Duzaar and Steffen [5–7]. The versions presented in Theorem 2 and 3 require a priori $C^1 \cap H^2_{2,\mathrm{loc}}$-regularity of the solution, which is, however, always satisfied in the application we have in mind later in Section 4.7, due to certain regularity results for obstacle problems, cp. Section 4.8. Our proof of Theorem 2 in Chapter 4.4 is self-contained and independent of the argument in Duzaar and Steffen [5–7]; it cannot be extended to H_2^1-subsolutions. The proof of Theorem 3 in Chapter 4.4 is reminiscent to Proposition 2.4 in Duzaar and Steffen [7] and uses the same type of test function argument. We also mention the geometric inclusion principle of Gulliver and Spruck [2] which uses strict energy minimality of the solution considered. In fact, pushing in a surface under an assumption on the boundary curvature similar to those in Theorems 2 and 3 of Section 4.4 saves energy, and hence energy minimizers cannot touch the boundary of the inclusion domain.

The following terminology due to P. Levy has become customary (see Nitsche [28], pp. 364, 671–672, [37], pp. 354, 373): A closed set $\mathcal{K}$ in $\mathbb{R}^3$ is said to be *H-convex* if for every point $P \in \partial\mathcal{K}$ there is a locally supporting minimal surface $\mathcal{M}$, i.e.: For any $P \in \mathcal{M}$ there is an $\epsilon > 0$ such that $\mathcal{K} \cap B_\epsilon(P)$ lies on one side of $\mathcal{M} \cap B_\epsilon(P)$.

If $\partial\mathcal{K}$ is a regular C^2-surface then H-convexity of $\mathcal{K}$ means that the mean curvature Λ of $\partial\mathcal{K}$ with respect to the inward normal is nonnegative.

4.9.2 The Isoperimetric Problem. Historical Remarks and References to the Literature

Among all closed curves of a given length, the circle encloses a domain of maximal area. This is the classical *isoperimetric property of the circle* which was already known in antiquity. The first transmitted proof of this property is due to Zenodorus who lived between 200 B.C. and 100 A.D. Concerning the history of the isoperimetric problem we refer to Gericke [1]. Of the later proofs we mention that of Galilei [1], pp. 57–60 who prompts Sagredo to say at the end of the discussion:

"Mà dove siamo trascorsi à ingolfarci nella Geometria ..."[3]

The problem became again popular through the work of Steiner who contributed many beautiful ideas to this and to related questions. Yet all of his proofs were imperfect as they only showed that no other curve than the circle can enclose maximal area. It remained open whether there is a curve of given perimeter whose interior maximizes area. The first rigorous proof of the isoperimetric property of the circle was given by Weierstrass in his lectures, and his student H.A. Schwarz established the isoperimetric property of the sphere, a much more difficult question. A beautiful discussion of the isoperimetric problem can be found in Blaschke's classic [3]: *Kreis und Kugel* (with a historical survey in §14).

Fig. 1. Rügen, an island in the Baltic Sea, furnishes an example of a planar domain whose area A is far less than $L^2/4\pi$, L being the length of its circumference. It shows how bold it is to draw conclusions about the area of a domain from the time it takes to sail around it

---

[3] It seems that Galileo was enthusiastic by rights as his reasoning (according to an oral communication by E. Giusti) can be turned into a proof that is correct by our standards.

Concerning references to the modern literature we refer to Nitsche [28], pp. 290–292, and particularly to Osserman's survey paper [19] that provides a thorough discussion of all pertaining results as well as a report on related questions.

The isoperimetric inequality for minimal surfaces of the type of the disk was first proved by Carleman [3] in 1921.

Beckenbach and Radó [1] proved in 1933: Let S be a surface in $\mathbb{R}^3$ with Gauss curvature K. *Then the inequality* $4\pi A \le L^2$ *holds for all simply connected domains* Ω *in* S *(*A = *area* S, L = *length* ∂S*) if and only if* K *is nonpositive.*

The simple connectivity of Ω is crucial as one immediately realizes by looking a long cylinders. Moreover, in the Beckenbach–Radó theorem it is essential that S is a regular surface, whereas in Carleman's theorem the minimal surface may have branch points. Note that in Theorems 1 and 2 of Section 4.5 the minimal surface is allowed to have arbitrarily many branch points.

It is still an open question whether the sharp isoperimetric inequality

$$A(\mathfrak{X}) \le \frac{1}{4\pi} L^2(\mathfrak{X}) \tag{1}$$

holds for any compact minimal surface $\mathfrak{X} : M \to \mathbb{R}^3$ with boundary, or if additional assumptions on $\mathfrak{X}$ are truly necessary for (1) to be true. It is, however, known that certain extra-assumptions suffice to ensure the validity of (1). For instance, Osserman and Schiffer [1] proved (1) for minimal surfaces $\mathfrak{X} : M \to \mathbb{R}^3$ defined on an annulus M, and Feinberg [1] showed that (1) also holds for annulus-type surfaces $\mathfrak{X} : M \to \mathbb{R}^n, n \ge 2$. The Osserman–Schiffer result implies that the sharp isoperimetric inequality also holds for minimal surfaces of the topological type of the Möbius strip, see Osserman [18]. The beautiful result of Theorem 3 of Section 4.5 was found by Li, Schoen, and Yau [1]. Amazingly it is strong enough to (essentially) imply the Osserman–Schiffer result. Other interesting conditions guaranteeing (1) were discovered by Alexander-Hoffmann-Osserman [1] and by Osserman [17].

A variant of the *linear isoperimetric inequality (21)* in Section 4.5 using the oscillation of a minimal surface X was pointed out by Nitsche [28]. Küster [3] showed that the radius of the smallest ball containing $X(B)$ leads to the optimal version of the inequality for which equality holds precisely for plane disks.

Concerning generalizations of the isoperimetric inequality to H-surfaces we refer e.g. to papers by Heinz and Hildebrandt [2], Heinz [11], and Kaul [2,3]. A survey of the entire field of geometric inequalities can be found in the treatise of Burago and Zalgaller [1]. B. White [3] showed that, for each integer $n > 1$, there is a smooth Jordan curve Γ in $\mathbb{R}^4$ such that $(1/n)\alpha(n\Gamma) < (1/k)(\alpha(k\Gamma)$ for $1 \le k < n$. Here $\alpha(k\Gamma)$ denotes the least area (counting multiplicities) of any oriented surface with boundary $k\Gamma$ (= k-fold multiple of Γ). In a different way, examples of this kind were somewhat earlier constructed by F. Morgan.

4.9.3 Experimental Proof of the Isoperimetric Inequality

There are two simple soap film experiments by means of which one can demonstrate the isoperimetric property of the circle. For instance, take a wire that has the shape of a plane curve, attach a handle to it, and dip it into a soap solution. On removing it from the liquid, a soap film spanning the wire will be formed. Then place a thin loop of thread onto the film and break the part of the soap film inside of the loop with a blunt tool. As the soap film wants to reduce its area, it will pull the thread tight into the shape of a circle (see Fig. 2). The soap film has minimal energy and therefore minimal area; hence the interior of the strained loop is maximizing area, and since the thread apparently has the form of a circle, we have an "experimental proof" of its isoperimetric property. Further experiments and results with soap films and threads will be described in Chapter 5.

Another experimental proof will be obtained by blowing a soap bubble between two parallel wetted glass plates. Let us begin with a bubble in the form of a hemisphere sitting on one of the plates. By blowing more air into the bubble, it will enlarge until it touches the other plate, whereupon it changes into a circular cylinder that meets both plates perpendicularly in circles (see Fig. 3). The cylinder has minimal area among all surfaces enclosing a fixed volume which touch both plates (a discussion of related mathematical questions can be found in papers by Athanassenas [1,2] and Vogel [1]), whence one concludes that the circle has minimal length among all closed curves bounding the same amount of area. But this "dual property" is equivalent to the isoperimetric property of the circle. This second experiment was apparently first described by Courant (see Courant and Robbins [1]).

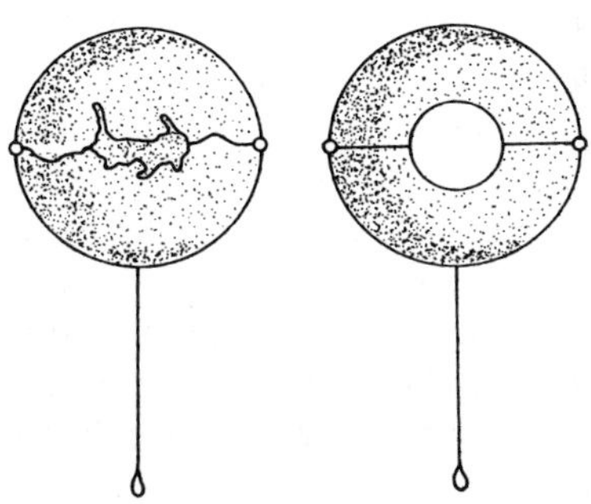

Fig. 2. Experimental proof of the isoperimetric inequality

4.9.4 Estimates for the Length of the Free Trace

The first estimate of this kind was derived by Hildebrandt and Nitsche [4]; an improved version of their result with the optimal constant 2 is due to

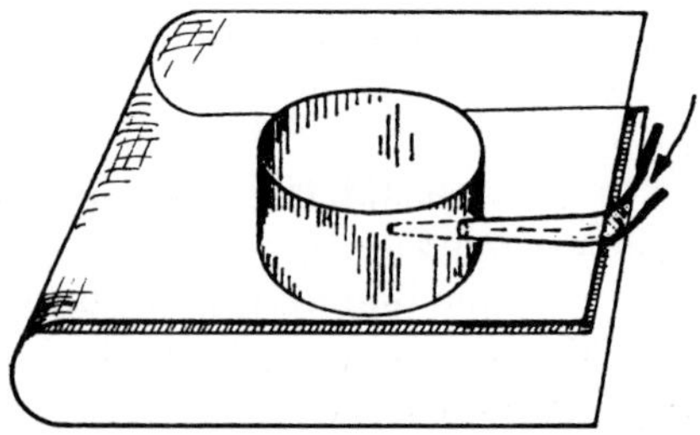

Fig. 3. Another experimental demonstration of this isoperimetric property of the circle

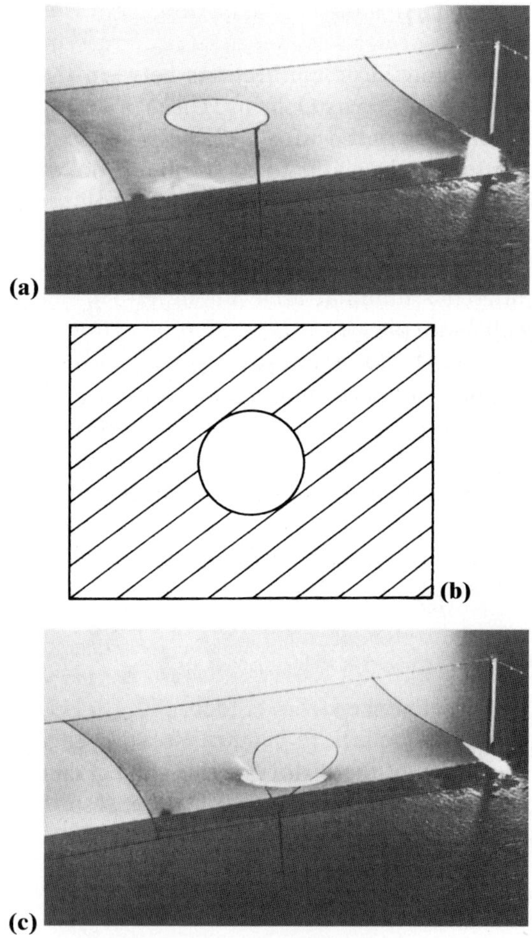

Fig. 4. (**a**), (**b**) Experimental proof of the isoperimetric property of the circle. (**c**) If the thread is pulled down, one obtains a curve of constant curvature (see Chapter 5). (**a**), (**c**) courtesy of Institut für Leichte Flächentragwerke, Stuttgart – Archive

Küster [2]. Finally Dziuk [8] removed the assumption that the minimal surface be free of branch points of odd order on the free boundary. We have presented this result as Theorem 2.

Using the idea of Hildebrandt and Nitsche, Ye [2] has stated estimates of the length of the free trace and of the area of a minimal surface with a partially free boundary in terms of the length of the fixed boundary, in case that the supporting surface is a strict graph (= λ-graph). Ye also provided the example described in Remark 4 which shows that the estimates of Section 4.6, Proposition 1 and Theorem 1, are in a sense optimal. Küster [2] contributed Remark 7, which shows that neither a bound on the Gauss curvature of the support surface S nor a bound on its mean curvature will imply an estimate such as stated in Theorem 2 of Section 4.6; instead, one needs bounds on both principal curvatures of S. Hence the *R-sphere condition* is really adequate in Theorem 2 and by no means artificial.

The partition problem was treated in the paper [2] of Grüter, Hildebrandt, and Nitsche. These authors derived boundary regularity for arbitrary stationary solutions as well as bounds on the length of the free trace such as stated in formulas (58)–(63) of Section 4.6.

We finally mention that an approach to estimates on the length of the free trace for area-minimizing solutions of free boundary problems can already be found in the fundamental work of H. Lewy [4].

Osserman [18] pointed out that there are close connections between the isoperimetric inequality and an inequality suggested by Gehring: *Given in* $\mathbb{R}^3$ *any closed Jordan curve* Γ *of length* $L(\Gamma)$ *which is linked with a closed set* Σ *such that* dist $(\Gamma, \Sigma) \geq r$, *then* $L(\Gamma) \geq 2\pi r$. Osserman was able to establish a proof of Gehring's inequality by means of the isoperimetric inequality. Generalizations to higher dimensions ($n > 3$) follow from work of White [1] and Almgren [7]. Other proofs and generalizations were given by Bombieri and Simon [1], Gage [1], and Gromov [1].

4.9.5 The Plateau Problem for H-Surfaces

In Sections 4.7 and 4.8 we have discussed the Plateau problem for H-surfaces in Euclidean space and in Riemannian manifolds. For $H = \text{const}$ this problem was first treated by E. Heinz [2], H. Werner [1,2], and S. Hildebrandt [4,7], and for variable H by Hildebrandt [5,6]. The Riemannian case was first studied by Hildebrandt and Kaul [1] and R. Gulliver [3]. Further pioneering work in this field is due to H. Wente [1–4,6–8], K. Steffen [1–6], Brezis and Coron [1,3], and M. Struwe [5,7]. The optimal results are due to K. Steffen [6] and Duzaar and Steffen [6].

We particularly mention the solution of Rellich's problem by the work of Brezis and Coron [1,3], M. Struwe [5,7], and K. Steffen [6].

In Section 4.7 (Theorems 3–6) we have outlined several regularity results for variational problems with obstacles due to S. Hildebrandt [12,13]; for similar results see Tomi [4]. We have added some important remarks to make these results accessible to applications for the existence procedure for H-surfaces, cp. Theorems 8 and 9. The existence result for H-surfaces in a closed ball is due to Hildebrandt [5,6]. Our proof given here is a slight modification of his argument. Theorem 9 was found independently and almost simultaneously by Gulliver and Spruck [1] and Hildebrandt [10].

A slight improvement of Gulliver and Spruck's [2] existence theorem for H-surfaces contained in arbitrary closed sets K with suitably curved boundaries is presented in Theorem 10. We have replaced their *pushing in* argument for minimizers by the geometric maximum principles *Enclosure Theorem* 2 and 3 of Section 4.4.

H. Wente [1–4] and K. Steffen [1–6] have initiated a completely different approach to prove existence theorems for H-surfaces by invoking the isoperimetric inequality in a suitable way. In his pioneering work, Wente [1] considered the energy functional for constant H,

$$E_H(x) = D(x) + 2HV,$$

where

$$V(x) = \frac{1}{3}\int_B X \cdot (X_u \wedge X_v)\, du\, dv$$

is the volume enclosed by the surface X and the cone over the boundary trace of X. Using the isoperimetric inequality in $\mathbb{R}^3$ he was able to prove lower semicontinuity of $E_H(\cdot)$ in a class of surfaces with suitably small Dirichlet integral. In a mayor achievement, Steffen [1–6] generalized and improved these results to variable H.

The following result holds:

Theorem 1. (Wente, Steffen). *Suppose that*

$$\sup_{\mathbb{R}^3} |H| \le c\sqrt{\frac{\pi}{A_\Gamma}},$$

where A_Γ is the infimum of area of all surfaces spanned by Γ and $c = \sqrt{2/3}$. Then there is an H-surface X bounded by Γ.

Clearly this theorem gives better existence results than the Theorems 6–9 in Section 4.7 for curves Γ which are of the shape of a curled and knotted rectangle of side lengths ϵ and $\frac{1}{\epsilon}$ spread over a large region of $\mathbb{R}^3$. Probably the optimal constant c in Theorem 1 is $c = 1$. According to Heinz [12] (see Section 4.7, Theorem 1), c cannot be larger than one, and a result by Struwe for constant H indicates that $c = 1$ is the best possible value. Using concepts from geometric measure theory, Steffen [3,4] introduced his notion of an H-volume, replacing the volume term V_H above, thereby obtaining several striking existence theorems under very natural conditions on the prescribed curvature H. A typical result is the following

Theorem 2. (Steffen [4]). *Suppose* $H : \mathbb{R}^3 \to \mathbb{R}$ *satisfies*

$$\int_{\mathbb{R}^3} |H|^3 \, dx < \frac{9\pi}{2}.$$

Then there is an H*-surface* X *bounded by* Γ *which is as regular as* H *(and* Γ*) permit. In particular, if* H *is continuous then* X *is of class* $C^{1,\alpha}(B, \mathbb{R}^3)$ *for every* $0 < \alpha < 1$*, and if* H *is locally Hölder continuous on* $\mathbb{R}^3$ *then* X *is of class* $C^{2,\alpha}$ *and solves the* H*-surface system in the classical sense.*

It may surprise that no condition on the boundary curve Γ is needed here. We remark that all results mentioned above possess suitable analogs for H-surfaces in three-dimensional manifolds M, see Hildebrandt and Kaul [1], Gulliver [3], Steffen [6], and Duzaar and Steffen [6,7]. Corresponding results hold also for H-surfaces which are restricted to lie in given sets K of $\mathbb{R}^3$, see Steffen [4] and Dierkes [2]. We mention in particular the survey articles by Steffen [6] and Duzaar and Steffen [6,7] for a thorough account of existence results for H-surfaces in three-manifolds which are not restricted to a coordinate patch.

Surfaces with prescribed mean curvature vector in manifolds of arbitrary dimensions were found by R. Gulliver [1].

The differential geometric background of Section 4.8 is taken from papers by Hildebrandt and Kaul [1], H. Karcher [6], and S. Hildebrandt [17].

Chapter 5

The Thread Problem

The problem to be studied in this chapter is another generalization of the isoperimetric problem which is related to minimal surfaces. Consider a fixed arc Γ with endpoints P_1 and P_2 connected by a movable arc Σ of fixed length. One may conceive Γ as a thin rigid wire, at the ends of which a thin inextensible thread Σ is fastened. Then the *thread problem* is to determine a minimal surface minimizing area among all surfaces bounded by the boundary configuration $\langle \Gamma, \Sigma \rangle$. The particular feature distinguishing this problem from the ordinary Plateau problem is the movability of the arc Σ.

In Section 5.1 we shall describe several variants of the thread problem, and we shall depict some experimental solutions. Most of these questions have not yet been treated mathematically; that is, no existence proof can be found in the literature. We shall state the mathematical formulation of the thread problem in the simplest case, and in Section 5.2 we shall outline the existence proof given by H.W. Alt for this case. The main difficulty to be overcome is that one can no longer preassign the topological type of the parameter domain on which the desired minimizer will be defined. The regularity of the movable part Σ of the boundary of the area-minimizing surface will be investigated in Section 5.3. The main result is that Σ is a regular real analytic arc of constant curvature.

5.1 Experiments and Examples. Mathematical Formulation of the Simplest Thread Problem

Imagine N points $P_1, P_2, \dots, P_N$ in $\mathbb{R}^3$ which are connected by k fixed arcs $\Gamma_1, \dots, \Gamma_k$ and by l movable arcs $\Sigma_1, \Sigma_2, \dots, \Sigma_l$ in such a way that the resulting configuration $\langle \Gamma, \Sigma \rangle := \langle \Gamma_1, \dots, \Gamma_k, \Sigma_1, \dots, \Sigma_l \rangle$ consists of n disjoint closed curves $C_1, C_2, \dots, C_n$ of finite length. The lengths of the arcs Σ_j are thought to be fixed. Experimentally we can realize the points $P_1, \dots, P_N$ as small holes in a plate or as endpoints of thin rods stuck in a plate. The arcs

U. Dierkes, S. Hildebrandt, A.J. Tromba, *Regularity of Minimal Surfaces*,
Grundlehren der mathematischen Wissenschaften 340,
DOI 10.1007/978-3-642-11700-8_5, © Springer-Verlag Berlin Heidelberg 2010

Γ_i are made of thin rigid wires, and the curves Σ_j can be realized by thin and essentially weightless synthetic fibres. Into such a boundary configuration we want to span a surface of minimal area, which can experimentally be achieved by dipping the array into a soap solution and then withdrawing it. This way a soap film will be generated which models a surface of minimal area within the configuration. The following figures show a few such experiments. One obtains particularly attractive and surprising results if all arcs are flexible, and one may very well assume that several of the threads Σ_j form closed loops which, by flexible connections, are attached to the ends of supporting rods. The resulting soap films will often be *multiply connected* minimal surfaces.

We may also conceive boundary configurations consisting of wires Γ_i, of threads Σ_j, and of supporting surfaces $S_1, \ldots, S_m$ on which parts of the boundary of the soap film are allowed to move freely.

Still different soap film experiments can be carried out by using threads as supporting ridges. This leads to a kind of mathematical questions which are to be viewed as *obstacle problems* with movable thin obstacles. Apparently such questions have not yet been treated.

We want to mention that thread experiments are used by architects to design light weight structures such as roofs and tents. Beautiful models are depicted in the publications of Frei Otto and collaborators (cf. Otto [1], Glaeser [1]).

Let us now consider the simplest case of a thread problem that was already mentioned in the introduction. Here we want to minimize area among all surfaces spanned in a boundary frame that consists of a fixed rectifiable Jordan arc Γ and of a movable curve Σ of given length L, having the same endpoints P_1 and P_2 as $\Gamma, P_1 \neq P_2$. We note that the thread experiment may lead to solutions which are no longer connected surfaces but disintegrate into several components, even if Γ is a smooth arc. One can even envision boundary configurations $\langle \Gamma, \Sigma \rangle$ for which the solution of the thread problem decomposes into countably many components since the movable arc Σ may in part adhere to the fixed arc Γ. The existence result to be described in the next section will take this phenomenon into account. We shall obtain solutions that are parametrized on a compact connected parameter domain B, the interior $\mathring{B}$ of which consists of at most countably many components.

Let us now specify *the mathematical setting of the thread problem* $\mathcal{P}(\Gamma, L)$ that will be solved in the following section.

Notational Convention. In Sections 5.1 and 5.2 we shall, *deviating from our usual notation, denote a disk of center z_0 and radius r by $B(z_0, r)$ instead of $B_r(z_0)$.*

An *admissible parameter domain for the thread problem* is defined to be a compact set B which can be represented in the form

$$(1) \qquad B = [-1, 1] \cup \bigcup_{\nu=1}^{\nu_B} B_\nu, \quad 1 \leq \nu_B \leq \infty.$$

Fig. 1a. Thread experiments. Courtesy of Institut für Leichte Flächentragwerke, Stuttgart – Archive

Here the sets B_ν with $\nu \in \mathbb{N}$ and $\nu \le \nu_B$ denote the closures of mutually disjoint disks $B(u_\nu, r_\nu), r_\nu > 0$ whose centers u_ν are contained in the open interval $\{u\colon -1 < u < 1\}$ on the real axis. Moreover, all disks B_ν are supposed to be contained in the unit disk $B(0,1)$.

Introducing the numbers a_ν and b_ν by

$$a_\nu := u_\nu - r_\nu, \quad b_\nu := u_\nu + r_\nu, \tag{2}$$

we then have

$$a_\nu, b_\nu \in [-1, 1]. \tag{3}$$

Let us denote the set of all admissible parameter domains B by $\mathcal{B}$.

For every $B \in \mathcal{B}$, we introduce the two mappings p_B^+ and $p_B^-\colon [-1,1] \to \partial B$ by

$$p_B^\pm(u) := \begin{cases} u & \\ & \text{if} \\ u \pm i\sqrt{r_\nu^2 - (u - u_\nu)^2} & \end{cases} \begin{matrix} u \in \partial B \cap [-1,1] \\ \\ |u - u_\nu| \le r_\nu. \end{matrix} \tag{4}$$

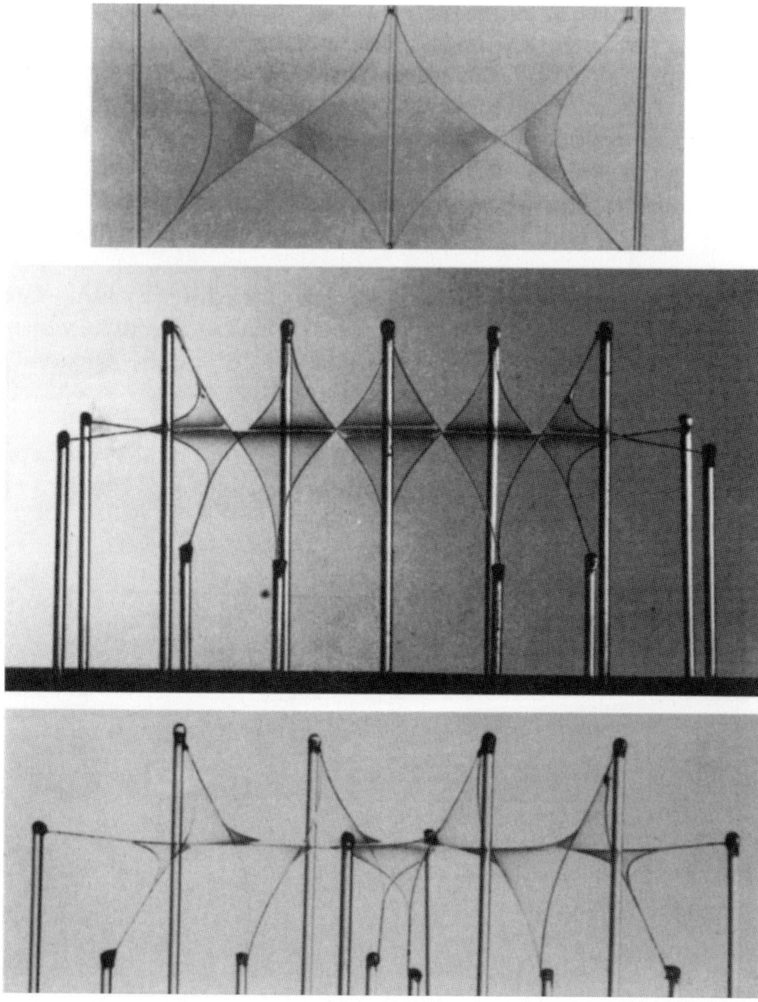

Fig. 1b. Thread experiments. Courtesy of Institut für Leichte Flächentragwerke, Stuttgart – Archive

Let c be a curve mapping a subinterval $I' = [\alpha, \beta]$ of $I = [-1, 1]$ into $\mathbb{R}^3$,

$$c \colon I' \to \mathbb{R}^3.$$

Then the length of c is given by

$$l(c, I') = \sup \sum_{j=1}^{n} |c(t_j) - c(t_{j-1})| \tag{5}$$

where the supremum is to be taken with respect to all possible decompositions $\alpha = t_0 < t_1 < t_2 < \cdots < t_n = \beta$ of I'.

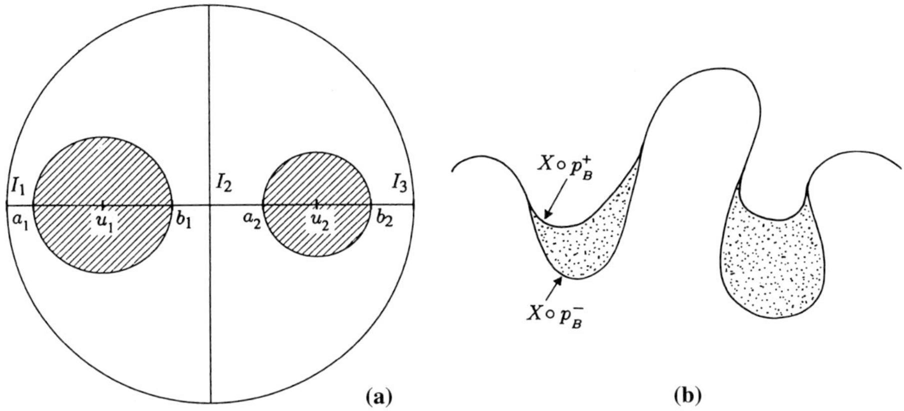

Fig. 2. (**a**) A parameter domain, and (**b**) a corresponding solution to the thread problem consisting of two components

If $I' = I$ we shall write

$$l(c) := l(c, I).$$

For any two intervals I_1 and I_2 in $\mathbb{R}$ we introduce the set $\mathcal{M}(I_1, I_2)$ of continuous, nondecreasing mappings $\theta\colon I_1 \to I_2$ of I_1 onto I_2, and we set $\mathcal{M}(I) := \mathcal{M}(I, I)$.

We observe that the length L of the movable curve Σ is bounded from below by the distance of its endpoints P_1 and P_2,

$$|P_1 - P_2| \le L. \tag{6}$$

Given a rectifiable Jordan curve Γ with endpoints P_1, P_2, and a number L satisfying $0 < |P_1 - P_2| < L$, we are now going to define the *set* $\mathcal{C}(\Gamma, L)$ *of admissible surfaces* X *for the thread problem* as follows:

Definition 1. *The set* $\mathcal{C}(\Gamma, L)$ *consists of the mappings* $X \in C^0(B, \mathbb{R}^3) \cap \mathring{H}_2^1(\mathring{B}, \mathbb{R}^3)$ *with* $B \in \mathcal{B}$ *which satisfy the following two conditions:*

(i) $l(X \circ p_B^+) \le L$;

(ii) *there exists some mapping* $\theta \in \mathcal{M}(I), I = [-1, 1]$, *such that* $\theta|_{\partial B \cap I} = \mathrm{id}|_{\partial B \cap I}$ *and* $X \circ p_B^- = \gamma \circ \theta$ *where* γ *denotes a fixed Lipschitz continuous representation of* Γ *which maps* I *bijectively onto* Γ.

In other words, a function X is admissible if it is parametrized on some domain $B \in \mathcal{B}$, if it is continuous and has a finite Dirichlet integral, if the length of the free part $X \circ p_B^+$ is less or equal to L, and if $X \circ p_B^-$ yields a weakly monotonic parametrization of Γ. Note that Γ and Σ may have one or more interior points in common, that is, Σ may in part adhere to Γ.

The *thread problem* $\mathcal{P}(\Gamma, L)$ now consists in finding some surface $X \in \mathcal{C}(\Gamma, L)$, defined on some parameter domain $B \in \mathcal{B}$, such that X minimizes the Dirichlet integral

(7) $$D(X, \mathring{B}) = \frac{1}{2} \int_{\mathring{B}} |\nabla X|^2 \, du \, dv$$

among all surfaces of $\mathcal{C}(\Gamma, L)$.

The solution of this problem will be carried out in two steps. First we shall single out a set $B \in \mathcal{B}$ which can serve as a parameter domain of a solution of $\mathcal{P}(\Gamma, L)$; this is the nonstandard part of the construction. We shall obtain such domains B as minimal elements with respect to inclusion. In a second step we shall construct a minimizing mapping X parametrized over B.

Let us now introduce the following three infima d, d^+, and d^-:

(8) $d = d(\Gamma, L) := \inf\{D(X, \mathring{B}) \colon X \in \mathcal{C}(\Gamma, L)\}$;

(9) $d^+ = d^+(\Gamma, L) := \inf\{D(X) \colon X \in \mathcal{C}(\Gamma, L), B = \overline{B}(0,1)\}$, where $D(X) := D(X, B(0,1))$;

(10) $d^- = d^-(\Gamma, L) := \inf\{\delta \colon \delta \text{ has the approximation property } (\mathcal{A})\}$.

The approximation property $(\mathcal{A})$ is defined as follows: *There exists some decreasing sequence of real numbers* $\lambda_n > 0$ *with* $\lambda_n \to 0$ *and a sequence of surfaces* $X_n \in \mathcal{C}(\Gamma, L + \lambda_n)$ *with parameter domains* $B_n \in \mathcal{B}$ *such that* $D(X_n, \mathring{B}_n) \to \delta$ *as* $n \to \infty$.

An obvious consequence of these definitions is the relation

(11) $$d^- \le d \le d^+.$$

We shall prove that

$$d^- = d = d^+$$

holds provided that we assume

$$|P_1 - P_2| < L.$$

In what follows we have to characterize a minimal parameter domain B among all domains in $\mathcal{B}$. To this end it will be convenient to single out a certain subclass $\mathcal{B}^*(\Gamma, L)$ of $\mathcal{B}$ which is defined as follows:

Definition 2. $\mathcal{B}^*(\Gamma, L)$ *is the class of admissible parameter domains* $B \in \mathcal{B}$ *with the following property: There exists a decreasing sequence of positive numbers* λ_n *with* $\lambda_n \to 0$ *and a sequence of surfaces* $X_n \in \mathcal{C}(\Gamma, L + \lambda_n)$, *parametrized over* B, *such that* $D(X_n, \mathring{B}) \to d^-$ *as* $n \to \infty$.

5.2 Existence of Solutions to the Thread Problem

Consider now the particular case $\mathcal{P}(\Gamma, L)$ of the thread problem that was formulated at the end of the previous section. Our main goal is the proof of the following existence result which is formulated as

Theorem 1. *Suppose that* $|P_1 - P_2| < L < l(\Gamma)$*. Then we obtain*

$$d^-(\Gamma, L) = d(\Gamma, L) = d^+(\Gamma, L).$$

Moreover, there exists an admissible parameter domain B *and a surface* $X \in \mathcal{C}(\Gamma, L)$ *parametrized over* B *such that*

$$D(X, \mathring{B}) = d(\Gamma, L).$$

This minimizer X *is a minimal surface, that is,* X *is of class* $C^2(\mathring{B}, \mathbb{R}^3)$ *and satisfies the equations*

$$\Delta X = 0,$$

$$|X_u|^2 = |X_v|^2, \quad \langle X_u, X_v \rangle = 0,$$

in $\mathring{B}$*, and furthermore, the free boundary of* X *is of maximal length, i.e.,*

$$l(X \circ p_B^+) = L.$$

The proof of this theorem is divided into two parts. The first one is concerned with the existence of a minimal parameter domain $B \in \mathcal{B}$. In the second part of our discussion we will show that such a parameter set B is the domain of a solution X for the thread problem $\mathcal{P}(\Gamma, L)$. This will be achieved by establishing the existence of a minimizing sequence $\{X_n\}$ whose elements are defined on B and converge to a solution X of $\mathcal{P}(\Gamma, L)$.

PART I. *Construction of a Minimal Parameter Set* $B \in \mathcal{B}$.

We begin our discussion with the following

Lemma 1. *Suppose that* X *is a surface of class* $\mathcal{C}(\Gamma, L)$ *which is defined on* $B \in \mathcal{B}$*, and let* ε *be an arbitrary positive number. Then there exists some* $X_\varepsilon \in \mathcal{C}(\Gamma, L + \varepsilon)$*, parametrized over* $\overline{B}(0,1)$*, such that*

$$|D(X_\varepsilon) - D(X, \mathring{B})| < \varepsilon.$$

(Recall that $D(X_\varepsilon)$ denotes the Dirichlet integral with the unit disk $B(0,1)$ as domain of integration.)

Proof. An admissible domain B is of the form given by formula (1) of Section 5.1. Since $l(\Gamma) < \infty$ and

$$D(X, \mathring{B}) = \sum_{\nu=1}^{\nu_B} D(X, \mathring{B}_\nu) < \infty,$$

we can find a number $\nu_0 \in \mathbb{N}$ such that

$$\sum_{\nu>\nu_0} D(X, \mathring{B}_\nu) < \varepsilon$$

and

$$\sum_{\nu>\nu_0} l(\gamma, [a_\nu, b_\nu]) < \varepsilon.$$

Set

$$B' := I \cup B_1 \cup B_2 \cup \cdots \cup B_{\nu_0}, \quad I = [-1, 1],$$

and

$$X_1(w) := \begin{cases} X(w) & \text{if } w \in B_1 \cup B_2 \cup \cdots \cup B_{\nu_0}, \\ \gamma(w) & \text{if } w \in \partial B' \cap [-1, 1]. \end{cases}$$

Then we infer $X_1 \in \mathcal{C}(\Gamma, L + \varepsilon)$ and

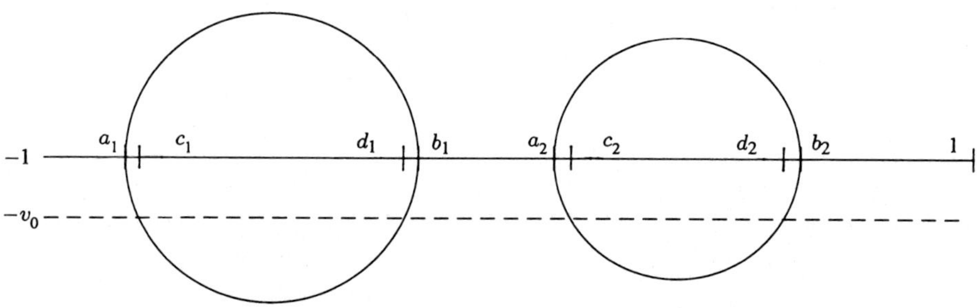

Fig. 1. A parameter domain with $\nu_0 = 2$, and the numbers $a_\nu, c_\nu, d_\nu, b_\nu$

(1) $$|D(X, \mathring{B}) - D(X_1, \mathring{B}')| \le \varepsilon.$$

For each v_0 with $0 < v_0 < \min\{r_1, r_2, \ldots, r_{\nu_0}\}$, there exist numbers c_ν, d_ν with $a_\nu < c_\nu < d_\nu < b_\nu$ such that $p^-_{B'}(c_\nu) = c_\nu - iv_0$, $p^-_{B'}(d_\nu) = d_\nu - iv_0$; cf. Fig. 1.

Now we choose v_0 so small that also the following conditions are fulfilled:

(i) $X_1(u - iv_0)$ is absolutely continuous with respect to $u \in \bigcup_{\nu \le \nu_0} [c_\nu, d_\nu]$ and has a square integrable first derivative;

(ii) $l(X_1 \circ p^-_{B'}, [a_\nu, c_\nu]) + l(X_1 \circ p^-_{B'}, [d_\nu, b_\nu]) \le \frac{\varepsilon}{2\nu_0}$.

For some arbitrary number $\delta > 0$, we define the set

$$\mathcal{D} = \mathcal{D}(\delta) := \overline{\mathcal{D}}_+ \cup \overline{\mathcal{D}}_- \cup \overline{\mathcal{Q}}$$

by

$$\begin{aligned} \mathcal{Q} &:= \{w = u + iv \colon |u| < 1, 0 < v < \delta\}, \\ \mathcal{D}_+ &:= \{w = u + iv \colon w - i(\delta + v_0) \in \mathring{B}' \text{ and } v > \delta\}, \\ \mathcal{D}_- &:= \{w = u + iv \colon w - iv_0 \in \mathring{B}' \text{ and } v < 0\}. \end{aligned}$$

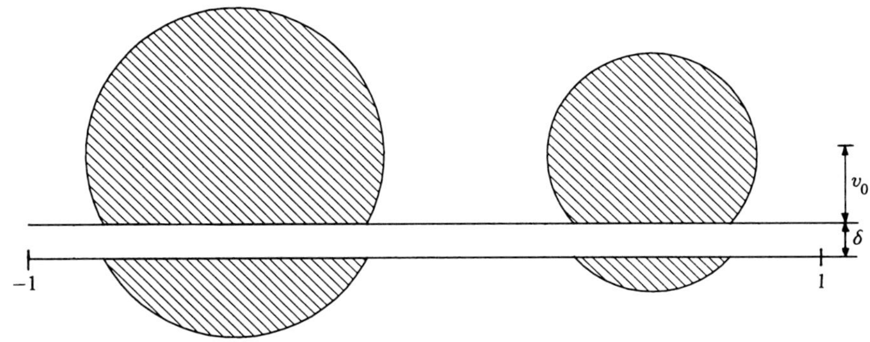

Fig. 2. The domain $\mathcal{D} = \mathcal{D}(\delta)$

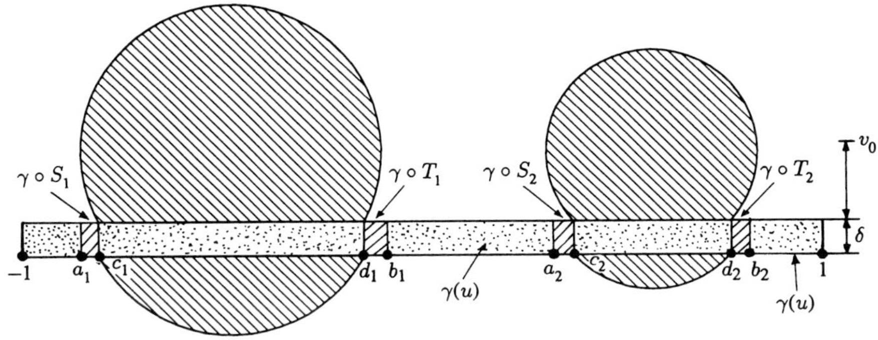

Fig. 3. The definition of X_2

We note that $\mathring{\mathcal{D}}$ is conformally equivalent to the unit disk. Thus, in order to prove the assertion of the lemma, we shall construct a suitable comparison function X_2 defined on $\mathcal{D}$. This function is defined as follows:

$$X_2(w) := \begin{cases} X_1(w - i(\delta + v_0)) & \text{if } w \in \overline{\mathcal{D}}_+, \\ X_1(w - iv_0) & \text{if } w \in \overline{\mathcal{D}}_-. \end{cases}$$

For $0 \le v \le \delta$, we set

$$X_2(w) := \begin{cases} X_1(u - iv_0) & u \in [c_\nu, d_\nu],\ 1 \le \nu \le \nu_0 \\ \gamma(u) & u \in [-1,1] \setminus \bigcup_{\nu \le \nu_0} (a_\nu, b_\nu) \\ & \text{if} \\ \gamma(S_\nu(u)) & u \in [a_\nu, c_\nu] \\ \gamma(T_\nu(u)) & u \in [d_\nu, b_\nu]. \end{cases}$$

Here S_ν is a linear mapping from $[a_\nu, c_\nu]$ onto $[a_\nu, \theta_1(c_\nu)]$, and T_ν is the linear map from $[d_\nu, b_\nu]$ onto $[\theta_1(d_\nu), b_\nu]$, where $\theta_1 \in \mathcal{M}(I)$ is the transformation $I \to I$ that corresponds to X_1. In other words, $\gamma \circ \theta_1 = X_1 \circ p_{\overline{B}'}$.

We infer from the construction that X_2 is of class $C^0(\overline{\mathcal{D}}) \cap H_2^1(\mathring{\mathcal{D}})$. Furthermore, we have

$$D(X_2, \mathring{\mathcal{D}}_+ \cup \mathring{\mathcal{D}}_-) = D(X_1, B') \tag{2}$$

and

$$\begin{aligned} D(X_2, \mathring{\mathcal{Q}}) &\le \frac{1}{2}\int_{\mathcal{Q}} |\nabla X_2|^2\, du\, dv = \frac{\delta}{2}\int_{-1}^{1} |D_u X_2|^2\, du \\ &\le \delta \sum_{\nu=1}^{\nu_0} \int_{c_\nu}^{d_\nu} |D_u X_1(u - iv_0)|^2\, du + \delta \int_{-1}^{1} |\dot\gamma(t)|^2\, dt. \end{aligned} \tag{2'}$$

We, moreover, note that the mapping

$$X_2 \colon \partial\mathcal{D} \cap \{\operatorname{Im} w \le 0\} \to \Gamma$$

is weakly monotonic, and from (ii) we derive the estimate

$$\begin{aligned} &l(X_2, \partial\mathcal{D} \cap \{\operatorname{Im} w > 0\}) \\ &\le l(X_1 \circ p_{B'}^+) + 2 \sum_{\nu \le \nu_0} \{l(X_1 \circ p_{B'}^-, [a_\nu, c_\nu]) + l(X \circ p_{B'}^-, [d_\nu, b_\nu])\} \\ &\le l(X_1 \circ p_{B'}^+) + \varepsilon \le L + 2\varepsilon. \end{aligned}$$

Now let $\tau \colon \overline{B}(0,1) \to \mathcal{D}$ be a conformal mapping of $B(0,1)$ onto $\mathring{\mathcal{D}}$, leaving the two points $w = \pm 1$ fixed. Then $X_\varepsilon := X_2 \circ \tau$ is of class $\mathcal{C}(\Gamma, L + 2\varepsilon)$, and we infer

$$\begin{aligned} |D(X_\varepsilon) - D(X, \mathring{B})| &= |D(X_2, \mathring{\mathcal{D}}) - D(X, \mathring{B})| \\ &\le |D(X_2, \mathring{\mathcal{D}}) - D(X_1, \mathring{B}')| + |D(X_1, \mathring{B}') - D(X, \mathring{B})| \\ &\le \delta \cdot \text{const} + \varepsilon, \end{aligned}$$

taking (1), (2) and (2′) into account. Since we can choose $\delta > 0$ arbitrarily small, the assertion of Lemma 1 is proved. □

The following result is an easy consequence of Lemma 1.

Proposition 1. *The class $\mathcal{C}(\Gamma, L)$ is nonvoid, and $\overline{B}(0,1) \in \mathcal{B}^*(\Gamma, L)$.*

Proof. Define

$$\gamma^*(e^{it}) := \begin{cases} \gamma(1) + \frac{t}{\pi}[\gamma(-1) - \gamma(1)] & \text{if } 0 \le t \le \pi \\ \gamma\left(-1 + 2\frac{t-\pi}{\pi}\right) & \text{if } \pi \le t \le 2\pi, \end{cases}$$

where $\gamma(-1) = P_1$ and $\gamma(1) = P_2$. Then $\gamma^* \colon \partial B \to \mathbb{R}^3$ is Lipschitz continuous, and a straight-forward computation shows that $X^*(w) := |w|\gamma^*(\frac{w}{|w|})$ is of class $\mathcal{C}(\Gamma, L)$. Hence $\mathcal{C}(\Gamma, L)$ is nonempty.

It follows from the definition of d^- that there is a sequence of surfaces $X_n \in \mathcal{C}(\Gamma, L + \frac{1}{n})$ parametrized on domains $B_n \in \mathcal{B}$ such that

$$|D(X_n, \overset{\circ}{B}_n) - d^-| < \frac{1}{n} \quad \text{for all } n \in \mathbb{N}.$$

By virtue of Lemma 1, we can choose a sequence of mappings $X_n^* \in \mathcal{C}(\Gamma, L+\frac{2}{n})$ which are parametrized over $B(0,1)$ and satisfy

$$|D(X_n^*) - D(X_n, \overset{\circ}{B}_n)| \le \frac{1}{n}, \qquad n = 1, 2, \ldots .$$

Thus we infer

$$|D(X_n^*) - d^-| \le \frac{2}{n}, \qquad n = 1, 2, \ldots ,$$

and it follows that $\overline{B}(0,1) \in \mathcal{B}^*(\Gamma, L)$. □

In the next lemma we prove the existence of sets $B \in \mathcal{B}^*(\Gamma, L)$ which are minimal with respect to an ordering of sets defined by inclusion.

Lemma 2. *Suppose that $L < l(\Gamma)$. Then any set of elements $B \in \mathcal{B}^*(\Gamma, L)$ which is totally ordered with respect to inclusion possesses an infimum in $\mathcal{B}^*(\Gamma, L)$.*

Proof. Let $\{B_\alpha^*\}_{\alpha \in A}$ be an arbitrary set of elements $B_\alpha^* \in \mathcal{B}^*(\Gamma, L)$ with the index set A which is totally ordered with respect to inclusion, and set

$$B := \bigcap_{\alpha \in A} B_\alpha^* .$$

We have to show that B is an element of $\mathcal{B}^*(\Gamma, L)$. The first step will be to prove

(i) $$\overset{\circ}{B} \ne \emptyset .$$

In fact, if $\overset{\circ}{B}$ were empty, we would have

$$I = \operatorname{clos} \left(\bigcup_{\alpha \in A} (I \setminus \overset{\circ}{B}{}_\alpha^*) \right), \qquad I := [-1, 1].$$

Then, for any partition

$$-1 = t_0 < t_1 < t_2 < \cdots < t_k = 1$$

of I, there exist numbers $t_j^n \in \bigcup_{\alpha \in A} (I \setminus \overset{\circ}{B}{}_\alpha^*)$ with $0 \le j \le k$ and $n \in \mathbb{N}$ such that

$$\lim_{n \to \infty} t_j^n = t_j \quad \text{for } j = 0, 1, \ldots, k.$$

Since the set $\{B^*_\alpha\}_{\alpha\in A}$ is totally ordered, we infer that, for every $n \in \mathbb{N}$, there exists an index $\alpha_n \in A$ such that $t^n_j \in I \setminus \mathring{B}^*_{\alpha_n}$ holds for all $j = 0, \ldots, k$. As all domains $B^*_{\alpha_n}$ are contained in $\mathcal{B}^*(\Gamma, L), n = 1, 2, \ldots$, there exist surfaces $X_n \in \mathcal{C}(\Gamma, L + \frac{1}{n})$ parametrized over $B^*_{\alpha_n}$. This implies

$$\sum_{j=1}^{k} |\gamma(t^n_j) - \gamma(t^n_{j-1})| = \sum_{j=1}^{k} |X_n(t^n_j) - X_n(t^n_{j-1})|$$
$$\le l(X_n \circ p^+_{B^*_{\alpha_n}}) \le L + \frac{1}{n} \to L \quad \text{as } n \to \infty.$$

Since

$$\lim_{n\to\infty} \sum_{j=1}^{k} |\gamma(t^n_j) - \gamma(t^n_{j-1})| = \sum_{j=1}^{k} |\gamma(t_j) - \gamma(t_{j-1})|,$$

we arrive at

$$\sum_{j=1}^{k} |\gamma(t_j) - \gamma(t_{j-1})| \le L.$$

As the partition $t_0, t_1, \ldots, t_k$ of I may be chosen arbitrarily, we conclude that

$$l(\Gamma) \le L$$

which contradicts our assumption $l(\Gamma) > L$.

Now we turn to the proof of

(ii) $$B \in \mathcal{B}^*(\Gamma, L).$$

We have to find surfaces defined on B whose Dirichlet integrals converge to d^-, and whose free boundaries (threads) exceed L only by an arbitrarily small amount.

First of all, for every $\varepsilon > 0$ there exists some $\nu_0 \in \mathbb{N}$ with $1 \le \nu_0 \le \nu_B$ such that

(3) $$\sum_{\nu\ge\nu_0} l(\gamma, [a_\nu, b_\nu]) \le \varepsilon.$$

For $\nu \ge \nu_0$ we define

$$Q_\nu := \{w = u + iv\colon a_\nu \le u \le b_\nu, |v| \le \varepsilon 2^{-\nu-1}\}$$

and choose conformal mappings $\tau_\nu\colon \mathring{B}_\nu \to \mathring{Q}_\nu$ from $\mathring{B}_\nu$ onto $\mathring{Q}_\nu$ with fixed points a_ν, b_ν. Here $B_1, B_2, \ldots$ denote the components of the domain B (cf. Section 5.1, (1)). Then the surfaces

$$X_\nu := \gamma(\operatorname{Re} \tau_\nu), \quad \nu \ge \nu_0,$$

are continuous and have the Dirichlet integrals

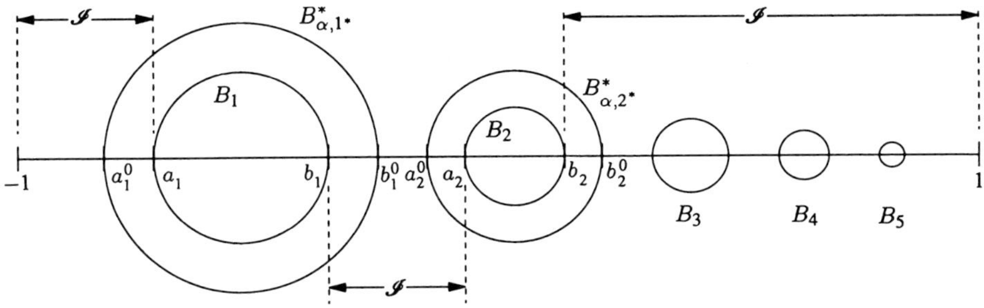

Fig. 4. The case $\nu_0 = 2$

$$(4)\quad D(X_\nu, \mathring{B}_\nu) = \frac{1}{2}\int_{B_\nu} |\nabla X_\nu|^2\,du\,dv$$
$$= \frac{1}{2}\int_{a_\nu}^{b_\nu}\int_{-\varepsilon 2^{-\nu-1}}^{\varepsilon 2^{-\nu-1}} |\dot\gamma(u)|^2\,du\,dv = \varepsilon 2^{-\nu-1}\int_{a_\nu}^{b_\nu} |\dot\gamma(u)|^2\,du.$$

Moreover, $X_\nu \circ p_{\overline{B}}$ is monotonic on $[\alpha_\nu, b_\nu]$.

For $\alpha \in A$ and $\nu \in \{1, 2, \dots, \nu_0\}$ there is a uniquely determined $\nu^* = \nu^*(\nu, \alpha)$ with $1 \le \nu^* \le \nu_{B^*_\alpha}$ such that $B_\nu \subset B^*_{\alpha,\nu^*}$. Here B^*_{α,ν^*} is the ν^*-th component of the domain B^*_α; cf. Section 5.1, (1). Since $\{B^*_\alpha\}_{\alpha\in A}$ is totally ordered, we infer from the definition of B that there is an index $\alpha_0 \in A$ such that the disks $\mathring{B}^*_{\alpha_0,\nu^*}$ are mutually disjoint, and that

$$(5)\qquad \int_{a_\nu^0}^{a_\nu} |\dot\gamma|\,dt + \int_{b_\nu}^{b_\nu^0} |\dot\gamma|\,dt \le \frac{\varepsilon}{\nu_0}$$

holds. Here a_ν and b_ν are the numbers associated with B which are defined in formula (2) of Section 5.1, and a_ν^0, b_ν^0 are the corresponding numbers for $B^*_{\alpha_0,\nu^*}$, i.e. $[a_\nu^0, b_\nu^0] := I \cap \overline{B}^*_{\alpha_0,\nu^*}$.

We have finitely many (at most $\nu_0 + 1$) open intervals $\mathcal{J} \subset I$ such that

$$(6)\qquad \int_{-1}^{1} |\dot\gamma|\,dt - \sum_{\nu\le\nu_0}\int_{a_\nu}^{b_\nu} |\dot\gamma|\,dt = \sum_{\{\mathcal{J}\}}\int_{\mathcal{J}} |\dot\gamma|\,dt.$$

For any such interval $\mathcal{J}$, there is a partition

$$t_0 < t_1 < t_2 < \dots < t_k, \quad t_j \in \mathcal{J},$$

such that

$$(7)\qquad \int_{\mathcal{J}} |\dot\gamma| dt \le \frac{\varepsilon}{\nu_0 + 1} + \sum_{j=1}^{k} |\gamma(t_j) - \gamma(t_{j-1})|.$$

Passing to a suitable refinement of this partition, we may also assume that there is a subset $\mathcal{K}$ of $\{1, 2, \dots, k\}$ with the following properties:

(I) $j \notin \mathcal{K}$ if and only if $(t_{j-1}, t_j) \subset (a_\nu, b_\nu)$ for some $\nu > \nu_0$;
(II) $j \in \mathcal{K}$ if and only if $[t_{j-1}, t_j] \subset \mathcal{I} \setminus \mathring{B}$.

Moreover we can choose an index $\alpha_0 \in A$ such that the following can be achieved:

The values $t_j \in \mathcal{I} \setminus \mathring{B}$ are replaced by values $t'_j \in \mathcal{I} \setminus \mathring{B}^*_{\alpha_0}$; all other values t_j remain unaltered and will be called t'_j; we have $t'_0 < t'_1 < \cdots < t'_k$ and

$$\sum_{j \in \mathcal{K}} |\gamma(t_j) - \gamma(t_{j-1})| \le \frac{\varepsilon}{\nu_0 + 1} + \sum_{j \in \mathcal{K}} |\gamma(t'_j) - \gamma(t'_{j-1})|. \tag{8}$$

We infer from (3), (7), and (8) that

$$\begin{aligned} \sum_{\{\mathcal{I}\}} \int_{\mathcal{I}} |\dot\gamma| \, dt &\le \varepsilon + \sum_{\nu > \nu_0} \int_{a_\nu}^{b_\nu} |\dot\gamma| \, dt + \sum_{\{\mathcal{I}\}} \sum_{j \in \mathcal{K}(\mathcal{I})} |\gamma(t_j) - \gamma(t_{j-1})| \\ &\le 3\varepsilon + \sum_{\{\mathcal{I}\}} \sum_{j \in \mathcal{K}(\mathcal{I})} |\gamma(t'_j) - \gamma(t'_{j-1})|. \end{aligned} \tag{9}$$

After these preparations, we proceed as follows: Since $B^*_{\alpha_0} \in \mathcal{B}^*(\Gamma, L)$, there is some surface $X \in \mathcal{C}(\Gamma, L + \varepsilon)$ defined on $B^*_{\alpha_0}$ such that

$$D(X, \mathring{B}^*_{\alpha_0}) \le d^- + \varepsilon. \tag{10}$$

Furthermore, for each ν with $1 \le \nu \le \nu_0$, there exists a conformal mapping $\tau_\nu \colon B_\nu \to B^*_{\alpha_0, \nu^*}$ such that $X \circ \tau_\nu \circ p_{B_\nu^-}$ furnishes a monotonic parametrization of that subarc of Γ which corresponds to $[a_\nu, b_\nu]$. Then $X'_\nu := X \circ \tau_\nu$ defines a continuous surface defined on B_ν satisfying

$$D(X'_\nu, \mathring{B}_\nu) = D(X, \mathring{B}^*_{\alpha_0, \nu^*}). \tag{11}$$

By virtue of (5) we obtain

$$l(X'_\nu \circ p^+_B, [a_\nu, b_\nu]) \le l(X \circ p^+_{B^*_{\alpha_0}}, [a^0_\alpha, b^0_\nu]) + \frac{\varepsilon}{\nu_0}. \tag{12}$$

Let us introduce the surface X_ε by

$$X_\varepsilon(w) := \begin{cases} X'_\nu(w) & \text{if } w \in B_\nu, 1 \le \nu \le \nu_0; \\ \gamma(w) & \phantom{\text{if }} w \in I \setminus \bigcup_{\nu=1}^{\nu_0} B_\nu. \end{cases}$$

Then we have $X_\varepsilon \in H^1_2(\mathring{B}, \mathbb{R}^3) \cap C^0(B, \mathbb{R}^3)$, and it follows from (4), (10) and (11) that

$$\begin{aligned} D(X_\varepsilon, \mathring{B}) &\le D(X, \mathring{B}^*_{\alpha_0}) + \sum_{\nu > \nu_0} \varepsilon 2^{-\nu-1} \int_{-1}^{1} |\dot\gamma|^2 \, dt \\ &\le d^-(\Gamma, L) + \varepsilon + \varepsilon \int_{-1}^{1} |\dot\gamma|^2 \, dt. \end{aligned} \tag{13}$$

The length of the movable part of the boundary of X_ε is estimated by

$$
\begin{aligned}
(14) \quad & l(X_\varepsilon \circ p_B^+) \\
& \leq l \sum_{\nu \leq \nu_0} l(X'_\nu \circ p_B^+, [a_\nu, b_\nu]) + \sum_{\{\mathcal{J}\}} \int_{\mathcal{J}} |\dot{\gamma}| \, dt \\
& \leq \varepsilon + \sum_{\nu \leq \nu_0} l(X \circ p^+_{B^*_{\alpha_0}}, [a^0_\nu, b^0_\nu]) + 3\varepsilon + \sum_{\{\mathcal{J}\}} \sum_{j \in \mathcal{K}} |\gamma(t'_j) - \gamma(t'_{j-1})| \\
& = 4\varepsilon + \sum_{\nu \leq \nu_0} l(X \circ p^+_{B^*_{\alpha_0}}, [a^0_\nu, b^0_\nu]) + \sum_{\{\mathcal{J}\}} \sum_{j \in \mathcal{K}} |X(t'_j) - X(t'_{j-1})| \\
& \leq 4\varepsilon + l(X \circ p^+_{B^*_{\alpha_0}}) \leq 5\varepsilon + L,
\end{aligned}
$$

on account of (12), (9) and of $X \in \mathcal{C}(\Gamma, L + \varepsilon)$.

The relations (13) and (14) yield $B \in \mathcal{B}^*(\Gamma, L)$. □

Applying Zorn's lemma we infer from this lemma that the following result holds true:

Proposition 2. *The set $\mathcal{B}^*(\Gamma, L)$ possesses minimal elements with respect to inclusion, provided that $L < l(\Gamma)$.*

PART II. *Existence of a Solution of $\mathcal{P}(\Gamma, L)$.*

Let $B \in \mathcal{B}^*(\Gamma, L)$ be a minimal element the existence of which was established in Proposition 2. We want to prove that B is the parameter domain of some minimizer X.

Lemma 3. *If X is a function of class $H_2^1(B(0,1), \mathbb{R}^3)$ with a trace $\xi \in L_2(\partial B(0,1), \mathbb{R}^3)$ on the circle $\partial B(0,1)$ which is of finite total variation $\int_{\partial B(0,1)} |d\xi|$, then the boundary values $\xi \colon \partial B(0,1) \to \mathbb{R}^3$ actually are continuous.*

The *proof* of this result is an immediate consequence of the Courant–Lebesgue lemma and has essentially been carried out in part (iii) of the proof of Proposition 3 in Section 4.7 of Vol. 1. In fact, we even know that ξ is absolutely continuous (see Theorem 1 of Section 4.7 of Vol. 1).

Now we turn to the crucial step in proving Theorem 1, which is to prove

Theorem 2. *Let $B \in \mathcal{B}^*(\Gamma, L)$ be a minimal parameter domain with respect to inclusion. Then there exists some $X \in \mathcal{C}(\Gamma, L)$, parametrized over B, such that*

$$D(X, \mathring{B}) = d^-(\Gamma, L) = d(\Gamma, L);$$

thus X is a solution of the minimum problem $\mathcal{P}(\Gamma, L)$.

Proof. Since $B \in \mathcal{B}^*(\Gamma, L)$, there is a sequence of surfaces $X_n \in \mathcal{C}(\Gamma, L + \frac{1}{n})$, $n \in \mathbb{N}$, satisfying

$$D(X_n, \mathring{B}) \le d^-(\Gamma, L) + \frac{1}{n} \le M$$

for some constant $M > 0$. Denote by $\theta_n \in \mathcal{M}(I)$ the mappings associated with X_n, i.e.,

$$\theta_n|_{\partial B \cap I} = \mathrm{id}|_{\partial B \cap I}, \quad X_n \circ p_{\overline{B}}^- = \gamma \circ \theta_n.$$

Moreover, let $u_\nu, r_\nu, a_\nu, b_\nu$ be the numbers corresponding to B, and let B_ν, $1 \le \nu \le \nu_B$, be the components of B (see Section 5.1, (1) and (2)). Applying suitable conformal reparametrizations, we can achieve that

$$\theta_n(u_\nu) = u_\nu \quad \text{for } n \in \mathbb{N} \text{ and } 1 \le \nu \le \nu_B.$$

We claim that *the mappings* $\theta_n|_{[a_\nu, b_\nu]}$, $n \in \mathbb{N}$, *are equicontinuous for every* $\nu \in \mathbb{N}$ *with* $\nu \le \nu_B$.

Otherwise we could find some $\varepsilon_0 > 0$, some $\nu \le \nu_B$ and two sequences $\{t_n\}, \{t_n'\}$ with $a_\nu \le t_n < t_n' \le b_\nu$, converging to some point $t_0 \in [a_\nu, b_\nu]$, such that

$$(15) \qquad |\theta_n(t_n) - \theta_n(t_n')| \ge \varepsilon_0 \quad \text{for all } n \in \mathbb{N}$$

is satisfied. (Actually, this would hold true for some subsequence of $\{\theta_n\}$. However, by renumbering this subsequence we could achieve that (15) is fulfilled.) We want to show that (15) leads to a contradiction. In order to do so, we distinguish the two cases (i) $a_\nu < t_0 < b_\nu$, and (ii) $t_0 = a_\nu$ or b_ν. Case (i) can be excluded by the discussion given in Chapter 4 of Vol. 1, where we have proved that the boundary values of a minimizing sequence for the ordinary Plateau problem are equicontinuous. By this reasoning we obtain that the functions $\gamma \circ \theta_n|_{[c_\nu, d_\nu]}$ are equicontinuous for every interval $[c_\nu, d_\nu] \subset (a_\nu, b_\nu)$. The injectivity of γ then implies that also the functions $\theta_n|_{[c_\nu, d_\nu]}$ are equicontinuous, which contradicts (15). In fact, there is some $n_0 \in \mathbb{N}$ such that

$$a_\nu < c_\nu \le t_n < t_n' \le d_\nu < b_\nu$$

holds for $n > n_0$ and for suitably chosen numbers c_ν and d_ν. Then it follows from (15) that

$$|\gamma(\theta_n(t_n)) - \gamma(\theta_n(t_n'))| \ge c(\varepsilon_0) > 0$$

for some fixed number $c(\varepsilon_0) > 0$ and for all $n > n_0$, which contradicts the equicontinuity of the sequence $\gamma \circ \theta_n|_{[c_\nu, d_\nu]}$. Thus case (i) cannot occur.

Now we want to exclude case (ii) as well.

It suffices to show that $t_0 = a_\nu$ is impossible since the case $t_0 = b_\nu$ can be handled analogously. Thus let us assume that $t_0 = a_\nu$.

We can choose sequences of numbers δ_n, r_n, and s_n' with $\delta_n \in (0, 1)$, $\delta_n \to 0$, $0 < r_n < \delta_n$, $t_n' < s_n' \le u_\nu$, $p_{\overline{B}}^-(s_n') \in \partial B(a_\nu, r_n)$, and with

$$\left\{ \int \left| \frac{\partial}{\partial \varphi} X_n(r_n, \varphi) \right| d\varphi \right\}^2 \le \frac{2\pi M}{\log 1/\delta_n}.$$

Here r, φ denote polar coordinates around a_ν, and the integral on the left-hand side is extended over the φ-interval in $[-\pi, \pi]$ corresponding to the arc in $B \cap \partial B(a_\nu, r_n)$ which contains $\varphi = 0$; cf. Section 4.4 of Vol. 1, Lemma 1.

There is a subsequence of $\{\theta_n(s_n')\}$ converging to some value u_0; renumbering this sequence we may assume that $\theta_n(s_n') \to u_0$ as $n \to \infty$. By virtue of (15) we have $u_0 \ge a_\nu + \varepsilon_0$.

Choose values s_n with $a_\nu \le s_n \le u_\nu$ and $\theta_n(s_n) = u_0$, and consider the two closed disks $\mathcal{D}_1$ and $\mathcal{D}_2$ defined by

$$\mathring{\mathcal{D}}_1 := B\left(\frac{a_\nu + u_0}{2}, \frac{u_0 - a_\nu}{2}\right), \quad \mathring{\mathcal{D}}_2 := B\left(\frac{u_0 + b_\nu}{2}, \frac{b_\nu - u_0}{2}\right).$$

Our aim is to define surfaces Y_n on $\mathcal{D}_1 \cup \mathcal{D}_2$ such that the surfaces $X_n^*: B^* \to \mathbb{R}^3$, given by

$$X_n^*(w) := \begin{cases} X_n(w) & \\ & \text{for} \\ Y_n(w) & \end{cases} \begin{matrix} w \in B \setminus B_\nu, \\ \\ w \in \mathcal{D}_1 \cup \mathcal{D}_2, \end{matrix} \tag{16}$$

$$B^* := (B \setminus B_\nu) \cup \mathcal{D}_1 \cup \mathcal{D}_2 \in \mathcal{B},$$

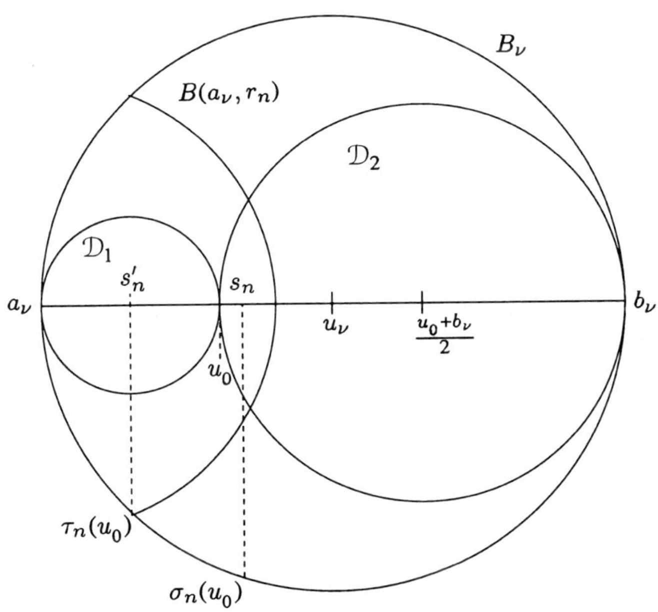

Fig. 5. The disks $\mathcal{D}_1, \mathcal{D}_2$, and B_ν

are of class $\mathcal{C}(\Gamma, L + \lambda_n)$ with $\lambda_n \to 0$ and satisfy

$$D(X_n, \mathring{B}^*) \to d^-(\Gamma, L)$$

as $n \to \infty$. This, clearly, would contradict the minimality of B, and therefore we would also have ruled out $t_0 = a_\nu$ (or b_ν), i.e. case (ii) cannot occur either.

Passing to a subsequence and then renumbering, we can achieve that either

$$s_n' \le s_n \quad \text{for all } n \in \mathbb{N}$$

or else

$$s_n \le s_n' \quad \text{for all } n \in \mathbb{N}$$

holds true. We only treat the first case; the second one can be dealt with in an analogous way.

Consider topological mappings

$$\tau_n \colon \mathcal{D}_1 \to \overline{B}(a_\nu, r_n) \cap B_\nu, \quad \sigma_n \colon \mathcal{D}_2 \to B_\nu \setminus B(a_\nu, r_n)$$

with $\tau_n(a_\nu) = a_\nu, \tau_n(u_0) = p_{\overline{B}}^-(s_n'), \sigma_n(b_\nu) = b_\nu, \sigma_n(u_0) = p_{\overline{B}}^-(s_n)$ such that $\mathring{\mathcal{D}}_1$ is conformally mapped onto $B(a_\nu, r_n) \cap \mathring{B}_\nu$ by τ_n, and that σ_n maps $\mathring{\mathcal{D}}_2$ conformally onto $\mathring{B}_\nu \setminus \overline{B}(a_\nu, r_n)$.

Note that

$$(X_n \circ \tau_n)(u_0) = X_n(p_{\overline{B}}^-(s_n')) = \gamma(\theta_n(s_n')) \to \gamma(u_0),$$
$$(X_n \circ \sigma_n)(u_0) = X_n(p_{\overline{B}}^-(s_n)) = \gamma(\theta_n(s_n)) \to \gamma(u_0).$$

If we had $(X_n \circ \tau_n)(u_0) = \gamma(u_0)$, we would simply define

$$Y_n := \begin{cases} X_n \circ \tau_n & \text{in } \mathcal{D}_1, \\ X_n \circ \sigma_n & \text{in } \mathcal{D}_2, \end{cases}$$

and the proof would be complete. As we only know $X_n \circ \tau_n(u_0) \to \gamma(u_0)$ as $n \to \infty$, we have to adjust the data correctly. The idea is the same as in the proof of Lemma 1: we have to fill in the missing parts of Γ, thereby slightly changing the Dirichlet integral and the length of the free boundary of X_n. This way we obtain from $X_n \circ \tau_n \colon \mathcal{D}_1 \to \mathbb{R}^3$ a new surface $(X_n \circ \tau_n)_{\delta_n} =: Z_n$ with $Z_n(u_0) = \gamma(u_0)$ such that

$$Y_n(w) := \begin{cases} Z_n(w) & \text{for } w \in \mathcal{D}_1, \\ X_n \circ \sigma_n(w) & \text{for } w \in \mathcal{D}_2 \end{cases}$$

satisfies both

$$D(Y_n, \mathring{\mathcal{D}}_1 \cup \mathring{\mathcal{D}}_2) \le D(X_n, \mathring{B}_\nu) + \delta_n$$

and

$$\begin{aligned} l(Y_n \circ p^+_{\mathcal{D}_1 \cup \mathcal{D}_2}, [a_\nu, b_\nu]) &\le l(X_n \circ p^+_B, [a_\nu, b_\nu]) + \delta_n + 2\int \left| \frac{\partial}{\partial\varphi} X_n(r_n, \varphi)\right| d\varphi \\ &\quad + 2l(\gamma, [\theta_n(s'_n), \theta_n(s_n)]) \\ &= l(X_n \circ p^+_B, [a_\nu, b_\nu]) + \lambda_n, \qquad \lim_{n\to\infty} \lambda_n = 0, \end{aligned}$$

and that the surfaces $X^*_n \colon B^* \to \mathbb{R}^3$ defined by (16) are of class $\mathcal{C}(\Gamma, L + \lambda_n)$, $\lambda_n \to 0$. This finishes the proof of equicontinuity of the mappings $\theta_n|_{[a_\nu,b_\nu]}$, $n \in \mathbb{N}$, for every $\nu \in \mathbb{N}$ with $1 \le \nu \le \nu_B$.

Now we can apply the reasoning of Chapter 4 of Vol. 1 to the sequence $\{X_n\}$ of surfaces $X_n \in \mathcal{C}(\Gamma, L+\frac{1}{n})$ which are defined on the minimal parameter domain $B \in \mathcal{B}^*(\Gamma, L)$ and satisfy

$$D(X_n, \mathring{B}) \le d^-(\Gamma, L) + \frac{1}{n} \le M \quad \text{for all } n \in \mathbb{N}. \tag{17}$$

From this inequality, together with

$$\sup_{\partial B} |X_n| \le M' \quad \text{for all } n \in \mathbb{N}$$

and some constant M' independent of n, we obtain that $\{X_n\}$ is a bounded sequence in $H^1_2(\mathring{B}, \mathbb{R}^3)$.

Passing to a suitable subsequence of X_n and renumbering it, we can assume that the sequence $\{X_n\}$ tends weakly in $H^1_2(\mathring{B}, \mathbb{R}^3)$ to some limit $X \in H^1_2(\mathring{B}, \mathbb{R}^3)$ such that X_n tends a.e. and also in the L_2-sense on every boundary ∂B_ν to the trace of X. By virtue of the equicontinuity result proved above we can assume that the mappings $\theta_n \in \mathcal{M}(I)$ associated with X_n tend uniformly on I to some limit $\theta \in \mathcal{M}(I)$ such that the relations

$$\theta|_{\partial B \cap I} = \mathrm{id}|_{\partial B \cap I}, \quad \xi \circ p^-_B = \gamma \circ \theta$$

hold true for some continuous, weakly monotonic mapping ξ from $p^-_B(I)$ onto Γ, with the property that ξ and X coincide a.e. on $p^-_B(I) \setminus I$. Thus we can use ξ to define X on $p^-_B(I)$ by setting $X(w) := \xi(w)$ for $w \in p^-_B(I)$, and we have $X \circ p^-_B = \gamma \circ \theta$.

Moreover, on account of Helly's selection theorem[1] and of the assumption $l(X_n \circ p^+_B) \le L + \frac{1}{n}$, we can assume that $X_n \circ p^+_B$ tends to $X \circ p^+_B$ everywhere on I, and that $l(X \circ p^+_B) \le L$.

By Lemma 3 we conclude that X has continuous boundary values on every ∂B_ν, and consequently X is continuous on ∂B.

Recall that Dirichlet's integral is weakly lower semicontinuous on $H^1_2(\mathring{B}, \mathbb{R}^3)$, that is, the weak convergence of X_n to X implies

$$D(X, \mathring{B}) \le \liminf_{n\to\infty} D(X_n, \mathring{B}).$$

[1] Cf. for instance Natanson [1], p. 250.

Then, by (17), we arrive at

$$D(X, \mathring{B}) \le d^-(\Gamma, L).$$

Consider now mappings $H_\nu \in C^0(B_\nu, \mathbb{R}^3) \cap H_2^1(\mathring{B}, \mathbb{R}^3)$ which are harmonic in B_ν and coincide with X on ∂B_ν. Then we also have

$$D(H_\nu, \mathring{B}_\nu) \le D(X, \mathring{B}_\nu).$$

Set

$$X^* := \begin{cases} X & \text{on } B \setminus \mathring{B}, \\ H_\nu & \text{on } B_\nu, \ \nu \in \mathbb{N}, 1 \le \nu \le \nu_B. \end{cases}$$

The surface X^* is of class $C^0(B, \mathbb{R}^3) \cap H_2^1(\mathring{B}, \mathbb{R}^3)$ and satisfies

$$D(X^*, \mathring{B}) \le D(X, \mathring{B}),$$

$$l(X^* \circ p_B^+) = l(X \circ p_B^+) \le L$$

and

$$X^* \circ p_B^- = \gamma \circ \theta, \quad \theta|_{\partial B \cap I} = \mathrm{id}|_{\partial B \cap I}.$$

Consequently we have $X^* \in \mathcal{C}(\Gamma, L)$, whence

$$d(\Gamma, L) \le D(X^*, \mathring{B}).$$

Thus we obtain

$$d^-(\Gamma, L) \le d(\Gamma, L) \le D(X^*, \mathring{B}) \le D(X, \mathring{B}) \le d^-(\Gamma, L),$$

and therefore

$$d(\Gamma, L) = d^-(\Gamma, L) = D(X^*, \mathring{B}) = D(X, \mathring{B})$$

which implies that $X = X^*$ holds, and that X is a solution of $\mathcal{P}(\Gamma, L)$. □

Theorem 3. *Suppose that $L < l(\Gamma)$. If $X \in \mathcal{C}(\Gamma, L)$ satisfies $D(X, \mathring{B}) = d(\Gamma, L)$, then X is a minimal surface, that is, X is nonconstant, the equations*

$$\begin{aligned} \Delta X &= 0, \\ |X_u|^2 &= |X_v|^2, \quad \langle X_u, X_v \rangle = 0 \end{aligned}$$

are satisfied in $\mathring{B}$, and it follows that

$$l(X \circ p_B^+) = L.$$

(That is, for any solution of $\mathcal{P}(\Gamma, L)$, the movable part of the boundary is taut.)

Proof. The minimal-surface property of X can be derived as in Chapter 4 of Vol. 1, since each of the mappings $X|_{B_\nu}$ solves a Plateau problem with respect to the boundary curve $\Gamma_\nu := X(\partial B_\nu)$. (Here B_ν denotes the disk-components of the parameter domain B of X.) Thus we only have to prove

$$l(X \circ p_B^+) = L.$$

Suppose that this inequality were not true. Then, because of $X \in \mathcal{C}(\Gamma, L)$, we would have

$$l(X \circ p_B^+) < L.$$

We recall that $\mathring{B}_1 = B(u_1, r_1)$, and we set $w_0 := u_1 + ir_1$. Then we can find some $r_0 \in (0, r_1)$ such that

$$l(X \circ p_B^+) + l(X, B_1 \cap \partial B(w_0, r_0)) < L.$$

Let τ be a topological mapping of $B_1 \setminus B(w_0, r_0)$ onto B_1 with $\tau(a_1) = a_1$ and $\tau(b_1) = b_1$ that maps the interior of B_1 conformally onto $B_1 \setminus \overline{B}(w_0, r_0)$. We use τ to define the comparison map $X^* \in \mathcal{C}(\Gamma, L)$ by defining

$$X^*(w) := \begin{cases} X(\tau(w)) & \\ & \text{for} \\ X(w) & \end{cases} \begin{matrix} w \in B_1, \\ \\ w \in B \setminus B_1. \end{matrix}$$

Then it follows that

$$d(\Gamma, L) \le D(X^*, \mathring{B}),$$

and because of

$$\begin{aligned} D(X^*, \mathring{B}) &= D(X, \mathring{B}) - D(X, \mathring{B}_1 \cap B(w_0, r_0)) \\ &= d(\Gamma, L) - D(X, \mathring{B}_1 \cap B(w_0, r_0)) \end{aligned}$$

we infer that $X|_{B(w_0, r_0)} = \text{const}$, whence $X|_{B_1} = \text{const}$, as $X|_{B_1}$ is harmonic and therefore real analytic. The relation $X|_{B_1} = \text{const}$ is a contradiction to $X(a_1) \ne X(b_1)$. □

Proposition 3. *If* $|P_1 - P_2| < L$ *then it follows that*

$$d^-(\Gamma, L) = d^+(\Gamma, L).$$

Proof. Case (i). Suppose that $L \ge l(\Gamma)$. Then we define the surface $Z\colon Q \to \mathbb{R}^3$ on $Q = \{u + iv\colon |u| \le 1, |v| \le \delta\}$ by setting $Z(u + iv) := \gamma(u)$. It follows that

$$D(Z, \mathring{Q}) = \delta \int_{-1}^{1} |\dot\gamma(u)|^2 \, du.$$

Consider a homeomorphism of $\overline{B}(0,1)$ onto Q which maps $B(0,1)$ conformally onto $\mathring{Q}$. Then $X := Z \circ \tau$ is of class $\mathcal{C}(\Gamma, L)$, and we have

$$d^+(\Gamma, L) \le D(X, B(0,1)) = D(Z, \mathring{Q}) = \delta \int_{-1}^{1} |\dot{\gamma}(u)|^2 \, du.$$

As we can make $\delta > 0$ arbitrarily small, it follows that $d^+(\Gamma, L) = 0$ whence

$$d(\Gamma, L) = d^-(\Gamma, L) = d^+(\Gamma, L) = 0.$$

Case (ii). Assume now that $l(\Gamma) > L$. By Theorem 2, there is some $X \in \mathcal{C}(\Gamma, L)$ such that

$$d^-(\Gamma, L) = d(\Gamma, L) = D(X, \mathring{B}),$$

where B is the parameter domain of X.

For given $\varepsilon > 0$ there exists some surface

$$X_\varepsilon \in \mathcal{C}(\Gamma, L + \varepsilon^2) \cap H_2^1(B(0,1), \mathbb{R}^3)$$

with

$$D(X_\varepsilon, B(0,1)) \le D(X, \mathring{B}) + \varepsilon = d^-(\Gamma, L) + \varepsilon,$$

if we take Lemma 1 into account.

Consider now the surface $X^* \in \mathcal{C}(\Gamma, |P_1 - P_2|) \cap H_2^1(B(0,1), \mathbb{R}^3)$ which was constructed in the proof of Proposition 1. We define the 1-parameter family of surfaces

$$X_\varepsilon^* := \varepsilon X^* + (1 - \varepsilon) X_\varepsilon, \quad 0 < \varepsilon \le L - |P_1 - P_2|.$$

Then we infer $X_\varepsilon^* \in \mathcal{C}(\Gamma, L_\varepsilon)$ where L_ε is estimated by

$$\begin{aligned} L_\varepsilon &\le \varepsilon |P_1 - P_2| + (1 - \varepsilon)(L + \varepsilon^2) \\ &= \varepsilon |P_1 - P_2| + L + \varepsilon^2 - \varepsilon L - \varepsilon^3 \le L - \varepsilon^3 < L. \end{aligned}$$

It follows that

$$d^+(\Gamma, L) \le D(X_\varepsilon^*) \quad \text{for } 0 < \varepsilon \le L - |P_1 - P_2|.$$

Furthermore we have

$$\begin{aligned} D(X_\varepsilon^*) &= \varepsilon^2 D(X^*) + \varepsilon(1 - \varepsilon) \int_B \langle \nabla X^*, \nabla X_\varepsilon \rangle \, du \, dv + (1 - \varepsilon)^2 D(X_\varepsilon) \\ &\le d^-(\Gamma, L) + \varepsilon K \end{aligned}$$

for some number $K > 0$ which does not depend on ε with

$$0 < \varepsilon \le L - |P_1 - P_2|.$$

Letting $\varepsilon \to +0$, we arrive at the inequality

$$d^+(\Gamma, L) \le d^-(\Gamma, L).$$

On the other hand, we have

$$d^-(\Gamma, L) \le d(\Gamma, L) \le d^+(\Gamma, L)$$

whence

$$d^-(\Gamma, L) = d(\Gamma, L) = d^+(\Gamma, L). \qquad \square$$

Theorem 4. *If X minimizes the Dirichlet integral $D(X, B)$ in the class $\mathcal{C}(\Gamma, L)$, then X also furnishes the minimum of the area functional $A(X, B)$ with $\mathcal{C}(\Gamma, L)$.*

Proof. This result can be derived from Morrey's lemma on ε-conformal mappings that we have described in Section 4.5 of Vol. 1. One can proceed in the same way as in the proof of Theorem 4 in Section 4.5 of Vol. 1. The proof can also be obtained by the method described in Section 4.10 of Vol. 1. □

5.3 Analyticity of the Movable Boundary

In this section we want to investigate the regularity of the movable part Σ of a solution X of the thread problem. Let us begin by considering a special case. We assume that Γ is a planar curve. By a projection argument it can easily be seen that X has to be contained in the plane E determined by Γ. In fact, if we assume without loss of generality that E is the plane $\{z = 0\}$, and that $X(w) = (x(w), y(w), z(w))$ is a solution of $\mathcal{P}(\Gamma, L)$, then also $X^*(w) := (x(w), y(w), 0)$ is a surface of class $\mathcal{C}(\Gamma, L)$, and we have

$$D(X^*, \mathring{B}) \le D(X, \mathring{B}).$$

The equality sign holds if and only if $D(z, \mathring{B}) = 0$, and $D(z, \mathring{B})$ vanishes if and only if $z(w) = 0$ holds for all $w \in \bigcup_{\nu=1}^{\nu_B} B_\nu$. As X is an absolute minimizer for the thread problem, there cannot be any surface in $\mathcal{C}(\Gamma, L)$ with a Dirichlet integral smaller than $D(X, \mathring{B})$. Thus we infer that $z(w) = 0$ on $\mathring{B}$. Since $z \in H_2^1(\mathring{B})$, we also have $z(w) = 0$ a.e. on $B \setminus I$. Finally, on $B \setminus \bigcup_{\nu=1}^{\nu_0} B_\nu$ the function $z(w)$ coincides with the z-component of Γ so that $z(w)$ vanishes identically on I and therefore on all of B.

Thus, X is in fact a planar surface, and by a classical result of analysis, every part of the movable curve Σ not attached to Γ must be a circular arc, that is, a regular real analytic curve of constant curvature.

It is the aim of this section to show that the same result holds true for any solution X of $\mathcal{P}(\Gamma, L)$, even if Γ is not a planar curve. As by-product of our investigation we shall also obtain that all free (i.e. nonattached) parts of Σ are asymptotic curves of constant geodesic curvature on X, and it can be proved that the curvature is the same for all free parts of Σ.

Clearly we can restrict our discussion of X to any part $X|_{B_\nu}$ where B_ν is an arbitrary disk-component of the parameter domain B of X. Thus we shall assume that X is a solution of a thread problem which is parametrized on a disk, say, the unit disk. For this reason we shall from now on abolish the notation of Sections 5.1 and 5.2 and, instead, return to another notation similar to that used in previous chapters. To be precise, we now denote by B the open disk

$$B = \{w = u + iv \colon |w| < 1\}$$

in the u, v-plane $\mathbb{C} \hat{=} \mathbb{R}^2$, and by C^+ and C^- its boundary parts

$$C^+ = \{w = u + iv\colon |w| = 1, v \geq 0\},$$
$$C^- = \{w = u + iv\colon |w| = 1, v \leq 0\}.$$

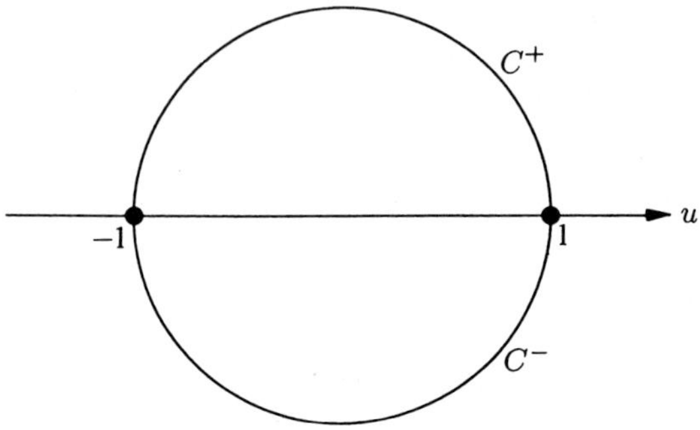

Fig. 1.

The set $\mathcal{C}(\Gamma, L)$ of comparison functions $X(w), w \in \overline{B}$, now consists of all surfaces of class $C^0(\overline{B}, \mathbb{R}^3) \cap H_2^1(B, \mathbb{R}^3)$ which map C^- in a weakly monotonic way onto a given rectifiable Jordan arc Γ, and whose total variation on C^+ is equal to a fixed number L,

(1) $$l(\Sigma) := \int_{C^+} |dX| = L.$$

Here Σ denotes the movable part $X\colon C^+ \to \mathbb{R}^3$ of the boundary of any $X \in \mathcal{C}(\Gamma, L)$. We assume that

(2) $$|P_1 - P_2| < L < l(\Gamma),$$

where P_1 and P_2 denote the endpoints of Γ, and $l(\Gamma)$ stands for the length of the fixed arc Γ.

Let $X \in \mathcal{C}(\Gamma, L)$ be a minimizer of the Dirichlet integral

$$D_B(X) = \frac{1}{2} \int_B |\nabla X|^2 \, du \, dv$$

among all surfaces in $\mathcal{C}(\Gamma, L)$. Such a minimizer will now be called *a solution of the thread problem* $\mathcal{P}(\Gamma, L)$. We already know that any such solution has to be a minimal surface. That is, the equations

$$\Delta X = 0,$$
$$|X_u|^2 = |X_v|^2, \quad \langle X_u, X_v \rangle = 0$$

hold true in B, and $X(w) \not\equiv \text{const}$ on B.

Now we state the main result of this section.

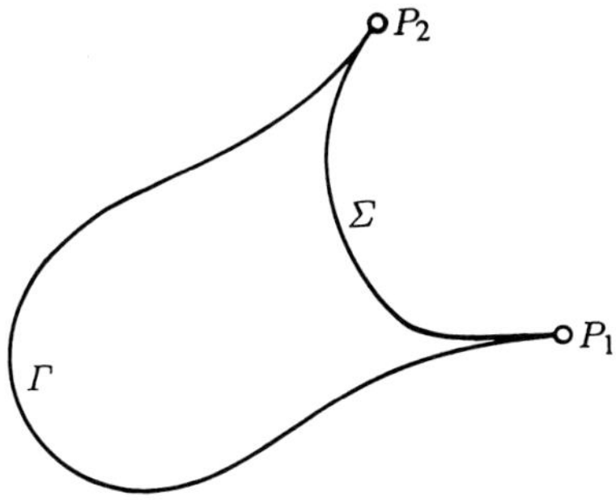

Fig. 2.

Theorem 1. *Let $X \in \mathcal{C}(\Gamma, L)$ be a minimal surface, that is, X satisfies*

$$\Delta X = 0 \quad \text{in } B \tag{3}$$

as well as the conformality relations. Introducing polar coordinates r, θ around the origin by $w = re^{i\theta}$, these relations can be written as

$$r^2|X_r|^2 = |X_\theta|^2, \quad \langle X_r, X_\theta \rangle = 0. \tag{4}$$

Moreover, suppose that X minimizes the Dirichlet integral within the class $\mathcal{C}(\Gamma, L)$. Then $X(w)$ can be continued analytically as a minimal surface across the arc C^+, and it has on C^+ no branch points of odd order nor any true branch points of even order. If, moreover, the boundary mapping $X: \partial B \to \mathbb{R}^3$ is assumed to be an embedding, then $X(w)$ has no false branch points of even order on C^+ either. Correspondingly, in this case, the free trace Σ defined by $X: C^+ \to \mathbb{R}^3$ is a regular, real analytic curve of constant curvature $\kappa \neq 0$.

For the following we recall some results on the boundary behaviour of minimal surfaces with a finite Dirichlet integral and with boundary values of bounded variation. The assumption $D_B(X) < \infty$ implies that $X(r,\theta) = X(re^{i\theta})$ possesses L^2-boundary values $X(1,\theta)$ on ∂B which are assumed in the L^2-sense as $r \to 1-0$. From $\int_0^{2\pi} |dX(1,\theta)| < \infty$ we conclude that $X(1,\theta)$ depends continuously on θ (cf. Lemma 3 of Section 5.2). More subtle results have been derived in Section 4.7 of Vol. 1. For the convenience of the reader, we collect the pertinent statements in the following lemma.

Lemma 1. *Let $X: B \to \mathbb{R}^3$ be a disk-type minimal surface, i.e. let (3) and (4) be satisfied, and denote by $X^*: B \to \mathbb{R}^3$ the adjoint minimal surface to X which, up to an additive constant, is uniquely determined by the equations*

$$X_r = \frac{1}{r} X^*_\theta, \quad \frac{1}{r} X_\theta = -X^*_r. \tag{5}$$

Assume that $D_B(X) < \infty$ and $\int_{\partial B} |dX| < \infty$. Then we have:

(i) *X and X^* are of class $C^0(\overline{B}, \mathbb{R}^3)$ and*

(6) $$D_B(X) = D_B(X^*), \quad \int_{\partial B} |dX| = \int_{\partial B} |dX^*|.$$

(ii) *The boundary values $X(1,\theta)$ and $X^*(1,\theta)$ are absolutely continuous functions of θ, and $X_\theta(r,\theta), X^*_\theta(r,\theta)$ tend in the L^2-sense to the derivatives $X_\theta(1,\theta), X^*_\theta(1,\theta)$ of the boundary values $X(1,\theta)$ and $X^*(1,\theta)$ respectively as $r \to 1-0$. Then, on account of (5), we deduce that also $X_r(r,\theta)$ and $X^*_r(r,\theta)$ converge in L^2 to boundary values as $r \to 1-0$, and we set*

$$X_r(1,\theta) = \lim_{r \to 1-0} X_r(r,\theta), \quad X^*_r(1,\theta) = \lim_{r \to 1-0} X^*_r(r,\theta).$$

It follows that a.e.

(7) $$X_r(1,\theta) = X^*_\theta(1,\theta), \quad X_\theta(1,r) = -X^*_r(1,\theta),$$

(8) $$|X_r(1,\theta)| = |X_\theta(1,\theta)|, \quad \langle X_r(1,\theta), X_\theta(1,\theta)\rangle = 0.$$

(iii) *If C is an open subarc of ∂B, and ξ is a test function of class $H^1_2(B,\mathbb{R}^3) \cap L_\infty(C,\mathbb{R}^3)$ with $\xi = 0$ on $\partial B \setminus C$, then*

(9) $$\int_B \langle \nabla X, \nabla \xi\rangle r\, dr\, d\theta = \int_C \langle X_r, \xi\rangle\, d\theta.$$

(iv) *If $X \not\equiv$ const on B, then $X_\theta(1,\theta)$ and $X^*_\theta(1,\theta)$ vanish at most on a subset of $[0,2\pi]$ of one-dimensional measure zero.*

Now we turn to the proof of Theorem 1 which we want to break up into three parts. In the first one we consider a *stationary version of the thread problem*; here the existence of a Lagrange multiplier is supposed. Thereafter we prove that every minimizer in $\mathcal{C}(\Gamma, L)$ is in fact a solution of the stationary problem by establishing the existence of a Lagrange multiplier, and in the third part we sketch how branch points can be excluded by using the minimum property.

Definition. *A minimal surface $X\colon B \to \mathbb{R}^3$ is said to be a* **stationary solution of the thread problem** *with respect to some open subarc C of ∂B if the following holds:*
(i) *$D_B(X) < \infty$, $\int_{\partial B} |dX| < \infty$;*
(ii) *there is a real number $\lambda \neq 0$ such that*

(10) $$\int_B \langle \nabla X, \nabla \xi\rangle r\, dr\, d\theta + \lambda \int_C \left\langle \frac{X_\theta}{|X_\theta|}, \xi_\theta \right\rangle d\theta = 0$$

holds for all $\xi \in C^1(\overline{B}, \mathbb{R}^3)$ with $\xi = 0$ on $\partial B \setminus C$.

Taking the identity (9) into account we arrive at

$$\int_C (\langle X_r, \xi\rangle + \lambda |X_\theta|^{-1}\langle X_\theta, \xi_\theta\rangle)\, d\theta = 0,$$

and (8) yields

$$\int_C \langle X_r, \xi\rangle\, d\theta = \int_C \langle X_\theta^*, \xi\rangle\, d\theta = -\int_C \langle X^*, \xi_\theta\rangle\, d\theta.$$

Thus (10) is equivalent to

$$\int_C \langle X^* - \lambda |X_\theta|^{-1} X_\theta, \xi_\theta\rangle\, d\theta = 0 \tag{11}$$
$$\text{for all } \xi \in C^1(\overline{B}, \mathbb{R}^3) \quad \text{with } \xi = 0 \text{ on } \partial B \setminus C.$$

DuBois–Reymond's lemma now implies that (11) – and therefore also (10) – is equivalent to the following property of X:

There exists a constant vector $P \in \mathbb{R}^3$ such that

$$X^* = \lambda |X_\theta|^{-1} X_\theta + P \quad \textit{a.e. on } C \tag{12}$$

holds.

We now prove

Theorem 2. *Let $X\colon B \to \mathbb{R}^3$ be a minimal surface which is a stationary solution of the thread problem with respect to the open arc $C \subset \partial B$. Then, for some $P \in \mathbb{R}^3$ and some $\lambda \in \mathbb{R}, \lambda \neq 0$, equation (12) is satisfied. Moreover, X and its adjoint X^* are real analytic on $B \cup C$, and X^* intersects the sphere*

$$S = \{Z \in \mathbb{R}^3 \colon |Z - P|^2 = \lambda^2\}$$

orthogonally along its free trace Σ^* defined by $X^*\colon C \to \mathbb{R}^3$. Both X and X^* have no boundary branch points of odd order on C. Finally, $\Sigma = X|_C$ has a representation $\mathfrak{X}(s), 0 < s < 1$, by its arc length s as parameter, which is of class C^2 and satisfies $|\dot{\mathfrak{X}}(s)| \equiv 1$, $|\ddot{\mathfrak{X}}(s)| \equiv \frac{1}{|\lambda|}$. Thus Σ represents a regular curve of constant curvature $\kappa = \frac{1}{|\lambda|}$.

Proof. As we have noticed, the assumption on X implies that (12) holds for some $P \in \mathbb{R}^3$ and some $\lambda \in \mathbb{R}, \lambda \neq 0$. Taking the continuity of $X^*(1, \theta)$ into account, we infer that

$$|X^* - P|^2 = \lambda^2 \quad \text{on } C. \tag{13}$$

In other words, the trace Σ^* lies on S. Moreover, equations (12) and (7) yield

(14) $$X^* - P = -\lambda |X_r^*|^{-1} \cdot X_r^* \quad \text{a.e. on } C.$$

Therefore the vector X_r^* is normal to S a.e. on C. Thus for almost all $w \in C$ the surface X^* has a tangent plane which meets S at a right angle. By the reasoning of Section 1.4 (cf. Theorem 1) we conclude that the adjoint surface X^* is a critical point of Dirichlet's integral within the boundary configuration $\langle \Gamma, S\rangle$ consisting of the arc $\Gamma^* = \{X^*(w)\colon w \in \partial B \setminus C\}$ and of the surface S. We can therefore apply Theorem 2′ of Section 2.8 to X^* and obtain that X^* can be continued analytically across C as a minimal surface. (Note that for this regularity theorem it is not necessary to assume that Γ^* be a Jordan arc which does not meet S except in his two endpoints.) By virtue of (5) we infer that both X and X^* are real analytic in $B \cup C$, as we have claimed.

We furthermore note that, because of (5), X and X^* have the same boundary branch points $w_0 \in C$. Since $X + iX^*$ is a nonconstant holomorphic mapping $U \to \mathbb{C}^3$ of some full neighbourhood U of each branch point $w_0 \in C$, we have the asymptotic formula

(15) $$X_w(w) = A(w - w_0)^\nu + O(|w - w_0|^{\nu+1}) \quad \text{as } w \to w_0,$$

for some integer $\nu \geq 1$ and some vector $A \neq 0$. Since $X_w^* = -iX_w$, we also have

(15′) $$X_w^*(w) = -iA(w - w_0)^\nu + O(|w - w_0|^{\nu+1}) \quad \text{as } w \to w_0.$$

That is, the order of w_0 as branch point of X equals its order as branch point of X^*. We moreover infer from (15) and (15′) that the boundary branch points of X and X^* are isolated. In addition, the conformality relations (5) imply $\langle A, A\rangle = 0$. Thus A is of the form $A = \frac{1}{2}(a - ib)$, where $a, b \in \mathbb{R}^3$, $|a| = |b| \neq 0$, $\langle a, b\rangle = 0$.

If $w = e^{i\theta}$ is not a branch point of X on C, we can define the unit tangent vectors

$$T(\theta) = \frac{X_\theta(1,\theta)}{|X_\theta(1,\theta)|}, \quad T^*(\theta) = \frac{X_\theta^*(1,\theta)}{|X_\theta^*(1,\theta)|}$$

of the curves Σ and Σ^* at $X(w)$ and $X^*(w)$, respectively.

Let $w_0 = e^{i\theta_0} \in C$ be a branch point of X (and of X^*). Then we infer from (15) and (15′) that the one-sided limits

$$T_\pm(\theta_0) = \lim_{\theta \to \theta_0 \pm 0} T(\theta), \quad T_\pm^*(\theta_0) = \lim_{\theta \to \theta_0 \pm 0} T^*(\theta)$$

exist. Moreover, we have

(16) $$T_+(\theta_0) = T_-(\theta_0), \quad T_+^*(\theta_0) = T_-^*(\theta_0)$$

if the order ν of the boundary branch point w_0 is *even*, whereas

(16′) $$T_+(\theta_0) = -T_-(\theta_0), \quad T_+^*(\theta_0) = -T_-^*(\theta_0)$$

if ν is *odd*.

We note that the limits $T_\pm(\theta_0), T^*_\pm(\theta_0)$ are unit vectors. Equation (12), on the other hand, yields that

$$T(\theta) = \frac{1}{\lambda}\{X^*(1,\theta) - P\} \tag{17}$$

holds for all θ satisfying $0 < |\theta - \theta_0| < \varepsilon$ where ε is a sufficiently small number and, moreover, the right-hand side depends continuously on $\theta \in (\theta_0 - \varepsilon, \theta_0 + \varepsilon)$. Therefore, $T_+(\theta_0) = T_-(\theta_0)$, and ν must be of even order. Hence X and also X^* can only have even order branch points on C, as we have claimed. If we define $T(\theta_0)$ by $T_+(\theta_0)$ at a branch point $w_0 = e^{i\theta_0} \in C$ of even order, we infer from (16) that $T(\theta)$ is a continuous function on C with $|T(\theta)| \equiv 1$, and (17) holds everywhere on C.

Suppose now that $C = \{e^{i\theta} : \theta_1 < \theta < \theta_2\}$ and set

$$l = \int_{\theta_1}^{\theta_2} |X_\theta(1,\theta)|\, d\theta = \int_{\theta_1}^{\theta_2} |X^*_\theta(1,\theta)|\, d\theta.$$

We furthermore introduce

$$s = s(\theta) = \int_{\theta_1}^{\theta} |X_\theta(1,\theta)|\, d\theta = \int_{\theta_1}^{\theta} |X^*_\theta(1,\theta)|\, d\theta,$$

$\theta_1 \le \theta \le \theta_2$, which is the arc length parameter of Σ as well as of Σ^*. Since $s'(\theta) = |X_\theta(1,\theta)| \ge 0$ has only isolated zeros, the function $s(\theta)$ can be inverted. Let $\theta(s), 0 \le s \le l$, be its (continuous) inverse. For $0 < s < l$ we introduce

$$t(s) = T(\theta(s)), \quad t^*(s) = T^*(\theta(s)),$$
$$\mathfrak{X}(s) = X(1,\theta(s)), \quad \mathfrak{X}^*(s) = X^*(1,\theta(s)).$$

So far, we only know that $\theta(s)$ is continuously differentiable in s-intervals corresponding to θ-intervals free of branch points. We already know that $t(s)$ and $t^*(s)$ are continuous for $0 < s < l$, and that $\dot{\mathfrak{X}}(s) = t(s), \dot{\mathfrak{X}}^*(s) = t^*(s)$ holds at values of s which do not correspond to branch points on C. Then a simple argument employing the mean value theorem yields that $\mathfrak{X}(s)$ and $\mathfrak{X}^*(s)$ are of class C^1 for $0 < s < l$, and that

$$\dot{\mathfrak{X}}(s) = t(s), \quad \dot{\mathfrak{X}}^*(s) = t^*(s) \quad \text{for } 0 < s < l. \tag{18}$$

(In these formulas as well as in the following ones, the dot denotes differentiation with respect to the arc length: $\dot{} = \frac{d}{ds}$.) Thus Σ and Σ^* are representations of regular curves of class C^1.

From (17) and (18) we derive the equation

$$t(s) = \frac{1}{\lambda}\{\mathfrak{X}^*(s) - P\} \tag{19}$$

and

$$(20) \qquad \dot{t}(s) = \frac{1}{\lambda} t^*(s)$$

for $0 < s < l$. Thus $\mathfrak{X}(s)$ is actually of class C^2 on $(0, l)$, $|\dot{t}(s)| = |\ddot{\mathfrak{X}}(s)| = 1/|\lambda|$. This means, Σ represents a regular C^2-curve of constant curvature $1/|\lambda|$. This concludes the proof of Theorem 2. □

The first of Frenet's equations yields

$$(21) \qquad \dot{t}(s) = \kappa n(s), \quad \kappa = \frac{1}{|\lambda|},$$

where $n(s)$ is the principal normal of the curve $\mathfrak{X}(s)$. On the other hand, differentiating (17) with respect to θ and employing (7) and (8), we arrive at

$$(22) \qquad \dot{t}(s) = \frac{1}{\lambda} \cdot \frac{X_r}{|X_r|}(1, \theta(s)).$$

Hence $n = \pm|X_r|^{-1} X_r$, and thus the normal curvature of Σ vanishes. Thus as a by-product of our discussion we obtain the following

Corollary 1. *Under the assumptions of Theorem 1 the free trace Σ of X is an asymptotic line of the surface X of constant geodesic curvature $\pm\kappa$.*

Remark 1. In general, stationary solutions of the thread problem will have boundary branch points of even order. In fact, one can easily construct examples of planar minimal surfaces $X^* : \overline{B} \to \mathbb{R}^3$ that satisfy (14) for some nonempty open subarc C of ∂B and have a branch point w_0 of second order on C. The adjoint surface X of $-X^*$ will then satisfy (12) or, equivalently, (10). Hence X is a stationary solution of a thread problem with respect to C that has a branch point of second order on C.

Next we come to the second part of the proof of Theorem 1. We shall prove that, for each solution of the real thread problem, there exists a Lagrange multiplier. This is not totally trivial since the applicability of the standard Lagrange multiplier theorem (which requires continuous differentiability of the involved functions) is not clear. The following result provides an appropriate substitute.

Lemma 2. *Let $\varphi(\varepsilon, t)$ and $\psi(\varepsilon, t)$ be real-valued functions of*

$$(\varepsilon, t) \in [-\varepsilon_0, \varepsilon_0] \times [-t_0, t_0], \quad \varepsilon_0 > 0, \; t_0 > 0,$$

which split in the form

$$\varphi(\varepsilon, t) = \varphi_0 + \varphi_1(\varepsilon) + \varphi_2(t), \quad \psi(\varepsilon, t) = \psi_0 + \psi_1(\varepsilon) + \psi_2(t).$$

Here it is assumed that φ_0 and ψ_0 are constant, and that

$$\varphi_1(0) = \varphi_2(0) = \psi_1(0) = \psi_2(0) = 0.$$

We also suppose that ψ_2 *is continuous on* $[-t_0, t_0]$, *that the derivatives* $\varphi_1'(0), \varphi_2'(0), \psi_1'(0), \psi_2'(0)$ *exist, and that* $\psi_2'(0) = 1$. *Finally, let the inequality* $\varphi(\varepsilon, t) \geq \varphi(0,0)$ *hold for all* (ε, t) *in* $[-\varepsilon_0, \varepsilon_0] \times [-t_0, t_0]$ *with* $\psi(\varepsilon, t) = \psi_0$.

Then the relation

$$\varphi_1'(0) + \lambda\psi_1'(0) = 0 \tag{23}$$

is satisfied for $\lambda = -\varphi_2'(0)$.

Proof. The assumptions imply that there is a function $\eta(t), -t_0 \leq t \leq t_0$, which satisfies

$$\lim_{t \to 0} \eta(t) = \eta(0) = 0$$

and

$$\psi_2(t) = t\{1 + \eta(t)\}.$$

Then we choose a number δ_0 with $0 < \delta_0 < \frac{t_0}{2}$ such that $|\eta(2t)| < \frac{1}{2}$ for $|t| < \delta_0$, and infer that

$$\psi_2(-2t) < -t < t < \psi_2(2t) \quad \text{for } t \in (0, \delta_0).$$

The continuity of ψ_2 now implies the relation

$$[-\delta, \delta] \subset \psi_2([-2\delta, 2\delta]) \quad \text{for all } \delta \in (0, \delta_0).$$

We also note that $\lim_{\varepsilon \to 0} \psi_1(\varepsilon) = 0$ holds. Therefore we can find a number ε_1 with $0 < \varepsilon_1 \leq \varepsilon_0$ such that $|\psi_1(\varepsilon)| < \delta_0$ is satisfied for each $\varepsilon \in [-\varepsilon_1, \varepsilon_1]$. Consequently there exists a real-valued function $\tau(\varepsilon), -\varepsilon_1 \leq \varepsilon \leq \varepsilon_1$, with the properties

$$\tau(0) = \lim_{\varepsilon \to 0} \tau(\varepsilon) = 0, \quad \psi_2(\tau(\varepsilon)) + \psi_1(\varepsilon) = 0,$$
$$|\tau(\varepsilon)| \leq 2|\psi_1(\varepsilon)| < t_0,$$

whence also $\psi(\varepsilon, \tau(\varepsilon)) = \psi_0$ for $-\varepsilon_1 \leq \varepsilon \leq \varepsilon_1$. From the identities

$$\frac{\tau(\varepsilon)}{\varepsilon} = \frac{\tau(\varepsilon) - \tau(0)}{\varepsilon} = -\frac{\psi_1(\varepsilon) - \psi_1(0)}{\varepsilon} \cdot \frac{1}{1 + \eta(\tau(\varepsilon))}$$

for $0 < |\varepsilon| \leq \varepsilon_1$ we infer that the function $\tau(\varepsilon)$ is differentiable at $\varepsilon = 0$, and that

$$\tau'(0) = \lim_{\varepsilon \to 0} \frac{\tau(\varepsilon)}{\varepsilon} = -\psi_1'(0). \tag{24}$$

Moreover, the minimum property

$$\varphi(\varepsilon, \tau(\varepsilon)) \geq \varphi(0,0) \quad \text{for } 0 < \varepsilon \leq \varepsilon_1$$

implies the inequality

$$(25)\qquad 0 \le \frac{\varphi_1(\varepsilon)}{\varepsilon} + \frac{\varphi_2(\tau(\varepsilon))}{\varepsilon}.$$

Suppose now that we would have $\tau(\varepsilon) \equiv 0$ on some interval $(0, \varepsilon']$, where $0 < \varepsilon' \le \varepsilon_1$. Then we obtain

$$0 \le \frac{\varphi_1(\varepsilon)}{\varepsilon} \quad \text{for } 0 < \varepsilon \le \varepsilon'$$

on account of (25), and therefore $\varphi_1'(0) \ge 0$. By virtue of (24) we furthermore have $\tau'(0) = 0$ and $\psi_1'(0) = 0$, whence

$$(26)\qquad 0 \le \varphi_1'(0) - \psi_1'(0)\varphi_2'(0).$$

If, on the other hand, there is no $\varepsilon' > 0$ such that $\tau(\varepsilon) \equiv 0$ on $(0, \varepsilon']$, then there exists a sequence of numbers $\varepsilon_2, \varepsilon_3, \varepsilon_4, \dots$ tending to zero, with $0 < \varepsilon_i \le \varepsilon'$ for $i \ge 2$ and $\tau(\varepsilon_i) \ne 0$. Set $\tau_i = \tau(\varepsilon_i)$. We then infer from (25) that

$$0 \le \frac{\varphi_1(\varepsilon_i)}{\varepsilon_i} + \frac{\varphi_2(\tau_i)}{\tau_i} \cdot \frac{\tau_i}{\varepsilon_i}, \qquad i = 2, 3, 4, \dots,$$

holds. For $i \to \infty$ we once again arrive at the inequality (26) which thus is established. Similarly we can verify the opposite inequality

$$0 \ge \varphi_1'(0) - \psi_1'(0)\varphi_2'(0),$$

and the Lemma is proved. □

In order to apply the previous lemma, we will introduce *the class* $\mathcal{F}(C^+)$ *of test functions* defined in the following way:

A function ζ *is said to be of class* $\mathcal{F}(C^+)$ *if it lies in* $C^1(\overline{B}, \mathbb{R}^3)$, *and if there are a point* $w_0 \in C^+$ *and a number* $r \in (0,1)$ *such that* $\partial B \cap \overline{B}_r(w_0)$ *is contained in the open arc* C^+ *and that* $\zeta(w) = 0$ *for all* $w \in \overline{B} \setminus B_{r/2}(w_0)$.

Lemma 3. *Suppose that* (2) *holds and that* X *is a mapping of class* $\mathcal{C}(\Gamma, L)$ *which satisfies the assumptions of Theorem* 1. *Then there exists some* $\zeta \in \mathcal{F}(C^+)$ *such that*

$$(27)\qquad \int_{C^+} |X_\theta|^{-1} \langle X_\theta, \zeta_\theta \rangle \, d\theta = 1.$$

Proof. It clearly suffices to establish the existence of some $\zeta \in \mathcal{F}(C^+)$ for which the integral in (27) is nonzero. To this end, let us suppose that the integral vanishes for all $\zeta \in \mathcal{F}(C^+)$. Then, by DuBois–Reymond's lemma, there would exist a unit vector $e \in \mathbb{R}^3$ such that

$$|X_\theta(1,\theta)|^{-1} X_\theta(1,\theta) = e$$

for almost all $\theta \in (0, \pi)$. Hence $X(C^+)$ would be contained in some straight line $\mathcal{L}$, and since $X: \partial B \to \mathbb{R}^3$ is a continuous mapping, $\mathcal{L}$ would have to be the straight line connecting the two points P_1 and P_2. Applying the reflection principle we could extend X analytically and as a minimal surface across C^+. Hence X is real analytic on $B \cup C^+$ and possesses at most denumerably many isolated branch points on C^+. Then we infer from the equation

$$X_\theta(1, \theta) = |X_\theta(1, \theta)| e \quad \text{for all } \theta \in (0, \pi)$$

that $X(1, \theta)$ yields a strictly monotonic mapping of $[0, \pi]$ onto the straight segment on $\mathcal{L}$ with the endpoints P_1 and P_2, whence we would get

$$L = \int_{C^+} |dX| = |P_1 - P_2|.$$

But this contradicts the assumption required in (2). □

Lemma 4. *Suppose that* (2) *holds and that* $X \in \mathcal{C}(\Gamma, L)$ *satisfies the assumptions of Theorem* 1. *Then* X *is a stationary solution of the thread problem with respect to the arc* $C^+ = \{e^{i\theta} : 0 < \theta < \pi\}$.

Proof. By Lemma 3 there is a test function $\zeta \in \mathcal{F}(C^+)$ such that (27) holds. By definition of $\mathcal{F}(C^+)$, there exist $w_0 \in C^+$ and $r \in (0, 1)$ such that $\zeta(w)$ vanishes for all $w \in \overline{B} \setminus B_{r/2}(w_0)$ and that the closed arc $\gamma := \partial B \cap \overline{B}_r(w_0)$ is contained in C^+. Then $C^+ \setminus \gamma$ consists of two non-empty open arcs C_1 and C_2. We first want to show that X *is a stationary solution of the thread problem with respect to* C_1 *as well as to* C_2. Since the reasoning will be the same for both arcs, it suffices to verify the assertion for, say, C_1.

Firstly, the assumptions of Theorem 1 imply that

$$D_B(X) < \infty, \quad \int_{\partial B} |dX| < \infty, \quad \text{and} \quad X(w) \not\equiv \text{const}.$$

Secondly we have to prove that

$$\int_B \langle \nabla X, \nabla \xi \rangle r \, dr \, d\theta + \lambda_1 \int_{C_1} \langle |X_\theta|^{-1} X_\theta, \xi_\theta \rangle \, d\theta = 0 \tag{28}$$

holds for some real number $\lambda_1 \neq 0$ and for all $\xi \in C^1(\overline{B}, \mathbb{R}^3)$ that vanish on $\partial B \setminus C_1$.

Clearly, it suffices to verify (28) for all $\xi \in C_c^1(B \cup C_1, \mathbb{R}^3)$. We shall, in fact, see that (28) only has to be established for an even smaller class of test functions. For this purpose, we choose some open disk B' with the property that $\partial B \cap B' = C_1$, and that $\Omega := B \cap B'$ does not meet the disk $B_{r/2}(w_0)$. By virtue of some appropriate partition of unity, each element $\xi \in C_c^1(B \cup C_1, \mathbb{R}^3)$ can be written as the sum $\xi = \xi_1 + \xi_2$ of a function $\xi_1 \in C_c^1(\Omega \cup C_1, \mathbb{R}^3)$ and of another function $\xi_2 \in C_c^1(B, \mathbb{R}^3)$. We now note that both integrals appearing in (28) vanish separately if ξ is of class $C_c^1(B, \mathbb{R}^3)$. Thus it remains to prove the following:

There is some number $\lambda_1 \neq 0$ *such that* (28) *holds for all test functions* $\xi \in C_c^1(\Omega \cup C_1, \mathbb{R}^3)$.

This will be achieved by employing Lemma 2. To this end we choose some arbitrary $\xi \in C_c^1(\Omega \cup C_1, \mathbb{R}^3)$ which in the sequel is thought to be fixed, and set

$$X_{\varepsilon,t} = X + \varepsilon\xi + t\zeta, \quad |\varepsilon| \leq \varepsilon_0, \ |t| \leq t_0$$

for some number $\varepsilon_0 > 0, t_0 > 0$. (At present, the subscripts ε and t indicate the dependence of the 2-parameter family $X_{\varepsilon,t}$ on the parameters ε and t and do, deviating from the previous way of notation, not stand for partial derivatives.)

Let us introduce the functions

$$\varphi(\varepsilon, t) := D_B(X_{\varepsilon,t}), \quad \psi(\varepsilon, t) := \int_{C+} |dX_{\varepsilon,t}| = \int_{C+} \left| \frac{d}{d\theta} X_{\varepsilon,t}(1, \theta) \right| d\theta$$

of $(\varepsilon, t) \in [-\varepsilon_0, \varepsilon_0] \times [-t_0, t_0]$. Then we have the representations

$$\varphi(\varepsilon, t) = \varphi_0 + \varphi_1(\varepsilon) + \varphi_2(t), \quad \psi(\varepsilon, t) = \psi_0 + \psi_1(\varepsilon) + \psi_2(t),$$

where we have set

$$\begin{aligned}
\varphi_0 &:= D_B(X), \quad \psi_0 := \int_{C+} |X_\theta(1, \theta)| \, d\theta, \\
\varphi_1(\varepsilon) &:= D_\Omega(X + \varepsilon\xi) - D_\Omega(X), \quad \varphi_2(t) := D_{\Omega_0}(X + t\zeta) - D_{\Omega_0}(X), \\
\Omega &:= B \cap B_{r/2}(w_0), \\
\psi_1(\varepsilon) &:= \int_{C_1} |X_\theta + \varepsilon\xi_\theta| \, d\theta - \int_{C_1} |X_\theta| \, d\theta, \\
\psi_2(t) &:= \int_\gamma |X_\theta + t\zeta_\theta| \, d\theta - \int_\gamma |X_\theta| \, d\theta.
\end{aligned}$$

(We now have once again used: $X_\theta = \frac{\partial}{\partial\theta} X$, etc.) The functions φ_1 and φ_2 are quadratic polynomials, and clearly

$$0 = \varphi_1(0) = \varphi_2(0) = \psi_1(0) = \psi_2(0).$$

Moreover, the function $\psi_2(t)$ is continuous on $[-t_0, t_0]$. We also claim that the derivatives $\psi_1'(0)$ and $\psi_2'(0)$ exist. In fact, the formula $a^2 - b^2 = (a + b)(a - b)$ yields

$$\frac{1}{\varepsilon}\{|X_\theta + \varepsilon\xi_\theta| - |X_\theta|\} = f(\varepsilon) + g(\varepsilon),$$

where

$$f(\varepsilon) = \frac{2\langle X_\theta, \xi_\theta \rangle}{|X_\theta + \varepsilon\xi_\theta| + |X_\theta|}, \quad g(\varepsilon) = \frac{\varepsilon|\xi_\theta|^2}{|X_\theta + \varepsilon\xi_\theta| + |X_\theta|}.$$

Hence we infer that

$$|f(\varepsilon)| \le 2|\xi_\theta|, \quad |g(\varepsilon)| \le |\xi_\theta| \quad \text{a.e. on } C^+$$

and for $|\varepsilon| > 0$. By Lebesgue's theorem on dominated convergence the derivatives $\psi_1'(0)$ and $\psi_2'(0)$ exist, and

$$(29)\quad \psi_1'(0) = \int_{C_1} |X_\theta|^{-1}\langle X_\theta, \xi_\theta\rangle\, d\theta, \quad \psi_2'(0) = \int_{C^+} |X_\theta|^{-1}\langle X_\theta, \zeta_\theta\rangle\, d\theta = 1.$$

Thus the assumptions of Lemma 2 are satisfied, and we obtain

$$\varphi_1'(0) + \lambda_1\psi_1'(0) = 0, \quad \text{where } \lambda_1 = -\varphi_2'(0).$$

On the other hand, we infer from (29) and from

$$\varphi_1(\varepsilon) = \varepsilon \int_\Omega \langle \nabla X, \nabla \xi\rangle r\, dr\, d\theta + \frac{\varepsilon^2}{2} D_\Omega(\xi)$$

that (28) is true for an arbitrarily chosen $\xi \in C_c^1(\Omega \cup C_1, \mathbb{R}^3)$, and hence (28) holds for all $\xi \in C^1(\overline{B}, \mathbb{R}^3)$ that vanish on $\partial B \setminus C_1$. Because of the equivalence of relations (10) and (12) we conclude that

$$(30)\qquad X^* = \lambda_1 |X_\theta|^{-1} X_\theta + P_1$$

holds a.e. on C_1 for some constant vector $P_1 \in \mathbb{R}^3$. If $\lambda_1 = 0$, we would get $X^* = P_1$; i.e. $X_\theta^*(1,\theta) = 0$ a.e on C_1, and this contradicts Lemma 1, (iv). Hence we have indeed $\lambda_1 \ne 0$, and it is proved that X is a stationary solution of the thread problem with respect to C_1 (and to C_2). By Theorem 2, the mappings X and X^* are real analytic on $B \cup C_1 \cup C_2$ and have at most isolated branch points.

In order to complete the proof of Lemma 4 we now assume w.l.o.g. that $C_1 = \{e^{i\theta} : 0 < \theta < \theta_1\}$ for some $\theta_1 \in (0, \pi)$. Then we introduce the two arcs

$$\gamma_1 = \{e^{i\theta} : 0 < \theta < \tfrac{1}{2}\theta_1\}, \quad \gamma_2 = \{e^{i\theta} : \tfrac{1}{2}\theta_1 < \theta < \pi\}.$$

Let us choose two disks B_1 and B_2 with centers outside of B such that $\gamma_1 = \partial B \cap B_1, \gamma_2 = \partial B \cap B_2$, and that the open sets $\Omega_1 = B \cap B_1$ and $\Omega_2 = B \cap B_2$ are disjoint. We claim that there is a function $\zeta_1 \in C_c^1(\Omega_1 \cup \gamma_1, \mathbb{R}^3)$ such that

$$\int_{\gamma_1} |X_\theta|^{-1} \left\langle X_\theta, \frac{\partial \zeta_1}{\partial \theta}\right\rangle d\theta = 1.$$

Otherwise we would have

$$|X_\theta|^{-1} X_\theta = \text{const} \quad \text{on } \gamma_1,$$

whence by (30) $X^*(1,\theta) = \text{const}$ for $0 < \theta < \frac{1}{2}\theta_1$, i.e. $X_r^* = X_\theta^* = 0$ on γ_1. This would be impossible since the branch points of X^* on γ_1 are isolated. In addition, we choose an arbitrary function $\xi \in C_c^1(\Omega_2 \cup \gamma_2, \mathbb{R}^3)$. Then we apply the previous reasoning to the 2-parameter family

$$X_{\varepsilon,t} = X + \varepsilon\xi + t\zeta_1, \quad |\varepsilon| \le \varepsilon_0, \ |t| \le t_0.$$

By the same arguments as before we can establish the existence of a constant vector $P \in \mathbb{R}^3$ and of a number $\lambda \in \mathbb{R}, \lambda \neq 0$, such that

$$X^* = \lambda |X_\theta|^{-1} X_\theta + P \tag{31}$$

holds on γ_2, and we also know that X and X^* are real analytic on $B \cup \gamma_2$. On the other hand, equation (30) is satisfied on C_1. Since

$$C_1 \cap \gamma_2 = \{e^{i\theta} : \tfrac{1}{2}\theta_1 < \theta < \theta_1\},$$

we may infer that $\lambda = \lambda_1$ and $P = P_1$. Thus we have proved that X and X^* are real analytic on $B \cup C^+$, and that (31) is satisfied on all of C^+. This in turn yields

$$\int_B \langle \nabla X, \nabla \xi \rangle r \, dr \, d\theta + \lambda \int_{C^+} \langle |X_\theta|^{-1} X_\theta, \xi_\theta \rangle \, d\theta = 0$$

for all $\xi \in C^1(\overline{B}, \mathbb{R}^3)$ with $\xi = 0$ on $\partial B \setminus C^+$, and Lemma 4 is proved. □

Resuming the results of Theorem 2 and of the Lemmata 2–4, we see that all assertions of Theorem 1 are proved, except for the claim that *Σ is a regular curve*. The proof of this fact will be sketched in the third and last part of our discussion. We shall proceed by proving that no minimizer X can have branch points of even order on C^+. Recall that branch points of odd order were already excluded in Theorem 2; they cannot even occur for stationary solutions of the thread problem. On the other hand, stationary solutions may very well possess branch points of even order, as we have noted in Remark 1. Thus we now really have to employ the minimizing property of X if we wish to exclude branch points of even order. In what follows we shall describe some of the main ideas that lead to the exclusion of true branch points of even order for minimizers X. For this we use some of the reasoning of Gulliver–Lesley and of Osserman [12]. The impossibility of false branch points of even order will not be discussed since we have already described the pertinent ideas in Section 1.9. For further information and for filling in all details we refer the reader to the Scholia of Chapter 6 (see Section 6.4).

It will be convenient to choose the parameter domain of any minimizer X as the semi-disk.

$$B = \{w = u + iv : |w| < 1, v > 0\},$$

and C^+, C^- will be replaced by

$$C = \{w = u + iv : |w| = 1, v \ge 0\}$$

and

$$I = \{u \in \mathbb{R} \colon |u| < 1\}.$$

We now assume that $X \colon C \to \mathbb{R}^3$ yields a monotonic parametrization of Γ, and $X \colon I \to \mathbb{R}^3$ describes the free trace of X, i.e., its movable part Σ of the boundary. It follows from the previous discussion that X can be continued analytically as a minimal surface across I. Let u_0 be an arbitrary branch point of even order for X with $u_0 \in I$. We want to show that the existence of such a branch point contradicts the minimizing property of X.

Without loss of generality we can assume that $u_0 = 0$ and that $X(0) = 0$ because we can always transform $u = u_0$ into $u = 0$ by a conformal self-mapping of B that keeps the points $u = \pm 1$ fixed, and $X(0) = 0$ can be achieved by a suitable translation of $\mathbb{R}^3$. Performing an appropriate rotation of $\mathbb{R}^3$, we can also accomplish the asymptotic representation

$$\begin{aligned} x(w) + iy(w) &= aw^{m+1} + O(|w|^{m+2}), \quad a \neq 0, \\ z(w) &= O(|w|^{m+2}) \end{aligned}$$

for the Cartesian coordinates $x(w), y(w), z(w)$ of $X(w)$ in the neighbourhood of $w = 0$, where a denotes some positive constant and $m = 2\nu, \nu \geq 1$, is the order of the branch point $w = 0$. By a suitable scaling it can also be arranged that

$$\begin{gathered} x(w) + iy(w) = w^{m+1} + O(|w|^{m+2}), \\ z(w) = O(|w|^{m+2}) \end{gathered}$$

holds true for $w \to 0$. Because of the power-series expansion of $X(w)$ at $w = 0$ we may write

$$\begin{aligned} x(w) + iy(w) &= w^{m+1} + \sigma(w), \\ z(w) &= \psi(w), \\ \nabla^k \sigma(w), \nabla^k \psi(w) &= O(|w|^{m+2-k}) \quad \text{for } 0 \leq k \leq 2 \end{aligned} \tag{32}$$

with $m = 2\nu > 0$.

We will now show that this representation can be simplified even further.

Lemma 5. *Let $X \colon B_R(0) \to \mathbb{R}^3$ be a minimal surface with the representation (32) at $w = 0$. Then there exist two neighbourhoods $\mathcal{U}, \mathcal{V}$ of 0 in $B_R(0)$, a function $\varphi \in C^2(\mathcal{V})$ with*

$$\nabla^k \varphi(w) = O(|w|^{m+2-k}) \quad \textit{for } 0 \leq k \leq 2$$

and a C^1-diffeomorphism $F \colon \mathcal{U} \to \mathcal{V}$ of $\mathcal{U}$ onto $\mathcal{V}$ such that the formulas

$$\begin{aligned} x(w) + iy(w) &= F^{m+1}(w), \\ z(w) &= \varphi(F(w)) \end{aligned} \tag{33}$$

hold true for $w \in \mathcal{U}$.

(Note that we use the complex notation $\omega = F(w) \in \mathbb{C}$; thus ω^{m+1} is the $(m+1)$-th power of ω.)

Proof. Define

$$F(w) := w\{1 + w^{-m-1}\sigma(w)\}^{1/(m+1)}$$

on a sufficiently small neighbourhood of $w = 0$. Because of $\sigma(w) = O(|w|^{m+2})$, this definition is meaningful if we choose the $(m+1)$-th root to be one at $w = 0$. Moreover, we have

$$\lim_{w \to 0} \frac{F(w)}{w} = 1.$$

Hence $\nabla F(0)$ exists, and $\nabla F(0) = \mathrm{id}$. Moreover, we have

$$(D_u + iD_v)F(w) = 1 + o(1) \quad \text{as } w \to 0,$$

whence

$$\nabla F(w) \to \nabla F(0) \quad \text{as } w \to 0,$$

and this implies $F \in C^1$. By the inverse function theorem, there exists a C^1-inverse f of F on a neighbourhood $\mathcal{V}$ of the origin; set $\mathcal{U} := f(\mathcal{V})$. Since $F \in C^2(\mathcal{U} \setminus \{0\})$, we see that $f \in C^2(\mathcal{V} \setminus \{0\})$, and it is not difficult to prove that

$$\nabla^2 F(w) = o(|w|^{-1}).$$

In order to be able to use the summation convention, we write $w = u + iv = u^1 + iu^2, u^1 = u, u^2 = v$. Then the identity

$$f^\alpha(F(w)) = u^\alpha, \quad \alpha = 1, 2,$$

implies

$$f^\alpha_{,\beta}(F(w))F^\beta_{,\gamma}(w) = \delta^\alpha_\gamma \quad \text{in } \mathcal{U},$$

that is,

$$f^\alpha_{,\beta}(\tilde{w})F^\beta_{,\gamma}(f(\tilde{w})) = \delta^\alpha_\gamma \quad \text{in } \mathcal{V}.$$

Moreover, we obtain

$$f^\alpha_{,\beta\sigma}F^\beta_{,\gamma}(f) + f^\alpha_{,\beta}F^\beta_{,\gamma\tau}(f)f^\tau_{,\sigma} = 0 \quad \text{in } \mathcal{V} \setminus \{0\}.$$

Multiplying this identity by $f^\gamma_{,\rho}$ we infer

$$f^\alpha_{,\rho\sigma} = -F^\beta_{,\gamma\tau}f^\tau_{,\sigma}f^\alpha_{,\beta}f^\gamma_{,\rho}$$

whence we derive that

$$\nabla^2 f(\tilde{w}) = o(|\tilde{w}|^{-1}) \quad \text{as } \tilde{w} \to 0, \ \tilde{w} = F(w).$$

Now we define $\varphi \colon \mathcal{V} \to \mathbb{R}$ by $\varphi(\tilde{w}) = \psi(f(\tilde{w}))$. Then φ is a well defined function of class $C^1(\mathcal{V}) \cap C^2(\mathcal{V} \setminus \{0\})$ which satisfies

(34) $$\varphi_{,\alpha} = \psi_{,\gamma} f^\gamma_{,\alpha} \quad \text{in } \mathcal{V}$$

and

(35) $$\varphi_{,\alpha\beta} = \psi_{,\gamma\rho} f^\rho_{,\beta} f^\gamma_{,\alpha} + \psi_{,\gamma} f^\gamma_{,\alpha\beta} \quad \text{in } \mathcal{V} \setminus \{0\}.$$

The assumptions of the lemma in conjunction with (34) imply that $\nabla\varphi = O(|w|^{m+1})$. Thus $\nabla^2\varphi(0)$ exists and is equal to zero. On the other hand, we infer from (35) that $\nabla^2\varphi(\tilde{w}) = O(|\tilde{w}|^m)$ holds. Altogether we arrive at $\varphi \in C^2(\mathcal{V})$, and the lemma is proved. □

Lemma 5 permits the introduction of a new independent variable $\tilde{w} = F(w) \in \mathcal{V}$ such that $X = (x, y, z)$ can be written as

(36) $$\begin{aligned} x(\tilde{w}) + iy(\tilde{w}) &= \tilde{w}^{m+1} \\ z(\tilde{w}) &= \varphi(\tilde{w}) \end{aligned} \qquad \text{for } \tilde{w} \in \mathcal{V},$$

where $\varphi \in C^2(\mathcal{V})$ and $\nabla^k\varphi(\tilde{w}) = O(|\tilde{w}|^{m+2-k})$ for $0 \le k \le 2$. (The reader will excuse the sloppy notation $X(\tilde{w})$ for the transformed surface; actually we should write $X(F^{-1}(\tilde{w}))$.)

Now we want to describe some local properties of the function φ which appears in the representation formula (36).

Lemma 6. *Let φ be the function that appears in* (36), *and let $w = u^1 + iu^2$. Then we obtain*

(37) $$D_\alpha \left\{ \frac{\varphi_{u^\alpha}}{\sqrt{1 + c^{-2}|\nabla\varphi|^2}} \right\} = 0 \quad \text{on } \mathcal{V},$$

where $c(w) := (m+1)|w|^m$, $w = u^1 + iu^2$.

(Here, we were even more careless and renamed $\tilde{w}$ as w. Thus the reader should bear in mind that $X(w)$ actually means the transformed surface $X(F^{-1}(\tilde{w}))$. The advantage of our sloppiness is that the following formulas become less cumbersome to read.)

Proof. From (36_1) we see that every point $p \in \mathcal{V} \setminus \{0\}$ has a neighbourhood $\mathcal{V}_1(p)$ which is mapped in a regular way onto a neighbourhood $\mathcal{V}_2$ in the x, y-plane. We write $x^1 = x, x^2 = y$. On $\mathcal{V}_2$ the function $\varphi(u^1, u^2)$ obtains a new representation $\psi(x^1, x^2)$, i.e.,

$$\varphi(u^1, u^2) = \psi(x^1, x^2).$$

As X is a minimal surface, we infer that

$$z = \psi(x^1, x^2)$$

provides a nonparametric representation of this minimal surface. Therefore $\psi(x^1,x^2)$ must satisfy the minimal surface equation

$$D_\alpha \left\{ \frac{\psi_{x^\alpha}}{\sqrt{1 + \psi^2_{x^1} + \psi^2_{x^2}}} \right\} = 0 \quad \text{in } \mathcal{V}_2.$$

From $\varphi(w) = \psi(\text{Re } w^m, \text{Im } w^m)$, we conclude by a straightforward computation that (37) holds in $\mathcal{V}_1$ and therefore also in $\mathcal{V} \setminus \{0\}$.

Now we claim that $\sqrt{1 + c^{-2}|\nabla\varphi|^2}$ is of class $C^1(\mathcal{V})$. In fact, let $\lambda_\alpha := c^{-1}\varphi_{u^\alpha}$. Then we see that $\lambda_\alpha(w) = O(|w|)$, whence we can extend $\lambda_\alpha(w)$ in a continuous way to $\mathcal{V}$ by setting $\lambda_\alpha(0) = 0$. It follows that $\lambda_\alpha(w)\lambda_\beta(w) = O(|w|^2)$, and therefore $\nabla(\lambda_\alpha\lambda_\beta)(0) = 0$. Finally we derive from

$$\lambda_{\alpha,\beta} = \varphi_{,\alpha\beta}\, c^{-1} - c^{-2}\varphi_{,\alpha}\, c_{,\beta} = O(1)$$

that $\nabla(\lambda_\alpha\lambda_\beta) = O(|w|)$, whence $\lambda_\alpha\lambda_\beta \in C^1(\mathcal{V})$. This concludes the proof of (37). □

It follows from the representation (36) that selfintersections of X occur at points which are images of points $w \in \mathcal{V}$ with $\varphi(w) = \varphi(\eta^j w)$ where η denotes some primitive $(m+1)$-th root of unity, and $j \not\equiv 0 \bmod (m+1)$. Note that $\varphi^*(w) := \varphi(\eta w)$ again satisfies (37). Hence the difference $\Phi(w) := \varphi(w) - \varphi^*(w)$ is a solution of a linear elliptic differential equation. To be precise, we have

Lemma 7. *The difference function* Φ *satisfies*

$$\{a_{\alpha\beta}(w)\Phi_{u^\alpha}\}_{u^\beta} = 0 \quad \text{in } \mathcal{V}, \tag{38}$$

where $a_{\alpha\beta}$ *is of class* $C^1(\mathcal{V})$ *and uniformly elliptic on* $\mathcal{V}$, *and* $a_{\alpha\beta}(0) = \delta_{\alpha\beta}$.

Proof. Set $T_\alpha(w, q) := q_\alpha / \sqrt{1 + c^{-2}(w)|q|^2}$ with $|q|^2 = q_\alpha q_\alpha$, and observe that

$$\begin{aligned} T_\alpha(w, \nabla\varphi^*) - T_\alpha(w, \nabla\varphi) &= \int_0^1 \frac{d}{dt} T_\alpha(w, t\nabla\varphi^* + (1-t)\nabla\varphi)\, dt \\ &= \left(\int_0^1 T_{\alpha, q_\beta}(w, t\nabla\varphi^* + (1-t)\nabla\varphi)\, dt \right) \Phi_{u^\beta}. \end{aligned}$$

Then one sees that the assertion follows for

$$a_{\alpha\beta}(w) := \int_0^1 T_{\alpha, q_\beta}(w, t\nabla\varphi^*(w) + (1-t)\nabla\varphi(w))\, dt. \qquad \square$$

It will be useful to obtain an asymptotic representation for the difference function Φ. This can be achieved by the technique of Hartman and Wintner (cf. Section 3.1), which yields the following *alternative*:

Either $\Phi(w) \equiv 0$, *or there exists some integer* $n \geq 1$ *and some number* $a \in \mathbb{C}$, $a \neq 0$, *such that*

$$\Phi_{u^1} - i\Phi_{u^2} = a w^{n-1} + \rho(w) \tag{39}$$

holds with $\rho(w) = o(|w|^{n-1})$ *as* $w \to 0$.

Integrating (39), we arrive at

$$\begin{aligned}\Phi(w) &= \operatorname{Re}\left\{\frac{a}{n}w^n\right\} + \sigma(w),\\ \sigma_{u^1}(w) - i\sigma_{u^2}(w) &= \rho(w), \quad \sigma(w) = o(|w|^n) \quad \text{as } w \to 0.\end{aligned} \tag{40}$$

Applying once again the reasoning used in the proof of Lemma 5 we obtain the existence of some diffeomorphism T defined on some open disk $B_R(0)$ such that

$$\Phi(w) = \operatorname{Re} T^n(w), \tag{41}$$

and that $T(0) = 0$ and $T'(0) \neq 0$ hold.

Then we derive from the alternative above the following result:

Proposition 1. *Let $X: B \cup I \to \mathbb{R}^3$ denote some solution of the thread problem, and suppose that $0 \in I$ is a branch point of X of order $m = 2\nu$. Furthermore, let φ, φ^*, Φ and T be the mappings which we have defined before. Then there exists some neighbourhood $\mathcal{V}_0$ of the origin 0 in $\mathbb{C} \hat{=} \mathbb{R}^2$ such that the following alternative holds true:*

(i) *Either $X|_{\mathcal{V}_0}$ can be reparametrized in such a way that it becomes an immersed surface,*

(ii) *or else, there exist two simple C^1-arcs $\gamma_1, \gamma_2 : [0, \varepsilon] \to \mathcal{V}_0 \cap \overline{B}$ with $\gamma_j(0) = 0$, $|\gamma_j'(0)| = 1, \gamma_1'(0) \neq \gamma_2'(0), X(\gamma_1(t)) = X(\gamma_2(t))$ for all $t \in [0, \varepsilon]$ and such that the vectors $X_u(\gamma_1(t)) \wedge X_v(\gamma_1(t))$ and $X_u(\gamma_2(t)) \wedge X_v(\gamma_2(t))$ are linearly independent for all $t \in [0, \varepsilon]$.*

Proof. Suppose first that $\Phi(w) \equiv 0$. Then, as in Lemma 5, we can show that (i) holds with $\mathcal{V}_0 = \mathcal{U}$. In fact, the system (36) assigns to each $\tilde{w} \in \mathcal{V}$ or to each $w \in \mathcal{U}$ a unique point $X(w) = (x^1(w), x^2(w), x^3(w))$, and the surface X may locally be written as $x^3 = \psi(x^1, x^2)$ with $\psi(x^1, x^2) = \varphi(\tilde{w}), \tilde{w} = F(w)$ and $\psi \in C^1$ since $D\varphi(\tilde{w}) = O(|\tilde{w}|^{m+1})$.

Now we want to settle the case $\Phi(w) \not\equiv 0$ using the expansion (40). We note that $n \geq m+2$ since $\Phi(w) = O(|w|^{m+2})$. Since $m = 2\nu \geq 2$, we find that $n \geq 4$. Define $\mathcal{V}_0 := F^{-1}(B_R(0))$ with a sufficiently small number $R > 0$, and consider the mapping $T \circ F: \mathcal{V}_0 \to \mathbb{C}$ which is conformal at the origin. Let $\zeta := T \circ F(w)$, and denote by $\mathcal{R}_j, 1 \leq j \leq 2n$, the $2n$ rays in the ζ-plane which emanate from $\zeta = 0$ and are defined by Re $\zeta^n = 0$. The rays $\mathcal{R}_j$ correspond to $2n$ curves γ_j in $\mathcal{V}_0$ via the mapping $T \circ F$. Moreover, since $n \geq 4$, at least one of the curves γ_j meets the positive real axis at an angle which is between 0 and $\pi/3$. We can assume that $\gamma_j(t)$ is such an arc, and we can also assume that t is the parameter of arc length along γ_1. Then we have

$$0 = \Phi(F \circ \gamma_1(t)) = \varphi(F \circ \gamma_1(t)) - \varphi(\eta F \circ \gamma_1(t)).$$

Setting $\gamma_2(t) := F^{-1} \circ (\eta F \circ \gamma_1(t))$, we arrive at $X \circ \gamma_1(t) = X \circ \gamma_2(t)$.

Moreover, because of conformality, $\gamma_2(t)$ hits the positive real axis under an angle which is strictly between $\pi/3$ and $\pi/3 + \frac{2\pi}{m+1} < \pi$. For sufficiently small $\varepsilon > 0$, the mappings γ_1 and γ_2 will map $[0, \varepsilon]$ into $\mathcal{V}_0 \cap \overline{B}$. Since Φ describes the difference of two branches of X and because of (41), it immediately follows that the two surface normals along γ_1 and γ_2 respectively are linearly independent. □

Let us now recall the definition of true and false branch points given in Section 1.9.

Definition. *The branch point $w = 0$ of the minimal surface $X(w)$ is called a* false branch point *if case* (i) *holds true; otherwise $w = 0$ is called a true branch point.*

Concerning true branch points, we shall prove:

Proposition 2. *If $X : \overline{B} \to \mathbb{R}^3$ is a solution of the thread problem, then there are no true branch points on the interval $I = \{u \in \mathbb{R} : |u| < 1\}$, which is mapped by X onto the movable boundary Σ.*

Proof. We first recall that X not only minimizes Dirichlet's integral within $\mathcal{C}(\Gamma, L)$ but also the area functional

$$A_B(X) = \int_B |X_u \wedge X_v| \, du \, dv;$$

cf. Theorem 4 of Section 5.2.

We may again assume that the true branch point $w \in I$ under consideration is the point $w = 0$.

Choose a neighbourhood $\mathcal{W}$ of 0 in $\mathbb{C}$ such that $\overline{\mathcal{W} \cap B}$ is diffeomorphic to $\overline{B}$. Suppose that the curves γ_1 and γ_2 first leave $\mathcal{W}$ at $\gamma_1(2\delta)$ and $\gamma_2(2\delta)$ transversally to $\partial\mathcal{W}$. Moreover, let $h : \overline{\mathcal{W} \cap B} \to \overline{B}$ be some C^1-diffeomorphism under consideration which, in addition, maps $\partial\mathcal{W} \cap B$ onto $\partial B \setminus \overline{I}$ and $\overline{\mathcal{W} \cap I}$ onto $\overline{I}$. Furthermore we may assume that

$$h \circ \gamma_1(t) = \frac{t}{\delta}\xi, \quad h \circ \gamma_2(t) = -\frac{t}{\delta}\overline{\xi}$$

for $0 \le t \le 2\delta$, where $\xi \in \mathbb{C}$ denotes some number with $|\xi| = \frac{1}{2}$.

It is now possible to construct a mapping $G : \overline{B} \to \overline{B}$ with the following properties:

(I) G is continuous and one-to-one on $\overline{B} \setminus [0, \frac{i}{2}]$;
(II) $G|_{\partial B} = \mathrm{id}|_{\partial B}$;
(III) For $\zeta \in \mathbb{C}$ with $\operatorname{Re} \zeta > 0$ and $0 \le t < 1$, the following relations are fulfilled:

$$\lim_{\zeta \to 0} G\left(\frac{i}{4}(1 \pm t) + \zeta\right) = (1 - t)\xi,$$

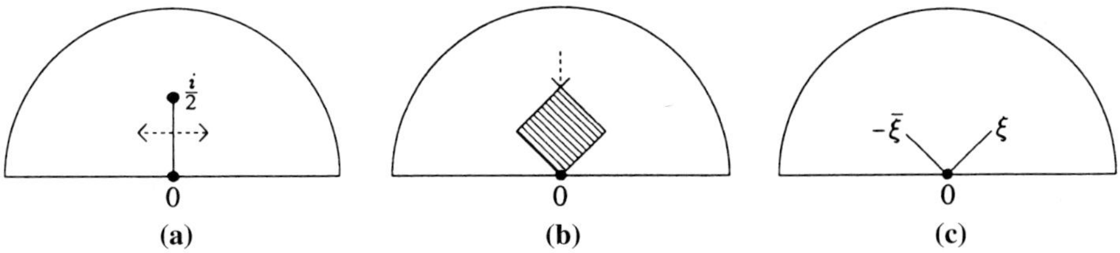

Fig. 3.

$$\lim_{\zeta \to 0} G\left(\frac{i}{4}(1 \pm t) - \zeta\right) = -(1-t)\bar{\xi};$$

(IV) G is piecewise C^1 and extends to a C^1-diffeomorphism on each edge of the slit $[0, \frac{i}{2}]$.

We refrain from constructing G explicitly by formulas; Fig. 3 describes the topological action of G.

Now we define a comparison function $X^* \colon \overline{B} \to \mathbb{R}^3$ by

$$X^*(w) := \begin{cases} X(w) & \text{for } w \in B \setminus \mathcal{W}, \\ (X \circ h^{-1} \circ G \circ h)(w) & \text{for } w \in \mathcal{W}. \end{cases}$$

It is clear that $X^* \in \mathcal{C}(\Gamma, L)$ and that $A_B(X) = A_B(X^*)$. Hence X^* minimizes $A_B(X)$ within $\mathcal{C}(\Gamma, L)$. This leads to a contradiction, since any point $w_0 \in \mathcal{W}$ satisfying $h(w_0) \in (0, \frac{i}{2}]$ possesses some neighbourhood which is mapped onto a surface with two portions intersecting along $X(\gamma_1)$. In view of (ii) this surface has an edge, and by "smoothing out" one can construct from X^* a new surface $X^{**} \in \mathcal{C}(\Gamma, L)$ with $A_B(X^{**}) < A_B(X^*) = A_B(X)$, a contradiction to the minimizing property of X. □

To exclude false branch points we assume that $X|_{\partial B}$ is an embedding of ∂B into $\mathbb{R}^3$. The pertinent reasoning is sketched in Section 4.7 of Vol. 1. A detailed discussion can be found in the paper of Gulliver, Osserman, and Royden [1].

By these remarks we conclude the proof of Theorem 1. □

5.4 Scholia

1. The existence of solutions of the thread problem in its simplest form was first proved by H.W. Alt [3]. Except for minor modifications we have presented Alt's existence proof in Section 5.2. Without any changes the proof can be carried over to 2-dimensional surfaces in $\mathbb{R}^N, N \geq 2$. A different proof has been given by K. Ecker [1], using methods of geometric measure theory; it even works for the analogue of the thread problem concerning n-dimensional

surfaces in $\mathbb{R}^N$. In the framework of integral currents, Ecker has proved the existence of a minimizer, the movable boundary of which has prescribed mass.

2. It seems to have been known for a long time that the *unattached* part of the movable boundary Σ consists of space curves of constant curvature; cf. van der Mensbrugghe [1], Otto [1]. A satisfactory proof was given by Nitsche [21] under the assumption that the free part of Σ is known to be regular and smooth; cf. also Nitsche [28], pp. 435–437 and pp. 706–707.

3. The first results concerning the boundary regularity of solutions for the thread problem were found by Nitsche [23–25]. He proved that the open components of the non-attached part of the movable boundary have a parametrization of class $C^{2,\alpha}$, for some $\alpha \in (0,1)$. Between branch points (the existence of which was not excluded by Nitsche) these parametrizations turn out to be of class C^∞.

The sharper regularity results, presented in Section 5.3, and their proofs are taken from Dierkes, Hildebrandt, and Lewy [1]. We have quite closely followed the presentation given in their paper.

4. By completely different techniques, K. Ecker [1] has established C^∞-regularity of the free part of the movable boundary Σ in the context of his integral-current solutions; the analyticity is in this case still an open question.

5. It is not known whether the *thread* of the solution constructed in Section 10.2 can have self-intersections; we are tempted to conjecture that this cannot occur. In the context of rectifiable flat chains modulo 2 this was in fact proved by R. Pilz [1]. He showed that the free boundary of a minimizer of this kind has no singular points in $\mathbb{R}^3 \setminus \Gamma$, Γ being the fixed part of the boundary.

6. Alt [3] has also proved that the movable arc Σ must always lift off Γ in a tangential way whenever it adheres to Γ in a subarc of positive length provided that Γ is supposed to be smooth.

7. As Alt [3] has pointed out, all pieces of the movable boundary Σ not attached to Γ have the same constant curvature κ. This can easily be proved by the reasoning given in the proofs of Lemmata 2–4 of Section 5.3.

8. In excluding branch points on the free parts of Σ we have used arguments of Gulliver and Lesley [1] and of Gulliver, Osserman, and Royden [1]. This part of our reasoning is restricted to $\mathbb{R}^3$ and cannot be carried over to $\mathbb{R}^n, n \geq 4$, according to an example by Federer [2].

9. A new existence proof for the thread problem was given by E. Kuwert in Section 4 of his Habilitationsschrift [5], pp. 51–52. This proof is a by-product of Kuwert's work on the minimization of Dirichlet's integral $D(X)$ among surfaces $X : B \to \mathbb{R}^n$ whose boundary curves $X\big|_{\partial B}$ represent a given homotopy class α of free loops in a closed configuration $S \subset \mathbb{R}^n$. We refer to the Scholia of Chapter 1 in this volume and to Kuwert [6,7].

10. Recently, the thread problem was anew studied by B.K. Stephens [1–3]. In Section 2 of [1], a new proof of Alt's theorem is given, and Section 3 presents two quantitative bounds on the nearness of minimizers to the wire in

the case that the thread length L is not much less than the length $\ell(\Gamma)$ of the wire Γ: *Suppose that* $\kappa_{\max}$ *is a bound on the curvature of* Γ, *and let* $0 < \lambda \ll 1$ *(relative to the* C^3*-data of* Γ*). Then there is a constant* $R(\Gamma, \lambda) > 0$ *with the following property: If* X *is a minimizer of* $\mathcal{P}(\Gamma, \ell(\Gamma) - \lambda)$, *then the image* M *of* X *lies in a "normal* $\mathcal{R}(\Gamma, \lambda)$*-neighbourhood of* Γ*" whose radius is estimated by*

$$R(\Gamma, \lambda) \le 2\lambda^{1/2}/(\pi\kappa_{\max})^{1/2} + o(\lambda^{1/2}),$$

and the area of M *is bounded by*

$$A(M) \le \lambda/\kappa_{\max} + o(\lambda^{1/2}).$$

Consider now the situation studied in Theorem 1 of Section 5.3 (cf. also Fig. 2 of 5.3), and as Stephens [2], call minimizers of this kind *"crescents"*. In [2], several geometric properties of *"near-wire crescents"* are proved. For instance, the representation of such a crescent X as a graph of a Lipschitz function f with $\mathrm{Lip}(f) \le \mathrm{const}(\Gamma) R^{1/12}$ is established if X lies in an R-tubular neighbourhood of Γ, $0 < R \ll 1$. The main tool is a sophisticated generalization of a result due to Radó (see Vol. 1, Section 4.9, Lemma 2), which Stephens calls "Free Radó Lemma", as it is an adjustment of the original Radó Lemma to the situation available in the thread-problem case.

Chapter 6

Branch Points

In $\mathbb{R}^3$ any solution of Plateau's problem minimizing Dirichlet's integral D or, equivalently, the area functional A, is an immersion in the sense that it has no interior branch points. This fact can easily be proved for planar boundaries as we have seen earlier, while the corresponding result in $\mathbb{R}^n$ is false for $n \geq 4$ according to Federer's counterexample. Therefore it remains to prove the assertion for *nonplanar* minimizers. Here we describe a new method, due to A. Tromba, to exclude interior branch points for nonplanar relative minimizers of Dirichlet's integral D. This method is based on the observation that one can compute any higher derivative of Dirichlet's integral in the direction of so-called *(interior) forced Jacobi fields*, using methods of complex analysis such as power series expansions and Cauchy's integral theorem as well as the residue theorem. These Jacobi fields lie in the kernel of the second variation of D; they also play a fundamental role in the index theory and the Morse theory of minimal surfaces.

We begin by calculating the first five derivatives of Dirichlet's integral in the direction of special types of forced Jacobi fields, thereby establishing that relative minimizers of D cannot have certain kinds of interior branch points. These introductory calculations will be carried out in Section 6.1, together with an outline of the variational procedure to be used in the sequel. These calculations are made transparent by shifting the branch point that is studied into the origin, and by bringing the minimal surface into a *normal form* with respect to the branch point $w = 0$ with an *order* n. Then also the *index* m of this branch point can be defined, with $m > n$. Furthermore, $w = 0$ is called an *exceptional branch point* if there is an integer $\kappa > 1$ such that $m + 1 = \kappa(n + 1)$. It turns out that Tromba's method works perfectly in excluding nonexceptional branch points of relative minimizers of D, while the exclusion of exceptional branch points only succeeds for absolute minimizers of the area A in $\mathcal{C}(\Gamma)$. Since the general investigation is quite lengthy, we only discuss one of the several general cases that are possible for nonexceptional

U. Dierkes, S. Hildebrandt, A.J. Tromba, *Regularity of Minimal Surfaces*, Grundlehren der mathematischen Wissenschaften 340,
DOI 10.1007/978-3-642-11700-8_6, © Springer-Verlag Berlin Heidelberg 2010

branch points (see Section 6.2). A comprehensive presentation of the method for all cases will be given in forthcoming work by A. Tromba.

In Section 6.1 it is described how the variations $\hat{Z}(t)$ of a minimal surface $\hat{X}$ are constructed by using interior forced Jacobi fields. This leads to the (rather weak) notion of a *weak minimizer of* D. Any absolute or weak relative minimizer of D in $\mathcal{C}(\Gamma)$ will be a weak D-minimizer, and the aim is to investigate whether such minimizers can have $w = 0$ as in interior branch point. This possibility is excluded if one can find an integer $L \geq 3$ and a variation $\hat{Z}(t)$ of $\hat{X}$, $|t| \ll 1$, such that $E(t) := D(\hat{Z}(t))$ satisfies

$$E^{(j)}(0) = 0 \quad \text{for } 1 \leq j \leq L-1, \quad E^{(L)}(0) < 0.$$

It will turn out that the existence of such an L depends on the order n and the index m of the branch point $w = 0$.

In Section 6.1, this idea is studied by investigating the third, fourth and fifth derivatives of $E(t)$ at $t = 0$. Here one meets fairly simple cases for testing the technique which show its efficiency. Furthermore, the difficulties are exhibited that will come up generally.

A case of general nature is treated in Section 6.2. Assuming that $n+1$ is even and $m+1$ is odd (whence $w = 0$ is nonexceptional) it will be seen that $E^{(m+1)}(0)$ can be made negative while $E^{(j)}(0) = 0$ for $1 \leq j \leq m$, and so $\hat{X}$ cannot be a weak minimizer of D.

In Section 6.3 we study boundary branch points of a minimal surface $\hat{X} \in \mathcal{C}(\Gamma)$ with a smooth boundary contour. In particular we show that $\hat{X}$ cannot be a minimizer of D in $\mathcal{C}(\Gamma)$ if it has a boundary branch point whose order n and index m satisfy the condition $2m-2 < 3n$ (Wienholtz's theorem).

Furthermore, in Sections 6.1 and 6.3 we exhibit geometric conditions which furnish bounds for the index of interior and boundary branch points. These estimates supplement the bounds on the order of branch points provided by the Gauss–Bonnet theorem.

6.1 The First Five Variations of Dirichlet's Integral, and Forced Jacobi Fields

In this chapter we take the point of view of Jesse Douglas and consider minimal surfaces as critical points of Dirichlet's integral within the class of harmonic surfaces $X : B \to \mathbb{R}^3$ that are continuous on the closure of the unit disk B and map $\partial B = S^1$ homeomorphically onto a closed Jordan curve Γ of $\mathbb{R}^3$. It will be assumed that *Γ is smooth of class C^∞ and nonplanar.* Then any minimal surface bounded by Γ will be a nonplanar surface of class $C^\infty(\overline{B}, \mathbb{R}^3)$, and so we shall be allowed to take directional derivatives (i.e. "variations") of any order of the Dirichlet integral along an arbitrary C^∞-smooth path through the minimal surface.

The first goal is to develop a *technique* which enables us to compute variations of any order of Dirichlet's integral, D, at an arbitrary minimal surface bounded by Γ, using complex analysis in form of Cauchy's integral theorem. This will be achieved by varying a given minimal surface via a one-parameter family of admissible harmonic mappings. Such harmonic variations will be generated by varying the boundary values of a given minimal surface in an admissible way and then extending the varied boundary values harmonically into B. From this point of view the admissible boundary maps $\partial B = S^1 \to \Gamma$ are the primary objects while their harmonic extensions $\overline{B} \to \mathbb{R}^3$ are of secondary nature. This calls for a change of notation: An admissible boundary map will be denoted by $X : \partial B \to \Gamma$, whereas $\hat{X}$ is the uniquely determined harmonic extension of X into B; i.e. $\hat{X} \in C^0(\overline{B}, \mathbb{R}^3) \cap C^2(B, \mathbb{R}^3)$ is the solution of

$$\triangle \hat{X} = 0 \quad \text{in } B, \quad \hat{X}(w) = X(w) \quad \text{for } w \in \partial B.$$

Instead of $\hat{X}$ we will occasionally write HX or $H(X)$ for this extension, and

$$D(\hat{X}) := \frac{1}{2} \int_B \nabla \hat{X} \cdot \nabla \hat{X} \, du \, dv$$

is its Dirichlet integral.

In the sequel the main idea is to vary the boundary values X of a given minimal surface $\hat{X}$ in direction of a so-called *forced Jacobi field*, as this restriction will enable us to evaluate the variations of D at X by means of Cauchy's integral theorem. In order to explain what forced Jacobi fields are we first collect a few useful formulas.

Let us begin with an arbitrary mapping $X \in C^\infty(\partial B, \mathbb{R}^n)$ and its harmonic extension $\hat{X} \in C^\infty(\overline{B}, \mathbb{R}^3)$. Then $\hat{X}$ is of the form

$$\hat{X}(w) = \operatorname{Re} f(w), \tag{1}$$

where f is holomorphic on B and can be written as

$$f = \hat{X} + i\hat{X}^* \quad \text{with } \hat{X}_u = \hat{X}^*_v \text{ and } \hat{X}_v = -\hat{X}^*_u. \tag{2}$$

We also note that

$$f'(w) = 2\hat{X}_w(w) = \hat{X}_u(w) - i\hat{X}_v(w) \quad \text{in } B. \tag{3}$$

Conversely, *if f is holomorphic in B and $\hat{X} = \operatorname{Re} f$ then f' and $\hat{X}_w$ are related by the formula $f' = 2\hat{X}_w$;* in particular, *$\hat{X}_w$ is holomorphic in B.* This simple, but basic fact will be used repeatedly in later computations.

Let us introduce polar coordinates r, θ about the origin by $w = re^{i\theta}$, and set $\hat{Y}(r, \theta) = \hat{X}(re^{i\theta})$. Then a straight-forward computation yields

$$iw\hat{X}_w(w)\Big|_{w=e^{i\theta}} = \frac{1}{2}\left[\hat{Y}_\theta(1,\theta) + i\hat{Y}_r(1,\theta)\right] \tag{4}$$

whence

(5) $$2\text{Re}\left\{iw\hat{X}_w(w)\right\}\Big|_{w=e^{i\theta}} = \hat{Y}_\theta(1,\theta) = \frac{\partial}{\partial\theta}X(e^{i\theta}) = Y_\theta(\theta)$$

since

$$\hat{Y}(1,\theta) = \hat{X}(e^{i\theta}) = X(e^{i\theta}) =: Y(\theta).$$

If $X \in C^\infty(S^1, \mathbb{R}^3)$ maps S^1 homeomorphically onto Γ then $Y_\theta(\theta)$ is tangent to Γ at $Y(\theta)$, i.e. $Y_\theta(\theta) \in T_{Y(\theta)}\Gamma$, and so the left-hand side of (5) is tangent to Γ.

Consider now a continuous function $\tau : \overline{B} \to \mathbb{C}$ that is meromorphic in B with finitely many poles in B, and that is real on ∂B. Then τ can be extended to a meromorphic function on an open set Ω with $\overline{B} \subset \Omega$, and τ is holomorphic in a strip containing ∂B. It follows from (5) that

(6) $$2\text{Re}\left\{iw\hat{X}_w(w)\tau(w)\right\}\Big|_{w=e^{i\theta}} = \tau(e^{i\theta})Y_\theta(\theta) \in T_{Y(\theta)}\Gamma.$$

Suppose now that $\hat{X}$ is a minimal surface with finitely many branch points in $\overline{B}$. These points are the zeros of the function $F(w) := \hat{X}_w(w)$ which is of class C^∞ on $\overline{B}$ and holomorphic in B. If $\tau(w)$ has its poles at most at the (interior) zeros of the function $wF(w)$, and if the order of any pole does not exceed the order of the corresponding zero of $wF(w)$, then the function $K(w) := iw\hat{X}_w(w)\tau(w)$ is holomorphic in B and of class $C^\infty(\overline{B}, \mathbb{R}^3)$. We call $\hat{h} :=$ Re K an **inner forced Jacobi field** $\hat{h} : \overline{B} \to \mathbb{R}^3$ **at** $\hat{X}$ with the **generator** τ.

In case that one wants to study boundary branch points of $\hat{X}$ it will be useful to admit factors $\tau(w)$ which are meromorphic on $\overline{B}$, real on ∂B, with poles at most at the zeros of $wF(w)$, the pole orders not exceeding the orders of the associated zeros of $wF(w)$. Then

(7) $$\hat{h} := \text{Re}\,K \quad \text{with } K(w) := iwF(w)\tau(w),\ w \in \overline{B},\ F := \hat{X}_w,$$

is said to be a (general) **forced Jacobi field** $\hat{h} : \overline{B} \to \mathbb{R}^3$ at the minimal surface $\hat{X}$, and τ is called the **generator** of $\hat{h}$.

The boundary values $\hat{h}|_{S^1}$ of a forced Jacobi field $\hat{h}$ are given by

(8) $$h(\theta) := \hat{h}(e^{i\theta}) = \text{Re}\,K(e^{i\theta}) = \frac{1}{2}\tau(e^{i\theta})Y_\theta(\theta), Y(\theta) := \hat{X}(\cos\theta, \sin\theta).$$

Using the asymptotic expansion of $F(w) = X_w(w)$ at a branch point $w_0 \in \overline{B}$ having the order $\lambda \in \mathbb{N}$, we obtain the factorization

(9) $$F(w) = (w - w_0)^\lambda G(w) \quad \text{with } G(w_0) \neq 0,$$

and, using Taylor's expansion in B or Taylor's formula on ∂B respectively, it follows that $G(w) = G(u, v)$ is a holomorphic function of w in B and a C^∞-function of $(u, v) \in \overline{B}$. It follows that *any forced Jacobi field* $\hat{h} : \overline{B} \to \mathbb{R}^3$ *is of class* $C^\infty(\overline{B}, \mathbb{R}^3)$ *and harmonic in* B.

Denote by $J(\hat{X})$ the linear space of forced Jacobi fields at $\hat{X}$, and let $J_0(\hat{X})$ be the linear subspace of inner forced Jacobi fields. The importance of $J(\hat{X})$ arises from the fact that *every forced Jacobi field* $\hat{h}$ *at* $\hat{X}$ *annihilates the second variation of* D, *i.e.*

$$\delta^2 D(\hat{X}, \hat{h}) = 0 \quad \textit{for all } \hat{h} \in J(\hat{X}).$$

This will be proved later in Section 6.3. In the present section we only deal with inner forced Jacobi fields, and so we only prove the weaker statement (cf. Proposition 1):

$$\delta^2 D(\hat{X}, \hat{h}) = 0 \quad \text{for all } \hat{h} \in J_0(X).$$

The existence of forced Jacobi fields arises from the group of conformal automorphisms of $\overline{B}$ and from the presence of branch points; the more branch points $\hat{X}$ has, and the higher their orders are, the more Jacobi fields appear—this explains the adjective 'forced'. To see the first statement we consider one-parameter families of conformal automorphisms $\varphi(\cdot, t)$, $|t| < \epsilon, \epsilon > 0$ of $\overline{B}$ with

$$(10) \qquad w \mapsto \varphi(w,t) = w + t\eta(w) + o(t) \text{ and } \varphi(w,0) = w, \dot{\varphi}(w,0) = \eta(w).$$

Type I:

$$\varphi_1(w,t) = e^{i\alpha(t)} w$$

with $\alpha(t) \in \mathbb{R}, \alpha(0) = 0, \dot{\alpha}(0) = a$. Then $\varphi_1(w,t) = w + tiwa + o(t)$, and so

$$\eta_1(w) = iwa \quad \text{with } a \in \mathbb{R}.$$

Type II:

$$\varphi_2(w,t) := \frac{w + i\beta(t)}{1 - i\beta(t)w}$$

with $\beta(t) \in \mathbb{R}, \beta(0) = 0, \dot{\beta}(0) = b$.
Then $\varphi_2(w,t) = w + t\eta_2(w) + o(t)$ with $\eta_2(w) = ib + ibw^2$, and so

$$\eta_2(w) = iw\left(\frac{b}{w} + bw\right) \quad \text{with } b \in \mathbb{R}.$$

Type III:

$$\varphi_3(w,t) := \frac{w - \gamma(t)}{1 - \gamma(t)w}$$

with $\gamma(t) \in \mathbb{R}, \gamma(0) = 0, \dot{\gamma}(0) = c$.
Then $\varphi_3(w,t) = w + t\eta_3(w) + o(t)$ with $\eta_3(w) = -c + cw^2$, whence

$$\eta_3(w) = iw\left(\frac{ic}{w} - icw\right).$$

We set

$$\tau_1(w) := a, \quad \tau_2(w) := b \cdot \left(\frac{1}{w} + w\right), \quad \tau_3(w) := c \cdot \left(\frac{i}{w} - iw\right), \tag{11}$$

with arbitrary constants $a, b, c \in \mathbb{R}$. For $w = e^{i\theta} \in \partial B$ we have

$$\tau_1(w) = a, \quad \tau_2(w) = 2b\cos\theta, \quad \tau_3(w) = -2c\sin\theta,$$

and so $\tau_j, j = 1, 2, 3$, are generators of the 'special' forced Jacobi field $\hat{h}_j := \operatorname{Re} K_j$, defined by

$$K_j(w) := iwF(w)\tau_j(w), \quad w \in \overline{B}, \ F := \hat{X}_w, \tag{12}$$

which are inner forced Jacobi fields for any minimal surface $\hat{X}$ *bounded by* Γ. If we vary $\hat{X}$ by means of $\varphi = \varphi_1, \varphi_2, \varphi_3$ with $\alpha := \operatorname{Re}\varphi$, $\beta := \operatorname{Im}\varphi$, i.e. $\varphi(w,t) = \alpha(u,v,t) + i\beta(u,v,t)$, setting

$$\hat{Z}(w,t) := \hat{X}(\varphi(w,t)) = \hat{X}(\alpha(u,v,t), \beta(u,v,t)),$$

we obtain

$$\begin{aligned}\frac{d}{dt}\hat{Z} &= \frac{d}{dt}\hat{X} \circ \varphi = \frac{d}{dt}\hat{X}(\alpha,\beta) = \hat{X}_u(\alpha,\beta)\dot{\alpha} + \hat{X}_v(\alpha,\beta)\dot{\beta} \\ &= 2\operatorname{Re}\hat{X}_w(\varphi)\dot{\varphi},\end{aligned}$$

and so

$$\frac{d}{dt}\hat{Z}\Big|_{t=0} = 2\operatorname{Re}\{\hat{X}_w\dot{\varphi}(0)\}.$$

For $\varphi = \varphi_j$ we have $\dot{\varphi}(0) = \eta_j$, hence

$$\frac{d}{dt}\hat{Z}(w,t)\Big|_{t=0} = 2\operatorname{Re}\{iw\hat{X}_w(w)\tau_j(w)\} = 2\hat{h}_j(w). \tag{13}$$

Let us now generate variations $\hat{Z}(t)$, $|t| \ll 1$, of a minimal surface $\hat{X}$ using any inner forced Jacobi field $\hat{h} \in J_0(\hat{X})$. We write $\hat{Z}(t) = \hat{Z}(\cdot, t)$ for the variation of $\hat{X}$ and $Z(t)$ for the variation of the boundary values X of $\hat{X}$, and start with the definition of $Z(t)$. Then $\hat{Z}(t)$ will be defined as the harmonic extension of $Z(t)$, i.e.

$$\hat{Z}(t) = H(Z(t)). \tag{14}$$

First we pick a smooth family $\gamma(t) = \gamma(\cdot, t), |t| < \delta$, of smooth mappings $\gamma(t) : \mathbb{R} \to \mathbb{R}$ with $\gamma(0) = \mathrm{id}_{\mathbb{R}}$ which are "shift periodic" with the period 2π, i.e.

$$\gamma(\theta, 0) = \theta \quad \text{and} \quad \gamma(\theta + 2\pi, t) = \gamma(\theta, t) + 2\pi \quad \text{for } \theta \in \mathbb{R}. \tag{15}$$

Setting $\sigma(\theta,t) := \gamma(\theta,t) - \theta$ we obtain

$$\gamma(\theta,t) = \theta + \sigma(\theta,t) \quad \text{with } \sigma(\theta,0) = 0 \text{ and } \sigma(\theta+2\pi,t) = \sigma(\theta,t)$$

and

$$\gamma_\theta(\theta,t) = 1 + \sigma_\theta(\theta,t) = 1 + \sigma_{\theta t}(\theta,0)t + o(t).$$

Choosing $\delta > 0$ sufficiently small it follows that

$$\gamma_\theta(\theta,t) > 0 \quad \text{for } (\theta,t) \in \mathbb{R} \times (-\delta,\delta).$$

Now we define the variation $\{Z(t)\}_{|t|<\delta}$ of X by

$$(16) \qquad Z(e^{i\theta},t) := X(e^{i\gamma(\theta,t)}) = \hat{X}(\cos\gamma(\theta,t), \sin\gamma(\theta,t)).$$

Then

$$\frac{\partial}{\partial t} Z(e^{i\theta},t) = \left[-\hat{X}_u(e^{i\gamma(\theta,t)})\sin\gamma(\theta,t) + \hat{X}_v(e^{i\gamma(\theta,t)})\cos\gamma(\theta,t)\right]\gamma_t(\theta,t).$$

By (4) we have

$$ie^{i\theta}\hat{X}_w(e^{i\theta}) = \frac{1}{2}\left[X_\theta(\theta) + i\hat{X}_r(1,\theta)\right]$$

if we somewhat sloppily write $\hat{X}(r,\theta)$ for $\hat{X}(re^{i\theta})$ and $X(\theta)$ for $\hat{X}(1,\theta) = X(e^{i\theta})$. This leads to

$$-\hat{X}_u(e^{i\gamma(\theta,t)})\sin\gamma(\theta,t) + \hat{X}_v(e^{i\gamma(\theta,t)})\cos\gamma(\theta,t) = X_\theta(\gamma(\theta,t))$$

whence

$$\frac{\partial}{\partial t} Z(e^{i\theta},t) = X_\theta(\gamma(\theta,t))\gamma_\theta(\theta,t) \cdot \frac{\gamma_t(\theta,t)}{\gamma_\theta(\theta,t)}.$$

On account of

$$(17) \qquad Z(\theta,t) := Z(e^{i\theta},t) = X(\gamma(\theta,t))$$

we have

$$Z_\theta(\theta,t) = X_\theta(\gamma(t,\theta)) \cdot \gamma_\theta(\theta,t),$$

and so it follows that

$$\frac{\partial}{\partial t} Z(e^{i\theta},t) = \frac{\partial}{\partial t} Z(\theta,t) = \frac{\partial}{\partial \theta} Z(\theta,t) \cdot \phi(\theta,t)$$

with

$$(18) \qquad \phi(\theta,t) := \frac{\gamma_t(\theta,t)}{\gamma_\theta(\theta,t)}.$$

Defining the family $\{\phi(t)\}_{|t|<\delta}$ of 2π-periodic functions $\phi(t) : \mathbb{R} \to \mathbb{R}$ by $\phi(t) := \phi(\cdot,t)$, we have

$$\frac{\partial}{\partial t} Z(t) = \phi(t) Z(t)_\theta =: h(t). \tag{19}$$

Now we consider the varied Dirichlet integral

$$E(t) := D(\hat{Z}(t)) = \frac{1}{2} \int_B \nabla \hat{Z}(t) \cdot \nabla \hat{Z}(t) \, du \, dv. \tag{20}$$

Then

$$\frac{d}{dt} E(t) = \int_B \nabla \hat{Z}(t) \cdot \nabla \frac{d}{dt} \hat{Z}(t) \, du \, dv.$$

Since the operations $\frac{d}{dt}$ and H (i.e. $\hat{Z}$) commute, we have

$$\frac{d}{dt} \hat{Z}(t) = H \left(\frac{d}{dt} Z(t) \right)$$

and therefore

$$\frac{d}{dt} E(t) = \int_B \nabla \hat{Z}(t) \cdot \nabla H \left(\frac{d}{dt} Z(t) \right) \, du \, dv.$$

Since $\triangle \hat{Z}(t) = 0$, an integration by parts leads to

$$\frac{d}{dt} E(t) = \int_0^{2\pi} \frac{\partial}{\partial r} \hat{Z}(t) \cdot h(t) \, d\theta \quad \text{with } h(t) = \frac{\partial}{\partial t} Z(t). \tag{21}$$

For brevity we write in the following computations $\hat{Z}$ instead of $\hat{Z}(t)$. We have

$$w \hat{Z}_w = \frac{1}{2} (\hat{Z}_r - i \hat{Z}_\theta)$$

if we write $\hat{Z}(r, \theta)$ for $\hat{Z}(w)|_{w = re^{i\theta}}$, cf. (4), and also

$$dw = iw \, d\theta \quad \text{for } w = e^{i\theta} \in \partial B.$$

Then on ∂B:

$$\begin{aligned} w \hat{Z}_w \cdot \hat{Z}_w \, dw &= i (w \hat{Z}_w) \cdot (w \hat{Z}_w) \, d\theta \\ &= \frac{i}{4} (\hat{Z}_r - i \hat{Z}_\theta) \cdot (\hat{Z}_r - i \hat{Z}_\theta) \, d\theta \\ &= \left[\frac{1}{2} \hat{Z}_r \cdot \hat{Z}_\theta - \frac{i}{4} (\hat{Z}_r \cdot \hat{Z}_r - \hat{Z}_\theta \cdot \hat{Z}_\theta) \right] d\theta, \end{aligned}$$

and so

$$2 \mathrm{Re} [w \hat{Z}_w \cdot \hat{Z}_w \phi \, dw] = \hat{Z}_r \cdot \hat{Z}_\theta \phi \, d\theta \quad \text{on } \partial B.$$

Furthermore, $\hat{Z}_\theta = Z_\theta$ on ∂B as well as $h = \phi Z_\theta$ (see (19)), and so (21) leads to the formula

$$\frac{d}{dt}E(t) = 2\text{Re}\int_{S^1} w\hat{Z}(t)_w \cdot \hat{Z}(t)_w \phi(t)\, dw, \tag{22}$$

where the closed curve S^1 is positively oriented. This formula will be the starting point for calculating all higher order derivatives $\frac{d^n}{dt^n}E(t)$ and, in particular, of $\frac{d^n}{dt^n}E(0) := \frac{d^n}{dt^n}E(t)\big|_{t=0}$. In order to evaluate the latter expressions for any n, it will be essential that we can choose $\phi(t)$ and any number of t-derivatives of $\phi(t)$ in an arbitrary way. This is indeed possible according to the following result:

Lemma 1. *By a suitable choice of $\gamma(\theta, t) = \theta + \sigma(\theta, t)$ with $\sigma \in C^\infty$ on $\mathbb{R} \times (-\delta, \delta), \sigma(\theta, 0) = 0$ and $\sigma(\theta + 2\pi, t) = \sigma(\theta, t)$ we can ensure that the variation of the boundary values of the minimal surface $\hat{X}$, defined by $Z(\theta, t) := X(\gamma(\theta, t))$, leads to "test functions" $\phi(\theta, t)$ in formula (22) such that the functions*

$$\phi_\nu(\theta) := \frac{\partial^\nu}{\partial t^\nu}\phi(\theta, t)\big|_{t=0}, \quad \nu = 0, 1, 2, \ldots, n,$$

can arbitrarily be prescribed as 2π-periodic functions of class C^∞.

Proof. Let us first check that, given $\phi_0, \phi_1, \ldots, \phi_n$, the computation of σ, and so of γ, can be carried out in a formal way. Consider the Fourier expansion of the function $\sigma(\theta, t)$ which is to be determined:

$$\sigma(\theta, t) = \frac{1}{2}a_0(t) + \sum_{k=1}^{\infty}[a_k(t)\cos k\theta + b_k(t)\sin k\theta]. \tag{23}$$

From $\sigma(\theta, 0) = 0$ it follows that

$$a_0(0) = a_k(0) = b_k(0) = 0 \quad \text{for } k \in \mathbb{N}.$$

Furthermore,

$$\sigma_\nu(\theta) := \frac{\partial^\nu}{\partial t^\nu}\sigma(\theta, 0) = \frac{1}{2}a_0^{(\nu)}(0) + \sum_{k=1}^{\infty}[a_k^{(\nu)}(0)\cos k\theta + b_k^{(\nu)}(0)\sin k\theta]. \tag{24}$$

Hence if $D_t^\nu \sigma(\theta, 0)$ are known for $\nu = 1, 2, \ldots, n$, one also knows all derivatives $D_\theta D_t^\nu \sigma(\theta, 0) = \sigma_\nu'(\theta)$ from the defining equation (18) for σ which amounts to

$$\phi(\theta, t) = \frac{\sigma_t(\theta, t)}{1 + \sigma_\theta(\theta, t)}.$$

By differentiation with respect to t we obtain

$$\begin{aligned}
\phi_t &= \frac{\sigma_{tt}}{1+\sigma_\theta} - \frac{\sigma_t\sigma_{\theta t}}{(1+\sigma_\theta)^2},\\
\phi_{tt} &= \frac{\sigma_{ttt}}{1+\sigma_\theta} - \frac{2\sigma_{tt}\sigma_{\theta t}}{(1+\sigma_\theta)^2} - \frac{\sigma_t\sigma_{\theta tt}}{(1+\sigma_\theta)^2} + \frac{2\sigma_t(\sigma_{t\theta})^2}{(1+\sigma_\theta)^3}
\end{aligned}$$

etc. Setting $t = 0$ and observing that $\sigma_\theta(\theta, 0) = 0$ it follows that

$$\begin{aligned}
\sigma_1 &= \phi_0 = \phi, \\
\sigma_2 &= \phi_1 + \sigma_1\sigma_1', \\
\sigma_3 &= \phi_2 + 2\sigma_2\sigma_1' + \sigma_1\sigma_2' - 2\sigma_1(\sigma_1')^2, \\
&\dots \\
\sigma_{\nu+1} &= \phi_\nu + f_\nu(\sigma_1, \dots, \sigma_\nu, \sigma_1', \dots, \sigma_\nu').
\end{aligned}$$

Here f_ν is a polynomial in the variables $\sigma_1, \dots, \sigma_\nu, \sigma_1', \dots, \sigma_\nu'$. This shows that, given $\phi_0, \phi_1, \dots, \phi_n$, we can successively determine $\sigma_1, \sigma_2, \dots, \sigma_{n+1}$. On account of (23) we then obtain

$$A_0^\nu := a_0^{(\nu)}(0), \quad A_k^\nu := a_k^{(\nu)}(0), \quad B_k^\nu := b_k^{(\nu)}(0) \quad \text{for } k \in \mathbb{N}.$$

Defining

$$a_k(t) := \sum_{\nu=1}^{n+1} \frac{1}{\nu!} A_k^\nu t^\nu, \quad b_k(t) := \sum_{\nu=1}^{n+1} \frac{1}{\nu!} B_k^\nu t^\nu,$$

equation (23) furnishes the function $\gamma(\theta, t) = \theta + \sigma(\theta, t)$ with the desired properties. Furthermore, the construction shows that this procedure leads to a C^∞-function σ that is 2π-periodic with respect to θ. □

Let us inspect a variation $\hat{Z}(t) = H(Z(t))$ of a minimal surface $\hat{X} \in C^\infty(\overline{B}, \mathbb{R}^3)$ as we have just discussed. It is the harmonic extension of a variation $Z(t)$ of the boundary values X of $\hat{X}$, given by (15) and (16). Clearly, $\hat{Z}(t)$ is not merely an "inner variation" of $\hat{X}$, generated as a reparametrization $\hat{X} \circ \sigma(t)$ with a perturbation $\sigma(t) = \mathrm{id}_{\overline{B}} + t\lambda + \cdots$ of the identity $\mathrm{id}_{\overline{B}}$ on $\overline{B}$, but the image $\hat{Z}(t)(B)$ will differ from the image $\hat{X}(B)$. Only the images $Z(t)(S^1)$ and $X(S^1)$ of the boundary $S^1 = \partial B$ will be the same set Σ, but described by different parametrizations $Z(t) : S^1 \to \Sigma$ and $X : S^1 \to \Sigma$.

Definition 1. *We call such a variation $\hat{Z}(t)$ a* **boundary preserving variation of** $\hat{X}$ *(for $|t| \ll 1$).*

Note: If $\hat{X} \in \mathcal{C}(\Gamma)$ then any boundary preserving variation $\hat{Z}(t)$ (with $|t| \ll 1$) lies in $\mathcal{C}(\Gamma)$.

Definition 2. *We say that $\hat{X}$* **is a weak relative minimizer of** D *(with respect to its own boundary) if $E(0) \le E(t)$ holds for any variation $E(t) = D(\hat{Z}(t))$ of D by an arbitrary boundary preserving variation $\hat{Z}(t)$ of $\hat{X}$ with $|t| \ll 1$.*

If $\hat{X} \in \mathcal{C}(\Gamma)$ is a weak relative minimizer of D in $\mathcal{C}(\Gamma)$ with respect to some C^k-norm on $\overline{B}$, then $\hat{X}$ clearly is a weak relative minimizer of D in the sense of Definition 2.

Let us return to formula (19) which states that

$$\frac{\partial}{\partial t} Z(t) = \phi(t) Z(t)_\theta .$$

According to (5) we have

$$Z(t)_\theta = 2\mathrm{Re}[iw \hat{Z}_w(w,t)]\big|_{w=e^{i\theta}},$$

and since ϕ is real-valued it follows that

(25) $$\frac{\partial}{\partial t} Z(\theta, t) = 2\mathrm{Re}[iw \hat{Z}_w(w,t)\phi(\theta,t)]\big|_{w=e^{i\theta}} .$$

Since $\frac{\partial}{\partial t}$ and the harmonic extension H commute we obtain

(26) $$\frac{\partial}{\partial t} \hat{Z}(t) = H\{2\mathrm{Re}[iw \hat{Z}(t)_w \phi(t)]\} \quad \text{in } \overline{B}$$

having for brevity dropped the w, except for the factor iw (as this would require a clumsy notation). Then, by

$$\frac{\partial}{\partial t}\frac{\partial}{\partial w} \hat{Z}(t) = \frac{\partial}{\partial w}\frac{\partial}{\partial t} \hat{Z}(t),$$

it follows that

(27) $$\frac{\partial}{\partial t} \hat{Z}(t)_w = \Big(H\{2\mathrm{Re}[iw \hat{Z}(t)_w \phi(t)]\}\Big)_w .$$

Now a straight-forward differentiation of (22) yields

(28) $$\begin{aligned}\frac{d^2}{dt^2} E(t) &= 4\mathrm{Re} \int_{S^1} w \left\{ \frac{\partial \hat{Z}(t)}{\partial t} \right\}_w \cdot \hat{Z}(t)_w \phi(t)\, dw \\ &\quad + 2\mathrm{Re} \int_{S^1} w \hat{Z}(t)_w \cdot \hat{Z}(t)_w \phi_t(t)\, dw .\end{aligned}$$

From (22) and (28) we obtain

Proposition 1. *Since $\hat{X} = \hat{Z}(0)$ is a minimal surface we have*

(29) $$\frac{dE}{dt}(0) = 0$$

and

(30) $$\frac{d^2 E}{dt^2}(0) = 4\mathrm{Re} \int_{S^1} w \left\{ \frac{\partial \hat{X}}{\partial t} \right\}_w \cdot \hat{X}_w \tau \, dw$$

with $\tau := \phi(0)$. If τ is the generator of an inner forced Jacobi field attached to $\hat{X}$, then

$$\frac{d^2E}{dt^2}(0) = 0. \tag{31}$$

This means that

$$\delta^2 D(\hat{X}, \hat{h}) = 0 \quad \textit{for all } \hat{h} \in J_0(\hat{X}), \tag{32}$$

i.e. for all inner forced Jacobi fields $\hat{h} = \text{Re}[iwX_w(w)\tau(w)]$.

Proof. We have $\hat{X}_w \cdot \hat{X}_w = 0$ since $\hat{X}$ is a minimal surface, and so (29) and (30) are proved. Secondly, $\hat{h}$ is holomorphic in B, as it is an inner forced Jacobi field, and the w-derivative of any harmonic mapping is holomorphic whence $\{\frac{\partial \hat{X}}{\partial t}\}_w$ is holomorphic in B. Thus the integrand of $\int_{S^1}(\dots)\,dw$ in (30) is holomorphic. Hence this integral vanishes, since Cauchy's integral theorem implies $\int_{\partial B_r(0)}(\dots)\,dw = 0$ for any $r \in (0,1)$ and then $\int_{S^1}(\dots)\,dw = \lim_{r\to 1-0}\int_{\partial B_r(0)}(\dots)\,dw = 0$ as the integrand $(\dots)$ is continuous (and even of class C^∞) on $\overline{B}$. □

Now we want to compute $\frac{d^3}{dt^3}E(t)$, and in particular $\frac{d^3E}{dt^3}(0)$ if $\tau = \phi(0)$ is the generator of an inner forced Jacobi field. Differentiating (28) it follows

$$\begin{aligned} \frac{d^3}{dt^3}E(t) &= 4\text{Re}\int_{S^1} w\left\{\frac{\partial \hat{Z}(t)}{\partial t}\right\}_w \cdot \left\{\frac{\partial \hat{Z}(t)}{\partial t}\right\}_w \phi(t)\,dw \\ &\quad + 4\text{Re}\int_{S^1} w\left\{\frac{\partial^2 \hat{Z}(t)}{\partial t^2}\right\}_w \cdot \hat{Z}(t)_w\phi(t)\,dw \\ &\quad + 8\text{Re}\int_{S^1} w\left\{\frac{\partial \hat{Z}(t)}{\partial t}\right\}_w \cdot \hat{Z}(t)_w\phi_t(t)\,dw \\ &\quad + 2\text{Re}\int_{S^1} w\hat{Z}(t)_w \cdot \hat{Z}(t)_w\phi_{tt}(t)\,dw. \end{aligned} \tag{33}$$

Proposition 2. *Since* $\hat{X} = \hat{Z}(0)$ *is a minimal surface we have*

$$\frac{d^3E}{dt^3}(0) = -4\text{Re}\int_{S^1} w^3\hat{X}_{ww}\cdot\hat{X}_{ww}\tau^3\,dw \tag{34}$$

if $\tau := \phi(0)$ *is the generator of an inner forced Jacobi field at* $\hat{X}$.

Proof. The fourth integral in (33) vanishes at $t = 0$ since

$$\hat{Z}(0)_w \cdot \hat{Z}(0)_w = X_w \cdot X_w = 0.$$

The integrand of the second integral in (33) is

$$\left\{\frac{\partial^2 \hat{Z}}{\partial t^2}(0)\right\}_w \cdot w\hat{X}_w\tau(w)$$

which is holomorphic in B since the w-derivative of a harmonic mapping is holomorphic and $\hat{h} = \mathrm{Re}[iw\hat{X}_w\tau]$ is an inner forced Jacobi field. So also the second integral in (33) vanishes on account of Cauchy's integral theorem. Next, using (27), we obtain

$$(35) \qquad \left\{\frac{\partial}{\partial t}\hat{Z}(t)\right\}_w\Big|_{t=0} = 2\frac{\partial}{\partial w}H\left\{\mathrm{Re}[iw\hat{X}_w\tau]\right\} = [iw\hat{X}_w\tau]_w.$$

This implies

$$\begin{aligned}&\left[w\left\{\frac{\partial}{\partial t}\hat{Z}(t)\right\}_w \cdot \hat{Z}(t)_w\right]\Big|_{t=0}\\ &\quad = w[iw\hat{X}_w\tau]_w \cdot \hat{X}_w\\ &\quad = iw\hat{X}_w \cdot \hat{X}_w\tau + iw^2\hat{X}_{ww} \cdot \hat{X}_w\tau + iw^2\hat{X}_w \cdot \hat{X}_w\tau_w = 0\end{aligned}$$

since $\hat{X}_w \cdot \hat{X}_w = 0$, which also yields $\hat{X}_{ww} \cdot \hat{X}_w = 0$. Thus

$$(36) \qquad \left[w\left\{\frac{\partial}{\partial t}\hat{Z}(t)\right\}_w \cdot \hat{Z}(t)_w\right]\Big|_{t=0} = 0$$

and so the third integral in (33) vanishes for $t = 0$. Finally, by (35),

$$\begin{aligned}&\left(\left\{\frac{\partial}{\partial t}\hat{Z}(t)\right\}_w \cdot \left\{\frac{\partial}{\partial t}\hat{Z}(t)\right\}_w\right)\Big|_{t=0}\\ &\quad = [iw\hat{X}_w\tau]_w \cdot [iwX_w\tau]_w\\ &\quad = [i\hat{X}_w\tau + iw\hat{X}_{ww}\tau + iw\hat{X}_w\tau_w] \cdot [i\hat{X}_w\tau + iw\hat{X}_{ww}\tau + iw\hat{X}_w\tau_w]\\ &\quad = -w^2\hat{X}_{ww} \cdot \hat{X}_{ww}\tau^2,\end{aligned}$$

using again $\hat{X}_w \cdot \hat{X}_w = 0$ and $\hat{X}_w \cdot \hat{X}_{ww} = 0$, i.e.

$$(37) \qquad \left(\left\{\frac{\partial}{\partial t}\hat{Z}(t)\right\}_w \cdot \left\{\frac{\partial}{\partial t}\hat{Z}(t)\right\}_w\right)\Big|_{t=0} = -w^2\hat{X}_{ww} \cdot \hat{X}_{ww}\tau^2.$$

Thus the first integral in (33) amounts to

$$-4\mathrm{Re}\int_{S^1} w^3\hat{X}_{ww} \cdot \hat{X}_{ww}\tau^3\,dw.$$

□

In order to simplify notation we drop the t in (33) and write

$$\begin{aligned}\frac{d^3}{dt^3}E = \mathrm{Re}\bigg[&4\int_{S^1} w\hat{Z}_{tw} \cdot \hat{Z}_{tw}\phi\,dw + 4\int_{S^1} w\hat{Z}_{ttw} \cdot \hat{Z}_w\phi\,dw\\ &+8\int_{S^1} w\hat{Z}_{tw} \cdot \hat{Z}_w\phi_t\,dw + 2\int_{S^1} w\hat{Z}_w \cdot \hat{Z}_w\phi_{tt}\,dw\bigg].\end{aligned}$$

Differentiation yields

$$\frac{d^4}{dt^4}E = \mathrm{Re}\left[12\int_{S^1} w\hat{Z}_{ttw}\cdot\hat{Z}_{tw}\phi\,dw + 4\int_{S^1} w\hat{Z}_{tttw}\cdot\hat{Z}_w\phi\,dw \right. \tag{38}$$
$$+ 12\int_{S^1} w\hat{Z}_{tw}\cdot\hat{Z}_{tw}\phi_t\,dw + 12\int_{S^1} w\hat{Z}_{ttw}\cdot\hat{Z}_w\phi_t\,dw$$
$$\left. +12\int_{S^1} w\hat{Z}_{tw}\cdot\hat{Z}_w\phi_{tt}\,dw + 2\int_{S^1} w\hat{Z}_w\cdot\hat{Z}_w\phi_{ttt}\,dw\right]$$
$$= \mathrm{Re}[I_1 + I_2 + I_3 + I_4 + I_5 + I_6].$$

We have $I_6(0) = 0$ since $\hat{Z}_w(0)\cdot\hat{Z}_w(0) = \hat{X}_w\cdot\hat{X}_w = 0$. Moreover, by Cauchy's theorem, $I_2(0) = 0$ since both $\hat{Z}_{tttw}\big|_{t=0} = [\hat{Z}_{ttt}(0)]_w$ and $w\hat{X}_w\tau$ are holomorphic. On account of (36) we also get $I_5(0) = 0$. Finally, taking (17) into account, we see that

$$I_3(0) = -12\int_{S^1} w^3\hat{X}_{ww}\cdot\hat{X}_{ww}\tau^2\phi_t(0)\,dw,$$

and we arrive at

Proposition 3. *Since $\hat{X} = \hat{Z}(0)$ is a minimal surface we have*

$$\frac{d^4E}{dt^4}(0) = 12\mathrm{Re}\int_{S^1}\hat{Z}_{ttw}(0)\cdot[w\hat{Z}_{tw}(0)\tau + w\hat{X}_w\phi_t(0)]\,dw \tag{39}$$
$$- 12\mathrm{Re}\int_{S^1} w^3\hat{X}_{ww}\cdot\hat{X}_{ww}\tau^2\phi_t(0)\,dw,$$

provided that $\tau = \phi(0)$ is the generator of an inner forced Jacobi field at $\hat{X}$.

Finally, as an exercise, we even compute $\frac{d^5E}{dt^5}(0)$. Differentiating (38) it follows that

$$\frac{d^5E}{dt^5} = \mathrm{Re}\sum_{j=1}^{9} I_j \tag{40}$$

with

$$I_1 := 16\int_{S^1} w\hat{Z}_{tttw}\cdot\hat{Z}_{tw}\phi\,dw, \quad I_2 := 12\int_{S^1} w\hat{Z}_{ttw}\cdot\hat{Z}_{ttw}\phi\,dw,$$
$$I_3 := 4\int_{S^1} w\hat{Z}_{ttttw}\cdot\hat{Z}_w\phi\,dw, \quad I_4 := 16\int_{S^1} w\hat{Z}_{tttw}\cdot\hat{Z}_w\phi_t\,dw,$$
$$I_5 := 48\int_{S^1} w\hat{Z}_{ttw}\cdot\hat{Z}_{tw}\phi_t\,dw, \quad I_6 := 24\int_{S^1} w\hat{Z}_{ttw}\cdot\hat{Z}_w\phi_{tt}\,dw,$$
$$I_7 := 24\int_{S^1} w\hat{Z}_{tw}\cdot\hat{Z}_{tw}\phi_{tt}\,dw, \quad I_8 := 16\int_{S^1} w\hat{Z}_{tw}\cdot\hat{Z}_w\phi_{ttt}\,dw,$$
$$I_9 := 2\int_{S^1} w\hat{Z}_w\cdot\hat{Z}_w\phi_{tttt}\,dw.$$

$I_3(0)$ vanishes by Cauchy's theorem since both $\hat{Z}_{tttt}(0)_w$ and $w\hat{X}_w\tau$ are holomorphic provided that $\tau = \phi(0)$ is the generator of a forced Jacobi field at $\hat{X}$. Furthermore, $I_8(0) = 0$ because of (36), and $\hat{X}_w \cdot \hat{X}_w = 0$ implies $I_9(0) = 0$. Thus we obtain by (37):

Proposition 4. *Since $\hat{X}$ is a minimal surface we have*

$$
\begin{aligned}
(41) \qquad \frac{d^5 E}{dt^5}(0) = {} & 16\mathrm{Re} \int_{S^1} \hat{Z}_{tttw}(0) \cdot [w\hat{Z}_{tw}(0)\tau + w\hat{X}_w\phi_t(0)]\, dw \\
& + 12\mathrm{Re} \int_{S^1} Z_{ttw}(0) \cdot [w\hat{Z}_{ttw}(0)\tau \\
& + 4w\hat{Z}_{tw}(0)\phi_t(0) + 2w\hat{X}_w\phi_{tt}(0)]\, dw \\
& - 24\mathrm{Re} \int_{S^1} w^3 \hat{X}_{ww} \cdot \hat{X}_{ww}\tau^2\phi_{tt}(0)\, dw
\end{aligned}
$$

provided that $\tau = \phi(0)$ is the generator of an inner forced Jacobi field at $\hat{X}$.

Note also that in (39) and (41) we can express $\hat{Z}_{tw}(0)$ by (35) which we write as

$$
(42) \qquad \hat{Z}_{tw}(0) = [iw\hat{X}_w\tau]_w.
$$

The values of $E''(0)$ and $E'''(0)$ in (30) and (34) depend only on $\tau = \phi(0)$ and not on any derivatives of $\phi(t)$ at $t = 0$; in this sense we say that $E''(0)$ and $E'''(0)$ are *intrinsic*. As we shall see later, this reflects important facts, namely: The Dirichlet integral D has an intrinsic second derivative d^2D, and an intrinsic third derivative d^3D in direction of forced Jacobi fields.

Let us try to show that a nonplanar weak relative minimizer $\hat{X}$ of D cannot have a branch point in $\overline{B}$. To achieve this goal, a somewhat naive approach would be to compute sufficiently many derivatives $E^{(j)}(0) := \frac{d^j E}{dt^j}(0)$ and to hope that one can find some first nonvanishing derivative, say, $E^{(L)}(0) \neq 0$, whereas $E^{(j)}(0) = 0$ for $j = 1, 2, \ldots, L-1$. Then Taylor's formula with Cauchy's remainder term yields

$$
E(t) = E(0) + \frac{1}{L!}E^{(L)}(\vartheta t)t^L \quad \text{for } |t| \ll 1,\ 0 < \vartheta < 1,
$$

that is,

$$
D(\hat{Z}(t)) = D(\hat{X}) + \frac{1}{L!}E^{(L)}(\vartheta t)t^L,
$$

and we infer for some t with $0 < |t| \ll 1$ that

(i) $D(\hat{Z}(t)) < D(\hat{X})$ if L odd $= 2\ell + 1 \geq 3$ and $E^{(2\ell+1)}(0) \neq 0$,

and

(ii) $D(\hat{Z}(t)) < D(\hat{X})$ if L even $= 2\ell \geq 4$ and $E^{(2\ell)}(0) < 0$.

Let us see under which assumption on $\hat{X}$ this approach works for $L = 3$. Note that an arbitrary branch point $w_0 \in B$ of a minimal surface $\hat{X}$ can be moved to the origin by means of a suitable conformal automorphism of $\overline{B}$. Hence it is sufficient for our purposes to show that a minimizer $\hat{X}$ of D in $\mathcal{C}(\Gamma)$ does not have $w = 0$ as a branch point. Therefore we shall from now on assume the following **normal form of a nonplanar minimal surface** $\hat{X}$ (cf. Vol. 1, Section 3.2):

$\hat{X}$ has $w = 0$ as a branch point of *order* n, i.e.

$$\hat{X}_w(w) = aw^n + o(w^n) \quad \text{as } w \to 0.$$

Choosing a suitable Cartesian coordinate system in $\mathbb{R}^3$ we may assume that $\hat{X}_w$ can be written as

$$(43) \quad \hat{X}_w(w) = (A_1 w^n + A_2 w^{n+1} + \cdots, R_m w^m + R_{m+1} w^{m+1} + \cdots), \quad m > n,$$

with $A_j \in \mathbb{C}^2, R_j \in \mathbb{C}, A_1 \neq 0$ and $R_m \neq 0$ for some integer m satisfying $m > n$; the number m is called *index* of the branch point $w = 0$ of $\hat{X}$ given in the normal form (43). Note that a surface $\hat{X}$ can also be brought into the normal form (43) (with $n = 0$) if $\hat{X}$ is regular at $w = 0$.

Lemma 2. *The normal form* (43) *satisfies*

$$(44) \quad \begin{aligned} &A_1 \cdot A_1 = 0, \quad A_k = \lambda_k \cdot A_1 \quad \textit{for } k = 1, 2, \ldots, 2(m-n), \\ &A_1 \cdot A_{2m-2n+1} = -\frac{1}{2} R_m^2, \end{aligned}$$

and therefore

$$(45) \qquad \hat{X}_{ww}(w) \cdot \hat{X}_{ww}(w) = (m-n)^2 R_m^2 w^{2m-2} + \cdots, \quad R_m \neq 0.$$

Proof. Equation (43) implies

$$\hat{X}_w(w) \cdot \hat{X}_w(w) = (w^{2n} p(w) + R_m^2 w^{2m}) + O(|w|^{2m+1}) \quad \text{as } w \to 0,$$

where $p(w)$ is a polynomial of degree 2ℓ in w with $\ell := m - n$ which is of the form

$$\begin{aligned} p(w) &= A_1 \cdot A_1 + 2A_1 \cdot A_2 w + (2A_1 \cdot A_3 + A_2 \cdot A_2) w^2 \\ &\quad + (2A_1 \cdot A_4 + 2A_2 \cdot A_3) w^3 + (2A_1 \cdot A_5 + 2A_2 \cdot A_4 + A_3 \cdot A_3) w^4 \\ &\quad + \cdots + (2A_1 \cdot A_{2\ell+1} + 2A_2 \cdot A_{2\ell} + \cdots + 2A_{\ell+2} \cdot A_\ell + A_{\ell+1} \cdot A_{\ell+1}) w^{2\ell} \\ &= c_0 + c_1 w + c_2 w^2 + \cdots + c_{2\ell} w^{2\ell}, \quad c_j \in \mathbb{C}. \end{aligned}$$

Since $\hat{X}_w \cdot \hat{X}_w = 0$ we obtain

$$c_0 = c_1 = \cdots = c_{2\ell-1} = 0, \quad c_{2\ell} + R_m^2 = 0.$$

Let $\langle A', A''\rangle := A' \cdot \overline{A''}$ be the Hermitian scalar product of two vectors $A', A'' \in \mathbb{C}^2$. The two equations $c_0 = 0$ and $c_1 = 0$ yield $A_1 \cdot A_1 = 0$ and $A_1 \cdot A_2 = 0$ which are equivalent to

$$\langle A_1, \overline{A}_1\rangle = 0 \quad \text{and} \quad \langle A_2, \overline{A}_1\rangle = 0.$$

Since $A_1 \neq 0$ and $\overline{A}_1 \neq 0$ this implies

$$A_2 = \lambda_2 A_1 \quad \text{for some } \lambda_2 \in \mathbb{C},$$

and so we also obtain

$$A_2 \cdot A_2 = \lambda_2^2 A_1 \cdot A_1 = 0.$$

On account of $c_2 = 0$ it follows $A_1 \cdot A_3 = 0$, and thus it follows

$$\langle A_1, \overline{A}_1\rangle = 0 \quad \text{and} \quad \langle A_3, \overline{A}_1\rangle = 0$$

whence

$$A_3 = \lambda_3 A_1 \quad \text{for some } \lambda_3 \in \mathbb{C},$$

and so

$$A_2 \cdot A_3 = \lambda_2\lambda_3 A_1 \cdot A_1 = 0.$$

Then $c_3 = 0$ yields $A_1 \cdot A_4 = 0$, therefore

$$\langle A_1, \overline{A}_1\rangle = 0 \quad \text{and} \quad \langle A_4, \overline{A}_1\rangle = 0;$$

consequently

$$A_4 = \lambda_4 A_1 \quad \text{for some } \lambda_4 \in \mathbb{C}.$$

In this way we proceed inductively using $c_0 = 0, \ldots, c_{2\ell-1} = 0$ and obtain $A_k = \lambda_k A_1$ for $k = 1, 2, \ldots, 2(m-n)$. Since $A_1 \cdot A_1 = 0$ it follows that

$$A_j \cdot A_k = 0 \quad \text{for } 1 \le j, k \le 2(m-n). \tag{46}$$

Then the equation $c_{2\ell} + R_m^2 = 0$ implies $2A_1 \cdot A_{2\ell+1} + R_m^2 = 0$, i.e.

$$A_1 \cdot A_{2(m-n)+1} = -\frac{1}{2} R_m^2. \tag{47}$$

Furthermore, from

$$\hat{X}_w(w) = (A_1 w^n + A_2 w^{n+1} + \cdots + A_{2m-2n+1} w^{2m-n} + \cdots, R_m w^m + \cdots)$$

we infer

$$\begin{aligned} \hat{X}_{ww}(w) \;=\; & (nA_1 w^{n-1} + \cdots + (2m-n) A_{2m-2n+1} w^{2m-n-1} \\ & + \cdots, m R_m w^{m-1} + \cdots). \end{aligned}$$

Then (46) implies

$$\hat{X}_{ww}(w) \cdot \hat{X}_{ww}(w) = [2n(2m-n) A_1 \cdot A_{2m-2n+1} + m^2 R_m^2] w^{2m-2} + \cdots,$$

and by (47) we arrive at

$$\hat{X}_{ww}(w) \cdot \hat{X}_{ww}(w) = [-n(2m-n) R_m^2 + m^2 R_m^2] w^{2m-2} + \cdots,$$

which is equivalent to (45). □

Theorem 1. (D. Wienholtz). *Let $\hat{X}$ be a minimal surface in normal form with a branch point at $w = 0$ which is of order n and index m, $n < m$, and suppose that $2m - 2 < 3n$ (or, equivalently, $2m + 2 \le 3(n+1)$). Then we can choose a generator τ of a forced Jacobi field $\hat{h}$ such that $E^{(3)}(0) < 0$, and so $\hat{X}$ is not a weak relative minimizer of D.*

Proof. Define the integer k by

$$k := (2m + 2) - 2(n + 1).$$

Because of $m > n$ and $2m - 2 < 3n$ it follows that

$$1 < k \le n + 1.$$

Let

$$\tau_0 := cw^{-n-1} + \bar{c}w^{n+1}, \quad \tau_1 := cw^{-k} + \bar{c}w^k, \quad c \in \mathbb{C},$$

and set

(i) $\tau := \tau_0$ if $k = n + 1$;
(ii) $\tau := \epsilon\tau_0 + \tau_1$, $\epsilon > 0$, if $k < n + 1$.

In both cases τ is a generator of a forced Jacobi field at $\hat{X}$, since $w\hat{X}_w(w)$ has a zero of order $n + 1$ at $w = 0$, and $\operatorname{Im} \tau = 0$ on ∂B. By (45) it follows for $w \in B$ that

$$w^3 \hat{X}_{ww}(w) \cdot \hat{X}_{ww}(w) = (m - n)^2 R_m^2 w^{2m+1} + \cdots,$$

where $+\cdots$ always stands for higher order terms of a convergent power series. In case (i) one has

$$\tau^3(w) = c^3 w^{-3(n+1)} + \cdots,$$

and so

$$w^3 \hat{X}_{ww}(w) \cdot \hat{X}_{ww}(w)\tau(w)^3 = (m - n)^2 R_m^2 c^3 w^{-1} + f(w),$$

where $f(w)$ is holomorphic in B and continuous on $\overline{B}$. Then formula (34) of Proposition 3 in conjunction with Cauchy's integral theorem yields

$$E^{(3)}(0) = -4\operatorname{Re}[2\pi i(m - n)^2 R_m^2 c^3] \quad \text{if } k = n + 1.$$

With a suitable choice of $c \in \mathbb{C}$ we can arrange for $E^{(3)}(0) < 0$ since $R_m \ne 0$ and $(m - n)^2 \ge 1$.

In case (ii) we write $w^3 \hat{X}_{ww} \cdot \hat{X}_{ww}$ as

$$w^3 \hat{X}_{ww}(w) \cdot \hat{X}_{ww}(w) = (m - n)^2 R_m^2 w^{2m+1} + f(w),$$

where

$$f(w) := w^{2m+2} \sum_{j=0}^{\infty} a_j w^j, \quad a_j \in \mathbb{C}.$$

From

$$\tau^3 = \epsilon^3 \tau_0^3 + 3\epsilon^2 \tau_0^2 \tau_1 + 3\epsilon \tau_0 \tau_1^2 + \tau_1^3$$

it follows that

$$g(w) := w^3 \hat{X}_{ww}(w) \cdot \hat{X}_{ww}(w) \tau^3(w)$$

is meromorphic in B, continuous in $\{w : \rho < |w| \leq 1\}$ for some $\rho \in (0,1)$, and its Laurent expansion at $w = 0$ has the residue

$$\operatorname{Res}_{w=0}(g) = 3\epsilon^2 c^3 (m-n)^2 R_m^2 + \epsilon^3 c^3 a_{n-k}, \quad 1 < k \leq n.$$

Cauchy's residue theorem together with formula (34) of Proposition 3 then imply

$$E^{(3)}(0) = -4\operatorname{Re}\{2\pi i [3\epsilon^2 c^3 (m-n)^2 R_m^2 + \epsilon^3 c^3 a_{n-k}]\} \quad \text{for } k < n+1.$$

By an appropriate choice of $c \in \mathbb{C}$ and ϵ with $0 < \epsilon < 1$ we can achieve that $E^{(3)}(0) < 0$ also in case (ii). □

The following definition will prove to be very useful.

Definition 3. *Let $\hat{X}$ be a minimal surface in normal form having $w = 0$ as a branch point of order n and of index m. Then $w = 0$ is called an* **exceptional branch point** *if $m + 1 = \kappa(n+1)$ for some $\kappa \in \mathbb{N}$; necessarily $\kappa > 1$.*

Remark 1. If $2m - 2 < 3n$, i.e. $2(m+1) \leq 3(n+1)$, then $w = 0$ is not exceptional, because $(m+1) = \kappa(n+1)$ with $\kappa > 1$ implies $2\kappa(n+1) \leq 3(n+1)$ and therefore $2\kappa \leq 3$ which is impossible for $\kappa \in \mathbb{N}$ with $\kappa > 1$.

Remark 2. Now we want to show that the notion "*$w = 0$ is an exceptional branch point*" is closely related to the notion "*$w = 0$ is a false branch point*". To this end we choose an arbitrary minimal surface $\hat{Z}(\zeta)$, $\zeta \in B$, in normal form without $\zeta = 0$ being a branch point, i.e. $\hat{Z} = \operatorname{Re} g$ where $g : B \to \mathbb{C}^3$ is holomorphic and of the form

$$g(\zeta) = \hat{Z}(0) + (B_0 \zeta + B_1 \zeta^2 + \cdots, C_\kappa \zeta^\kappa + \cdots), \quad B_0 \neq 0, \ C_\kappa \neq 0, \ \kappa > 1.$$

Consider a conformal mapping $w \mapsto \zeta = \varphi(w)$ from B into B with $\varphi(0) = 0$ which is provided by a holomorphic function

$$\varphi(w) = aw + \cdots, \quad a \neq 0, \ w \in B.$$

Then $\hat{X}(w) := \operatorname{Re} f(w)$ with $f(w) := g(\varphi^{n+1}(w))$, $w \in B$, is a minimal surface $\hat{X} : B \to \mathbb{R}^3$ such that $\hat{X}(0) = \hat{Z}(0)$ and

$$f(w) = \hat{X}(0) + (a^{n+1} B_0 w^{n+1} + \cdots, \ a^{\kappa(n+1)} C_\kappa w^{\kappa(n+1)} + \cdots).$$

Thus we obtain for $\hat{X}_w = \frac{1}{2} f'$ that

$$\hat{X}_w(w) = (A_1 w^n + \cdots, \ R_m w^m + \cdots), \quad A_1 \neq 0, \ R_m \neq 0,$$

and so $\hat{X}(w)$, $w \in B$, is a minimal surface in normal form which has the branch point $w = 0$ of order n and index $m := \kappa(n+1) - 1$, whence $w = 0$ is *exceptional.* Clearly $\hat{X}$ is obtained from the minimal immersion $\hat{Z}(\zeta)$ as a *false branch point* by setting $\hat{X} := \hat{Z} \circ \varphi^{n+1}$. As the "false parametrization" $\hat{X}$ of the regular surface $\mathcal{S} := \hat{Z}(B)$ is produced by an analytic expression $\zeta = \varphi^{n+1}(w)$ we call $w = 0$ an "*analytic false branch point*".

In Remark 1 we have noted that $w = 0$ cannot be "exceptional" if $2m - n < 3n$, and so it cannot be an "analytic false branch point".

It will be useful to have a **characterization of the nonexceptional branch points**, the proof of which is left to the reader.

Lemma 3. *The branch point $w = 0$ is nonexceptional if and only if one of the following two conditions is satisfied:*

(i) *There is an even integer L with*

$$(L-1)(n+1) < 2(m+1) < L(n+1). \tag{48}$$

(ii) *There is an odd integer L with*

$$(L-1)(n+1) < 2(m+1) \leq L(n+1). \tag{49}$$

We say that $w = 0$ satisfies condition (T_L) if either (48) *with L even or* (49) *with L odd holds.*

In Theorem 1 it was shown that $E^{(3)}(0)$ can be made negative if $2m - 2 < 3n$. Therefore *we shall now assume that* $2m - 2 \geq 3n$. It takes some experience to realize that the right approach to success lies in separating the two cases "$w = 0$ *is nonexceptional*" and "$w = 0$ *is exceptional*". Instead one might guess that the right generalization of Wienholtz's theorem consists in considering the cases

$$(C_L) \qquad (L-1)n \leq 2m - 2 < Ln, \quad L \in \mathbb{N} \text{ with } L \geq 3$$

and hoping that one can prove

$$E^{(j)}(0) = 0 \quad \text{for } 1 \leq j \leq L-1, \quad E^{(L)}(0) < 0$$

using appropriate choices of forced Jacobi fields in varying the minimal surface $\hat{X}$. Unfortunately this is not the case. To see what happens we study the two cases

$$(C_4) \qquad 3n \leq 2m - 2 < 4n$$

and

$$4n \le 2m - 2 < 5n \tag{C_5}$$

by computing $E^{(4)}(0)$ in the first case and $E^{(5)}(0)$ in the second one. We begin by treating special cases of (C_4) and (C_5), where we can proceed in a similar way as before with $E^{(3)}(0)$ for $2n \le 2m - 2 < 3n$.

The case (C_4) with $2m - 2 = 4p, p \in \mathbb{N}$.

Proposition 5. *If $w\hat{Z}_{tw}(0)\tau + w\hat{X}_w\phi_t(0)$ is holomorphic, then*

$$E^{(4)}(0) = -12\text{Re}\int_{S^1} w^3 \hat{X}_{ww} \cdot \hat{X}_{ww}\tau^2\phi_t(0)\,dw. \tag{50}$$

Proof. Since $\hat{Z}_{ttw}(0)$ is holomorphic in B, the integrand of the first integral in (39) is holomorphic, and so this integral vanishes. □

Remark 3. In case (C_4) with $2m - 2 = 4p$ the branch point $w = 0$ is nonexceptional. To see this we note that $p < n$ whence

$$2m + 2 = 4(p + 1) < 4(n + 1)$$

and therefore

$$n + 1 < m + 1 < 2(n + 1).$$

Also note that $n = 1, 2, 3$ are not possible since $n = 1$ would imply $p < 1$; $n = 2$ would mean $p = 1$ whence $6 = 3n \le 4p = 4$; and $n = 3$ would imply $p \le 2$, and so $9 = 3n \le 4p = 8$. Finally $3n \le 4p$ and $n \ge 4$ yields $p \ge 3$.

Theorem 2. *If $3n \le 2m - 2 = 4p < 4n$ for some $p \in \mathbb{N}$, then one can find a variation $\hat{Z}(t)$ of $\hat{X}$ such that $E^{(4)}(0) < 0$, whereas $E^{(j)}(0) = 0$ for $j = 1, 2, 3$.*

Proof. First we want to choose $\tau = \phi(0)$ and $\phi_t(0)$ in such a way that the assumption of Proposition 5 is satisfied. To this end, set

$$\tau(w) := (a - ib)w^{-p-1} + (a + ib)w^{p+1},$$

which clearly is a generator of a forced Jacobi field. By (43) we get

$$\begin{aligned} &w\hat{X}_w(w)\tau(w) \\ &\quad= (a - ib)(A_1 w^{n-p} + A_2 w^{n-p+1} + \cdots + A_{2m-2n+1}w^{2m-n-p} + \cdots, \\ &\qquad R_m w^{m-p} + \cdots) + (a + ib)(A_1 w^{n+p+2} + \cdots, R_m w^{m+p+2} + \cdots). \end{aligned}$$

By (35) it follows

$$\begin{aligned} &w\hat{Z}_{tw}(w, 0)\tau(w) = w[iw\hat{X}_w(w)\tau(w)]_w\tau(w) \\ &\quad= i(a - ib)^2((n - p)A_1 w^{n-2p-1} + (n - p + 1)A_2 w^{n-2p} + \cdots \\ &\qquad + (2m - n - p)A_{2m-2n+1}w^{2m-n-2p-1} + \cdots, (m - p)R_m w^{m-2p-1} + \cdots). \end{aligned}$$

Note that $2m-2=4p$ implies $m-2p-1=0$, whence $n-2p-1<0$ because of $m>n$, but $2m-n-2p-1=(m-2p-1)+(m-n)=m-n>0$. Thus the third component above has no pole, while the first (vectorial) component has a pole at least in the first term, but no pole anymore from the $(2m-2n+1)$-th term on. These poles will be removed by adding $w\hat{X}_w\phi_t(0)$ to $w\hat{Z}_{tw}(0)\tau$ with an appropriately chosen value of $\phi_t(0)$. We set $\phi_t(0)=\psi_1+\cdots+\psi_s$ where $\psi_1,\ldots,\psi_s$ are defined inductively. First set

$$\begin{aligned}\psi_1(w) := &-i(n-p)(a-ib)^2w^{-2p-2}\\ &+i(n-p)(a+ib)^2w^{2p+2}.\end{aligned}$$

Now $w\hat{Z}_{tw}(0)\tau+w\hat{X}_w\psi_1$ has no pole associated to A_1 while the poles associated to A_k, $1<k\leq s$, are of the same order as before. Then we choose ψ_2 so that there is no pole associated to A_2, etc. The number s is the index of the last term $(n-p+s)A_{s+1}w^{n-2p+s-1}$ where $n-2p+s-1$ is ≥ 0 and $\leq 2m-2n$. Note that

$$w\hat{X}_w(w)=(A_1w^{n+1}+A_2w^{n+2}+\cdots+A_{2m-2n+1}w^{2m-n+1}+\cdots, R_mw^{m+1}+\cdots)$$

and

$$A_1\cdot A_k=0 \quad \text{for } k=1,2,\ldots,2m-2n.$$

Therefore, $w\hat{X}_w\phi_t(0)=w\hat{X}_w\cdot[\psi_1+\psi_2+\cdots+\psi_s]$ removes all poles from $w\hat{Z}_{tw}(0)\tau$ and creates no new poles. Consequently $w\hat{Z}_{tw}(0)\tau+w\hat{X}_w\phi_t(0)$ is holomorphic, and so we have

$$E^{(4)}(0)=-12\mathrm{Re}\int_{S^1} w^3\hat{X}_{ww}\cdot\hat{X}_{ww}\tau^2\phi_t(0)\,dw.$$

Formula (45) yields

$$w^3\hat{X}_{ww}(w)\cdot\hat{X}_{ww}(w)=(m-n)^2R_m^2w^{2m+1}+\cdots.$$

The leading term in $\phi_t(0)$ is that of ψ_1, and

$$\psi_1(w)=-i(n-p)(a-ib)^2w^{-2p-2}+\cdots.$$

Furthermore,

$$\tau^2(w)=(a-ib)^2w^{-2p-2}+\cdots,$$

and so

$$\tau^2(w)\phi_t(w,0)=-i(a-ib)^4(n-p)w^{-4p-4}+\cdots.$$

Noticing that $2m+1=(2m+2)-1=4(p+1)-1$, and setting

$$\kappa := 12(m-n)^2(n-p)>0$$

we obtain

$$E^{(4)}(0) = \kappa \mathrm{Re}\left[i(a-ib)^4 R_m^2 \int_{S^1} \frac{dw}{w} \right] = -2\pi\kappa \mathrm{Re}[(a-ib)^4 R_m^2]$$

and an appropriate choice of a and b yields $E^{(4)}(0) < 0$. Finally we note that $E^{(2)}(0) = 0$ and $E^{(3)}(0) = 0$ for the above choice of $\hat{Z}(t)$. The first statement follows from Proposition 1. To verify the second, we recall formula (34) from Proposition 2:

$$E^{(3)}(0) = -4\mathrm{Re} \int_{S^1} w^3 \hat{X}_{ww} \cdot \hat{X}_{ww} \tau^3 \, dw.$$

From the preceding computations it follows that

$$w^3 \hat{X}_{ww}(w) \cdot \hat{X}_{ww}(w) \tau^3(w) = (m-n)^2 R_m^2 (a-ib)^3 w^{2m+1-3(p+1)} + \cdots,$$

and, by assumption, $2m-2 = 4p$, whence

$$2m+1-3(p+1) = 4p+3-3(p+1) = p > 1;$$

therefore $E^{(3)}(0) = 0$. □

Remark 4. Under the special assumption that $2m-2 = 4p$ we were able to carry out the program outlined above for $L = 4$. However, applying the method from Theorem 2 to cases when $2m-2 \not\equiv 0 \mod 4$ one will get nowhere. Instead, trying another approach similar to that used in the proof of Theorem 1, one is able to handle the case (C_4) under the additional assumption $2m-2 \equiv 2 \mod 4$ by considering the next higher derivative, namely $E^{(5)}(0)$ instead of $E^{(4)}(0)$, cf. Theorem 4 stated later on. This seems to shatter the hope that one can always make $E^{(L)}(0)$ negative, with $E^{(j)}(0) = 0$ for $1 \le j \le L-1$, if (C_L) is satisfied. In fact, by studying assumption (C_5) we shall realize that (C_L) is probably not the appropriate classification for developing methods that in general lead to our goal. Rather, the case (C_5) will show us that one should distinguish between the cases "exceptional" and "nonexceptional" using the classification given in Lemma 3 to reach this purpose.

Let us mention that, assuming (C_4),the branch point $w = 0$ is nonexceptional according to Lemma 3, since $3n \le 2m-2 < 4n$ implies

$$3(n+1) < 3n+4 \le 2m+2 < 4(n+1).$$

Let us now turn to the *investigation of* (C_5) *by means of the fifth derivative* $E^{(5)}(0)$.

Lemma 4. *If* $f(w) := w\hat{Z}_{tw}(0)\tau + w\hat{X}_w \phi_t(0)$ *is holomorphic, then*

$$\begin{aligned} &\hat{Z}_{ttw}(0) = \{iw[iw\hat{X}_w\tau]_w\tau + iw\hat{X}_w\phi_t(0)\}_w, \\ &\hat{Z}_{ttw}(0) \cdot \hat{X}_w = -\hat{Z}_{tw}(0) \cdot \hat{Z}_{tw}(0) = w^2 \hat{X}_{ww} \cdot \hat{X}_{ww} \tau^2. \end{aligned} \tag{51}$$

Proof. By (27) we have

$$\hat{Z}_{tw} = \{2H[\mathrm{Re}(iw\hat{Z}_w\phi)]\}_w$$

whence

$$\hat{Z}_{ttw} = \{2H[\mathrm{Re}(iw\hat{Z}_{tw}\phi + iw\hat{Z}_w\phi_t)]\}_w$$

and therefore

$$\begin{aligned}\hat{Z}_{ttw}(0) &= \{2H[\mathrm{Re}(if)]\}_w = \{if\}_w \\ &= \{iw\hat{Z}_{tw}(0)\tau + iw\hat{X}_w\phi_t(0)\}_w.\end{aligned}$$

By (35),

$$\hat{Z}_{tw}(0) = [iw\hat{X}_w\tau]_w,$$

and so

$$\hat{Z}_{ttw}(0) = \{iw[iw\hat{X}_w\tau]_w\tau + iw\hat{X}_w\phi_t(0)\}_w.$$

It follows that

$$Z_{ttw}(0)\cdot\hat{X}_w = \{iw[i\hat{X}_w\tau + iw\hat{X}_{ww}\tau + iw\hat{X}_w\tau_w]\tau + iw\hat{X}_w\phi_t(0)\}_w\cdot\hat{X}_w.$$

From $\hat{X}_w\cdot\hat{X}_w = 0$ one obtains $\hat{X}_w\cdot\hat{X}_{ww} = 0$, and then

$$\hat{X}_{www}\cdot\hat{X}_w = -\hat{X}_{ww}\cdot\hat{X}_{ww}.$$

This leads to

$$\begin{aligned}\hat{Z}_{ttw}(0)\cdot\hat{X}_w &= -w^2\hat{X}_{www}\cdot\hat{X}_w\tau^2 \\ &= w^2\hat{X}_{ww}\cdot\hat{X}_{ww}\tau^2 = -\hat{Z}_{tw}(0)\cdot\hat{Z}_{tw}(0),\end{aligned}$$

taking (37) into account. □

Proposition 4 and Lemma 4 imply

Proposition 6. *If* $f(w) := w\hat{Z}_{tw}(0)\tau + w\hat{X}_w\phi_t(0)$ *is holomorphic, then*

(52) $$E^{(5)}(0) = 12\mathrm{Re}\int_{S^1}[w\hat{Z}_{ttw}(0)\cdot\hat{Z}_{ttw}(0)\tau + 4w\hat{Z}_{ttw}(0)\cdot\hat{Z}_{tw}(0)\phi_t(0)]\,dw.$$

We are now going to discuss the envisioned program for the case (C_5) using the simplified form (52) for the fifth derivative $E^{(5)}(0)$. It will be useful to distinguish several subcases of (C_5):

(a) $5n \le 2m+2$,
(b) $5n > 2m+2$.

In case (a) we have $5n \le 2m + 2 < 5n + 4$, that is,

$$2m + 2 = 5n + \alpha, \quad 0 \le \alpha \le 3.$$

Therefore (a) consists of the four subcases

$$2m - 5n = 0, 1, -1, -2. \tag{53}$$

In case (b) we have $5n > 2m + 2$, and (C_5) implies $2m + 2 \ge n + 4$, whence $5n > n + 4$, and so we have $n > 1$ in case (b).

Case (a) allows an easy treatment based on the following representation of $2m + 2$ which we apply successively for $\alpha = 0, 1, 2, 3$ to deal with the four cases (53). We write

$$\alpha(n + 1) + \beta n = 2m + 2$$

with $\alpha := 2m + 2 - 5n$, $\beta := 5 - \alpha$ where $0 \le \alpha \le 3$ and $\beta \ge 2$. Then we choose

$$\tau := \tau_0 + \epsilon\tau_1, \quad \epsilon > 0,$$

where

$$\tau_0 := cw^{-n} + \overline{c}w^n, \quad \tau_1 := cw^{-n-1} + \overline{c}w^{n+1}, \quad c \in \mathbb{C}.$$

With an appropriate choice of $\phi_t(0)$ we obtain by an elimination procedure similar to the one used in the proof of Theorem 2 that $f := w\hat{Z}_{tw}(0)\tau + w\hat{X}_w\phi_t(0)$ is holomorphic. Here and in the sequel we omit the lengthy computations and merely state the results. As f is holomorphic one can use formula (52) for $E^{(5)}(0)$; we investigate the four different cases of (53) separately, but note that always

$$E^{(j)}(0) = 0, \quad j = 1, \dots, 4.$$

(I) $\mathbf{2m - 5n = 0}$, $1 \le n \le 4$. Only (i) $n = 2$ and (ii) $n = 4$ are possible. This leads to

(i) $n = 2$, $m = 5$, $(m + 1) = 2(n + 1)$, i.e. $w = 0$ is exceptional;
(ii) $n = 4$, $m = 10$, hence $m + 1 \not\equiv 0 \mod (n + 1)$, and so $w = 0$ is not exceptional.

For (i) we obtain $E^{(5)}(0) = 0 + o(\epsilon)$, whereas (ii) yields

$$E^{(5)}(0) = 12\text{Re}[2\pi i \cdot 336 \cdot \epsilon^2 \cdot c^5 R_m^2] + o(\epsilon^2)$$

which can be made negative by appropriate choice of c. Thus the method is inconclusive for (i), but gives the desired result for (ii).

(II) $\mathbf{2m - 5n = 1}$, $1 \le n \le 4$. Then n necessarily either (i) $n = 1$ or (ii) $n = 3$. Here,

(i) $n = 1,\ m = 3,\ m + 1 = 2(n + 1)$, i.e. $w = 0$ is exceptional;
(ii) $n = 3,\ m = 8$, and $m+1 \not\equiv 0 \bmod (n+1)$, hence $w = 0$ is not exceptional.

For (i) it follows that $E^{(5)}(0) = 0 + o(\epsilon^3)$, i.e. the method is inconclusive, while for (ii) one gets

$$E^{(5)}(0) = 12 \cdot \mathrm{Re}[2\pi i \cdot 250 \cdot \epsilon^3 \cdot c^5 R_m^2] + o(\epsilon^3),$$

and so $E^{(5)}(0) < 0$ for a suitable choice of c.

(III) $2m - 5n = -1$, $1 \leq n \leq 4$. Then either (i) $n = 1$ or (ii) $n = 3$, i.e.

(i) $n = 1,\ m = 2$, and so $m+1 \not\equiv 0 \bmod (n+1)$, i.e. $w = 0$ is not exceptional.
(ii) $n = 3,\ m = 7$, whence $m + 1 = 2(n + 1)$, i.e. $w = 0$ is exceptional.

For (i) we have $2m - 2 < 3n$, and this case was already dealt with in the positive sense by using $E^{(3)}(0)$, cf. Theorem 1. For (ii) the method is again inconclusive since one obtains

$$E^{(5)}(0) = 0 + o(\epsilon).$$

(IV) $2m - 5n = -2$, $1 \leq n \leq 4$. Then either (i) $n = 2$ or (ii) $n = 4$, that is,

(i) $n = 2,\ m = 4$, whence $m+1 \not\equiv 0 \bmod (n+1)$, i.e. $w = 0$ is not exceptional.
(ii) $n = 4,\ m = 9$, and so $m + 1 = 2(n + 1)$, i.e. $w = 0$ is exceptional.

In case (i) we have $3n = 2m - 2 < 4n$, i.e. condition (C_4) holds, and this case will be tackled by Theorem 4, to be stated later on. Case (ii) leads to $E^{(5)}(0) = 0 + o(1)$ as $\epsilon \to 0$ which is once again inconclusive.

Conclusion. *The method is inconclusive in all of the exceptional cases. In the nonexceptional cases it either leads to the positive result $E^{(5)}(0) < 0$ for appropriate choice of c, or one can apply the cases (C_3) or (C_4), and here one obtains the desired results $E^{(3)}(0) < 0$ or $E^{(4)}(0) < 0$ respectively (see Theorems* 1 *and* 4*).*

Now we turn to the case (b). We first note that (C_5) together with (b) implies $4(n + 1) \leq 2m + 2 < 5n$. Hence either (i) $2(n + 1) = m + 1$, or (ii) $4(n + 1) < 2m + 2 < 5n$. Therefore, $w = 0$ is exceptional in case (i) and nonexceptional in case (ii). Furthermore we have

$$2m + 2 = 4n + k \quad \text{with} \ \leq k < n,$$

where $k = 4$ is the case (i) and $4 < k < n$ is the case (ii).

In order to treat the case (b) which in some sense is the "*general subcase*" of (C_5) we use

$$\tau := c \cdot (\epsilon w^{-n} + w^{-k}) + \bar{c} \cdot (\epsilon w^n + w^k).$$

Choosing $\phi_t(0)$ appropriately we achieve that f is holomorphic, and so $E^{(5)}(0)$ is given by (52). Moreover, $E^{(j)}(0) = 0$ for $1 \le j \le 4$. It turns out that

$$E^{(5)}(0) = 12 \cdot \text{Re}[2\pi i c^3 \epsilon^4 \gamma R_m^2] + o(\epsilon^4), \quad \epsilon > 0,$$

with

$$\gamma = (m-n)(k-4)^2 \left[\frac{5}{4}n + \frac{5}{8}(k-2)\right]$$

and $\gamma = 0$ in case (i), whereas $\gamma > 0$ in case (ii).

Thus the following result is established:

Theorem 3. *Suppose that* (C_5) *and* (b) *hold, hence* $4n + 4 \le 2m + 2 < 5n$. *This implies* $2m + 2 = 4n + k$ *with* $4 \le k < n$. *For* $k = 4$ *the branch point* $w = 0$ *is exceptional, and the method is nonconclusive. If, however,* $4 < k < n$, *then* $\tau = \phi(0)$ *and* $\phi_t(0)$ *can be chosen in such a way that* $E^{(5)}(0) < 0$ *and* $E^{(j)}(0) = 0$ *for* $j = 1, \dots, 4$.

Next, we want to prove that the remaining cases of (C_4) lead to a conclusive result also for the remaining possibility $2m - 2 \ne 4p$ for some $p \in \mathbb{N}$ with $1 \le p < n$. Because of $3n \le 2m - 2 < 4n$ we can write $2m - 2 = 4p + k$ with $0 < k < 4$ (the case $k = 0$ was treated before). Since k must be even, we are left with $k = 2$, and we recall that $w = 0$ is a nonexceptional branch point in the case (C_4).

Theorem 4. *Suppose that* $3n \le 2m - 2 = 4p + 2 < 4n$ *with* $1 \le p < n$ *holds (this is the subcase of* (C_4) *that was not treated in Theorem 2). Then* $\tau = \phi(0)$ *and* $\phi_t(0)$ *can be chosen in such a way that*

$$E^{(j)}(0) = 0 \quad \textit{for } j = 1, \dots, 4, \qquad E^{(5)}(0) < 0.$$

Proof. This follows with

$$\tau := c(w^{-k} + \epsilon w^{-p-1}) + \bar{c} \cdot (w^k + \epsilon w^{p+1}), \quad \epsilon > 0.$$

Then $E^{(j)}(0) = 0$ for $1 \le j \le 4$ and

$$E^{(5)}(0) = 12 \cdot \text{Re}[2\pi i c^5 \epsilon^4 R_m^2 \gamma] + o(\epsilon^4),$$

where

$$\begin{aligned}\gamma := {} & (m-n)^2(m-2p-1)^2 + 4(m-n)^2(m-2p-1)(m-k-p) \\ & -8(n-p)(m-p)(m-n)(m-k-p) \\ & -4(m-n)(m-2p-1)[(n-p)(m-p+1) + (m-p)(2n-p-k+1)].\end{aligned}$$

Since $4(p+1) + k = 2m + 2$ and

$$5(p+1) = 4p + k + p + (5-k) = (2m-2) + 3 + p \ge 2m + 2$$

one can prove that $\gamma < 0$. Thus one can make $E^{(5)}(0) < 0$ for a suitable choice of c. □

Let us return to the case (C_4): $3n \le 2m - 2 < 4n$ which splits into the two subcases $2m - 2 \equiv 0 \bmod 4$ and $2m - 2 \equiv 2 \bmod 4$. The first one was dealt with by $E^{(4)}(0)$, cf. Theorem 2, the second by $E^{(5)}(0)$, see Theorem 4. Combining both results we obtain

Theorem 5. *Let $\hat{X}$ be a minimal surface in normal form having the branch point $w = 0$ with the order n and the index m such that (C_4) holds. Then $\hat{X}$ cannot be a weak minimizer of D.*

We want to give a new proof of this result which combines both cases into a single one. Note first that $3n \le 2m - 2 < 4n$ is equivalent to $3(n+1) + 1 \le 2m + 2 < 4(n+1) = 3(n+1) + n + 1$. Therefore $w = 0$ is not exceptional, and

(54) $$2m + 2 = 3(n+1) + r, \quad 1 \le r \le n.$$

The new approach consists in choosing the generator $\tau = \phi(0)$ as

(55) $$\tau = \tau_0 + \tau_1 \quad \text{with } \tau_0 := \epsilon c w^{-n-1} + \epsilon \bar{c} w^{n+1}, \; \tau_1 := c w^{-r} + \bar{c} w^r, \; c \in \mathbb{C}.$$

We need the following auxiliary result:

Lemma 5. *For any $\nu \in \mathbb{N}$ and $a \in \mathbb{C}$ we have*

(56) $$\{2H[\mathrm{Re}(aw^{-\nu})]\}_w = \nu \bar{a} w^{\nu-1} \quad \textit{on } \overline{B}.$$

Proof. On S^1 one has $w^{-\nu} = \overline{w}^\nu$ whence

$$aw^{-\nu} = a\overline{w}^\nu = \overline{\bar{a}w^\nu} \quad \text{on } S^1$$

and therefore

$$\mathrm{Re}(aw^{-\nu}) = \mathrm{Re}(\bar{a}w^\nu) \quad \text{on } S^1.$$

Consequently

$$2H[\mathrm{Re}(aw^{-\nu})] = 2H[\mathrm{Re}(\bar{a}w^\nu)] \quad \text{on } \overline{B}.$$

This implies

$$\{2H[\mathrm{Re}(aw^{-\nu})]\}_w = \{2H[\mathrm{Re}(\bar{a}w^\nu)]\}_w \quad \text{on } \overline{B}.$$

Finally, since $\bar{a}w^\nu$ is holomorphic in $\mathbb{C}$, it follows that

$$\{2H[\mathrm{Re}(\bar{a}w^\nu)]\}_w = \frac{d}{dw}(\bar{a}w^\nu) = \nu \bar{a} w^{\nu-1} \quad \text{on } \overline{B}. \qquad \square$$

Now we calculate $E^{(4)}(0)$ using the formulae (37) and (39):

(57) $$E^{(4)}(0) = 12\mathrm{Re} \int_{S^1} \hat{Z}_{ttw}(0) \cdot [w\hat{Z}_{tw}(0)\tau + w\hat{X}_w \phi_t(0)]\, dw + 12\mathrm{Re} \int_{S^1} w\hat{Z}_{tw}(0) \cdot \hat{Z}_{tw}(0)\phi_t(0)\, dw.$$

From

$$w\hat{X}_w = (A_1 w^{n+1} + \cdots + A_{2m-2n+1} w^{2m-n+1} + \cdots , R_m w^{m+1} + \cdots)$$

it follows that

$$\begin{aligned} & w\hat{X}_w \tau \\ & = c\epsilon(A_1 + \cdots + A_{2m-2n+1} w^{2m-2n} + \cdots , R_m w^{m-n} + \cdots) \\ & \quad + c(A_1 w^{n+1-r} + \cdots + A_{2m-2n+1} w^{2m-n-r+1} + \cdots , R_m w^{m+1-r} + \cdots) \\ & \quad + g(w), \quad g(w) := w\hat{X}_w(w) \cdot [\epsilon\bar{c} w^{n+1} + \bar{c} w^r]. \end{aligned}$$

The expression $g(w)$ is "better" than the sum $T_1 + T_2$ of the first two terms T_1, T_2 on the right-hand side of this equation, in the sense that it is built in a similar way as $T_1 + T_2$ except that it is less singular. In the sequel this phenomenon will appear repeatedly, and so we shall always use a notation similar to the following:

$$w\hat{X}_w \tau = T_1 + T_2 + \langle \text{better} \rangle.$$

This sloppy notation will not do any harm since in the end we shall see that each of the two integrands in (57) possesses exactly one term of order w^{-1} as w-terms of least order, and no expression labelled "better" is contributing to them.

Using (35) one obtains

$$\begin{aligned} \hat{Z}_{tw}(0) = \; & ic\epsilon(A_2 + \cdots + (2m-2n) A_{2m-2n+1} w^{2m-2n-1} + \cdots , \\ & (m-n) R_m w^{m-n-1} + \cdots) \\ & + ic((n+1-r) A_1 w^{n-r} + \cdots \\ & + (2m-n+1-r) A_{2m-2n+1} w^{2m-n-r} \\ & + \cdots , (m+1-r) R_m w^{m-r} + \cdots) + \langle \text{better} \rangle. \end{aligned}$$

This implies

$$\begin{aligned} w\hat{Z}_{tw}(0)\tau = \; & ic^2\epsilon^2(A_2 w^{-n} + \cdots + (2m-2n) A_{2m-2n+1} w^{2m-3n-1} + \cdots , \\ & (m-n) R_m w^{m-2n-1} + \cdots) \\ & + ic^2\epsilon((n+1-r) A_1 w^{-r} + \cdots \\ & + (2m-n+1-r) A_{2m-2n+1} w^{2m-2n-r} \\ & + \cdots , (m+1-r) R_m w^{m-n-r} + \cdots) + \langle \text{better} \rangle. \end{aligned}$$

Recall that $A_k = \lambda_k A_1$ for $k = 1, \ldots, 2m-2n$. In order to remove all poles in the first two components of

$$f := w\hat{Z}_{tw}(0)\tau + w\hat{X}_w \phi_t(0)$$

one chooses $\phi_t(0)$ in a fashion similar to that used in the proof of Theorem 2:

$$\phi_t(0) := -ic^2\lambda_2\epsilon^2 w^{-2n-1} - ic^2\epsilon(n+1-r)w^{-n-1-r} + \cdots .$$

Then

$$\begin{aligned} f = {} & ic^2\epsilon^2(\cdots(2m-2n)A_{2m-2n+1}w^{2m-3n-1} \\ & + \cdots, (m-n)R_m w^{m-2n-1} + \cdots) \\ & + ic^2\epsilon(\cdots(2m-n+1-r)A_{2m-2n+1}w^{2m-2n-r} \\ & + \cdots, (m-n)R_m w^{m-n-r} + \cdots) \\ & + \langle\text{better}\rangle. \end{aligned}$$

Here and in the sequel, $\cdots$ stand for non-pole terms with coefficients A_j with $j \le 2m-2n$.

The first two components of f (i.e. the expressions before the commata) are holomorphic; the worst pole in the third component is the term with the power w^{m-2n-1}; note that

$$\gamma := m - 2n - 1 = \frac{1}{2}[(2m+2) - 4(n+1)] < 0.$$

Thus Lemma 5 yields

$$\{H[\mathrm{Re}(R_m w^\gamma)]\}_w = -\gamma \overline{R}_m w^{-\gamma-1}.$$

Using a formula established in the proof of Lemma 4 one obtains

$$\begin{aligned} \hat{Z}_{ttw}(0) = {} & -c^2\epsilon^2(\cdots(2m-2n)(2m-3n-1)A_{2m-2n+1}w^{2m-3n-2}, \\ & (m-n)(2n+1-m)\overline{R}_m w^{2n-m} + \cdots) \\ & - c^2\epsilon(\cdots(2m-n)(2m-2n-r)A_{2m-2n+1}w^{2m-2n-r-1} + \cdots, \\ & (m-n)(m-n-r)R_m w^{m-n-r-1}) + \langle\text{better}\rangle. \end{aligned}$$

It follows that

$$\begin{aligned} & \hat{Z}_{ttw}(0) \cdot [w\hat{Z}_{tw}(0)\tau + w\hat{X}_w\phi_t(0)] \\ & \quad = \{-ic^4\epsilon^3(m-n)^2(m-n-r)R_m^2 w^{-1} + \cdots\} + o(\epsilon^3) \end{aligned}$$

since

$$(58) \qquad 2m - 3n - r - 2 = (2m+2) - [3(n+1) + r] - 1 = -1.$$

A straight-forward calculation shows

$$\begin{aligned} & w\hat{Z}_{tw}(0) \cdot \hat{Z}_{tw}(0)\phi_t(0) \\ & \quad = \{ic^4\epsilon^3(m-n)^2(n+1-r)R_m^2 w^{-1} + \cdots\} + o(\epsilon^3). \end{aligned}$$

Thus one obtains by (57) that

$$E^{(4)}(0) = 12\epsilon^3 \mathrm{Re} \int_{S^1} ikc^4 R_m^2 \frac{dw}{w} + o(\epsilon^3)$$

with

$$k := (m-n)^2(n+1-r) - (m-n)^2(m-n-r).$$

Since

$$m - n - r = \frac{1}{2}\{(2m+2) - 2(n+1) - 2r\} = \frac{1}{2}(n+1-r)$$

it follows that

$$k = \frac{1}{2}(m-n)^2(n+1-r) > 0.$$

Hence, by suitable choice of $c \in \mathbb{C}$ one can achieve that $E^{(4)}(0) < 0$, while $E^{(j)}(0) = 0$ for $j = 1, 2, 3$. This concludes the new proof of Theorem 5. □

Finally we want to show that it often is possible to estimate the index m of an interior branch point w_0 of a minimal surface $\hat{X} \in \mathcal{C}(\Gamma)$ with the aid of a geometric condition on its boundary contour Γ. Following an idea by J.C.C. Nitsche, we use Radó's lemma for this purpose (cf. Vol. 1, Section 4.9), which states the following. *If $f \in C^0(\overline{B})$ is harmonic in B, $f(w) \not\equiv 0$ in B, and $\nabla^j f(w_0) = 0$ at $w_0 \in B$ for $j = 0, 1, \ldots, m$, then f has at least $2(m+1)$ different zeros on ∂B.*

We can assume that the minimal surface $\hat{X}$ is transformed into the normal form with respect to the branch point $w_0 = 0$ having the index m. If the contour Γ is nonplanar, then $X^3(w) \not\equiv X_0^3 := X^3(0)$, whence $m < \infty$ and

$$X^3(w) = X_0^3 + \mathrm{Re}[cw^{m+1} + O(w^{m+2})] \quad \text{for } w \to 0$$

with $c \in \mathbb{C} \setminus \{0\}$. Hence $f := X^3 - X_0^3$ satisfies the assumptions of Radó's lemma, and therefore f has at least $2(m+1)$ different zeros on ∂B. Hence the plane $\Pi := \{(x^1, x^2, x^3) \in \mathbb{R}^3 : x^3 = X_0^3\}$ intersects Γ in at least $2(m+1)$ different points. If $m = \infty$ then even $\Gamma \subset \Pi$, and so we obtain:

Proposition 7. *If the minimal surface $\hat{X} \in \mathcal{C}(\Gamma)$ possesses a branch point $w_0 \in B$ with the index m, then there is a plane Π in $\mathbb{R}^3$ which intersects Γ in at least $2(m+1)$ different points. Consequently, if every plane in $\mathbb{R}^3$ intersects Γ in at most k different points, then the index m is bounded by*

$$2m + 2 \leq k.$$

This result motivates the following

Definition 4. *The* **cut number** *$c(\Gamma)$ of a closed Jordan curve Γ in $\mathbb{R}^3$ is the supremum of the number of intersection points of Γ with any (affine) plane Π in $\mathbb{R}^3$, i.e.*

(59) $$c(\Gamma) := \sup\{\sharp(\Gamma \cap \Pi) : \ \Pi = \textit{affine plane in } \mathbb{R}^3\}.$$

It is easy to see that

(60) $$4 \le c(\Gamma) \le \infty,$$

and for any nonplanar, real analytic, closed Jordan curve the cut number $c(\Gamma)$ *is finite.*

We can rephrase the second statement of Proposition 7 as follows:

Proposition 8. *The index* m *of any interior branch point of a minimal surface* $\hat{X} \in \mathcal{C}(\Gamma)$ *is bounded by*

(61) $$2m + 2 \le c(\Gamma).$$

If n is the order and m the index of some branch point, then $1 \le n < m$. On the other hand, $c(\Gamma) = 4$ implies $m \le 1$, and $c(\Gamma) = 6$ yields $m \le 2$. Thus we obtain

Corollary 1. (i) *If* $c(\Gamma) = 4$ *then every minimal surface* $\hat{X} \in \mathcal{C}(\Gamma)$ *is free of interior branch points.*

(ii) *If* $c(\Gamma) = 6$ *then any minimal surface* $\hat{X} \in \mathcal{C}(\Gamma)$ *has at most simple interior branch points of index two; if* $\hat{X}$ *has an interior branch point, it cannot be a weak minimizer of* D *in* $\mathcal{C}(\Gamma)$.

Proof. (i) follows from $1 \le n < m \le 1$, which is impossible. (ii) $1 \le n < m \le 2$ implies $n = 1$ and $m = 2$ for an interior branch point w_0 of $\hat{X}$, whence $2n \le 2m - 2 < 3$. Thus condition (C_3) is satisfied, and therefore the last assertion follows from Theorem 1. □

Corollary 2. *Let* $\hat{X} \in \mathcal{C}(\Gamma)$ *be a minimal surface with an interior branch point of order* n, *and suppose that the cut number of* Γ *satisfies* $c(\Gamma) \le 4n+3$. *Then* $\hat{X}$ *is not a weak minimizer of* D *in* $\mathcal{C}(\Gamma)$.

Proof. By (61) we have

$$2m + 2 \le 4n + 3;$$

hence either

$$2n + 4 \le 2m + 2 < 3n + 4 \Leftrightarrow 2n \le 2m - 2 < 3n$$

or

$$3n + 4 \le 2m + 2 < 4n + 4 \Leftrightarrow 3n \le 2m - 2 < 4n$$

hold true, i.e. either (C_3) or (C_4) are fulfilled. In the first case the assertion follows from Theorem 1, in the second from Theorem 5. □

6.2 The Theorem for $n+1$ Even and $m+1$ Odd

In this section we want to show that *a (nonplanar) weak relative minimizer* $\hat{X}$ *of Dirichlet's integral* D *that is given in the normal form cannot have* $w=0$ *as a branch point if its order* n *is odd and its index* m *is even.* Note that such a branch point is **nonexceptional** since $n+1$ cannot be a divisor of $m+1$. We shall give the proof only under the assumptions $n \geq 3$ since $n=1$ is easily dealt with by a method presented in a forthcoming book by A. Tromba. (Moreover it would suffice to treat the case $m \geq 6$ since $2m-2 < 3n$ is already treated by the Wienholtz theorem. So $2m \geq 3n+2 \geq 11$, i.e. $m \geq 6$ since m is even.)

The Strategy of the Proof

The strategy to find the first nonvanishing derivative of $E(t)$ at $t=0$ that can be made negative consists in the following four steps:

(I) Guess the candidate L for which $E^{(L)}(0) < 0$ can be achieved with a suitable choice of the generator $\tau = \phi(0)$.
(II) Select $D_t^\beta \phi(0)$, $\beta \geq 1$, so that the lower order derivatives $E^{(j)}(0), j = 1, 2, \ldots, L-1$ vanish, ($D_t^\beta := \frac{\partial^\beta}{\partial t^\beta}$).
(III) Prove that

$$E^{(L)}(0) = \operatorname{Re} \int_{S^1} c^L k R_m^2 \frac{dw}{w} = \operatorname{Re}\{2\pi i c^L k R_m^2\},$$

where $c \neq 0$ is a complex number which can be chosen arbitrarily, and $k \in \mathbb{C}$ is to the computed.
(IV) Show that $k \neq 0$.

Remark 1. In order to achieve (II) one tries to choose $D_t^\beta \phi(0)$, $\beta \geq 1$, in such a way that the integrands of $E^{(j)}(0)$ for $j < L$ are free of any poles and, therefore, free of first-order poles. To see that this strategy is advisable, let us consider the case $L=5$; then we have to achieve $E^{(4)}(0) = 0$. Recall that $E^{(4)}(0)$ consists of two terms, one of which has the form

$$I := 12 \operatorname{Re} \int_{S^1} \{2H[\operatorname{Re} if]\}_w f \, dw,$$

where

$$f := w[iw\hat{X}_w \tau]_w \tau + w\hat{X}_w \phi_t(0).$$

Assume that f had poles, say,

$$f(w) = g(w) + h(w), \quad g(w) = \sum_{j \geq 1} a_j w^{-j}, \; h = \text{holomorphic in } B,$$

and $h \in C^0(\overline{B})$. Then, by Lemma 5 of Section 6.1,

$$\{2H[\mathrm{Re}\, if]\}_w(w) = g^*(w) + h'(w), \quad g^*(w) := -i\sum_{j\geq 1} j\overline{a}_j w^{j-1}.$$

Thus, $I = 12 \cdot \{I_1 + I_2 + I_3\}$, with

$$I_1 := \mathrm{Re}\int_{S^1} g^* g\, dw, \quad I_2 := \mathrm{Re}\int_{S^1} h' g\, dw, \quad I_3 := \mathrm{Re}\int_{S^1} (g^* h + h' h)\, dw.$$

The worst term is I_1; one obtains

$$I_1 = \mathrm{Re}\int_{S^1} \sum_{j,\ell\geq 1} (-ij\overline{a}_j w^{j-1} a_\ell w^{-\ell})\, dw = 2\pi \sum_{j\geq 1} j|a_j|^2 > 0$$

and $I_3 = 0$. Hence, in order to achieve $I = 0$, one would have to balance I_2 against $I_1 > 0$ which seems to be pretty hopeless.

Let us now apply the "strategy" to prove

Theorem 1. *Let $\hat{X}$ be a nonplanar minimal surface in normal form that has $w = 0$ as a branch point of odd order $n \geq 3$ and of even index $m \geq 4$. Then, by a suitable choice of $\tau = \phi(0)$ and $D_t^\beta \phi(0)$, one can achieve that*

$$E^{(m+1)}(0) < 0 \quad \textit{and} \quad E^{(j)}(0) = 0 \quad \textit{for } 1 \leq j \leq m.$$

Proof. Set $N := L - 1$, $M := L - (\alpha + \beta + 1) = N - (\alpha + \beta)$, hence $L - 1 = \alpha + \beta + M$. By Leibniz's formula,

$$D_t^N\{[\hat{Z}_w \cdot \hat{Z}_w]\phi\} = \sum_{\alpha=0}^{N-\beta} \sum_{\beta=0}^{N} \frac{N!}{\alpha!\beta!(N-\beta-\alpha)!} (D_t^{N-\beta-\alpha}\hat{Z}_w) \cdot (D_t^\alpha \hat{Z}_w) D_t^\beta \phi.$$

Since

$$D_t E(t) = 2\mathrm{Re}\int_{S^1} w\hat{Z}(t)_w \cdot \hat{Z}(t)_w \phi(t)\, dw,$$

we can use Leibniz's formula to compute $E^{(L)}(t)$ from

$$E^{(L)}(t) = 2\mathrm{Re}\int_{S^1} w D_t^N\{[\hat{Z}_w(t) \cdot \hat{Z}_w(t)]\phi(t)\}\, dw.$$

We choose $L := m + 1$; then $L \geq 5$ as we have assumed $m \geq 4$. It follows that

$$E^{(L)}(0) = J_1 + J_2 + J_3, \tag{1}$$

where the terms J_1, J_2, J_3 are defined as follows: Set

$$T^{\alpha,\beta} := w(D_t^\alpha \hat{Z}(0))_w D_t^\beta \phi(0). \tag{2}$$

Then,

$$(3)\ J_1 := 4\,\mathrm{Re}\int_{S^1}[D_t^{L-1}\hat{Z}(0)]_w\cdot(w\hat{X}_w\tau)\,dw$$
$$+\,4\cdot(L-1)\ \mathrm{Re}\int_{S^1}[D_t^{L-2}\hat{Z}(0)]_w f\,dw$$
$$+\,4\sum_{M>\frac{1}{2}(L-1)}^{L-3}\frac{(L-1)!}{M!(L-M-1)!}\,\mathrm{Re}\int_{S^1}[D_t^M\hat{Z}(0)]_w\cdot g_{L-M-1}\,dw,$$

$$f := T^{1,0}+T^{0,1} = w[\hat{Z}_t(0)]_w\tau + w\hat{X}_w\phi_t(0),$$
$$g_\nu := \sum_{\alpha+\beta=\nu} c^\nu_{\alpha\beta}T^{\alpha,\beta}\quad\text{with } c^\nu_{\alpha\beta} := \frac{\nu!}{\alpha!\beta!};$$

$$(4)\qquad J_2 := \sum_{M=2}^{\frac{1}{2}(L-1)}\frac{2(L-1)!}{M!M!}\,\mathrm{Re}\int_{S^1}[D_t^M\hat{Z}(0)]_w\cdot h_M\,dw$$
$$+\,2(L-1)(L-2)\,\mathrm{Re}\int_{S^1}[\hat{Z}_t(0)]_w\cdot T^{1,L-3}\,dw,$$

$$h_M := \sum_{\alpha=0}^{M}\psi(M,\alpha)\frac{M!}{\alpha!(L-1-M-\alpha)!}T^{\alpha,L-1-M-\alpha},$$
$$\psi(M,\alpha) := 1\quad\text{for } \alpha = M,\qquad \psi(M,\alpha) := 2\quad\text{for } \alpha\neq M;$$

$$(5)\qquad J_3 := 4(L-1)\,\mathrm{Re}\int_{S^1}w\hat{Z}_{tw}(0)\cdot\hat{X}_w D_t^{L-2}\phi(0)\,dw$$
$$+\,2\mathrm{Re}\int_{S^1}w\hat{X}_w\cdot\hat{X}_w D_t^{L-1}\phi(0)\,dw.$$

We have $J_3 = 0$ since $\hat{X}_w\cdot\hat{X}_w = 0$ and $\hat{Z}_{tw}(0)\cdot\hat{X}_w = 0$ on account of formula (36) in 6.1.

Now we proceed as follows:

Step 1. We choose $\tau = \phi(0)$ and $D_t^\beta\phi(0)$ for $\beta\geq 1$ in such a way that f and g_{L-M-1} are holomorphic. Then the integrands of the three integrals in J_1 are holomorphic because all w-derivatives $[D_t^j\hat{Z}(0)]_w$ of the harmonic functions $D_t^j\hat{Z}(t)$ are holomorphic. Then it follows that $J_1 = 0$, and thus we have

$$(6)\qquad E^{(L)}(0) = J_2.$$

Step 2. Then it will be shown that $E^{(L)}(0)$ reduces to the single term

$$(7)\qquad E^{(L)}(0) = \frac{2\cdot m!}{(\frac{m}{2})!\,(\frac{m}{2})!}\,\mathrm{Re}\int_{S^1}w[D_t^{m/2}\hat{Z}(0)]_w\cdot[D_t^{m/2}\hat{Z}(0)]_w\tau\,dw$$

which can be calculated explicitly; it will be shown that

$$E^{(L)}(0) = \frac{2 \cdot m!}{(\frac{m}{2})!\,(\frac{m}{2})!} \mathrm{Re}(2\pi i \cdot \kappa \cdot R_m^2), \tag{8}$$

where κ is the number

$$\kappa := i^{L-1}(a - ib)^L (m-1)^2 (m-3)^2 \cdots 3^2 \cdot 1^2 \tag{9}$$

if the generator $\tau = \phi(0)$ is chosen as

$$\tau(w) := (a - ib)w^{-2} + (a + ib)w^2. \tag{10}$$

For a suitable choice of $(a - ib)$ one obtains $E^{(L)}(0) < 0$. Furthermore the construction will yield $E^{(j)}(0) = 0$ for $1 \le j \le L - 1$.

Before we carry out this program for general $n \ge 3$, $m \ge 4$, $n =$ odd, $m =$ even, we explain the procedure for the simplest possible case: $n = 3$ and $m = 4$.

From the normal form for $\hat{X}_w$ with the order n and the index m of the branch point $w = 0$ we obtain

$$w\hat{X}_w = (A_1 w^{n+1} + \cdots + A_{2m-2n+1} w^{2m-n+1} + \cdots, R_m w^{m+1} + \cdots). \tag{11}$$

Choosing τ according to (10) it follows from

$$[\hat{Z}_t(0)]_w = (iw\hat{X}_w\tau)_w$$

that

$$\begin{aligned} [\hat{Z}_t(0)]_w &= (a - ib)(i(n-1)A_1 w^{n-2} + inA_2 w^{n-1} + \cdots \\ &\quad + i(2m-n-1)A_{2m-2n+1} w^{2m-n-2}, i(m-1)R_m w^{m-2} + \cdots) \\ &\quad + \langle \text{better} \rangle. \end{aligned} \tag{12}$$

Here, $\langle$better$\rangle$ stands again for terms that are similarly built as those in the preceding expression but whose w-powers attached to corresponding coefficients are of higher order. Then

$$\begin{aligned} w[\hat{Z}_t(0)]_w\tau &= (a - ib)^2(i(n-1)A_1 w^{n-3} + inA_2 w^{n-2} + \cdots \\ &\quad + i(2m-n-1)A_{2m-2n+1} w^{2m-n-3} + \cdots, i(m-1)R_m w^{m-3} + \cdots) \\ &\quad + \langle \text{better} \rangle. \end{aligned} \tag{13}$$

Since this term is holomorphic we have the freedom to set $\phi_t(0) = 0$. Then $f(w) = w\hat{Z}_{tw}(0)\tau + w\hat{X}_w\phi_t(0)$ is holomorphic, and Proposition 6 in Section 6.1 yields

$$E^{(5)}(0) = 12\mathrm{Re}\int_{S^1} w\hat{Z}_{ttw}(0)\cdot\hat{Z}_{ttw}(0)\tau\,dw. \tag{14}$$

(This follows of course also from the general formulas stated above.)

From formula (51) of Lemma 4 in Section 6.1 we get

$$\hat{Z}_{ttw}(0) = \{iw[iw\hat{X}_w\tau]_w\tau\}_w = i\{w\hat{Z}_{tw}(0)\tau\}_w,$$

and so

$$\begin{aligned}\hat{Z}_{ttw}(0) &= -(a-ib)^2((n-1)(n-3)A_1 w^{n-4}+\cdots\\ &\quad + (2m-n-1)(2m-n-3)A_{2m-2n+1}w^{2m-n-4}\\ &\quad +\cdots,(m-1)(m-3)R_m w^{m-4}+\cdots) + \langle\text{better}\rangle.\end{aligned} \tag{15}$$

Since $n-3=0$ and $m=4$, this leads to

$$\hat{Z}_{ttw}(0)\cdot\hat{Z}_{ttw}(0) = (a-ib)^4(m-1)^2(m-3)^2R_m^2+\cdots, \tag{16}$$

and by (14) we obtain for $L=m+1=5$:

$$\begin{aligned}E^{(L)}(0) &= E^{(5)}(0) = 12\cdot\mathrm{Re}\int_{S^1}(a-ib)^5(m-1)^2(m-3)^2R_m^2\frac{dw}{w}\\ &= 12\cdot\mathrm{Re}[2\pi i(a-ib)^5(m-1)^2(m-3)^2R_m^2], \quad m=4.\end{aligned} \tag{17}$$

Now we turn to the **general case** of an odd $n\ge 3$ and an even index $m\ge 4$.

Step 1. *The pole removal technique to make the expressions f and g_{L-M-1} in the integral J_1 holomorphic.*

We have already seen that $f(w)$ is holomorphic if we set $\phi_t(0)=0$. In fact, we set

$$D_t^\beta\phi(0) = 0 \quad \text{for } 1\le\beta\le\frac{n-1}{2} \text{ and for } \beta>\frac{1}{2}(L-3) \tag{18}$$

and prove the following

Lemma 1. *By the pole-removing technique we can inductively choose $D_t^\beta\phi(0)$ for $\beta\le\frac{1}{2}(L-3)$ such that g_ν is holomorphic for $\nu=0,1,\dots,\frac{1}{2}(L-3)$. Then the derivative $[D_t^\gamma\hat{Z}(0)]_w$ is not only holomorphic, but can be obtained in the form*

$$[D_t^\gamma\hat{Z}(0)]_w = \{ig_{\gamma-1}\}_w \quad \textit{for } \gamma=1,2,\dots,\frac{1}{2}(L-1). \tag{19}$$

Suppose this result were proved. Since in J_1 there appear only g_ν with $\nu=L-M-1$ where $\frac{1}{2}(L-1)<M\le L-3$, i.e. $2\le\nu\le\frac{1}{2}(L-3)$, all integrands in J_1 were indeed holomorphic, and so $J_1=0$. Thus it remains to prove Lemma 1.

Proof of Lemma 1. By definition we have

(20) $$g_\nu = \sum_{\alpha+\beta=\nu} c^\nu_{\alpha\beta} T^{\alpha,\beta}, \quad T^{\alpha,\beta} := w[D_t^\alpha \hat{Z}(0)]_w D_t^\beta \phi(0),$$

and $\phi(0) = \tau$.

The expressions $w[D_t^\alpha \hat{Z}(0)]_w \tau$ have no pole for $\alpha \le \frac{n-1}{2}$, and we make the important observation that there are numbers c, c' such that

$$w[D_t^{\frac{n-1}{2}} \hat{Z}(0)]_w \tau = (cA_1 + \cdots, c' R_m w^{m-n} + \cdots).$$

Thus, a pole in $w[D_t^\alpha \hat{Z}(0)]_w \tau$ may arise at first for $\alpha = \frac{1}{2}(n+1)$; then we have, say

(21) $$w[D_t^{\frac{1}{2}(n+1)} \hat{Z}(0)]_w \tau = (cA_2 w^{-1} + \cdots, c' R_m w^{m-n-2} + \cdots).$$

This requires a nonzero $D_t^{\frac{n+1}{2}} \phi(0)$ in case that $cA_2 \ne 0$ if we want to make $g_{\frac{1}{2}(n+1)}$ pole-free. Now we go on and discuss the pole removal for $\nu = \frac{1}{2}(n+3), \frac{1}{2}(n+5), \dots, \frac{1}{2}(L-3)$.

Observation 1. Since m is even, n is odd, and $m > n$, we have

(22) $$m = n + (2k+1), \quad k = 0, 1, 2, \dots,$$

and therefore

(23) $$\frac{1}{2}(L-3) = \frac{1}{2}(m-2) = \frac{1}{2}(n+2k-1).$$

Thus, for $m = n+1$, all g_ν with $2 \le \nu \le \frac{1}{2}(L-3)$ are pole-free if we set $D_t^\beta \phi(0) = 0$ for all $\beta \ge 1$; cf. (18). For $m = n+3$, we have to choose $D_t^\beta \phi(0)$ appropriately for $\beta = \frac{1}{2}(n+1)$ while the other $D_t^\beta \phi(0)$ are taken to be zero. For $m = n+5$, we must also choose $D_t^\beta \phi(0)$ appropriately for $\beta = \frac{1}{2}(n+3)$ whereas the other $D_t^\beta \phi(0)$ are set to be zero. In this way we proceed inductively and choose $D_t^\beta \phi(0)$ in a suitable way for $\beta = \frac{1}{2}(n+1), \frac{1}{2}(n+3), \dots, \frac{1}{2}(n+2k-1)$ in case that $m = n+2k+1$ while all other $D_t^\beta \phi(0)$ are taken to be zero according to (18).

Observation 2. The pole-removal procedure would only stop for some g_ν with $\frac{1}{2}(n+1) \le \nu \le \frac{1}{2}(L-3)$ if the w-power attached to $A_{2m-2n+1}$ became negative. We have to check that this does not happen for $\nu \le \frac{1}{2}(L-3)$. Since at the α-th stage in defining $[D_t^\alpha \hat{Z}(0)]_w$ the w-powers have been reduced by 2α, we must check that the terms $T^{\alpha,\beta}$ have no poles connected with $A_{2m-2n+1}$ if $\alpha+\beta \le \frac{1}{2}(L-3)$. Looking first only at $T^{\alpha,0} = w[D_t^\alpha \hat{Z}(0)]_w \tau$ for $\alpha \le \frac{1}{2}(L-3)$, we must have

$$2m - n - 2\alpha = 2m - n + 1 - 2(\alpha + 1) \geq 0 \quad \text{for } \alpha \leq \frac{1}{2}(L-3),$$

which is true since

$$2m - n + 1 - 2 \cdot \frac{1}{2}(L-1) = m - n + 1 > 0.$$

We must also check that during the process no pole is introduced into the third complex component. Again we first look at $T^{\alpha,0}$ for $\alpha \leq \frac{1}{2}(L-3)$. Then the order of the w-power at the R_m-term is

$$m - 2\alpha - 1 = (m+1) - 2(\alpha+1) \geq (m+1) - (L-1) = 1,$$

and so there is no pole.

Let us now look at the pole-removal procedure. For $m = n+1$ all g_ν with $2 \leq \nu \leq \frac{1}{2}(L-3)$ are pole-free if we assume (18). If $m = n+3$ we have to make $g_{\frac{1}{2}(n+1)}$ pole-free. To this end it suffices to choose $D_t^{\frac{1}{2}(n+1)}\phi(0)$ appropriately; it need have a pole at most of order $(n+2)$ in order to remove a possible pole of $T^{\alpha,0}, \alpha = \frac{1}{2}(n+1)$, cf. (21).

If $m = n+5$, we have to choose $D_t^\beta \phi(0)$ appropriately for $\beta = \frac{1}{2}(n+1)$ and $\beta = \frac{1}{2}(n+3)$. The derivative $D_t^{\frac{1}{2}(n+1)}\phi(0)$ will be taken as before, while $D_t^{\frac{1}{2}(n+3)}\phi(0)$ is to be chosen in such a way that

$$g_{\frac{1}{2}(n+3)} = T^{\frac{1}{2}(n+3),0} + T^{1,\frac{1}{2}(n+1)} + T^{0,\frac{1}{2}(n+3)}$$

becomes holomorphic. Since

$$\begin{aligned} T^{1,\frac{1}{2}(n+1)} &= w[\hat{Z}_t(0)]_w D_t^{\frac{1}{2}(n+1)}\phi(0) \\ &= (i(n-1)(a-ib)A_1 w^{n-1} + \cdots, \\ &\quad\; i(m-1)(a-ib)R_m w^{m-1} + \cdots) D_t^{\frac{1}{2}(n+1)}\phi(0) \\ &= (cA_1 w^{-3} + \cdots, c' R_m w^{m-n-3} + \cdots) \end{aligned}$$

with some constants c, c', the derivative $D_t^{\frac{1}{2}(n+3)}\phi(0)$ in

$$T^{0,\frac{1}{2}(n+3)} = w\hat{X}_w D_t^{\frac{1}{2}(n+3)}\phi(0)$$

should have a pole of order $n+4$, while a pole of lower order than $n+4$ is needed to remove a possible singularity in the first term $T^{\frac{1}{2}(n+3),0} = w[D_t^{\frac{1}{2}(n+3)}\hat{Z}(0)]_w\tau$.

In this way we can proceed inductively choosing the poles of $D_t^\beta\phi(0)$ always at most of order

$$(24) \qquad n + 2\left(\beta - \frac{n-1}{2}\right) = 2\beta + 1 \quad \text{for } \frac{1}{2}(n+1) \leq \beta \leq \frac{1}{2}(L-3).$$

This is the crucial estimate on the order of the pole of $D_t^\beta\phi(0)$ in order to ensure that these derivatives play no role in the final calculations.

Observation 3. Consider the last complex component of

$$g_{\frac{1}{2}(n+1)} = w[D_t^{\frac{1}{2}(n+1)}\hat{Z}(0)]_w\tau + w\hat{X}_w D_t^{\frac{1}{2}(n+1)}\phi(0).$$

The lowest w-power attached to R_m in the first term is $1+m-(n+1)-2 = m-n-2 \geq 1$ (since in this case $m \geq n+3$ according to Observation 1). The lowest w-power associated to R_m in the second term is $1+m-(n+2) = m-n-1 > m-n-2$. Continuing inductively we see that the lowest w-power attached to R_m in any g_ν arises from $\tau = \phi(0)$ and not from any $D_t^\beta\phi(0)$. □

This ends the proof of Step 1, and we have found that $E^{(L)}(0) = J_2$. Now we come to

Step 2. The integral J_2 is a linear combination of the real parts of the integrals

$$(25) \qquad I_{\alpha\gamma\beta} := \int_{S^1} w[D_t^\alpha \hat{Z}(0)]_w \cdot [D_t^\gamma \hat{Z}(0)]_w D_t^\beta \phi(0)\, dw,$$

where $1 \leq \alpha, \gamma \leq \frac{1}{2}(L-1)$ and $\beta = (L-1) - \alpha - \gamma$. Then we have

$$(26) \qquad \beta = 0 \quad \text{if and only if} \quad \alpha = \gamma = \frac{1}{2}(L-1) = \frac{m}{2}.$$

This implies

$$(27) \qquad J_2 = \frac{2 \cdot m!}{(\frac{m}{2})!(\frac{m}{2})!}\mathrm{Re}\int_{S^1} w[D_t^{\frac{m}{2}}\hat{Z}(0)]_w \cdot [D_t^{\frac{m}{2}}\hat{Z}(0)]_w\tau\, dw$$

because of the following

Lemma 2. *We have*

$$(28) \qquad I_{\alpha\gamma\beta} = 0 \quad \textit{for } 1 \leq \alpha, \gamma \leq \frac{1}{2}(L-1) \textit{ and } 1 \leq \beta = m - \alpha - \gamma.$$

Proof. Let us first show that the product of the last complex components of $[D_t^\alpha \hat{Z}(0)]_w$ and $[D_t^\gamma \hat{Z}(0)]_w$ and of $wD_t^\beta\phi(0)$ have a zero integral. In fact, this product has the form

$$\begin{aligned}
&\mathrm{const}(wR_m w^{m-2\alpha} \cdot R_m w^{m-2\gamma} + \cdots)(w^{-2\beta-1} + \cdots) \\
&= \mathrm{const} R_m^2 w^{1+2m-2(\alpha+\beta+\gamma)-1} + \cdots = \mathrm{const} \cdot R_m^2 + \cdots
\end{aligned}$$

since $\alpha + \beta + \gamma = L - 1 = m$.

The same holds true for the scalar product of the first two complex components, multiplied by $wD_t^\beta\phi(0)$. To see this we assume without loss of generality that $\alpha \geq \gamma$. Denote by $P^{\alpha\gamma}$ the expression

$$P^{\alpha\gamma} := w[C_1^\alpha \cdot C_1^\gamma + C_2^\alpha \cdot C_2^\gamma],$$

where C_1^α, C_2^α and C_1^γ, C_2^γ are the first two complex components of $[D_t^\alpha \hat{Z}(0)]_w$ and $[D_t^\gamma \hat{Z}(0)]_w$ respectively.

Case 1. If $2\gamma \le 2\alpha < n$ then

$$\begin{aligned} P^{\alpha\gamma} &= w(\text{const}A_j w^{n-2\alpha} + \cdots + \text{const}A_{2m-2n+1} w^{2m-n-2\alpha} + \cdots) \\ &\quad \cdot (\text{const}A_\ell w^{n-2\gamma} + \cdots + \text{const}A_{2m-2n+1} w^{2m-n+\gamma} + \cdots) \end{aligned}$$

with $j, \ell < 2m-2n+1$.

Case 2. If $2\gamma < n < 2\alpha$ then

$$\begin{aligned} P^{\alpha\gamma} &= w(\text{const}A_j + \cdots + \text{const}A_{2m-2n+1} w^{2m-n} + \cdots) \\ &\quad \cdot (\text{const}A_\ell w^{n-2\gamma} + \cdots + \text{const}A_{2m-2n+1} w^{2m-n-2\gamma} + \cdots) \end{aligned}$$

with $j, \ell < 2m-2n+1$.

Case 3. If $n < 2\alpha$ and $n < 2\gamma$ then

$$\begin{aligned} P^{\alpha\gamma} &= w(\text{const}A_j + \cdots + \text{const}A_{2m-2n+1} w^{2m-n-2\alpha} + \cdots) \\ &\quad \cdot (\text{const}A_\ell + \cdots + \text{const}A_{2m-2n+1} w^{2m-n-2\gamma} + \cdots). \end{aligned}$$

Let $\mu(\alpha,\gamma)$ be the lowest w-power appearing in $P^{\alpha\gamma} D_t^\beta \phi(0)$. Recalling $\alpha + \beta + \gamma = m$ we obtain the following results:

Case 1.

$$\begin{aligned} \mu(\alpha,\gamma) &= 1 + 2m - 2\gamma - 2\alpha - 2\beta - 1 \\ &= 2 + 2m - 2(\alpha+\beta+\gamma+1) \\ &= 2 + 2m - 2(m+1) = 0. \end{aligned}$$

Case 2. $\mu(\alpha,\gamma)$ is either zero as in Case 1, or

$$\begin{aligned} \mu(\alpha,\gamma) &= 1 + 2m - n - 2\gamma - 2\beta - 1 \\ &= 2 + 2m - n - 2(\gamma+\beta+1) \\ &= 2 + 2m - n - 2(m+1-\alpha) = 2\alpha - n > 0. \end{aligned}$$

Case 3. As in Case 2 we have $\mu(\alpha,\gamma) > 0$.

This proves $I_{\alpha\gamma\beta} = 0$ for $1 \le \alpha, \gamma \le \frac{m}{2}$ and $1 \le \beta = m - \alpha - \gamma$, which yields Lemma 2. □

Thus we have arrived at (27), and a straight-forward computation leads to (8) and (9); so the proof of Theorem 1 is complete. □

6.3 Boundary Branch Points

In this section we first show that Dirichlet's integral possesses intrinsic second and third derivatives at a minimal surface $\hat{X}$ on the tangent space $T_X M$ of $M := H^2(\partial B, \mathbb{R}^n)$ of $X = \hat{X}|_{\partial B}$ on the space $J(\hat{X})$ of forced Jacobi fields for $\hat{X}$. These results will also be used in Vol. 3, Chapters 5 and 6. In particular it will be seen that $J(\hat{X})$ is a subspace of the kernel of the Hessian $D^2E(X)$ of Dirichlet's integral $E(X)$ defined in (1) below, and an interesting formula (see (16)) for the second variation of Dirichlet's integral is derived.

Secondly we prove that, for a sufficiently smooth contour Γ in $\mathbb{R}^3$, not only the *order*, but also the *index* of a boundary branch point of a minimal surface $X \in \mathcal{C}(\Gamma)$ can be estimated in terms of the *total curvature* of Γ if curvature and torsion of Γ are nowhere zero.

Finally we prove Wienholtz's theorem, which states a condition under which a minimizer for Plateau's problem cannot possess a boundary branch point. In particular we show: If n is the order and m the index of a boundary branch point of $\hat{X}$ such that $2m-2 < 3n$ (equivalently $2m+2 \leq 3(n+1)$) then $\hat{X}$ cannot be a minimizer of Dirichlet's integral or of area. The key idea of the proof will be to recompute the third derivative of Dirichlet's integral, D, in an intrinsic way on $J(\hat{X})$, thereby showing that the formula for $E^{(3)}(0) = \frac{d^3}{dt^3} D(\hat{Z}(t))\big|_{t=0}$ derived in Section 6.1 is valid in the presence of boundary branch points as well.

Towards these goals, we first show that if the boundary contour $\Gamma \subset \mathbb{R}^n$ is of class $D^{\boldsymbol{r}+7}, \boldsymbol{r} \geq 3$, the space $\mathcal{H}^{5/2}_\Gamma(\overline{B}, \mathbb{R}^n)$ of harmonic surfaces from $\overline{B}$ into $\mathbb{R}^n$, mapping $S^1 = \partial B$ to Γ, is a $C^{\boldsymbol{r}}$ manifold, in fact, a $C^{\boldsymbol{r}}$-submanifold of the space $\mathcal{H}^{5/2}(\overline{B}, \mathbb{R}^n)$ of harmonic mappings from $\overline{B}$ into $\mathbb{R}^n$. Instead of the dimension $n = 3$ we do this for arbitrary dimension n, since this result is necessary for the index theorem to be derived in Chapter 5 of Vol. 3. Here it is essential that we operate in the context of a manifold since the third derivative of any real-valued C^3-smooth function is seen to be well defined as a trilinear form on the kernel of the Hessian of this function at any critical point. As in Chapters 5 and 6 of Vol. 3 we shall use the symbol D for the total derivative or the Fréchet derivative. Therefore we need another notation for Dirichlet's integral; instead of D we employ the symbol E and consider E as a function of boundary values $X : S^1 \to \mathbb{R}^n$ (instead of their harmonic extension $\hat{X}$), i.e.

$$(1) \qquad E(X) := \frac{1}{2} \int_B (\hat{X}_u \cdot \hat{X}_u + \hat{X}_v \cdot \hat{X}_v) \, du \, dv \quad \text{for } X \in H^{1/2}(S^1, \mathbb{R}^n).$$

It is a well-known fact that $\mathbb{R}^n$ carries a $C^{\boldsymbol{r}+6}$-Riemannian metric g with respect to which Γ is totally geodesic, i.e. any g-geodesic $\sigma : (-1, 1) \to \mathbb{R}^n$ with $\sigma(0) \in \Gamma$ and $\sigma'(0) \in T_{\sigma(0)}\Gamma$ remains on Γ. Let $(p, v) \mapsto \exp_p v$ denote the exponential map of g; it is of class $C^{\boldsymbol{r}+4}$. Via harmonic extension we identify the space

$$M := H^2(S^1, \Gamma)$$

of H^2-maps from S^1 to Γ with the space $\mathcal{H}^{5/2}_\Gamma(\overline{B}, \mathbb{R}^n)$. In order to show that M is a submanifold of $H^2(S^1, \mathbb{R}^n)$ we need to identify the tangent space $T_X M$ for $X \in H^2(S^1, \Gamma)$. (In Vol. 3, Chapters 5 and 6, we shall denote M by $\mathcal{N}_\alpha$ if Γ is given by $\Gamma = \alpha(S^1)$.)

Definition 1. *We define the tangent space $T_X M$ of M at $X \in H^2(S^1, \Gamma)$ as*

$$T_X M := \{Y \in H^2(S^1, \mathbb{R}^n) : Y(e^{i\theta}) \in T_{X(e^{i\theta})}\Gamma,\ \theta \in \mathbb{R}\}.$$

Clearly $T_X M$ is a Hilbert subspace of $H^2(S^1, \mathbb{R}^n)$. Our goal is to show that the map

$$\Phi(Y)(s) := \exp_{X(s)} Y(s), \quad s = e^{i\theta},$$

is a local C^r-diffeomorphism about the zero $0 \in H^2(S^1, \mathbb{R}^n)$ mapping a neighbourhood of zero in $T_X M$ onto a neighbourhood of X in M. Towards this goal we have:

Theorem 1. *If $\varphi \in C^{r+3}(\mathbb{R}^n, \mathbb{R}^n)$, then $\Phi : H^2(S^1, \mathbb{R}^n) \to H^2(S^1, \mathbb{R}^n)$ defined by $\Phi(Y) := \varphi \circ Y$ is of class C^r. Furthermore,*

$$D^m \Phi_Y(\lambda_1, \dots, \lambda_m)(s) = D^m \varphi_{Y(s)}(\lambda_1(0), \dots, \lambda_m(s)) \quad \textit{for } 0 \le m \le r.$$

The proof of this theorem will be a consequence of the following

Lemma 1. *Let $\mathcal{L}^m(\mathbb{R}^n, \mathbb{R}^n)$ be the space of m-linear maps from $\mathbb{R}^n$ into $\mathbb{R}^n$, and suppose that $f \in C^3(\mathbb{R}^n, \mathcal{L}^m(\mathbb{R}^n, \mathbb{R}^n))$. Then the map $F : H^2(S^1, \mathbb{R}^n) \to \mathcal{L}^m(H^2(S^1, \mathbb{R}^n), H^2(S^1, \mathbb{R}^n))$ defined by*

$$Y \mapsto F(Y)(\lambda_1, \dots, \lambda_m)(s) := f(Y(s))(\lambda_1(s), \dots, \lambda_m(s))$$

is continuous. Moreover, if $f \in C^4$ then $F \in C^1$, and the derivative of $Y \mapsto F(Y)$ is

$$\lambda \mapsto df(Y(s))(\lambda(s), \lambda_1(s), \dots, \lambda_m(s)).$$

Proof. Recall that $H^2(S^1, \mathbb{R}^n)$ is continuously and compactly embedded into $C^1(S^1, \mathbb{R}^n)$. Assume for simplicity that

$$\|\lambda_j\|_{H^2} \le 1, \quad \|Y\|_{H^2} < 2, \quad \|\tilde{Y}\|_{H^2} < 2,$$

and consider the difference

$$[F(Y) - F(\tilde{Y})](\lambda_1, \dots, \lambda_m)(s) = [f(Y(s)) - f(\tilde{Y}(s))](\lambda_1(s), \dots, \lambda_m(s)).$$

Then

$$
\begin{aligned}
&\frac{d}{ds}[F(Y) - F(\tilde{Y})](\lambda_1, \ldots, \lambda_m)(s) \\
&\quad = df(Y(s))(Y'(s))(\lambda_1(s), \ldots, \lambda_m(s)) - df(\tilde{Y}(s))(\tilde{Y}'(s))(\lambda_1(s), \ldots, \lambda_m(s)) \\
&\qquad + \sum_{j=1}^{m} [f(Y(s)) - f(\tilde{Y}(s))](\lambda_1(s), \ldots, \lambda_{j-1}(s), \lambda_j'(s), \lambda_{j+1}(s), \ldots, \lambda_m(s)) \\
&\quad = df(Y(s))(Y'(s) - \tilde{Y}'(s))(\lambda_1(s), \ldots, \lambda_m(s)) \\
&\qquad + [df(Y(s) - df(\tilde{Y}(s))](\tilde{Y}'(s))(\lambda_1(s), \ldots, \lambda_m(s)) \\
&\qquad + \sum_{j=1}^{m} [f(Y(s)) - f(\tilde{Y}(s))](\lambda_1(s), \ldots, \lambda_j'(s), \ldots, \lambda_m(s)).
\end{aligned}
$$

Since f is Lipschitz continuous, we have

$$
\begin{aligned}
\sup_s |f(Y(s)) - f(\tilde{Y}(s))| &\leq \text{const} \sup_s |Y(s) - \tilde{Y}(s)| \\
&\leq \text{const} \, \|Y - \tilde{Y}\|_{H^1},
\end{aligned}
$$

and therefore

$$
\begin{aligned}
&\left| \sum_{j=1}^{m} [f(Y(s)) - f(\tilde{Y}(s))](\lambda_1(s), \ldots, \lambda_j'(s), \ldots, \lambda_m(s)) \right| \\
&\quad \leq \text{const} \sum_{j=1}^{m} \|Y - \tilde{Y}\|_{H^1} |\lambda_j'(s)|,
\end{aligned}
$$

from which it follows that

$$
\left\| \sum_{j=1}^{m} [f(Y) - f(\tilde{Y})](\lambda_1, \ldots, \lambda_j', \ldots, \lambda_m) \right\|_{L^2} \leq \text{const} \, \|Y - \tilde{Y}\|_{H^2}.
$$

Furthermore, the Lipschitz continuity of df implies

$$
\begin{aligned}
\|df(Y)(Y' - \tilde{Y}')(\lambda_1, \ldots, \lambda_m)\|_{L^2} &\leq \text{const} \, \|Y - \tilde{Y}\|_{H^2}, \\
\|df(Y) - df(\tilde{Y})](\tilde{Y}')(\lambda_1, \ldots, \lambda_m)\|_{L^2} &\leq \text{const} \, \|Y - \tilde{Y}\|_{H^2}.
\end{aligned}
$$

Summarizing these estimates we obtain

$$
\left\| \frac{d}{ds}[F(Y) - \tilde{F}(Y)](\lambda_1, \ldots, \lambda_m) \right\|_{L^2} \leq \text{const} \, \|Y - \tilde{Y}\|_{H^2}.
$$

In the same manner we infer

$$
\left\| \frac{d^2}{ds^2}[F(Y) - \tilde{F}(Y)](\lambda_1, \ldots, \lambda_m) \right\|_{L^2} \leq \text{const} \, \|Y - \tilde{Y}\|_{H^2},
$$

since f, df, and $d^2 f$ are Lipschitz continuous, using

$$
\begin{aligned}
&\frac{d^2}{ds^2}[F(Y) - F(\tilde{Y})](\lambda_1, \dots, \lambda_m)(s) \\
&\quad = d^2 f(Y(s))(Y'(s))(Y'(s) - \tilde{Y}'(s))(\lambda_1(s), \dots, \lambda_m(s)) \\
&\qquad + df(Y(s))(Y''(s) - \tilde{Y}''(s))(\lambda_1(s), \dots, \lambda_m(s)) \\
&\qquad + \sum_{j=1}^m df(Y(s))(Y'(s) - \tilde{Y}'(s))(\lambda_1(s), \dots, \lambda_j'(s), \dots, \lambda_m(s)) \\
&\qquad + [d^2 f(Y(s))(Y'(s)) - d^2 f(\tilde{Y}(s))(\tilde{Y}'(s))](\tilde{Y}'(s))(\lambda_1(s), \dots, \lambda_m(s)) \\
&\qquad + [df(Y(s)) - df(\tilde{Y}(s))](\tilde{Y}''(s))(\lambda_1(s), \dots, \lambda_m(s)) \\
&\qquad + \sum_{j=1}^m [df(Y(s)) - df(\tilde{Y}(s))](Y'(s))(\lambda_1(s), \dots, \lambda_j'(s), \dots, \lambda_m(s)) \\
&\qquad + \sum_{j=1}^m [f(Y(s)) - f(\tilde{Y}(s))](\lambda_1(s), \dots, \lambda_j''(s), \dots, \lambda_m(s)) \\
&\qquad + \sum_{j,k=1, j<k}^m [f(Y(s)) - f(\tilde{Y}(s))](\lambda_1(s), \dots, \lambda_j'(s), \dots, \lambda_k'(s), \dots, \lambda_m(s)) \\
&\qquad + \sum_{j=1}^m [df(Y(s))(Y'(s)) - df(\tilde{Y})(\tilde{Y}'(s))](\lambda_1(s), \dots, \lambda_j'(s), \dots, \lambda_m(s)).
\end{aligned}
$$

The estimates above prove that F maps $H^2(S^1, \mathbb{R}^n)$ continuously into the space

$$\mathfrak{L}^m(H^2(S^1, \mathbb{R}^n), H^2(S^1, \mathbb{R}^m)).$$

If $f \in C^4$ then $df \in C^3$ and $d^2 f \in C^2$, and Taylor's theorem yields

$$f(u + h) - f(u) - df(u)h = r(u, h)(h, h),$$

where

$$r(u, h)(h, h) := \int_0^1 (1 - t)[d^2 f(u + th) - d^2 f(u)](h, h)\, dt.$$

Since f is in C^4 we obtain

$$\|r(u, h)(h, h)\|_{H^2} \le \text{const}\, \|h\|_{H^2}^2 \quad \text{for } \|h\|_{H^2} \le 1.$$

This shows that the mapping F is differentiable, and its derivative $DF(Y)$ at $Y \in H^2(S^1, \mathbb{R}^n)$ is given by

$$(DF(Y)h)(s) = df(Y(s))h(s).$$

Since $df \in C^3$, the first part of the lemma yields $DF \in C^0$. □

Proof of Theorem 1. Applying Lemma 1 to $f = d^m \varphi$ succesively to $m = 0, 1, \dots, \boldsymbol{r} - 1$, we infer that $D\Phi, D^2\Phi, \dots, D^{\boldsymbol{r}}\Phi$ exist and are continuous. □

Theorem 2. *$M = H^2(S^1, \Gamma)$ is a C^r-submanifold of $H^2(S^1, \mathbb{R}^n)$.*

Proof. Since $H^2(S^1, \mathbb{R}^n) \subset C^1(S, \mathbb{R}^n)$, the set M is closed in $H^2(S^1, \mathbb{R}^n)$. Consider the map $Y \mapsto \Phi(Y)$ defined by

$$\Phi(Y)(s) := \exp_{X(s)} Y(s) \quad \text{for } X \in H^2(S^1, \Gamma),$$

which is of class C^r by virtue of Theorem 1.

Since $\Phi(0)$ is the identity map, the inverse function theorem implies that Φ is a local C^r-diffeomorphism about 0. Moreover, as the Riemannian metric g is totally geodesic with respect to Γ, we see that Φ maps $T_X M$ into M. Since Φ is also locally invertible, it provides a coordinate chart for M as a submanifold of $H^2(S^1, \mathbb{R}^n)$. □

Before we can apply the preceding results to Plateau's problem we need an abstract functional analytic reasoning which shows that a C^3-function $E : M \to \mathbb{R}$ on a C^r-smooth submanifold M of a Hilbert space $\mathcal{H}, \boldsymbol{r} \geq 3$, possesses intrinsic first, second, and third order derivatives for any critical point x of E (i.e. $DE(x) = 0$). To prove this we need a few prerequisites.

By $E \in C^3(M)$ we mean that E extends to a C^3-map on a neighbourhood of every point $x \in M$. Equivalently we can use coordinate charts as follows. From the definition of a submanifold it follows that about each point $x \in M$ there is a C^r-diffeomorphism $\rho : \mathcal{V} \to \mathcal{V}'$ from a neighbourhood $\mathcal{V}$ of x in $\mathcal{H}$ onto a neighbourhood $\mathcal{V}'$ of 0 in $\mathcal{H}$ with $\rho(x) = 0$ such that $\rho(\mathcal{V} \cap M)$ is an open subset of a fixed subspace $\mathcal{H}_0$ of $\mathcal{H}$. Then "$E \in C^3(M)$" means that $E \circ \psi$ is of class C^3 for any such chart $(\rho, \mathcal{V})$ where ψ is the inverse of ρ. For $x \in M$ with the image $0 = \rho(x)$ we define the tangent space $T_x M$ of M at x by

$$T_x M := D\psi(0)[\mathcal{H}_0] \subset \mathcal{H},$$

i.e. as the image of $\mathcal{H}_0$ under the mapping provided by the derivative $D\psi(0)$. This definition of $T_x M$ does not depend on the choice of the chart $(\rho, \mathcal{V})$.

As each $h \in T_x M$ can be written as $h = D\psi(0)\tilde{h}$ with $\tilde{h} \in \mathcal{H}_0$, we define

$$DE(x)h := D(E \circ \psi)(0)\tilde{h},$$

which again can be shown to be independent of the choice of the chart.

A point $x \in M$ is a critical point of $E : M \to \mathbb{R}$ if $DE(x) = 0$. At a critical point x of E there is a well-defined bilinear form

$$D^2 E(x) : T_x M \times T_x M \to \mathbb{R}$$

defined by

$$D^2 E(x)(h, k) := D^2(E \circ \psi)(0)(\tilde{h}, \tilde{k}) \quad \text{for}$$

$$h = D\psi(0)\tilde{h}, \quad k = D\psi(0)\tilde{k}; \quad \tilde{h}, \tilde{k} \in \mathcal{H}_0.$$

This is the Hessian (bilinear form), which again does not depend on the choice of the chart $(\rho, \mathcal{V})$, as we will shortly show. Surprisingly, there is also a third intrinsic derivative $D^3E(x)$, but this is intrinsically defined only on the kernel K_x of $D^2E(x)$, i.e. on

$$K_x := \{h \in T_xM : D^2E(x)(h,k) = 0 \text{ for all } k \in T_xM\}.$$

Let us state this formally as

Theorem 3. *At a critical point x of $E \in C^3(M)$ there is an intrinsically defined*[1] *second derivative $D^2E(x) : T_xM \times T_xM \to \mathbb{R}$, and a third derivative $D^3E(x) : K_x \times K_x \times K_x \to \mathbb{R}$ defined as a trilinear map on the kernel K_x of $D^2E(x)$.*

To prove this we have to show that, with respect to any transition map $\varphi : U \to U$ on $U \subset M$ fixing the critical point $x \in U$ of E, the second and third derivative of $E \circ \varphi$ depend only on the first derivative of φ and are independent of $D^2\varphi(x)$ and $D^3\varphi(x)$. Since we may choose the critical point x as the origin 0, the theorem is a consequence of the following

Lemma 2. *Let U be an open subset of a Hilbert space and suppose that $0 \in U$ is a critical point of $E \in C^3(U)$. Assume also that K is the kernel of the Hessian of E at 0 and $\varphi : U \to U$ is a C^3-diffeomorphism of U onto itself with $\varphi(0) = 0$. Then*

$$D^2(E\circ\varphi)(0)(k_1,k_2) = D^2E(0)(D\varphi(0)k_1, D\varphi(0)k_2),$$

and furthermore, if $D\varphi(0)k_j \in K, j = 1,2,3,$ then

$$D^3(E\circ\varphi)(0)(k_1,k_2,k_3) = D^3E(0)(D\varphi(0)k_1, D\varphi(0)k_2, D\varphi(0)k_3).$$

Proof. Repeatedly using the chain rule we see that

(i) $D(E\circ\varphi)(x)(h) = DE(\varphi(x))D\varphi(x)h,$

(ii) $$D^2(E\circ\varphi)(x)(h,k) = D^2E(\varphi(x))(D\varphi(x)h, D\varphi(x)k) + DE(\varphi(x))D^2\varphi(x)(h,k).$$

(iii) $$\begin{aligned} D^3(E\circ\varphi)(x)(h,k,\ell) &= D^3E(\varphi(x))(D\varphi(x)h, D\varphi(x)k, D\varphi(x)\ell) \\ &+ D^2E(\varphi(x))(D^2\varphi(x)(h,\ell), D\varphi(x)k) \\ &+ D^2E(\varphi(x))(D\varphi(x)h, D^2\varphi(x)(k,\ell)) \\ &+ D^2E(\varphi(x))(D^2\varphi(x)(h,k), D\varphi(x)\ell) \\ &+ DE(\varphi(x))D^3\varphi(x)(h,k,\ell). \end{aligned}$$

[1] An intrinsic derivative $D^*f(x)$ of a map $f : M \to \mathbb{R}$ on a subspace σ of the tangent space T_xM is an r-linear form $\sigma^+ \to \mathbb{R}$ of $\sigma^r = \sigma \times \cdots \times \sigma$ which is defined independently of the choice of any coordinate chart.

Set $k_1 := h, k_2 := k, k_3 := \ell$ and note that $DE(0) = 0$. Then the first assertion follows from (ii) and $\varphi(0) = 0$. The second claim is a consequence of (iii) noting that $\varphi(0) = 0$, $DE(0) = 0$, and by assumption $D\varphi(0)k_j \in K$, $1 \le j \le 3$. □

Now we shall apply the preceding result to Dirichlet's integral $E : H^2(S^1, \mathbb{R}^n) \to \mathbb{R}$ defined by (1). Recall the assumption $\Gamma \in C^{\boldsymbol{r}+7}, \boldsymbol{r} \ge 3$. By Theorem 2 it follows that $M := H^2(S^1, \Gamma)$ is a $C^{\boldsymbol{r}}$-submanifold of $H^2(S^1, \mathbb{R}^n)$, and since $E : H^2(S^1, \mathbb{R}^n) \to \mathbb{R}$ is of class C^∞, it follows immediately that the restriction $E|M$ is of class $C^{\boldsymbol{r}}$. Let us simply write E instead of $E|M$, i.e. we view E as a function of class $C^r(M)$.

We wish now to calculate the intrinsic third derivative in the direction of certain specific elements of the kernel of $D^2E(X) : T_XM \times T_XM \to \mathbb{R}$, namely the forced Jacobi fields, in the case that $X \in H^2(S^1, \Gamma)$ is a minimal surface. By the results of Chapter 2 we know that $\hat{X} \in C^{\boldsymbol{r}+6,\alpha}(\overline{B}, \mathbb{R}^n)$ and therefore also $X \in C^{\boldsymbol{r}+6,\alpha}(S^1, \mathbb{R}^n)$ for all $\alpha \in (0,1)$.

Besides assuming that $\Gamma \in C^{\boldsymbol{r}+7}$ we make another standing assumption on Γ, namely that the total curvature $\int_\Gamma \kappa\, ds$ of Γ satisfies

(2) $$\int_\Gamma \kappa\, ds \le \frac{1}{3}\pi \boldsymbol{r},$$

which implies $\boldsymbol{r} \ge 6$. Then the generalized Gauss–Bonnet formula (19) of Section 2.11 implies

$$2\pi \sum_{w_j \in B} \nu(w_j) + \pi \sum_{\zeta_k \in \partial B} \nu(\zeta_k) + 2\pi \le \frac{1}{3}\pi \boldsymbol{r},$$

where $\nu(w_j)$ are the orders of the interior branch points w_j of a (branched) minimal surface $\hat{X} \in \mathcal{C}(\Gamma)$, and $\nu(\zeta_k)$ are the orders of its boundary branch points, $k = 1, \dots, q$. Suppose that $q \ge 1$. Then

(3) $$\nu(\zeta_k) \le \boldsymbol{r}/3 - 2.$$

Recall the definition of a *forced Jacobi field* of a minimal surface $\hat{X} : \overline{B} \to \mathbb{R}^3$ which we now generalize to a minimal surface $\hat{X} : \overline{B} \to \mathbb{R}^n$ with $n \ge 3$ which has the interior branch points $w_1, \dots, w_p$ and the boundary branch points $\zeta_1, \dots, \zeta_q$. The *generator* τ of a forced Jacobi field $\hat{Y}$ for $\hat{X}$ is a meromorphic function on $\overline{B}$ with poles possibly at $w = 0$ and at the branch points of $\hat{X}$ whose orders are at most $\nu(w_j)$ at $w_j \ne 0$, $\nu(0) + 1$ at $w = 0$, $\nu(\zeta_j)$ at ζ_j, and which is real on ∂B. Then the *forced Jacobi field* $\hat{Y}$ of $\hat{X}$ with the generator τ is a mapping $\hat{Y} : \overline{B} \to \mathbb{R}^n$ of the form

$$\hat{Y} = 2\beta \operatorname{Re}(iw\hat{X}_w\tau) \quad \text{with } \beta \in \mathbb{R},$$

and

$$Y = \beta X_\theta \tau|_{S^1} : S^1 \to \mathbb{R}^n$$

are its boundary values. From the regularity of $\hat{X}$ and (3) we infer as in 6.1 that certainly $Y \in H^2(S^1, \mathbb{R}^n)$, $\hat{Y}_w \in C^0(\overline{B}, \mathbb{R}^n)$, and clearly $Y \in T_X M$. The space of forced Jacobi fields of $\hat{X}$ is denoted by $J(\hat{X})$.

We shall show that the forced Jacobi fields are in the kernel of the Hessian of $E : M \to \mathbb{R}$, and we will compute the second and third derivative of E in these directions. In Chapter 5 of Vol. 3 we shall describe how the forced Jacobi fields were discovered.

Computation of D^2E and D^3E.
Let $\Omega(p) : \mathbb{R}^n \to T_p\Gamma$ be the C^{r+6}-smooth orthogonal projection of $\mathbb{R}^n$ onto the tangent space $T_p\Gamma$ for $p \in \Gamma$. We extend $\Omega(p)$ to a C^{r+6}-smooth mapping $p \mapsto \Omega(p)$ from $\mathbb{R}^n$ into $\mathfrak{L}(\mathbb{R}^n, \mathbb{R}^n)$. We then can write the first derivative of E at $X \in M = H^2(S^1, \Gamma)$ as

$$(4) \qquad DE(X) = \int_{S^1} \langle \Omega(X)\hat{X}_r, h\rangle \, d\theta, \quad \hat{X}_r = \text{radial derivative of } \hat{X}.$$

A slight generalization of Theorem 1 yields that $X \to \Omega(X)$ belongs to $C^r(M, H^2(S^1, \mathfrak{L}(\mathbb{R}^n, \mathbb{R}^n)))$, $M = H^2(S^1, \Gamma)$, if we take Theorem 2 into account. Clearly, *X is a critical point of E if and only if*

$$(5) \qquad \Omega(X)\hat{X}_r = 0.$$

$\hat{X}$ will be a solution to Plateau's problem if X is also a monotonic map from S^1 onto Γ.

The derivative of $\Omega(X)\hat{X}_r$ is given by

$$(6) \qquad h \mapsto \Omega(X)\hat{h}_r + D\Omega(X)h[\hat{X}_r],$$

and so the Hessian of E is

$$(7) \qquad D^2E(X)(h,k) = \int_{S^1} \langle \Omega(X)\hat{h}_r + D\Omega(X)h[\hat{X}_r], k\rangle \, d\theta.$$

It follows that *the kernel of* (6) *is just the kernel of the Hessian $D^2E(X)$ of E at X.*

Claim: *The forced Jacobi fields of X lie in the kernel of $D^2E(X)$.* To see this we first note that

$$(8) \qquad |X_\theta|^2 \Omega(X)m = \langle m, X_\theta\rangle X_\theta \quad \text{for } m \in \mathbb{R}^n.$$

Differentiating this in direction of a tangent vector $h \in T_X M$, $M = H^2(S^1, \Gamma)$, we obtain

$$(9) \qquad \begin{aligned} &2\langle X_\theta, h_\theta\rangle \Omega(X)[m] + |X_\theta|^2 D\Omega(X)(h)[m] \\ &= \langle m, h_\theta\rangle X_\theta + \langle m, X_\theta\rangle h_\theta. \end{aligned}$$

Thus the kernel of (6) is the kernel of

$$h \mapsto |X_\theta|^{-2}\{\langle \hat{X}_r, h_\theta\rangle X_\theta + \langle \hat{X}_r, X_\theta\rangle h_\theta - 2\langle X_\theta, h_\theta\rangle \Omega(X)\hat{X}_r\} + \Omega(X)\hat{h}_r.$$

From (5) we infer

$$\langle \hat{X}_r, X_\theta\rangle = 0 \quad \text{and} \quad \Omega(X)\hat{X}_r = 0,$$

and (8) yields

$$\Omega(X)\hat{h}_r = |X_\theta|^{-2}\langle \hat{h}_r, X_\theta\rangle X_\theta.$$

Thus h is in the kernel of (6) if and only if

$$|X_\theta|^{-2}\{\langle \hat{X}_r, h_\theta\rangle X_\theta + \langle X_\theta, \hat{h}_r\rangle X_\theta\} = 0$$

that is, if and only if

$$(10) \qquad \langle \hat{X}_r, h_\theta\rangle + \langle X_\theta, \hat{h}_r\rangle = 0,$$

since the zeros of $X_\theta(\theta)$ are isolated because of the asymptotic expansion of $\hat{X}_w$ at branch points $w_0 \in \overline{B}$.

On $S^1 = \partial B$ we have

$$iw\hat{X}_w = \frac{1}{2}(X_\theta + i\hat{X}_r), \quad iw\hat{h}_w = \frac{1}{2}(h_\theta + i\hat{h}_r),$$

implying that

$$(11) \qquad \langle \hat{X}_r, h_\theta\rangle + \langle X_\theta, \hat{h}_r\rangle = -4\,\mathrm{Im}\{w^2\langle \hat{X}_w, \hat{h}_w\rangle\}.$$

If $\hat{h}$ is a forced Jacobi field we have

$$h = \beta X_\theta \tau|_{S^1} \quad \text{and} \quad \hat{h} = 2\,\mathrm{Re}(\beta i w \hat{X}_w \tau)$$

with $\beta \in \mathbb{R}$ and τ the generator of $\hat{h}$. Since $w\hat{X}_w\tau$ is holomorphic on $\overline{B}$, it follows

$$\hat{h}_w = \beta[iw\hat{X}_w\tau]_w.$$

Hence, if $w \in \overline{B}$ is not a branch point of $\hat{X}$, we obtain

$$\hat{h}_w(w) = \beta[i\hat{X}_w(w)\tau + iw\hat{X}_{ww}(w)\tau(w) + iw\hat{X}_w(w)\tau_w(w)].$$

On the other hand, a minimal surface $\hat{X}$ satisfies

$$\langle \hat{X}_w, \hat{X}_w\rangle = 0$$

and therefore also

$$\langle \hat{X}_w(w), \hat{h}_w(w)\rangle = 0$$

if $w \in \overline{B}$ is not a branch point of $\hat{X}$, and by continuity of $\hat{h}_w$ on $\overline{B}$ it follows

$$(12) \qquad \langle \hat{X}_w, \hat{h}_w\rangle = 0 \quad \text{if } \hat{h} \in J(\hat{X}).$$

From (10), (11) and (12) we infer that for a forced Jacobi field $\hat{h}$ its boundary values h lie in the kernel of (6) and therefore in the kernel K_X of the Hessian $D^2E(X)$. This proves the claim, and we have established

Proposition 1. *If $\hat{X}$ is a minimal surface with $X \in M = H^2(S^1, \Gamma)$ then the boundary values h of any $\hat{h} \in J(\hat{X})$ lie in the kernel K_X of the Hessian $D^2E(X)$ of E at X, that is, $h \in T_X M$ and*

$$D^2E(X)(h,k) = 0 \quad \textit{for all } k \in T_X M.$$

Remark 1. We would like to point out that $D^2E(X)$ has been defined for *branched* minimal surfaces without making normal variations of $\hat{X}$.

Before we compute $D^3E(X)$ we give a geometric interpretation of

$$D^2E(X)(h,h) = \delta^2 E(X,h),$$

i.e. of the second variation of E at X in direction of $h \in T_X M$. An integration by parts yields

$$\begin{aligned} (13) \qquad \int_B \nabla\hat{h} \cdot \nabla\hat{h}\, du\, dv &= \int_{S^1} \langle \hat{h}_r, h\rangle\, d\theta - \int_B \langle \Delta\hat{h}, \hat{h}\rangle\, du\, dv \\ &= \int_{S^1} \langle \hat{h}_r, h\rangle\, d\theta \end{aligned}$$

since $\Delta\hat{h} = 0$. Away from branch points on S^1 we set

$$h = aX_\theta \quad \text{and} \quad b = \langle \hat{h}_r, X_\theta\rangle.$$

By (8) we have

$$\Omega(X)\hat{h}_r = |X_\theta|^{-2}\langle \hat{h}_r, X_\theta\rangle X_\theta,$$

and so

$$\langle h, \Omega(X)\hat{h}_r\rangle = \langle aX_\theta, bX_\theta\rangle |X_\theta|^{-2} = ab = \langle \hat{h}_r, aX_\theta\rangle = \langle \hat{h}_r, h\rangle$$

and by continuity it follows

$$\langle \hat{h}_r, h\rangle = \langle h, \Omega(X)\hat{h}_r\rangle \quad \text{on } S^1.$$

On account of (7) and (13) it follows that

$$(14) \qquad D^2E(X)(h,h) = \int_B |\nabla\hat{h}|^2\, du\, dv + \int_{S^1} \langle h, D\Omega(X)h[\hat{X}_r]\rangle\, d\theta.$$

In order to simplify the boundary term we return to (9) where we insert $m = \hat{X}_r$. Since $\langle \hat{X}_r, X_\theta\rangle = 0$ we have $\Omega(X)\hat{X}_r = 0$ on S^1, and so two terms in (9) vanish. We are left with

$$D\Omega(X)h[\hat{X}_r] = |X_\theta|^{-2}\langle \hat{X}_r, h_\theta\rangle X_\theta.$$

Since $h = aX_\theta$ (away from branch points), we have

$$h_\theta = aX_{\theta\theta} + a_\theta X_\theta$$

whence

$$\langle \hat{X}_r, h_\theta \rangle = a\langle \hat{X}_r, X_{\theta\theta} \rangle.$$

This implies

$$\begin{aligned}
&\langle h, D\Omega(X)h[\hat{X}_r] \rangle \\
&\quad = |X_\theta|^{-2} \langle aX_\theta, a\langle \hat{X}_r, X_{\theta\theta} \rangle X_\theta \rangle = a^2 \langle \hat{X}_r, X_{\theta\theta} \rangle \\
&\quad = |h|^2 |X_\theta|^{-2} \langle \hat{X}_r, X_{\theta\theta} \rangle = |h|^2 k_g,
\end{aligned}$$

where

(15) $$k_g := |X_\theta|^{-2} \langle \hat{X}_r, X_{\theta\theta} \rangle$$

is the *signed geodesic curvature of* Γ *in the minimal surface* $\hat{X}$, *i.e. the interior product of the curvature vector of* Γ *with the unit vector* $|\hat{X}_r|^{-1}\hat{X}_r$, since $|X_\theta| = |\hat{X}_r|$ on S^1.

Thus we infer from (14) the following result which was independently obtained by R. Böhme and A. Tromba:

Proposition 2. *If* $\hat{X}$ *is a minimal surface with* $X \in M = H^2(S^1, \Gamma)$ *then, for any* $h \in T_X M$, *we obtain*

(16) $$D^2E(X)(h,h) = \int_B |\nabla \hat{h}|^2 \, du \, dv + \int_{S^1} k_g |h|^2 \, d\theta,$$

where k_g *is the signed geodesic curvature* (15) *of the boundary contour* Γ *in the minimal surface* $\hat{X}$.

Now we proceed to compute the intrinsic third derivative $D^3E(X)$. Let us return to formula (9) which will be differentiated in direction of a vector $k \in T_X M$. This yields

$$\begin{aligned}
&2\langle h_\theta, k_\theta \rangle \Omega(X)m + 2\langle X_\theta, h_\theta \rangle D\Omega(X)[k]m \\
&\qquad + 2\langle X_\theta, k_\theta \rangle D\Omega(X)[h]m + |X_\theta|^2 D^2\Omega(X)(h,k)m \\
&\quad = \langle m, h_\theta \rangle k_\theta + \langle m, k_\theta \rangle h_\theta.
\end{aligned}$$

Choosing $m := \hat{X}_r$ we see that

$$\begin{aligned}
&2\langle X_\theta, h_\theta \rangle D\Omega(X)(k)[\hat{X}_r] + 2\langle X_\theta, k_\theta \rangle D\Omega(X)(h)[\hat{X}_r] \\
&\qquad + |X_\theta|^2 D^2\Omega(X)(h,k)[\hat{X}_r] = \langle \hat{X}_r, h_\theta \rangle k_\theta + \langle \hat{X}_r, k_\theta \rangle h_\theta.
\end{aligned}$$

By (7) we may write for h, k in the kernel of $D^2E(X)$ (and therefore in the kernel of (6))

(17) $$D\Omega(X)(h)[\hat{X}_r] = -\Omega(X)\hat{h}_r, \quad D\Omega(X)(k)[\hat{X}_r] = -\Omega(X)\hat{k}_r,$$

then obtaining

$$
\begin{aligned}
(18)\qquad & -2\langle X_\theta, h_\theta\rangle \Omega(X)\hat{k}_r - 2\langle X_\theta, k_\theta\rangle \Omega(X)\hat{h}_r \\
& + |X_\theta|^2 D^2\Omega(X)(h,k)[\hat{X}_r] = \langle \hat{X}_r, h_\theta\rangle k_\theta + \langle \hat{X}_r, k_\theta\rangle h_\theta .
\end{aligned}
$$

Setting in (9) $m = \hat{k}_r$ we get

$$
\begin{aligned}
(19)\qquad & 2\langle X_\theta, h_\theta\rangle \Omega(X)\hat{k}_r + |X_\theta|^2 D\Omega(X)[h]\hat{k}_r \\
& = \langle \hat{k}_r, h_\theta\rangle X_\theta + \langle \hat{k}_r, X_\theta\rangle h_\theta .
\end{aligned}
$$

Commuting h and k it follows also

$$
\begin{aligned}
(20)\qquad & 2\langle X_\theta, k_\theta\rangle \Omega(X)\hat{h}_r + |X_\theta|^2 D\Omega(X)[k]\hat{h}_r \\
& = \langle \hat{h}_r, k_\theta\rangle X_\theta + \langle \hat{h}_r, X_\theta\rangle k_\theta .
\end{aligned}
$$

Adding (19) and (20) to (18) we see that

$$
\begin{aligned}
(21)\qquad & |X_\theta|^2 D^2\Omega(X)(h,k)\hat{X}_r + |X_\theta|^2 D\Omega(X)[h]\hat{k}_r + |X_\theta|^2 D\Omega(X)[k]\hat{h}_r \\
& = \langle \hat{X}_r, h_\theta\rangle k_\theta + \langle \hat{X}_r, k_\theta\rangle h_\theta + \langle \hat{h}_r, k_\theta\rangle X_\theta + \langle \hat{h}_r, X_\theta\rangle k_\theta \\
& \quad + \langle \hat{k}_r, h_\theta\rangle X_\theta + \langle \hat{k}_r, X_\theta\rangle h_\theta .
\end{aligned}
$$

By (10) we have

$$
\langle X_\theta, \hat{h}_r\rangle = -\langle \hat{X}_r, h_\theta\rangle \quad \text{and} \quad \langle X_\theta, \hat{k}_r\rangle = -\langle \hat{X}_r, k_\theta\rangle .
$$

Therefore (21) reduces to

$$
\begin{aligned}
(22)\qquad & |X_\theta|^2 \{D^2\Omega(X)(h,k)\hat{X}_r + D\Omega(X)[h]\hat{k}_r + D\Omega(X)[k]\hat{h}_r\} \\
& = \{\langle \hat{h}_r, k_\theta\rangle + \langle \hat{k}_r, h_\theta\rangle\} X_\theta .
\end{aligned}
$$

Suppose now that h, k, ℓ lie in the space $J(\hat{X})$ of forced Jacobi fields. By (7) we have

$$
(22')\qquad D^2E(X)(h,\ell) = \int_{S^1} \langle D\Omega(X)h[\hat{X}_r] + \Omega(X)\hat{h}_r, \ell\rangle \, d\theta .
$$

Differentiating this in direction of k it follows

$$
\begin{aligned}
(23)\qquad & D^3E(X)(h,\ell,k) \\
& = \int_{S^1} \langle D^2\Omega(X)(h,k)[\hat{X}_r] + D\Omega[h]\hat{k}_r + D\Omega(X)[k]\hat{h}_r, \ell\rangle \, d\theta ,
\end{aligned}
$$

which by (22) yields

$$
\begin{aligned}
(24)\qquad & D^3E(X)(h,\ell,k) \\
& = \int_{S^1} \{\langle \hat{h}_r, k_\theta\rangle + \langle \hat{k}_r, h_\theta\rangle\} |X_\theta|^{-2} \langle X_\theta, \ell\rangle \, d\theta .
\end{aligned}
$$

Actually there are two more terms on the right-hand side of (24) which come from the derivatives ℓ' and h' of ℓ and h. We have to shows that these terms are zero if ℓ and h are forced Jacobi fields. The additional ℓ'-term is

$$\int_{S^1} \langle D\Omega(X)h[\hat{X}_r] + \Omega(X)\hat{h}_r, \ell'\rangle \, d\theta.$$

It vanishes since

$$D\Omega(X)h[\hat{X}_r] + \Omega(X)\hat{h}_r = 0,$$

as h is a forced Jacobi field.
The second additional term becomes

$$\int_{S^1} \langle h', (\widehat{\lambda X_\theta})_r - (\lambda \hat{X}_r)_\theta\rangle \, d\theta$$

if we write $\ell = \lambda X_\theta = \mathrm{Re}\{\lambda i w \hat{X}_w\}$ and integrate by parts. But ℓ is holomorphic in B. and so the Cauchy–Riemann equations yield

$$-\frac{\partial}{\partial\theta}(\widehat{\lambda X_\theta}) + \frac{\partial}{\partial r}(\widehat{\lambda X_\theta}) = 0.$$

This equation extends to the boundary $S^1 = \partial B$, and so the second additional term vanishes too.

The two expressions (23) and (24) yield the intrinsic third derivative of E at X. We synonymously write

$$\begin{aligned} \frac{\partial E}{\partial h}(X) &= DE(X)h, \\ \frac{\partial^2 E}{\partial h \partial k}(X) &= D^2E(X)(h,k), \\ \frac{\partial^3 E(X)}{\partial h \partial \ell \partial k} &= D^3E(X)(h,\ell,k). \end{aligned} \tag{25}$$

Suppose that $h, k, \ell \in J(\hat{X})$ have the generators τ, ρ, λ; we shall write τ, ρ, λ also for the boundary values $\tau|_{S^1}, \rho|_{S^1}, \lambda|_{S^1}$:

$$\begin{aligned} h(\theta) &= \tau(\theta)X_\theta(\theta), \quad \text{so} & \hat{h}(w) &= 2\mathrm{Re}(iw\tau(w)\hat{X}_w(w)), \\ k(\theta) &= \rho(\theta)X_\theta(\theta), & \hat{k}(w) &= 2\mathrm{Re}(iw\rho(w)\hat{X}_w(w)), \\ \ell(\theta) &= \lambda(\theta)X_\theta(\theta), & \hat{\ell}(w) &= 2\mathrm{Re}(iw\lambda(w)\hat{X}_w(w)). \end{aligned} \tag{26}$$

Then (24) becomes

$$D^3E(X)(h,\ell,k) = \int_{S^1} \{\langle \hat{h}_r, k_\theta\rangle + \langle \hat{k}_r, h_\theta\rangle\}\lambda(\theta)\, d\theta. \tag{27}$$

On S^1 we have $d\theta = \frac{dw}{iw}$ and

$$2w\hat{h}_w = \hat{h}_r - ih_\theta, \quad 2w\hat{k}_w = \hat{k}_r - ik_\theta$$

whence

$$\langle \hat{h}_r, k_\theta \rangle + \langle \hat{k}_r, h_\theta \rangle = -4\,\mathrm{Im}(w^2 \hat{h}_w \hat{k}_w).$$

Furthermore,

$$\begin{aligned} \hat{h}_w &= (iw\hat{X}_w\tau)_w = i(w\tau\hat{X}_{ww} + \hat{X}_w\tau + w\hat{X}_w\tau_w), \\ \hat{k}_w &= (iw\hat{X}_w\rho)_w = i(w\rho\hat{X}_{ww} + \hat{X}_w\rho + w\hat{X}_w\rho_w). \end{aligned}$$

Since $\hat{X}_w \cdot \hat{X}_w = 0$ and $\hat{X}_w \cdot \hat{X}_{ww} = 0$ it follows that

$$w^2 \hat{h}_w \hat{k}_w = -w^4 \tau\rho \hat{X}_{ww} \cdot \hat{X}_{ww}$$

and consequently

$$\langle \hat{h}_r, k_\theta \rangle + \langle \hat{k}_r, h_\theta \rangle = 4\,\mathrm{Im}(w^4 \tau\rho \hat{X}_{ww} \cdot \hat{X}_{ww}).$$

This implies

$$\begin{aligned} D^3E(X)(h, \ell, k) &= 4 \int_{S^1} \mathrm{Im}(w^4 \tau\rho \hat{X}_{ww} \cdot \hat{X}_{ww})\lambda\, d\theta \\ &= 4\,\mathrm{Im} \int_{S^1} w^4 \tau\rho\lambda \hat{X}_{ww} \cdot \hat{X}_{ww}\, d\theta \\ &= 4\,\mathrm{Im} \int_{S^1} w^4 \tau\rho\lambda \hat{X}_{ww} \cdot \hat{X}_{ww} \frac{dw}{iw}, \end{aligned}$$

and we arrive at

$$\begin{aligned} D^3E(X)(h, \ell, k) &= -4\,\mathrm{Re} \int_{S^1} w^3 \tau\rho\lambda \hat{X}_{ww} \cdot \hat{X}_{ww}\, dw \\ &= 4 \int_{S^1} \mathrm{Im}(w^4 \tau\rho\lambda \hat{X}_{ww} \cdot \hat{X}_{ww})\, d\theta. \end{aligned} \tag{28}$$

It follows from (23) that the right-hand side of (28) is the integral of a continuous function. If we wish to apply the residue theorem to evaluate the integral in (28) we have to get a better grip to the integrand. To this end we impose an *additional standing assumption: $n = 3$, i.e. we consider boundary contours only in $\mathbb{R}^3$.*

First we wish to understand what the generators τ of forced Jacobi fields for a minimal surface $\hat{X}$ with a boundary branch point $w_0 \in S^1$ are. By means of a rotation we can move w_0 to the point $w = 1$. Thus we make the following further standing **assumption:**

$\hat{X} \in \mathcal{C}(\Gamma)$ is a minimal surface in the unit disk B with the boundary branch point $w = 1$ of order n, and the boundary contour $\Gamma \in C^2$ has a total curvature $\kappa(\Gamma) := \int_\Gamma \kappa(s)\, ds$ satisfying $3\kappa(\Gamma) \le \tau\boldsymbol{r}$. It is also assumed that

$\Gamma \in C^{r+7}, r \geq 2$, which implies $\hat{X} \in C^{r+6,\beta}(\overline{B}, \mathbb{R}^3)$, $0 < \beta < 1$, and $n \leq r/3 - 2$.

It is easy to verify that

$$\tau(w) := \beta \left(i \frac{w+1}{w-1} \right)^{\ell}, \quad \beta \in \mathbb{R}, \tag{29}$$

is a meromorphic function on $\overline{B}$ with a pole of order ℓ at $w = 1$ such that $\tau(w) \in \mathbb{R}$ for $w \in S^1 \setminus \{1\}$. If $\ell \leq n$ then $\hat{X}_w(w)\tau(w)$ is holomorphic in B and at least continuous on $\overline{B}$ since we have the asymptotic expansion

$$\begin{aligned} \hat{X}_w(w) &= a(w-1)^n + o(|w-1|^n) \quad \text{as } w \to 1, \ w \in \overline{B} \setminus \{1\} \\ &\text{with } a \in \mathbb{C}^3, \ a \neq 0, \ \text{and } a \cdot a = 0. \end{aligned} \tag{30}$$

Thus τ generates a forced Jacobi field for $\hat{X}$. Consider the conformal mapping $\varphi : \overline{B} \setminus \{-1\} \to \overline{\mathcal{H}}$, defined by

$$w \mapsto z = \varphi(w) := -i \frac{w-1}{w+1}, \quad w \in \overline{B} \setminus \{-1\}, \tag{31}$$

which maps $B = \{w \in \mathbb{C} : |w| < 1\}$ onto the upper halfplane

$$\mathcal{H} := \{z \in \mathbb{C} : \operatorname{Im} z > 0\}$$

and takes $S^1 \setminus \{-1\}$ onto the real line $\mathbb{R}$ such that $\varphi(1) = 0, \varphi(i) = 1, \varphi(-1) = \infty$. The inverse $\psi := \varphi^{-1}$ is given by

$$z \mapsto w = \psi(z) := \frac{1+iz}{1-iz}. \tag{32}$$

We write $z = x + iy$ with $x = \operatorname{Re} z$ and $y = \operatorname{Im} z$, while $w = u + iv, u = \operatorname{Re} w, v = \operatorname{Im} w$. From (31) we infer

$$\frac{1}{z} = i \frac{w+1}{w-1}$$

and so

$$\sigma := \tau \circ \psi = \frac{\beta}{z^{\ell}}. \tag{33}$$

Transforming the minimal surface $\hat{X}(w)$ to the new parameter z, we obtain

$$\hat{Y}(z) := \hat{X}(\psi(z)) \tag{34}$$

which has the branch point $z = 0$ on $\mathbb{R} = \partial\mathcal{H}$ with the asymptotic expansion

$$\begin{aligned} &\hat{Y}_z(z) = bz^n + o(|z|^n) \quad \text{as } z \to 0, \ z \in \overline{\mathcal{H}} \setminus \{0\} \\ &b \in \mathbb{C}^3 \setminus \{0\}, \quad b \cdot b = 0. \end{aligned}$$

Choosing a suitable coordinate system in $\mathbb{R}^3$ we may assume that $\hat{Y}_z(z)$ can be written in the normal form

$$\hat{Y}_z(z) = \tilde{A}_1 z^n + o(z^n) \tag{35}$$

with $\tilde{A}_1 = (a_1 + ib_1)$; $a_1, b_1 \in \mathbb{R}^3$, $|a_1|^2 = |b_1|^2 \neq 0$; $a_1 \cdot b_1 = 0$, $a_1 = (n+1)\alpha e_1$, $e_1 = (1,0,0)$, $\alpha > 0$, where a_1, b_1 span the tangent space to $\hat{X}$ at $X(1)$. Let us recall that the order of any boundary branch point is even; thus we can set

$$n = 2\nu \quad \text{with } \nu \in \mathbb{N}. \tag{36}$$

Now we wish to write $\hat{Y}_z$ in the more specific form

$$\hat{Y}_z(z) = (A_1 z^n + \cdots + A_{m-n+1} z^m + O(|z|^{m+1}), \quad R_m z^m + O(|z|^{m+1})) \tag{37}$$

with

$$R_m \neq 0. \tag{38}$$

By Taylor's theorem and (35) we can achieve (37) for any $m \in \mathbb{N}$ with $m > n$ and such that $\hat{Y} \in C^{m+2}(\overline{\mathcal{H}}, \mathbb{R}^3)$.

However, it is not at all a priori obvious that one can achieve also (38). This fact is ensured by the following

Proposition 3. *Suppose that $\hat{Y} \in C^{3n+6}(\overline{\mathcal{H}}, \mathbb{R}^3)$ and that both the torsion $\boldsymbol{\tau}$ and the curvature $\boldsymbol{\kappa}$ of Γ are nonzero. Then there is an $m \in \mathbb{N}$ with $n+1 < m+1 \leq 3(n+1)$ such that*

$$\hat{Y}_z^3(z) = R_m z^m + O(|z|^{m+1}) \quad \textit{for } |z| \ll 1 \textit{ and } R_m \neq 0. \tag{39}$$

Proof. Otherwise we have

$$\hat{Y}_z^3(z) = O(|z|^{3n+3}). \tag{40}$$

Let $\gamma(s) = (\gamma_1(s), \gamma_2(s), \gamma_3(s))$ be the local representation of Γ with respect to its arc-length parameter s such that $\gamma(0) = \hat{Y}(0)$ and $\gamma'(0) = e_1$. By (35) and (36) we have

$$\hat{Y}_x(x,0) = (n+1)\alpha e_1 x^n + O(x^{n+1}), \quad n = 2\nu,$$

and so s and x are related by $s = \sigma(x)$ with

$$\sigma'(x) = |Y_x(x)| = [(n+1)\alpha x^n + O(x^{n+1})],$$

whence

$$\sigma(x) = \alpha x^{n+1} + O(x^{n+2}) \quad \text{as } x \to 0. \tag{41}$$

Then $Y(x) = \gamma(\sigma(x))$ for $|x| \ll 1$, and therefore the third component Y^3 of Y is given by

$$Y^3(x) = \gamma_3(\sigma(x)) = \gamma_3(\alpha x^{n+1} + O(x^{n+2})) \quad \text{for } x \to 0.$$

Because of (40) we have $Y_x^3(x) = O(x^{3n+3})$ as $x \to 0$, which implies

$$Y^3(x) = O(x^{3n+4}) \quad \text{as } x \to 0. \tag{42}$$

On the other hand

$$\gamma(s) = \gamma'(0)s + O(s^2) \quad \text{as } s \to 0.$$

Consequently

$$Y^3(x) = \gamma_3'(0)\alpha x^{n+1} + O(x^{n+2}) \quad \text{as } x \to 0.$$

On account of (42) and $\alpha > 0$ it follows $\gamma_3'(0) = 0$. Thus we can write

$$\gamma_3(s) = \frac{1}{2}\gamma_3''(0)s^2 + O(s^3) \quad \text{as } s \to 0,$$

which implies

$$Y^3(x) = \frac{1}{2}\gamma_3''(0)\alpha^2 x^{2n+2} + O(x^{2n+3}) \quad \text{as } x \to 0.$$

By (42) and $\alpha > 0$ we obtain $\gamma_3''(0) = 0$, and we have

$$\gamma_3(s) = \frac{1}{6}\gamma_3'''(0)s^3 + O(s^4) \quad \text{as } s \to 0.$$

Hence,

$$Y^3(x) = \frac{1}{6}\gamma_3'''(0)\alpha^3 x^{3n+3} + O(x^{3n+4}) \quad \text{as } x \to 0,$$

and then $\gamma_3'''(0) = 0$ on account of (42) and $\alpha > 0$. Thus we have found

$$\gamma_3'(0) = 0, \quad \gamma_3''(0) = 0, \quad \gamma_3'''(0) = 0,$$

and so the three vectors $\gamma'(0), \gamma''(0), \gamma'''(0)$ are linearly dependent. This will contradict our assumption $\kappa(s) \neq 0$ and $\tau(s) \neq 0$. To see this we introduce the Frenet triple $\boldsymbol{T}(s), \boldsymbol{N}(s), \boldsymbol{B}(s)$ of the curve Γ satisfying $\boldsymbol{T} = \gamma', \boldsymbol{T}' = \gamma'', \boldsymbol{T}'' = \gamma'''$, and

$$\begin{aligned}
\boldsymbol{T}' &= \qquad\quad \boldsymbol{\kappa N} \\
\boldsymbol{N}' &= -\boldsymbol{\kappa T} \qquad\qquad +\boldsymbol{\tau B} \\
\boldsymbol{B}' &= \qquad\quad -\boldsymbol{\tau N}.
\end{aligned}$$

Then $\boldsymbol{T}_3(0) = 0, \boldsymbol{T}_3'(0) = 0, \boldsymbol{T}_3''(0) = 0$, and from $\boldsymbol{T}' = \boldsymbol{\kappa N}$ and $\boldsymbol{\kappa} \neq 0$ it follows that $\boldsymbol{N}_3(0) = 0$. Since

$$\boldsymbol{N}' = \left(\frac{1}{\boldsymbol{\kappa}}\right)' \boldsymbol{T}' + \frac{1}{\boldsymbol{\kappa}}\boldsymbol{T}''$$

we obtain $\boldsymbol{N}_3'(0) = 0$ whence $\boldsymbol{\tau}(0)\boldsymbol{B}_3(0) = 0$. Because of $\boldsymbol{\tau} \neq 0$ it follows that $\boldsymbol{B}_3(0) = 0$, and so $\boldsymbol{T}(0), \boldsymbol{N}(0), \boldsymbol{B}(0)$ are linearly dependent. This is a contradiction since $(\boldsymbol{T}, \boldsymbol{N}, \boldsymbol{B})$ is an orthonormal frame, hence the assumption (40) is impossible. □

Remark 2. Note that $n \le \boldsymbol{r}/3 - 2$ implies $3n + 6 \le \boldsymbol{r} < \boldsymbol{r} + 7$. Thus the assumption $\hat{Y} \in C^{3n+6}(\overline{\mathcal{H}}, \mathbb{R}^3)$ is certainly satisfied if we assume $3\kappa(\Gamma) \le \pi \boldsymbol{r}$ and $\Gamma \in C^{\boldsymbol{r}+7}$. Thus we have a lower bound on $\boldsymbol{r}$ and upper bounds on n and m. We call the number m in (39) with $n < m < 3n + 3$ the *index* of the boundary branch point $z = 0$ of $\hat{Y}$, or of the boundary branch point $w = 1$ of $\hat{X}$.

Assumption. *In what follows we assume that the assumptions and therefore also the conclusions of Proposition 3 are satisfied.*

Proposition 4. *If $m + 1 \not\equiv 0 \bmod (n + 1)$ (i.e. if $z = 0$ is not an exceptional branch point of $\hat{Y}$) then the coefficient R_m in (39) satisfies*

$$\operatorname{Re} R_m = 0, \tag{43}$$

i.e. R_m is purely imaginary, and therefore

$$R_m^2 < 0 \tag{44}$$

since $R_m \neq 0$. If we write (39) in the form

$$Y_z^3(z) = R_m z^m + R_{m+1} z^{m+1} + R_{m+2} z^{m+2} + o(|z|^{m+2}) \quad \textit{for } |z| \ll 1 \tag{45}$$

and if $2m - 2 < 3n$, then we in addition obtain that

$$\operatorname{Re} R_{m+1} = 0 \quad \textit{and, if } n > 2, \textit{ also} \quad \operatorname{Re} R_{m+2} = 0. \tag{46}$$

Finally, independent of any assumption on m, we have

$$A_j = \mu_j A_1, \quad j = 1, \dots, \min\{n + 1, 2m - 2n\}, \ \textit{with } \mu_j \in \mathbb{R} \tag{47}$$

for the coefficients A_j in the expansion (37).

Remark 3. The relations (47) are in some sense a strengthening of the equations

$$A_j = \lambda_j A_1, \quad j = 1, \dots, 2m - 2n, \text{ with } \lambda_j \in \mathbb{C}$$

which hold at an interior branch point $w = 0$ of a minimal surface $\hat{X}$ in normal form.

Proof of Proposition 4. (i) From (45) we infer

$$Y^3(x) = \operatorname{Re}\left(\frac{R_m}{m+1} x^{m+1} + \frac{R_{m+1}}{m+2} x^{m+2} + \frac{R_{m+2}}{m+3} x^{m+3} + o(x^{m+3}) \right) \tag{48}$$

for $x \to 0$.

On the other hand,

$$Y^3(x) = \gamma_3(\alpha x^{n+1} + o(x^{n+1}))$$

and $\gamma(0) = 0, \gamma'(0) = e_3$ whence also $\gamma_3(0) = \gamma_3'(0) = 0$. As pointed out before it is then impossible that both $\gamma_3''(0) = 0$ and $\gamma_3'''(0) = 0$ because this would imply that $\boldsymbol{T}(0), \boldsymbol{N}(0), \boldsymbol{B}(0)$ are linearly dependent. Thus we obtain

$$\gamma_3(s) = \frac{1}{k!}\gamma^{(k)}(0)s^k + O(s^{k+1}) \quad \text{as } s \to 0, \gamma^{(k)}(0) \neq 0,$$

for $k = 2$ or $k = 3$. Therefore

$$Y^3(x) = \frac{1}{k!}\gamma_3^{(k)}(0)\alpha^k x^{k(n+1)} + o(x^{k(n+1)}) \quad \text{as } x \to 0. \tag{49}$$

Comparing (48) and (49) it follows that $\operatorname{Re} R_m \neq 0$ implies $m + 1 = k(n+1)$ for $k = 2$ or $k = 3$, which is excluded by assumption. Thus $\operatorname{Re} R_m = 0$, and we have

$$\begin{aligned} Y^3(x) &= \operatorname{Re}\left(\frac{R_{m+1}}{m+2}x^{m+2} + \frac{R_{m+2}}{m+3}x^{m+3} + o(x^{m+3})\right) \\ &= \frac{1}{k!}\gamma^k(0)\alpha^k x^{k(n+1)} + o(x^{k(n+1)}) \quad \text{as } x \to 0. \end{aligned} \tag{50}$$

Suppose now that $2m - 2 < 3n$, which is equivalent to

$$2m \leq 3n \tag{51}$$

since n is even, and so

$$m + 2 < m + 3 \leq \frac{3}{2}n + 3 < 3(n+1).$$

Thus, for $k = 3$, equation (50) can only hold if

$$\operatorname{Re} R_{m+1} = 0 \quad \text{and} \quad \operatorname{Re} R_{m+2} = 0.$$

Furthermore, (51) yields also

$$m + 2 < m + 3 \leq \frac{3}{2}n + 3 = (2n+2) + \left(1 - \frac{n}{2}\right)\begin{cases} = 2n+2 \\ < 2n+2 \end{cases} \text{and} \quad \begin{matrix} n = 2 \\ n > 2. \end{matrix}$$

Hence it follows in this case that always $\operatorname{Re} R_{m+1} = 0$ while $\operatorname{Re} R_{m+2} = 0$ holds for $n > 2$.

(ii) From $Y_x(x) = 2\operatorname{Re}\hat{Y}_z(x, 0)$, (50) and (37) it follows that

$$Y_x(x) = 2\operatorname{Re}(A_1 x^n + \cdots + A_{n+1}x^{2n} + o(x^{2n}), o(x^{2n}))$$

whence

$$Y(x) = 2\operatorname{Re}\left(\frac{A_1}{n+1}x^{n+1} + \cdots + \frac{A_{n+1}}{2n+1}x^{2n+1} + o(x^{2n+1}), o(x^{2n+1})\right).$$

Furthermore,

$$\gamma(s) = e_1 s + O(s^2) \quad \text{as } s \to 0$$

and

$$\sigma(x) = b_1 x^{n+1} + \cdots + b_{n+1} x^{2n+1} + o(x^{2n+1}) \quad \text{as } x \to 0$$

with $b_1, \ldots, b_{n+1} \in \mathbb{R}$, $\alpha e_1 = b_1 e_1 = \frac{2}{n+1} \operatorname{Re} A_1$. Then

$$Y(x) = \gamma(\sigma(x)) = (b_1 x^{n+1} + \cdots + b_{n+1} x^{2n+1}) e_1 + O(x^{2n+2}).$$

Comparing the coefficients we get

$$2 \operatorname{Re} A_j = (n+j) b_j e_1 \quad \text{with } \alpha = b_1 > 0 \text{ for } 1 \le j \le n+1.$$

Then $\operatorname{Re} A_j = \frac{(n+j)b_j}{(n+1)\alpha} \operatorname{Re} A_1$, and so

$$\operatorname{Re} A_j = \mu_j \operatorname{Re} A_1 \quad \text{for } j = 2, \ldots, n+1$$

with

$$\mu_j := \frac{n+j}{n+1} \frac{b_j}{\alpha}, \quad 2 \le j \le n+1.$$

Set $A_j := a_j + i b_j$; $a_j := \operatorname{Re} A_j$, $b_j := \operatorname{Im} A_j \in \mathbb{R}^n$. We know from 6.1 that $A_j = \lambda_j A_1$ for $j = 1, \ldots, 2m-2n$ with $\lambda_j \in \mathbb{C}$ hence

$$a_j = (\operatorname{Re} \lambda_j) a_1 - (\operatorname{Im} \lambda_j) b_1 \quad \text{for } 2 \le j \le 2m-2n$$

and

$$a_j = \mu_j a_1 \quad \text{for } 2 \le j \le n+1.$$

From $|\hat{Y}_x| = |\hat{Y}_y|$ it follows that $|b_1| = |a_1| = \frac{n+1}{2}\alpha > 0$, and $\hat{Y}_x \cdot \hat{Y}_y = 0$ yields $a_1 \cdot b_1 = 0$; thus we obtain $\operatorname{Im} \lambda_j = 0$ for $j = 2, \ldots, n+1$ whence $\lambda_j = \mu_j \in \mathbb{R}$ and $A_j = \mu_j A_1$ for $1 \le j \le \min\{n+1, 2m-2n\}$. □

Let us now return to formula (28) for $D^3 E(X)(h, k, \ell)$ in the direction of forced Jacobi fields (with the boundary values) h, k, ℓ; note that (28) is symmetric in h, k, ℓ. We already know that (28) is the integral of a continuous function; but we need to understand (28) at a level where we can apply the residue theorem. To this end we consider the conformal mapping (31) defined by

$$w \mapsto z = \varphi(w) := -i \frac{w-1}{w+1}, \quad w \in \overline{B} \setminus \{-1\},$$

which has the derivative

$$\varphi'(w) = \frac{-2i}{(w+1)^2}. \tag{52}$$

Using the inverse

$$z \mapsto w = \psi(z) := \frac{1+iz}{1-iz}$$

we obtain

$$\varphi'(\psi(z)) = \frac{-i}{2}(1 - iz)^2, \tag{53}$$

or sloppily

$$\frac{dz}{dw} = -\frac{i}{2}(1 - iz)^2.$$

From (34) we get $\hat{X}(w) = \hat{Y}(\varphi(w))$, whence

$$\hat{X}_{ww} = \hat{Y}_{zz}(\varphi)(\varphi')^2 + \hat{Y}_z(\varphi)\varphi''.$$

From $\hat{Y}_z \cdot \hat{Y}_z = 0$ it follows $\hat{Y}_z \cdot \hat{Y}_{zz} = 0$, and then

$$\hat{X}_{ww} \cdot \hat{X}_{ww} = \hat{Y}_{zz}(\varphi) \cdot \hat{Y}_{zz}(\varphi)(\varphi')^4, \tag{54}$$

which we sloppily write

$$\hat{X}_{ww} \cdot \hat{X}_{ww} = \hat{Y}_{zz} \cdot \hat{Y}_{zz} \left(\frac{dz}{dw}\right)^4.$$

Lemma 3. *Assuming* $2m - 2 < 3n$ *(i.e.* $2m \leq 3n$*) we obtain the Taylor expansion*

$$(\hat{Y}_{zz} \cdot \hat{Y}_{zz})(z) = \sum_{j=0}^{s} Q_j z^{2m-2+j} + R(z) \tag{55}$$

with $s := (3n - 1) - (2m - 2) = (3n - 2m) + 1 \geq 1$, $R(z) = O(z^{3n})$, *where* $Q_0 := (m - n)^2 R_m^2 < 0$ *and* $\operatorname{Im} Q_j = 0$ *for* $0 \leq j \leq s$.

Proof. From $2m - 2 < 3n$ we infer $2m \leq 3n$ since n is even. Thus $s \geq 1$ and $2m - 2n + 1 \leq n + 1$. Consider the Taylor expansion

$$\hat{Y}_z(z) = (A_1 z^n + A_2 z^{n+1} + \cdots, R_m z^m + R_{m+1} z^{m+1} + \cdots),$$

where "$+\cdots$" indicates further z-powers plus a remainder term. As for interior branch points we have

$$A_1 \cdot A_{2m-2n+1} = -R_m^2/2 \tag{56}$$

and

$$A_2 \cdot A_{2m-2n+1} + A_1 \cdot A_{2m-2n+2} = -R_m R_{m+1}. \tag{57}$$

By (44) we have $R_m^2 < 0$ whence $A_1 \cdot A_{2m-2n+1} \in \mathbb{R}$. Since $2 \leq 2m - 2n \leq n$ it follows $A_2 = \mu_2 A_1$ with $\mu_2 \in \mathbb{R}$ on account of (47). Then (56) implies $A_2 \cdot A_{2m-2n+1} \in \mathbb{R}$, and furthermore $R_m R_{m+1} \in \mathbb{R}$ in virtue of (44) and (46). Then (57) yields $A_1 \cdot A_{2m-2n+2} \in \mathbb{R}$, and we arrive at

$$\hat{Y}_{zz}(z) \cdot \hat{Y}_{zz}(z) = Q_0 z^{2m-2} + Q_1 z^{2m-1} + \cdots$$

with $Q_0 = (m-n)^2 R_m^2$, (see Section 6.1), and $Q_0 < 0$ as well as $Q_1 \in \mathbb{R}$, since Q_1 is a real linear combination of $A_1 \cdot A_{2m-2n+2}$, $A_2 \cdot A_{2m-2n+1}$, and $R_m R_{m+1}$. Suppose now that $s = 3n-2m+1 > 1$. In order to show $\operatorname{Im} Q_j = 0$ for $2 \leq j \leq s$, we note that by (54)

$$\tau\rho\lambda w^4 \hat{X}_{ww} \cdot \hat{X}_{ww} = \tau\rho\lambda \hat{Y}_{zz} \cdot \hat{Y}_{zz} \left(w \frac{dz}{dw} \right)^4 ,$$

where τ, ρ, λ are generators of forced Jacobi fields with the pole $w = 1$. Furthermore, by (52),

$$w \frac{dz}{dw} = \frac{-2iw}{(w+1)^2} = \frac{1+z^2}{2i}. \tag{58}$$

Thus

$$\operatorname{Im}(\tau\rho\lambda w^4 \hat{X}_{ww} \cdot \hat{X}_{ww}) = \frac{1}{16} \operatorname{Im}[\tau\rho\lambda(1+z^2)^4 \hat{Y}_{zz} \cdot \hat{Y}_{zz}]. \tag{59}$$

By (28) the left-hand side of (59) is a continuous function on S^1, and thus the right-hand side must be continuous in a neighbourhood of 0 in $\overline{\mathcal{H}}$ for all generators τ, ρ, λ of forced Jacobi fields $\hat{h}, \hat{k}, \hat{l}$ with poles at $w = 1$.

Suppose now that not all Q_j with $2 \leq j \leq s$ are real, $s = (3n-1)-(2m-2)$, and let J be the smallest of the indices $j \in \{2, \ldots, s\}$ with the property that $\operatorname{Im} Q_j \neq 0$. Then we choose λ, ρ, τ such that the sum of their pole orders at $w = 1$ equals $(J+1)+(2m-2) \leq 3n$. Transforming λ, ρ, τ from w to z it follows for $z = x \in \mathbb{R} = \partial\mathcal{H}$ that

$$\begin{aligned} &\operatorname{Im}[\tau\rho\lambda(1+z^2)^4 \hat{Y}_{zz} \cdot \hat{Y}_{zz}]\Big|_{z=x\in\mathbb{R}} \\ &\quad = (1+x^2)^4 \beta_1 (\operatorname{Im} Q_J) \frac{1}{x} + \langle \text{terms continuous in } x \rangle, \end{aligned} \tag{60}$$

$\beta_1 \in \mathbb{R} \setminus \{0\}$. This is clearly not a continuous function unless $\operatorname{Im} Q_J = 0$, a contradiction, therefore no such J exists. □

Now we want to evaluate the integral in (28) by applying the residue theorem. To this and we state

Proposition 5. *Let τ be given by (29), and consider the function*

$$f(w) := \tau(w)^4 w^4 \hat{X}_{ww}(w) \cdot \hat{X}_{ww}(x), \quad w \in \overline{B}, \tag{61}$$

which has a continuous imaginary part on $S^1 = \partial B$. Then there is a meromorphic function $g(w)$ on $\overline{B}$ with a pole only at $w = 1$ such that

(i) $\operatorname{Im}[f(w) - g(w)] = 0$ *for* $w \in S^1 = \partial B$;

(ii) $f - g$ *is continous on* $\overline{B}$.

Proof. Setting $w = \psi(z) = (1+iz)/(1-iz)$ we obtain

$$f(\psi(z)) = \frac{1}{16}\tau(\psi(z))^3(1+z^2)^4\hat{Y}_{zz}(z)\cdot\hat{Y}_{zz}(z).$$

By (55) of Lemma 3 we see that, in a neighbourhood of $z=0$ in $\mathcal{H}$, we can write the right-hand side as

$$\sum_{j=0}^{s}\sum_{\ell_j}\tilde{\beta}_j\tilde{Q}_j z^{-l_j} + G(z)$$

with $\tilde{\beta}_j \in \mathbb{R}, \tilde{Q}_j \in \mathbb{R}$, $0 < l_j \le (3n-1)-(2m-2) = s$, and a continuous term $G(z)$. Set

$$\tilde{g}(z) := \sum_{j=0}^{s}\sum_{l_j=1}^{s}\tilde{\beta}_j\tilde{Q}_j z^{-l_j} \quad \text{for } z \in \overline{\mathcal{H}}\setminus\{0\}$$

and

$$g(w) := \tilde{g}(\varphi(w)) = \sum_{j=0}^{s}\sum_{l_j=1}^{s}\tilde{\beta}_j\tilde{Q}_j\left(i\frac{w+1}{w-1}\right)^{l_j}.$$

Clearly f and g satisfy (i) and (ii). □

Corollary 1. *We have*

$$\int_{S^1}[f(w)-g(w)]\,d\theta = -2\pi\,\mathrm{res}_{w=0}\frac{g(w)}{w}. \tag{62}$$

Proof. For $w = e^{i\theta} \in S^1$ we have $d\theta = dw/(iw)$, whence

$$\begin{aligned}\int_{S^1}[f(w)-g(w)]\,d\theta &= \int_{S^1}[f(w)-g(w)]\frac{dw}{iw}\\ &= 2\pi\,\mathrm{res}_{w=0}\left\{\frac{f(w)-g(w)}{w}\right\}\\ &= -2\pi\,\mathrm{res}_{w=0}\left\{\frac{g(w)}{w}\right\}\end{aligned}$$

since $f(w)/w$ is holomorphic at $w=0$. □

Since $\mathrm{Im}\, g = 0$ on S^1, we obtain

Corollary 2. *We have*

$$\mathrm{Im}\int_{S^1} f(w)\,d\theta = 2\pi\,\mathrm{Im}\,\mathrm{res}_{w=0}\left\{\frac{g(w)}{w}\right\}. \tag{63}$$

Furthermore we have

$$\begin{aligned}-4\mathrm{Re}\{w^3\tau^3\hat{X}_{ww}\cdot\hat{X}_{ww}\,dw\} &= (-4)\mathrm{Re}\{iw^4\tau^3\hat{X}_{ww}\cdot\hat{X}_{ww}\}\,d\theta\\ &= 4\mathrm{Im}\{w^4\tau^3\hat{X}_{ww}\cdot\hat{X}_{ww}\}\,d\theta = 4\,\mathrm{Im}\,f(w)\,d\theta.\end{aligned}$$

Then (28) and Corollary 2 imply

$$D^3E(X)(h,h,h) = -8\pi\,\mathrm{Im}\,\mathrm{res}_{w=0}\left\{\frac{g(w)}{w}\right\}. \tag{64}$$

Remark 4. We note the following slight, but very useful generalization of the three preceding results. Namely, if $\hat{X}$ has other boundary branch points than $w = 1$ we are allowed to change τ by an additive term having poles of first order at these branch points. Then Proposition 5 as well as Corollaries 1 and 2 also hold for the new f defined by (61) and the modified τ. This observation is used in order to ensure that the forced Jacobi field $\hat{h}$ generated by τ produces a variation $\hat{Z}(t)$, $|t| \ll 1$, of $\hat{X}$ which is monotonic on $\partial B = S^1$.

Now we turn to evaluation of $D^3E(X)(h,h,h)$ using formula (64). We distinguish three possible cases: There is an $l \in N$ such that

(i) $2m - 1 = 3l$; then l is odd;

(ii) $2m - 2 = 3l$; in this case l is even;

(iii) $2m = 3l$; here l is again even.

Since $2m \le 3n$ it follows $l < n$ for (i) and (ii), whereas $l \le n$ in case (iii).

Case (i). Choose τ as

$$\begin{aligned}\tau &:= \beta\tau_1 + \epsilon\tau^* \quad \text{and} \quad \beta > 0,\ \epsilon > 0, \quad \text{and}\\ \tau_1 &= \left(i\frac{w+1}{w-1}\right)^l = \frac{1}{z^l}, \quad w \in \overline{B}\setminus\{1\},\end{aligned} \tag{65}$$

$w = \psi(z), w \in \overline{B}\setminus\{-1\}, z \in \overline{\mathcal{H}}\setminus\{0\}$. We will choose τ^* as a meromorphic function that has poles of order 1 at the boundary branch points different from $w = 1$ or $z = 0$ respectively. Then close to $w = 1$ or $z = 0$ respectively we have

$$\begin{aligned}\tau^3w^4\hat{X}_{ww}\cdot\hat{X}_{ww} &= \frac{1}{16}\tau^3(1+z^2)^4\hat{Y}_{zz}\cdot\hat{Y}_{zz}\\ &\overset{(55)}{=} \frac{\beta^3}{16}(m-n)^2R_m^2\frac{1}{z} + G(z) + O(\epsilon)\end{aligned}$$

with a continuous $G(z)$.

Choose

$$g(w) = \frac{\beta^3}{16}(m-n)^2R_m^2\left(i\frac{w+1}{w-1}\right)$$

and let $\hat{h}(w) = \text{Re}(iw\hat{X}_w(w)\tau(w))$ be the forced Jacobi field generated by $\tau, h := \hat{h}|_{S^1}$. Then by Proposition 5 and the Corollaries 1, 2 we obtain

$$\text{(66)} \quad D^3E(X)(h,h,h) = -\frac{8\pi}{16}\beta^3(m-n)^2R_m^2 \,\text{Im}\left\{\text{res}_{w=1}\frac{i}{w}\left(\frac{w+1}{w-1}\right)\right\}$$
$$= \frac{1}{2}\pi\beta^3(m-n)^2R_m^2 + O(\epsilon).$$

Since $R_m^2 < 0$ this yields for $0 < \epsilon \ll 1$ that

$$D^3E(X)(h,h,h) < 0.$$

Case (ii). Here we have $3l = 2m - 2 < 3n$ whence $l < n$. Since both l and n are even we obtain $l+1 < n$ whence $n > 2$. Moreover, $2m-1 = 2(l+1)+(l-1)$. Set

$$\text{(67)} \qquad \tau := \epsilon\tau_1 + \beta\tau_2 + \epsilon^3\tau^*, \quad \beta > 0,\ \epsilon > 0,$$
$$\tau_1 := \left(i\frac{w+1}{w-1}\right)^{l+1}, \ \tau_2 := \left(i\frac{w+1}{w-1}\right)^{l-1}, \ \tau^* \text{ as in Case 1.}$$

Note also that both $l+1$ and $l-1$ are odd. We then have that

$$\tau^3 = \beta^3\tau_2^3 + 3\beta^2\tau_2^2\tau_1\epsilon + 3\beta\epsilon^2\tau_1^2\tau_2 + O(\epsilon^3)$$
$$= \beta^3 z^{-2m+5} + 3\beta^2 z^{-2m+3} + 3\beta\epsilon^2 z^{-2m+1} + O(\epsilon^3)$$

for z close to zero, but this does not add a contribution to (64).

By the same procedure as in Case 1 we find for $\hat{h} = \text{Re}(iw\hat{X}_w\tau)$ that

$$\text{(69)} \qquad D^3E(X)(h,h,h) = \frac{3}{2}\pi\epsilon^2\beta(m-n)^2R_m^2 + O(\epsilon^3),$$

which implies

$$D^3E(X)(h,h,h) < 0 \quad \text{for } 0 < \epsilon \ll 1.$$

Case (iii). Now we have $2m = 3l, l =$ even. We have two subcases.

(a) If $l = n$ we write $2m - 1 = 2l + (l-1)$ and set

$$\text{(70)} \qquad \tau_1 := \left(i\frac{w+1}{w-1}\right)^{l-1}, \quad \tau_2 := \left(i\frac{w+1}{w-1}\right)^{l},$$
$$\tau := \beta\tau_1 + \epsilon\tau_2 + \epsilon^3\tau^*, \quad \beta > 0,\ \epsilon^3 > 0.$$

(b) If $l < n$ we write $2m - 1 = 2(l-1) + (l+1)$ and set

$$\text{(71)} \qquad \tau_1 := \left(i\frac{w+1}{w-1}\right)^{l+1}, \quad \tau_2 := \left(i\frac{w+1}{w-1}\right)^{l-1},$$
$$\tau := \epsilon\tau_1 + \beta\tau_2 + \epsilon^3\tau^*.$$

Then our now established procedure yields

(72)
$$D^3E(X)(h,h,h) = \begin{cases} \frac{1}{2}3\pi\beta\epsilon^2(m-n)^2R_m^2 + O(\epsilon^3) & \text{in Subcase (a)}, \\ \frac{1}{2}3\pi\beta^2\epsilon(m-n)^2R_m^2 + O(\epsilon^2) & \text{in Subcase (b)}. \end{cases}$$

This again implies $D^3E(X)(h,h,h) < 0$ for $0 < \epsilon \ll 1$ and $\hat{h} = \text{Re}(iw\hat{X}_w\tau)$.

Remark 5. The choice of τ^* has to be carried out in such a way that the variation $Z(t)$ of X produced by $\hat{h} = \text{Re}(iw\hat{X}_w\tau)$ furnishes a monotonic mapping of $\partial B = S^1$ onto the boundary contour Γ. The details on how this can be achieved by the formulae (65), (67), (70) and (71) can be found in the thesis of D. Wienholtz [2]. The complete proof is technically quite involved and will here be omitted. We just sketch the intuitive idea underlying the proof; we shall argue only locally, identifying Γ with its tangent line, and writing $\Gamma \hat{=} \mathbb{R}$. The boundary values $Y(x)$, $x \in \mathbb{R}$, of our minimal surface $\hat{Y}(z)$ are then interpreted as a mapping $Y : \mathbb{R} \to \mathbb{R}$ with $\hat{Y}(0) = 0$ where $z = 0$ is the boundary branch point of Y which we consider. Then we have

$$Y(x) = \frac{1}{n+1}a_n x^{n+1} + o(x^{n+1}) \quad \text{as } x \to 0$$

with $a_n > 0$ which shows that $Y(x)$ is (locally) monotone. Suppose now that $\tau(x) = \beta x^{-k}, k < n$, k odd, $\beta > 0$. Now define a one-parameter family $Z(t)$ of variations

$$Z(x,t) = Z(t)(x) = Y(x) + tY_x(x)\tau(x).$$

Then

$$\frac{\partial}{\partial x}Z(x,t) = a_n x^n + \beta t(n-k)x^{n-k-1} + o(x^n) + to(x^{n-k-1})$$

and we have

$$\frac{\partial}{\partial x}Z(x,t) > 0 \quad \text{for } 0 < |x| \ll 1 \text{ and } t > 0$$

since $[n-(k+1)]$ is even; thus $Z(x,t)$ is monotonic in x for $|x| \ll 1$ and $t > 0$.

In the actual proof one defines a variation

$$\tilde{Z}(t) := Y + tY_x\tau,$$

and then, using either the normal bundle projection for Γ or an exponential map, we project $\tilde{Z}(t)$ onto Γ, which defines $Z(t)$. The technical difficulty lies in showing that this variation remains monotonic near the branch point 0 for all $t \geq 0$. Global monotonicity for small $t \geq 0$ will follow from the compactness of S^1.

In conclusion we have

Theorem 4. (D. Wienholtz). *If $\hat{X}$ is a minimal surface in $\mathcal{C}(\Gamma)$ with $\Gamma \in C^{r+7}, 3\int_\Gamma \kappa\, ds \le \pi\boldsymbol{r}$, having a boundary branch point of order n and index m satisfying the Wienholtz condition $2m - 2 < 3n$, then X cannot be an $H^2(S^1, \mathbb{R}^3)$-minimizer for Dirichlet's integral $E(X)$ defined by* (1)*, and thus $\hat{X}$ cannot be an $H^{5/2}(\overline{B}, \mathbb{R}^3)$-minimizer of area.*

Remark 6. There remains the question if one can use higher derivatives of E to show that minimizers $\hat{X}$ cannot have boundary branch points if Γ is taken to be sufficiently smooth. Proposition 3 implies that for this purpose it would suffice to consider at most seven derivatives of E if one assumes nonvanishing curvature and torsion of Γ. Focussing on nonexceptional branch points, merely six derivatives of E would suffice.

The exceptional case is even more challenging since we no longer have $\mathrm{Re}\, R_m = 0$.

6.4 Scholia

The solution of Plateau's problem presented by J. Douglas [12] and T. Radó [17] was achieved by a – very natural – redefinition of the *notion of a minimal surface* $X : \Omega \to \mathbb{R}^3$ which is also used in our book[2]: Such a surface is a harmonic and conformally parametrized mapping; but it is not assumed to be an immersion. Consequently X may possess branch points, and thus some authors speak of "branched immersions". This raises the question whether or not Plateau's problem always has a solution which is immersed, i.e. regular in the sense of differential geometry. Certainly there exist minimal surfaces with branch points; but one might conjecture that area minimizing solutions of Plateau's problem are free of (interior) branch points. To be specific, let Γ be a closed, rectifiable Jordan curve in $\mathbb{R}^3$, and denote by $\mathcal{C}(\Gamma)$ the class of disk-type surfaces $X : B \to \mathbb{R}^3$ bounded by Γ which was defined in Vol. 1, Section 4.2. Then one may ask: *Suppose that $X \in \mathcal{C}(\Gamma)$ is a disk-type minimal surface $X : \overline{B} \to \mathbb{R}^3$ which minimizes both A and D in $\mathcal{C}(\Gamma)$. Does X have branch points in B (or in $\overline{B}$)?*

Radó [17], pp. 791–795 gave a first answer to this question for some special classes of boundary contours Γ, using the following result:

If $X_w(w)$ vanishes at some point $w_0 \in B$ then any plane through the point $P_0 := X(w_0)$ intersects Γ in at least four distinct points.

This observation has the following interesting consequence: *Suppose that there is a straight line $\mathcal{L}$ in $\mathbb{R}^3$ such that any plane through $\mathcal{L}$ intersects Γ in at most two distinct points. Then any minimal surface $X \in \mathcal{C}(\Gamma)$ has no branch points in B*. In fact, for $P_0 \notin \mathcal{L}$, the plane Π determined by P_0 and $\mathcal{L}$

[2] We now denote a minimal surface by X and no longer by $\hat{X}$, i.e. we no longer emphasize the difference between a surface $\hat{X}$ and its boundary values X.

meets Γ in at most two points, and for $P_0 \in \mathfrak{L}$ there are infinitely many such planes.

In particular: *If Γ has a simply covered star-shaped image under a (central or parallel) projection upon some plane Π_0, then any minimal surface $X \in \mathcal{C}(\Gamma)$ is free of branch points in B.*

Somewhat later, Douglas [15], pp. 733, 739, 753 thought that he had found a contour Γ with the property that any minimal surface $X \in \mathcal{C}(\Gamma)$ is branched, namely a curve whose orthogonal projection onto the x^1, x^2-plane is a certain closed curve with a double point. Radó [21], p. 109 commented on this assertion as follows: A curve Γ with this x^1, x^2-projection can be chosen in such a way that its x^1, x^3-projection is a simply covered star-shaped curve in the x^1, x^3-plane; thus no minimal surface in $\mathcal{C}(\Gamma)$ has a branch point.

In 1941, Courant [11] believed to have found a contour Γ for which some minimizer of Dirichlet's integral in $\mathcal{C}(\Gamma)$ has an interior branch point. This assertion is not correct, as Osserman [12], p. 567 pointed out in 1970. Moreover, in [12] he described an ingenious line of argumentation which seemed to exclude interior branch points for area minimizing solutions of Plateau's problem. For this purpose he distinguished between *true* and *false* branch points (cf. Osserman, [15], p. 154, Definition 6; and, more vaguely, [12], p. 558): A branch point is false, if the image of some neighbourhood of the branch point lies on a regularly embedded minimal surface; otherwise it is a true branch point. Osserman's treatment of the false branch points is incomplete,but contains essential ideas used by later authors, while his exclusion of true branch points is essentially complete (see also W.F. Pohl [1], Gulliver, Osserman, and Royden [1], p. 751, D. Wienholtz [1], p. 2). The principal ideas of Osserman in dealing with true branch points w_0 are the following: First, the geometric behaviour of the minimal surface X in the neighbourhood of w_0 is studied, yielding the existence of branch lines. Then a remarkable discontinuous parameter transformation G is introduced such that $\tilde{X} := X \circ G$ lies again in $\mathcal{C}(\Gamma)$ and has the same area as X, but in addition $\tilde{X}$ has a wedge, and so its area can be reduced by "smoothing out" the wedge. Osserman's definition of G is somewhat sloppy, but K. Steffen has kindly pointed out to us how this can be remedied and that the construction of the area reducing surface can rigorously be carried out.

Osserman's paper [12] was the decisive break-through in excluding true branch points for area minimizing minimal surfaces in $\mathbb{R}^3$, and it inspired the succeeding papers by R. Gulliver [2] and H.W. Alt [1,2], which even tackled the more difficult branch point problem for H-surfaces and for minimal surfaces in a Riemannian manifold (Gulliver). Nearly simultaneously, both authors published proofs of the assertion that area minimizing minimal surfaces in $\mathcal{C}(\Gamma)$ possess no interior branch points (and of the analogous statement for H-surfaces).

Gulliver's reasoning runs as follows: Let us assume that $w_0 = 0$ is an interior branch point of the minimal surface $X \in \overline{\mathcal{C}}(\Gamma), X : \overline{B} \to \mathbb{R}^3$. Then there is a neighbourhood $V \Subset B$ of 0 in which two oriented Jordan arcs

$\gamma_1, \gamma_2 \in C^1([0,1], B)$ exist with $\gamma_1(0) = \gamma_2(0) = 0, |\gamma_j'(0)| = 1, \gamma_1'(0) \neq \gamma_2'(0), X(\gamma_1(t)) \equiv X(\gamma_2(t))$, and such that $(X_u \wedge X_v)(\gamma_1(t)), (X_u \wedge X_v)(\gamma_2(t))$ are linearly independent for $0 < t \leq 1$. One can assume that ∂V is smooth, and that γ_1, γ_2 meet ∂V transversally at distinct points $\gamma_1(\epsilon), \gamma_2(\epsilon), 0 < \epsilon < 1$. Then there is a homeomorphism $F : \overline{B}_\epsilon \to \overline{V}$ with $F(it) = \gamma_1(t), F(-it) = \gamma_2(t)$ for $0 \leq t \leq \epsilon$, and $F \in C^2(\overline{B}_\epsilon \setminus \{0\})$ where $B_\epsilon := B_\epsilon(0) = \{w \in \mathbb{C} : |w| < \epsilon\}$. Define a discontinuous map $G : \overline{B}_\epsilon \to \overline{B}_\epsilon$ such that $\{it : 0 < t \leq 1\}$ and $\{-it : 0 < t \leq 1\}$ are mapped to i and $-i$ respectively; $\pm\epsilon/2$ are taken to zero; on the segments of discontinuity $[-\epsilon/2, 0]$ and $[0, \epsilon/2]$ are each given two linear mappings by limiting values under approach from the two sides; G is continuous on a neighbourhood of ∂B_ϵ with $G|_{\partial B_\epsilon} = \mathrm{id}_{\partial B_\epsilon}$; and G is conformal on each component of $B_\epsilon \setminus I_\epsilon \setminus$imaginary axis, where I_ϵ is the interval $[-\epsilon/2, \epsilon/2]$ on the real axis. Thus $X \circ F \circ G$ is continuous and piecewise C^2. Now define

$$\overline{X}(w) := \begin{cases} (X \circ F \circ G \circ F^{-1})(w) & \text{for } w \in V, \\ X(w) & \text{for } w \in \overline{B} \setminus V. \end{cases}$$

Then $\overline{X}$ is continuous and piecewise C^2, and $\overline{X} \in \mathcal{C}(\Gamma)$. The metric

$$ds^2 := \langle d\overline{X}, d\overline{X} \rangle = a\,du^2 + 2b\,du\,dv + c\,dv^2,$$

$$a := |\overline{X}_u|^2, \quad b := \langle \overline{X}_u, \overline{X}_v \rangle, \quad c := |\overline{X}_v|^2,$$

induced on B by pulling back the metric induced from $\mathbb{R}^3$ along $\overline{X}$ has bounded, piecewise smooth coefficients. "It follows from the uniformization theorem of Morrey ([1], Theorem 3) that there exists $T : B \to B$ with L^2 second derivatives, which is almost everywhere conformal from B with its usual metric to B with its induced metric, and T may be extended to a homeomorphism $\overline{B} \to \overline{B}$".

Now define $\tilde{X} := \overline{X} \circ T$; then $\tilde{X} \in \mathcal{C}(\Gamma), A(\tilde{X}) = A(X)$, and $\langle \tilde{X}_w, \tilde{X}_w \rangle = 0$ a.e. on B, and consequently

$$\inf_{\mathcal{C}(\Gamma)} D = \inf_{\mathcal{C}(\Gamma)} A = D(X) = A(X) = A(\tilde{X}) = D(\tilde{X}).$$

Thus $\tilde{X}$ is D-minimizing, and so its surface normal $\tilde{N}$ is continuous on B. On the other hand, the sets $\overline{X}(B)$ and $\tilde{X}(B)$ are the same, and so $\tilde{X}(B)$ has an edge, whence $\tilde{N}$ cannot be continuous, a contradiction.

This reasoning requires two comments. First, D. Wienholtz in his Diploma thesis [1], p. 3 (published as [2]), noted that Gulliver's discontinuous map $G : \overline{B}_\epsilon \to \overline{B}_\epsilon$ does not exist, since its existence contradicts Schwarz's reflection principle. A remedy of this deficiency is to set up another definition of G or T, such as used in Alt [1], pp. 360–361, or in Steffen and Wente [1], p. 218, or by a modification of the definition of G as in Gulliver and Lesley [1], p. 24.

Secondly, the application of one of Morrey's uniformization theorems from [1] is not immediately justified, as Theorem 3 of §2 requires besides $a, b, c \in L^\infty(B)$ the assumption

$$(*) \qquad ac - b^2 = 1,$$

and Theorem 3 of Moorey's §4 demands the existence of constants $\lambda_1, \lambda_2 \in \mathbb{R}$ with $0 < \lambda_1 \le \lambda_2$ such that

$$(**) \qquad \lambda_1[\xi^2 + \eta^2] \le a(w)\xi^2 + 2b(w)\xi\eta + c(w)\eta^2 \le \lambda_2[\xi^2 + \eta^2]$$

for all $(\xi, \eta) \in \mathbb{R}^2$ and for almost all $w \in B$. However, $\overline{X}(w) \equiv X(w)$ on $B \setminus V$, and X might have another branch point $w_0' \in B \setminus V$; then $a(w_0') = b(w_0') = c(w_0') = 0$, and so neither (*) nor (**) were satisfied.

This difficulty is overcome by assuring that $\overline{X}$ is quasiconformal in the sense that

$$|\overline{X}_u|^2 + |\overline{X}_v|^2 \le \kappa |\overline{X}_u \wedge \overline{X}_v| \quad \text{(a.e. on } B)$$

holds for some constant $\kappa > 0$. Then it follows

$$a, |b|, c \le \kappa\sqrt{ac - b^2},$$

and thus the quadratic form

$$d\sigma^2 := \alpha\, du^2 + 2\beta\, du\, dv + \gamma\, dv^2$$

with

$$\alpha := \frac{a}{\sqrt{ac - b^2}}, \quad \beta := \frac{b}{\sqrt{ac - b^2}}, \quad \gamma := \frac{c}{\sqrt{ac - b^2}}$$

satisfies $|\alpha|, |\beta|, |\gamma| \le \kappa$ and $\alpha\gamma - \beta^2 = 1$. Hence one can apply Morrey's first uniformization theorem (as quoted above), obtaining a homeomorphism T from $\overline{B}$ onto $\overline{B}$ with $T, T^{-1} \in H_2^1(B, B)$ such that the pull-back $T^*\, d\sigma^2$ is a multiple of the Euclidean metric ds_e^2, i.e.

$$T^*\, d\sigma^2 = \lambda\, ds_e^2$$

whence

$$T^*\, ds^2 = \tilde{\lambda}\, ds_e^2$$

with $\tilde{\lambda} := \lambda\sqrt{\tilde{a}\tilde{c} - \tilde{b}^2}$, $\tilde{a} := a \circ T$, $\tilde{b} := b \circ T$, $\tilde{c} := c \circ T$.

Now one can proceed for $\tilde{X} := \overline{X} \circ T$ as above. Alt's method to exclude true branch points (worked out in detail by D. Wienholtz [1,2]) eventually uses the same contradiction argument as Gulliver, namely to derive the existence of an energy minimizer $\tilde{X} \in \mathcal{C}(\Gamma)$ with a discontinuous normal $\tilde{N}$. The construction of $\tilde{X}$ is different from Gulliver's approach. Alt defines a new surface $\overline{X}$ on B_ϵ which is quasiconformal, and by reparametrization a new surface $\tilde{X} = \overline{X} \circ \tau$ is obtained which is energy minimizing with respect to its boundary values. Here Morrey's lemma on ϵ-conformal mappings is used as well as an elaboration of Lemma 9.3.3 in Morrey [8].

The nonexistence of false branch points for solutions X of Plateau's problem was proved by R. Gulliver [2], H.W. Alt [2], and then by Gulliver, Osserman, and Royden in their fundamental 1973-paper [1]. Here one only needs that $X|_{\partial B}$ is $1-1$, and this observation is used by Alt as well as by Gulliver, Osserman, and Royden, while Gulliver also employs the minimizing property of X. K. Steffen pointed out to us that Osserman's original paper [12] already contains significant contributions to the problem of excluding false branch points, and it even is satisfactory if, for some reason, an inner point of X cannot lie on the boundary curve Γ, say, if Γ lies on the surface of a convex body.

Furthermore, in Section 6 of their paper, Gulliver, Osserman, and Royden proved a rather general result on branched surfaces $X : \overline{B} \to \mathbb{R}^n$, $n \geq 2$, such that $X|_{\partial B}$ is injective, which implies the following: *A minimal surface $X \in \mathcal{C}(\Gamma)$ has no false boundary branch points* (see [1], pp. 799–809, in particular Theorem 6.16).

In 1973, R. Gulliver and F.D. Lesley [1] published the following result which we cite in a slightly weaker form: *If Γ is a real analytic and regular contour in $\mathbb{R}^3$, then any area minimizing minimal surface in $\mathcal{C}(\Gamma)$ has no boundary branch points.*

To prove this result they extend a minimizer X across the boundary of the parameter domain B as a minimal surface, so that a branch point w_0 on ∂B can be treated as an inner point. Then the same analysis of X in a small neighbourhood of w_0 can be carried out, and w_0 is either seen to be false or true. To exclude the possibility of a true branch point, they apply the method from Gulliver's paper [2], except that a new discontinuous "Osserman-type" mapping G is described, which is appropriate for this situation. In a different way, true boundary branch points for analytic Γ were excluded by B. White [24], see below.

The elimination of the possibility of false branch points in the Gulliver–Lesley paper is achieved by using results from the theory of "*branched immersions*", created by Gulliver, Osserman, and Royden.

The theory of branched immersions was extended by Gulliver [4,5,7] in such a way that it applies to surfaces of higher topological type (minimal surfaces and H-surfaces in a Riemannian manifold).

K. Steffen and H. Wente [1] showed in 1978 that minimizers of

$$E_Q(X) := \int_B \left[\frac{1}{2}|\nabla X|^2 + Q(X) \cdot (X_u \wedge X_v)\right] du\, dv$$

in $\mathcal{C}(\Gamma)$ subject to a volume constraint $V(X) = \text{const}$ with

$$V(X) := \frac{1}{3}\int_B X \cdot (X_u \wedge X_v)\, du\, dv$$

have no interior branch points. While their treatment of true branch points essentially follows Osserman [12], they simplified, in their special situation, the discussion of false branch points by Gulliver, Osserman, and Royden [1] and Gulliver [4].

In 1980, Beeson [2] showed that a minimal surface in $\mathcal{C}(\Gamma)$, given by a local Weierstrass representation, cannot have a true interior branch point if it is a C^1-local minimizer of D in $\mathcal{C}(\Gamma)$. (According to D. Wienholtz, Beeson's proof does not work for C^k-local minimizers with $k \geq 2$.) Motivated by the discovery of forced Jacobi fields, Beeson achieved this result by arguing that some first non-vanishing derivative must be negative.

Later on, in 1994, M. Micaleff and B. White [1] excluded the existence of true interior branch points for area minimizing minimal surfaces in a Riemannian 3-manifold, and in 1997, B. White [24] proved that an area minimizing minimal surface $X : \overline{B} \to \mathbb{R}^n, n \geq 3$, cannot have a true branch point on any part of ∂B which is mapped by X onto a real analytic portion of Γ, even if $n \geq 4$. This is quite surprising as X may have interior branch points if $n \geq 4$ (Federer's examples). However, White pointed out that, for any $k < \infty$, one can find C^k-curves Γ in $\mathbb{R}^4$ that bound area minimizing disk-type minimal surfaces with true boundary branch points, and Gulliver [11] found a C^∞-curve in $\mathbb{R}^6$ bounding an area minimizer with a true boundary branch point.

It is a major open question to decide whether or not an area minimizing minimal surface of disk-type in $\mathbb{R}^3$ *can have a boundary branch point assuming that it is bounded by a (regular)* C^k- *or* C^∞-*contour* Γ, *rather than by an analytic one.*

We furthermore mention the paper of H.W. Alt and F. Tomi [1] where the nonexistence of branch points for minimizers to certain free boundary problems is proved (see also Section 1.9 of this volume, Theorem 5), and the work of R. Gulliver and F. Tomi [1] where the absence of interior branch points for minimizers of higher genus is established. Specifically, they showed that such a minimizer $X : M \to N$ cannot possess false branch points if X induces an isomorphism on fundamental groups.

In 1977–81, R. Böhme and A. Tromba [1,2] showed that, *generically*, every smooth Jordan curve in $\mathbb{R}^n$, $n \geq 4$, bounds only immersed minimal surfaces, and admits only simple interior branch points for $n = 3$, but no boundary branch points. "Generic" means that there is an open and dense subset in the space of all sufficiently smooth $\alpha : S^1 \to \mathbb{R}^n$ defining a Jordan curve Γ, for which subset the assertion holds. This result is based on the Böhme–Tromba index theory, which is presented in Vol. 3.

A completely new method to exclude the existence of branch points for *minimal surfaces in* $\mathbb{R}^3$ *which are weak relative minimizers of* D was developed by A.J. Tromba [11] in 1993 by deriving an *intrinsic third derivative of* D *in direction of forced Jacobi fields.* He showed that if $X \in \mathcal{C}(\Gamma)$ has only simple interior branch points satisfying a *Schüffler condition* (a condition which by K. Schüffler [2] had been identified as generic), then the third variation of D can be made negative, while the first and second derivatives are zero, and so X cannot be a weak relative minimizer of D in $\mathcal{C}(\Gamma)$. D. Wienholtz in his Doctoral thesis [3] generalized Tromba's method to interior and boundary branch points of arbitrary order, satisfying a "Schüffler-type condition", by computing the third derivative of D in suitable directions generated by forced Jacobi fields. This work of Tromba and Wienholtz is described in Sections 6.1 and 6.3. We note that Wienholtz's results also refer to boundary branch points of minimal surfaces in $\mathbb{R}^n, n \geq 3$, but they do not apply to Gulliver's $\mathbb{R}^6$-example (see Wienholtz [3], p. 244). In forthcoming work by Tromba it will be shown how the ideas presented in Sections 6.1 and 6.2 can be used to exclude interior branch points for absolute minimizers of A in $\mathcal{C}(\Gamma)$.

Bibliography

This bibliography is a selection from the references collected in Vol. 1. Here we only cite authors whose results are relevant for the topics discussed in this volume.

We also refer the reader to the following Lecture notes:

MSG: Minimal submanifolds and geodesics. Proceedings of the Japan–United States Seminar on Minimal Submanifolds, including Geodesics, Tokyo, 1977. Kagai Publications, Tokyo, 1978

SDG: Seminar on differential geometry, edited by S.T. Yau, Ann. Math. Studies **102**, Princeton, 1982

SMS: Seminar on minimal submanifolds, edited by Enrico Bombieri. Ann. Math. Studies **103**, Princeton, 1983

TVMA: Théorie des variétés minimales et applications. Séminaire Palaiseau, Oct. 1983–June 1984. Astérisque **154–155** (1987)

GACG: Geometric analysis and computer graphics. Proceedings of the Conference on Differential Geometry, Calculus of Variations and Computer Graphics, edited by P. Concus, R. Finn, D.A. Hoffman. Math. Sci. Res. Inst. **17**. Springer, New York, 1991

GTMS: Global theory of minimal surfaces, edited by D. Hoffman. Proceedings of the Clay Mathematics Institute 2001 Summer School, MSRI, Berkeley, June 25–27, 2001. Clay Math. Proceedings **2**, Am. Math. Soc., Providence, 2005

We also mention the following report by H. Rosenberg which appeared in May 1992:

Some recent developments in the theory of properly embedded minimal surfaces in $\mathbb{R}^3$. Séminaire Bourbaki **34**, Exp. No. 759 (1992), 73 pp.

Furthermore, we refer to:

EMS: Encyclopaedia of Math. Sciences **90**, Geometry V, Minimal surfaces (ed. R. Osserman), Springer, 1997. This volume contains the following reports:

I. D. Hoffman, H. Karcher, Complete embedded minimal surfaces of finite total curvature.

II. H. Fujimoto, Nevanlinna theory and minimal surfaces.

III. S. Hildebrandt, Boundary value problems for minimal surfaces.

IV. L. Simon, The minimal surface equation.

U. Dierkes, S. Hildebrandt, A.J. Tromba, *Regularity of Minimal Surfaces*,
Grundlehren der mathematischen Wissenschaften 340,
DOI 10.1007/978-3-642-11700-8, © Springer-Verlag Berlin Heidelberg 2010

At last we quote the predecessor of the present treatise:

DHKW: Dierkes, U., Hildebrandt, S., Küster, A., Wohlrab, O. Minimal surfaces I, II. Grundlehren Math. Wiss. **295**, **296**. Springer, Berlin, 1992

Abresch, U.

1. Constant mean curvature tori in terms of elliptic functions. J. Reine Angew. Math. **374**, 169–192 (1987)

Acerbi, E., Fusco, N.

1. Semicontinuity problems in the calculus of variations. Arch. Ration. Mech. Anal. **86**, 125–145 (1986)

Adams, R.A.

1. Sobolev spaces. Academic Press, New York, 1975

Agmon, S., Douglis, A., Nirenberg, L.

1. Estimates near the boundary for solutions of elliptic differential equations satisfying general boundary conditions I. Commun. Pure Appl. Math. **12**, 623–727 (1959)
2. Estimates near the boundary for solutions of elliptic differential equations satisfying general boundary conditions II. Commun. Pure Appl. Math. **17**, 35–92 (1964)

Alexander, H., Hoffman, D., Osserman, R.

1. Area estimates for submanifolds of Euclidean space. INDAM Symp. Math. **14**, 445–455 (1974)

Alexander, H., Osserman, R.

1. Area bounds for various classes of surfaces. Am. J. Math. **97**, 753–769 (1975)

Alexandroff, P., Hopf, H.

1. Topologie **1**. Springer, Berlin, 1935

Allard, W.K.

1. On the first variation of a varifold. Ann. Math. **95**, 417–491 (1972)
2. On the first variation of a manifold: boundary behavior. Ann. Math. **101**, 418–446 (1975)

Almgren, F.J.

1. Some interior regularity theorems for minimal surfaces and an extension of Bernstein's theorem. Ann. Math. **84**, 277–292 (1966)
2. Plateau's problem. An invitation to varifold geometry. Benjamin, New York, 1966
3. Existence and regularity almost everywhere of solutions to elliptic variational problems with constraints. Mem. Am. Math. Soc. **4**(165) (1976)
4. The theory of varifolds; a variational calculus in the large for the k-dimensional area integrand. Mimeographed notes, Princeton, 1965
5. Minimal surface forms. Math. Intell. **4**, 164–172 (1982)
6. Q-valued functions minimizing Dirichlet's integral and the regularity of area minimizing rectifiable currents up to codimension two. Bull. Am. Math. Soc. **8**, 327–328 (1983) and typoscript, 3 vols., Princeton University
7. Optimal isoperimetric inequalities. Indiana Univ. Math. J. **35**, 451–547 (1986)

Almgren, F.J., Simon, L.

1. Existence of embedded solutions of Plateau's problem. Ann. Sc. Norm. Super. Pisa, Cl. Sci. **6**, 447–495 (1979)

Alt, H.W.

1. Verzweigungspunkte von H-Flächen. I. Math. Z. **127**, 333–362 (1972)
2. Verzweigungspunkte von H-Flächen. II. Math. Ann. **201**, 33–55 (1973)
3. Die Existenz einer Minimalfläche mit freiem Rand vorgeschriebener Länge. Arch. Ration. Mech. Anal. **51**, 304–320 (1973)

Alt, H.W., Tomi, F.

1. Regularity and finiteness of solutions to the free boundary problem for minimal surfaces. Math. Z. **189**, 227–237 (1985)

Athanassenas, M.

1. Ein Variationsproblem für Flächen konstanter mittlerer Krümmung. Diplomarbeit, Bonn, 1985
2. A variational problem for constant mean curvature surfaces with free boundary. J. Reine Angew. Math. **377**, 97–107 (1987)

Beckenbach, E.F.

1. The area and boundary of minimal surfaces. Ann. Math. **33**, 658–664 (1932)
2. Minimal surfaces in euclidean n-space. Am. J. Math. **55**, 458–468 (1933)
3. Bloch's theorem for minimal surfaces. Bull. Am. Math. Soc. **39**, 450–456 (1933)
4. Remarks on the problem of Plateau. Duke Math. J. **3**, 676–681 (1937)
5. On a theorem of Fejér and Riesz. J. Lond. Math. Soc. **13**, 82–86 (1938)
6. Functions having subharmonic logarithms. Duke Math. J. **8**, 393–400 (1941)
7. The analytic prolongation of a minimal surface. Duke Math. J. **9**, 109–111 (1942)
8. Painlevé's theorem and the analytic prolongation of minimal surfaces. Ann. Math. **46**, 667–673 (1945)
9. Some convexity properties of surfaces of negative curvature. Am. Math. Mon. **55**, 285–301 (1948)
10. An introduction to the theory of meromorphic minimal surfaces. In: Proc. Symp. Pure Math. **11**, pp. 36–65. Am. Math. Soc., Providence, 1968

Beckenbach, E.F., Radó, T.

1. Subharmonic functions and minimal surfaces. Trans. Am. Math. Soc. **35**, 648–661 (1933)
2. Subharmonic functions and surfaces of negative curvature. Trans. Am. Math. Soc. **35**, 662–664 (1933)

Beeson, M.

1. The behavior of a minimal surface in a corner. Arch. Ration. Mech. Anal. **65**, 379–393 (1977)
2. On interior branch points of minimal surfaces. Math. Z. **171**, 133–154 (1980)
3. Some results on finiteness in Plateau's problem, part I. Math. Z. **175**, 103–123 (1980)
4. Some results on finiteness in Plateau's problem, part II. Math. Z. **181**, 1–30 (1980)
5. The 6π theorem about minimal surfaces. Pac. J. Math. **117**, 17–25 (1985)

Bemelmans, J., Dierkes, U.

1. On a singular variational integral with linear growth. I: existence and regularity of minimizers. Arch. Ration. Mech. Anal. **100**, 83–103 (1987)

Bernstein, S.

1. Sur les surfaces définies au moyen de leur courbure moyenne ou totale. Ann. Sci. Ecole Norm. Super. (3) **27**, 233–256 (1910)
2. Sur la généralisation du problème de Dirichlet. Math. Ann. **69**, 82–136 (1910)
3. Sur les équations du calcul des variations. Ann. Sci. Ecole Norm. Super. (3) **29**, 431–485 (1912)
4. Sur un théorème de géométrie et ses applications aux équations aux dérivées partielles du type eliptique. Comm. de la Soc. Math. de Kharkov (2-ème sér.) **15**, 38–45 (1915–1917) [Translation in Math. Z. **26**, 551–558 (1927) under the title: Über ein geometrisches Theorem und seine Anwendung auf die partiellen Differentialgleichungen vom elliptischen Typus.]

Bers, L., John, F., Schechter M.

1. Partial differential equations. Interscience, New York, 1964

Bethuel, F.
1. The approximation problem for Sobolev mappings between manifolds. Acta Math. **167**, 167–201 (1991)
2. Un résultat de régularité pour les solutions de l'equation des surfaces à courbure moyenne prescrite. C. R. Acad. Sci. Paris **314**, 1003–1007 (1992)

Bethuel, F., Rey, O.
1. Multiple solutions to the Plateau problem for nonconstant mean curvature. Duke Math. J. **73**, 593–646 (1994)

Bianchi, L.
1. Lezioni di geometria differenziale. Spoerri, Pisa, 1894 [German translation: Vorlesungen über Differentialgeometrie, Teubner, Leipzig, 1899]
2. Lezioni di geometria differenziale. E. Spoerri, Pisa, 1903

Blaschke, W.
1. Vorlesungen über Differentialgeometrie. I. Elementare Differentialgeometrie. II. Affine Differentialgeometrie (zusammen mit K. Reidemeister). III. Differentialgeometrie der Kreise und Kugeln (zusammen mit G. Thomsen). Springer, Berlin, 1921, 1923, 1929
2. Einführung in die Differentialgeometrie. Springer, Berlin, 1950
3. Kreis und Kugel. Veit und Co., Leipzig, 1916

Blaschke, W., Leichtweiß, K.
1. Elementare Differentialgeometrie, 5th edn. Springer, Berlin, 1973

Bliss, G.E.
1. Calculus of variations. Open Court Publ. Co., La Salle, 1925

Böhme, R.
1. Die Zusammenhangskomponenten der Lösungen analytischer Plateauprobleme. Math. Z. **133**, 31–40 (1973)
2. Stabilität von Minimalflächen gegen Störung der Randkurve. Habilitationsschrift, Göttingen, 1974
3. Die Jacobifelder zu Minimalflächen im $\mathbb{R}^3$. Manuscr. Math. **16**, 51–73 (1975)
4. Stability of minimal surfaces. In: Function theoretic methods for partial differential equations. Proc. Internat. Sympos. Darmstadt, 1976. Lect. Notes Math. **561**, pp. 123–137. Springer, Berlin, 1976
5. Über Stabilität und Isoliertheit der Lösungen des klassischen Plateau-Problems. Math. Z. **158**, 211–243 (1978)
6. New results on the classical problem of Plateau on the existence of many solutions. Séminaire Bourbaki **34**, Exp. No. 579, 1–20 (1981/1982)

Böhme, R., Hildebrandt, S., Tausch, E.
1. The two-dimensional analogue of the catenary. Pac. J. Math. **88**, 247–278 (1980)

Böhme, R., Tomi, F.
1. Zur Struktur der Lösungsmenge des Plateauproblems. Math. Z. **133**, 1–29 (1973)

Böhme, R., Tromba, A.J.
1. The number of solutions to the classical Plateau problem is generically finite. Bull. Am. Math. Soc. **83**, 1043–1044 (1977)
2. The index theorem for classical minimal surfaces. Ann. Math. **113**, 447–499 (1981)

Bokowski, J., Sperner, E.
1. Zerlegung konvexer Körper durch minimale Trennflächen. J. Reine Angew. Math. **311/312**, 80–100 (1979)

Bol, G.
1. Isoperimetrische Ungleichungen für Bereiche auf Flächen. Jahresber. Dtsch. Math.-Ver. **51**, 219–257 (1941)

Bolza, O.
1. Vorlesungen über Variationsrechnung. Teubner, Leipzig, 1909
2. Gauss und die Variationsrechnung. In: C.F. Gauss, Werke. Bd. X.2, Abh. 5, Springer, Berlin, 1922–1923

Bombieri, E.
1. Lecture Séminaire Bourbaki, February 1969
2. Theory of minimal surfaces and a counterexample to the Bernstein conjecture in high dimensions. Lectures, Courant Institute, New York University, 1970
3. Recent progress in the theory of minimal surfaces. Enseign. Math. **25**, 1–8 (1979)
4. An introduction to minimal currents and parametric variational problems. Mathematical Reports **2**, Part 3. Harwood, London, 1985

Bombieri, E., de Giorgi, E., Giusti, E.
1. Minimal cones and the Bernstein problem. Invent. Math. **7**, 243–268 (1969)

Bombieri, E., de Giorgi, E., Miranda, M.
1. Una maggioriazione a priori relativa alle ipersurfici minimali non parametriche. Arch. Ration. Mech. Anal. **32**, 255–267 (1969)

Bombieri, E., Giusti, E.
1. Harnack inequality for elliptic differential equations on minimal surfaces. Invent. Math. **15**, 24–46 (1971/1972)
2. Local estimates for the gradient of non-parametric surfaces of prescribed mean curvature. Commun. Pure Appl. Math. **26**, 381–394 (1973)

Bombieri, E., Simon, L.
1. On the Gehring link problem. In: Bombieri, E. (ed.) Seminar on minimal submanifolds. Ann. Math. Stud. **103**, pp. 271–274. Am. Math. Soc., Princeton, 1983

Bonnesen, T., Fenchel, W.
1. Theorie der konvexen Körper. Chelsea, New York, 1948

Bonnet, O.
1. Mémoire sur la théorie générale des surfaces. J. Ecole Polytech. Cahier **32**, 131 (1848)
2. Deuxième note sur les surfaces à lignes de courbure sphérique. C. R. Acad. Sci. Paris **36**, 389–391 (1853)
3. Troisième note sur les surfaces à lignes de courbure planes et sphériques. C. R. Acad. Sci. Paris **36**, 585–587 (1853)
4. Note sur la théorie générale des surface. C. R. Acad. Sci. Paris **37**, 529–532 (1853)
5. Observations sur les surfaces minima. C. R. Acad. Sci. Paris **41**, 1057–1058 (1855)
6. Nouvelles remarques sur les surfaces à aire minima. C. R. Acad. Sci. Paris **42**, 532–535 (1856)
7. Mémoire sur l'emploi d'un nouveau système de variables dans l'étude des surfaces courbes. J. Math. Pures Appl. (2), 153–266 (1860)
8. Sur la détermination des fonctions arbitraires qui entrent dans l'équations intégrale des surfaces à aire minima. C. R. Acad. Sci. Paris **40**, 1107–1110 (1865)
9. Sur la surface réglée minima. Bull. Sci. Math. (2) **9**, 14–15 (1885)

Brakke, K.A.
1. Minimal surfaces, corners, and wires. J. Geom. Anal. **2**, 11–36 (1992)
2. The surface evolver. Expo. Math. **1**, 141–165 (1992)

Brézis, H.R., Coron, M.
1. Sur la conjecture de Rellich pour les surfaces à courbure moyenne prescribe. C. R. Acad. Sci. Paris, Ser. I **295**, 615–618 (1982)
2. Large solutions for harmonic maps in two dimensions. Commun. Math. Phys. **92**, 203–215 (1983)
3. Multiple solutions of H-systems and Rellich's conjecture. Commun. Pure Appl. Math. **37**, 149–187 (1984)
4. Convergence of solutions of H-systems or how to blow bubbles. Arch. Ration. Mech. Anal. **89**(1), 21–56 (1985)

Brézis, H.R., Kinderlehrer, D.
1. The smoothness of solutions to nonlinear variational inequalities. Indiana Univ. Math. J. **23**, 831–844 (1974)

Burago, Y.D.
1. Note on the isoperimetric inequality on two dimensional surfaces. Sib. Mat. Zh. **14**, 666–668 (1973)

Burago, Y.D., Zalgaller, V.A.
1. Geometric inequalities. Grundlehren Math. Wiss. **285**. Springer, Berlin, 1988

Bürger, W., Kuwert, E.
1. Area-minimizing disks with free boundary and prescribed enclosed volume. J. Reine Angew. Math. **621**, 1–27 (2008)

Caffarelli, L.A., Nirenberg, L., Spruck, J.
1. On a form of Bernstein's theorem. In: Analyse mathématique et applications, pp. 55–56. Gauthier-Villars, Paris, 1988

Caffarelli, L.A., Rivière, N.M.
1. Smoothness and analyticity of free boundaries in variational inequalities. Ann. Sc. Norm. Super. Pisa, Cl. Sci. **3**, 289–310 (1976)

Calabi, E.
1. Minimal immersions of surfaces in euclidean space. J. Differ. Geom. **1**, 111–125 (1967)
2. Quelques applications de l'analyse complex aux surfaces d'aire minima. In: Rossi, H. (ed.) Topics in complex manifolds. Les Presses de l'Université de Montréal, Montréal, 1968
3. Examples of Bernstein problems for some nonlinear equations. In: Global analysis. Proc. Symp. Pure Math., pp. 223–230. Am. Math. Soc., Providence, 1968

Callahan, M.J., Hoffman, D.A., Hoffman, J.T.
1. Computer graphics tools for the study of minimal surfaces. Commun. ACM **31**, 648–661 (1988)

Carathéodory, C.
1. Über die Begrenzung einfach zusammenhängender Gebiete. Math. Ann. **73**, 323–370 (1913)
2. Über die gegenseitige Beziehung der Ränder bei der konformen Abbildung des Inneren einer Jordankurve auf einen Kreis. Math. Ann. **73**, 305–320 (1913)
3. Variationsrechnung und partielle Differentialgleichungen erster Ordnung. Teubner, Leipzig, 1935
4. Conformal representation. Cambridge University Press, Cambridge, 1952
5. Theory of functions of a complex variable, Vols. 1 and 2. Chelsea, New York, 1954

Carleman, T.
1. Über eine isoperimetrische Aufgabe und ihre physikalischen Anwendungen. Math. Z. **3**, 1–7 (1919)

2. Sur la répresentation conforme des domaines multiplement connexes. C. R. Acad. Sci. Paris **168**, 843–845 (1919)
3. Zur Theorie der Minimalflächen. Math. Z. **9**, 154–160 (1921)

Chavel, I.
1. On A. Hurwitz' method in isoperimetric inequalities. Proc. Am. Math. Soc. **71**, 275–279 (1978)

Chen, C.C.
1. A characterization of the catenoid. An. Acad. Bras. Ciênc. **51**, 1–3 (1979)
2. Complete minimal surfaces with total curvature -2π. Bull. Braz. Math. Soc. **10**, 71–76 (1979)
3. Elliptic functions and non-existence of complete minimal surfaces of certain type. Proc. Am. Math. Soc. **79**, 289–293 (1980)
4. Total curvature and topological structure of complete minimal surfaces. IME-USP **8**, 1980
5. On the image of the generalized Gauss map of a complete minimal surface in $\mathbb{R}^4$. Pac. J. Math. **102**, 9–14 (1982)
6. The generalized curvature ellipses and minimal surfaces. Bull. Inst. Math. Acad. Sin. **11**, 329–336 (1983)

Chen, Y.-W.
1. Branch points, poles and planar points of minimal surfaces in $\mathbb{R}^3$. Ann. Math. **49**, 790–806 (1948)
2. Existence of minimal surfaces with a pole at infinity and condition of transversality on the surface of a cylinder. Trans. Am. Math. Soc. **65**, 331–347 (1949)
3. Discontinuity and representations of minimal surface solutions. In: Proc. Conf. on Minimal Surface Solutions, pp. 115–138. University of Maryland Press, College Park, 1956

Chern, S.S.
1. Topics in differential geometry. The Institute for Advanced Study, Princeton, 1951
2. An elementary proof of the existence of isothermal parameters on a surface. Proc. Am. Math. Soc. **6**, 771–782 (1955)
3. Differentiable manifolds. Lecture Notes, University of Chicago, 1959
4. On the curvatures of a piece of hypersurface in Euclidean space. Abh. Math. Semin. Univ. Hamb. **29**, 77–91 (1965)
5. Minimal surfaces in a Euclidean space of N dimensions. In: Differential and combinatorial topology, a symposium in honor of Marston Morse, pp. 187–198. Princeton University Press, Princeton, 1965
6. Simple proofs of two theorems on minimal surfaces. Enseign. Math. **15**, 53–61 (1969)
7. Differential geometry; its past and its future. In: Actes, congrès Intern. Math. **I**, pp. 41–53, 1970
8. On the minimal immersions of the two-sphere in a space of constant curvature. In: Problems in analysis, pp. 27–40. Princeton University Press, Princeton, 1970

Chern, S.S., Hartman, P., Wintner, A.
1. On isothermic coordinates. Comment. Math. Helv. **28**, 301–309 (1954)

Cheung, L.F.
1. Communications to the authors.
2. Geometric properties of non-compact stable constant mean curvature surfaces. Thesis. Mathematisches Institut der Universität Bonn, Bonn, 1990

Chicco, M.
1. Principio di massimo forte per sottosoluzioni di equazioni ellittiche di tipo variazionale. Boll. Unione Mat. Ital. (3) **22**, 368–372 (1967)

Choe, J.

1. The isoperimetric inequality for a minimal surface with radially connected boundary. Preprint No. 00908-89, Math. Sci. Res. Inst. Berkeley, 1988
2. Index, vision number, and stability of complete minimal surfaces. Preprint No. 02008-89, Math. Sci. Res. Inst. Berkeley, 1988
3. On the existence and regularity of fundamental domains with least boundary area. J. Differ. Geom. **29**, 623–663 (1989)
4. On the analytic reflection of a minimal surface. Preprint No. 102, SFB 256, Univ. Bonn, 1990
5. The isoperimetric inequality for a minimal surface with radially connected boundary. Ann. Sc. Norm. Super. Pisa, Cl. Sci. **17**, 583–593 (1990)
6. The isoperimetric inequality for minimal surfaces in a Riemannian manifold. J. Reine Angew. Math. **506**, 205–214 (1999)
7. Isoperimetric inequalities of minimal submanifolds; cf. GTMS (2005)

Clarenz, U.

1. Sätze über Extremalen zu parametrischen Funktionalen. Bonner Math. Schriften **322**. Mathematisches Institut der Universität Bonn, Bonn, 1999
2. Enclosure theorems for extremals of elliptic parametric functionals. Calc. Var. Partial Differ. Equ. **15**, 313–324 (2002)

Clarenz, U., von der Mosel, H.

1. Compactness theorem and an isoperimetric inequality for critical points of elliptic parametric functionals. Calc. Var. Partial Differ. Equ. **12**, 85–107 (2000)

Cohn-Vossen, S.

1. Kürzeste Wege and Totalkrümmung auf Flächen. Compos. Math. **2**, 69–133 (1935)

Courant, R.

1. Über direkte Methoden bei Variations- und Randwertproblemen. Jahresber. Dtsch. Math.-Ver. **97**, 90–117 (1925)
2. Über direkte Methoden in der Variationsrechnung und über verwandte Fragen. Math. Ann. **97**, 711–736 (1927)
3. Neue Bemerkungen zum Dirichletschen Prinzip. J. Reine Angew. Math. **165**, 248–256 (1931)
4. On the problem of Plateau. Proc. Natl. Acad. Sci. USA **22**, 367–372 (1936)
5. Plateau's problem and Dirichlet's Principle. Ann. Math. **38**, 679–724 (1937)
6. The existence of a minimal surface of least area bounded by prescribed Jordan arcs and prescribed surfaces. Proc. Natl. Acad. Sci. USA **24**, 97–101 (1938)
7. Remarks on Plateau's and Douglas' problem. Proc. Natl. Acad. Sci. USA **24**, 519–523 (1938)
8. Conformal mapping of multiply connected domains. Duke Math. J. **5**, 314–823 (1939)
9. The existence of minimal surfaces of given topological structure under prescribed boundary conditions. Acta Math. **72**, 51–98 (1940)
10. Soap film experiments with minimal surfaces. Am. Math. Mon. **47**, 168–174 (1940)
11. On a generalized form of Plateau's problem. Trans. Am. Math. Soc. **50**, 40–47 (1941)
12. Critical points and unstable minimal surfaces. Proc. Natl. Acad. Sci. USA **27**, 51–57 (1941)
13. On the first variation of the Dirichlet–Douglas integral and on the method of gradients. Proc. Natl. Acad. Sci. USA **27**, 242–248 (1941)
14. On Plateau's problem with free boundaries. Proc. Natl. Acad. Sci. USA **31**, 242–246 (1945)
15. Dirichlet's principle, conformal mapping, and minimal surfaces. Interscience, New York, 1950

Courant, R., Davids, N.
1. Minimal surfaces spanning closed manifolds. Proc. Natl. Acad. Sci. USA **26**, 194–199 (1940)

Courant, R., Hilbert, D.
1. Methoden der mathematischen Physik, vol. 2. Springer, Berlin, 1937
2. Methods of mathematical physics I. Interscience, New York, 1953
3. Methods of mathematical physics II. Interscience, New York, 1962

Courant, R., Hurwitz, A.
1. Funktionentheorie, 1st edn. Springer, Berlin, 1922 and 1929 (third edition)

Courant, R., Manel, B., Shiffman, M.
1. A general theorem on conformal mapping of multiply connected domains. Proc. Natl. Acad. Sci. USA **26**, 503–507 (1940)

Courant, R., Robbins, H.
1. What is mathematics? Oxford University Press, London, 1941

Croke, C.B.
1. Some isoperimetric inequalities and eigenvalue estimates. Ann. Sci. Ecole Norm. Super. **13**, 419–435 (1980)
2. A sharp four dimensional isoperimetric inequality. Comment. Math. Helv. **59**, 187–192 (1984)

Croke, C.B., Weinstein, A.
1. Closed curves on convex hypersurfaces and periods of nonlinear oscillations. Invent. Math. **64**, 199–202 (1981)

Davids, N.
1. Minimal surfaces spanning closed manifolds and having prescribed topological position. Am. J. Math. **64**, 348–362 (1942)

De Giorgi, E.
1. Frontiere orientate di misura minima. Seminario Mat. Scuola Norm. Sup. Pisa 1–56, 1961
2. Una extensione del teorema di Bernstein. Ann. Sc. Norm. Super. Pisa, Cl. Sci. **19**, 79–85 (1965)

De Giorgi, E., Stampacchia, G.
1. Sulle singolarità eliminabili delle ipersuperficie minimali. Atti Accad. Naz. Lincei, VIII Ser., Rend. Cl. Sci. Fis. Mat. Nat. **38**, 352–357 (1965)

Delaunay, C.
1. Sur la surface de revolution dont la courbure moyenne est constante. J. Math. Pures Appl. **6**, 309–315 (1841)

Desideri, L.
1. Desideri, L. Problème de Plateau, équations fuchsiennes et problème de Riemann-Hilbert. Thèse de Doctorat, Université Paris VII, 2009

Dierkes, U.
1. Singuläre Variationsprobleme und Hindernisprobleme. Bonner Math. Schriften **155**. Mathematisches Institut der Universität Bonn, Bonn, 1984
2. Plateau's problem for surfaces of prescribed mean curvature in given regions. Manuscr. Math. **56**, 313–331 (1986)
3. An inclusion principle for a two-dimensional obstacle problem. Preprint No. 772, SFB 72, Bonn

4. A geometric maximum principle, Plateau's problem for surfaces of prescribed mean curvature, and the two-dimensional analogue of the catenary. In: Hildebrandt, S., Leis, R. (eds.) Partial differential equations and calculus of variations. Lect. Notes Math. **1357**, pp. 116–141. Springer, Berlin, 1988
5. Minimal hypercones and $C^{0,1/2}$-minimizers for a singular variational problem. Indiana Univ. Math. J. **37**, 841–863 (1988)
6. A geometric maximum principle for surfaces of prescribed mean curvature in Riemannian manifolds. Z. Anal. Ihre Anwend. **8**(2), 97–102 (1989)
7. Boundary regularity for solutions of a singular variational problem with linear growth. Arch. Ration. Mech. Anal. **105**(4), 285–298 (1989)
8. A classification of minimal cones in $\mathbb{R}^n \times \mathbb{R}^+$ and a counterexample to interior regularity of energy minimizing functions. Manuscr. Math. **63**, 173–192 (1989)
9. Singuläre Lösungen gewisser mehrdimensionaler Variationsprobleme. Habilitationsschrift, Saarbrücken, 1989
10. On the non-existence of energy stable minimal cones. Ann. Inst. Henri Poincaré, Anal. Non Linéaire **7**, 589–601 (1990)
11. Maximum principles and nonexistence results for minimal submanifolds. Manuscr. Math. **69**, 203–218 (1990)
12. A Bernstein result for energy minimizing hypersurfaces. Calc. Var. Partial Differ. Equ. **1**, 37–54 (1993)
13. Curvature estimates for minimal hypersurfaces in singular spaces. Invent. Math. **122**, 453–473 (1995)
14. Singular minimal surfaces. In: Hildebrandt, S., Karcher, H. (eds.) Geometric analysis and nonlinear partial differential equations, pp. 177–193. Springer, Berlin, 2003
15. On the regularity of solutions for a singular variational problem. Math. Z. **225**, 657–670 (1997)

Dierkes, U., Hildebrandt, S., Küster, A., Wohlrab, O.
1. Minimal surfaces I, II. Grundlehren der math. Wiss. **295** & **296**. Springer, Berlin, 1992

Dierkes, U., Hildebrandt, S., Lewy, H.
1. On the analyticity of minimal surfaces at movable boundaries of prescribed length. J. Reine Angew. Math. **379**, 100–114 (1987)

Dierkes, U., Huisken, G.
1. The N-dimensional analogue of the catenary: existence and non-existence. Pac. J. Math. **141**, 47–54 (1990)
2. The N-dimensional analogue of the catenary: prescribed area. In: Jost, J. (ed.) Calculus of variations and geometric analysis, pp. 1–13. International Press, New York, 1996

Dierkes U., Schwab, D.
1. Maximum principles for submanifolds of arbitrary codimension and bounded mean curvature. Calc. Var. Partial Differ. Equ. **22**, 173–184 (2005)

Dinghas, A.
1. Über Minimalabbildungen von Gebieten der komplexen Ebene in einen Hilbert-Raum. Jahresber. Dtsch. Math.-Ver. **67**, 43–48 (1964)
2. Über einen allgemeinen Verzerrungssatz für beschränkte Minimalflächen. Jahresber. Dtsch. Math.-Ver. **69**, 152–160 (1967)

do Carmo, M.
1. Differential geometry of curves and surfaces. Prentice-Hall, Englewood Cliffs, 1976
2. Stability of minimal submanifolds. In: Global differential geometry and global analysis. Lect. Notes Math. **838**. Springer, Berlin, 1981
3. Riemannian geometry. Birkhäuser, Boston, 1992

Dombrowski, P.
1. Krümmungsgrößen gleichungsdefinierter Untermannigfaltigkeiten Riemannscher Mannigfaltigkeiten. Math. Nachr. **38**, 133–180 (1968)

2. 150 years after Gauss, "Disquisitiones generales circa superficies curvas". Astérisque **62**, 1979 (cf. also: Differentialgeometrie—150 Jahre nach den "Disquisitiones generales circa superficies curvas" von Carl Friedrich Gauß. Abh. Braunschweig. Wiss. Ges. **27**, 63–101 (1977))
3. Differentialgeometrie. In: Festschrift zum Jubiläum der Dtsch. Math. Ver., pp. 323–360. Vieweg, Braunschweig, 1990

Douglas, J.
1. Reduction of the problem of Plateau to an integral equation. Bull. Am. Math. Soc. **33**, 143–144 (1927)
2. Reduction to integral equations of the problem of Plateau for the case of two contours. Bull. Am. Math. Soc. **33**, 259 (1927)
3. Reduction of the problem of Plateau to the minimization of a certain functional. Bull. Am. Math. Soc. **34**, 405 (1928)
4. A method of numerical solution of the problem of Plateau. Ann. Math. **29**, 180–188 (1928)
5. Solution of the problem of Plateau. Bull. Am. Math. Soc. **35**, 292 (1929)
6. Various forms of the fundamental functional in the problem of Plateau and its relation to the area functional. Bull. Am. Math. Soc. **36**, 49–50 (1930)
7. Solution of the problem of Plateau for any rectifiable contour in n-dimensional euclidean space. Bull. Am. Math. Soc. **36**, 189 (1930)
8. Solution of the problem of Plateau when the contour is an arbitrary Jordan curve in n-dimensional euclidean space. I. Bull. Am. Math. Soc. **36**, 189–190 (1930); II, Bull. Am. Math. Soc. **36**, 190 (1930)
9. The problem of Plateau and the theorem of Osgood–Carathéodory on the conformal mapping of Jordan regions. Bull. Am. Math. Soc. **36**, 190–191 (1930)
10. A general formulation of the problem of Plateau. Bull. Am. Math. Soc. **36**, 50 (1930)
11. The mapping theorem of Koebe and the problem of Plateau. J. Math. Phys. **10**, 106–130 (1930–31)
12. Solution of the problem of Plateau. Trans. Am. Math. Soc. **33**, 263–321 (1931)
13. The problem of Plateau for two contours. J. Math. Phys. **10**, 315–359 (1931)
14. The least area property of the minimal surface determined by an arbitrary Jordan contour. Proc. Natl. Acad. Sci. USA **17**, 211–216 (1931)
15. One-sided minimal surfaces with a given boundary. Trans. Am. Math. Soc. **34**, 731–756 (1932)
16. Seven theorems in the problem of Plateau. Proc. Natl. Acad. Sci. USA **18**, 83–85 (1932)
17. The problem of Plateau. Bull. Am. Math. Soc. **39**, 227–251 (1933)
18. An analytic closed space curve which bounds no orientable surface of finite area. Proc. Natl. Acad. Sci. USA **19**, 448–451 (1933)
19. A Jordan space curve which bounds no finite simply connected area. Proc. Natl. Acad. Sci. USA **19**, 269–271 (1933)
20. Crescent-shaped minimal surfaces. Proc. Natl. Acad. Sci. USA **19**, 192–199 (1933)
21. A Jordan curve no arc of which can form part of a contour which bounds a finite area. Ann. Math. **35**, 100–104 (1934)
22. Minimal surfaces general topological structure with any finite number of assigned boundaries. J. Math. Phys. **15**, 105–123 (1936)
23. Some new results in the problem of Plateau. J. Math. Phys. **15**, 55–64 (1936)
24. Remarks on Riemann's doctoral dissertation. Proc. Natl. Acad. Sci. USA **24**, 297–302 (1938)
25. Minimal surfaces of higher topological structure. Proc. Natl. Acad. Sci. USA **24**, 343–353 (1938)
26. Green's function and the problem of Plateau. Proc. Natl. Acad. Sci. USA **24**, 353–360 (1938)

27. The most general form of the problem of Plateau. Proc. Natl. Acad. Sci. USA **24**, 360–364 (1938)
28. Minimal surfaces of higher topological structure. Ann. Math. **40**, 205–298 (1939)
29. The higher topological form of Plateau's problem. Ann. Sc. Norm. Super. Pisa, Cl. Sci. **8**, 195–218 (1939)
30. Green's function and the problem of Plateau. Am. J. Math. **61**, 545–589 (1939)
31. The most general form of the problem of Plateau. Am. J. Math. **61**, 590–608 (1939)

Douglas, J., Franklin, P.
1. A step-polygon of a denumerable infinity of sides which bounds no finite area. Proc. Natl. Acad. Sci. USA **19**, 188–191 (1933)

Dubrovin, B.A., Fomenko, A.T., Novikov, S.P.
1. Modern geometry—methods and applications I, II. Springer, Berlin, 1984 and 1985

Duzaar, F.
1. Ein teilweise freies Randwertproblem für Ströme vorgeschriebener mittlerer Krümmung. Habilitationsschrift, Düsseldorf (1990)
2. Existence and regularity of hypersurfaces with prescribed mean curvature and a free boundary. J. Reine Angew. Math. **457**, 23–43 (1994)
3. On the existence of surfaces with prescribed mean curvature and boundary in higher dimensions. Ann. Inst. Henri Poincaré, Anal. Non Linéaire **10**, 191–214 (1993)
4. Boundary regularity for area minimizing currents with prescribed volume. J. Geom. Anal. **7**, 585–592 (1997)

Duzaar, F., Fuchs, M.
1. On the existence of integral currents with prescribed mean curvature. Manuscr. Math. **67**, 41–67 (1990)
2. On the existence of integral currents with constant mean curvature. Rend. Semin. Mat. Univ. Padova **85**, 79–103 (1991)
3. Einige Bemerkungen über die Existenz orientierter Mannigfaltigkeiten mit vorgeschriebener mittlerer Krümmungsform. Z. Anal. Ihre Anwend. **10**, 525–534 (1991)
4. A general existence theorem for integral currents with prescribed mean curvature form. Boll. Unione Mat. Ital. **6-B**, 901–912 (1992)

Duzaar, F., Kuwert, E.
1. Minimization of conformally invariant energies in homotopy classes. Calc. Var. Partial Differ. Equ. **6**, 285–313 (1998)

Duzaar, F., Steffen, K.
1. Area minimizing hypersurfaces with prescribed volume and boundary. Math. Z. **209**, 581–618 (1992)
2. λ minimizing currents. Manuscr. Math. **80**, 403–447 (1993)
3. Boundary regularity for minimizing currents with prescribed mean curvature. Calc. Var. Partial Differ. Equ. **1**, 355–406 (1993)
4. Comparison principles for hypersurfaces of prescribed mean curvature. J. Reine Angew. Math. **457**, 71–83 (1994)
5. Existence of hypersurfaces with prescribed mean curvature in Riemannian manifolds. Indiana Univ. Math. J. **45**, 1045–1093 (1996)
6. The Plateau problem for parametric surfaces with prescribed mean curvature. In: Jost, J. (ed.) Geometric analysis and the calculus of variations, pp. 13–70. International Press, Cambridge, 1996
7. Parametric surfaces of least H-energy in a Riemannian manifold. Math. Ann. **314**, 197–244 (1999)

Dziuk, G.

1. Das Verhalten von Lösungen semilinearer elliptischer Systeme an Ecken eines Gebietes. Math. Z. **159**, 89–100 (1978)
2. Das Verhalten von Flächen beschränkter mittlerer Krümmung an C^1-Randkurven. Nachr. Akad. Wiss. Gött. II. Math.-Phys. Kl., 21–28 (1979)
3. On the boundary behavior of partially free minimal surfaces. Manuscr. Math. **35**, 105–123 (1981)
4. Über quasilineare elliptische Systeme mit isothermen Parametern an Ecken der Randkurve. Analysis **1**, 63–81 (1981)
5. Über die Stetigkeit teilweise freier Minimalflächen. Manuscr. Math. **36**, 241–251 (1981)
6. Über die Glattheit des freien Randes bei Minimalflächen. Habilitationsschrift, Aachen, 1982
7. C^2-Regularity for partially free minimal surfaces. Math. Z. **189**, 71–79 (1985)
8. On the length of the free boundary of a minimal surface. Control Cybern. **14**, 161–170 (1985)
9. Finite elements for the Beltrami operator on arbitrary surfaces. In: Hildebrandt, S., Leis, R. (eds.) Partial differential equations and calculus of variations. Lect. Notes Math. **1357**, pp. 142–155. Springer, Berlin, 1988
10. An algorithm for evolutionary surfaces. Preprint, SFB 256, Report No. 5, Bonn, 1989
11. Branch points of polygonally bounded minimal surfaces. Analysis **5**, 275–286 (1985)

Ecker, K.

1. Area-minimizing integral currents with movable boundary parts of prescribed mass. Ann. Inst. Henri Poincaré, Anal. Non Linéaire **6**, 261–293 (1989)
2. Local techniques for mean curvature flow. In: Proc. of Conference on Theoretical and Numerical Aspects of Geometric Variational Problems. Proc. Centre Math. Appl. Aust. Nat. Univ. **26**, pp. 107–119. Australian National University Press, Canberra, 1991
3. Regularity theory for mean curvature flow. Birkhäuser, Basel, 2004

Ecker, K., Huisken, G.

1. A Bernstein result for minimal graphs of controlled growth. J. Differ. Geom. **31**, 337–400 (1990)
2. Interior curvature estimates for hypersurfaces of prescribed mean curvature. Ann. Inst. Henri Poincaré, Anal. Non Linéaire **6**, 251–260 (1989)

Eells, J.

1. Minimal graphs. Manuscr. Math. **28**, 101–108 (1979)

Eells, J., Lemaire, L.

1. A report on harmonic maps. Bull. Lond. Math. Soc. **10**, 1–68 (1978)
2. On the construction of harmonic and holomorphic maps between surfaces. Math. Ann. **252**, 27–52 (1980)
3. Deformations of metrics and associated harmonic maps. Proc. Indian Acad. Sci. **90**, 33–45 (1981)
4. Selected topics in harmonic maps. CBMS–NSF Regional Conf. Ser. Appl. Math. **50**. SIAM, Philadelphia, 1983
5. Another report on harmonic maps. Bull. Lond. Math. Soc. **20**, 385–524 (1988)

Eells, J., Sampson, J.H.

1. Harmonic mappings of Riemannian manifolds. Am. J. Math. **86**, 109–160 (1964)

Eisenhart, L.P.

1. A treatise on the differential geometry of curves and surfaces. Ginn, Boston, 1909
2. An introduction to differential geometry. Princeton University Press, Princeton, 1947
3. Riemannian geometry, 5th edn. Princeton University Press, Princeton, 1964

Ekholm, T., White, B., Wienholtz, D.
1. Embeddedness of minimal surfaces with total boundary curvature at most 4π. Ann. Math. **155**, 209–234 (2002)

Eschenburg, J.H.
1. Maximum principle for hypersurfaces. Manuscr. Math. **64**, 55–75 (1989)

Eschenburg, J.H., Jost, J.
1. Differentialgeometrie und Minimalflächen, 256 pp. Springer, Berlin, 2007

Eschenburg, J.H., Tribuzy, R.
1. Branch points of conformal mappings of surfaces. Math. Ann. **279**, 621–633 (1988)

Euler, L.
1. Recherches sur la courbure de surfaces. Mém. Acad. Sci. Berl. **16**, 119–143 (1767)
2. De solidis quorum superficiem in planum explicare licet. Novi Comment. Acad. Sci. Petropol **16**, 3–34 (1772)
3. (i) De repraesentatione superficiei sphaericae super plano, (ii) De projectione geographica superficiei sphaericae. Acta Acad. Sci. Petropol. 1777: **I**, 107–132 & 133–142 (1778)
4. Commentationes geometricae. Opera omnia, Series prima, vols. 28 and 29, Lausanne 1955 and 1956
5. Methodus inveniendi lineas curvas maximi minimive proprietate gaudentes sive solutio problematis isoperimetrici latissimo sensu accepti. Lausannae et Genevae, Bousquet et Socios, 1744 (Opera omnia, Series I, vol. 24)
6. De insigni paradoxo quod in analysi maximorum et minimorum occurrit. Mém. Acad. Imp. Sci. St.-Pétersbg. **3**, 16–25 (1811) (Opera omnia, Ser. I, vol. 25, 286–292)

Federer, H.
1. Geometric measure theory. Grundlehren Math. Wiss. Springer, Berlin, 1969
2. Some theorems on integral currents. Trans. Am. Math. Soc. **117**, 43–67 (1965)
3. The singular sets of area minimizing rectifiable currents with codimension one and of area minimizing flat chains modulo two with arbitrary codimension. Bull. Am. Math. Soc. **76**, 767–771 (1970)

Federer, H., Fleming, W.H.
1. Normal and integral currents. Ann. Math. **72**, 458–520 (1960)

Feinberg, J.M.
1. The isoperimetric inequality for doubly-connected minimal surfaces in $\mathbb{R}^3$. J. Anal. Math. **32**, 249–278 (1977)
2. Some Wirtinger-like inequalities. SIAM J. Math. Anal. **10**, 1258–1271 (1979)

Ferus, D., Karcher, H.
1. , Karcher, H. Non rotational minimal spheres and minimizing cones. Comment. Math. Helv. **60**, 247–269 (1985)

Fiala, F.
1. Le problème des isoperimètres sur les surfaces ouvertes à courbure positive. Comment. Math. Helv. **13**, 293–346 (1941)

Finn, R.
1. Isolated singularities of solutions of non-linear partial differential equations. Trans. Am. Math. Soc. **75**, 383–404 (1953)
2. A property of minimal surfaces. Proc. Natl. Acad. Sci. USA **39**, 197–201 (1953)
3. On equations of minimal surface type. Ann. Math. (2) **60**, 397–416 (1954)
4. On a problem of minimal surface type, with application to elliptic partial differential equations. Arch. Ration. Mech. Anal. **3**, 789–799 (1954)
5. Growth properties of solutions of non-linear elliptic equations. Commun. Pure Appl. Math. **9**, 415–423 (1956)

6. On partial differential equations (whose solutions admit no isolated singularities). Scr. Math. **26**, 107–115 (1961)
7. Remarks on my paper "On equations of minimal surface type". Ann. Math. (2) **80**, 158–159 (1964)
8. New estimates for equations of minimal surface type. Arch. Ration. Mech. Anal. **14**, 337–375 (1963)
9. Remarks relevant to minimal surfaces and to surfaces of prescribed mean curvature. J. Anal. Math. **14**, 139–160 (1965)
10. On a class of conformal metrics, with application to differential geometry in the large. Comment. Math. Helv. **40**, 1–30 (1965)
11. Equilibrium capillary surfaces. Springer, New York, 1986
12. The Gauß curvature of an H-graph. Nachr. Ges. Wiss. Gött., Math.-Phys.-Kl. **2** (1987)

Finn, R., Osserman, R.
1. On the Gauss curvature of non-parametric minimal surfaces. J. Anal. Math. **12**, 351–364 (1964)

Frehse, J.
1. On the regularity of the solution of a second order variational inequality. Boll. Unione Mat. Ital. (4) **6**, 312–315 (1972)
2. Two dimensional variational problems with thin obstacles. Math. Z. **143**, 279–288 (1975)
3. On Signorini's problem and variational problems with thin obstacles. Ann. Sc. Norm. Super. Pisa, Cl. Sci., Ser. IV **4**, 343–363 (1977)
4. Un problème variationnel bidimensionnel possédant des extremales bornées et discontinues. C. R. Acad. Sci. Paris, Sér. A **289**, 751–753 (1979)

Gage, M.
1. A proof of Gehring's linked spheres conjecture. Duke Math. J. **47**, 615–620 (1980)

Galilei, G.
1. Discorsi e dimostrazioni matematiche intorno à due nuove scienze. Elsevier, Leyden, 1638

Galle, A.
1. Über die geodätischen Arbeiten von Gauß. In: Gauß, Werke, Bd. 11.2, 165 pp. Akad. Wiss. Göttingen (1924–1929)

Gallot, S.
1. Isoperimetric inequalities based on integral norms of Ricci curvature. Astérisque **157–158**, 191–216 (1988)
2. Inégalités isopérimétriques et analytiques sur les variétés Riemanniennes. Astérisque **163–164**, 31–91 (1988)

Garnier, R.
1. Solution du problème de Riemann pour les systèmes différentiels linéaires du second ordre. Ann. Sci. Ecole Norm. Super. (3) **43**, 177–307 (1926)
2. Le problème de Plateau. Ann. Sci. Ecole Norm. Super. (3) **45**, 53–144 (1928)
3. Sur une théorème de Schwarz. Comment. Math. Helv. **25**, 140–172 (1951)
4. Sur le problème de Plateau pour un quadrilatère variable qui peut acquérir un point double. Ann. Mat. Pura Appl. (4) **58**, 1–34 (1962)
5. Sur le problème de Plateau pour les quadrilatères gauches ayant un sommet à l'infini. J. Math. Pures Appl. (9) **41**, 241–271 (1962)

Garofalo, N., Nhieu, D.-M.
1. Isoperimetric and Sobolev inequalities for Carnot–Carathéodory spaces and the existence of minimal surfaces. Commun. Pure Appl. Math. **49**, 1081–1144 (1996)

Gauß, C.F.

1. Allgemeine Auflösung der Aufgabe, die Theile einer gegebnen Fläche auf einer andern gegebnen Fläche so abzubilden, daß die Abbildung dem Abgebildeten in den kleinsten Theilen ähnlich wird. Astronomische Abhandlungen herausgeg. von H.C. Schumacher, Drittes Heft, Altona (1825)
2. Werke, Band 4 (Wahrscheinlichkeitsrechnung und Geometrie). Band 8 (Nachträge zu Band 1–4), Akad. Wiss. Göttingen, 1880 und 1900
3. Disquisitiones generales circa superficies curvas. Gött. Nachr. **6**, 99–146 (1828). German Translation: Allgemeine Flächentheorie. Herausgeg. von A. Wangerin, Ostwald's Klassiker, Engelmann, Leipzig, 1905 (cf. also Dombrowski [2], and General investigations of curved surfaces. Raven Press, New York, 1965)

Gergonne, J.D.

1. Questions proposées/résolues. Ann. Math. Pure Appl. **7**, 68, 99–100, **156**, 143–147 (1816)

Gerhardt, C.

1. Regularity of solutions of nonlinear variational inequalities. Arch. Ration. Mech. Anal. **52**, 389–393 (1973)

Gericke, H.

1. Zur Geschichte des isoperimetrischen Problems. Math. Semesterber. **29**, 160–187 (1982)

Geveci, T.

1. On differentiability of minimal surfaces at a boundary point. Proc. Am. Math. Soc. **28**, 213–218 (1971)

Giaquinta, M.

1. Multiple integrals in the calculus of variations and nonlinear elliptic systems. Ann. Math. Stud. **105**. Princeton University Press, Princeton, 1983
2. On the Dirichlet problem for surfaces of prescribed mean curvature. Manuscr. Math. **12**, 73–86 (1974)

Giaquinta, M., Hildebrandt, S.

1. Calculus of variations, vols. I, II. Grundlehren Math. Wiss. **310** & **311**. Springer, Berlin, 1996, 2nd edn. 2004

Giaquinta, M., Martinazzi, L.

1. An introduction to the regularity theory for elliptic systems, harmonic maps and minimal graphs. Scuola Normale Superiore, Pisa, 2005

Giaquinta, M., Modica, G., Souček, J.

1. Cartesian currents in the calculus of variations, vols. I, II. Ergebnisse Math. Grenzgeb. **37** & **38**, 3rd edn. Springer, Berlin, 1998

Giaquinta, M., Pepe, L.

1. Esistenza e regolarità per il problema dell'area minima con ostacolo in n variabili. Ann. Sc. Norm. Super. Pisa, Cl. Sci. **25**, 481–507 (1971)

Gilbarg, D., Trudinger, N.S.

1. Elliptic partial differential equations of second order. Grundlehren Math. Wiss. **224**. Springer, Berlin, 1977. 2nd edn. 1983

Giusti, E.

1. Superficie minime cartesiane con ostacoli discontinui. Arch. Ration. Mech. Anal. **40**, 251–267 (1971)
2. Nonparametric minimal surfaces with discontinuous and thin obstacles. Arch. Ration. Mech. Anal. **49**, 41–56 (1972)
3. Boundary behavior of non-parametric minimal surfaces. Indiana Univ. Math. J. **22**, 435–444 (1972)

4. Minimal surfaces and functions of bounded variation. Birkhäuser, Boston, 1984
5. Harmonic mappings and minimal immersions. Lect. Notes Math. **1161**. Springer, Berlin, 1985
6. On the regularity of the solution to a mixed boundary value problem for the non-homogeneous minimal surface equation. Boll. U. M. I. **11**, 349–374
7. Boundary value problems for non-parametric surfaces of prescribed mean curvature. Ann. Sc. Norm. Super. Pisa, Cl. Sci., Sér. 4 **3**, 501–548 (1976)

Glaeser, L.
1. The work of Frei Otto and his team 1955–1976. Information of the Institut für leichte Flächentragwerke (IL, Institute for Lightweight Structures), Stuttgart, 1978

Goldhorn, K.H.
1. Flächen beschränkter mittlerer Krümmung in einer dreidimensionalen Riemannschen Mannigfaltigkeit. Manuscr. Math. **8**, 189–207 (1973)

Goldhorn, K., Hildebrandt, S.
1. Zum Randverhalten der Lösungen gewisser zweidimensionaler Variationsprobleme mit freien Randbedingungen. Math. Z. **118**, 241–253 (1970)

Goldschmidt, B.
1. Determinatio superficiei minimae rotatione curvae data duo puncta jungentis circa datum axem ortae. Dissertation, Göttingen, 1831

Gonzales, E., Massari, U., Tamanini, I.
1. On the regularity of boundaries of sets minimizing perimeter with a volume constraint. Indiana Univ. Math. J. **32**, 25–37 (1983)

Gornik, K.
1. Zum Regularitätsverhalten parametrischer elliptischer Variationsprobleme mit Ungleichungen als Nebenbedingungen. Thesis. Bonner Math. Schriften **80**. Mathematisches Institut der Universität Bonn, Bonn, 1975
2. Ein Stetigkeitssatz für Variationsprobleme mit Ungleichungen als Nebenbedingung. Math. Z. **152**, 89–97 (1976)
3. Ein Differenzierbarkeitssatz für Lösungen zweidimensionaler Variationsprobleme mit "zweischaligem Hindernis". Arch. Ration. Mech. Anal. **64**, 127–135 (1977)

Greenberg, M.J.
1. Lectures on algebraic topology. Benjamin, New York, 1967

Greenberg, M., Harper, J.
1. Algebraic topology: a first course. Benjamin-Cummings, Reading, 1981

Gromoll, D., Meyer, W.
1. On differentiable functions with isolated critical points. Topology **8**, 361–369 (1969)

Gromoll, D., Klingenberg, W., Meyer, W.
1. Riemannsche Geometrie im Großen. Lect. Notes Math. **55**. Springer, Berlin, 1968

Gromov, M.
1. Filling Riemannian manifolds. J. Differ. Geom. **18**, 1–14 (1983)

Grüter, M.
1. Über die Regularität schwacher Lösungen des Systems $\Delta x = 2H(x)x_u \wedge x_v$. Dissertation, Düseldorf, 1979
2. Regularity of weak H-surfaces. J. Reine Angew. Math. **329**, 1–15 (1981)
3. A note on variational integrals which are conformally invariant. Preprint 502, SFB 72, Bonn, 1982
4. Conformally invariant variational integrals and the removability of isolated singularities. Manuscr. Math. **47**, 85–104 (1984)

5. Regularität von minimierenden Strömen bei einer freien Randwertbedingung. Habilitationsschrift, Düsseldorf, 1985
6. Regularity results for minimizing currents with a free boundary. J. Reine Angew. Math. **375/376**, 307–325 (1987)
7. Eine Bemerkung zur Regularität stationärer Punkte von konform invarianten Variationsintegralen. Manuscr. Math. **55**, 451–453 (1986)
8. The monotonicity formula in geometric measure theory, and an application to partially free boundary problems. In: Hildebrandt, S., Leis, R. (eds.) Partial differential equations and calculus of variations. Lect. Notes Math. **1357**, pp. 238–255. Springer, Berlin, 1988
9. Boundary regularity for solutions of a partitioning problem. Arch. Ration. Mech. Anal. **97**(3), 261–270 (1987)
10. Optimal regularity for codimension-one minimal surfaces with a free boundary. Manuscr. Math. **58**, 295–343 (1987)
11. Free boundaries in geometric measure theory and applications. In: Concus, P., Finn, R. (eds.) Variational methods for free surface interfaces. Springer, Berlin, 1987

Grüter, M., Hildebrandt, S., Nitsche, J.C.C.

1. On the boundary behavior of minimal surfaces with a free boundary which are not minima of the area. Manuscr. Math. **35**, 387–410 (1981)
2. Regularity for surfaces of constant mean curvature with free boundaries. Acta Math. **156**, 119–152 (1986)

Grüter, M., Jost, J.

1. On embedded minimal disks in convex bodies. Ann. Inst. Henri Poincaré, Anal. Non Linéaire **3**, 345–390 (1986)
2. Allard-type regularity results for varifolds with free boundaries. Ann. Sc. Norm. Super. Pisa, Cl. Sci., Ser. IV **13**(1), 129–169 (1986)

Günther, P.

1. Einige Vergleichssätze über das Volumenelement eines Riemannschen Raumes. Publ. Math. (Debr.) **7**, 258–287 (1960)

Gulliver, R.

1. Existence of surfaces with prescribed mean curvature vector. Math. Z. **131**, 117–140 (1973)
2. Regularity of minimizing surfaces of prescribed mean curvature. Ann. Math. **97**, 275–305 (1973)
3. The Plateau problem for surfaces of prescribed mean curvature in a Riemannian manifold. J. Differ. Geom. **8**, 317–330 (1973)
4. Branched immersions of surfaces and reduction of topological type, I. Math. Z. **145**, 267–288 (1975)
5. Finiteness of the ramified set for branched immersions of surfaces. Pac. J. Math. **64**, 153–165 (1976)
6. Removability of singular points on surfaces of bounded mean curvature. J. Differ. Geom. **11**, 345–350 (1976)
7. Branched immersions of surfaces and reduction of topological type, II. Math. Ann. **230**, 25–48 (1977)
8. Representation of surfaces near a branched minimal surface. In: Minimal submanifolds and Geodesics, pp. 31–42. Kaigai Publications, Tokyo, 1978
9. Index and total curvature of complete minimal surfaces. Proc. Symp. Pure Math. **44**, 207–212 (1986)
10. Minimal surfaces of finite index in manifolds of positive scalar curvature. In: Hildebrandt, S., Kinderlehrer, D., Miranda, M. (eds.) Calculus of variations and partial differential equations. Lect. Notes Math. **1340**, pp. 115–122. Springer, Berlin, 1988

11. A minimal surface with an atypical boundary branch point. In: Differential geometry. Pitman Monographs Surveys Pure Appl. Math. **52**, pp. 211–228. Longman, Harlow, 1991
12. Convergence of minimal submanifolds to a singular variety. Preprint No. 76, SFB 256, Bonn, 1989
13. On the non-existence of a hypersurface of prescribed mean curvature with a given boundary. Manuscr. Math. **11**, 15–39 (1974)
14. Necessary conditions for submanifolds and currents with prescribed mean curvature vector. In: Bombieri, E. (ed.) Seminar on minimal submanifolds. Ann. Math. Stud. **103**. Princeton University Press, Princeton, 1983
15. Minimal surfaces of finite index in manifolds of positive scalar curvature. In: Hildebrandt, S., Kinderlehrer, D., Miranda, M. (eds.) Lect. Notes Math. **1340**, pp. 115–122. Springer, Berlin, 1988

Gulliver, R., Hildebrandt, S.
1. Boundary configurations spanning continua of minimal surfaces. Manuscr. Math. **54**, 323–347 (1986)

Gulliver, R., Lawson, H.B.
1. The structure of stable minimal surfaces near a singularity. Proc. Symp. Pure Math. **44**, 213–237 (1986)
2. The structure of stable minimal hypersurfaces near a singularity. Proc. Symp. Pure Math. **44**, 213–237 (1986)

Gulliver, R., Lesley, F.D.
1. On boundary branch points of minimizing surfaces. Arch. Ration. Mech. Anal. **52**, 20–25 (1973)

Gulliver, R., Osserman, R., Royden, H.L.
1. A theory of branched immersions of surfaces. Am. J. Math. **95**, 750–812 (1973)

Gulliver, R., Scott, P.
1. Least area surfaces can have excess triple points. Topology **26**, 345–359 (1987)

Gulliver, R.D., Spruck, J.
1. The Plateau problem for surfaces of prescribed mean curvature in a cylinder. Invent. Math. **13**, 169–178 (1971)
2. Existence theorems for parametric surfaces of prescribed mean curvature. Indiana Univ. Math. J. **22**, 445–472 (1972)
3. On embedded minimal surfaces. Ann. Math. **103**, 331–347 (1976), with a correction in Ann. Math. **109**, 407–412 (1979)
4. Surfaces of constant mean curvature which have a simple projection. Math. Z. **129**, 95–107 (1972)

Gulliver, R., Tomi, F.
1. On false branch points of incompressible branched immersions. Manuscr. Math. **63**, 293–302 (1989)

Hahn, J., Polthier, K.
1. Bilder aus der Differentialgeometrie. Kalender 1987, Computergraphiken. Vieweg, Braunschweig, 1987

Hall, P.
1. Topological properties of minimal surfaces. Thesis, Warwick, 1983
2. Two topological examples in minimal surfaces theory. J. Differ. Geom. **19**, 475–481 (1984)
3. On Sasaki's inequality for a branched minimal disc. Preprint, 1985

4. A Picard theorem with an application to minimal surfaces. Trans. Am. Math. Soc. **314**, 597–603 (1989); II. Trans. Am. Math. Soc. **325**, 597–603 (1991)

Hardt, R.
1. Topological properties of subanalytic sets. Trans. Am. Math. Soc. **211**, 57–70 (1975)
2. An introduction to geometric measure theory. Lecture Notes, Melbourne University, 1979

Hardt, R., Simon, L.
1. Boundary regularity and embedded minimal solutions for the oriented Plateau problem. Ann. Math. **110**, 439–486 (1979)

Hardy, G.H., Littlewood, J.E., Pólya, G.
1. Inequalities., 2nd edn. Cambridge University Press, Cambridge, 1952

Harth, F.P.
1. Minimalflächen mit freiem Rand in Riemannschen Mannigfaltigkeiten. Manuscr. Math. **7**, 35–54 (1972)
2. Zur Regularität von H-Flächen mit freiem Rand. Math. Z. **150**, 71–74 (1976)

Hartman, P.
1. On homotopic harmonic maps. Can. J. Math. **29**, 673–687 (1987)

Hartman, P., Wintner, A.
1. On the local behavior of solutions of nonparabolic partial differential equations. Am. J. Math. **75**, 449–476 (1953)

Harvey, R., Lawson, B.
1. On boundaries of complex analytic varieties. I: Ann. Math. (2) **102**, 233–290 (1975); II: Ann. Math. (2) **106**, 213–238 (1977)
2. Extending minimal varieties. Invent. Math. **28**, 209–226 (1975)
3. Calibrated foliations. Am. J. Math. **103**, 411–435 (1981)
4. Calibrated geometries. Acta Math. **148**, 47–157 (1982)

Heinz, E.
1. Über die Lösungen der Minimalflächengleichung. Nachr. Akad. Wiss. Gött., Math.-Phys. Kl., 51–56 (1952)
2. Über die Existenz einer Fläche konstanter mittlerer Krümmung bei vorgegebener Berandung. Math. Ann. **127**, 258–287 (1954)
3. Über die Eindeutigkeit beim Cauchyschen Anfangswertproblem einer elliptischen Differentialgleichung 2, Ordnung. Nachr. Akad. Wiss. Gött., Math.-Phys. Kl., 1–12 (1955)
4. On the existence problem for surfaces of constant mean curvature. Commun. Pure Appl. Math. **9**, 467–470 (1956)
5. On certain nonlinear elliptic differential equations and univalent mappings. J. Anal. Math. **5**, 197–272 (1956/57)
6. On one-to-one harmonic mappings. Pac. J. Math. **9**, 101–105 (1959)
7. Existence theorems for one-to-one mappings associated with elliptic systems of second order I. J. Anal. Math. **15**, 325–352 (1962)
8. Über das Nichtverschwinden der Funktionaldeterminante bei einer Klasse eineindeutiger Abbildungen. Math. Z. **105**, 87–89 (1968)
9. Zur Abschätzung der Funktionaldeterminante bei einer Klasse topologischer Abbildungen. Nachr. Akad. Wiss. Gött., Math.-Phys. Kl., 183–197 (1968)
10. Ein Regularitätssatz für Flächen beschränkter mittlerer Krümmung. Nachr. Akad. Wiss. Gött. Math.-Phys. Kl., 2B **12**, 107–118 (1969)
11. An inequality of isoperimetric type for surfaces of constant mean curvature. Arch. Ration. Mech. Anal. **33**, 155–168 (1969)
12. On the nonexistence of a surface of constant mean curvature with finite area and prescribed rectifiable boundary. Arch. Ration. Mech. Anal. **35**, 249–252 (1969)

13. On surfaces of constant mean curvature with polygonal boundaries. Arch. Ration. Mech. Anal. **36**, 335–347 (1970)
14. Unstable surfaces of constant mean curvature. Arch. Ration. Mech. Anal. **38**, 257–267 (1970)
15. Über des Randverhalten quasilinearer elliptischer Systeme mit isothermen Parametern. Math. Z. **113**, 99–105 (1970)
16. Interior gradient estimates for surfaces $z = f(x, y)$ with prescribed mean curvature. J. Differ. Geom. **5**, 149–157 (1971)
17. Elementare Bemerkung zur isoperimetrischen Ungleichung im $\mathbb{R}^3$. Math. Z. **132**, 319–322 (1973)
18. Ein Regularitätssatz für schwache Lösungen nichtlinearer elliptischer Systeme. Nachr. Akad. Wiss Gött., Math.-Phys. Kl. 1–13 (1975)
19. Über die analytische Abhängigkeit der Lösungen eines linearen elliptischen Randwertproblems von Parametern. Nachr. Akad. Wiss. Gött., Math.-Phys. Kl. 1–12 (1979)
20. Über eine Verallgemeinerung des Plateauschen Problems. Manuscr. Math. **28**, 81–88 (1979)
21. Ein mit der Theorie der Minimalflächen zusammenhängendes Variationsproblem. Nachr. Akad. Wiss. Gött., Math.-Phys. Kl. 25–35 (1980)
22. Minimalflächen mit polygonalem Rand. Math. Z. **183**, 547–564 (1983)
23. Zum Marx-Shiffmanschen Variationsproblem. J. Reine Angew. Math. **344**, 196–200 (1983)
24. An estimate for the total number of branch points of quasi-minimal surfaces. Analysis **5**, 383–390 (1985)
25. Zum Plateauschen Problem für Polygone. In: Knobloch, E., Louhivaara, I.S., Winkler, J. (eds.) Zum Werk Leonhard Eulers. Vorträge des Euler-Kolloquiums im Mai 1983 in Berlin, pp. 197–204. Birkhäuser, Basel, 1984
26. Über Flächen mit eineindeutiger Projektion auf eine Ebene, deren Krümmungen durch Ungleichungen eingeschränkt sind. Math. Ann. **129**, 451–454 (1955)

Heinz, E., Hildebrandt, S.
1. Some remarks on minimal surfaces in Riemannian manifolds. Commun. Pure Appl. Math. **23**, 371–377 (1970)
2. On the number of branch points of surfaces of bounded mean curvature. J. Differ. Geom. **4**, 227–235 (1970)

Heinz, E., Tomi, F.
1. Zu einem Satz von S., Hildebrandt über das Randverhalten von Minimalflächen. Math. Z. **111**, 372–386 (1969)

Helein, F.
1. Régularité des applications faiblement harmoniques entre une surface et une varieté riemannienne. C. R. Acad. Sci. Paris, Ser. I **312**, 591–596 (1991)
2. Harmonic maps, conservation laws and moving frames. Cambridge University Press, Cambridge, 2002

Hewitt, E., Stromberg, K.
1. Real and abstract analysis. Springer, Berlin, 1965

Hilbert, D., Cohn-Vossen, S.
1. Anschauliche Geometrie. Springer, Berlin, 1932

Hicks, N.J.
1. Notes on differential geometry. Van Nostrand, Princeton, 1965

Hildebrandt, S.
1. Über das Randverhalten von Minimalflächen. Math. Ann. **165**, 1–18 (1966)

2. Über Minimalflächen mit freiem Rand. Math. Z. **95**, 1–19 (1967)
3. Boundary behavior of minimal surfaces. Arch. Ration. Mech. Anal. **35**, 47–82 (1969)
4. Über Flächen konstanter mittlerer Krümmung. Math. Z. **112**, 107–144 (1969)
5. Randwertprobleme für Flächen mit vorgeschriebener mittlerer Krümmung und Anwendungen auf die Kapillaritätstheorie, I. Fest vorgegebener Rand. Math. Z. **112**, 205–213 (1969)
6. Randwertprobleme für Flächen mit vorgeschriebener mittlerer Krümmung und Anwendungen auf die Kapillaritätstheorie, II. Freie Ränder. Arch. Ration. Mech. Anal. **39**, 275–293 (1970)
7. On the Plateau problem for surfaces of constant mean curvature. Commun. Pure Appl. Math. **23**, 97–114 (1970)
8. Einige Bemerkungen über Flächen beschränkter mittlerer Krümmung. Math. Z. **115**, 169–178 (1970)
9. Ein einfacher Beweis für die Regularität der Lösungen gewisser zweidimensionaler Variationsprobleme unter freien Randbedingungen. Math. Ann. **194**, 316–331 (1971)
10. Über einen neuen Existenzsatz für Flächen vorgeschriebener mittlerer Krümmung. Math. Z. **119**, 267–272 (1971)
11. Maximum principles for minimal surfaces and for surfaces of continuous mean curvature. Math. Z. **128**, 253–269 (1972)
12. On the regularity of solutions of two-dimensional variational problems with obstructions. Commun. Pure Appl. Math. **25**, 479–496 (1972)
13. Interior $C^{1+\alpha}$-regularity of solutions of two-dimensional variational problems with obstacles. Math. Z. **131**, 233–240 (1973)
14. Liouville's theorem for harmonic mappings and an approach to Bernstein theorems. In: Seminar on differential geometry, pp. 107–131. Princeton University Press, Princeton, 1982
15. Nonlinear elliptic systems and harmonic mappings. In: Proc. of the 1980 Beijing Symposium on Differential Geometry and Differential Equations **1**, pp. 481–615. Science Press, Beijing, 1982
16. Minimal surfaces with free boundaries. Miniconference on P.D.E., Canberra, C.M.A., A.N.U. Preprint, August 1985
17. Harmonic mappings of Riemannian manifolds. In: Giusti, E. (ed.) Harmonic mappings and minimal immersions. Lect. Notes Math. **1161**, pp. 1–117. Springer, Berlin, 1985
18. Boundary value problems for minimal surfaces. In: Osserman, R. (ed.) Geometry V, Minimal surfaces. Encycl. Math. Sci. **90**, pp. 153–237. Springer, Berlin, 1997
19. Analysis 2. Springer, Berlin, 2003
20. On Dirichlet's principle and Poincaré's méthode de balayage. Math. Nachr. **278**, 141–144 (2005)

Hildebrandt, S., Jäger, W.

1. On the regularity of surfaces with prescribed mean curvature at a free boundary. Math. Z. **118**, 289–308 (1970)

Hildebrandt, S., Jost, J., Widman, K.O.

1. Harmonic mappings and minimal submanifolds. Invent. Math. **62**, 269–298 (1980)

Hildebrandt, S., Kaul, H.

1. Two-dimensional variational problems with obstructions, and Plateau's problem for H-surfaces in a Riemannian manifold. Commun. Pure Appl. Math. **25**, 187–223 (1972)

Hildebrandt, S., Kaul, H., Widman, K.-O.

1. An existence theorem for harmonic mappings of Riemannian manifolds. Acta Math. **138**, 1–16 (1977)

Hildebrandt, S., Nitsche, J.C.C.
1. Minimal surfaces with free boundaries. Acta Math. **143**, 251–272 (1979)
2. Optimal boundary regularity for minimal surfaces with a free boundary. Manuscr. Math. **33**, 357–364 (1981)
3. A uniqueness theorem for surfaces of least area with partially free boundaries on obstacles. Arch. Ration. Mech. Anal. **79**, 189–218 (1982)
4. Geometric properties of minimal surfaces with free boundaries. Math. Z. **184**, 497–509 (1983)

Hildebrandt, S., Sauvigny, F.
1. Embeddedness and uniqueness of minimal surfaces solving a partially free boundary value problem. J. Reine Angew. Math. **422**, 69–89 (1991)
2. On one-to-one harmonic mappings and minimal surfaces. Nachr. Akad. Wiss. Gött., Math.-Phys. Kl. **3**, 73–93 (1992)
3. Uniqueness of stable minimal surfaces with partially free boundaries. J. Math. Soc. Jpn. **47**, 423–440 (1995)
4. Minimal surfaces in a wedge, I. Asymptotic expansions. Calc. Var. Partial Differ. Equ. **5**, 99–115 (1997)
5. Minimal surfaces in a wedge, II. The edge-creeping phenomenon. Arch. Math. **69**, 164–176 (1997)
6. Minimal surfaces in a wedge, III. Existence of graph solutions and some uniqueness results. J. Reine Angew. Math. **514**, 71–101 (1999)
7. Minimal surfaces in a wedge, IV. Hölder estimates of the Gauss map and a Bernstein theorem. Calc. Var. Partial Differ. Equ. **8**, 71–90 (1999)
8. An energy estimate for the difference of solutions for the n-dimensional equation with prescribed mean curvature and removable singularities. Analysis **29**, 141–154 (2009)
9. Relative minimizers of energy are relative minimizers of area. Calc. Var. **37**, 475–483 (2010)

Hildebrandt, S., Tromba, A.J.
1. Mathematics and optimal form. Scientific American Library. W.H. Freeman, New York, 1985. [French transl.: Mathématiques et formes optimales. Pour la Science. Diff. Belin, Paris, 1986. German transl.: Panoptimum, Spektrum der Wissenschaft, Heidelberg, 1987. Dutch transl.: Architectuur in de Natuur, Wetenschappl. Bibliotheek, Natuur en Techniek, Maastricht/Brussel, 1989. Spanish transl.: Matematica y formas optimas. Prensa Cientifica, Barcelona, 1990. Japanese transl. 1995]
2. The parsimonious Universe. Shape and form in the natural world. Springer, New York, 1996 [German transl.: Kugel, Kreis und Seifenblasen. Optimale Formen in Geometrie und Natur. Birkhäuser, Basel, 1996. Italian transl. 2005]
3. On the branch point index of minimal surfaces. Arch. Math. **92**, 493–500 (2009)

Hildebrandt, S., von der Mosel, H.
1. On two-dimensional parametric variational problems. Calc. Var. Partial Differ. Equ. **9**, 249–267 (1999)
2. Plateau's problem for parametric double integrals: Part I. Existence and regularity in the interior. Commun. Pure Appl. Math. **56**, 926–955 (2003)
3. The partially free boundary problem for parametric double integrals. In: Nonlinear problems in mathematical physics and related topics I. Internat. Math. Ser. **I**, pp. 145–165. Kluwer/Plenum, London, 2002
4. Plateau's problem for parametric double integrals: Part II. Regularity at the boundary. J. Reine Angew. Math. **565**, 207–233 (2003)
5. Dominance functions for parametric Lagrangians. In: Geometric analysis and nonlinear partial differential equations, pp. 297–326. Springer, Berlin, 2003

6. On Lichtenstein's theorem about globally conformal mappings. Calc. Var. Partial Differ. Equ. **23**, 415–424 (2005)
7. Conformal representation of surfaces, and Plateau's problem for Cartan functionals. Riv. Mat. Univ. Parma (7) **4***, 1–43 (2005)
8. Conformal mapping of multiply connected Riemann domains by a variational approach. Adv. Calc. Var. **2**, 137–183 (2009)

Hildebrandt, S., Wente, H.C.
1. Variational problems with obstacles and a volume constraint. Math. Z. **135**, 55–68 (1973)

Hildebrandt, S., Widman, K.-O.
1. Some regularity results for quasilinear elliptic systems of second order. Math. Z. **142**, 67–86 (1975)

Hoffman, D.A.
1. The discovery of new embedded minimal surfaces: elliptic functions; symmetry; computer graphics. In: Proceedings of the Berlin Conference on Global Differential Geometry, Berlin, 1984
2. Embedded minimal surfaces, computer graphics and elliptic functions. In: Lect. Notes Math. **1156**, pp. 204–215. Springer, Berlin, 1985
3. The computer-aided discovery of new embedded minimal surfaces. Math. Intell. **9**, 8–21 (1987)
4. The construction of families of embedded minimal surfaces. In: Concus, P., Finn, R. (eds.) Variational methods for free surface interfaces, pp. 25–36. Springer, Berlin, 1987
5. New examples of singly-periodic minimal surfaces and their qualitative behavior. Contemp. Math. **101**, 97–106 (1989)
6. Natural minimal surfaces via theory and computation (videotape). Science Television, New York, Dec. 1990
7. Computing minimal surfaces; cf. GTMS 2005

Hoffman, D.A., Karcher, H.
1. Complete embedded minimal surfaces of finite total curvature. In: Geometry V, Minimal surfaces. Encycl. Math. Sci. **90**, pp. 5–93. Springer, Berlin, 1997

Hoffman, D.A., Karcher, H., Rosenberg, H.
1. Embedded minimal annuli in $\mathbb{R}^3$ bounded by a pair of straight lines. Comment. Math. Helv. **66**, 599–617 (1991)

Hoffman, D.A., Karcher, H., Wei, F.
1. Adding handles to the helicoid. Bull., New Ser., Am. Math. Soc. **29**, 77–84 (1993)
2. The genus one helicoid and the minimal surfaces that led to its discovery. In: Uhlenbeck, K. (ed.) Global analysis and modern mathematics, pp. 119–170. Publish or Perish Press, Berkeley, 1993

Hoffman, D.A., Meeks, W.H.
1. A complete embedded minimal surface in $\mathbb{R}^3$ with genus one and three ends. J. Differ. Geom. **21**, 109–127 (1985)
2. Complete embedded minimal surfaces of finite total curvature. Bull. Am. Math. Soc. **12**, 134–136 (1985)
3. Properties of properly embedded minimal surfaces of finite topology. Bull. Am. Math. Soc. **17**, 296–300 (1987)
4. The strong halfspace theorem for minimal surfaces. Invent. Math. 373–377 (1990)
5. A variational approach to the existence of complete embedded minimal surfaces. Duke Math. J. **57**, 877–894 (1988)
6. The asymptotic behavior of properly embedded minimal surfaces of finite topology. J. Am. Math. Soc. **2**, 667–682 (1989)

7. One parameter families of embedded minimal surfaces. Preprint, 1989
8. Properly embedded minimal surfaces of finite topology. Ann. Math. (2) **131**, 1–34 (1990)
9. Limits of minimal surfaces and Scherk's fifth surface. Arch. Ration. Mech. Anal. **111**, 181–195 (1990)
10. The global theory of embedded minimal surfaces. Preprint
11. Minimal surfaces based on the catenoid. Am. Math. Mon. **97**, 702–731 (1990)

Hoffman, D.A., Osserman, R.
1. The geometry of the generalized Gauss map. Mem. Am. Math. Soc. **236**, 1980
2. The area of the generalized Gaussian image and the stability of minimal surfaces in S^n and $\mathbb{R}^n$. Math. Ann. **260**, 437–452 (1982)
3. The Gauss map of surfaces in $\mathbb{R}^n$. J. Differ. Geom. **18**, 733–754 (1983)
4. The Gauss map of surfaces in $\mathbb{R}^3$ and $\mathbb{R}^4$. Proc. Lond. Math. Soc. (3) **50**, 27–56 (1985)

Hoffman, D.A., Osserman, R., Schoen, R.
1. On the Gauss map of complete surfaces of constant mean curvature in $\mathbb{R}^3$ and $\mathbb{R}^4$. Comment. Math. Helv. **57**, 519–531 (1982)

Hoffmann, D.A., Spruck, J.
1. Sobolev and isoperimetric inequalities for Riemannian submanifolds. Commun. Pure Appl. Math. **28**, 715–727 (1974)

Hoffman, D.A., Wohlgemuth, M.
1. New embedded periodic minimal surfaces of Riemann-type. Preprint

Hohrein, J.
1. Existence of unstable minimal surfaces of higher genus in manifolds of nonpositive curvature. Thesis. Universität Heidelberg, 1994

Hopf, E.
1. On an inequality for minimal surfaces $z = z(x, y)$. J. Ration. Mech. Anal. **2**, 519–522, 801–802 (1953)
2. Bemerkungen zu einem Satz von S. Bernstein aus der Theorie der elliptischen Differentialgleichungen. Math. Z. **29**, 744–745 (1929)
3. On S. Bernstein's theorem on surfaces $z(x, y)$ of non-positive curvature. Proc. Am. Math. Soc. **1**, 80–85 (1950)
4. Kleine Bemerkung zur Theorie der elliptischen Differentialgleichungen. J. Reine Angew. Math. **165**, 50–51 (1931)
5. A theorem on the accessability of boundary parts of an open point set. Proc. Am. Math. Soc. **1**, 76–79 (1950)

Hopf, H.
1. Differential geometry in the large. Stanford Lecture Notes, 1955. Reprint: Lect. Notes Math. **1000**, 2nd edn. Springer, Berlin, 1989

Hopf, H., Rinow, W.
1. Über den Begriff der vollständigen differentialgeometrischen Fläche. Comment. Math. Helv. **3**, 209–225 (1931)

Hsiung, C.C.
1. Isoperimetric inequalities for two-dimensional Riemannian manifolds with boundary. Ann. Math. (2) **73**, 213–220 (1961)

Hubbard, J.H.
1. On the convex hull genus of space curves. Topology **19**, 203–208 (1980)

Huber, A.
1. On subharmonic functions and differential geometry in the large. Comment. Math. Helv. **32**, 13–72 (1957)

Imbusch, C., Struwe, M.
1. Variational principles for minimal surfaces. In: Escher, J., Simonetti, G. (eds.) Progress in Nonlinear Differential Equations and Their Applications. **35**, pp. 477–498. Birkhäuser, Basel, 1999

Jäger, W.
1. Behavior of minimal surfaces with free boundaries. Commun. Pure Appl. Math. **23**, 803–818 (1970)
2. Ein Maximumprinzip für ein System nichtlinearer Differentialgleichungen. Nachr. Akad. Wiss. Gött., Math.-Phys. Kl., 157–164 (1976)
3. Das Randverhalten von Flächen beschränkter mittlerer Krümmung bei $C^{1,\alpha}$-Rändern. Nachr. Akad. Wiss. Gött., Math.-Phys. Kl. **5**, 45–54 (1977)

Jakob, R.
1. Instabile Extremalen des Shiffman-Funktionals. Bonner Math. Schr. **362**, 1–103 (2003)
2. Unstable extremal surfaces of the "Shiffman-functional". Calc. Var. Partial Differ. Equ. **21**, 401–427 (2004)
3. H-surface-index-formula. Ann. Inst. Henri Poincaré, Anal. Non Linéaire **22**, 557–578 (2005)
4. A "quasi maximum principle" for $\mathfrak{J}$-surfaces. Ann. Inst. Henri Poincaré, Anal. Non Linéaire **24**, 549–561 (2007)
5. Unstable extremal surfaces of the "Shiffman functional" spanning rectifiable boundary curves. Calc. Var. Partial Differ. Equ. **28**, 383–409 (2007)
6. Boundary branch points of minimal surfaces spanning extreme polygons. Results Math. **55**, 87–100 (2009)
7. Mollified and classical Green functions on the unit disc. Preprint, Duisburger Math. Schriftenreihe Nr. 625 (2006)
8. Schwarz operators of minimal surfaces spanning polygonal boundary curves. Calc. Var. Partial Differ. Equ. **30**, 467–476 (2007)
9. Finiteness of the set of solutions of Plateau's problem for polygonal boundary curves. Habilitationsschrift, Duisburg, 2008
10. Local boundedness of the number of solutions of Plateau's problem for polygonal boundary curves. Ann. Glob. Anal. Geom. **33**, 231–244 (2008)

Jakobowski, N.
1. Multiple surfaces of prescribed mean curvature. Math. Z. **217**, 497–512 (1994)
2. A result on large surfaces of prescribed mean curvature in a Riemannian manifold. Calc. Var. Partial Differ. Equ. **5**, 85–97 (1997)

Jenkins, H., Serrin, J.
1. Variational problems of minimal surfaces type. I. Arch. Ration. Mech. Anal. **12**, 185–212 (1963)
2. Variational problems of minimal surface type. II: boundary value problems for the minimal surface equation. Arch. Ration. Mech. Anal. **21**, 321–342 (1965/1966)
3. The Dirichlet problem for the minimal surface equation in higher dimensions. J. Reine Angew. Math. **229**, 170–187 (1968)
4. Variational problems of minimal surface type. III. The Dirichlet problem with infinite data. Arch. Ration. Mech. Anal. **29**, 304–322 (1968)

Joachimsthal, F.
1. Demonstrationes theorematum ad superficies curvas spectantium. J. Reine Angew. Math. **30**, 347–350 (1846)

John, F.

1. Partial differential equations, 4th edn. Springer, New York, 1982

Jorge, L.P.M., Meeks, W.H.

1. The topology of complete minimal surfaces of finite total Gaussian curvature. Topology **22**, 203–221 (1983)

Jorge, L.P., Tomi, F.

1. The barrier principle for submanifolds of arbitrary codimension. Ann. Glob. Anal. Geom. **24**, 261–267 (2003)

Jorge, L.P.M., Xavier, F.

1. On the existence of a complete bounded minimal surface in $\mathbb{R}^3$. Bull. Braz. Math. Soc. **10**, 171–183 (1979)
2. A complete minimal surface in $\mathbb{R}^3$ between two parallel planes. Ann. Math. **112**, 203–206 (1980)

Jörgens, K.

1. Über die Lösungen der Differentialgleichung $rt - s^2 = 1$. Math. Ann. **127**, 130–134 (1954)
2. Harmonische Abbildungen und die Differentialgleichung $rt - s^2 = 1$. Math. Ann. **129**, 330–344 (1955)

Jost, J.

1. Univalency of harmonic mappings between surfaces. J. Reine Angew. Math. **342**, 141–153 (1981)
2. The Dirichlet problem for harmonic maps from a surface with boundary onto a 2-sphere with non-constant boundary values. J. Differ. Geom. **19**, 393–401 (1984)
3. Harmonic maps between surfaces. Lect. Notes Math. **1062**. Springer, Berlin, 1984
4. Harmonic mappings between Riemannian manifolds. Proc. CMA **4**. ANU-Press, Canberra, 1984
5. A note on harmonic maps between surfaces. Ann. Inst. Henri Poincaré, Anal. Non Linéaire **2**, 397–405 (1985)
6. Conformal mappings and the Plateau–Douglas problem in Riemannian manifolds. J. Reine Angew. Math. **359**, 37–54 (1985)
7. Lectures on harmonic maps (with applications to conformal mappings and minimal surfaces). In: Lect. Notes Math. **1161**, pp. 118–192. Springer, Berlin, 1985
8. On the regularity of minimal surfaces with free boundaries in a Riemannian manifold. Manuscr. Math. **56**, 279–291 (1986)
9. Existence results for embedded minimal surfaces of controlled topological type. I. Ann. Sc. Norm Super. Pisa, Cl. Sci. (Ser. IV) **13**, 15–50 (1986): II. Ann. Sc. Norm. Super. Pisa, Cl. Sci. (Ser. IV) **13**, 401–426 (1986); III. Ann. Sc. Norm. Super. Pisa, Cl. Sci. (Ser. IV) **14**, 165–167 (1987)
10. On the existence of embedded minimal surfaces of higher genus with free boundaries in Riemannian manifolds. In: Concus, P., Finn, R. (eds.) Variational methods for free surface interfaces, pp. 65–75. Springer, New York, 1987
11. Two-dimensional geometric variational problems. In: Proc. Int. Congr. Math. 1986, Berkeley, pp. 1094–1100. Am. Math. Soc., Providence, 1987
12. Continuity of minimal surfaces with piecewise smooth boundary. Math. Ann. **276**, 599–614 (1987)
13. Embedded minimal disks with a free boundary on a polyhedron in $\mathbb{R}^3$. Math. Z. **199**, 311–320 (1988)
14. Das Existenzproblem für Minimalflächen. Jahresber. Dtsch. Math.-Ver. **90**, 1–32 (1988)

15. Embedded minimal surfaces in manifolds diffeomorphic to the three dimensional ball or sphere. J. Differ. Geom. **30**, 555–577 (1989)
16. Bosonic strings: A mathematical treatment. International Press/Am. Math. Soc., Sommerville, 2001
17. Two-dimensional geometric variational problems. Wiley-Interscience, Chichester, 1991
18. Riemannian geometry and geometric analysis. Universitext. Springer, Berlin, 1995
19. Eine geometrische Bemerkung zu Sätzen, die ein Dirichletproblem lösen. Manuscr. Math. **32**, 51–57 (1980)

Jost, J., Schoen, R.

1. On the existence of harmonic diffeomorphisms between surfaces. Invent. Math. **66**, 353–359 (1982)

Jost, J., Struwe, M.

1. Morse–Conley theory for minimal surfaces of varying topological type. Invent. Math. **102**, 465–499 (1990)

Jost, J., Xin, Y.L.

1. Bernstein type theorems for higher codimension. Calc. Var. Partial Differ. Equ. **9**, 277–296 (1999)
2. A Bernstein theorem for special Lagrangian graphs. Calc. Var. Partial Differ. Equ. **5**, 299–312 (2002)

Jost, J., Xin, Y.L., Yang, L.

1. The regularity of harmonic maps into spheres and applications to Bernstein problems. Preprint, 2009

Kapouleas, N.

1. Compact constant mean curvature surfaces in Euclidean three-space. J. Differ. Geom. **33**, 683–715 (1991)
2. Constructions of minimal surfaces by gluing minimal immersions; cf. GTMS 2005

Karcher, H.

1. Embedded minimal surfaces derived from Scherk's examples. Manuscr. Math. **62**, 83–114 (1988)
2. The triply periodic minimal surfaces of Alan Schoen and their constant mean curvature companions. Manuscr. Math. **64**, 291–357 (1989)
3. Construction of minimal surfaces. Surveys in Geometry 1989/90, University of Tokyo, 1989. Also: Vorlesungsreihe Nr. 12, SFB 256, Bonn, 1989
4. Eingebettete Minimalflächen und ihre Riemannschen Flächen. Jahresber. Dtsch. Math.-Ver. **101**, 72–96 (1999)
5. Introduction to conjugate Plateau constructions; cf. GTMS 2005
6. Riemannian center of mass and mollifier smoothing. Commun. Pure Appl. Math. **30**, 509–541 (1977)

Karcher, H., Pinkall, U., Sterling, J.

1. New minimal surfaces in S^3. J. Differ. Geom. **28**, 169–185 (1988)

Karcher, H., Polthier, K.

1. Construction of triply periodic minimal surfaces. Philos. Trans. R. Soc. Lond. A **354**, 2077–2104 (1996)

Kaul, H.

1. Ein Einschließungssatz für H-Flächen in Riemannschen Mannigfaltigkeiten. Manuscr. Math. **5**, 103–112 (1971)
2. Remarks on the isoperimetric inequality for multiply connected H-surfaces. Math. Z. **128**, 271–276 (1972)

3. Isoperimetrische Ungleichung und Gauß–Bonnet Formel für H-Flächen in Riemannschen Mannigfaltigkeiten. Arch. Ration. Mech. Anal. **45**, 194–221 (1972)
4. Eindeutigkeit von Lösungen elliptischer Systeme. Vorlesungsreihe SFB 72 No. 4, Analysis-Seminar SS 1980, Bonn, 1980

Keiper, J.B.
1. The axially symmetric n-tectum. Preprint, 1989

Kellogg, O.D.
1. Harmonic functions and Green's integrals. Trans. Am. Math. Soc. **13**, 109–132 (1912)
2. On the derivatives of harmonic functions on the boundary. Trans. Am. Math. Soc. **33**, 486–510 (1931)

Kenmotsu, K.
1. On minimal immersions of $\mathbb{R}^2$ into S^n. J. Math. Soc. Jpn. **28**, 182–191 (1976)
2. Weierstrass formula for surfaces of prescribed mean curvature. Math. Ann. **245**, 89–99 (1979)
3. Minimal surfaces with constant curvature in four dimensional space forms. Proc. Am. Math. Soc. **89**, 131–138 (1983)

Kerékjártó, B. von
1. Vorlesungen über Topologie. Springer, Berlin, 1923

Kim, H.
1. Unstable minimal surfaces of annulus type in manifolds. Dissertation, Saarbrücken, 2003

Kinderlehrer, D.
1. The boundary regularity of minimal surfaces. Ann. Sc. Norm. Super. Pisa, Cl. Sci. **23**, 711–744 (1969)
2. Minimal surfaces whose boundaries contain spikes. J. Math. Mech. **19**, 829–853 (1970)
3. The coincidence set of solutions of certain variational inequalities. Arch. Ration. Mech. Anal. **40**, 231–250 (1971)
4. The regularity of minimal surfaces defined over slit domains. Pac. J. Math. **37**, 109–117 (1971)
5. Variational inequalities with lower dimensional obstacles. Isr. J. Math. **10**, 330–348 (1971)
6. How a minimal surface leaves an obstacle. Acta Math. **130**, 221–242 (1973)
7. The free boundary determined by the solution of a differential equation. Indiana Univ. Math. J. **25**, 195–208 (1976)

Kinderlehrer, D., Nirenberg, L., Spruck, J.
1. Regularity in elliptic free boundary problems. I. J. Anal. Math. **34**, 86–119 (1978); II. Ann. Sc. Norm. Super. Pisa, Cl. Sci. **6**, 637–683 (1979)

Kinderlehrer, D., Stampacchia, G.
1. An introduction to variational inequalities and their applications. Academic Press, New York, 1980

Klein, F.
1. Vorlesungen über die Entwicklung der Mathematik im 19. Jahrhundert, Teil 1 und 2. Springer, Berlin, 1926, 1927

Klingenberg, W.
1. A course on differential geometry. Springer, Berlin, 1978. Translated by D. Hoffman, 2nd edn., 1983

Klotz, T., Osserman, R.
1. On complete surfaces in E^3 with constant mean curvature. Comment. Math. Helv. **41**, 313–318 (1966–1967)

Klotz, T., Sario, L.
1. Existence of complete minimal surfaces of arbitrary connectivity and genus. Proc. Natl. Acad. Sci. USA **54**, 42–44 (1965)
2. Gaussian mapping of arbitrary minimal surfaces. J. Anal. Math. **17**, 209–217 (1966)

Klotz-Milnor, T.
1. Harmonically immersed surfaces. J. Differ. Geom. **14**, 205–214 (1979)

Kneser, H.
1. Lösung der Aufgabe 41. Jahresber. Dtsch. Math.-Ver. **35**, 123–124 (1926)
2. Die kleinste Bedeckungszahl innerhalb einer Klasse von Flächenabbildungen. Math. Ann. **103**, 347–358 (1930)

Kobayashi, S., Nomizu, K.
1. Foundations of differential geometry **II**. Interscience, New York, 1969

Koebe, P.
1. Über die konforme Abbildung mehrfach zusammenhängender Bereiche, insbesondere solcher Bereiche, deren Begrenzung von Kreisen gebildet wird. Jahresber. Dtsch. Math.-Ver. **15** (1906)
2. Abhandlungen zur Theorie der konformen Abbildung. I. Die Kreisabbildung des allgemeinen einfach und zweifach zusammenhängenden schlichten Bereichs und die Ränderzuordnung bei konformer Abbildung. J. Reine Angew. Math. **145**, 177–223 (1915)

Koiso, M.
1. On the finite solvability of Plateau's problem for extreme curves. Osaka J. Math. **20**, 177–183 (1983)
2. On the stability of minimal surfaces in $\mathbb{R}^3$. J. Math. Soc. Jpn. **36**, 523–541 (1984)
3. The stability and the Gauss map of minimal surfaces in $\mathbb{R}^3$. In: Lect. Notes Math. **1090**, pp. 77–92. Springer, Berlin, 1984
4. On the non-uniqueness for minimal surfaces in $\mathbb{R}^3$. Proc. Diff. Geom., Sendai, 1989
5. Function theoretic and functional analytic methods for minimal surfaces. Surveys in Geometry 1989/90. Minimal surfaces, Tokyo, 1989
6. The uniqueness for minimal surfaces in S^3. Manuscr. Math. **63**, 193–207 (1989)
7. On the space of minimal surfaces with boundaries. Osaka J. Math. **20**, 911–921 (1983)

Korevaar, N., Kusner, R., Solomon, B.
1. The structure of complete embedded minimal surfaces of constant mean curvature. J. Differ. Geom. **30**, 465–503 (1989)

Korn, A.
1. Über Minimalflächen, deren Randkurven wenig von ebenen Kurven abweichen. Abh. K. Preuss. Akad. Wiss. Phys.-Math. Kl. **II**, 1–37 (1909)
2. Zwei Anwendungen der Methode der sukzessiven Annäherungen. In: Schwarz-Festschrift, pp. 215–229. Springer, Berlin, 1914

Kühnel, W.
1. Zur Totalkrümmung vollständiger Flächen. Vorlesungsreihe des SFB 256, Universität Bonn, No. **5**, 98–101 (1988)
2. Differentialgeometrie. Vieweg, Wiesbaden, 1999

Küster, A.
1. Zweidimensionale Variationsprobleme mit Hindernissen und völlig freien Randbedingungen. Thesis, Mathematisches Institut der Universität Bonn, Bonn, 1983
2. An optimal estimate of the free boundary of a minimal surface. J. Reine Angew. Math. **349**, 55–62 (1984)
3. On the linear isoperimetric inequality. Manuscr. Math. **53**, 255–259 (1985)

Kuwert, E.

1. Der Minimalflächenbeweis des Positive Mass Theorem. Vorlesungsreihe des SFB 256, No. 14, Bonn, 1990
2. Embedded solutions for the exterior minimal surface problems. Manuscr. Math. **70**, 51–65 (1990)
3. On solutions of the exterior Dirichlet problem for the minimal surface equation. Ann. Inst. Henri Poincaré, Anal. Non Linéaire **10**, 445–451 (1993)
4. A bound for minimal graphs with a normal at infinity. Calc. Var. Partial Differ. Equ. **1**, 407–416 (1993)
5. Area-minimizing immersions of the disk with boundary in a given homotopy class. Habilitationsschrift, Universität Bonn, 1995
6. Weak limits in the free boundary problem for immersions of the disk which minimize a conformally invariant functional. In: Jost, J. (ed.) Geometric analysis and the calculus of variations, pp. 203–215. International Press, Somerville, 1996
7. A compactness result for loops with an $H^{1/2}$-bound. J. Reine Angew. Math. **505**, 1–22 (1998)
8. Minimizing the energy of maps from a surface into a 2-sphere with prescribed degree and boundary values. Manuscr. Math. **83**, 31–38 (1994)
9. Harmonic maps between flat surfaces with conical singularities. Math. Z. **221**, 421–436 (1996)

Ladyzhenskaya, O.A., Uraltseva, N.N.

1. Quasilinear elliptic equations and variational problems with several independent variables. Usp. Mat. Nauk **16**, 19–90 (1961) (in Russian)
2. Linear and quasilinear elliptic equations. Academic Press, New York, 1968

Lawlor, G., Morgan, F.

1. Minimizing cones and networks: immiscible fluids, norms, and calibrations. Preprint, 1991

Lawson Jr., H.B.

1. Local rigidity theorem for minimal hypersurfaces. Ann. Math. **89**, 187–197 (1969)
2. The global behavior of minimal surfaces in S^n. Ann. Math. **92**, 224–237 (1970)
3. Compact minimal surfaces in S^3. In: Global analysis. Proc. Symp. Pure Math. **15**, pp. 275–282. Am. Math. Soc., Providence, 1970
4. Complete minimal surfaces in S^3. Ann. Math. **92**, 335–374 (1970)
5. The unknottedness of minimal embeddings. Invent. Math. **11**, 183–187 (1970)
6. Lectures on minimal submanifolds. Publish or Perish Press, Berkeley, 1971
7. Some intrinsic characterizations of minimal surfaces. J. Anal. Math. **24**, 151–161 (1971)
8. The equivariant Plateau problem and interior regularity. Trans. Am. Math. Soc. **173**, 231–249 (1972)
9. Minimal varieties in real and complex geometry. University of Montreal Press, Montreal, 1973
10. Surfaces minimales et la construction de Calabi–Penrose. Sémin. Bourbaki 36e année **624**, 1–15 (1983/1984). Astérisque **121–122**, 197–211 (1985)

Lemaire, L.

1. Boundary value problem for harmonic and minimal maps of surfaces into manifolds. Ann. Sc. Norm. Super. Pisa, Cl. Sci. (4) **8**, 91–103 (1982)
2. Applications harmoniques de surfaces Riemanniennes. J. Differ. Geom. **13**, 51–78 (1978)

Lesley, F.D.

1. Differentiability of minimal surfaces on the boundary. Pac. J. Math. **37**, 123–140 (1971)

Lévy, P.

1. Surfaces minima et corps convexes moyenne. C. R. Acad. Sci. Paris, Ser. A–B **223**, 881–883 (1946)
2. Exemples de contours pour lesquels le problème de Plateau a 3 ou $2p+1$ solutions. C. R. Acad. Sci. Paris **224**, 325–327 (1947)
3. Le problème de Plateau. Mathematica **23**, 1–45 (1947)

Lewy, H.

1. A priori limitations for solutions of Monge–Ampère equations, I, II. Trans. Am. Math. Soc. **37**, 417–434 (1935); **41**, 365–374 (1937)
2. Aspects of the calculus of variations. University of California Press, Berkeley, 1939 (Notes by J.W. Green)
3. On the nonvanishing of the Jacobian in certain one-to-one mappings. Bull. Am. Math. Soc. **42**, 689–692 (1936)
4. On minimal surfaces with partially free boundary. Commun. Pure Appl. Math. **4**, 1–13 (1951)
5. On the boundary behavior of minimal surfaces. Proc. Natl. Acad. Sci. USA **37**, 103–110 (1951)
6. On a variational problem with inequalities on the boundary. J. Math. Mech. **17**, 861–884 (1968)
7. On the non-vanishing of the Jacobian of a homeomorphism by harmonic gradients. Ann. Math. (2) **88**, 518–529 (1968)
8. About the Hessian of a spherical harmonic. Am. J. Math. **91**, 505–507 (1969)
9. On the coincidence set in variational inequalities. J. Differ. Geom. **6**, 497–501 (1972)
10. Über die Darstellung ebener Kurven mit Doppelpunkten. Nachr. Akad. Wiss. Gött., Math.-Phys. Kl. 109–130 (1981)

Lewy, H., Stampacchia, G.

1. On the regularity of the solution of a variational inequality. Commun. Pure Appl. Math. **22**, 153–188 (1969)
2. On existence and smoothness of solutions of some non-coercive variational inequalities. Arch. Ration. Mech. Anal. **41**, 241–253 (1971)

Li, P., Schoen, R., Yau, S.T.

1. On the isoperimetric inequality for minimal surfaces. Ann. Sc. Norm. Super. Pisa Cl. Sci., Ser. IV **XI.2**, 237–244 (1984)

Lichtenstein, L.

1. Neuere Entwicklung der Potentialtheorie, Konforme Abbildung. In: Encykl. Math. Wiss. **II** C 3, pp. 177–377. B.G. Teubner, Leipzig, 1909–1921
2. Beweis des Satzes, daß jedes hinreichend kleine, im wesentlichen stetig gekrümmte, singularitätenfreie Flächenstück auf einen Teil einer Ebene zusammenhängend und in den kleinsten Teilen ähnlich abgebildet wird. Abh. Königl. Preuss. Akad. Wiss. Berlin, Phys.-Math. Kl., Anhang, Abh. VI, 1–49 (1911)
3. Zur Theorie der konformen Abbildung. Konforme Abbildung nichtanalytischer singularitätenfreier Flächenstücke auf ebene Gebiete. Bull. Acad. Sci. Cracovie, Cl. Sci. Math. Nat. A, 192–217 (1916)
4. Neuere Entwicklung der Theorie partieller Differentialgleichungen zweiter Ordnung vom elliptischen Typus. In: Encykl. Math. Wiss. 2.3.2, pp. 1277–1334. B.G. Teubner, Leipzig, 1923–1927 (completed 1924)
5. Über einige Hilfssätze der Potentialtheorie, IV. Sitzungsber. Sächs. Akad. Wiss. Leipz., Math.-Nat. Wiss. Kl. **82**, 265–344 (1930)

Lin, F.-H.
1. Uniqueness and Nonuniqueness of the Plateau problem. Indiana Univ. Math. J. **36**(4), 843–848 (1987)
2. Regularity for a class of parametric obstacle problems. Dissertation, University of Minnesota, 1985
3. Plateau's problem for H-convex curves. Manuscr. Math. **58**, 491–511 (1987)
4. Estimates for surfaces which are stationary for an elliptic parametric integral. CMA report R28–86 (1986)

Lindelöf, L.
1. Théorie des surfaces de révolution à courbure moyenne constante. Acta Soc. Sci. Fenn. **7**, 345–372 (1863)
2. Sur les limites entre lesquelles le caténoïde est une surface minima. Acta Soc. Sci. Fenn. **9**, 353–360 (1871); also Math. Ann. **2**, 160–166 (1870)

Lipkin, L.J.
1. A free boundary problem for parametric integrals of the calculus of variations. Rend. Circ. Mat. Palermo (2) **17**, 33–67 (1968)

Longinetti, M.
1. On minimal surfaces bounded by two convex curves in parallel planes. Publ. dell'Ist. di Anal. Glob. & Appl. No. 12, Firenze, 1985

Luckhaus, S.
1. The Douglas-problem for surfaces of prescribed mean curvature. Preprint No. 234, SFB 72, Bonn, 1978

Lyusternik, L., Shnirelman, L.
1. Méthodes topologiques dans les problèmes variationnels. Actualités scient. et indust. **188**. Herman, Paris, 1934
2. Functional topology and abstract variational theory. Trans. Am. Math. Soc. **35**, 716–733 (1933)

Martín, F., Nadirashvili, N.
1. A Jordan curve spanned by a complete minimal surface. Arch. Ration. Mech. Anal. **184**, 285–301 (2007)

Marx, I.
1. On the classification of unstable minimal surfaces with polygonal boundaries. Commun. Pure Appl. Math. **8**, 235–244 (1955)

Massari, U., Miranda, M.
1. Minimal surfaces of codimension one. North-Holland Math. Stud. **91**. North-Holland, Amsterdam, 1984

Massey, W.
1. Algebraic topology: an introduction. Brace & World, Harcourt, 1967

McShane, E.J.
1. Parametrizations of saddle surfaces with applications to the problem of Plateau. Trans. Am. Math. Soc. **35**, 716–733 (1933)
2. On the analytic nature of surfaces of least area. Ann. Math. (2) **35**, 456–473 (1934)

Meeks, W.H.
1. The conformal structure and geometry of triply periodic minimal surfaces in $\mathbb{R}^3$. Ph.D. thesis, Berkeley, 1975
2. The conformal structure and geometry of triply periodic minimal surfaces in $\mathbb{R}^3$. Bull. Am. Math. Soc. **83**, 134–136 (1977)
3. Lectures on Plateau's problem. Escola de Geometria Differencial, Universidade Federal do Cearà (Brazil), de 17 a 28 de Julho de 1978

4. The classification of complete minimal surfaces in $\mathbb{R}^3$ with total curvature greater than -8π. Duke Math. J. **48**, 523–535 (1981)
5. Uniqueness theorems for minimal surfaces. Ill. J. Math. **25**, 318–336 (1981)
6. A survey of the geometric results in the classical theory of minimal surfaces. Bull. Braz. Math. Soc. **12**, 29–86 (1981)
7. The topological uniqueness of minimal surfaces in three-dimensional Euclidean space. Topology **20**, 389–410 (1981)
8. Recent progress on the geometry of surfaces in $\mathbb{R}^3$ and on the use of computer graphics as a research tool. In: Proceedings of the International Congress of Math., pp. 551–559. Berkeley, 1987
9. The topology and geometry of embedded surfaces of constant mean curvature. J. Differ. Geom. **27**, 539–552 (1988)
10. Regularity of the Albanese map for nonorientable surfaces. J. Differ. Geom. **29**, 345–352 (1989)
11. The geometry of triply-periodic minimal surfaces. Indiana Univ. Math. J. **39**, 877–936 (1990)
12. Global problems in classical minimal surface theory; cf. GTMS 2005

Meeks, W.H., Rosenberg, H.

1. The global theory of doubly periodic minimal surfaces. Invent. Math. **97**, 351–379 (1989)
2. The maximum principle at infinity for minimal surfaces in flat three-manifolds. Comment. Math. Helv. **69**, 255–270 (1990)
3. The geometry and conformal structure of properly embedded minimal surfaces of finite topology in $\mathbb{R}^3$. Invent. Math. **114**, 625–639 (1993)
4. Minimal surfaces of finite topology; cf. GTMS 2005

Meeks, W.H., Simon, L., Yau, S.T.

1. The existence of embedded minimal surfaces, exotic spheres and positive Ricci curvature. Ann. Math. **116**, 221–259 (1982)
2. Embedded minimal surfaces, exotic spheres, and manifolds with positive Ricci curvature. Ann. Math. **116**, 621–659 (1982)

Meeks, W.H., White, B.

1. Minimal surfaces bounding two convex curves in parallel planes. Comment. Math. Helv. **66**, 263–278 (1991)
2. The space of minimal annuli bounded by an extremal pair of planar curves. Commun. Anal. Geom. **1**, 415–437 (1993)

Meeks, W.H., Yau, S.-T.

1. Topology of three-manifolds and the embedding problems in minimal surface theory. Ann. Math. **112**, 441–484 (1980)
2. The equivariant Dehn's lemma and loop theorem. Comment. Math. Helv. **56**, 225–239 (1981)
3. The classical Plateau problem and the topology of three-dimensional manifolds. Topology **21**, 409–440 (1982)
4. The existence of embedded minimal surfaces and the problem of uniqueness. Math. Z. **179**, 151–168 (1982)
5. The equivariant loop theorem for three-dimensional manifolds and a review of existence theorems for minimal surfaces. In: The Smith conjecture, pp. 153–163. Academic Press, New York, 1984
6. The topological uniqueness theorem for minimal surfaces of finite type. Topology **31**, 305–316 (1992)

Mese, C.

1. Minimal surfaces and harmonic maps into singular geometry; cf. GTMS 2005

Micaleff, M.J., White, B.
1. The structure of branch points in minimal surfaces and in pseudoholomorphic curves. Ann. Math. **139**, 35–85 (1994)

Michael, F.H., Simon, L.M.
1. Sobolev and mean value inequalities on generalized submanifolds of $\mathbb{R}^n$. Commun. Pure Appl. Math. **26**, 361–379 (1973)

Mickle, E.J.
1. A remark on a theorem of Serge Bernstein. Proc. Am. Math. Soc. **1**, 86–89 (1950)

Miersemann, E.
1. Zur Regularität verallgemeinerter Lösungen von quasilinearen Differentialgleichungen in Gebieten mit Ecken. Z. Anal. Anwend. (4) **1**, 59–71 (1982)
2. Zur Gleichung der Fläche mit gegebener mittlerer Krümmung in zweidimensionalen eckigen Gebieten. Math. Nachr. **110**, 231–241 (1983)
3. Zur gemischten Randwertaufgabe für die Minimalflächengleichung. Math. Nachr. **115**, 125–136 (1984)

Milnor, J.
1. Morse theory. Ann. Math. Stud. **51**. Princeton University Press, Princeton, 1963
2. Topology from the differentiable view point. University Press of Virginia, Charlottesville, 1965

Minding, F.
1. Bemerkung über die Abwickelung krummer Linien auf Flächen. J. Reine Angew. Math. **6**, 159–161 (1830)
2. Zur Theorie der Curven kürzesten Umringes, bei gegebenem Flächeninhalt, auf krummen Flächen. J. Reine Angew. Math. **86**, 279–289 (1879)

Miranda, C.
1. Sul problema misto per le equazioni lineari ellitiche. Ann. Mat. Pura Appl. **39**, 279–303 (1955)

Miranda, M.
1. Sulle singolarità eliminabili delle soluzioni della equazione delle ipersurficie minimale. Ann. Scuola Norm. Pisa, Ser. IV A, 129–132 (1977)
2. Disuguaglianze di Sobolev sulle ipersuperfici minimali. Rend. Semin. Mat. Univ. Padova **38**, 69–79 (1967)
3. Una maggiorazione integrale per le curvature delle ipersuperfici minimali. Rend. Semin. Mat. Univ. Padova **38**, 91–107 (1967)
4. Some remarks about a free boundary type problem. In: Ericksen, Kinderlehrer (eds.) Liquid crystals. IMA Vol. Math. Appl. **5**. Springer, Berlin, 1987

Morgan, F.
1. A smooth curve in $\mathbb{R}^4$ bounding a continuum of area minimizing surfaces. Duke Math. J. **43**, 867–870 (1976)
2. Almost every curve in $\mathbb{R}^3$ bounds a unique area minimizing surface. Invent. Math. **45**, 253–297 (1978)
3. A smooth curve in $\mathbb{R}^3$ bounding a continuum of minimal manifolds. Arch. Ration. Mech. Anal. **75**, 193–197 (1980)
4. On the singular structure of two-dimensional area minimizing surfaces in $\mathbb{R}^n$. Math. Ann. **261**, 101–110 (1982)
5. On finiteness of the number of stable minimal hypersurfaces with a fixed boundary. Bull. Am. Math. Soc. **13**, 133–136 (1985)
6. Geometric measure theory: A beginner's guide. Academic Press, San Diego, 1988; 3rd edn., 2000

7. Clusters minimizing area plus length of singular curves. Math. Ann. **299**, 697–714 (1994)
8. Regularity of isoperimetric hypersurfaces in Riemannian manifolds. Trans. Am. Math. Soc. **355**, 5041–5052 (2003)

Morgan, F., Ritoré, M.
1. Geometric measure theory and the proof of the double bubble conjecture; cf. GTMS 2005

Morrey, C. B.
1. On the solutions of quasi-linear elliptic partial differential equations. Trans. Am. Math. Soc. **43**, 126–166 (1938)
2. Multiple integral problems in the calculus of variations and related topics. Univ. Calif. Publ. Math., New Ser. **1**(1), 1–130 (1943)
3. The problem of Plateau on a Riemannian manifold. Ann. Math. (2) **49**, 807–851 (1948)
4. Second order elliptic systems of differential equations. In: Contributions to the theory of partial differential equations. Ann. Math. Stud. **33**, pp. 101–160. Princeton University Press, Princeton, 1954
5. On the analyticity of the solutions of analytic non-linear elliptic systems of partial differential equations. Am. J. Math. **80**, I. 198–218, II. 219–234 (1958)
6. Multiple integral problems in the calculus of variations and related topics. Ann. Sc. Norm. Super. Pisa, Cl. Sci. **14**, 1–61 (1960)
7. The higher dimensional Plateau problem on a Riemannian manifold. Proc. Natl. Acad. Sci. USA **54**, 1029–1035 (1965)
8. Multiple integrals in the calculus of variations. Grundlehren Math. Wiss. **130**. Springer, Berlin, 1966
9. The parametric variational problem for double integrals. Commun. Pure Appl. Math. **14**, 569–575 (1961)

Morrey, C.B., Nirenberg, L.
1. On the analyticity of the solutions of linear elliptic systems of partial differential equations. Commun. Pure Appl. Math. **10**, 271–290 (1957)

Morse, M.
1. The calculus of variations in the large. Am. Math. Soc. Colloquium Publication **18**, 1934
2. Functional topology and abstract variational theory. Ann. Math. **38**, 386–449 (1937)

Morse, M., Tompkins, C.B.
1. Existence of minimal surfaces of general critical type. Ann. Math. **40**, 443–472 (1939); correction in **42**, 331 (1941)
2. Existence of minimal surfaces of general critical type. Proc. Natl. Acad. Sci. USA **25**, 153–158 (1939)
3. The continuity of the area of harmonic surfaces as a function of the boundary representation. Am. J. Math. **63**, 825–838 (1941)
4. Unstable minimal surfaces of higher topological structure. Duke Math. J. **8**, 350–375 (1941)
5. Minimal surfaces of non-minimum type by a new mode of approximation. Ann. Math. (2) **42**, 62–72 (1941)

Moser, J.
1. A new proof of De Giorgi's theorem concerning the regularity problem for elliptic differential equations. Commun. Pure Appl. Math. **13**, 457–468 (1960)
2. On Harnack's theorem for elliptic differential equations. Commun. Pure Appl. Math. **14**, 577–591 (1961)

Müller, F.
1. Analyticity of solutions for semilinear elliptic systems of second order. Calc. Var. Partial Differ. Equ. **15**, 257–288 (2002)

2. On the continuations for solutions for elliptic equations in two variables. Ann. Inst. Henri Poincaré, Anal. Non Linéaire **19**, 745–776 (2002)
3. On the analytic continuation of H-surfaces across the free boundary. Analysis **22**, 201–218 (2002)
4. Hölder continuity of surfaces with bounded mean curvature at corners where Plateau boundaries meet free boundaries. Calc. Var. Partial Differ. Equ. **24**, 283–288 (2005)
5. On stable surfaces of prescribed mean curvature with partially free boundaries. Analysis **26**, 289–308 (2005)
6. A priori bounds for surfaces with prescribed mean curvature and partially free boundaries. Calc. Var. Partial Differ. Equ. **24**, 471–489 (2006)
7. On the regularity of H-surfaces with free boundaries on a smooth support manifold. Analysis **28**, 401–419 (2009)
8. Investigations on the regularity of surfaces with prescribed mean curvature and partially free boundaries. Habilitationsschrift, BTU Cottbus, 2007
9. Growth estimates for the gradient of an H-surface near singular points of the boundary configuration. Z. Anal. Anwend. **28**, 87–102 (2009)
10. The asymptotic behaviour of surfaces with prescribed mean curvature near meeting points of fixed and free boundaries. Ann. Sc. Norm. Super. Pisa, Cl. Sci. (5) **6**, 529–559 (2007)
11. On the behaviour of free H-surfaces near singular points of the support surface. Adv. Calc. Var. **1**, 345–378 (2008)

Müller, F., Schikorra, A.
1. Boundary regularity via Uhlenbeck–Rivière decomposition. Analysis **29**, 199–220 (2009)

Müller, F., Winklmann, S.
1. Projectability and uniqueness of F-stable immersions with partially free boundaries. Pac. J. Math. **230**, 409–426 (2007)

Müntz, C.H.
1. Zum Randwertproblem der partiellen Differentialgleichung der Minimalflächen. J. Reine Angew. Math. **139**, 52–79 (1911)
2. Die Lösung des Plateauschen Problems über konvexen Bereichen. Math. Ann. **94**, 53–96 (1925)
3. Zum Plateauschen Problem. Erwiderung auf die vorstehende Note des Herrn Radó. Math. Ann. **96**, 597–600 (1927)

Nadirashvili, N.
1. Hadamard's and Calabi–Yau's conjectures on negatively curved and minimal surfaces. Invent. Math. **126**, 457–465 (1996)

Nakauchi, N.
1. Multiply connected minimal surfaces and the geometric annulus theorem. J. Math. Soc. Jpn. **37**, 17–39 (1985)

Natanson, I.P.
1. Theorie der Funktionen einer reellen Veränderlichen. Akademie-Verlag, Berlin, 1975

Nehring, G.
1. Embedded minimal annuli solving an exterior problem. Calc. Var. Partial Differ. Equ. **2**, 373–388 (1994)

Nirenberg, L.
1. Remarks on strongly elliptic partial differential equations. Commun. Pure Appl. Math. **8**, 648–674 (1955)
2. On elliptic partial differential equations. Ann. Sc. Norm. Super. Pisa, Cl. Sci., Ser. 3 **13**, 115–162 (1959)

Nitsche, J.C.C.

1. Über eine mit der Minimalflächengleichung zusammenhängende analytische Funktion und den Bernsteinschen Satz. Arch. Math. **7**, 417–419 (1956)
2. Elementary proof of Bernstein's theorem on minimal surfaces. Ann Math. (2) **66**, 543–544 (1957)
3. A uniqueness theorem of Bernstein's type for minimal surfaces in cylindrical coordinates. J. Math. Mech. **6**, 859–864 (1957)
4. A characterization of the catenoid. J. Math. Mech. **11**, 293–302 (1962)
5. Über die Ausdehnung gewisser zweifach zusammenhängender Minimalflächen. Math. Ann. **149**, 144–149 (1963)
6. Review of Sasaki Sasaki [1]. Math. Rev. **25**, Nr. 492 (1963)
7. A supplement to the condition of J. Douglas. Rend. Circ. Mat. Palermo (2) **13**, 192–198 (1964)
8. A necessary criterion for the existence of certain minimal surfaces. J. Math. Mech. **13**, 659–666 (1964)
9. On the non-solvability of Dirichlet's problem for the minimal surface equation. J. Math. Mech. **14**, 779–788 (1965)
10. The isoperimetric inequality for multiply connected minimal surfaces. Math. Ann. **160**, 370–375 (1965)
11. On new results in the theory of minimal surfaces. Bull. Am. Math. Soc. **71**, 195–270 (1965)
12. Über ein verallgemeinertes Dirichletsches Problem für die Minimalflächengleichung und hebbare Unstetigkeiten ihrer Lösungen. Math. Ann. **158**, 203–214 (1965)
13. Ein Einschließungssatz für Minimalflächen. Math. Ann. **165**, 71–75 (1966)
14. Contours bounding at least three solutions of Plateau's problem. Arch. Ration. Mech. Anal. **30**, 1–11 (1968)
15. Note on the non-existence of minimal surfaces. Proc. Am. Math. Soc. **19**, 1303–1305 (1968)
16. The boundary behavior of minimal surfaces—Kellogg's theorem and branch points on the boundary. Invent. Math. **8**, 313–333 (1969)
17. Concerning the isolated character of solutions of Plateau's problem. Math. Z. **109**, 393–411 (1969)
18. A variational problem with inequalities as boundary conditions. Bull. Am. Math. Soc. **75**, 450–452 (1969)
19. Variational problems with inequalities as boundary conditions, or, how to fashion a cheap hat for Giacometti's brother. Arch. Ration. Mech. Anal. **35**, 83–113 (1969)
20. Concerning my paper on the boundary behavior of minimal surfaces. Invent. Math. **9**, 270 (1970)
21. An isoperimetric property of surfaces with movable boundaries. Am. Math. Mon. **77**, 359–362 (1970)
22. Minimal surfaces with partially free boundary. Least area property and Hölder continuity for boundaries satisfying a chord-arc condition. Arch. Ration. Mech. Anal. **39**, 131–145 (1970)
23. Minimal surfaces with movable boundaries. Bull. Am. Math. Soc. **77**, 746–751 (1971)
24. The regularity of minimal surfaces on the movable parts of their boundaries. Indiana Univ. Math. J. **21**, 505–513 (1971)
25. On the boundary regularity of surfaces of least area in euclidean space. In: Continuum mechanics and related problems in analysis, pp. 375–377. Nauka, Moscow, 1972
26. A new uniqueness theorem for minimal surfaces. Arch. Ration. Mech. Anal. **52**, 319–329 (1973)
27. Plateau problems and their modern ramifications. Am. Math. Mon. **81**, 945–968 (1974)

28. Vorlesungen über Minimalflächen. Grundlehren Math. Wiss. **199**. Springer, Berlin, 1975
29. Non-uniqueness for Plateau's problem. A bifurcation process. Ann. Acad. Sci. Fenn., Ser. A 1 Math. **2**, 361–373 (1976)
30. The regularity of the trace for minimal surfaces. Ann. Sc. Norm. Super. Pisa, Cl. Sci., Ser. IV **3**, 139–155 (1976)
31. Contours bounding at most finitely many solutions of Plateau's problem. In: Complex analysis and its applications, dedicated to I.N. Vekua. Nauka, Moscow, 1978
32. Uniqueness and non-uniqueness for Plateau's problem—one of the last major questions. In: Minimal submanifolds and geodesics. Proceedings of the Japan–United States Seminar on Minimal Submanifolds, including Geodesics, Tokyo, 1977, pp. 143–161. Kagai Publications, Tokyo, 1978
33. The higher regularity of liquid edges in aggregates of minimal surfaces. Nachr. Akad. Wiss. Gött. Math.-Phys. Kl., 2B **2**, 31–51 (1978)
34. Minimal surfaces and partial differential equations. In: MAA Studies in Mathematics **23**, pp. 69–142. Math. Assoc. Am., Washington, 1982
35. Stationary partitioning of convex bodies. Arch. Ration. Mech. Anal. **89**, 1–19 (1985); corrigendum in Arch. Ration. Mech. Anal. **95**, 389 (1986)
36. Nonparametric solutions of Plateau's problem need not minimize area. Analysis **8**, 69–72 (1988)
37. Lectures on minimal surfaces, vol. 1: Introduction, fundamentals, geometry and basic boundary problems. Cambridge University Press, Cambridge, 1989
38. On an estimate for the curvature of minimal surfaces $z = z(x, y)$. J. Math. Mech. **7**, 767–769 (1958)

Nitsche, J.C.C., Leavitt, J.

1. Numerical estimates for minimal surfaces. Math. Ann. **180**, 170–174 (1969)

Osserman, R.

1. Proof of a conjecture of Nirenberg. Commun. Pure Appl. Math. **12**, 229–232 (1959)
2. On the Gauss curvature of minimal surfaces. Trans. Am. Math. Soc. **96**, 115–128 (1960)
3. Minimal surfaces in the large. Comment. Math. Helv. **35**, 65–76 (1961)
4. On complete minimal surfaces. Arch. Ration. Mech. Anal. **13**, 392–404 (1963)
5. Global properties of minimal surfaces in E^3 and E^n. Ann. Math. (2) **80**, 340–364 (1964)
6. Some geometric properties of polynomial surfaces. Comment. Math. Helv. **37**, 214–220 (1962–63)
7. Global properties of classical minimal surfaces. Duke Math. J. **32**, 565–573 (1965)
8. Le théorème de Bernstein pour des systèmes. C. R. Acad. Sci. Paris, Sér. A **262**, 571–574 (1966)
9. Minimal surfaces. Usp. Mat. Nauk **22**, 56–136 (1967) (in Russian)
10. A survey of minimal surfaces. Van Nostrand, New York, 1969
11. Minimal varieties. Bull. Am. Math. Soc. **75**, 1092–1120 (1969)
12. A proof of the regularity everywhere of the classical solution to Plateau's problem. Ann. Math. (2) **91**, 550–569 (1970)
13. Some properties of solutions to the minimal surface system for arbitrary codimension. In: Global analysis. Proc. Symp. Pure Math. **15**, pp. 283–291. Am. Math. Soc., Providence, 1970
14. On the convex hull property of immersed manifolds. J. Differ. Geom. **6**, 267–270 (1971)
15. Branched immersions of surfaces. In: Symposia Mathematica of Istituto Nazionale di Alta Matematica Roma **10**, pp. 141–158. Academic Press, London, 1972
16. On Bers' theorem on isolated singularities. Indiana Univ. Math. J. **23**, 337–342 (1973)

17. Isoperimetric and related inequalities. In: Proc. Symp. Pure Math. **27**, pp. 207–215. Am. Math. Soc., Providence, 1975
18. Some remarks on the isoperimetric inequality and a problem of Gehring. J. Anal. Math. **30**, 404–410 (1976)
19. The isoperimetric inequality. Bull. Am. Math. Soc. **84**, 1182–1238 (1978)
20. Properties of solutions to the minimal surface equation in higher codimension. In: Minimal submanifolds and geodesics. Proceedings of the Japan–United States Seminar on Minimal Submanifolds, including Geodesics, Tokyo, 1977, pp. 163–172. Kagai Publications, Tokyo, 1978
21. Minimal surfaces, Gauss maps, total curvature, eigenvalue estimates, and stability. In: The Chern Symposium, 1979, pp. 199–227. Springer, Berlin, 1980
22. The total curvature of algebraic surfaces. In: Contributions to analysis and geometry, pp. 249–257. John Hopkins University Press, Baltimore, 1982
23. The minimal surface equation. In: Seminar on nonlinear partial differential equations. Math. Sci. Res. Inst. Publ. **2**, pp. 237–259. Springer, Berlin, 1984
24. Minimal surfaces in $\mathbb{R}^3$. In: Chern, S.S. (ed.) Global differential geometry. MAA Studies in Mathematics **27**, pp. 73–98. Math. Assoc. Am., Washington, 1990
25. Riemann surfaces of class A. Trans. Am. Math. Soc. **82**, 217–245 (1956)

Osserman, R., Schiffer, M.

1. Doubly connected minimal surfaces. Arch. Ration. Mech. Anal. **58**, 285–307 (1974/75)

Otsuki, T.

1. Minimal hypersurfaces in a Riemannian manifold of constant curvature. Am. J. Math. **92**, 145–173 (1970)

Otto, F., Editor

1. Zugbeanspruchte Konstruktionen, Bd. 1 und 2. Ullstein Fachverlag, Berlin Frankfurt/M. Wien, 1962 und 1966

Painlevé, P.

1. Sur la théorie de la représentation conforme. C. R. Acad. Sci. Paris **112**, 653–657 (1891)

Pilz, R.

1. On the thread problem for minimal surfaces. Calc. Var. Partial Differ. Equ. **5**, 117–136 (1997)

Pinkall, U., Polthier, K.

1. Computing discrete minimal surfaces and their conjugates. Exp. Math. **2**, 15–36 (1993)

Pitts, J.

1. Existence and regularity of minimal surfaces on Riemannian manifolds. Ann. Math. Stud. **27**. Princeton University Press, Princeton, 1981

Pitts, J., Rubinstein, H.

1. Existence of minimal surfaces of bounded topological type in 3-manifolds. In: Proceedings of the Centre for Mathematical Analysis **10**. Australian National University, Canberra, Australia, 1987

Plateau, J.A.F.

1. Statique expérimentale et théoretique des liquides soumis aux seule forces moléculaires, vols. I, II. Gauthier-Villars, Paris, 1873

Pohl, W.F.

1. Review of Osserman [12]. Math Reviews **42**, 1, Nr. 979 (1972)

Polthier, K.

1. Neue Minimalflächen in $\mathbb{H}^3$. Diplomarbeit, Bonn, 1989
2. Geometric data for triply periodic minimal surfaces in spaces of constant curvature. Preprint, SFB 256, Report No. 4, Bonn, 1989

3. Bilder aus der Differentialgeometrie. Kalender 1989. Vieweg, Braunschweig, 1989
4. Geometric data for triply periodic minimal surfaces in spaces of constant curvature. In: Concus, P., Finn, R., Hoffman, D. (eds.) Geometric analysis and computer graphics, pp. 141–145. Springer, New York, 1991
5. Computational aspects and discrete minimal surfaces. In: Hoffman, D. (ed.) Proc. Clay Summerschool on Minimal Surfaces. Am. Math. Soc., Providence, 2002
6. Unstable periodic discrete minimal surfaces. In: Hildebrandt, S., Karcher, H. (eds.) Geometric analysis and nonlinear partial differential equations, pp. 129–145. Springer, Berlin, 2003
7. Computational aspects of discrete minimal surfaces; cf. GTMS 2005

Polthier, K., Wohlgemuth, M.
1. Bilder aus der Differentialgeometrie. Kalender 1988. Computergraphiken. Vieweg, Braunschweig, 1988

Protter, M., Weinberger, H.
1. Maximum principles in differential equations. Prentice-Hall, Englewood, 1967

Quien, N.
1. Über die endliche Lösbarkeit des Plateau-Problems in Riemannschen Mannigfaltigkeiten. Manuscr. Math. **39**, 313–338 (1982)

Quien, N., Tomi, F.
1. Nearly planar Jordan curves spanning a given number of minimal immersions of the disc. Arch. Math. **44**, 456–460 (1985)

Radó, T.
1. Über die Fundamentalabbildung schlichter Gebiete. Acta Litt. Sci. Univ. Szeged. 240–251(1923)
2. Über den analytischen Charakter der Minimalflächen. Math. Z. **24**, 321–327 (1925)
3. Bemerkung über die Differentialgleichungen zweidimensionaler Variationsprobleme. Acta Litt. Sci. Univ. Szeged. 147–156 (1925)
4. Über den Begriff der Riemannschen Fläche. Acta Litt. Sci. Univ. Szeged. 101–121 (1925)
5. Aufgabe 41. Jahresber. Dtsch. Math.-Ver. **35**, 49 (1926)
6. Geometrische Betrachtungen über zweidimensionale reguläre Variationsprobleme. Acta Litt. Sci. Univ. Szeged. 228–253 (1926)
7. Sur le calcul de l'aire des surface courbes. Fundam. Math. **10**, 197–210 (1926)
8. Das Hilbertsche Theorem über den analytischen Charakter der Lösungen der partiellen Differentialgleichungen zweiter Ordnung. Math. Z. **25**, 514–589 (1926)
9. Bemerkung über das Doppelintegral $\iint(1 + p^2 + q^2)^{1/2}\,dx\,dy$. Math. Z. **26**, 408–416 (1927)
10. Zu einem Satz von S. Bernstein über Minimalflächen im Großen. Math. Z. **26**, 559–565 (1927)
11. Bemerkung zur Arbeit von Herrn Ch. H. Müntz über das Plateausche Problem. Math. Ann. **96**, 587–596 (1927)
12. Sur l'aire des surfaces courbes. Acta Litt. Sci. Univ. Szeged. **3**, 131–169 (1927)
13. Über das Flächenmaß rektifizierbarer Flächen. Math. Ann. **100**, 445–479 (1928)
14. Bemerkung über die konformen Abbildungen konvexer Gebiete. Math. Ann. **102**, 428–429 (1929)
15. Über zweidimensionale reguläre Variationsprobleme der Form $\iint F(p,q)\,dx\,dy =$ Minimum. Math. Ann. **101**, 620–632 (1929)
16. Some remarks on the problem of Plateau. Proc. Natl. Acad. Sci. USA **16**, 242–248 (1930)
17. The problem of the least area and the problem of Plateau. Math. Z. **32**, 763–796 (1930)

18. On Plateau's problem. Ann. Math. (2) **31**, 457–469 (1930)
19. On the functional of Mr. Douglas. Ann. Math. (2) **32**, 785–803 (1931)
20. Contributions to the theory of minimal surfaces. Acta Sci. Math. Univ. Szeged. **6**, 1–20 (1932)
21. On the problem of Plateau. Ergebnisse der Math. Band 2. Springer, Berlin, 1933
22. An iterative process in the problem of Plateau. Trans. Am. Math. Soc. **35**, 869–887 (1933)
23. Length and area. Am. Math. Soc. Colloq. Publ. **30**. Am. Math. Soc., Providence, 1948

Radó, T., Reicheldorfer, P.
1. Note on an inequality of Steiner. Bull. Am. Math. Soc. **47**, 102–108 (1941)

Reade, M.
1. Analogue of a theorem of F. and M. Riesz for minimal surfaces. J. Math. Soc. Jpn. **8**, 177–179 (1956)

Reid, C.
1. Courant. Springer, Berlin, 1976

Reid, W.T.
1. The isoperimetric inequality and associated boundary problems. J. Math. Mech. **8**, 897–905 (1959)

Reifenberg, E.R.
1. Solution of the Plateau problem for m-dimensional surfaces of varying topological type. Acta Math. **104**, 1–92 (1960)
2. An epiperimetric inequality related to the analyticity of minimal surfaces. Ann. Math. **80**, 1–14 (1964)
3. On the analyticity of minimal surfaces. Ann. Math. **80**, 15–21 (1964)

Reilly, R.C.
1. Extrinsic rigidity theorems for compact submanifolds of the sphere. J. Differ. Geom. **4**, 487–497 (1970)

Riemann, B.
1. Gesammelte mathematische Werke. B.G. Teubner, Leipzig, 1876 (1. Auflage), 1892 (2. Auflage) und Nachträge, 1902
2. Über die Fläche vom kleinsten Inhalt bei gegebener Begrenzung. Abh. K. Ges. Wiss. Gött., Math. Kl. **13**, 3–52 (1867) (K. Hattendorff, ed.)

Riesz, F.
1. Über die Randwerte einer analytischen Funktion. Math. Z. **18**, 87–95 (1923)

Riesz, F., Riesz, M.
1. Über die Randwerte einer analytischen Funktion. In: Comptes rendus du 4. Congr. des Math. Scand. Stockh., pp. 27–44, 1916

Ripoll, J., Tomi, F.
1. Maximum principles for minimal surfaces having noncompact boundary and a uniqueness theorem for the helicoid. Manuscr. Math. **87**, 417–434 (1995)

Ritter F.
1. Solution of Schwarz' problem concerning minimal surfaces. Rev. Univ. Nac. Tucumán **1**, 40–62 (1940)

Rivière, T.
1. Conservation laws for conformally invariant variational problems. Invent. Math. **168**, 1–22 (2007)

Rivière, T., Struwe, M.
1. Partial regularity for harmonic maps and related problems. Commun. Pure Appl. Math. **61**, 451–463 (2008)

Ruchert, H.
1. Ein Eindeutigkeitssatz für Flächen konstanter mittlerer Krümmung. Arch. Math. **33**, 91–104 (1979)
2. A uniqueness result for Enneper's minimal surface. Indiana Univ. Math. J. **30**, 427–431 (1981)

Rudin, W.
1. Real and complex analysis. Tata McGraw-Hill, New Delhi, 1966
2. Functional analysis. McGraw-Hill, New York, 1973

Ruh, E.A., Vilms, J.
1. The tension field of the Gauss map. Trans. Am. Math. Soc. **149**, 569–573 (1970)

Rummler, H.
1. Quelques notions simples en géométrie riemannienne et leurs applications aux feuilletages compacts. Comment. Math. Helv. **54**, 224–239 (1979)

Sacks, J., Uhlenbeck, K.
1. The existence of minimal immersions of two-spheres. Ann. Math. **113**, 1–24 (1981)
2. Minimal immersions of closed Riemann surfaces. Trans. Am. Math. Soc. **271**, 639–652 (1982)

Sasaki, S.
1. On the total curvature of a closed curve. Jpn. J. Math. **29**, 118–125 (1959)

Sauvigny, F.
1. Flächen vorgeschriebener mittlerer Krümmung mit eineindeutiger Projektion auf eine Ebene. Dissertation, Göttingen, 1981
2. Flächen vorgeschriebener mittlerer Krümmung mit eineindeutiger Projektion auf eine Ebene. Math. Z. **180**, 41–67 (1982)
3. Ein Eindeutigkeitssatz für Minimalflächen im $\mathbb{R}^p$ mit polygonalem Rand. J. Reine Angew. Math. **358**, 92–96 (1985)
4. On the Morse index of minimal surfaces in $\mathbb{R}^p$ with polygonal boundaries. Manuscr. Math. **53**, 167–197 (1985)
5. Die zweite Variation von Minimalflächen im $\mathbb{R}^p$ mit polygonalem Rand. Math. Z. **189**, 167–184 (1985)
6. On the total number of branch points of quasi-minimal surfaces bounded by a polygon. Analysis **8**, 297–304 (1988)
7. A-priori-Abschätzungen der Hauptkrümmungen für Immersionen vom Mittleren-Krümmungs-Typ mittels Uniformisierung und Sätze vom Bernstein-Typ. Habilitationsschrift, Göttingen, 1989
8. A priori estimates of the principal curvatures for immersions of prescribed mean curvature and theorems of Bernstein type. Math. Z. **205**, 567–582 (1990)
9. Curvature estimates for immersions of minimal surfaces type via uniformization and theorems of Bernstein type. Manuscr. Math. **67**, 69–97 (1990)
10. On immersions of constant mean curvature: compactness results and finiteness theorems for Plateau's problem. Arch. Ration. Mech. Anal. **110**, 125–140 (1990)
11. A new proof for the gradient estimate for graphs of prescribed mean curvature. Manuscr. Math. **74**, 83–86 (1992)
12. Uniqueness of Plateau's problem for certain contours with a one-to-one, nonconvex projection onto a plane. In: Jost, J. (ed.) Geometric analysis and the calculus of variations, pp. 297–314. International Press, Somerville, 1996

13. Introduction to isothermal parameters into a Riemannian metric by the continuity method. Analysis **19**, 235–243 (1999)
14. Global $C^{2+\alpha}$-estimates for conformal maps. In: Hildebrandt, S., Karcher, H. (eds.) Geometric analysis and nonlinear partial differential equations, pp. 105–115. Springer, Berlin, 2003
15. Partielle Differentialgleichungen der Geometrie und der Physik. Bd. 1: Grundlagen und Integraldarstellungen. Springer, Berlin, 2004. Bd. 2: Funktionalanalytische Lösungsmethoden. Springer, Berlin, 2005
16. Partial equations. Vol. 1: Foundations and integral representation. Vol. 2: Functional analytic methods. Springer Universitext, Berlin, 2006
17. Un problème aux limites mixte des surfaces minimales avec une multiple projection plane et le dessin optimal des escaliers tournants. To appear 2010/2011 in Analyse Non Linéaire

Schauder, J.
1. Potentialtheoretische Untersuchungen. I. Math. Z. **33**, 602–640 (1931)
2. Über lineare elliptische Differentialgleichungen zweiter Ordnung. Math. Z. **38**, 257–282 (1934)

Schikorra, A.
1. A remark on gauge transformations and the moving frame method. Ann. Inst. Henri Poincaré, Anal. Non Linéaire **27**, 503–515 (2010)

Schlesinger, L.
1. Handbuch der Theorie der linearen Differentialgleichungen. Band I, II.1, II.2. Teubner, Leipzig, 1895
2. Über die Lösungen gewisser linearer Differentialgleichungen als Funktionen der singulären Punkte. J. Reine Angew. Math. **129**, 287–294 (1905)
3. Vorlesungen über lineare Differentialgleichungen. Teubner, Leipzig, 1908
4. Einführung in die Theorie der gewöhnlichen Differentialgleichungen auf funktionentheoretischer Grundlage. de Gruyter, Berlin, 1922

Schneider, R.
1. A note on branch points of minimal surfaces. Proc. Am. Math. Soc. **17**, 1254–1257 (1966)
2. Ein Eindeutigkeitssatz zum Plateauschen Problem. Preprint, 1969 (unpublished)

Schoen, R.
1. A remark on minimal hypercones. Proc. Natl. Acad. Sci. USA **79**, 4523–4524 (1982)
2. Estimates for stable minimal surfaces in three dimensional manifolds. In: Seminar on minimal submanifolds. Ann. Math. Stud. **103**, pp. 111–126. Princeton University Press, Princeton, 1983
3. Uniqueness, symmetry, and embedded minimal surfaces. J. Differ. Geom. **18**, 791–809 (1983)

Schoen, R., Simon, L.
1. Regularity of stable minimal hypersurfaces. Commun. Pure Appl. Math. **34**, 741–797 (1981)
2. Regularity of simply connected surfaces with quasiconformal Gauss map. In: Seminar on minimal submanifolds. Ann. Math. Stud. **103**, pp. 127–145. Princeton University Press, Princeton, 1983

Schoen, R., Simon, L., Yau, S.-T.
1. Curvature estimates for minimal hypersurfaces. Acta Math. **134**, 275–288 (1974)

Schoen, R., Yau, S.-T.
1. On univalent harmonic maps between surfaces. Invent. Math. **44**, 265–278 (1978)
2. Existence of incompressible minimal surfaces and the topology of three-dimensional manifolds with non-negative scalar curvature. Ann. Math. **110**, 127–142 (1979)

3. On the proof of the positive mass conjecture in general relativity. Commun. Math. Phys. **65**, 45–76 (1979)
4. Compact group, actions and the topology of manifolds with non-positive curvature. Topology **18**, 361–380 (1979)
5. Proof of the mass theorem II. Commun. Math. Phys. **79**, 231–260 (1981)

Schubert, H.
1. Topologie. Teubner, Stuttgart, 1971

Schüffler, K.
1. Stabilität mehrfach zusammenhängender Minimalflächen. Manuscr. Math. **30**, 163–197 (1979)
2. Isoliertheit und Stabilität von Flächen konstanter mittlerer Krümmung. Manuscr. Math. **40**, 1–15 (1982)
3. Jacobifelder zu Flächen konstanter mittlerer Krümmung. Arch. Math. **41**, 176–182 (1983)
4. Eine globalanalytische Behandlung des Douglas'schen Problems. Manuscr. Math. **48**, 189–226 (1984)
5. Zur Fredholmtheorie des Riemann–Hilbert-Operators. Arch. Math. **47**, 359–366 (1986)
6. Function theory and index theory for minimal surfaces of genus 1. Arch. Math. **48**, Part I: 250–266, II: 343–352, III: 446–457 (1987)
7. On holomorphic functions on Riemann surfaces and the Riemann–Hilbert problem. Analysis **9**, 283–296 (1989)
8. Minimalflächen auf Möbius-Bändern. Z. Anal. Anwend. **9**, 503–517 (1990)

Schüffler, K., Tomi, F.
1. Ein Indexsatz für Flächen konstanter mittlerer Krümmung. Math. Z. **182**, 245–258 (1983)

Schulz, F.
1. Regularity theory for quasilinear elliptic systems and Monge–Ampère equations in two dimensions. Lect. Notes Math. **1445**. Springer, Berlin, 1990

Schwab, D.
1. Hypersurfaces of prescribed mean curvature in central projection I, II. Arch. Math. **82**, 245–262 (2004) and Arch. Math. 84, 171–182 (2005)
2. Interior regularity of conical capillary surfaces. Preprint, Universität Duisburg-Essen, 2005

Schwarz, H.A.
1. Fortgesetzte Untersuchungen über spezielle Minimalflächen. Monatsberichte der Königlichen Akad. Wiss. Berlin, 3–27 (1872). Gesammelte Math. Abhandlungen I, 126–148 (1890)
2. Gesammelte Mathematische Abhandlungen, Band I und II. Springer, Berlin, 1890
3. Zur Theorie der Minimalflächen, deren Begrenzung aus geradlinigen Strecken besteht. Sitzungsber. K. Preuß. Akad. Wiss. Berl., Phys.-Math. Kl. 1237–1266 (1894)

Seidel, W.
1. Über die Ränderzuordnung bei konformen Abbildungen. Math. Ann. **104**, 183–243 (1931)

Seifert, H.
1. Minimalflächen von vorgegebener topologischer Gestalt. Sitzungsber. Heidelberg Akad. Wiss., Math.-Nat. Wiss. Kl. 5–16 (1974)

Seifert, H., Threlfall, W.
1. Lehrbuch der Topologie. Teubner, Leipzig, 1934. Reprint: Chelsea, New York
2. Variationsrechnung im Großen. Teubner, Leipzig, 1938

Serrin, J.
1. A priori estimates for solutions of the minimal surface equation. Arch. Ration. Mech. Anal. **14**, 376–383 (1963)
2. Removable singularities of elliptic equations, II. Arch. Ration. Mech. Anal. **20**, 163–169 (1965)
3. The Dirichlet problem for quasilinear equations with many independent variables. Proc. Natl. Acad. Sci. USA **58**, 1829–1835 (1967)
4. The problem of Dirichlet for quasilinear elliptic equations with many independent variables. Philos. Trans. R. Soc. Lond., Ser. A **264**, 413–496 (1969)
5. On surfaces of constant mean curvature which span a given space curve. Math. Z. **112**, 77–88 (1969)

Shiffman, M.
1. The Plateau problem for minimal surfaces which are relative minima. Ann. Math. (2) **39**, 309–315 (1938)
2. The Plateau problem for non-relative minima. Ann. Math. (2) **40**, 834–854 (1939)
3. The Plateau problem for minimal surfaces of arbitrary topological structure. Am. J. Math. **61**, 853–882 (1939)
4. Unstable minimal surfaces with any rectifiable boundary. Proc. Natl. Acad. Sci. USA **28**, 103–108 (1942)
5. Unstable minimal surfaces with several boundaries. Ann. Math. (2) **43**, 197–222 (1942)
6. On the isoperimetric inequality for saddle surfaces with singularities. In: Studies and essays presented to R. Courant, pp. 383–394. Interscience, New York, 1948
7. On surfaces of stationary area bounded by two circles, or convex curves, in parallel planes. Ann. Math. **63**, 77–90 (1956)
8. Instability for double integral problems in the calculus of variations. Ann. Math. **45**, 543–576 (1944)

Simader, C.G.
1. Equivalence of weak Dirichlet's principle, the method of weak solutions, and Perron's method towards classical solutions of Dirichlet's problem for harmonic functions. Math. Nachr. **279**, 415–430 (2006)

Simões, P.
1. On a class of minimal cones in $\mathbb{R}^n$. Bull. Am. Math. Soc. **80**, 488–489 (1974)

Simon, L.
1. Remarks on curvature estimates for minimal hypersurfaces. Duke Math. J. **43**, 545–553 (1976)
2. A Hölder estimate for quasiconformal maps between surfaces in Euclidean space. Acta Math. **139**, 19–51 (1977)
3. On a theorem of de Giorgi and Stampacchia. Math. Z. **155**, 199–204 (1977)
4. On some extensions of Bernstein's theorem. Math. Z. **154**, 265–273 (1977)
5. Equations of mean curvature type in 2 independent variables. Pac. J. Math. **69**, 245–268 (1977)
6. Isolated singularities of minimal surfaces. In: Proc. Centre Math. Anal. **1**, pp. 70–100. Australian National University, Centre for Mathematical Analysis, Canberra, Australia, 1982
7. Asymptotics for a class of nonlinear evolution equations with applications to geometric problems. Ann. Math. (2) **118**, 525–571 (1983)
8. Lectures on geometric measure theory. In: Proc. Centre Math. Anal. **3**. Australian National University, Centre for Mathematical Analysis, Canberra, Australia, 1984
9. Survey lectures on minimal submanifolds. In: Seminar on minimal submanifolds. Ann. Math. Stud. **103**, pp. 3–52. Princeton University Press, Princeton, 1983

10. Asymptotic behaviour of minimal graphs over exterior domains. Ann. Inst. Henri Poincaré, Anal. Non Linéaire **4**, 231–242 (1987)
11. Entire solution of the minimal surface equation. J. Differ. Geom. **30**, 643–688 (1989)
12. A strict maximum principle for area minimizing hypersurfaces. J. Differ. Geom. **26**, 327–335 (1987)
13. Regularity of capillary surfaces over domains with corners. Pac. J. Math. **88**(2), 363–377 (1980)
14. Asymptotics for exterior solutions of quasilinear elliptic equations. In: Berrick, Loo, Wang (eds.) Proceedings of the Pacific Rim Geometry Conference, University of Singapore, 1994. de Gruyter, Berlin, 1997
15. Asymptotic behaviour of minimal submanifolds and harmonic maps. Research report CMA, R51-84, Australian National University, Canberra
16. Growth properties for exterior solutions of quasilinear elliptic equations. Research report CMA, R17-89, Australian National University, Canberra
17. The minimal surface equation. In: Osserman, R. (ed.) Geometry V, Minimal surfaces. Encycl. Math. Sci. **90**, pp. 239–266. Springer, Berlin, 1997
18. Lower growth estimates for solutions of the minimal surface equation. In preparation

Simon, L., Smith, F.

1. On the existence of embedded minimal 2-spheres in the 3-sphere, endowed with an arbitrary metric. Published in the thesis of F. Smith, Melbourne University, 1983

Simons, J.

1. Minimal varieties in Riemannian manifolds. Ann. Math. (2) **88**, 62–105 (1968)
2. Minimal cones, Plateau's problem, and the Bernstein conjecture. Proc. Natl. Acad. Sci. USA **58**, 410–411 (1967)

Sinclair, E.

1. On the minimum surface of revolution in the case of one variable end point. Ann. Math. (2) **8**, 177–188 (1906–1907)
2. The absolute minimum in the problem of the surface of revolution of minimum area. Ann. Math. (2) **9**, 151–155 (1907–1908)
3. Concerning a compound discontinuous solution in the problem of the surface of revolution of minimum area. Ann. Math. (2) **10**, 55–80 (1908–1909)

Smyth, B.

1. Stationary minimal surfaces with boundary on a simplex. Invent. Math. **76**, 411–420 (1984)

Söllner, M.

1. Über die Struktur der Lösungsmenge des globalen Plateau-Problems bei Flächen konstanter mittlerer Krümmung. Dissertation, Bochum, 1982
2. Plateau's problem for surfaces of constant mean curvature from a global point of view. Manuscr. Math. **43**, 191–217 (1983)

Solomon, B.

1. On the Gauss map of an area-minimizing hypersurface. J. Differ. Geom. **19**, 221–232 (1984)

Spanier, E.H.

1. Algebraic topology. McGraw-Hill, New York, 1966

Spivak, M.

1. A comprehensive introduction to differential geometry. 5 vols., 2nd edn. Publish or Perish, Berkeley, 1979

Springer, G.

1. Introduction to Riemann surfaces. Addison-Wesley, Reading, 1957

Spruck, J.
1. Infinite boundary value problems for surfaces of constant mean curvature. Arch. Ration. Mech. Anal. **49**, 1–31 (1972)
2. Gauss curvature estimates for surfaces of constant mean curvature. Commun. Pure Appl. Math. **27**, 547–557 (1974)
3. Remarks on the stability of minimal submanifolds of $\mathbb{R}^n$. Math. Z. **144**, 169–174 (1975)

Stäckel, P.
1. Gauß als Geometer. In: Gauß, Werke Bd. 10.2
2. Über bedingte Biegungen krummer Flächen. Jahresber. Dtsch. Math.-Ver. **1**, 70 (1890–91)

Steffen, K.
1. Flächen konstanter mittlerer Krümmung mit vorgegebenem Volumen oder Flächeninhalt. Arch. Ration. Mech. Anal. **49**, 99–128 (1972)
2. Ein verbesserter Existenzsatz für Flächen konstanter mittlerer Krümmung. Manuscr. Math. **6**, 105–139 (1972)
3. Isoperimetric inequalities and the problem of Plateau. Math. Ann. **222**, 97–144 (1976)
4. On the existence of surfaces with prescribed mean curvature and boundary. Math. Z. **146**, 113–135 (1976)
5. On the nonuniqueness of surfaces with prescribed constant mean curvature spanning a given contour. Arch. Ration. Mech. Anal. **94**, 101–122 (1986)
6. Parametric surfaces of prescribed mean curvature. In: Lect. Notes Math. **1713**, pp. 211–265. Springer, Berlin, 1996

Steffen, K., Wente, H.
1. The non-existence of branch points in solutions to certain classes of Plateau type variational problems. Math. Z. **163**, 211–238 (1978)

Stein, E.M.
1. Singular integrals and differentiability properties of functions. Princeton University Press, Princeton, 1970

Stenius, E.
1. Ueber Minimalflächen, deren Begrenzung von zwei Geraden und einer Fläche gebildet wird. Druckerei d. Finn. Litt.-Ges., Helsingfors, 1892

Stephens, B.K.
1. Existence of thread-wire minimizers, with quantitative estimate. Preprint, 2008
2. Near-wire thread-wire surfaces: Lipschitz regularity and localization. Preprint, 2008
3. Local dominates global: C^1 regularity of near-wire thread wire minimizers. Preprint, 2008
4. Small $C^{1,1}$ thread-wire surfaces and a finiteness conjecture. Preprint, 2008

Stone, A.
1. On the isoperimetric inequality on a minimal surface. Calc. Var. Partial Differ. Equ. **17**, 369–391 (2003)

Ströhmer, G.
1. Instabile Minimalflächen in Riemannschen Mannigfaltigkeiten nichtpositiver Schnittkrümmung. J. Reine Angew. Math. **315**, 16–39 (1980)
2. Instabile Flächen vorgeschriebener mittlerer Krümmung. Math. Z. **174**, 119–133 (1980)
3. Instabile Minimalflächen mit halbfreiem Rand. Analysis **2**, 315–335 (1982)
4. Instabile Lösungen der Eulerschen Gleichungen gewisser Variationsprobleme. Arch. Ration. Mech. Anal. **79**, 219–239 (1982)
5. Instabile Lösungen der Eulerschen Gleichungen gewisser Variationsprobleme. Arch. Ration. Mech. Anal. **79**, 219–239 (1982)

Struik, D.J.
1. Lectures on classical differential geometry. Addison-Wesley, Reading, 1950

Struwe, M.
1. Multiple solutions of differential equations without the Palais–Smale condition. Math. Ann. **261**, 399–412 (1982)
2. Quasilinear elliptic eigenvalue problems. Comment. Math. Helv. **58**, 509–527 (1983)
3. On a free boundary problem for minimal surfaces. Invent. Math. **75**, 547–560 (1984)
4. On a critical point theory for minimal surfaces spanning a wire in $\mathbb{R}^n$. J. Reine Angew. Math. **349**, 1–23 (1984)
5. Large H-surfaces via the mountain-pass-lemma. Math. Ann. **270**, 441–459 (1985)
6. On the evolution of harmonic mappings. Comment. Math. Helv. **60**, 558–581 (1985)
7. Nonuniqueness in the Plateau problem for surfaces of constant mean curvature. Arch. Ration. Mech. Anal. **93**, 135–157 (1986)
8. A Morse theory for annulus-type minimal surfaces. J. Reine Angew. Math. **368**, 1–27 (1986)
9. The existence of surfaces of constant mean curvature with free boundaries. Acta Math. **160**, 19–64 (1988)
10. Heat flow methods for harmonic maps of surfaces and applications to free boundary problems. In: Cardoso, Figueiredo, Iório-Lopes (eds.) Partial differential equations. Lect. Notes Math. **1324**, pp. 293–319. Springer, Berlin, 1988
11. Plateau's problem and the calculus of variations. Ann. Math. Stud. **35**. Princeton University Press, Princeton, 1988
12. Applications of variational methods to problems in the geometry of surfaces. In: Hildebrandt, S., Leis, R. (eds.) Partial differential equations and calculus of variations. Lect. Notes Math. **1357**, pp. 359–378. Springer, Berlin, 1988
13. Variational methods. Applications to nonlinear partial differential equations and Hamiltonian systems. Ergeb. Math. Grenzgebiete **34**. Springer, Berlin, 1996
14. Multiple solutions to the Dirichlet problem for the equation of prescribed mean curvature. In: Rabihowitz, P.H., Zehnder, E. (eds.) Analysis et cetera. pp. 639–666. Academic Press, Boston, 1990
15. Minimal surfaces of higher genus and general critical type. In: Chang et al. (eds.) Proceedings of Int. Conf. on Microlocal and Nonlinear Analysis, Nankai Institute. World Scientific, Singapore, 1992
16. Das Plateausche Problem. Jahresber. Dtsch. Math.-Ver. **96**, 101–116 (1994)

Strzelecki, P.
1. A new proof of regularity of weak solutions of the H-surface equation. Calc. Var. Partial Differ. Equ. **16**, 227–242 (2003)

Tausch, E.
1. A class of variational problems with linear growth. Math. Z. **164**, 159–178 (1978)
2. The n-dimensional least area problem for boundaries on a convex cone. Arch. Ration. Mech. Anal. **75**, 407–416 (1981)

Thompson, D'Arcy W.
1. On growth and form, abridged edn. Cambridge University Press, Cambridge, 1969

Thomson, W.
1. On the division of space with minimum partitional area. Acta Math. **11**, 121–134 (1887/88)

Toda, M.
1. On the existence of H-surfaces into Riemannian manifolds. Calc. Var. Partial Differ. Equ. **5**, 55–83 (1997)

Tolksdorf P.

1. A parametric variational principle for minimal surfaces of varying topological type. J. Reine Angew. Math. **345**, 16–49 (1984)
2. On minimal surfaces with free boundaries in given homotopy classes. Ann. Inst. Henri Poincaré, Anal. Non Linéaire **2**, 157–165 (1985)

Tomi, F.

1. Ein einfacher Beweis eines Regularitätssatzes für schwache Lösungen gewisser elliptischer Systeme. Math. Z. **112**, 214–218 (1969)
2. Ein teilweise freies Randwertproblem für Flächen vorgeschriebener mittlerer Krümmung. Math. Z. **115**, 104–112 (1970)
3. Minimal surfaces and surfaces of prescribed mean curvature spanned over obstacles. Math. Ann. **190**, 248–264 (1971)
4. Variationsprobleme vom Dirichlet-Typ mit einer Ungleichung als Nebenbedingung. Math. Z. **128**, 43–74 (1972)
5. A perturbation theorem for surfaces of constant mean curvature. Math. Z. **141**, 253–264 (1975)
6. On the local uniqueness of the problem of least area. Arch. Ration. Mech. Anal. **52**, 312–318 (1973)
7. Bemerkungen zum Regularitätsproblem der Gleichung vorgeschriebener mittlerer Krümmung. Math. Z. **132**, 323–326 (1973)
8. On the finite solvability of Plateau's problem. In: Lect. Notes Math. **597**, pp. 679–695. Springer, Berlin, 1977
9. Plateau's problem for embedded minimal surfaces of the type of the disc. Arch. Math. **31**, 374–381 (1978)
10. A finiteness result in the free boundary value problem for minimal surfaces. Ann. Inst. Henri Poincaré, Anal. Non Linéaire **3**, 331–343 (1986)
11. Plateau's problem for minimal surfaces with a catenoidal end. Arch. Math. **59**, 165–173 (1978)
12. Über elliptische Differentialgleichungen 4. Ordnung mit einer starken Nichtlinearität. Gött. Nachr. **3**, 33–42 (1976)
13. Plateau's problem for infinite contours. Analysis **29**, 155–167 (2009)

Tomi, F., Tromba, A.J.

1. Extreme curves bound an embedded minimal surface of disk type. Math. Z. **158**, 137–145 (1978)
2. On the structure of the set of curves bounding minimal surfaces of prescribed degeneracy. J. Reine Angew. Math. **316**, 31–43 (1980)
3. On Plateau's problem for minimal surfaces of higher genus in $\mathbb{R}^3$. Bull. Am. Math. Soc. **13**, 169–171 (1985)
4. A geometric proof of the Mumford compactness theorem. In: Chern, S.S. (ed.) Proc. of the DD7 Symposium on Partial Differential Equations, Nankai University, 1986. Lect. Notes Math. **1306**, pp. 174–181. Springer, Berlin, 1986
5. Existence theorems for minimal surfaces of non-zero genus spanning a contour. Mem. Am. Math. Soc. **71** (1988). [Appeared previously as preprint No. 5, Heidelberg, 1987 under the title "On Plateau's problem for minimal surfaces of prescribed topological type."]
6. The index theorem for higher genus minimal surfaces. Mem. Am. Math. Soc. **560**, 78 pp., 1995

Tomi, F., Ye, R.

1. The exterior Plateau problem. Math. Z. **205**, 233–245 (1990)

Tonelli, L.

1. Sul problema di Plateau, I & II. Rend. R. Accad. dei Lincei **24**, 333–339, 393–398 (1936) (cf. also: Opere scelte, Vol. III, 328–341)

Toponogov, W.A.

1. An isoperimetric inequality for surfaces whose Gauss curvature is bounded from above. Sib. Mat. Zh. **10**, 144–157 (1967) (in Russian). [Engl. translation in Sib. Math. J. **10**, 104–113 (1969).]

Toth, G.

1. Harmonic and minimal maps with applications in geometry and physics. Ellis Horwood, Chichester, 1984

Tromba, A.J.

1. Some theorems on Fredholm maps. Proc. Am. Math. Soc. **34**, 578–585 (1972)
2. Almost Riemannian structures on Banach manifolds, the Morse lemma, and the Darboux theorem. Can. J. Math. **28**, 640–652 (1976)
3. On the numbers of solutions to Plateau's problem. Bull. Am. Math. Soc. **82** (1976)
4. The set of curves of uniqueness for Plateau's problem has a dense interior. Lect. Notes Math. **597**. Springer, Berlin, 1977
5. On the number of simply connected minimal surfaces spanning a curve. Mem. Am. Math. Soc. No. 194, **12** (1977)
6. The Morse–Sard–Brown theorem for functionals and the problem of Plateau. Am. J. Math. **99**, 1251–1256 (1977)
7. The Euler characteristic of vector fields on Banach manifolds and a globalization of Leray–Schauder degree. Adv. Math. **28**, 148–173 (1978)
8. On Plateau's problem for minimal surfaces of higher genus in $\mathbb{R}^n$. Preprint No. 580, SFB 72, Bonn, 1983
9. A sufficient condition for a critical point of a functional to be a minimum and its application to Plateau's problem. Math. Ann. **263**, 303–312 (1983)
10. Degree theory on oriented infinite dimensional varieties and the Morse number of minimal surfaces spanning a curve in $\mathbb{R}^n$. Part II: $n = 3$. Manuscr. Math. **48**, 139–161 (1984)
11. Degree theory on oriented infinite dimensional varieties and the Morse number of minimal surfaces spanning a curve in $\mathbb{R}^n$. Part I: $n \geq 4$. Trans. Am. Math. Soc. **290**, 385–413 (1985)
12. On the Morse number of embedded and non-embedded minimal immersions spanning wires on the boundary of special bodies in $\mathbb{R}^3$. Math. Z. **188**, 149–170 (1985)
13. On a natural algebraic affine connection on the space of almost complex structures and the curvature of Teichmüller space with respect to its Weil–Petersson metric. Manuscr. Math. **56**, 475–497 (1986)
14. New results in the classical Plateau problem. In: Proc. Int. Congress Math. Berkeley, 1986
15. On an energy function for the Weil–Petersson metric on Teichmüller space. Manuscr. Math. **59**, 249–260 (1987)
16. A proof of the Douglas theorem on the existence of disc-like minimal surfaces spanning Jordan contours in $\mathbb{R}^n$. Astérisque **154–155**, 39–50 (1987)
17. Global analysis and Teichmüller theory. In: Tromba, A. (ed.) Seminar on new results in nonlinear partial differential equations. Aspects of Mathematics **10**. Vieweg, Braunschweig, 1987
18. Open problems in the degree theory for disc minimal surfaces spanning a curve in $\mathbb{R}^3$. In: Hildebrandt, S., Leis, R. (eds.) Partial differential equations and calculus of variations. Lect. Notes Math. **357**, pp. 379–401. Springer, Berlin, 1988

19. Existence theorems for minimal surfaces of non-zero genus spanning a contour. Mem. Am. Math. Soc. No. 382, vol. **1** (1988)
20. Seminar on new results in nonlinear partial differential equations. Aspects of Mathematics **E10**. Vieweg, Braunschweig, 1987
21. Intrinsic third derivatives for Plateau's problem and the Morse inequalities for disc minimal surfaces in $\mathbb{R}^3$. Calc. Var. Partial Differ. Equ. **1**, 335–353 (1993)
22. Dirichlet's energy on Teichmüller's moduli space and the Nielsen realization problem. Math. Z. **222**, 451–464 (1996)
23. On the Levi form for Dirichlet's energy on Teichmüller's moduli space. Appendix E in Tromba [19]
24. Teichmüller theory in Riemannian geometry. Lect. Notes Math. Birkhäuser, Basel, 1992 (based on notes taken by J. Denzler, ETH Zürich)
25. The Morse theory of two-dimensional closed branched minimal surfaces and their generic non-degeneracy in Riemann manifolds. Calc. Var. Partial Differ. Equ. **10**, 135–170 (2000)
26. A general approach to Morse theory. J. Differ. Geom. **11**, 47–85 (1977)
27. Fredholm vector fields and a transversality theorem. J. Funct. Anal. (1976)
28. Dirichlet's energy on Teichmüller's moduli space is strictly pluri-subharmonic. In: Jost, J. (ed.) Geometric analysis and calculus of variations, pp. 315–341. International Press, Somerville, 1996
29. Smale and nonlinear analysis: A personal perspective. In: Hirsch, M.W., Marsden, J.E., Shub, M. (eds.) From topology to computation. Proceedings of the Smalefest, pp. 481–492. Springer, New York, 1993

Tsuji, M.

1. On a theorem of F. and M. Riesz. Proc. Imp. Acad. (Tokyo) **18**, 172–175 (1942)
2. Potential theory in modern function theory. Maruzen, Tokyo, 1959

Turowski, G.

1. Nichtparametrische Minimalflächen vom Typ des Kreisrings und ihr Verhalten längs Kanten der Stützfläche. Thesis, Bonn, 1997. Bonner Math. Schr. **307**. Mathematisches Institut der Universität Bonn, Bonn, 1998
2. Existence of doubly connected minimal graphs in singular boundary configurations. Asymptot. Anal. **23**, 239–256 (2000)
3. Behaviour of doubly connected minimal surfaces at the edges of the support surface. Arch. Math. **77**, 278–288 (2001)

Uhlenbeck, K.

1. Closed minimal surfaces in hyperbolic 3-manifolds. In: Seminar on minimal submanifolds. Ann. Math. Stud. **103**, pp. 147–168. Princeton University Press, Princeton, 1983

Umehara, M., Yamada, K.

1. Complete surfaces of constant mean curvature -1 in the hyperbolic 3-space. Ann. Math. **137**, 611–638 (1993)
2. A parametrization of the Weierstrass formulae minimal surfaces in $\mathbb{R}^3$ into the hyperbolic 3-space. J. Reine Angew. Math. **432**, 93–116 (1992)

Van der Mensbrugghe, G.

1. Sur la tension des lames liquides. Bull. Acad. R. Sci. Brux. (2) **22**, 270–276, 308–328 (1866)
2. Discussion et réalisation expérimentale d'une surface paritulière à courbure moyenne nulle. Bull. Acad. R. Sci. Brux. (2) **21**, 552–566 (1866)
3. Sur la tension des lames liquide. 2me note. Bull. Acad. R. Sci. Brux. (2) **23**, 448–465 (1867)

Vekua, I.N.
1. Generalized analytic functions. Pergamon Press, Oxford, 1962
2. Verallgemeinerte analytische Funktionen. Akademie-Verlag, Berlin, 1963

Vogel, T.I.
1. Stability of a drop trapped between two parallel planes. Preliminary Report, Texas A&M University, 1985

Walter R.
1. Explicit examples to the H-problem of Heinz Hopf. Geom. Dedic. **23**, 187–213 (1987)
2. Constant mean curvature tori with spherical curvature lines in noneuclidean geometry. Manuscr. Math. **63**, 343–363 (1989)

Warschawski, S.E.
1. Über einen Satz von O.D. Kellogg. Nachr. Akad. Wiss. Gött., II. Math. Phys. Kl., 73–86 (1932)
2. Über das Randverhalten der Abbildungsfunktion bei konformer Abbildung. Math. Z. **35**, 321–456 (1932)
3. On the higher derivatives at the boundary in conformal mapping. Trans. Am. Math. Soc. **38**, 310–340 (1935)
4. On a theorem of L. Lichtenstein. Pac. J. Math. **5**, 835–839 (1955)
5. On the differentiability at the boundary in conformal mapping. Proc. Am. Math. Soc. **12**, 614–620 (1961)
6. Boundary derivatives of minimal surfaces. Arch. Ration. Mech. Anal. **38**, 241–256 (1970)

Weierstraß, K.
1. Mathematische Werke **3**. Mayer & Müller, Berlin, 1903
2. Fortsetzung der Untersuchung über die Minimalflächen. Monatsber. K. Akad. Wiss., 855–856, December 1866 und Mathematische Werke **3**, pp. 219–220. Mayer & Müller, Berlin, 1903
3. Über eine besondere Gattung von Minimalflächen. Monatsber. K. Akad. Wiss., 511–518, August 1887 und Math. Werke **3**, pp. 241–247. Mayer & Müller, Berlin, 1903
4. Analytische Bestimmung einfach zusammenhängender Minimalflächen, deren Begrenzung aus geradlinigen, ganz im endlichen liegenden Strecken besteht. In: Math. Werke **3**, pp. 221–238. Mayer & Müller, Berlin, 1903
5. Untersuchungen über die Flächen, deren mittlere Krümmung überall gleich Null ist. In: Math. Werke **3**, pp. 39–52. Mayer & Müller, Berlin, 1903

Weingarten, J.
1. Ueber eine Klasse aufeinander abwickelbarer Flächen. J. Reine Angew. Math. **59**, 382–393 (1861)
2. Ueber die durch eine Gleichung der Form $X + Y + Z = 0$ darstellbaren Minimalflächen. Nachr. K. Ges. Wiss. Univ. Gött. 272–275 (1887)
3. Ueber particuläre Integrale der Differentialgleichung $\partial^2 V/\partial x^2 + \partial^2 V/\partial y^2 + \partial^2 V/\partial z^2 = 0$ und eine mit der Theorie der Minimalflächen zusammenhängende Gattung von Flüssigkeitsbewegungen. Nachr. K. Ges. Wiss. Univ. Gött. 313–335 (1890)

Wente, H.
1. An existence theorem for surfaces of constant mean curvature. J. Math. Anal. Appl. **26**, 318–344 (1969)
2. A general existence theorem for surfaces of constant mean curvature. Math. Z. **120**, 277–288 (1971)
3. An existence theorem for surfaces in equilibrium satisfying a volume constraint. Arch. Ration. Mech. Anal. **50**, 139–158 (1973)
4. The Dirichlet problem with a volume constraint. Manuscr. Math. **11**, 141–157 (1974)

5. The differential equation $\Delta x = 2Hx_u \wedge x_v$ with vanishing boundary values. Proc. Am. Math. Soc. **50**, 131–137 (1975)
6. The Plateau problem for symmetric surfaces. Arch. Ration. Mech. Anal. **60**, 149–169 (1976)
7. Large solutions to the volume constrained Plateau problem. Arch. Ration. Mech. Anal. **75**, 59–77 (1980)
8. Counterexample to a question of H. Hopf. Pac. J. Math. **121**, 193–243 (1986)
9. Twisted tori of constant mean curvature in $\mathbb{R}^3$. In: Tromba, A.J. (ed.) Seminar on new results in non-linear partial differential equations, Max-Planck-Institut für Mathematik, pp. 1–36. Vieweg, Braunschweig, 1987
10. A note on the stability theorem of J.L. Barbosa and M. do Carmo for closed surfaces of constant mean curvature. Pac. J. Math. **147**, 375–379 (1991)
11. The Plateau problem for boundary curves with connectors. In: Jost, J. (ed.) Geometric analysis and the calculus of variations, pp. 343–359. International Press, Cambridge, 1996
12. Constant mean curvature immersions of Enneper type. Mem. Am. Math. Soc. Nr. 47, 83 pp., 1992

Werner, H.

1. Das Problem von Douglas für Flächen konstanter mittlerer Krümmung. Math. Ann. **133**, 303–319 (1957)
2. The existence of surfaces of constant mean curvature with arbitrary Jordan curves as assigned boundary. Proc. Am. Math. Soc. **11**, 63–70 (1960)

White, B.

1. Existence of least area mappings of N-dimensional domains. Ann. Math. **118**, 179–185 (1983)
2. Tangent cones to two-dimensional area-minimizing currents are unique. Duke Math. J. **50**, 143–160 (1983)
3. The least area bounded by multiples of a curve. Proc. Am. Math. Soc. **90**, 230–232 (1984)
4. Mappings that minimize area in their homotopy classes. J. Differ. Geom. **20**, 433–446 (1984)
5. Generic regularity of unoriented two-dimensional area minimizing surfaces. Ann. Math. **121**, 595–603 (1985)
6. Homotopy classes in Sobolev spaces and energy minimizing maps. Bull., New Ser., Am. Math. Soc. **13**, 166–168 (1985)
7. Infima of energy functionals in homotopy classes of mappings. J. Differ. Geom. **23**, 127–142 (1986)
8. The space of m-dimensional surfaces that are stationary for a parametric integrand. Indiana Univ. Math. J. **30**, 567–602 (1987)
9. Curvature estimates and compactness theorems in 3-manifolds for surfaces that are stationary for parametric elliptic functionals. Invent. Math. **88**, 243–256 (1987)
10. Complete surfaces of finite total curvature. J. Differ. Geom. **26**, 315–316 (1987). Correction: J. Differ. Geom. **28**, 359–360 (1988)
11. Homotopy classes in Sobolev spaces and the existence of energy minimizing maps. Acta Math. **160**, 1–17 (1988)
12. Some recent developments in differential geometry. Math. Intel. **11**, 41–47 (1989)
13. New applications of mapping degrees to minimal surface theory. J. Differ. Geom. **29**, 143–162 (1989)
14. A new proof for the compactness theorem for integral currents. Comment. Math. Helv. **64**, 207–220 (1989)

15. Every metric of positive Ricci curvature on S^3 admits a minimal embedded torus. Bull. Am. Math. Soc. **21**, 71–75 (1989)
16. Existence of smooth embedded surfaces of prescribed topological type that minimize parametric even elliptic functionals on three-manifolds. J. Differ. Geom. **33**, 413–443 (1991)
17. On the topological type of minimal submanifolds. Topology **31**, 445–448 (1992)
18. The space of minimal submanifolds for varying Riemannian metrics. Indiana Univ. Math. J. **40**, 161–200 (1991)
19. Regularity of singular sets for Plateau-type problems. Preprint
20. The space of m-dimensional surfaces that are stationary for a parametric elliptic integral. Indiana Univ. Math. J. **36**, 567–602 (1987)
21. The bridge principle for stable minimal surfaces. Calc. Var. Partial Differ. Equ. **2**, 405–425 (1994)
22. The bridge principle for unstable and for singular minimal surfaces. Commun. Anal. Geom. **2**, 513–532 (1994)
23. Half of Enneper's surface minimizes area. In: Jost, J. (ed.) Geometric analysis and the calculus of variations, pp. 361–367. International Press, Somerville, 1996
24. Classical area minimizing surfaces with real-analytic boundaries. Acta Math. **179**, 295–305 (1997)
25. The space of minimal submanifolds for varying Riemannian metrics. Indiana Univ. Math. J. **40**, 161–200 (1991)

Whitehead, G.W.
1. Elements of homotopy theory. Springer, Berlin, 1978

Whittemore, J.K.
1. The isoperimetrical problem on any surface. Ann. Math. (2) **2**, 175–178 (1900–1901)
2. Minimal surfaces applicable to surfaces of revolution. Ann. Math. (2) **19**, 1–20 (1917–1918)
3. Spiral minimal surfaces. Trans. Am. Math. Soc. **19**, 315–330 (1918)
4. Associate minimal surfaces. Am. J. Math. **40**, 87–96 (1918)
5. Minimal surfaces containing straight lines. Ann. Math. (2) **22**, 217–225 (1921)

Widman, K.-O.
1. On the boundary behavior of solutions to a class of elliptic partial differential equations. Ark. Mat. **6**, 485–533 (1966)
2. Inequalities for the Green function of the gradient of solutions of elliptic differential equations. Math. Scand. **21**, 17–37 (1967)
3. Hölder continuity of solutions of elliptic systems. Manuscr. Math. **5**, 299–308 (1971)

Wienholtz, D.
1. Der Ausschluß von eigentlichen Verzweigungspunkten bei Minimalflächen vom Typ der Kreisscheibe. Diplomarbeit, Universität München. SFB 256, Universität Bonn, Lecture notes No. 37, 1996
2. Zum Ausschluß von Randverzweigungspunkten bei Minimalflächen. Bonner Math. Schr. **298**. Mathematisches Institut der Universität Bonn, Bonn, 1997
3. A method to exclude branch points of minimal surfaces. Calc. Var. Partial Differ. Equ. **7**, 219–247 (1998)

Wigley, N.M.
1. Development of the mapping function at a corner. Pac. J. Math. **15**, 1435–1461 (1965)

Williams, G.
1. The Dirichlet problem for the minimal surface equation with Lipschitz continuous boundary data. J. Reine Angew. Math. **354**, 123–140 (1984)

Winklmann, S.

1. Enclosure theorems for generalized mean curvature flows. Calc. Var. Partial Differ. Equ. **16**, 439–447 (2003)
2. Integral curvature estimates for F-stable hypersurfaces. Calc. Var. Partial Differ. Equ. **23**, 391–414 (2005)
3. Pointwise curvature estimates for F-stable hypersurfaces. Ann. Inst. Henri Poincaré, Anal. Non Linéaire **22**, 543–555 (2005)
4. Estimates for stable hypersurfaces of prescribed F-mean curvature. Manuscr. Math. **118**, 485–499 (2005)

Wohlgemuth, M.

1. Abelsche Minimalflächen. Diplomarbeit, Bonn, 1988
2. Higher genus minimal surfaces by growing handles out of a catenoid. Manuscr. Math. **70**, 397–428 (1991)

Wohlrab, O.

1. Einschließungssätze für Minimalflächen und Flächen mit vorgegebener mittlerer Krümmung. Bonner Math. Schr. **138**. Mathematisches Institut der Universität Bonn, Bonn, 1982
2. Zur numerischen Behandlung von parametrischen Minimalflächen mit halbfreien Rändern. Dissertation, Bonn, 1985
3. Die Berechnung und graphische Darstellung von Randwertproblemen für Minimalflächen. In: Jürgens, H., Saupe, D. (eds.) Visualisierung in Mathematik und Naturwissenschaften, Springer, Berlin, 1989

Wood, J.C.

1. Singularities of harmonic maps and applications of the Gauss–Bonnet formula. Am. J. Math. **99**, 1329–1344 (1977)

Yau, S.T.

1. Some function-theoretic properties of complete Riemannian manifolds and their applications to geometry. Indiana Univ. Math. J. **25**, 659–670 (1976)
2. Problem section. In: Seminar on differential geometry. Ann. Math. Stud. **102**, pp. 669–706. Princeton University Press, Princeton, 1982
3. Survey on partial differential equations in differential geometry. In: Seminar on differential geometry. Ann Math. Stud. **102**, pp. 3–72. Princeton University Press, Princeton, 1982
4. Minimal surfaces and their role in differential geometry. In: Global Riemannian geometry, pp. 99–103. Horwood, Chichester, 1984
5. Nonlinear analysis in geometry. Monogr. 33, Enseign. Math. 5–54 (1986)

Ye, R.

1. Randregularität von Minimalflächen. Diplomarbeit, Bonn, 1984
2. A priori estimates for minimal surfaces with free boundary which are not minima of the area. Manuscr. Math. **58**, 95–107 (1987)
3. On minimal surfaces of higher topology. Preprint, Stanford, 1988
4. Regularity of a minimal surface at its free boundary. Math. Z. **198**, 261–275 (1988)
5. Existence, regularity and finiteness of minimal surfaces with free boundary. SFB 256 preprint, No. 1, Bonn, 1987
6. On the existence of area-minimizing surfaces with free boundary. Math. Z. **206**, 321–331 (1991)
7. Construction of embedded area-minimizing surfaces via a topological type induction scheme. Calc. Var. Partial Differ. Equ. **19**, 391–420 (2004)

Young, L.C.

1. On the isoperimetric ratio for a harmonic surface. Proc. Lond. Math. Soc. (2) **49**, 396–408 (1947)
2. Some new methods in two-dimensional variational problems with special reference to minimal surfaces. Commun. Pure Appl. Math. **9**, 625–632 (1956)

Zeidler, E.

1. Applied functional analysis. Main principles and their applications. Appl. Math. Sci. **109**. Springer, New York, 1995

Ziemer, W.

1. Weakly differentiable functions. Sobolev spaces and functions of bounded variation. Graduate Texts Math. **120**. Springer, Berlin, 1989

Index

U. Dierkes, S. Hildebrandt, A.J. Tromba, *Regularity of Minimal Surfaces*,
Grundlehren der mathematischen Wissenschaften 340,
DOI 10.1007/978-3-642-11700-8, © Springer-Verlag Berlin Heidelberg 2010

Printing and Binding: Stürtz GmbH, Würzburg